Methods in Enzymology

Volume XXX
NUCLEIC ACIDS AND PROTEINS
Part F

METHODS IN ENZYMOLOGY

EDITORS-IN-CHIEF

Sidney P. Colowick   Nathan O. Kaplan

Methods in Enzymology

Volume XXX

Nucleic Acids and Protein Synthesis

Part F

EDITED BY

Kivie Moldave

DEPARTMENT OF BIOLOGICAL CHEMISTRY
CALIFORNIA COLLEGE OF MEDICINE
UNIVERSITY OF CALIFORNIA
IRVINE, CALIFORNIA

Lawrence Grossman

GRADUATE DEPARTMENT OF BIOCHEMISTRY
BRANDEIS UNIVERSITY
WALTHAM, MASSACHUSETTS

1974

ACADEMIC PRESS New York and London
A Subsidiary of Harcourt Brace Jovanovich, Publishers

ACADEMIC PRESS, INC.
111 Fifth Avenue, New York, New York 10003

United Kingdom Edition published by
ACADEMIC PRESS, INC. (LONDON) LTD.
24/28 Oval Road, London NW1

LIBRARY OF CONGRESS CATALOG CARD NUMBER: 74-3577

PRINTED IN THE UNITED STATES OF AMERICA

Table of Contents

Section I. Initiation Factors in Protein Synthesis

Section II. Elongation Factors in Protein Synthesis

Section III. Termination Factors in Protein Synthesis

Section IV. Ribosome Structure and Function

Section V. Messenger RNA and Protein Synthesizing Systems

Contributors to Volume XXX

Article numbers are in parentheses following the names of contributors.
Affiliations listed are current.

W. FRENCH ANDERSON (12, 13, 14, 17, 58), *Section on Molecular Hematology, National Heart and Lung Institute, National Institutes of Health, Bethesda, Maryland*

R. BAIERLEIN (33), *Department of Physics, Wesleyan University, Middletown, Connecticut*

J. P. G. BALLESTA (23), *Instituto de Biología Celular, C.S.I.C., Madrid, Spain*

DAVID BALTIMORE (66), *Department of Biology, Massachusetts Institute of Technology, Cambridge, Massachusetts*

MARIANO BARBACID (41), *Instituto de Biología Celular, C.S.I.C., Madrid, Spain*

E. BATTANER (29), *Instituto de Biología Celular, C.S.I.C., Madrid, Spain*

MIGUEL BEATO (65), *Institute of Cancer Research, College of Physicians and Surgeons, Columbia University, New York, New York*

A. L. BEAUDET (30), *Department of Internal Medicine, Baylor College of Medicine, Houston, Texas*

E. L. BENEDETTI (32), *Institut de Biologie Moléculaire, Universite Paris VII, Paris, France*

ANTON J. M. BERNS (63), *Department of Biochemistry, University of Nijmegen, Nijmegen, The Netherlands*

T. A. BICKLE (52), *Department of Biological Chemistry, School of Medicine, University of California, Davis, California*

LAWRENCE BITTE (54), *Department of Biochemistry, University of Oregon Medical School, Portland, Oregon*

HANS BLOEMENDAL (32, 63), *Department of Biochemistry, University of Nijmegen, Nijmegen, The Netherlands*

JAMES W. BODLEY (24), *Department of Biochemistry, University of Minnesota, Minneapolis, Minnesota*

W. S. BONT (32), *Department of Biochemistry, The Netherlands Cancer Institute, Amsterdam, The Netherlands*

GEORGE BRAWERMAN (56), *Department of Biochemistry and Pharmacology, Tufts University School of Medicine, Boston, Massachusetts*

NATHAN BROT (24), *Department of Biochemistry, Roche Institute of Molecular Biology, Nutley, New Jersey*

DONALD D. BROWN (60), *Department of Embryology, Carnegie Institute of Washington, Washington, D.C.*

GLENN E. BROWN (37), *Department of Molecular Biology and Biochemistry, School of Biological Sciences, University of California, Irvine, California*

L. CARRASCO (29), *Instituto de Biología Celular, C.S.I.C., Madrid, Spain*

LINDA M. CASHION (16), *Department of Molecular Biology and Biochemistry, School of Biological Sciences, University of California, Irvine, California*

C. T. CASKEY (30), *Department of Medicine, Baylor College of Medicine, Houston, Texas*

YANG C. CHEN (15), *Department of Chemistry, University of Nebraska, Lincoln, Nebraska*

J. M. CIMADEVILLA (21), *Clayton Foundation Biochemical Institute, Department of Chemistry, University of Texas, Austin, Texas*

JOHN M. CLARK, JR. (69), *Department of Biochemistry, University of Illinois, Urbana, Illinois*

RONALD G. CRYSTAL (12), *Section on Molecular Hematology, National Heart and Lung Institute, National Institutes of Health, Bethesda, Maryland*

BERNARD D. DAVIS (10), *Bacterial Physiology Unit, Harvard Medical School, Boston, Massachusetts*

xi

GERALDINE L. DETTMAN (16), *Department of Radiological Sciences, California College of Medicine, University of California, Irvine, California*

RALPH F. DI CAMELLI (36), *Department of Biochemistry, The University of Chicago, Chicago, Illinois*

DOV EFRON (68), *The Institute for Cancer Research, Philadelphia, Pennsylvania*

CHARLES EIL (55), *Department of Biochemistry, The University of Chicago, Chicago, Illinois*

DAVID ELSON (40), *Biochemistry Department, The Weizmann Institute of Science, Rehovot, Israel*

NORTON A. ELSON (12), *Section on Molecular Hematology, National Heart and Lung Institute, National Institutes of Health, Bethesda, Maryland*

V. ERDMANN (53), *Institute for Enzyme Research, University of Wisconsin, Madison, Wisconsin*

STEPHEN FAHNESTOCK (47, 53), *Department of Biochemistry, Althouse Laboratory, Pennsylvania State University, University Park, Pennsylvania*

JOHN L. FAKUNDING (3), *Department of Biological Chemistry, School of Medicine, University of California, Davis, California*

AMY K. FALVEY (58), *Section on Molecular Hematology, National Heart and Lung Institute, National Institutes of Health, Bethesda, Maryland*

PHILIP FEIGELSON (65), *Institute of Cancer Research, College of Physicians and Surgeons, Columbia University, New York, New York*

L. P. GAVRILOVA (43), *Institute of Protein Research, Academy of Sciences of the USSR, Moscow, USSR*

CARMEN GONZALES (59), *Department of Biological Sciences, Stanford University, Stanford, California*

MICHAEL GOTTLIEB (10), *Bacterial Physiology Unit, Harvard Medical School, Boston, Massachusetts*

HERMANN GRAF (13), *Section on Molecular Hematology, National Heart and Lung Institute, National Institutes of Health, Bethesda, Maryland*

YORAM GRONER (7, 8), *Biochemistry Department, Weizmann Institute of Science, Rehovot, Israel*

NABA K. GUPTA (15), *Department of Chemistry, University of Nebraska, Lincoln, Nebraska*

MARY G. HAMILTON (38, 51), *Sloan-Kettering Institute for Cancer Research, New York, New York*

B. HARDESTY (21), *Clayton Foundation Biochemical Institute, Department of Chemistry, The University of Texas, Austin, Texas*

EDWARD HERBERT (67a, 67b, 67c, 67d), *Department of Chemistry, University of Oregon, Eugene, Oregon*

C. F. HEREDIA (27), *Instituto de Enzimología del C.S.I.C., Facultad de Medicina de la Universidad Autónoma, Madrid, Spain*

PETER HERRLICH (61), *Max-Planck-Institut für Molekulare Genetik, Berlin-Dahlem, Germany*

JOHN W. B. HERSHEY (3), *Department of Biological Chemistry, School of Medicine, University of California, Davis, California*

STUART M. HEYWOOD (62), *Genetics and Cell Biology Section, University of Connecticut, Storrs, Connecticut*

G. A. HOWARD (50), *Department of Biological Chemistry, School of Medicine, University of California, Davis, California*

A. A. INFANTE (33), *Department of Biology, Wesleyan University, Middletown, Connecticut*

JOEL L. IVEY (67a), *Department of Chemistry, University of Oregon, Eugene, Oregon*

DAVID KABAT (54), *Department of Biochemistry, University of Oregon Medical School, Portland, Oregon*

MARY L. KIELY (59), *Department of Biological Sciences, Stanford University, Stanford, California*

WILLIAM H. KLEIN (69), *Department of*

Biochemistry, University of Illinois, Urbana, Illinois

ALFRED J. KOLB (37), *Department of Molecular Biology and Biochemistry, School of Biological Sciences, University of California, Irvine, California*

DAVID P. LEADER (18), *Institute of Biochemistry, University of Glasgow, Glasgow, Scotland*

BERNARD LEBLEU (57), *Department of Molecular Biology, University of Brussels, Brussels, Belgium*

SYLVIA LEE-HUANG (4, 6), *Department of Biochemistry, School of Medicine, New York University Medical Center, New York, New York*

FRITZ LIPMANN (26), *The Rockefeller University, New York, New York*

HARVEY F. LODISH (66), *Department of Biology, Massachusetts Institute of Technology, Cambridge, Massachusetts*

NICOLETTE H. LUBSEN (10), *Bacterial Physiology Unit, Harvard Medical School, Boston, Massachusetts*

MAXSON MCDOWELL (66), *Department of Microbiology and Immunology, Duke University Medical Center, Durham, North Carolina*

G. STANLEY MCKNIGHT (64), *Department of Biological Sciences, Stanford University, Stanford, California*

ABRAHAM MARCUS (11, 68), *The Institute for Cancer Research, Philadelphia, Pennsylvania*

JOHN R. MENNINGER (31), *Department of Zoology, The University of Iowa, Iowa City, Iowa*

WILLIAM MERRICK (13), *Section on Molecular Hematology, National Heart and Lung Institute, National Institutes of Health, Bethesda, Maryland*

DAVID LEE MILLER (22), *Department of Biochemistry, Roche Institute of Molecular Biology, Nutley, New Jersey*

M. J. MILLER (1, 2), *Laboratory of Molecular Biology, University of Sherbrooke School of Medicine, Sherbrooke, Quebec, Canada*

RUTH MISKIN (40), *Biochemistry De-*

partment, The Weizmann Institute of Science, Rehovot, Israel

JUAN MODOLELL (9), *Instituto de Biología Celular, C.S.I.C., Madrid, Spain*

HELMUT NEUMANN (47), *Hoffmann-La Roche Co., Basel, Switzerland*

ARTHUR W. NIENHUIS (58), *Section on Molecular Hematology, National Heart and Lung Institute, National Institutes of Health, Bethesda, Maryland*

M. NOMURA (53), *Institute for Enzyme Research, University of Wisconsin, Madison, Wisconsin*

SEVERO OCHOA (4, 5, 6, 20), *Department of Biochemistry, School of Medicine, New York University Medical Center, New York, New York*

RAFAEL PALACIOS (59), *Department of Biological Sciences, Stanford University, Stanford, California*

SIDNEY PESTKA (27, 42, 44, 45, 46), *Department of Biochemistry, Roche Institute of Molecular Biology, Nutley, New Jersey*

MARY L. PETERMANN (34), *The Sloan-Kettering Institute for Cancer Research, New York, New York*

DANTE PICCIANO (17), *Section on Molecular Hematology, National Heart and Lung Institute, National Institutes of Health, Bethesda, Maryland*

PHILIP M. PRICHARD (14), *Section on Molecular Hematology, National Heart and Lung Institute, National Institutes of Health, Bethesda, Maryland*

MICHEL REVEL (7, 8), *Biochemistry Department, Weizmann Institute of Science, Rehovot, Israel*

ROBERT E. RHOADS (64), *Department of Biological Sciences, Stanford, University, Stanford, California*

ALEXANDER RICH (47), *Department of Biology, Massachusetts Institute of Technology, Cambridge, Massachusetts*

DIETMAR RICHTER (25, 26), *Max-Planck-Institut für Molekulare Genetik, Berlin-Dahlem, Germany*

ARTHUR W. ROURKE (62), *Genetics and*

Cell Biology Section, University of Connecticut, Storrs, Connecticut

STEVEN SABOL (5), Department of Biochemistry, School of Medicine, New York University Medical Center, New York, New York

A. SANDOVAL (27), Instituto de Enzimología del C.S.I.C., Facultad de Medicina de la Universidad Autónoma, Madrid, Spain

ROBERT T. SCHIMKE (59, 64), Department of Biological Sciences, Stanford University, Stanford, California

GUNTHER SCHUTZ (65), Max-Planck-Institut für Molekulare Genetik, Berlin-Dahlem, Germany

MANFRED SCHWEIGER (61), Max-Planck-Institut für Molekulare Genetik, Berlin-Dahlem, Germany

SAMARENDA N. SEAL (11), The Institute for Cancer Research, Philadelphia, Pennsylvania

CORINNE C. SHERTON (36, 49), Department of Biochemistry, The University of Chicago, Chicago, Illinois

MELVIN V. SIMPSON (39), Department of Biochemistry, State University of New York, Stony Brook, New York

A. S. SPIRIN (43), A. N. Bakh Institute of Biochemistry, Academy of Sciences of the USSR, Moscow, USSR

WENDELL M. STANLEY, JR. (16, 37), Department of Molecular Biology and Biochemistry, School of Biological Sciences, University of California, Irvine, California

DREW SULLIVAN (59), Department of Biological Sciences, Stanford University, Stanford, California

YOSHIAKI SUZUKI (60), Laboratory of Radiation Research, National Institute of Health, Tokyo, Japan

PAUL S. SYPHERD (35), Department of Medical Microbiology, California College of Medicine, University of California, Irvine, California

W. P. TATE (30), Department of Internal Medicine, Baylor College of Medicine, Houston, Texas

JOHN M. TAYLOR (59), Department of

Biological Sciences, Stanford University, Stanford, California

A. TORAÑO (27), Instituto de Enzimología del C.S.I.C., Facultad de Medicina de la Universidad Autónoma, Madrid, Spain

JOLINDA A. TRAUGH (3), Department of Biological Chemistry, School of Medicine, University of California, Davis, California

ROBERT R. TRAUT (3, 50, 52), Department of Biological Chemistry, School of Medicine, University of California, Davis, California

C. VAZQUEZ (2), Laboratory of Molecular Biology, University of Sherbrooke School of Medicine, Sherbrooke, Quebec, Canada

DAVID VAZQUEZ (29, 41), Instituto de Biología Celular, C.S.I.C., Madrid, Spain

LYDIA VILLA-KOMAROFF (66), Department of Biology, Massachusetts Institute of Technology, Cambridge, Massachusetts

ZVI VOGEL (40), Biochemistry Department, The Weizmann Institute of Science, Rehovot, Israel

A. J. WAHBA (1, 2), Laboratory of Molecular Biology, University of Sherbrooke School of Medicine, Sherbrooke, Quebec, Canada

CHARLINE WALKER (31), Department of Zoology, The University of Iowa, Iowa City, Iowa

DONALD P. WEEKS (11, 68), The Institute for Cancer Research, Philadelphia, Pennsylvania

HERBERT WEISSBACH (22, 24), Department of Biochemistry, Roche Institute of Molecular Biology, Nutley, New Jersey

RICHARD E. H. WETTENHALL (19), Queen Elizabeth Medical Center, Birmingham, England

PRAPON WILAIRAT (67c), Department of Chemistry, University of Oregon, Eugene, Oregon

JOHN W. WIREMAN (35), Department of

Microbiology, University of Minnesota Medical School, Minneapolis,Minnesota

H. G. WITTMANN (48), *Max-Planck-Institut für Molekulare Genetik, Berlin-Dahlem, Germany*

CHARLES L. WOODLEY (15), *Department of Chemistry, University of Nebraska, Lincoln, Nebraska*

WILLIAM R. WOODWARD (67a, 67b, 67c, 67d), *Department of Chemistry, University of Oregon, Eugene, Oregon*

IRA G. WOOL (18, 19, 36, 49, 55), *Department of Biochemistry, University of Chicago, Chicago, Illinois*

ADA ZAMIR (40), *Biochemistry Department, The Weizmann Institute of Science, Rehovot, Israel*

MICHAEL ZASLOFF (20), *Department of Biochemistry, School of Medicine, New York University Medical Center, New York, New York*

Preface

The introduction of two additional volumes dealing with nucleic acids and protein synthesis (Volume XXIX, Part E and Volume XXX, Part F) attests to the remarkable progress that continues to be made in these fields of research.

In Volume XXIX details are provided for the isolation, purification, and properties of both DNA-directed and RNA-directed DNA polymerases derived from a variety of prokaryotic and eukaryotic organisms. The isolation of those ancillary proteins which can influence the rate and extent of some polymerization reactions is also described. In addition, the preparation of cells with ultrapermeability properties allowing for the direct assessment to DNA polymerase activities *in situ* is documented. A variety of currently available techniques for the sequence determinations of DNA and analyses of repeating DNA sequences are described in great detail. One section deals with the preparation, resolution, and characterization of tRNA's, of some derivatives of tRNA, and of some enzymes that use tRNA as a substrate.

This volume deals with the various prokaryotic and eukaryotic systems that can carry out protein synthesis and/or intermediary reactions involved in this process. A portion of the volume is devoted to some new and improved methods for examining the roles of initiation, elongation, and termination factors. A fairly extensive section describes the preparation, physical and biological characterization, and the protein compositional analysis of ribosomes and their corresponding subunits; assays for individual reactions catalyzed by these particles are also included. The isolation of several messenger RNA's and the preparation of a number of biological systems capable of *de novo* synthesis of complete, identifiable proteins *in vitro,* particularly from eukaryotic cells, are also described.

The methods presented in these two new volumes reflect the most recent advances in the methodology with which problems in molecular biology are currently under investigation; we hope that they will be found equally useful.

We thank the many colleagues who have so generously acknowledged the influence of these volumes and the reliability of the methods. The credit belongs to the numerous authors who have contributed so ably. We also wish to acknowledge the valuable assistance and cooperation of the very capable staff of Academic Press.

KIVIE MOLDAVE
LAWRENCE GROSSMAN

METHODS IN ENZYMOLOGY

EDITED BY

Sidney P. Colowick and Nathan O. Kaplan

VANDERBILT UNIVERSITY
SCHOOL OF MEDICINE
NASHVILLE, TENNESSEE

DEPARTMENT OF CHEMISTRY
UNIVERSITY OF CALIFORNIA
AT SAN DIEGO
LA JOLLA, CALIFORNIA

I. Preparation and Assay of Enzymes
II. Preparation and Assay of Enzymes
III. Preparation and Assay of Substrates
IV. Special Techniques for the Enzymologist
V. Preparation and Assay of Enzymes
VI. Preparation and Assay of Enzymes (*Continued*)
Preparation and Assay of Substrates
Special Techniques
VII. Cumulative Subject Index

METHODS IN ENZYMOLOGY

EDITORS-IN-CHIEF

Sidney P. Colowick   Nathan O. Kaplan

Section I

Initiation Factors in Protein Synthesis

[1] Chain Initiation Factors from *Escherichia coli*

By A. J. Wahba *and* M. J. Miller

The first step in protein synthesis in bacteria and eukaryotes is the formation of a chain initiation complex. In bacteria this complex is located at the peptidyl site on the ribosomes and contains fMet-tRNA$_f$ bound to an AUG codon in the messenger RNA molecule. Three protein factors, IF1, IF2, and IF3 are required for the formation of a stable initiation complex and for maximal rates of amino acid incorporation with natural mRNA.[1,2] All three factors are readily isolated from the 1 M NH$_4$Cl ribosomal wash. In the first section of this article, we describe the assay methods for each chain initiation factor. The isolation and purification of the three factors from *E. coli* are presented in the second section.

Materials

Buffer A: 1.0 M NH$_4$Cl; 20 mM Tris·HCl, pH 7.8; 10 mM Mg acetate; 1.0 mM dithiothreitol (DTT)

Buffer B: 0.5 M NH$_4$Cl; 20 mM Tris·HCl, pH 7.8; 10 mM Mg acetate; 1.0 mM DTT; 50% glycerol

Buffer C: 1.0 M NH$_4$Cl; 20 mM Tris·HCl, pH 7.6; 0.2 mM Mg acetate; 1.0 mM DTT; 1 mg of bovine serum albumin per milliliter; 5% glycerol

Buffer D: 1.0 M NH$_4$Cl; 50 mM Tris·HCl, pH 7.2; 5 mM Mg acetate

Buffer E: 60 mM NH$_4$Cl; 50 mM Tris·HCl, pH 7.8; 5 mM Mg acetate

Buffer F: 20 mM Tris·HCl, pH 7.8; 10 mM Mg acetate; 10 mM 2-mercaptoethanol

Buffer G: 1.0 M NH$_4$Cl; 20 mM Tris·HCl, pH 7.8; 10 mM Mg acetate; 10 mM 2-mercaptoethanol

Buffer H: 5 mM phosphate·Tris, pH 7.5; 5% glycerol

Buffer I: 10 mM Tris·HCl, pH 7.4

Buffer J: 0.2 M NH$_4$Cl; 10 mM Tris·HCl, pH 7.4

Buffer K: 30 mM NH$_4$Cl; 20 mM Tris·HCl, pH 7.8; 10 mM Mg acetate, 10 mM 2-mercaptoethanol

[1] A. J. Wahba, Y.-B. Chae, K. Iwasaki, R. Mazumder, M. J. Miller, S. Sabol, and M. A. G. Sillero, *Cold Spring Harbor Symp. Quant. Biol.* 34, 285 (1969).

[2] A. J. Wahba, K. Iwasaki, M. J. Miller, S. Sabol, M. A. G. Sillero, and C. Vasquez, *Cold Spring Harbor Symp. Quant. Biol.* 34, 291 (1969).

Buffer L: 20 mM NH$_4$Cl; 20 mM Tris·HCl, pH 7.6; 0.2 mM Mg acetate; 1.0 mM DTT; 5% glycerol

Buffer M: 6 M urea; 20 mM NH$_4$Cl; 20 mM Tris·HCl, pH 7.6; 0.2 mM Mg acetate; 1.0 mM DTT; 5% glycerol

Buffer N: 1.0 M NH$_4$Cl; 20 mM Tris·HCl, pH 7.6; 0.2 mM Mg acetate; 1.0 mM DTT; 5% glycerol

Buffer O: 0.15 M phosphate·Tris, pH 7.5; 5% glycerol

Buffer P: 0.75 M phosphate·Tris, pH 7.5; 5% glycerol

E. coli Q13 and MRE 600, grown to late log phase[3] and frozen in 500-g lots

E. coli W tRNA; ammonium sulfate (ultrapure); urea (ultrapure), from Schwarz/Mann

U-[^{14}C]methionine (spec. act. 221 Ci/mole); U-[^{14}C]lysine (spec. act. 255 Ci/mole); U-[^{14}C]phenylalanine (spec. act. 384 Ci/mole); Liquifluor, from New England Nuclear Corp.

ApUpG, from Miles Laboratories

GTP, from P. L. Biochemicals, Inc.

Leucovorin (^{5}N-formyltetrahydrofolic acid), from American Cyanamid Co.

Dithiothreitol (Cleland's reagent); creatine phosphate, from Calbiochem

Creatine phosphokinase, from Sigma Chemical Company

Alumina, levigated, from Fisher

DNase I, RNase free, from Worthington Biochemical Corp.

Phosphocellulose (P-11, 7.4 meq/g); DEAE-cellulose (DE-52, 1.0 meq/g); carboxymethyl cellulose (CM-23, 0.6 meq/g); from Whatman

Millipore filters (HA 0.45 μm), from Millipore Corp.

Preparation of Reagents

E. coli W tRNA is charged with [^{14}C]methionine, specific activity 221 Ci/mole, in the presence of ^{5}N-formyltetrahydrofolic acid and a dialyzed E. coli MRE 600 S150 fraction. The RNA of R17 coliphage is prepared by phenol extraction and ethanol precipitation.[4] Stock solutions of 10 M urea are prepared and deionized immediately before use by the procedure of Duesberg and Rueckert.[5] For the preparation of 1 M phosphate·Tris stock solution, 1 M orthophosphoric acid is neutralized at 0° by the addition of solid Tris base to pH 7.5.

Ribosomes and S150 fractions are prepared from freshly harvested

[3] I. Haruna and S. Spiegelman, Proc. Nat. Acad. Sci. U.S. **54**, 579 (1965).
[4] C. Weissmann and G. Feix, Proc. Nat. Acad. Sci. U.S. **55**, 1264 (1966).
[5] P. H. Duesberg and R. R. Rueckert, Anal. Biochem. **11**, 342 (1965).

E. coli Q13 or MRE 600 cells as previously described.[6] Initiation factors are removed by gently stirring the ribosomes for 4 hours at 5° in buffer A. The ribosomes are pelleted at 4° by centrifugation overnight at 40,000 rpm. The pellets are suspended in the same buffer, washed again for 4 hours at 5°, and concentrated by centrifugation at 5° and 65,000 rpm for 2 hours (Spinco, 65 Ti rotor). Ribosomes are suspended in buffer B and stored at −80° at a concentration of 1000 A_{260} units/ml. Preparations of ribosomes stored in this manner are stable for 4 months.

Polyacrylamide Disc Gel Electrophoresis of IF2 and IF3

Analytical disc gel electrophoresis and determination of molecular weights is performed by the method of Weber and Osborn.[7] Samples of protein at a concentration of 0.05–0.1 mg/ml are incubated at 37° for 2 hours in 10 mM sodium phosphate buffer, pH 7.0, containing 1% SDS and 1% 2-mercaptoethanol. The samples are then dialyzed overnight against 10 mM sodium phosphate buffer, pH 7.0, containing 0.1% SDS and 0.1% 2-mercaptoethanol. Electrophoresis is performed using 10% acrylamide gels at 8 mA per gel constant current for 5–6 hours. Protein bands are stained at room temperature for 18 hours in 0.25% Coomassie brilliant blue in 10% acetic acid and 50% methanol, and destained in 7.5% acetic acid. For the determination of the molecular weights of IF2-α and IF2-β, phosphorylase A, catalase, and ovalbumin are employed as markers. Ovalbumin, aldolase, β-lactoglobulin, and lysozyme are used as markers in the determination of the molecular weight of IF3. Densitometer tracing of polyacrylamide gels is carried out on a Gilford 240 spectrophotometer equipped with a model 2410 linear transport.

Chain Initiation Factor Assays

Determination of IF2 Activity

Principle. Initiation factor IF2 stimulates the binding of fMet-tRNA to *E. coli* ribosomes in the presence of IF1 and IF3 with AUG or natural mRNA's as templates.[1,2] The 70 S initiation complex which is formed may be detected either by the Millipore filtration procedure of Nirenberg and Leder,[8] or by sucrose density gradient centrifugation.[2] The former procedure is readily adaptable to the routine assay of large numbers of samples during purification, whereas the latter is primarily used for the detection of IF2 activity when coliphage RNA is the template. Highly purified IF1

[6] K. Iwasaki, S. Sabol, A. J. Wahba, and S. Ochoa, *Arch. Biochem. Biophys.* **125**, 542 (1968).
[7] K. Weber and M. Osborn, *J. Biol. Chem.* **244**, 4406 (1969).
[8] M. W. Nirenberg and P. Leder, *Science* **145**, 1399 (1964).

and IF3 are required for both procedures, especially during the final stages of purification. With purified IF2, a 16-fold stimulation of fMet-tRNA binding is observed by the addition of IF1 and IF3. For the determination of specific activity, the assay must be linearly dependent on the amount of IF2 added (usually in the range of 0.02–0.2 μg of purified IF2).

IF2 activity in column eluates containing 6 M urea can be measured directly by the AUG-dependent binding assay. Although the final concentration of urea in the assay may be as high as 1.2 M, a 10-fold stimulation of fMet-tRNA binding by IF2 is obtained. By this means, the elution of IF2 from DEAE-cellulose or phosphocellulose columns containing urea is readily followed. The specific activity of IF2 is determined after removal of urea from the samples by dialysis.

Procedure A. AUG-Dependent Binding of f[^{14}C]Met-tRNA to Ribosomes

The standard assay contains, in a volume of 50 μl, 0.10 M NH$_4$Cl; 50 mM Tris·HCl buffer, pH 7.2; 5 mM magnesium acetate; 1.0 mM DTT; 0.2 mM GTP; 1 nmole of AUG; 18 pmoles of f[^{14}C]Met-tRNA (221 Ci/mole); 1 A_{260} unit of ribosomes; 2.0 μg of IF1; 0.03–0.3 μg of IF2; and 0.65–1.04 μg of IF3. Samples of IF2 are diluted in buffer C and incubated at 25° for 5 minutes before addition to the assay mixture. Incubation is then carried out for 15 minutes at 25°, and the reaction is terminated by addition of 1 ml of ice cold buffer D. The solution is then filtered through a Millipore filter[8]; the filter is washed three times with 1-ml aliquots of the same buffer and dried. The radioactivity retained by the filter is measured in a Packard Tri-Carb liquid scintillation spectrometer. A unit of IF2 activity is defined as 1 nmole of f[^{14}C]Met-tRNA bound under standard assay conditions. Specific activities are expressed as units of activity per milligram of protein.

Procedure B. R17 RNA-Dependent Binding of f[^{14}C]Met-tRNA to Ribosomes

The standard assay contains, in a volume of 0.125 ml, 60 mM NH$_4$Cl; 50 mM Tris·HCl buffer, pH 7.8; 5 mM magnesium acetate; 1.0 mM DTT; 0.2 mM GTP; 64 pmoles of R17 RNA; 34 pmoles of f[^{14}C]Met-tRNA; 5 A_{260} units of ribosomes; 2.0 μg of IF1; 0.25–0.6 μg of IF2; 2.6 μg of IF3. After incubation for 15 minutes at 37°, 0.1-ml aliquots are layered on 5-ml linear (5 to 20%) sucrose gradients in buffer E. The gradients are centrifuged at 5° for 130 minutes at 38,000 rpm (Spinco, SW 50L rotor). Aliquots (0.16 ml) are then collected dropwise and diluted with 1.0 ml of water. Absorbance of each aliquot is measured at 260 nm, and radioactivity is determined in a Packard Tri-Carb scintillation spectrometer with Bray's solution as scintillation fluid.[9]

Determination of IF1 Activity

The assay for IF1 utilizes the same components as Procedure A for IF2 with the following exceptions. Each reaction mixture contains approximately 1.5–3.0 μg of crude IF2 (Purification of IF2, Procedure B). Under these conditions the amount of fMet-tRNA bound with IF2 alone is quite low and 5- to 10-fold stimulation of fMet-tRNA binding is observed upon addition of IF1. Incubation is carried out for 5 minutes at 25°, and complex formation is detected as described under Procedure A for IF2.

Determination of IF3 Activity

Principle. Initiation factor IF3 exhibits a stimulatory effect on the following reactions: (a) AUG-dependent binding of fMet-tRNA to ribosomes[2,10]; (b) R17 RNA-dependent binding of fMet-tRNA to ribosomes[2,10]; (c) dissociation of 70 S ribosomal particles into 30 S and 50 S subunits[2,11,12]; (d) translation of natural messenger RNA such as R17, MS2, or T4 RNA[2,6,13]; (e) poly(U)-directed polyphenylalanine synthesis.[2,13]

During purification, reaction (e) is used for the routine assay of column fractions. Reaction (d) may be also used for detection of IF3 activity with coliphage RNA as template.

Procedure A. Poly(U)-Dependent Phenylalanine Incorporation

The assay contains, in a volume of 0.125 ml, 0.1 M NH$_4$Cl; 63 mM Tris·HCl, pH 7.8; 18 mM magnesium acetate; 16 mM 2-mercaptoethanol; 1.3 mM ATP; 0.3 mM GTP; 17 mM creatine phosphate; 8 μg of creatine phosphokinase; 0.2–0.3 mg of *E. coli* S150 fraction (dialyzed); 4 μg poly(U); 8 A_{260} units of ribosomes; 0.3–1.3 μg of IF3; 770 μg *E. coli* W tRNA; 0.1 mM [^{14}C]phenylalanine (spec. act. 10 Ci/mole). After incubation for 20 minutes at 37°, the reaction is stopped by the addition of 3 ml of 5% trichloroacetic acid. The samples are then heated for 15 minutes at 90°. Precipitated material is collected on Millipore filters and is washed with approximately 6 ml of 5% trichloroacetic acid. Radioactivity is measured in a Packard Tri-Carb scintillation spectrometer with Liquifluor as solvent. The Mg^{2+} dependence of the assay is illustrated in Fig. 1.

[9] G. A. Bray, *Anal. Biochem.* 1, 279 (1960).
[10] M. J. Miller and A. J. Wahba, *J. Biol. Chem.* 248, 1084 (1973).
[11] A. R. Subramanian and B. D. Davis, *Nature (London)* 228, 1273 (1970).
[12] R. Kaempfer, *Proc. Nat. Acad. Sci. U.S.* 68, 2458 (1971).
[13] N. Schiff, M. J. Miller, and A. J. Wahba, unpublished observations (1971).

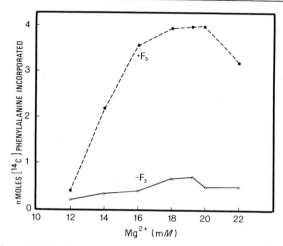

FIG. 1. Effect of IF3 on the translation of poly(U). Phenylalanine incorporation is determined as described in Procedure A (Determination of IF3 Activity). Blank values in the absence of poly(U), but with and without IF3, are subtracted from the values shown.

Procedure B. R17 RNA-Dependent Lysine Incorporation

The assay contains, in a volume of 0.125 ml, 60 mM NH$_4$Cl; 50 mM Tris·HCl buffer, pH 7.8; 12 mM magnesium acetate; 16 mM 2-mercapto-ethanol; 1.3 mM ATP; 0.3 mM GTP; 17 mM creatine phosphate; 8 μg creatine phosphokinase; 125 μg $E.$ $coli$ W tRNA; 80 μg R17 RNA; 5 A_{260} units of ribosomes; 1.5 μg of IF1; 40–80 μg of crude IF2 (Purification of IF2, Procedure B); 1.0–5.0 μg of IF3; 0.3 mg of $E.$ $coli$ S150 fraction (dialyzed); 0.1 mM [^{14}C]lysine (spec. act. 10 Ci/mole); and the remaining (unlabeled) 19 amino acids, each 0.1 mM. After incubation for 20 minutes at 37°, the reaction is stopped and radioactivity incorporated into protein is determined as described in Procedure A for IF3.

Isolation of Chain Initiation Factors

Purification of IF1

Procedure A[1]

Step 1. Preparation of 1.0 M NH$_4$Cl Wash. All operations are performed at 0–4° unless otherwise stated. Two kilograms of frozen $E.$ $coli$ Q13 or MRE 600 cells are thawed and suspended in buffer F. The cells are disrupted in a Manton-Gaulin continuous-flow homogenizer at 8000 psi. DNase I (20 mg) is added, and the suspension is stirred for 10 minutes

and then centrifuged (Sorvall, GSA rotor) at 12,000 rpm for 60 minutes. The supernatant is collected and stored at $-20°$. For each preparation of pure IF1, a total of 1200 ml of S30 extract is thawed and the ribosomes are pelleted by centrifugation for 195 minutes at 50,000 rpm (Spinco, 60 Ti rotor). The ribosomal pellets are suspended in 1.0 liter of buffer G and stirred overnight. The solution is centrifuged again for 195 minutes at 50,000 rpm (Spinco, 60 Ti rotor), and the supernatant is stored at 0–5°. Approximately 15 g of protein is recovered at this step.

Step 2. Ammonium Sulfate Fractionation. The ammonium chloride wash is fractionated by stepwise ammonium sulfate precipitation. The fraction between 0.55 and 0.70 $(NH_4)_2SO_4$ saturation is suspended in a minimal volume of buffer H and dialyzed overnight against this buffer. A total of 4.95 g of protein is obtained at this step.

Step 3. DEAE-Cellulose Chromatography. The above fraction containing approximately 5 g of protein is diluted to a final volume of 300 ml in buffer H and applied to a column (2.2 × 66 cm) of DEAE-cellulose previously equilibrated with buffer H. The column is washed with buffer H at a flow rate of 2.0 ml per minute until the A_{280} of the effluent reaches the base line and all unadsorbed material is eluted. Fractions of 20 ml are collected, and those containing IF1 activity are pooled (see IF1 Assay), yielding 195 mg of protein.

Step 4. Heating of IF1. The solution from step 3 is heated with shaking in a 67° water bath until the temperature of the solution reaches 65°. It is kept at 65° for 5 minutes and then rapidly cooled to 0°. Precipitated proteins are removed by centrifugation at 15,000 rpm for 20 minutes, and the supernatant is used as the source of IF1, yielding 155 mg of protein. This step inactivates about 60% of an enzymatic activity which hydrolyzes the ester linkage of fMet-tRNA. The elution pattern of this contaminating activity from DEAE-cellulose and its heat lability resemble the properties of an enzyme previously described.[14,15]

Step 5. Carboxymethyl Cellulose Chromatography. The total solution from step 4 is applied to a column (1.5 × 36 cm) of carboxymethyl cellulose previously equilibrated with buffer I. The column is washed with this buffer at a flow rate of 1 ml per minute until the A_{280} of the effluent is close to zero. A linear gradient of 0 to 0.35 M NH_4Cl in a total volume of 1500 ml of buffer I is applied. Fractions of 12 ml are collected, and those containing IF1 activity are pooled, yielding 8.05 mg of protein. IF1 is eluted from carboxymethyl cellulose at approximately 0.1 M NH_4Cl.

Step 6. Phosphocellulose Chromatography. The concentration of NH_4Cl

[14] F. Cuzin, N. Kretchmer, R. E. Greenberg, R. Hurwitz, and F. Chapeville, *Proc. Nat. Acad. Sci. U.S.* 58, 2079 (1967).
[15] Z. Vogel, A. Zamir, and D. Elson, *Proc. Nat. Acad. Sci. U.S.* 61, 701 (1968).

in the solution from step 5 is raised to 0.2 M NH$_4$Cl, and the final volume is applied to a column (0.9 × 23 cm) of phosphocellulose equilibrated with buffer J. The column is washed with buffer J at a rate of 0.6 ml per minute until the A_{280} of the effluent reaches the base line. Then a linear gradient of 0.2 to 0.7 M NH$_4$Cl in a total volume of 200 ml in buffer I is applied to the column. Fractions of 3.2 ml are collected and IF1 activity is eluted at approximately 0.35 M NH$_4$Cl. At this step, 5 mg of pure IF1 can be recovered.

Concentration of IF1. Diluted samples of IF1 can be concentrated up to 0.80 mg/ml by stepwise elution from a small phosphocellulose column (about 1 ml bed volume) equilibrated with buffer J. The solution containing IF1 activity, from step 5 or 6, is adjusted to a concentration of 0.2 M NH$_4$Cl and applied to the column with a flow rate of 0.1 ml per minute. After washing with buffer J (about 10 ml), IF1 is eluted with 0.7 M NH$_4$Cl in buffer I. Concentrated solutions of IF1 can be stored at 4° for 3 months with very little loss of activity.

Discussion. A summary of the purification procedure is given in Table I. Starting from step 3, 20-fold purification is obtained with a recovery of 47% of the initial activity. In steps 1 and 2, IF1 activity cannot be determined accurately because of the presence of IF2 and IF3. Step 6 IF1 has an $A_{280}:A_{260}$ ratio of 1.8 and gives one band on SDS-polyacrylamide gel

TABLE I

PURIFICATION OF IF1[a]

Purification step	Total protein[c] (mg)	Total activity (units)[d]	Specific activity (units/mg)	Yield (%)
Step 1. NH$_4$Cl wash of ribosomes[b]	15,360	—	—	—
Step 2. 55–70% (NH$_4$)$_2$SO$_4$ fractionation[b]	4,950	—	—	—
Step 3. DEAE-cellulose chromatography	195	55.5	0.28	100
Step 4. Heating 5 minutes at 65°	155	46.9	0.30	84.5
Step 5. Carboxymethyl cellulose chromatography	8	33.6	4.18	60
Step 6. Phosphocellulose chromatography	5	26.2	5.24	47

[a] From 2 kg of frozen *Escherichia coli* Q13 cells.

[b] IF1 activity cannot be determined accurately in this step due to the presence of IF2 and IF3.

[c] Protein is determined by the method of O. H. Lowry, N. J. Rosebrough, A. L. Farr, and R. J. Randall [*J. Biol. Chem.* **193**, 265 (1951)].

[d] One unit = 1.0 nmole of f[^{14}C]Met-tRNA bound under standard assay conditions (5 minutes, 25°).

electrophoresis.[7] The protein concentration determined by the Lowry method[16] is approximately 3 times higher than that estimated by ultraviolet absorption. The average molecular weight of IF1, as determined by the meniscus depletion method of Yphantis[17] is 9207. For this molecular weight determination, samples of step 6 IF1 are dialyzed overnight against 10 mM KCl, 0.1 M NH$_4$Cl, pH 2.4 (adjusted with HCl). Under these conditions, IF1 can be recovered without appreciable loss of activity.

Procedure B

Step 1. NH$_4$Cl Wash. The 1.0 M NH$_4$Cl ribosomal wash is prepared as in step 1 of Procedure A.

Step 2. Ammonium Sulfate Fractionation. IF1 is isolated from the 35–80% (NH$_4$)$_2$SO$_4$ fraction of the 1.0 M NH$_4$Cl wash. This fraction is dialyzed against buffer H overnight before storage at −80°. All operations are performed at 0–4° unless otherwise stated.

Step 3. DEAE-Cellulose Chromatography. Approximately 900–1200 mg of protein from step 2, at a concentration of 12–15 mg per ml, are applied to a column (2.6 × 100 cm) of DEAE-cellulose, previously equilibrated with buffer H. The column is washed with 1 liter of buffer H and IF1 activity is detected in the void volume. Fractions containing IF1 activity are pooled and heated for 5 minutes at 65°. Precipitated proteins are removed by centrifugation at 15,000 rpm for 20 minutes, and the supernatant is used as the source of IF1.

Discussion. IF1 is very easily isolated in this manner and may be used in the IF2 (Procedures A and B) and IF3 (Procedure B) assays. Preparations of IF1 may be stored at 4° for 4 weeks with 50% loss of activity.

Purification of IF2

Procedure A

Principle. A purification procedure for IF2 will be described utilizing ion exchange chromatography in 6 M urea.[10] The advantages of this method are increased yields of IF2 and the resolution of two distinct IF2 proteins, designated IF2-α and IF2-β.

Step 1. Preparation of 1.0 M NH$_4$Cl Wash. The purification of IF2 is outlined in Fig. 2. All procedures are carried out at 0–5° unless otherwise indicated. Frozen *E. coli* MRE 600 cells (500 g) are ground with 1 kg of alumina and suspended in buffer K. Debris and alumina are removed by centrifugation for 40 minutes at 12,000 rpm (Sorvall, GSA rotor). The

[16] O. H. Lowry, N. J. Rosebrough, A. L. Farr, and R. J. Randall, *J. Biol. Chem.* **193**, 265 (1951).

[17] D. A. Yphantis, *Biochemistry* **3**, 297 (1964).

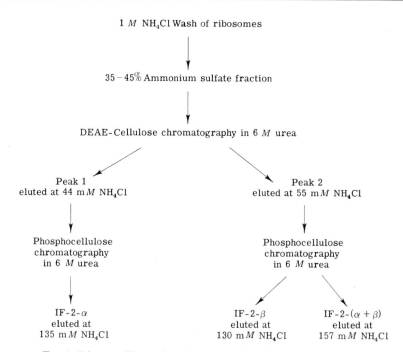

FIG. 2. Diagram illustrating the preparation of IF2-α and IF2-β.

viscous supernatant is treated with DNase I (final concentration 3 μg/ml) for 15 minutes at 5°, and the solution is clarified by centrifugation at 12,000 rpm for 40 minutes (Sorvall, GSA rotor). Ribosomes are prepared by centrifugation of the extract for 120 minutes at 60,000 rpm (Spinco, 60 Ti rotor), and the ribosomal pellets are washed overnight by gentle stirring in buffer G. The ribosomal wash is retained as the source of initiation fractors after removal of the ribosomes by centrifugation (120 minutes, Spinco, 60 Ti rotor).

Step 2. Ammonium Sulfate Fractionation. The supernatant (1 M NH₄Cl ribosomal wash) is precipitated with (NH₄)₂SO₄ to 35% saturation (19.4 g per 100 ml of solution), stirred for 20 minutes and centrifuged for 20 minutes at 12,000 rpm (Sorvall, GSA rotor). The precipitate is discarded and the supernatant brought to 45% saturation with ammonium sulfate (6.4 g per 100 ml) and stirred for 20 minutes. The precipitate is collected by centrifugation, dissolved in buffer L, dialyzed overnight against the same buffer, and then stored at −80°.

Step 3. DEAE-Cellulose Chromatography. The 35–45% (NH₄)₂SO₄ fraction from 1.5 kg of cells (513 mg of protein) is applied to a column (2.5 × 100 cm) of DEAE-cellulose previously equilibrated with buffer L.

The column is first washed with 1 liter of buffer L at a flow rate of 1 ml per minute, then with 1 liter of buffer M. No IF2 activity is eluted at this stage. A linear gradient of 20 to 100 M NH$_4$Cl in a total volume of 2 liters of buffer M is then applied to the column at a flow rate of 1 ml per minute. Ten-milliliter fractions are collected, and every third fraction is assayed for activity. IF2 is found in two peaks. The first DEAE peak (containing IF2-α) eluates at 44 mM NH$_4$Cl and the second DEAE peak (containing IF2-β and an unresolved mixture of IF2-α and β) elutes at 55 mM NH$_4$Cl.

Step 4. Phosphocellulose Chromatography. The fractions of each DEAE peak are combined and diluted 2-fold with buffer M (without NH$_4$Cl) to reduce the final concentration of NH$_4$Cl to approximately 20 mM. Each pool of IF2 is then applied, at a flow rate of 0.25 ml per minute to a column (0.9 × 30 cm) of phosphocellulose previously equilibrated with buffer M. After washing the column with 2 volumes of buffer M, IF2-α (from the first DEAE pool) is eluted from phosphocellulose by a linear gradient of 0.02 to 0.20 M NH$_4$Cl in buffer M (total volume 200 ml). IF2-β and a mixture of IF2-α and β (from the second DEAE pool) are eluted by a linear gradient of 0.02 to 0.30 M NH$_4$Cl in buffer M (total volume 200 ml). Five-milliliter fractions are collected and assayed for activity. IF2-α elutes at 135 mM NH$_4$Cl and IF2-β at 130 mM NH$_4$Cl. The unresolved mixture of IF2-α and β elutes at 157 mM NH$_4$Cl.

Concentration of IF2. Phosphocellulose fractions are diluted 2-fold with buffer M (without NH$_4$Cl) and are applied to a 2-ml column of phosphocellulose previously equilibrated with buffer M. The column is washed with buffer M until the A_{280} of the effluent is close to zero. IF2 is eluted with buffer M containing 0.5 M NH$_4$Cl.

Discussion. A summary of the purification procedure is given in Table II. Phosphocellulose fractions having IF2 activity (step 4 IF2) as well as concentrated IF2 are stored in 50% glycerol after dialysis against buffer N, which contains 1 M NH$_4$Cl. Factor activity is rapidly lost if samples are diluted in a buffer with a concentration of NH$_4$Cl less than 0.25 M. In this purification procedure (NH$_4$)$_2$SO$_4$ precipitation of IF2 is avoided since this invariably leads to loss of activity. An overall yield of 26% is obtained at step 4. Of the total IF2, approximately 75% is present as IF2-α and 25% as IF2-β. Both species can be stored at $-80°$ for 8 months without appreciable loss of activity.

The homogeneity of IF2-α and IF2-β may be determined by densitometer tracings of SDS-polyacrylamide gels. The purity of IF2-α is estimated to be 81.8% and that of IF2-β, 73.4%. Step 4 IF2-α and β have an $A_{280}:A_{260}$ ratio of 1.28 and 1.39, respectively. The protein concentration determined by the Lowry method[16] is from 2 to 3 times higher than that estimated by ultraviolet absorption for both species of IF2. The mo-

TABLE II
PURIFICATION OF IF2[a]

Purification step	Total protein[d] (mg)	Total activity (units)[e]	Specific activity (units/mg)	Yield (%)	Purification (fold)
Step 1. NH₄Cl wash of ribosomes	4546	2182	0.48	100	1
Step 2. 35–45% (NH₄)₂SO₄ fractionation	513	1430	2.80	65.5	5.8
Step 3. DEAE-cellulose chromatography					
Peak 1	22.2	1650	74.3	75.6	154
Peak 2	50.8	564	11.1	25.8	23.1
Step 4. Phosphocellulose chromatography 1[b] Phosphocellulose chromatography 2[c]	5.10	361	70.7	16.5	147
Peak 1	0.66	77.2	117.0	3.54	243
Peak 2	2.16	143.0	65.8	6.56	136

[a] From 1.5 kg of frozen *Escherichia coli* MRE 600 cells.
[b] The active fractions from DEAE peak 1 are applied to this column.
[c] The active fractions from DEAE peak 2 are applied to this column.
[d] Protein is determined by the method of O. H. Lowry, N. J. Rosebrough, A. L. Farr, and R. J. Randall [*J. Biol. Chem.* **193**, 265 (1951)].
[e] One unit = 1.0 nmole of f[¹⁴C]Met-tRNA bound under standard assay conditions (15 minutes, 25°).

lecular weight of IF2-α is 98,000 and that of IF2-β is 83,000 (Fig. 3), as measured by SDS-gel electrophoresis.[7]

In the presence of IF1 and IF3, both IF2-α and IF2-β promote the binding of fMet-tRNA to ribosomes with either AUG, GUG, or R17 RNA as messenger, and the activities of both species are additive.[10] They are, however, inactive in promoting the translation of natural mRNA. A crude preparation of IF2 (Procedure B) stimulates phage RNA-dependent amino acid incorporation in the presence of IF1 and IF3.

Procedure B

Step 1. Ammonium Sulfate Fractionation. The 35–80% (NH₄)₂SO₄ fraction of the 1 M NH₄Cl ribosomal wash is used as a source of crude IF2. This fraction is dialyzed overnight against buffer K and stored at −80°.

Step 2. DEAE-Cellulose Chromatography. Approximately 1 g of protein from step 1 is applied to a column (2.6 × 100 cm) of DEAE-cellulose previously equilibrated with buffer L (containing 10 mM 2-mercaptoethanol as a substitute for DTT). The column is washed with buffer L until

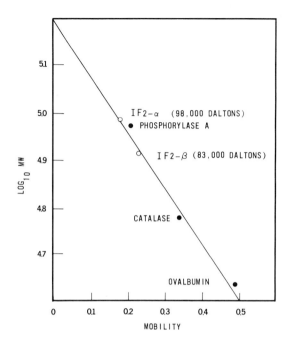

FIG. 3. Molecular weight determination of IF2-α and IF2-β by SDS-polyacrylamide gel electrophoresis. Electrophoresis of IF2 and calculation of molecular weights are performed as described by K. Weber and M. Osborn [*J. Biol. Chem.* **244**, 4406 (1969)].

the A_{280} of the effluent is close to zero. A linear gradient of 20 mM to 0.35 M NH$_4$Cl in a total volume of 2 liters of buffer L is then applied to the column at a flow rate of 1 ml per minute. IF2 activity is recovered in two peaks at approximately 0.2 M NH$_4$Cl, and the combined fractions are concentrated by precipitation with (NH$_4$)$_2$SO$_4$ to 80% saturation. The precipitate is dialyzed overnight against buffer N and stored at $-20°$. Fractions stored in this manner are stable for several months.

Discussion. IF2 may also be isolated from the 35–80% (NH$_4$)$_2$SO$_4$ fraction which is applied to a DEAE-cellulose column previously equilibrated with buffer H (Procedure B, Purification of IF1). A linear gradient of 20 mM to 0.35 mM NH$_4$Cl in buffer H will elute IF2 activity from the column.

Purification of IF3

Step 1. Preparation of 1.0 M NH$_4$Cl Wash. All procedures are carried out at 0–5°. The ribosomal wash is prepared as in step 1, Procedure A (Purification of IF2).

Step 2. Ammonium Sulfate Fractionation. The ribosomal wash is frac-

TABLE III
PURIFICATION OF IF3[a]

Purification step	Total protein[b] (mg)	Poly(U)-dependent Phe incorporation			R17 RNA-dependent Lys incorporation		
		Total activity (units)[c]	Specific activity (units/mg)	Yield (%)	Total activity (units)[d]	Specific activity (units/mg)	Yield (%)
Step 1. 1 M NH₄Cl wash	3005	54,054	17.9	100	5,051	1.68	100
Step 2. 55–75% (NH₄)₂SO₄ fractionation	744	55,843	75	103	3,809	5.12	75
Step 3. DEAE-cellulose chromatography	111	36,218	558	67	4,287	38.5	84
Step 4. Phosphocellulose chromatography 1	7.5	18,133	2418	34	2,231	297.4	44
Step 5. Phosphocellulose chromatography 2	3.5	13,292	3826	25	1,204	344	23

[a] From 900 g of frozen Escherichia coli MRE 600 cells.
[b] Protein is determined by the method of O. H. Lowry, N. J. Rosebrough, A. L. Farr, and R. J. Randall [J. Biol. Chem. 193, 265 (1951)].
[c] One unit = 1.0 nmole of [¹⁴C]phenylalanine incorporated under standard assay conditions (20 minutes, 37°).
[d] One unit = 1.0 nmole of [¹⁴C]lysine incorporated under standard assay conditions (20 minutes, 37°).

tionated with $(NH_4)_2SO_4$. The precipitate between 55 and 75% saturation contains approximately 80% of the total IF3 activity. This fraction is dialyzed against buffer H overnight and stored at $-80°$.

Step 3. DEAE-Cellulose Chromatography. The $(NH_4)_2SO_4$ fraction from step 2 (600–800 mg) is diluted with buffer H to a concentration of 8.5 mg/ml of protein and applied to a column (2.6 × 100 cm) of DEAE-cellulose previously equilibrated with buffer H. The column is then washed with the same buffer until the A_{280} of the effluent is close to zero. IF3 is eluted with buffer O.

Step 4. Phosphocellulose Chromatography. The fractions containing IF3 activity are pooled and applied to a column (1.5 × 30 cm) of phosphocellulose previously equilibrated with buffer O. The column is washed with buffer O until the A_{280} of the effluent is approximately zero and IF3 is eluted with buffer P. The fractions containing IF3 activity are pooled, dialyzed against buffer O and applied to a second column (0.8 × 6 cm) of phosphocellulose previously equilibrated with the same buffer. IF3 is eluted with buffer P, and the fractions containing activity are combined and stored at $4°$.

Discussion. The use of a second phosphocellulose column results in the

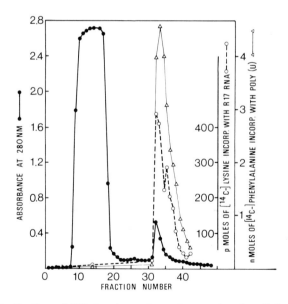

FIG. 4. Purification of IF3 by chromatography on phosphocellulose. Each fraction is assayed by poly(U)-dependent [^{14}C]phenylalanine incorporation (△—△) and R17 RNA-directed [^{14}C]lysine incorporation (○- - -○) into acid-insoluble protein as described in Procedures A and B (Determination of IF3 Activity). Absorbance is measured at 280 nm (●—●).

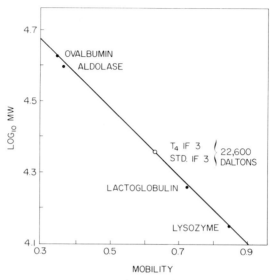

FIG. 5. Molecular weight determination by SDS-polyacrylamide gel electrophoresis of standard IF3 and IF3 from T4-infected *Escherichia coli* MRE 600 cells.

removal of a large amount of inactive protein in the void volume. Throughout the purification procedure (Table III) IF3 activity in column fractions is determined by the poly(U) and R17 assays. The elution pattern of IF3 from phosphocellulose is illustrated in Fig. 4. IF3 activity as detected by amino acid incorporation with poly(U) coincides with that detected by the R17 assay.

The preparation of IF3 obtained after step 5 has an $A_{280}:A_{260}$ ratio of 1.7. A sample of purified IF3 migrates as one major band on SDS-polyacrylamide gel electrophoresis, and the molecular weight is estimated to be approximately 22,600 (Fig. 5). The above purification procedure can also be used for the preparation of IF3 from T4-infected *E. coli* MRE 600 cells.

Samples of step 5 IF3 can be stored at 4° in buffer P (750 m*M* phosphate·Tris, pH 7.5) for several months without appreciable loss of activity. Before using IF3 in an assay, tRNA (1.2 mg/ml) is added, and samples are dialyzed overnight against buffer H. Addition of tRNA is necessary to prevent inactivation of IF3 in solutions with low phosphate concentrations.

[2] Electron Microscopic Studies of Chain Initiation Complexes and Polysomes

By C. VASQUEZ, M. J. MILLER, and A. J. WAHBA

Formation of the chain initiation complex (mRNA·ribosome·fMet-tRNA) which purified *E. coli* ribosomes and dependence of this reaction on the chain initiation factors was studied by centrifugation in sucrose density gradients and by electron microscopy. The three chain initiation factors are required for maximum binding of natural mRNA and fMet-tRNA to the 70 S region. A complex involving mRNA and fMet-tRNA may also be formed on the 30 S ribosomal subunits. Both sucrose density gradient analyses and electron microscopic observations indicate the presence of one primary initiation site on intact phage RNA.[1]

A system low in nuclease activity was utilized for the translation of phage RNA. Electron microscopy of polysomes formed with MS2 RNA reveals the attachment of 30–40 ribosomes to the intact RNA filament.[1]

Reagents

Parlodion, 0.75%, in isoamyl acetate
Formaldehyde solution, 8%, containing 2 mM $MgCl_2$ and 0.2 M
    potassium phosphate, pH 7.0
Tris·HCl buffer, 10 mM, containing 2 mM $MgCl_2$, pH 7.5
Uranyl acetate, 0.1 mM, in acetone
Ethanol

Formation of Chain Initiation Complex

The samples contain, in a volume of 0.125 ml, 60 mM NH_4Cl; 50 mM Tris·HCl buffer, pH 7.8; 5 mM magnesium acetate; 1.0 mM DTT; 0.2 mM GTP; 64 pmoles of R17 RNA (labeled or unlabeled); 34 pmoles of f[^{14}C]Met-tRNA; 5 A_{260} units of purified *E. coli* MRE 600 ribosomes; and purified initiation factors IF1, IF2, and IF3 as specified in Procedure B (Determination of IF2 Activity, this volume [1]). After incubation for 15 minutes at 37°, 0.1-ml aliquots are layered on 5-ml linear (5–20%) sucrose gradients in a solution containing 50 mM Tris·HCl, pH 7.8, 60 mM NH_4Cl, and 6 mM magnesium acetate. The gradients are centrifuged at 3° for 130 minutes (38,000 rpm) in a Spinco SW 39 rotor. Fractions (0.16 ml) are collected dropwise and diluted with 1.0 ml of water. Ab-

[1] A. J. Wahba, K. Iwasaki, M. J. Miller, S. Sabol, M. A. G. Sillero, and C. Vasquez, *Cold Spring Harbor Symp. Quant. Biol.* **34**, 291 (1969).

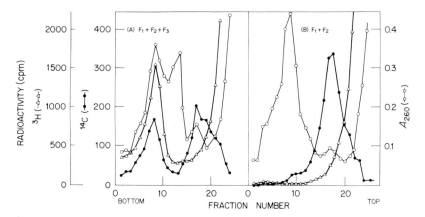

FIG. 1. Formation of chain initiation complex with R17 RNA, *Escherichia coli* MRE 600 ribosomes and fMet-tRNA. The basic composition of the samples and incubation mixtures is as described in the text. (A) With IF1, IF2, and IF3; (B) with IF1 and IF2. A_{260}, —○—○—; ^{14}C-labeled R17 RNA, —●—●—; f[3H]Met-tRNA, —△—△—.

sorbancy is measured at 260 nm and radioactivity determined in a Packard Tri-Carb scintillation spectrometer with Bray's solution as scintillation fluid. Formation of a chain initiation complex is determined by the appearance of radioactivity (from fMet-tRNA or R17 RNA) in the 70 S region (Fig. 1A). Samples taken from the 70 S region are used for electron microscopy.

In Vitro Polysome Formation with Phage RNA

The reaction mixture for the translation of MS2 RNA contains, in a volume of 0.125 ml, 65 mM Tris·HCl, pH 7.8; 70 mM NH$_4$Cl; 14 mM magnesium acetate; 18 mM 2-mercaptoethanol; 1.3 mM ATP; 0.3 mM GTP; 17 mM creatine phosphate; 3.1 μg creatine phosphokinase; 125 μg *E. coli* W tRNA; 1 A_{260} unit MS2 RNA; 40 μg dialyzed *E. coli* Q13 S150 fraction; 180 μg supernatant fraction from *Lactobacillus arabinosus;* 0.1 mM [14C]lysine (spec. act. 10 Ci/mole); 0.1 mM each of the remaining (unlabeled) 19 amino acids; 14 μg IF1, 25 μg IF2; and 7.5 A_{260} units of 0.5 M NH$_4$Cl washed *E. coli* Q13 ribosomes. IF1 and IF2 and 0.5 M NH$_4$Cl-washed ribosomes are prepared as previously described.[2] *L. arabinosus* supernatant is prepared from the 35–60% (NH$_4$)$_2$SO$_4$ fraction of the high speed supernatant fluid (S150).[3] This fraction is dissolved in buffer

[2] M. Salas, M. J. Miller, A. J. Wahba, and S. Ochoa, *Proc. Nat. Acad. Sci. U.S.* **57**, 1865 (1967).

[3] M. Salas, M. A. Smith, W. M. Stanley, Jr., A. J. Wahba, and S. Ochoa, *J. Biol. Chem.* **240**, 3988 (1965).

A (20 mM Tris·HCl, pH 7.8; 10 mM magnesium acetate; and 10 mM 2-mercaptoethanol), and dialyzed overnight at 4° against the same buffer.

After incubation for 20 minutes at 37°, 0.1-ml aliquots of the reaction mixtures are layered on 5 ml linear (5–20%) sucrose gradients in a solution containing 10 mM Tris·HCl, pH 7.8, 6 mM NH$_4$Cl, and 10 mM magnesium acetate. The gradients are centrifuged at 3° for 100–120 minutes (38,000 rpm) in a Spinco SW 39 rotor. The sucrose solution is siphoned out of each tube and 0.1 ml of buffer A is added to dissolve the pellet at the bottom of the tube. The samples are then used for electron microscopy.

Procedure

A clean glass slide is covered with a collodion film of uniform thickness[4] (0.75% Parlodion in isoamyl acetate), and the film is allowed to dry slowly for 45 minutes. By means of an evaporating unit, a very thin layer of carbon (75 Å) is next deposited on the slide. The carbon-collodion film is cut with a razor blade into small squares 2 mm in size, which are then floated on a clean water surface. The squares are mounted from below on top of clean platinum specimen mounts (Siemens type).

Mounting the Specimen

Freshly collected samples from the sucrose gradients are fixed at 4° in equal volumes of 8% formaldehyde solution, containing 2 mM MgCl$_2$ and 0.2 M potassium phosphate, pH 7.0.[5] After 1 hour of fixation, the samples are diluted to a final concentration of 3 μg ribosomal RNA per milliliter in 0.01 M Tris·HCl buffer, pH 7.5, containing 2 mM MgCl$_2$.

Samples of the fixed and diluted material are deposited on grids by the droplet technique and dried by slowly blotting the surface of the grid with filter paper. This process stretches and orients most of the polysomes and initiation complexes.

Staining and Decoration

The preparations are washed for 10 seconds in water, stained for 60 seconds with uranyl acetate in acetone, and then rinsed with ethanol for 5 seconds.[6] During these steps, precautions are taken so that only the surface of the grids touches the solution. After the grids are dried, they are placed in an oven at 180° for 10 minutes in order to remove the collodion. Where no staining is applied, the preparations are first rinsed in water and

[4] D. E. Bradley, *in* "Techniques for Electron Microscopy" (D. H. Kay, ed.), p. 58. Davis, Philadelphia, 1965.
[5] E. Shelton and E. L. Kuff, *J. Mol. Biol.* **22**, 23 (1966).
[6] C. N. Gordon and A. K. Kleinschmidt, *Biochim. Biophys. Acta* **155**, 305 (1968).

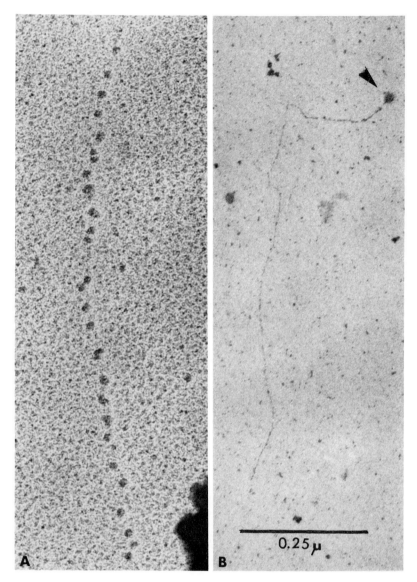

FIG. 2. (A) *In vitro* polysome formation with *Escherichia coli* 30 S subunits and MS2 RNA. Polysomes collected by sucrose density gradients are fixed for 30 minutes in formaldehyde. Uranium shadow. (B) *In vitro* formation of the chain initiation complex with R17 RNA. Samples collected by sucrose density gradients are fixed for 60 minutes in formaldehyde. The arrow points to the 70 S ribosome attached to the end of the RNA strand. Uranyl acetate staining. A, B, ×140,000.

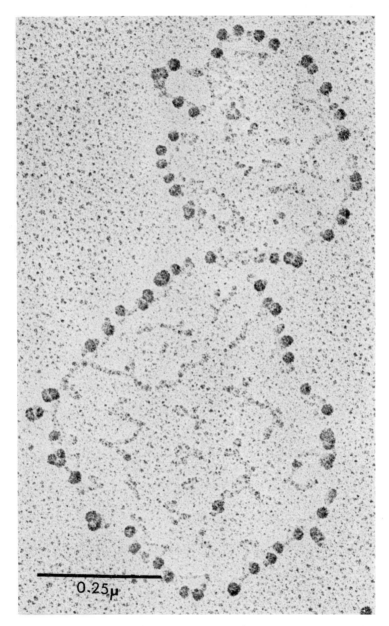

FIG. 3. *In vitro* polysome formation with 70 ribosomes and MS2 RNA. Polysomes collected by sucrose density gradients are fixed for 90 minutes in formaldehyde. Two cross-linked polysomes are observed. Uranium shadow. ×140,000.

then in ethanol. The grids are later rotary shadowed in a vacuum chamber (greater than 10^{-4} mm Hg) with premelted uranium.[7]

Electron Microscopy

Electron micrographs are taken at 20,000 × magnifications or higher, and several grids of each preparation are carefully checked. If a short fixation time is used, only 30 S subunits are observed attached to mRNA (Fig. 2A). The 70 S subunits may be seen only after 60 minutes of fixation (Fig. 2B) when the concentration of ribosomal RNA in the samples is greater than 3 μg/ml. The macromolecules tend to aggregate during the streaking of the grids (Fig. 3). Spreading or diffusion techniques do not yield promising results because dilution of the samples disrupts the complexes.

[7] T. H. Dunnebacke and A. K. Kleinschmidt, *Z. Naturforsch.* B **22**, 159 (1967).

[3] Purification and Phosphorylation of Initiation Factor IF2[1]

By JOHN L. FAKUNDING, JOLINDA A. TRAUGH, ROBERT R. TRAUT, and JOHN W. B. HERSHEY

Studies on the mechanism of action of initiation factors in protein synthesis are greatly facilitated by the availability of pure, radioactively labeled factors. One approach to their procurement is the purification of labeled factors from cells grown with radioactive amino acids or sulfate; [³H]IF1[2] and [³⁵S]IF3[3] have been prepared in this way. A second approach is the large-scale purification of initiation factors followed by chemical or enzymatic modification of the factor with a radioactive chemical group which does not alter its biological activity. Reductive alkylation with radioactive formaldehyde and sodium borohydride has been used to label IF3[4] and

[1] Supported by research grants from the American Cancer Society (NP 70), American Heart Association (69 461), Cancer Research Funds of the University of California, and the United States Public Health Service (GM 17924). R.T. is an Established Investigator of the American Heart Association, and J.T. is a recipient of a U.S.P.H.S. Postdoctoral Fellowship (GM 505590).
[2] J. W. B. Hershey, K. F. Dewey, and R. E. Thach, *Nature (London)* **222**, 944 (1969).
[3] S. Sabol and S. Ochoa, *Nature (London) New Biol.* **234**, 233 (1971).
[4] C. L. Pon, S. M. Friedman, and C. Gualerzi, *Mol. Gen. Genet.* **116**, 192 (1972).

IF1.[5] The observation from this laboratory,[6,7] that IF2 can be phosphorylated by a protein kinase from rabbit skeletal muscle, has provided a convenient way to label pure IF2 with ^{32}P. Phosphorylation has been shown to have no detectable effect on the activity of the factor.[7] The procedures described here are first, a simple, rapid method for the preparation of IF2 from *Escherichia coli* strain MRE600 and then, the preparation of [^{32}P]IF2 with skeletal muscle protein kinase.

Materials

Buffer A: 10 mM Tris·HCl, pH 7.4; 22 mM NH$_4$Cl; 10.5 mM Mg acetate; 0.5 mM EDTA; 1 mM dithiothreitol (DTT)

Buffer B: 10 mM Tris·HCl, pH 7.4; 500 mM NH$_4$Cl; 21 mM Mg acetate; 1 mM EDTA; 1 mM DTT

Buffer C: 10 mM K phosphate, pH 7.5; 7 mM β-mercaptoethanol; 5% glycerol (v/v)

Ribosomes, 70 S, prepared and washed twice with 1 M NH$_4$Cl buffer according to the procedure of Ohta *et al.*[8]

Protein kinase catalytic subunit, a gift of the laboratory of Dr. Edwin Krebs, prepared from rabbit skeletal muscle according to the method of Reimann *et al.*[9]

[H^3]fMet-tRNA, from *E. coli* B tRNA (General Biochemicals, Inc.), charged with [H^3]methionine (Schwarz/Mann, spec. act. 2800 Ci/mole) according to the procedure of Hershey and Thach[10]

[γ-^{32}P]ATP, prepared by a modification of the method of Glynn and Chappell[11] described by Reimann *et al.*[12]

DNase, RNase free, from Worthington Biochemical Corp.

Dithiothreitol (DTT) from Pierce Chemical Company

Yeast extract, from Difco

Glass beads, Superbrite 100, from 3M Company

Toluene scintillation fluid, prepared by mixing 4 g of PPO and 0.05 g of POPOP in 1 liter of toluene

Phosphocellulose, P-11 (7.4 meq/g), from Whatman

DEAE-Sephadex, A-50 (3.5 meq/g), from Pharmacia

[5] K. Johnston and J. W. B. Hershey, unpublished results.
[6] J. A. Traugh and R. R. Traut, *Biochemistry* 11, 2508 (1972).
[7] J. L. Fakunding, J. A. Traugh, R. R. Traut, and J. W. B. Hershey, *J. Biol. Chem.* 247, 6365 (1972).
[8] T. Ohta, S. Sarkar, and R. E. Thach, *Proc. Nat. Acad. Sci. U.S.* 58, 1638 (1967).
[9] E. M. Reimann, C. O. Brostrom, J. D. Corbin, C. A. King, and E. G. Krebs, *Biochem. Biophys. Res. Commun.* 42, 187 (1971).
[10] J. W. B. Hershey and R. E. Thach, *Proc. Nat. Acad. Sci. U.S.* 57, 759 (1967).
[11] I. M. Glynn and J. B. Chappell, *Biochem. J.* 90, 147 (1964).
[12] E. M. Reimann, D. A. Walsh, and E. G. Krebs, *J. Biol. Chem.* 246, 1986 (1971).

Isolation and Purification of IF2

E. coli Growth Conditions. *E. coli* strain MRE600 was grown in a New Brunswick Model F-130 fermenter in 100-liter batches in a buffered, highly enriched medium (per liter of water: 20 g of glucose, 10 g of yeast extract, 34 g KH_2PO_4, 9 g KOH, final pH 6.8).[13] Ten liters of exponentially growing cells were added to the medium, and growth was carried out at 37° with vigorous stirring (200 rpm) and aeration (8 ft^3/min). Foaming was suppressed by the addition of 20 ml of Antifoam A. Growth was stopped in mid-log phase ($A_{550} = 5$ in a Gilford 2400 S spectrophotometer) by the addition of excess ice to the medium. The temperature was reduced to 0° during a period of 25 minutes. The chilled cells were harvested by centrifugation in a Sharples continuous flow centrifuge, washed once by resuspension in 10 mM Tris·HCl, pH 7.4, and 10 mM Mg acetate, and pelleted in liter bottles at 8000 rpm in a Lourdes centrifuge. The cells, 800–900 g wet weight per 100-liter batch, were frozen and stored at −70°.

Isolation of Crude Initiation Factors. Frozen *E. coli* cells (500 g) were combined with glass beads (1250 g) and buffer A (500 ml) in a Waring Blendor (capacity, 1 gal) and homogenized for 30-sec intervals at medium speed. After each interval, powdered dry ice was added to maintain the temperature below 5°. Homogenization was for a total of 20 intervals or 10 minutes, and then 200 μg DNase were added and the cell extract was centrifuged in the Sorvall GSA rotor at 5000 rpm for 5 minutes. The supernatant fractions were combined and saved, and the beads were washed by resuspension in 300 ml of buffer A and again centrifuged. All supernatants were combined and centrifuged in the Sorvall SS34 rotor at 16,000 rpm for 20 minutes. The clear supernatant (total volume, about 600 ml) was carefully withdrawn and divided into 12 portions. Each portion was layered over a 25 ml cushion of 10% sucrose with buffer A in 75-ml polycarbonate tubes, and the contents were centrifuged for 8 hours at 35,000 rpm in Beckman 42 fixed-angle rotors. The supernatant (S100) was decanted and the ribosome pellets were suspended in 300 ml of buffer A. In order to remove the ribosome-bound initiation factors, the suspension was brought to the composition of buffer B by the addition of 1 M Tris·HCl, pH 7.4, 4 M NH₄Cl and 1 M Mg acetate. The ribosome solution, now in high salt, was stirred for 6 hours, layered in 75-ml polycarbonate tubes over 25 ml cushions of 20% sucrose with buffer B, and centrifuged for 10 hours as described above. The upper 80% of the clear supernatant, the ribosomal wash fraction which contains most of the initiation factors, was carefully removed.

[13] P. B. Moore, R. R. Traut, H. Noller, P. Pearson, and H. Delius, *J. Mol. Biol.* 31, 441 (1968).

Ammonium Sulfate Fractionation. The ribosomal wash (approximately 540 ml) was brought to 30% saturation with the slow dropwise addition of saturated (4°) ammonium sulfate during a period of 1 to 2 hours. The suspension was held at 4° for 2 hours, and was clarified by centrifugation at 10,000 rpm for 20 minutes in the Sorvall GSA rotor. The precipitate was discarded, and the clear supernatant was brought to 50% saturation as described above. The second precipitate, which contains the bulk of the IF2, was recovered by centrifugation. It can be stored for a few weeks at −20° with little loss of IF2 activity. The supernatant of the 50% ammonium sulfate fraction contains crude IF1 and IF3.

Phosphocellulose Column Chromatography. The protein precipitates from the 30–50% saturated ammonium sulfate fractions derived from 1 kg of cells were dissolved in 100 ml of buffer C and dialyzed overnight against 100 mM KCl in buffer C. A small amount of precipitate was removed by centrifugation, and the clear protein solution (about 1.2 g) was loaded onto a phosphocellulose column. The column (1.5 × 60 cm; 100 ml bed volume) was packed under 4 feet of hydrostatic pressure and equilibrated with 100 mM KCl in buffer C. Unadsorbed protein was removed by washing with 200 ml of 100 mM KCl in buffer C and the adsorbed protein was eluted with a 2-liter linear salt gradient from 100 to 600 mM KCl in buffer C, at a flow rate of 10 ml per hour; 20-ml fractions were collected. The protein concentration was determined by absorbancy at 280 nm, and salt concentration was determined from measurements of conductivity. IF2 activity was assayed by its stimulation of ApUpG-dependent [^{3}H]fMet-tRNA binding to ribosomes; 3-μl aliquots were employed in the assay, which has already been described in detail in this series.[14] Protein purity in the IF2 region was monitored by sodium dodecyl sulfate (SDS) poly-acrylamide gel electrophoresis.[15]

The protein elution profile, IF2 activity profile, and polyacrylamide gel patterns are shown in Fig. 1. IF2 activities eluted at a salt concentration of 200 to 300 mM KCl as two peaks. Fractions from the first peak contained predominantly a protein of molecular weight 82,000 (IF2b). The second, broader peak contained a mixture of both IF2b and a protein of molecular weight 91,000 (IF2a); the later eluting fractions of the peak contained IF2a free of IF2b. To facilitate the subsequent purification and separation of the two forms of IF2, the activity peaks were subdivided into three fractions: fraction 1 (tubes 26–30) which contained about 20 mg of protein; fraction 2 (tubes 31–36), about 25 mg of protein; and fraction 3 (tubes

[14] J. W. B. Hershey, E. Remold-O'Donnell, D. Kolakofsky, K. F. Dewey, and R. E. Thach, this series, Vol. 20, p. 235.
[15] R. R. Traut, *J. Mol. Biol.* 21, 571 (1966).

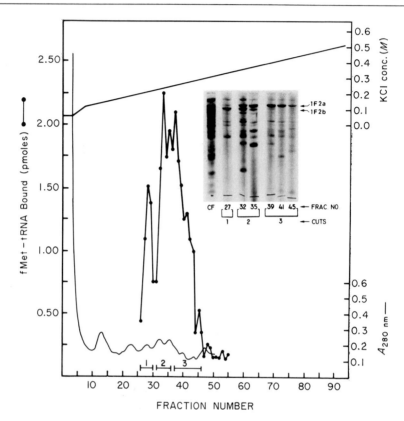

FIG. 1. Chromatography of IF2 on phosphocellulose.

37–46), about 35 mg of protein. Each pooled fraction was further purified by DEAE-Sephadex column chromatography.

DEAE-Sephadex Column Chromatography. DEAE-Sephadex A-50 was prepared according to the method of Moldave[16] and equilibrated with buffer C containing 100 mM KCl. A column (2 × 60 cm; 180 ml bed volume) was poured under a 40-cm hydrostatic pressure head and was washed with the buffer above. The procedure was the same for each of the three pooled fractions from the phosphocellulose column; that for fraction 3 is described here. The IF2 sample (fraction 3, above; 35 mg of protein) was diluted with 400 ml of buffer C to reduce the KCl concentration to 100 mM, and was applied to the column. The adsorbed protein was eluted with a 2-liter linear salt gradient, 100 to 400 mM KCl in buffer C, at about 10 ml per hour. Fractions of 20 ml were collected and analyzed as described above for the phosphocellulose eluate. The protein and IF2 activity profiles are shown in Fig. 2. IF2a elutes sharply at approximately 300 mM

[16] K. Moldave, W. Galasinski, and P. Rao, this series, Vol. 20, p. 337 (1971).

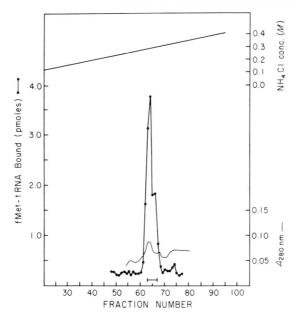

FIG. 2. Chromatography of IF2 on DEAE-cellulose.

KCl (fractions 63–67) as a homogeneous protein (yield, 15 mg). When fraction 1 from the phosphocellulose column was similarly treated, IF2b eluted at about 320 mM KCl and was nearly homogeneous (yield, 8 mg). The purity of the IF2 preparations thus obtained was determined by SDS gel electrophoresis (Fig. 3). The yields of IF2a and IF2b are quite high. Loss of activity during the various preparative procedures seems to be due mainly to adsorption of IF2 protein to glass.[14] Treatment of all glassware with silicone (e.g., Sigmacote) inhibits this adsorption and thus improves the yield. Dilute solutions are avoided when possible, and all procedures (except column elution) are carried out as rapidly as possible.

Concentration of IF2 Protein Solutions. Dilute solutions of IF2 were concentrated by adsorption and elution from a small phosphocellulose column. Solutions containing up to 20 mg protein were first diluted with buffer C to reduce the KCl concentration to 100 mM, and then applied to a tightly packed phosphocellulose column (0.9 × 8 cm; bed volume 5 ml). The adsorbed protein was eluted slowly with 500 mM KCl in buffer C, and 1-ml fractions were collected. Most of the protein eluted in 2 or 3 fractions.

Phosphorylation of IF2 with Skeletal Muscle Kinase

Conditions for Phosphorylation. Preliminary assays for optimal ratios of protein kinase and IF2 were routinely carried out prior to the prepara-

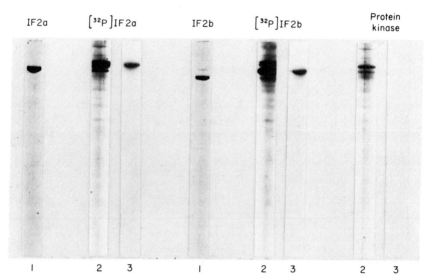

FIG. 3. Polyacrylamide gel electrophoresis of IF2 preparations; 10% SDS poly-acrylamide gel electrophoresis, staining with Coomassie blue, and autoradiography were carried out as described by J. L. Fakunding, J. A. Traugh, R. R. Traut, and J. W. B. Hershey [*J. Biol. Chem.* **247**, 6365 (1972)]. (1) Stained purified IF2; (2) stained analysis of the total reaction mixtures described for the phosphorylation of IF2; (3) autoradiography of 2. Control gels, showing the protein kinase preparation alone, are also shown.

tion of larger amounts of [^{32}P]IF2. Each assay mixture contained in 70 μl: 15 mM K phosphate, pH 7.0; 2 mM Mg acetate; 0.2 mM [γ-^{32}P]ATP (spec. act. 200–2000 Ci/mole); 10 μg of pure IF2a or IF2b; and skeletal muscle kinase in amounts ranging from 0.2 to 5 μg protein, depending on the specific activity of the preparation. The reaction was initiated by the addition of the protein kinase and was incubated for 25 minutes at 30°. Aliquots of 50 μl were pipetted onto 2-cm squares of Whatman ET 31 filter paper and immediately placed in cold 10% trichloroacetic acid (TCA) for 15 minutes with slow stirring. The samples were then washed three times in cold 5% TCA for 15 minutes, 5 minutes in 95% ethanol, and 5 minutes in diethyl ether. The squares were dried and the radioactivity counted in toluene scintillation fluid.

Preparative Phosphorylation of IF2a. The optimal conditions for phosphorylation were determined, then pure IF2a (300 μg) was incubated under the conditions described above. After 25 minutes at 30°, the reaction mixture was diluted with 1 volume of cold water and was applied to a small phosphocellulose column (0.4 × 3 cm; bed volume 0.5 ml) equilibrated with 50 mM KCl in buffer C. The column was washed with

10 ml of the same buffer, and the [^{32}P]IF2a was eluted with 500 mM KCl in buffer C. Fractions of 0.25 ml were collected and the radioactive protein was localized by counting 5-μl aliquots. The fractions containing [^{32}P]IF2a were combined and stored at $-70°$. Protein concentration was determined by the method of Lowry et al.[17] The radioactive phosphate was incorporated only into protein corresponding to IF2, as shown by autoradiography of SDS polyacrylamide gels (Fig. 3). Under optimal conditions nearly 1 mole of phosphate was incorporated per mole of IF2a or IF2b protein. The phosphate has been shown to be attached to a serine residue in both forms.[7]

Summary

Methods are described for the purification to homogeneity, as judged by SDS gel electrophoresis, of two forms of IF2: IF2a with a molecular weight of 91,000 and IF2b with a molecular weight of 82,000. Multiple forms of IF2 have been observed previously,[18-21] but the purification of both forms to near homogeneity has been reported only recently.[22,23] The two forms are equally active, are substrates for protein kinase, and retain activity following phosphorylation. A procedure is given for the *in vitro* labeling of IF2 with [^{32}P]phosphate. The [^{32}P]IF2 has been used to study IF2 binding to 30 S and 70 S ribosomes and its subsequent release during the process of initiation of protein synthesis.[24]

[17] O. H. Lowry, N. J. Rosebrough, A. L. Farr, and R. J. Randall, *J. Biol. Chem.* **193**, 265 (1951).
[18] M. Herzberg, J. C. Lelong, and M. Revel, *J. Mol. Biol.* **44**, 297 (1969).
[19] D. Kolakofsky, K. Dewey, and R. E. Thach, *Nature (London)* **223**, 694 (1969).
[20] J. C. Lelong, M. Grunberg-Manago, J. Dondon, D. Gros, and F. Gros, *Nature (London)* **226**, 505 (1970).
[21] P. S. Rudland, W. A. Whybrow, and B. F. C. Clark, *Nature (London) New Biol.* **231**, 76 (1971).
[22] J. L. Fakunding, J. A. Traugh, R. R. Traut, and J. W. B. Hershey, *Fed. Proc., Fed. Amer. Soc. Exp. Biol.* **31**, 410 (1972).
[23] M. Miller and A. Wahba, *J. Biol. Chem.* **248**, 1084 (1973).
[24] J. Fakunding and J. W. B. Hershey, *J. Biol. Chem.* **248**, 4206 (1973).

[4] Preparation and Properties of Crystalline Initiation Factor 1 (IF1) from *Escherichia coli*

By SYLVIA LEE-HUANG and SEVERO OCHOA

Two protein factors, IF1 and IF2, are implicated in formation of an initiation complex in *E. coli* systems with the trinucleoside diphosphate

ApUpG (AUG) as messenger.[1,2] With natural messengers, such as coli-phage RNA (e.g., MS2 or Q_β RNA), there is an additional requirement for a third initiation factor, IF3.[3] These factors are loosely associated to the ribosomes from which they can be extracted with salt solutions, e.g., 1.0 M NH_4Cl.

IF2 promotes the binding of the initiator, fMet-tRNA$_f$, to the 30 S ribosomal subunits to the exclusion of all other aminoacyl-tRNA's. IF2 is the initiation counterpart of the chain elongation factor EFT which pro-motes the ribosomal binding of all aminoacyl-tRNA's other than the initiator.

The function of IF1 is less clear. Whereas this protein is inactive by itself, IF2 has considerable activity, in the absence of IF1, when the initia-tion complex is formed at 0°. However, at 25° the amount of complex ob-tained with IF2 alone is small and is increased severalfold when IF1 is also present. These observations suggested that IF1 serves to stabilize the initia-tion complex[4] (see also Mazumder[5]) a function consistent with the notion that this protein, rather than an initiation factor, is an easily dissociable protein which forms part of the 30 S initiation site.[4] The remarkable sim-ilarity of amino acid composition of IF1 and the majority of the ribosomal proteins,[6] and recent studies with ribosomes and initiation factors from the bacterium *Caulobacter crescentus*,[7] support this view.

Assay

Principle. The assay of IF1 is based on the observation that in the presence of IF2, at low Mg^{2+} concentrations, IF1 markedly stimulates the binding of fMet-tRNA$_f$ to 1.0 M NH_4Cl-washed ribosomes at 25° with the trinucleoside diphosphate ApUpG (AUG) as messenger. In the absence of IF2, IF1 is totally inactive.[8] The ribosomal binding of f[^{14}C]Met-tRNA$_f$ is measured by the Millipore filter assay of Nirenberg and Leder.[9]

[1] M. Salas, M. B. Hille, J. A. Last, A. J. Wahba, and S. Ochoa, *Proc. Nat. Acad. Sci. U.S.* **57**, 387 (1967).
[2] J. S. Anderson, M. S. Bretscher, B. F. C. Clark, and K. A. Marcker, *Nature (London)* **215**, 490 (1967).
[3] S. Lee-Huang and S. Ochoa, this volume [6]; S. Sabol and S. Ochoa, this volume [5].
[4] Y.-B. Chae, R. Mazumder, and S. Ochoa, *Proc. Nat. Acad. Sci. U.S.* **63**, 828 (1969).
[5] R. Mazumder, *FEBS (Fed. Eur. Biochem. Soc.) Lett.* **18**, 64 (1971).
[6] S. Lee-Huang, M. A. G. Sillero, and S. Ochoa, *Eur. J. Biochem.* **18**, 536 (1971).
[7] S. Leffler and W. Szer, manuscript in preparation.
[8] Y.-B. Chae, R. Mazumder, and S. Ochoa, *Proc. Nat. Acad. Sci. U.S.* **62**, 1181 (1969).
[9] M. W. Nirenberg and P. Leder, *Science* **145**, 1399 (1964).

Reagents. The reaction mixtures, in a final volume of 0.05 ml, have the following composition:

Tris·HCl buffer, pH 7.2, 50 mM
NH$_4$Cl, 180 mM
Magnesium acetate, 3 mM
Dithiothreitol (DTT), 1 mM
GTP, disodium salt (P-L Biochemicals), 0.2 mM
AUG (Miles Laboratories), 0.025 A_{260} unit
f[^{14}C]Met-tRNA, 22 pmoles ($\sim$13,000 cpm)
E. coli Q13 unfractionated ribosomes (1.0 M NH$_4$Cl-washed), 2 A_{260} units[10]
Purified IF2 (step 5 or 6 of Chae *et al.*[8]), 2–3 μg
IF1, not over 0.007 unit

Procedure. The above components are mixed at 0° in the order listed and the samples are incubated for 5 minutes at 25°. They are then chilled, followed by the addition of 2 ml of a buffer containing 50 mM Tris·HCl, pH 7.2, 150 mM NH$_4$Cl, 3 mM magnesium acetate, and the bound radioactivity is measured by the Millipore procedure. Under the conditions of the assay the net amount of fMet-tRNA$_f$ bound is proportional to the concentration of IF1 up to about 0.5 μg of pure factor/0.05 ml. Specific activity is expressed as units per milligram of protein. One unit is defined as the amount of IF1 which stimulates binding to the extent of 1 nmole of fMet-tRNA$_f$ in 5 minutes at 25°. Protein is determined by the method of Lowry *et al.*[11]

Preparation of Assay Components

The preparation of 1.0 M NH$_4$Cl-washed ribosomes is described elsewhere in this volume.[10]

Crude f[^{14}C]Met-tRNA$_f$. The methionine-accepting species of tRNA (tRNA$_f^{Met}$ and tRNA$_M^{Met}$) present in preparations of total *E. coli* tRNA are charged with [^{14}C]methionine and the [^{14}C]Met-tRNA$_f$ species is formylated using *E. coli* high speed supernatant as a source of Met-tRNA synthetase and transformylase, and N^5-formyltetrahydrofolic acid (Leucovorin) as the formyl donor.

Reagents. The reaction mixture, in a total volume of 5 ml, has the following composition:

Tris·HCl buffer, pH 7.8, 58 mM
NH$_4$Cl, 30 mM

---

[10] For preparation, see this volume [6].
[11] O. H. Lowry, N. J. Rosebrough, A. L. Farr, and R. J. Randall, *J. Biol. Chem.* 193, 265 (1951).

Magnesium acetate, 15 mM
β-Mercaptoethanol, 16 mM
ATP, disodium salt (Sigma), 1.3 mM
Phosphocreatine (Sigma), 17 mM
Creatine kinase (Worthington), 320 μg
E. coli W tRNA (Schwarz/Mann), 14 mg
[14C]Methionine (New England Nuclear), 200–233 mCi/mmole, 110–215 nmoles
Leucovorin (American Cyanamid Company), 300 μg
E. coli high speed supernatant[10]

Procedure. The above components are mixed at 0° in the order listed and the mixture (5 ml) is incubated for 20 minutes at 37° followed by chilling. After addition of 0.5 ml of 2.0 M sodium acetate buffer, pH 5.3, and 5.5 ml of freshly distilled 80% phenol (equilibrated with the pH 5.3 acetate buffer), the phases are mixed by bubbling nitrogen and separated by low speed centrifugation at room temperature. The phenol phase is reextracted with 6 ml of 0.2 M acetate buffer, pH 5.3, and residual phenol in the pooled aqueous phases is removed by ether extraction (six times, each with 2 volumes of ether cooled to −20°). The RNA is precipitated by adding 3 volumes of ethanol at −20° and the precipitate recovered by centrifugation after standing at −20° for at least 3 hours. The precipitated RNA is redissolved in 1 ml of the 0.2 M acetate buffer, reprecipitated, and collected as above. The pellet is dried under a stream of nitrogen, dissolved in 0.2 M acetate buffer, pH 5.3, and dialyzed for 4 hours against the same buffer to remove any traces of [14C]methionine. Sephadex G-25 filtration may be alternatively used for this purpose for all of the RNA is eluted in the void volume while contaminants of low molecular weight are retarded.

Purification Procedure

IF1 is extracted, along with IF2, IF3, and other proteins, in the 1.0 M NH$_4$Cl ribosomal wash from which it is purified by ammonium sulfate fractionation, DEAE-cellulose chromatography, and chromatography on cation exchanges. Whereas IF2 precipitates between 0.30 and 0.45, IF1 precipitates along with IF3 between 0.55 and 0.70 ammonium sulfate saturation. Contrary to IF3, IF1 is not retained by DEAE-cellulose. All operations are carried out at 0–4° unless otherwise stated.

Materials

DNase (Worthington), RNase-free
DEAE-cellulose (Schleicher and Schuell), 0.97 meq/g

Carboxymethyl cellulose (Whatman CM 23), 0.6 meq/g
Cellulose phosphate (Whatman P-11), 7.4 meq/g
Buffer A: 20 mM Tris·HCl, pH 7.8, 10 mM magnesium acetate, 1 mM DTT
Buffer B: Buffer A containing 1.0 M NH$_i$Cl
Buffer C: 10 mM Tris·HCl, pH 7.4
Buffer D: Buffer C containing 0.1 M NH$_i$Cl

The pH of buffers was adjusted at 25°.

Growth of Cells. E. coli Q13 or MRE600 is grown at 37° in a vat fermenter on 180 liters of medium containing 0.01 M potassium phosphate buffer, pH 7.3, 5 mM MgSO$_4$, 0.5% glucose, 0.8% casamino acids, 0.05% yeast extract. The cells are harvested under refrigeration at mid-log phase and kept frozen until used.

Step 1. Ribosomal Wash. The frozen cells are thawed overnight in the cold room and suspended in one volume of buffer A. DNase is added to the suspension in the proportion of 1 μg/ml and the mixture is stirred gently for 15 minutes to a homogeneous paste. The cells are disrupted twice in a French decompression press at 8000 psi and the suspension is centrifuged for 1 hour at 9000 rpm (13,200 g) in a Sorvall centrifuge to remove cell debris. The supernatant is centrifuged in rotor No. 21 of the Spinco Model L centrifuge for 1 hour at 17,000 rpm (30,000 g) and the resulting supernatant (S20) is centrifuged once more in the Ti 60 rotor for 4.5 hours at 50,000 rpm. The ribosomal pellet thus obtained is washed with buffer B using 400–500 ml per kilogram of cells. The suspension is stirred gently for 20–24 hours, and the supernatant is recovered by centrifugation (Ti 60 rotor, Spinco Model L4 centrifuge) for 4 hours at 50,000 rpm. The ratio of absorbance at 280:260 nm of the supernatant is about 0.6.

Step 2. Ammonium Sulfate Fractionation. To the ribosomal wash is added 19.4 g of solid ammonium sulfate per 100 ml (0.3 saturation). After stirring for 15 minutes at 0° and further standing for 10 minutes the precipitate is removed by centrifugation and discarded. The supernatant is brought to 0.45 saturation with ammonium sulfate by the addition of 5.7 g/100 ml. The precipitate between 0.3 and 0.45 ammonium sulfate saturation, containing the bulk of the IF2 and the interference factors iα and iβ[12,13] is saved and used for purification of these factors. The supernatant is brought to 0.55 saturation by the addition of 5.9 g of ammonium sulfate per 100 ml. The precipitate is discarded, and the

[12] Y. Groner, Y. Pollard, H. Berissi, and M. Revel, *Nature* (*London*) *New Biol.* **239**, 16 (1972).
[13] S. Lee-Huang and S. Ochoa, *Biochem. Biophys. Res. Commun.* **49**, 371 (1972).

supernatant is made 0.7 saturated with ammonium sulfate by further addition of 9.3 g of ammonium sulfate per 100 ml. The precipitate, which contains the bulk of the IF1 and IF3, is dissolved in buffer C (10 ml) and dialyzed for 6–8 hours against 2 liters of the same buffer, which is changed three or four times during this period.

Step 3. DEAE-Cellulose Chromatography. The dialyzed solution from step 2 is applied to a column (2.5 × 76 cm) of DEAE-cellulose equilibrated with buffer C and the column washed with this buffer, at the rate of 3 ml/min, until the absorbance at 280 nm of the effluent is negligible. IF1 is not retained by DEAE-cellulose. Fractions of 25 ml are collected, and fractions 16 to 25, containing IF1 activity, are pooled. This step largely separates IF1 from IF3.

Step 4. Heating. The solution from step 3 (about 250 ml) is placed in a water bath at 67° with constant swirling; the temperature of the solution rises to 65° within 3 minutes. It is kept at 65° for 5 minutes, during which time a small amount of precipitate appears. After quickly cooling to 5° in an ice bath, the mixture is filtered through Whatman No. 1 filter paper and the precipitate on the filter is washed with a small amount (about 20 ml) of buffer C. The filtrate and the wash are combined. This step leads to a small loss of IF1 with little or no change in specific activity. It is used to remove contaminating N-substituted aminoacyl-tRNA hydrolase.[14]

Step 5. Carboxymethyl Cellulose Chromatography. The solution from step 4 is applied to a column (0.9 × 50 cm) of carboxymethyl cellulose previously equilibrated with buffer C. The column is washed with this buffer at a flow rate of 1 ml/min until the absorbance at 280 nm of the effluent is nil. It is then eluted with 300 ml of a linear gradient of NH_4Cl (0 to 0.35 M) in buffer C. Fractions (2.5 ml) are collected. IF1 activity usually begins to emerge from the column at about 0.13–0.14 M NH_4Cl. Fractions 50–60 encompassing a peak of IF1 activity are pooled.

Step 6. Cellulose Phosphate Chromatography. The solution from step 5 is diluted with buffer C, to give a final concentration of NH_4Cl of 0.1 M. This solution is applied to a column (0.35 × 20 cm) of cellulose phosphate equilibrated with buffer D at a flow rate of 6 ml/hr until the absorbance at 280 nm of the effluent is negligible. IF1 is then eluted with 50 ml of a linear gradient of NH_4Cl (0.1 to 0.5 M) at a flow rate of 0.1 ml/min. Fractions of 1 ml are collected. IF1 is eluted between fractions 50 and 60 (0.24 to 0.3 M NH_4Cl). As seen in Fig. 1, IF1 activity coincides with a sharply symmetrical protein peak. IF1 can also be eluted by stepwise increase of the NH_4Cl concentration (0.15, 0.2,

[14] Z. Vogel, A. Zamir, and D. Elson, *Proc. Nat. Acad. Sci. U.S.* **61**, 701 (1968).

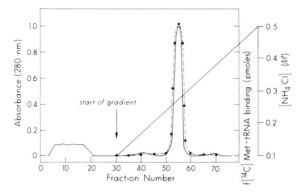

FIG. 1. Chromatography of IF1 on phosphocellulose. Step 6 of purification. It corresponds to the preparation summarized in Table I. ——, absorbance at 280 nm; ●- - -●, stimulation of f[^{14}C]Met-tRNA binding.

0.25, 0.3, and 0.35 M NH$_4$Cl) and the bulk of IF1 activity appears in the 0.25 M NH$_4$Cl fractions as a homogeneous peak. The solution is concentrated by partial lyophilization to a protein concentration of 2–3 mg/ml and the concentrated solution dialyzed against distilled water (500 ml) with four to five changes for a total of not over 3 hours) and lyophilized to dryness. The lyophilized material, stored at −20°, is stable for periods up to 5 months. Step 6 IF1 can also be stored in solution in the presence of 0.25 to 0.3 M NH$_4$Cl at refrigerator temperatures.

The absorbance ratio at 280:260 nm of step 6 IF1 is about 1.8 indicating freedom from nucleic acid. A summary of a typical purification run is given in Table I. The assay of IF1 is not reliable at steps 1 and 2 owing to the presence of IF2. For this reason, no activity data

TABLE I

PURIFICATION OF IF1[a]

Step	Volume (ml)	Protein (mg)	Activity (units)	Specific activity (units/mg protein)
1. 1.0 M NH$_4$Cl wash of ribosomes	745	8563	—	—
2. (NH$_4$)$_2$SO$_4$ fractionation	30	2660	—	—
3. DEAE-cellulose chromatography	250	151	94	0.62
4. Heating	230	122	81	0.66
5. Carboxymethyl cellulose chromatography	27	11	66	6.0
6. Phosphocellulose chromatography	10	4.2	59	14.0

[a] From 2.8 kg of *Escherichia coli* Q13 cells.

TABLE II

SPECIFIC ACTIVITY OF N-SUBSTITUTED AMINOACYL-tRNA HYDROLASE
AT VARIOUS STEPS OF IF1 PURIFICATION

IF1 purification step	Hydrolase specific activity[a]
1	0.05
2	0.09
3	0.98
4	0.14
5	0.12
6	0

[a] Specific activity in nanomoles of N-acetylphenylalanine-tRNA hydrolyzed (15 minutes at 25°) per milligram of protein.

have been entered in the corresponding columns of Table I. The IF1 specific activity at step 6 varied from 14 to 15.2 in six preparations. The yield calculated from step 3 taken as 100, varied from 47 to 63%. From Table II it may be seen that N-substituted aminoacyl-tRNA hydrolase (assayed with N-acetyl[^{14}C]Phe-tRNA as substitute) is actually purified along with IF1 up to step 3 and sharply decreases thereafter. Step 6 IF1 is free of this contaminant.

Crystallization. Lyophilized step 6 IF1 is dissolved in buffer D, at a concentration of 2–3 mg/ml. The solution is dialyzed at 4° successively against 10 mM, 6 mM, and 3 mM Tris·HCl, pH 7.2, 7.4 and 7.7, respectively. During the last dialysis the solution becomes slightly turbid. It is transferred to a test tube and allowed to stand at 4° for one or more weeks. Once a crop of crystals is obtained, seeding leads to crystallization in 1–3 days. The crystals are small and irregular in shape[6] and seem to have a hexagonal prismatic or pyramidal structure. The yield of crystalline material is about 65% and its specific activity is identical to that of step 6 IF1.

Properties

The molecular weight of IF1, determined by three independent methods (sedimentation equilibrium, dodecyl sulfate gel electrophoresis, and gel filtration on Sephadex G-50), is 9400 ± 100 in agreement with the value of 9000 obtained by gel filtration by Hershey et al.[15] The small size of IF1 may be responsible for the losses one observes upon dialysis of very dilute solutions of the factor, a fact also noticed by Hershey et al.[15] However, solutions containing 2–3 mg of IF1 per milliliter can be

[15] J. W. B. Hershey, K. F. Dewey, and R. E. Thach, *Nature* (*London*) **222**, 944 (1969).

dialyzed with little or no loss. IF1 molecules aggregate at high protein concentration.

IF1 contains all twenty of the common amino acids, with only one residue each of histidine, proline, cysteine, methionine, and tryptophan, and has alanine and lysine, respectively, at the amino- and carboxy-terminal ends. The amino acid composition of IF1 is remarkably similar to that of the majority of E. coli ribosomal proteins.[16]

[16] E. Kaltschmidt, M. Dzionara, D. Donner, and H. G. Wittmann, Mol. Gen. Genet. 100, 364 (1967).

[5] Preparation of Radioactive Initiation Factor 3

By STEVEN SABOL and SEVERO OCHOA

The polypeptide chain initiation factor 3 (IF3)[1] of Escherichia coli is a basic protein of molecular weight approximately 21,000 which is isolated, along with IF1 and IF2, from the ribosomes by a high-salt wash.[2-4] IF3 is required for formation of initiation complexes with natural messenger RNA molecules such as bacteriophage mRNA,[2,3] and it promotes the dissociation of ribosomal subunits under appropriate conditions.[5,6]

Labeled IF3 of high specific radioactivity, purity, and functional activity is useful in studies of its mechanism of action. Sulfur-labeled IF3 appropriate for such studies can be purified from cells grown in [35S] sulfate.[7,8] This approach is advantageous because it involves no chemical modification which might alter the functional properties of the factor. The preparation of [3H]IF1 has been described in this series by Hershey et al.[9]

[1] By agreement at a conference on protein biosynthesis held at the Fogarty International Center, National Institutes of Health, Bethesda, Maryland, in November 1971 the terms IF1, IF2, and IF3 replace the following terms, respectively: F_1, F_2, and F_3 (ref. 2), A, C, and B (ref. 3), and FI, FIII, and FII (ref. 4).

[2] K. Iwasaki, S. Sabol, A. Wahba, and S. Ochoa, Arch. Biochem. Biophys. 125, 542 (1968).

[3] M. Revel, H. Greenshpan, and M. Herzberg, this series, Vol. 20, p. 261.

[4] J. Dubnoff and U. Maitra, this series, Vol. 20, p. 248.

[5] S. Sabol, M. A. G. Sillero, K. Iwasaki, and S. Ochoa, Nature (London) 228, 1269 (1970).

[6] A. Subramanian and B. D. Davis, Nature (London) 228, 1273 (1970).

[7] S. Sabol and S. Ochoa, Nature (London) New Biol. 234, 233 (1971).

[8] S. Sabol, D. Meier, and S. Ochoa, Eur. J. Biochem. 33, 332 (1973).

[9] J. W. B. Hershey, E. Remold-O'Donnell, D. Kolakofsky, K. F. Dewey, and R. E. Thach, this series, Vol. 20, p. 235.

Other methods tried for labeling IF3 were enzymatic iodination with radioactive iodine and phosphorylation with γ-labeled [^{32}P]ATP in the presence of protein kinase. However, IF3 was not phosphorylated, and iodination inactivated the factor. Active ^{14}CH$_3$-IF3 has been prepared by reductive alkylation.[10]

Assay Method

Principle. A highly reproducible assay for IF3 is based on its stimulation of RNA phage RNA-dependent amino acid polymerization in the presence of initiation factor-free ribosomes, IF1, and IF2.[5]

Reagents. The reaction mixture, total volume 0.125 ml, has the following composition:

Tris·HCl, pH 7.4, 60 mM; NH$_4$Cl, 70 mM; magnesium acetate, 12 mM

2-Mercaptoethanol, 16 mM

ATP, disodium salt, 1.3 mM (P-L Labs)

GTP, disodium salt, 0.3 mM (P-L Labs)

Creatine phosphate, 17 mM (Calbiochem)

Creatine phosphokinase, 3 μg (Sigma)

L-[^{3}H]Leucine, 250 mCi/mmole, 0.04 mM (New England Nuclear) 19 nonradioactive amino acids, grade A, 0.1 mM each (Calbiochem)

E. coli W tRNA, 50 μg (Schwarz-Mann)

E. coli Q13 high-speed supernatant protein, 0.3 mg, prepared according to Iwasaki *et al.*[2]

E. coli Q13 or MRE600 1 M NH$_4$Cl-washed ribosomes, 6 A_{260} units, prepared according to Iwasaki *et al.*[2] with the exceptions that the two high-salt washings are carried out for 12 hours and 6 hours, respectively, and the DEAE-cellulose step is omitted. For storage, the ribosomes are suspended in 50% glycerol, 0.25 M NH$_4$Cl, 5 mM magnesium acetate, 10 mM Tris·HCl, pH 7.4, 1 mM DTT to yield a concentration of 1000 A_{260} units/ ml. They are then stored at $-70°$ with indefinite retention of activity.

IF1, CM-cellulose step of Lee-Huang *et al.*[11] 2 μg (saturating) (see also this volume [4])

IF2, DEAE-cellulose step of Salas *et al.*[12] 20 μg (saturating)

[10] C. L. Pon, S. M. Friedman, and C. Gualerzi, *Mol. Gen. Genet.* 116, 192 (1972).

[11] S. Lee-Huang, M. A. G. Sillero, and S. Ochoa, *Eur. J. Biochem.* 18, 536 (1971).

[12] M. Salas, M. J. Miller, A. J. Wahba, and S. Ochoa, *Proc. Nat. Acad. Sci. U.S.* 57, 1865 (1967).

[^{35}S]IF3, 0–1 μg

MS2 RNA, 1 A_{260} unit, prepared according to Weissmann and Feix[13]

Procedure. The reagents above are mixed at 0° in the order listed. Incubation is for 20 minutes at 37°. The reaction is stopped by adding 2 ml of 5% trichloroacetic acid (TCA). The samples are then heated at 90° for 15 minutes, then chilled and filtered through 2.5 cm Millipore filters. The filters are dried under a lamp, immersed in 15 ml of toluene containing Omnifluor (New England Nuclear), and counted in a scintillation counter set for simultaneous counting of ^{3}H and ^{35}S. The ratio of counts per minute per picomole of leucine incorporated is determined empirically by comparing identical assay samples containing unlabeled IF3, one with [^{3}H]leucine and one with [^{14}C]leucine of known specific activity. The contamination of ^{35}S radioactivity into the ^{3}H channel is determined from control samples containing [^{35}S]protein lacking IF3 activity. Assays of IF3 fractions cruder than the phosphocellulose fractions are unreliable because of the large amount of contaminating [^{35}S]protein.

One unit of activity incorporates 1 nmole of leucine into TCA-insoluble material in 20 minutes under the conditions described. The assay is linear up to 0.5 unit/assay.

Purification Procedure

The procedure is similar to that of Sabol *et al.,*[5] but modified to reduce losses noted in small-scale IF3 preparations. No effort is made to separate multiple IF3 species[14,15] (see this volume [6]). The growth of cells and steps 1 and 2 are preferably conducted in a laboratory restricted to radioisotope work. Buffer pH values are adjusted at 20°.

Materials

H$_2^{35}$SO$_4$, carrier free (New England Nuclear Co.)

DEAE-cellulose, Whatman DE-52, 1.0 meq/g, washed in 0.5 M KOH for 2 hours before equilibration

Phosphocellulose, Whatman P-11, 7.4 meq/g, soaked in 0.1 M KOH for 5 minutes and in 4 M NH$_4$Cl for 5 hours before equilibration

Sephadex G-75 (Pharmacia)

[13] C. Weissmann and G. Feix, *Proc. Nat. Acad. Sci. U.S.* **55**, 1264 (1966).
[14] M. Revel, H. Aviv, Y. Groner, and Y. Pollack, *FEBS (Fed. Eur. Biochem. Soc.) Lett.* **9**, 213 (1970).
[15] S. Lee-Huang and S. Ochoa, *Nature (London) New Biol.* **234**, 236 (1971).

Buffer A: 20 mM NH$_4$Cl, 10 mM magnesium acetate, 20 mM
Tris·HCl buffer pH 7.4, 10 mM 2-mercaptoethanol

Buffer B: 1 M NH$_4$Cl, 10 mM magnesium acetate, 20 mM Tris·HCl
buffer pH 7.4, 2 mM dithiothreitol (DTT)

Buffer C: 10 mM NH$_4$HCO$_3$, 5 mM Tris·HCl buffer pH 7.4,
1 mM 2-mercaptoethanol

Buffer D: 0.15 M NH$_4$HCO$_3$, 5 mM Tris·HCl buffer pH 7.4,
1 mM 2-mercaptoethanol

Buffer E: 6 M urea (deionized by passage of stock 10 M solution
over Dowex AG501-X8 mixed bed resin), 0.25 M NH$_4$Cl, 50
mM Tris·HCl buffer pH 7.4, 2 mM DTT

Buffer F: 0.25 M Tris·HCl buffer pH 7.4, 10% glycerol

Growth of Radioactive Cells. E. *coli* MRE600 cells are grown in a
carboy containing 10 liters of a medium consisting (per liter) of 8 g of
D-glucose (autoclaved separately), 11.2 g of K$_2$HPO$_4$ (3H$_2$O), 3 g of
KH$_2$PO$_4$, 1 g of NH$_4$Cl, 0.5 g of NaCl, 0.05 g of MgSO$_4$ (7 H$_2$O),
10 mCi of H$_2$^{35}SO$_4$ (total 100 mCi, added separately), 0.02 g of magnesium acetate, 0.0002 g of CaCl$_2$, 0.0006 g of FeCl$_3$, 0.005 g of thymine, 0.005 g of adenine, 0.005 g of uracil, and 0.04 g of each of 18
amino acids, grade A (Calbiochem) omitting cysteine and methionine.
The sulfate concentration used is slightly higher than that required to
limit the extent of growth. Growth is conducted at 37° with vigorous
forced aeration through a gas dispersion tube, and the pH is maintained
at 7.0 during the later stages of growth by addition of KOH. Anti-foam
A (Dow Corning) is added as needed. All operations are carried out
under a fume hood. Growth is stopped when the culture is in late log
phase; waiting until stationary phase may result in reduced factor yield.
Thus, when the culture reaches an OD$_{550 \, nm}$ of 1.4–1.5, ice is quickly
added to it and to the surrounding water bath. The cells are harvested
at 10,000 rpm for 20 minutes; the yield is about 40 g of wet-packed
cells containing 70–75% of the input radioactivity.

All subsequent operations are conducted at 0–4°. The radioactivity of
^{35}S-labeled protein is counted with an efficiency of 90% in Bray's
solution or Triton X-100 (Baker)–toluene–water (3:6:1) containing
Omnifluor.

Step 1. 1 M NH$_4$Cl Wash. The wet-packed cells (40 g) are lysed by
grinding with twice their weight of levigated Alumina (Fisher) and 250 μg
of RNase-free DNase (Worthington), and extracted with 80 ml of buffer
A. After centrifugation at 16,000 rpm for 30 minutes, the pellets are
reground with 20 ml of buffer A and the suspension is centrifuged as
before. The supernatants are pooled and centrifuged at 38,000 rpm for
7 hours. The resultant pellets are stirred with 50 ml of buffer B for 18

hours and the suspension is centrifuged again at 38,000 rpm. The resultant supernatant fraction is retained.

Step 2. *Ammonium Sulfate Precipitation.* To the NH_4Cl wash is added 29.1 g $(NH_4)_2SO_4$ per 100 ml. After 15 minutes the precipitate is centrifuged and discarded. To the supernatant is added 15.9 g per 100 ml more $(NH_4)_2SO_4$, and the precipitate is centrifuged, dissolved in buffer C, and dialyzed against several changes of buffer C. Buffers containing NH_4HCO_3 are employed in steps 2–4 because of the lyophilization after step 4.

Step 3. *DEAE-Cellulose Chromatography.* The dialyzed protein from step 2 is clarified by low-speed centrifugation, diluted to less than 5 mg/ml, and applied to a column, 0.8 × 20 cm, of DEAE-cellulose equilibrated with buffer C. One-milliliter fractions are collected at a rate of 0.25 ml/min. The column is washed with buffer C (which elutes IF1) and when the radioactivity of the effluent becomes negligible, buffer D is applied. Protein usually appears in a prominent leading peak and several trailing peaks. Fractions comprising the leading peak only are pooled.

Step 4. *Phosphocellulose Chromatography.* For steps 4 and 5 all glassware is precoated with 1% Siliclad (Clay-Adams). The pooled fractions from step 3 are applied to a phosphocellulose column, 0.5 × 9 cm, equilibrated with buffer D. One-milliliter fractions are collected at a rate of 0.2 ml/min. The column is washed with buffer D until all unadsorbed protein has been eluted; then it is washed successively with 15-ml portions of 0.30, 0.50, and 0.75 M NH_4HCO_3 in 5 mM Tris, 1 mM mercaptoethanol, pH adjusted to 7.4 (20°) with HCl. IF3 activity is found mainly or exclusively in the protein eluting at 0.50 M NH_4HCO_3. The 4–5 fractions comprising the peak, which contain 5% of the ^{35}S radioactivity of the step 3 pool, are pooled and frozen in an acetone–dry ice bath. Concentration is achieved by lyophilization to dryness, then redissolution and relyophilization once. The resultant solid material is dissolved in buffer E. If the volume is greater than 0.25 ml, further concentration can be achieved by placing the protein solution in a dialysis bag and suspending it overnight in dry Sephadex G-200 at 4°.

Step 5. *Gel Filtration in 6 M Urea.* The purpose of this step is to remove minor contaminants slightly larger and smaller than IF3. Urea prevents adsorption of IF3 to other proteins and to glass. The step 4 material (0.25 ml) is applied to a Sephadex G-75 column, 0.6 × 62 cm (void volume 5.0 ml), equilibrated with buffer E. Fractions (0.25 ml) are collected at a rate of 0.1 ml/min. Upon development of the column with buffer E, a single peak of ^{35}S radioactivity is eluted with its center at approximately 0.7 ml past the void volume. The peak of IF3 activity coincides with the ^{35}S peak, the ratio of IF3 activity to radioactivity being highest in the central fractions. The central fractions are pooled and dialyzed exhaustively vs. buffer F. The resultant solution should contain 75–100 µg protein, determined by the Lowry method[16] using bovine serum albumin as standard, with an approximate specific radioactivity of 25,000 cpm/µg.

Properties

The specific activity of [35S]IF3 prepared as described is 400–500 units/mg in the authors' assay system, but the value depends strongly on the activity of the critical assay reagents. Analysis by polyacrylamide gel electrophoresis[17] in 8 M urea followed by autoradiography[18] shows a single band of Coomassie-Brilliant Blue stained material and a single corresponding band of radioactivity.

The tenacity with which [35S]IF3 protein in dilute solutions adsorbs to glass, siliconized glass, and various plastics is a source of experimental difficulties. Many inert solutes were tested for their ability to prevent this binding, but only denaturing agents such as urea were effective.

The binding of [35S]IF3 to E. coli 30 S ribosomal subunits can be easily detected by sucrose gradient analysis.[7,8] The binding is slightly stronger at 4° than at 24° and plateaus at 1 mole bound per mole subunit. It is independent of the magnesium ion concentration between 5 and 15 mM and is slightly stimulated by the presence of IF1 and IF2. Binding to the 50 S subunit or the 70 S couple is not detected by sucrose gradient analysis. At low magnesium ion concentrations (e.g., 5 mM), interaction of labeled IF3 with the 70 S couples causes their dissociation into subunits, and the factor becomes bound to the newly formed 30 S subunits in an approximately 1:1 molar ratio. If 50 S subunits are added to 30 S·IF3 complexes at high magnesium ion concentration (e.g., 15 mM), the factor is released during or after the association of subunits. Labeled IF3 does not bind to 0.7 M KCl-washed active 40 S subunits of the brine shrimp *Artemia salina*. Binding to nucleic acids such as MS2 RNA and 16 S ribosomal RNA can be detected, but its significance is not established.

The labeled IF3 is highly active in promoting the formation of the initiation complex consisting of both ribosomal subunits, MS2 RNA, and fMet-tRNA$_f$. However, while it is required for this reaction, it is not bound to the 70 S initiation complex, indicating that it is released when the complex is formed. Thus, IF3 undergoes a cycle of association to and dissociation from the ribosomes during protein biosynthesis.

[16] O. H. Lowry, N. J. Rosebrough, A. L. Farr, and R. J. Randall, *J. Biol. Chem.* **193**, 265 (1951).
[17] R. A. Reisfeld, U. J. Lewis, and D. E. Williams, *Nature (London)* **195**, 281 (1962).
[18] J. V. Maizel, Jr., *in* "Methods in Virology" (K. Maramorosch and H. Koprowski, eds.), Vol. V, p. 180. Academic Press, New York, 1971.

[6] Purification of Two Messenger-Discriminating Species of Initiation Factor 3 (IF3) from *Escherichia coli*

By SYLVIA LEE-HUANG and SEVERO OCHOA

IF3 has a dual role in protein synthesis: (a) it is involved in formation of the polypeptide chain initiation complex,[1,2] and (b) it can promote the dissociation of 70 S ribosomes to 50 S and 30 S subunits.[3-5] Both activities are a consequence of the binding of IF3 to the 30 S subunit[6-8] from which it is released when the 70 S complex is formed.[6-8] Because IF3 is required largely for initiation with natural but not with synthetic messengers,[1,2] the factor may recognize mRNA initiation sites and direct the binding thereto of 30 S ribosomal subunits. Changing mRNA recognition patterns following infection of *E. coli* with phage T4 (for literature see Lee-Huang and Ochoa[9]) suggested the existence of multiple forms of IF3 capable of messenger or cistron selection. Virtually homogeneous but nonselecting preparations of IF3[3] have been resolved into two messenger-selecting molecular species[9,10]: one (IF3α) with high selectivity toward coliphage (e.g., MS2), *E. coli,* and early T4 RNA, and the other (IF3β) selecting for late T4 RNA. The unresolved IF3 will be referred to as IF3αβ.

Assay

The routine assay of IF3 (assay 1) is based on its requirement for translation of natural mRNA by 1.0 M NH$_4$Cl-washed *E. coli* ribosomes in the presence of excess IF1 and IF2.[3] IF3α is assayed with MS2 RNA and IF3β with late T4 RNA as messenger.[9,10] IF3 can also be assayed (assay 2) using the requirement of the factor for ribosomal binding of fMet-tRNA$_f$ when directed by natural mRNA.[3]

[1] K. Iwasaki, S. Sabol, A. J. Wahba, and S. Ochoa, *Arch. Biochem. Biophys.* **125**, 542 (1968).
[2] M. Revel, J. C. Lelong, G. Brawerman, and F. Gros, *Nature (London)* **219**, 1016 (1968).
[3] S. Sabol, M. A. G. Sillero, K. Iwasaki, and S. Ochoa, *Nature (London)* **228**, 1269 (1970).
[4] A. R. Subramanian and B. D. Davis, *Nature (London)* **228**, 1273 (1968).
[5] J. Dubnoff and U. Maitra, *Proc. Nat. Acad. Sci. U.S.* **68**, 318 (1971).
[6] S. Sabol and S. Ochoa, *Nature (London) New Biol.* **234**, 233 (1971).
[7] S. Sabol, D. Meier, and S. Ochoa, *Eur. J. Biochem.* **33**, 332 (1973).
[8] C. L. Pon, S. M. Friedman, and C. Gualerzi, *Mol. Gen. Genet.* **116**, 192 (1972).
[9] S. Lee-Huang and S. Ochoa, *Nature (London) New Biol.* **234**, 236 (1971).
[10] S. Lee-Huang and S. Ochoa, *Archives of Biochem. and Biophys.* **156**, 84–96 (1973).

Assay 1

Principle. Translation of MS2 or late T4 RNA is measured as the incorporation of [^{14}C]lysine into hot trichloroacetic acid (TCA)-insoluble material.

Reagents. The reaction mixtures, in a total volume of 0.125 ml, have the following composition:

> Tris·HCl buffer, pH 7.8, 60 mM
> ATP, disodium salt (Sigma), 1.3 mM
> GTP, disodium salt (P-L Labs), 0.3 mM
> Phosphocreatine (Sigma), 17 mM
> Creatine kinase (Worthington), 3 μg
> Magnesium acetate, 12 mM
> Dithiothreitol (DTT), 8 mM
> *E. coli* W tRNA (Schwarz/Mann), 50 μg
> [^{14}C]lysine (New England Nuclear), 10 mCi/mmole, 0.1 mM
> Nineteen nonlabeled amino acids (Sigma), 0.1 mM each
> *E. coli* Q13 or MRE 600 ribosomes (1.0 M NH$_4$Cl-washed), 6 A_{260} units
> *E. coli* MRE 600 high speed supernatant protein (see below under preparation of ribosomes), 0.3 mg
> NH$_4$Cl, 70 mM
> Homogeneous IF1,[11] 1 μg
> Purified IF2, 2.4 μg, prepared through DEAE-cellulose step of Salas *et al.*[12] and further purified by chromatography on phosphocellulose in buffer A (see below) using 0.5 M NH$_4$Cl in buffer A for elution
> MS$_2$ or late T4 RNA, 40 μg
> IF3, 0.1–0.3 unit

Procedure. The above components are mixed at 0° in the order listed. Incubation is for 20 minutes at 37°. The amount of [^{14}C]lysine incorporated into hot TCA-insoluble material is determined as in earlier work.[1] One unit of activity is the amount of IF3 promoting the incorporation of 1 nmole of lysine in 20 minutes under the conditions described. The assay is linear up to about 0.5 unit/assay.

Assay 2

Principle. The IF3-dependent ribosomal binding of f[^{14}C]Met-tRNA in the presence of excess IF1 and IF2, with either MS2 RNA or late T4

[11] S. Lee-Huang, M. A. G. Sillero, and S. Ochoa, *Eur. J. Biochem.*, **18**, 536 (1971). See also S. Lee-Huang and S. Ochoa, this volume [4].

[12] M. Salas, M. J. Miller, A. J. Wahba, and S. Ochoa, *Proc. Nat. Acad. Sci. U.S.* **57**, 1865 (1967).

RNA as messenger, is measured by the Millipore filter assay of Nirenberg and Leder.[13]

Reagents. The reaction mixtures, in a total volume of 0.05 ml, have the following composition:

Tris·HCl buffer, pH 7.2, 100 mM
NH$_4$Cl, 55 mM
Magnesium acetate, 5.5 mM
DTT, 1 mM
GTP, disodium salt (P-L Labs), 0.2 mM
f[^{14}C]Met-tRNA,[11] 12 pmoles (7200 cpm)
E. coli Q13 unfractionated ribosomes (1.0 M NH$_4$Cl-washed), 3 A_{260} units
MS2 or late T4 RNA (kept for 3 minutes at 37° in 0.1 mM EDTA before use), 40 μg
Homogeneous IF1,[11] 1 μg
Purified IF2, 2.4 μg
IF3, 0.6–1.0 μg

Procedure. The above components are mixed at 0° in the order listed. After incubation for 10 minutes at 37°, the samples are chilled, followed by the addition of 2 ml of a buffer containing 100 mM Tris·HCl, pH 7.2, 55 mM NH$_4$Cl, and 5.5 mM magnesium acetate, and the bound radioactivity is measured by the Millipore procedure.

Preparation of Assay Components

Buffers

Buffer A: Tris·HCl, pH 7.8, 10 mM; DTT, 1 mM; magnesium acetate, 10 mM
Buffer B: As buffer A but with 20 mM Tris·HCl, pH 7.8
Buffer C: Buffer A containing 1.0 M NH$_4$Cl
Buffer D: Tris·HCl, pH 7.8, 10 mM; NH$_4$Cl, 250 mM; magnesium acetate, 5 mM; glycerol, 50% (v/v)

Ribosomes. All operations are conducted at 0–4°. Freshly grown *E. coli* Q13 or *E. coli* MRE 600 cells, harvested at early log phase (60 g wet weight), are disrupted by grinding in a mortar with 120 g of Alumina (Sigma, type 305). The mixture is stirred into 120 ml of buffer A and incubated with 1 μg of DNase/ml (Worthington, RNase-free) for 30 minutes at 4°. The resulting suspension is centrifuged for 30 minutes at 30,000 g in the Sorvall B-2 centrifuge. The sediment of Alumina and cell debris is discarded, and the supernatant is recentrifuged as above to re-

[13] M. W. Nirenberg and P. Leder, *Science* 145, 1399 (1964).

move any remaining cell debris. The supernatant is centrifuged for 3 hours at 50,000 rpm (160,000 g) in rotor No. 65 of the Spinco Model L4 centrifuge. The upper two-thirds of the high speed supernatant is collected and immediately dialyzed against buffer B. This supernatant is used in the translation assay (assay 1) as a source of aminoacyl-tRNA synthetases, chain elongation, and chain termination factors. The ribosomal pellet is washed with buffer C (1.0 M NH$_4$Cl wash) with gentle stirring overnight, and the suspension is centrifuged at 30,000 g for 30 minutes; the sediment is discarded and the supernatant is centrifuged at 160,000 g for 3 hours. The washing is repeated two more times with stirring in buffer C overnight or just for a few hours. The brown layer on top of the ribosomal pellet is carefully removed and discarded each time. The transparent, colorless pellet is finally suspended in buffer D to a concentration of 1200 A_{260} units/ml and stored at $-20°$.

MS2 RNA. MS2 RNA is prepared according to Weissmann and Feix.[14] The RNA may be purified further by sucrose density gradient centrifugation. The gradients are prepared with sterilized buffer containing 10 mM Tris·HCl, pH 7.4, 2 mM MgCl$_2$, and 0.1 M NaCl, and the RNA is dissolved in the same buffer to a concentration of 150 A_{260} units/ml (6 mg/ml). One milliliter of this solution is layered on 50 ml of a 10–30% linear sucrose gradient in the same buffer and centrifuged for 16 hours at 23,000 rpm in the Spinco SW 25.2 rotor (60,000 g). After centrifugation the bottom of the tube is punctured and 0.5 ml fractions are collected in sterilized test tubes. One milliliter of ethanol at $-20°$ is immediately added to each tube, and the fractions are stored at $-20°$. Each fraction is analyzed by centrifugation through a linear 10–30% sucrose gradient using an Isco gradient analyzer (5 mm light path flow cell). The fractions that contain RNA sedimenting as a symmetric sharp peak at 27 S are pooled and stored at $-20°$ in 70% ethanol. When required, the RNA is collected by centrifugation and dissolved in sterilized water before use.

Late T4 RNA. Late T4 RNA is prepared from *E. coli* BA cells 15 minutes after infection with T4 phage by the procedure of Bautz and Hall[15] with some modifications. Freshly harvested T4-infected cells (15 g) are suspended in 300 ml of a buffer containing 10 mM Tris·HCl, pH 7.4, and 10 mM MgCl$_2$. The homogeneous suspension is divided into 50-ml portions, frozen in methanol–dry ice, and stored frozen. When required, the suspension is thawed and the cells are lysed with lysozyme (Nutritional Biochemicals, 0.6 mg/ml) at 20° for 3 minutes, followed by DNase (Worthington, RNase free, 10 μg/ml) for another 3 minutes. The pH is

[14] C. Weissmann and G. Feix, *Proc. Nat. Acad. Sci. U.S.* **55**, 1264 (1966).
[15] K. E. F. Bautz and B. D. Hall, *Proc. Nat. Acad. Sci. U.S.* **48**, 400 (1962).

adjusted to 5.0 with 0.02 M acetic acid, sodium dodecyl sulfate (SDS) is added to 5 mg/ml, and the solution is incubated at 37° for 4 minutes. After phenol extraction (20 minutes at room temperature with two volumes of redistilled phenol, previously equilibrated with 50 mM acetate buffer, pH 5.2, containing 10 mM $MgCl_2$), the phases are separated by centrifugation at 16,000 g for 30 minutes. Traces of phenol remaining in the aqueous layer are removed by extraction six times with an equal volume of ether. The RNA is then precipitated with two volumes of ethanol at $-20°$; the precipitate is dried and dissolved in 10 mM Tris·HCl, pH 7.4, containing 10 mM $MgCl_2$, and adjusted to 1.0 M NaCl. The precipitate formed in 1.0 M NaCl at 4° overnight is dissolved in water, dialyzed, and further purified on BD-cellulose (Schwarz/Mann). The RNA sample is adjusted to a concentration of 5 mg/ml in 0.02 M Tris·HCl (pH 7.2), 1.2 M NaCl, and applied to a column (0.9 × 90 cm) of BD-cellulose at room temperature. The column is washed with the same buffer until the A_{260} of the effluent becomes 0.01 or less. Late T4 RNA is eluted from the column with 1.5 M NaCl containing 15% ethanol. It is precipitated with ethanol and stored at $-20°$ in 70% ethanol.

Purification Procedure

The procedure follows essentially that of Sabol *et al.*[3] through the phosphocellulose chromatography step. Fractions enriched in IF3$\alpha\beta$ activity are pooled and further fractionated by phosphocellulose chromatography in 6.0 M urea. Fractions enriched in IF3α and IF3β activity are separately pooled and refractionated in the same way. Buffer pH values are adjusted at 25°. Protein is determined by the Lowry method.[16]

Materials

DNase (Worthington), RNase-free
DEAE-cellulose (Schleicher and Schuell), 0.97 meq/g
Cellulose phosphate (Whatman, P-11), 7.4 meq/g
Buffer A: 10 mM Tris·HCl, pH 7.4, 1 mM magnesium acetate, 1 mM DTT, 10% glycerol
Buffer B: Buffer A containing 0.25 M NH$_4$Cl
Buffer C: Buffer A containing 0.5 M NH$_4$Cl
Buffer D: Buffer A containing 6.0 M urea

Growth of Cells. *E. coli* MRE 600 is grown at 37° in a vat fermenter on 180 liters of medium containing 10 mM potassium phosphate buffer,

---

[16] O. H. Lowry, N. J. Rosebrough, A. L. Farr, and R. J. Randall, *J. Biol. Chem.* **193**, 265 (1951).

pH 7.3, 5 mM MgSO$_4$, 0.5% glucose, 0.8% casamino acids, 0.05% yeast extract. The cells are harvested under refrigeration at mid-log phase. All operations are conducted at 0–4° unless otherwise stated.

Step 1. 1.0 M NH$_4$Cl Ribosomal Wash. This is carried out as described for the purification of IF1.[11]

Step 2. Ammonium Sulfate Fractionation. This is also carried out as in the IF1 purification except that the fraction taken is between 0.45–0.7 ammonium sulfate saturation.

Step 3. DEAE-Cellulose Chromatography. The ammonium sulfate fraction (from a preparation starting with 1.59 kg of cells) is dissolved in 30 ml of buffer A and dialyzed against the same buffer for 8 hours with four 3-liter changes of buffer. Any precipitated protein is removed by centrifugation. The clear supernatant solution is applied to a column (2.5 × 179 cm) of DEAE-cellulose, made from two connected columns (2.5 × 92 cm and 2.5 × 87 cm each), previously equilibrated with buffer A. The column is washed with the same buffer until the A_{280} of the effluent is zero. IF3 is eluted with buffer B. Fractions of 15.8 ml are collected. Fractions 41 through 50 containing the bulk of IF3 activity are pooled. This step results in about 3-fold purification with 70% yield.

Step 4. Cellulose Phosphate Chromatography. The solution from the above step is applied to a column (2 × 50 cm) of cellulose phosphate (7.4 meq/g), previously equilibrated with buffer B, and the column is washed with the same buffer until the A_{280} of the effluent is nil. IF3 activity is eluted with buffer C. Fractions of 5 ml are collected and fractions 35–60 are pooled to give 125 ml of solution containing 30 mg of protein. This step gives about 15-fold purification with an average 65% yield.

Step 5. Cellulose Phosphate Chromatography in 6.0 M Urea. The solu-

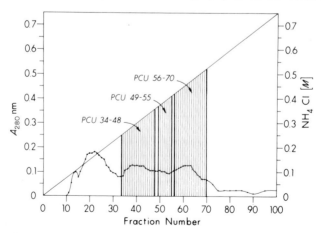

FIG. 1. Chromatography of IF3 on phosphocellulose in 6.0 M urea. This corresponds to step 5 in the table. Details are described in the text.

tion from step 4 is concentrated to 25 ml by lyophilization. It is then dialyzed overnight against buffer D with four 1-liter changes. The dialyzed sample is applied to a column (0.9 × 15 cm) of cellulose phosphate equilibrated with buffer D. The column is washed with the same buffer until the refractive index of the effluent matches that of the buffer (1.4109). A linear gradient of NH_4Cl from 0 to 0.75 M in buffer D (200 ml) is started; 2-ml fractions are collected. They are dialyzed against buffer A and assayed immediately.

Three main regions of IF3 activity are pooled (Fig. 1), namely fractions 34 through 48 (PCU 34–48), 49 through 55 (PCU 49–55), and 56 through 70 (PCU 56–70). The table shows that PCU 34–48 is enriched in IF3α whereas PCU 56–70 is enriched in IF3β activity. PCU 49–55 has similar IF3α and β activities. Each pool is concentrated by lyophilization to one-tenth of its original volume. PCU 34–48 and PCU 56–60 are separately dialyzed against buffer D prior to rechromatography. PCU 49–55 is stored at −20°.

Step 6. Rechromatography on Cellulose Phosphate in 6.0 M Urea. The

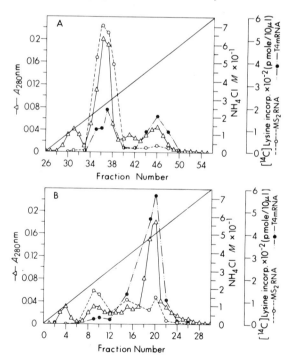

FIG. 2. Rechromatography of IF3 fractions on phosphocellulose in 6.0 M urea. (A) Rechromatography of fractions 34–48 (step 6a). (B) Rechromatography of fractions 56–70 (step 6b). Fractions 34–48 and 56–70 are from step 5 (see table).

ISOLATION OF MESSENGER DISCRIMINATING SPECIES OF IF3

Step	Volume (ml)	Protein (mg)	Units[b]		Specific activity[c]		Ratio T4L:MS2
			MS2	T4L	MS2	T4L	
1. NH$_4$Cl wash[a]	960	9600	8750	8400	0.9	0.9	1.0
2. (NH$_4$)$_2$SO$_4$ fractionation	50	3050	5200	5500	1.7	1.8	1.1
3. DEAE-cellulose chromatography	158	725	3660	3880	5.1	5.4	0.9
4. Phosphocellulose chromatography	125	30	2210	2690	73.6	89.5	1.2
5. Phosphocellulose chromatography in 6.0 M urea							
Fractions 34–48	30	3	840	441	280	147	0.5
49–55	10	0.8	121	107	151	134	0.9
56–70	30	3.4	312	780	92	230	2.5
6a. Phosphocellulose–urea rechromatography of step 5 fractions 34–48							
Fractions *34–37*	1	0.9	397	98	*440*	*109*	*0.25*
38–39	1	0.1	53	22	530	220	0.4
40–41	1	0.1	32	26	320	260	0.8
42–43	1	0.1	9	42	90	420	4.7
44–46	1	0.16	12	69	75	430	5.8
47–53	1	0.1	9	41	90	410	4.5
6b. Phosphocellulose–urea rechromatography of step 5 fractions 56–70							
Fractions 8–10	0.8	0.1	45	12	450	116	0.3
11–14	0.8	0.1	23	9	230	90	0.4
15–16	0.8	0.1	45	53	450	530	1.2
17–20	0.8	0.8	53	384	*66*	*480*	*7.3*
21–25	0.8	0.1	9	52	90	520	5.8

[a] From 1.59 kg of frozen *Escherichia coli* MRE600 cells.
[b] One unit = 1.0 nmole of [^{14}C]lysine incorporated under standard assay conditions.
[c] Units per milligram of protein.

dialyzed PCU 34–48 and PCU 56–70 pools from the previous step are separately rechromatographed on phosphocellulose in 6.0 M urea, as in step 5, using a 0.5 × 27 cm column and the same NH_4Cl gradient (60 ml); 2-ml fractions are collected. A peak of protein enriched in IF3α activity is eluted between 0.2 and 0.32 NH_4Cl (Fig. 2A) and one enriched in IF3β activity between 0.45 and 0.55 M NH_4Cl (Fig. 2B). Fractions are pooled as shown in the table, dialyzed, assayed, and concentrated as above. They are stored at −20° in buffer C; no loss of activity was observed over a 3-month period.

A summary of the isolation procedure is given in the table. This method results in about 500-fold purification of IF3α activity with about 4.5% yield (step 6a, fractions 34–47) and approximately the same degree of purification and the same yield of IF3β activity (step 6b, fraction 17–20). The T4L/MS2 discrimination ratio[17] of these two sets of fractions is about 30. Both of these preparations are virtually homogeneous by three criteria: (a) disc electrophoresis on polyacrylamide gels, (b) SDS gel electrophoresis, and (c) disc gel electrofocusing.

Properties

IF3α and IF3β have the same ribosome dissociation activity. They differ slightly in molecular weight (23,500 and 21,500 for the α and β species, respectively, as determined by SDS gel electrophoresis). Using the same method a value of about 21,000 was reported for unresolved IF3.[3,5] Filtration of IF3 through Sephadex G-100 columns in buffers of low ionic strength results in formation of aggregates. This can be avoided by the addition of SDS and DTT or by the use of buffers containing 0.5–1.0 M NH_4Cl. The molecular weights calculated from gel filtration data are 23,000 for IF3α and 21,000 for IF3β.

Although selective for two broad classes of messengers (see introduction) IF3α and IF3β do not seem to discriminate between cistrons of a polycistronic messenger such as MS2 RNA, or, if they do, they do not differ in this respect.[10] Since early and late T4 RNA's are translated[18] from the light and heavy strands of T4 DNA, respectively, predominant recognition of early T4 RNA by IF3α and of late T4 RNA by IF3β could mean that each IF3 species recognizes mRNA transcribed from complementary strands of DNA.

---

[17] The T4L:MS2 (late T4 RNA:MS2 RNA) discrimination ratio of step 6b fractions 17–20:step 6a fractions 34–37 is the quotient of their respective T4L:MS2 translation ratios, i.e. 7.3:0.25 = 29 (see last column of the table).

[18] A. Guha and W. Szybalski, *Virology* **34**, 608 (1968).

[7] Selection of Ribosome Binding Sites on mRNA in *Escherichia coli* by Cistron-Specific IF3 Subfractions, and Interference Factors

By Michel Revel and Yoram Groner

Initiation factor IF3 activity (defined by complementation with IF1 and IF2)[1] can be separated in subfractions differing by their relative activity for various mRNA templates.[2-6] In many cases, it can be shown that the template specificity of these subfractions results from the presence of other protein factors, having no IF3 activity by themselves, but which interfere with IF3 messenger selection activity.[6-8] Purified IF3 is active for both MS2 and T4 mRNA and interference factors are defined and assayed by their mRNA and cistron-specific stimulatory or inhibitory effects on this standard IF3 activity. Several IF3-interference activities have been detected already.[9] In this section we shall describe: (1) the methods used to separate IF3 into the different subfractions and to demonstrate the presence of interference factors in these subfractions, (2) the techniques used to study specific ribosome binding to various cistrons initiation sites.

Assay of IF3 Activity

IF3 is assayed by measuring the translation of natural mRNA with high salt-washed ribosomes, high speed supernatant, and purified initiation factors IF1 and IF2.[1] Messenger RNA discrimination is determined by comparing the overall translation of several mRNA templates or the translation of the different cistrons of the same mRNA.

[1] M. Revel, H. Greenshpan, and M. Herzberg, this series, Vol. 20, p. 261.

[2] M. Revel, H. Aviv, Y. Groner, and Y. Pollack, *FEBS* (*Fed. Eur. Biochem. Soc.*) *Lett.* 9, 213 (1970).

[3] H. Berissi, Y. Groner, and M. Revel, *Nature* (*London*) *New Biol.* 234, 44 (1971).

[4] S. Lee-Huang and S. Ochoa, *Nature* (*London*) *New Biol.* 234, 236 (1971).

[5] M. Yoshida and P. S. Rudland, *J. Mol. Biol.* 68, 465 (1972).

[6] Y. Groner, Y. Pollack, H. Berissi, and M. Revel, *FEBS* (*Fed. Eur. Biochem. Soc.*) *Lett.* 21, 223 (1972).

[7] Y. Groner, Y. Pollack, H. Berissi, and M. Revel, *Nature* (*London*) *New Biol.* 239, 16 (1972).

[8] S. Lee-Huang and S. Ochoa, *Biochem. Biophys. Res. Commun.* 49, 371 (1972).

[9] M. Revel, Y. Pollack, Y. Groner, R. Scheps, H. Inouye, H. Berissi, and H. Zeller, *Ribosomes: Structure, Function and Biogenesis, FEBS Symp.* 27, 261 (1972).

Materials

TEMPLATES

MS2 RNA (native), prepared as described by Gesteland and Spahr[10] or unfolded MS2 RNA, treated with 1 M formaldehyde at 37° as described by Lodish[11]

T4 mRNA prepared as by Salser *et al.*[12] (late T4 mRNA)

RNA transcribed by *E. coli* RNA-polymerase from T4 DNA, prepared as described previously[1] (early T4 mRNA)

RIBOSOMES AND FACTORS

Ribosomes washed twice with 2 M NH$_4$Cl, 30 mM Tris·HCl, pH 7.5, 10 mM MgCl$_2$, 14 mM β-mercaptoethanol[1]; high speed supernatant[1]; initiation factor IF1 and IF2 purified by standard procedures[1]; all from *E. coli* MRE 600

Procedures

1. Overall Amino Acid Incorporation. The assay mixture contained in 0.065 ml: 50 mM Tris·HCl, pH 7.5, 50 mM NH$_4$Cl, 13 mM MgCl$_2$, 2 mM ATP, 1 mM GTP, 5 mM phosphoenolpyruvate, 2 μg pyruvate kinase, 0.1 mM CTP and UTP, 0.15 mM 19 L-amino acids (minus valine), 6 mM dithiothreitol, 0.5 μg of formyltetrahydrofolate, 1.5 nmoles [^{14}C]valine (50 μCi/μmole), 100 μg high salt-washed ribosomes, 5 μl of supernatant, 5 μg of IF1, 15 μg of IF2, 25 μg of tRNA, 20 μg of native MS2 RNA, or 100 μg of T4 mRNA (extracted from *E. coli* 20 minutes after T4 infection at 30°[12]). Preparations containing IF3 or interference factors are added in parallel to reactions containing MS2 RNA or T4 mRNA. After 30 minutes' incubation at 37°, the incorporation of valine into hot trichloroacetic acid-insoluble material is measured by the filter paper disk method.[1]

2. Study of the Cistron Specificity of IF3 Activity by Measure of N-formyl [^{35}S]methionine Incorporation into the Specific Aminoterminal Peptides of Each Product. For these experiments, MS2 RNA treated with 1 M formaldehyde at 37°[11] is used, to allow independent initiation at the three cistrons. Protein synthesis is carried out as above except that [^{35}S]-formylmethionyl-tRNA (2 × 10^5 cpm, 25 μg, free of unformylated species[1]) is used as source of label. After 30 minutes at 37°, ribonuclease A (120 μg/ml) and 10 mM EDTA are added; after 1 hour, proteins are acid-precipitated, washed, and digested with trypsin and chymotrypsin as

---

[10] R. Gesteland and P. F. Spahr, *Biochem. Biophys. Res. Commun.* **41**, 1267 (1970).

[11] H. Lodish, *J. Mol. Biol.* **56**, 689 (1970).

[12] W. Salser, R. Gesteland, and M. A. Bolle, *Nature (London)* **215**, 588 (1967).

described by Lodish.[13,14] Analysis of the peptides is carried out by electrophoresis at pH 1.9[13] or by a two-dimensional system of electrophoresis at pH 3.5 and chromatography in isoamylalcohol–pyridine–water (35:35:30).[14] The effect of adding various IF3 fractions to IF1 and IF2 on the synthesis of each peptide is measured. T4 mRNA can also be used as template; in this case the two-dimensional system should be used.

Methods for the Fractionation of IF3 Activity and the Isolation of Interference Factors

Materials

Buffer I: 10 mM Tris·HCl, pH 7.5; 10 mM MgCl$_2$; 60 mM NH$_4$Cl; 7 mM β-mercaptoethanol; 5% glycerol (v/v)

Buffer II: 1 M NH$_4$Cl, 30 mM Tris·HCl, pH 7.5; 0.2 mM MgCl$_2$; 14 mM β-mercaptoethanol; 5% glycerol (v/v)

Buffer III: 20 mM potassium phosphate buffer, pH 7.2; 0.2 mM MgCl$_2$; 7 mM β-mercaptoethanol; 5% glycerol (v/v)

Buffer IV: 30 mM potassium phosphate buffer, pH 7.2; 20 mM NH$_4$Cl; 0.2 mM MgCl$_2$; 7 mM β-mercaptoethanol; 6 M urea (deionized by passing through an Amberlite ion exchange column)

DEAE-cellulose, Serva, 0.65 meq/g

DEAE-Sephadex A-50, Pharmacia

Phosphocellulose, Whatman P-11, 7.4 meq/g

DNase, Worthington

Bacteria: *E. coli* MRE 600 is grown in a 300-liter fermentor with 4% potassium phosphate, pH 7, 0.8% glucose, and 0.6% yeast extract (Difco) at 37° to mid-log phase. Culture is progressively withdrawn, cooled and centrifuged; the cells are washed in buffer I and stored at −20°.

Step 1. Preparation of Ribosome Wash Proteins. Cells (5 kg) are suspended in 5 liters of buffer I with 1 μg of DNase per milliliter and passed twice through a continuous-flow French press (Manton-Gaulin homogenizer, Everett, Massachusetts) at 7000 psi (temperature kept below 10°). Batches of 700 g are processed individually. S-30 is prepared by centrifugation at 30,000 g for 30 minutes. Ribosomes are sedimented in Spinco rotor R30 at 80,000 g for 16 hours; resuspended overnight with stirring in 2.1 liters of buffer II at 4°, and centrifuged at 250,000 g for 2.5 hours in Spinco rotor 60 Ti. The supernatant is precipitated with ammonium sulfate

[13] H. Lodish, *Nature (London)* **224**, 867 (1969).
[14] H. Lodish, *Nature (London)* **220**, 345 (1968).

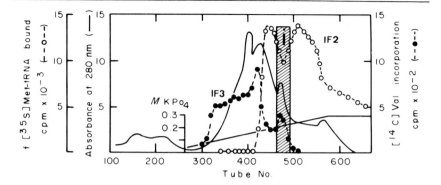

FIG. 1. DEAE-cellulose chromatography of crude ribosome wash protein. IF3 activity for T4 mRNA translation and IF2 activity for fMet-tRNA binding to ribosomes (M. Revel, H. Greenshpan, and M. Herzberg, this series, Vol. 20, p. 261) were assayed; for details, see text.

at 80% saturation, in order not to eliminate any IF3 subfraction. The proteins are dissolved in buffer III at a concentration of 10 mg/ml and dialyzed against the same buffer.

Step 2. DEAE-Cellulose Chromatography. The ribosome wash proteins (50 mg) are applied to a 6.6 × 78 cm column of DEAE cellulose equilibrated in buffer III. After washing the effluent proteins (containing IF1), IF3 is eluted with a 6-liter gradient of potassium phosphate from 20 mM to 300 mM in buffer III. Each fraction (20 ml) is assayed for IF3 activity in the translation of T4 mRNA. A typical elution profile is shown in Fig. 1. Elution profile of IF2[1] is also indicated.

Step 3. DEAE-Sephadex Chromatography. This is the step in which most of the fractionation into mRNA-specific species is achieved. Moreover, the use of DEAE-Sephadex after DEAE-cellulose separated very well IF3 activity from the bulk of the proteins. Two alternative methods have been used. (A) All the fractions having IF3 activity in the DEAE cellulose eluate are pooled and chromatographed on DEAE-Sephadex, or (B) different areas of the peak of IF3 on DEAE-cellulose are pooled separately and submitted each to DEAE-Sephadex chromatography.

Table I describes the results of a preparation from 1.2 kg *E. coli* MRE 600 using method A. The total IF3 from the DEAE-cellulose dialyzed against buffer III (1500 mg of protein) has been rechromatographed on a 1.5 × 34 cm column of DEAE-Sephadex A-50 with a 300-ml gradient of NH$_4$Cl from 0.02 to 0.3 M in buffer III (flow rate 8 ml/cm^2 per hour). Each fraction (5 ml) is assayed for its activity to stimulate translation of MS2 RNA and T4 mRNA. No discrete peaks are usually observed, and the eluent fractions are pooled according to their T4:MS2 translation discrimination ratio. In this way, one obtains an early eluting fraction

TABLE I

FRACTIONATION OF IF3 ACTIVITY ON DEAE-SEPHADEX

Nomenclature	NH₄Cl Molarity of elution (M)	Translation ratio T4 mRNA: MS2 RNA	Translation ratio MS2 coat: MS2 synthetase	Specific activity on T4 mRNA (pmoles valine per μg protein)
IF3-B1	0.05	1.9	0.6	32
IF3-B2	0.10	1.3	3.0	86
IF3-B3	0.15	3.8	1.0	70
IF3-B4	0.20	31.0	—	21

Assay for the translation of T4 mRNA and MS2 RNA and for the measure of MS2 coat and synthetase production is detailed in the text. From Y. Groner, Y. Pollack, H. Berissi, and M. Revel, *FEBS (Fed. Eur. Biochem. Soc.) Lett.* **21,** 223 (1972).

(IF3-B1) with a relatively higher translation of T4 mRNA than the next fraction (IF3-B2) which has the highest relative activity for MS2 RNA. The first fraction translates much more actively the MS2 synthetase cistron (using unfolded MS2 RNA as template) than does the second fraction (IF3-B2) which is quite specific for MS2 coat cistron (Table I). These early eluted fractions are well separated from the bulk of the protein. As shown by electrophoresis on polyacrylamide gels, IF3-B2 seems to be close to homogeneity (Fig. 2A).

The IF3 subfractions eluting later from the DEAE-Sephadex column have a much higher T4:MS2 translation ratio and the last IF3 peak (IF3-B4) is almost T4 mRNA specific. These fractions are, however, far from purity at this stage.

In larger preparations, it is preferable to use method B, in which the various regions of the IF3 peak from the DEAE cellulose are applied separately to DEAE-Sephadex columns. Figure 3 shows the chromatography of fractions 396–420 of Fig. 1 on a 2.8 × 25 cm DEAE-Sephadex A-50 equilibrated in buffer III. Elution is carried out with an 800-ml gradient of NH₄Cl from 20 mM to 300 mM, and 5-ml fractions are collected. The typical elution profile in Fig. 3 shows a main peak of IF3, active for both MS2 and T4 mRNA and two small peaks, one eluting earlier and the other eluting later than the main peak, and which both appear T4 mRNA specific. As shown below these mRNA discriminating fractions correspond to combinations between IF3 and interference factors, which can be separated in the further purification steps.

Step 4. Phosphocellulose. The next step for the subfractions obtained on DEAE-Sephadex is chromatography on phosphocellulose which gives

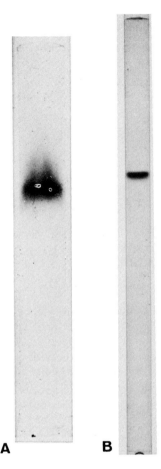

FIG. 2. Electrophoresis of purified IF3 on polyacrylamide gels. (A) 8 M urea polyacrylamide gel at pH 4.5; 43 μg IF3-B2 as in Table I. Run toward the cathode at the bottom of the gel. (B) SDS-polyacrylamide (15%) stacking gel [U. K. Laemmli, *Nature* (*London*) **227**, 680 (1970)]; 4.5 μg IF3 after chromatography through phosphocellulose with 6 M urea.

usually a very good purification since most proteins adsorbed on DEAE-Sephadex are not retained on phosphocellulose while IF3 is strongly retained. The use of this adsorbant for the purification of IF3 and mRNA discriminating interference factors is given here.

Isolation of IF3-Interference Factor i Complex

Figure 4 shows the purification of the T4 mRNA specific IF3-subfraction eluted very early from DEAE-Sephadex (IF3-B1 similar to fractions

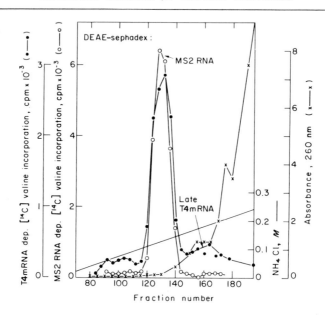

FIG. 3. Fractionation of T4 mRNA-specific IF3 on DEAE-Sephadex. For details
see text.

80–110 of Fig. 3). This fraction (18 mg protein) is chromatographed on
a 1.2 × 20 cm column of phosphocellulose equilibrated in buffer III.
Elution is carried out with a gradient of 100 ml NH_4Cl from 20 mM to
0.6 M, and 3-ml fractions are collected. A rather purified IF3 subfraction
is obtained, which translates T4 mRNA but has no activity with native
MS2 RNA. However, further analysis of this mRNA discriminating fraction shows that it is a complex between IF3 and a small portion of inter-

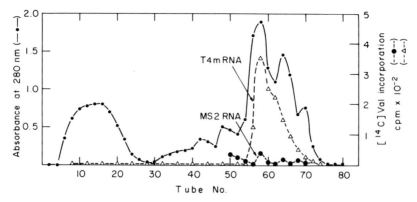

FIG. 4. Purification of T4 mRNA specific IF3 subfraction on phosphocellulose.
This fraction represents an IF3-i complex. For details, see text.

ference factor i (most of which is found free of IF3, as described in detail elsewhere[15]). The presence of factor i is demonstrated (1) by the cross-reaction of this IF3 subfraction with anti-i-serum[15]; (2) by the fact that this IF3-i complex sediments faster than IF3 on a glycerol gradient[6,7]; (3) by SDS-polyacrylamide gel electrophoresis, which shows the presence of both factor i and IF3 in this subfraction.[6] It is indeed this interference factor which is responsible for the absence of activity in the translation of native MS2 RNA. As expected from the properties of factor i,[7,15] the IF3-i complex is active on MS2 RNA which has been unfolded by treatment with formaldehyde. Factor i can be separated from this IF3-i complex by treatment with 6 M urea in buffer IV and rechromatography of this subfraction on a phosphocellulose column with 6 M urea (buffer IV). Under these conditions the IF3-i complex dissociates, and since free factor i is not retained on phosphocellulose,[15] it comes out in the column effluent separated from IF3. The IF3, freed of factor i, is eluted by 0.4 M NH$_4$Cl and, after this step, is active, like standard IF3, for both native MS2 RNA and T4 mRNA translation.

Isolation of Interference Factors i-β and i-γ

Figure 5 shows the analysis of the other T4 mRNA-specific IF3 subfraction from the DEAE-Sephadex step shown in Fig. 3 (fractions 150–170). This preparation (160 mg protein) is chromatographed on a 1.2 × 20 cm column of phosphocellulose equilibrated with 0.1 M NH$_4$Cl in buffer III, using a 150-ml gradient of NH$_4$Cl from 0.1 M to 1 M. Fractions of 3 ml each are collected. IF3, active for both T4 mRNA and MS2 RNA, is separated from an interference activity (designated factor i-γ), which when added to IF3 inhibits MS2 RNA translation more than that of T4 mRNA. This interference activity differs from factor i (or i-α)[8] by the fact that it inhibits also the translation of early T4 mRNA, T7 mRNA and formaldehyde-treated MS2 RNA, which are on the contrary stimulated by factor i.[7,15] Yet another interference factor (factor i-β),[8] is obtained from this fraction as shown by the chromatography of Fig. 5 (fractions 45–50).

Complete Separation of IF3 from Interference Factors

To obtain IF3 entirely free of interference factors the main IF3 fractions (as fraction 120–140 of the DEAE-Sephadex column of Fig. 3) is further purified by chromatography on phosphocellulose with 6 M urea (buffer IV). The column is washed with buffer IV and then IF3 is eluted with a gradient of NH$_4$Cl from 0.1 M to 0.6 M in buffer IV. Electrophoresis on SDS–polyacrylamide stacking gels shows (Fig. 2B) that IF3 is homogeneous at this stage.

---

[15] Y. Groner and M. Revel, this volume [8].

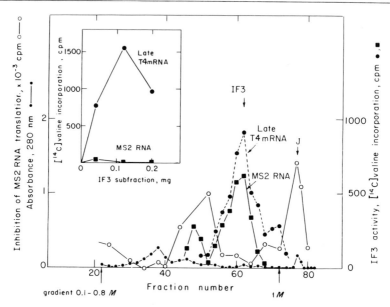

FIG. 5. Separation of IF3 and interference factor on phosphocellulose. The inset shows the IF3 activity of the fraction loaded on the column. During purification, IF3 activity for MS2 RNA reappears and is separated from interference factor i-β and i-γ.

Other Methods

Hydroxyapatite column chromatography has also been used to fractionate IF3.[2,4]

Methods to Study the Selection of Ribosomal Binding Site

Several methods are suitable for the study of the effects of IF3 subfractions and of interference factors (in combination with purified IF3) on the selection by ribosomes of the different initiation sites on mRNA. First, the ribosome attachment site may be deduced from an analysis of the protein products synthesized. This is best carried out by analysis of [35S]formylmethionine incorporation into the aminoterminal peptides.[6,7] More simply, the initial dipeptides formed may be identified and compared to the known sequence of the protein products; this has been successfully applied in the case of RNA phages.[11] However, the most direct method is the study of ribosome binding to radioactive mRNA and of the nucleotide sequences protected by the ribosomes against nuclease degradation.[3,7,16] This last method will be detailed here in the case of MS2 RNA.

[16] J. A. Steitz, *Nature* (*London*) **224**, 957 (1969).

Materials

SSC: 0.15 *M* NaCl–15 m*M* Na citrate, pH 7.0

Buffer V: 50 m*M* Tris·HCl, pH 7.5; 100 m*M* NH₄Cl; 8 m*M* Mg acetate; 1 m*M* dithiothreitol

DEAE cellulose, Serva, 0.68 meq/g

Formaldehyde solution, Fisher Scientific Company

Pancreatic ribonuclease A, Worthington; T1 ribonuclease, Calbiochem

Selectron BA85 membrane filters, Schleicher and Schuell

Materials for thin-layer chromatography: Cellulose MN 300 DEAE and cellulose MN 300 HR, Macherey and Nagel

Electrophoresis strip CA, Selectron, Schleicher and Schuell

Ribonucleic acid from yeast, BDH

[³²P]*MS2 RNA.* Very highly labeled [³²P]MS2 RNA is prepared by growing *E. coli* 58-161 in low phosphate medium [Bacto-peptone (Difco), 10 g; NaCl, 10 g; glucose, 1 g; CaCl₂, 0.22 g; casaminoacids (vitamin free), 1 g per liter of medium] to 5×10^8 cells/ml at 37° and infecting with MS2 at a multiplicity of 0.01. After 1 minute, 10 mCi of carrier free ³²P are added per 50 ml of culture, and vigorous aeration is maintained for 5 hours. Chloroform (0.01 ml) is added, followed by 1.5 ml of 0.5 *M* EDTA and 5 mg of lysozyme. After 1 hour at 22°, the lysate is centrifuged at 10,000 *g* for 10 minutes, and the supernatant is applied to a 2 × 3.5 cm column of DEAE-cellulose[10] equilibrated with 1 × SSC, at a flow rate of 30–40 ml/hr. The column is washed with 1 × SSC, and the phage is eluted with small aliquots of 4 × SSC. The most radioactive fractions are centrifuged at 140,000 *g* for 2 hours, and the pelleted phage resuspended in 0.1 *M* NaCl, 10 m*M* Na acetate, pH 5. After three extractions with water saturated phenol, the aqueous phase is extracted with ether and the RNA precipitated with 3 volumes of ethanol at −20°, for 1 hour. Before use, the RNA is dissolved in water, after drying. The specific activities obtained are about 2×10^6 cpm per microgram of RNA.

Unfolding of MS2 RNA

To measure the effect of factors on ribosome distribution among the three initiation sites, it is necessary to use RNA in which these sites are independently available. For this purpose the secondary structure of the phage RNA has to be disrupted. This is achieved either by using [³²P]RNA stored in concentrated aqueous solution for about a week, which ensures sufficient radiolysis,[16] or RNA treated with formaldehyde in the following way.[11] About 1 mg RNA is incubated in 1 ml of 10 m*M* sodium phosphate buffer pH 7.7, 0.2 *M* NaCl, 1 *M* formaldehyde for 15 minutes at 37°, followed by precipitation with 2.5 volumes of ethanol at $-20°$ for 15 minutes, washing with alcohol and with ether, and thoroughly drying in vacuum before dissolving in water.

Ribosome Binding to [^{32}P]MS2 RNA

Reaction mixtures in 0.2 ml contain 50 mM Tris·HCl pH 7.5, 100 mM NH$_4$Cl, 8 mM Mg acetate, 1 mM dithiothreitol, 0.3 mM GTP, 500 μg of high salt-washed ribosomes, 400 μg of tRNA (charged with methionine under formylating conditions and discharged to leave only fmet-tRNA) 50 μg of IF1, 150 μg of IF2, and 75 μg of [^{32}P]MS2 RNA (1.5×10^8 cpm). The effect of adding IF3 or an interference factor may be tested under these conditions. After 10 minutes at 37°, ribonuclease A (5 μg/ml) is added and incubation continued 10 minutes at 23°. The mixture is then

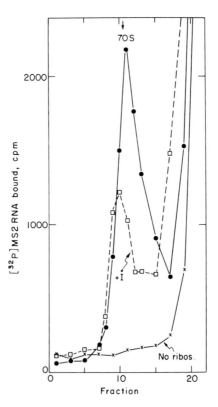

Fig. 6. Isolation of ribosome-protected [^{32}P]MS2 RNA fragments by sedimentation on a glycerol gradient as described in the text. The 70 S peak of RNase-resistant MS2 RNA appears only when ribosomes are present. The inhibitory effect of interference factor i on this reaction is illustrated.

centrifuged on a 5–20% glycerol gradient (25 ml) in buffer V for 13 hours at 17,000 rpm in a Spinco SW 25 rotor. A total of 35 fractions are collected, and each one is filtered through Selectron BA 85 membranes. Filters are washed with 10 ml of buffer V and counted while wet in a Nuclear Chicago gas flow counter. A typical radioactivity profile of the ribosome protected [^{32}P]RNA fragments is shown in Fig. 6. Filters corresponding to the 70 S peak are pooled in 4 ml 8 M urea, 1 M NaCl, 0.1 M Tris·HCl, pH 7.4, and 1 mM EDTA and agitated 3 hours at 4°. The [^{32}P]RNA eluted is precipitated with 3 volumes ethanol, dissolved in 0.5 ml water, extracted three times with water-saturated phenol, 5 times with ether, and lyophilized (extractions are conveniently carried out in small polypropylene conical tubes from Eppendorf).

Fingerprinting of Protected Fragments. The classical methods of Brownlee and Sanger[17] are used. Routinely it is enough to analyze the T1 RNase oligonucleotides originating from the three binding sites of MS2 RNA (Fig. 7). The lyophilized fragments (about 10^6 cpm) are dissolved in 5 μl containing 0.5 μg T1 RNase, 10 mM Tris·HCl pH 7.5, 1 mM EDTA. After incubation at 37° for 30 minutes, the reaction mixture is applied at about 15 cm from one end of a cellulose acetate electrophoresis strip (3 × 55 cm) wetted with 7 M urea in 5% acetic acid pH 3.5. Electrophoresis is carried out with the same buffer at 4500 V for 50 minutes, in a Savant high voltage electrophoresis apparatus, filled with Versol. The oligonucleotides are then blotted onto a DEAE-cellulose (1 part DEAE-cellulose to 7.5 parts cellulose) thin-layer glass plate (20 × 40 cm) and submitted to ascending homochromatography[18] with homomixture B (30% yeast RNA dialyzed against water, 30 ml, plus 200 ml 8 M urea, deionized by passing through an Amberlite ion exchanger). The plate is preheated and sprayed with water before chromatography. Autoradiography on Kodak Royal Blue X-ray film for 1–2 days gives the picture shown in Fig. 7. Areas of the thin layer containing the oligonucleotides are lifted up with a razor blade and counted by liquid scintillation.

The identity of each oligonucleotide should be verified by further pancreatic ribonuclease A digestion. (The thin-layer areas are washed with ethanol, and the oligonucleotides are extracted with aliquots of 30% triethylamine–carbonate buffer pH 9.5, desiccated and digested with pancreatic ribonuclease as above, and then separated by electrophoresis on DEAE-cellulose paper at pH 3.5.[17])

The radioactivity in each oligonucleotide is calculated as percent of the total radioactivity protected by the ribosomes. The values obtained in the

[17] F. Sanger and G. G. Brownlee, this series, Vol. 12A, p. 361.
[18] G. G. Brownlee and F. Sanger, *Eur. J. Biochem.* 11, 395 (1969).

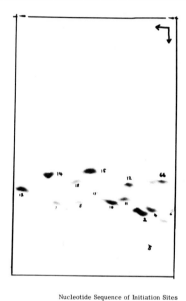

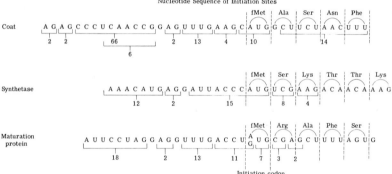

Nucleotide Sequence of Initiation Sites

FIG. 7. Fingerprints of the T1 oligonucleotides from ribosome binding sites of MS2 RNA. For experimental procedure see text. Numbers under each spot correspond to the oligonucleotides indicated in the sequence, as first published by J. A. Steitz [*Nature (London)* **224**, 957 (1969)].

presence or absence of IF3 or one of the interference factors are compared, and the variation produced by the factor studied in each oligonucleotide is calculated. Some of the results are presented in Table II.

Results. By this technique it is possible to show that protein factors influence the selection of ribosomal binding sites on mRNA. In the case of MS2 RNA, purified IF3-B2 fraction ·(of Table I) increases selectively ribosome binding to the coat initiation site. No stimulation of ribosome binding to the other cistron is seen. In contrast, interference factor i (added together with IF3) decreases ribosome binding to the coat protein

TABLE II

EFFECTS OF IF3 AND INTERFERENCE FACTOR i ON RIBOSOME BINDING TO THE
INITIATION SITES OF THE THREE CISTRONS OF MS2 RNA[a]

| Factor assayed | Maturation cistron (%) | Effect on ribosome binding (%)[b] | |
		Coat protein cistron	Synthetase cistron
IF3	−10	+130	−10
Factor i	−25	−50	+80

[a] Computed from published data with IF3-B2 [H. Berissi, Y. Groner, and M. Revel, *Nature (London) New Biol.* **234,** 46 (1971)], and factor i [Y. Groner, Y. Pollack, H. Berissi, and M. Revel, *Nature (London) New Biol.* **239,** 16 (1972)]. The data express the variation in percent of oligonucleotides from each binding site protected by ribosomes with IF1 + IF2 + IF3 versus IF1 + IF2 and with IF1 + IF2 + IF3 + factor i versus IF1 + IF2 + IF3. The values represent an average of the variations of the following oligonucleotides (Fig. 7): maturation cistron: 18, 11, 7; coat cistron: 14, 10, 66, 6; synthetase: 15, 12, 8; common maturation and coat: 13.
[b] Percent of control without the factor.

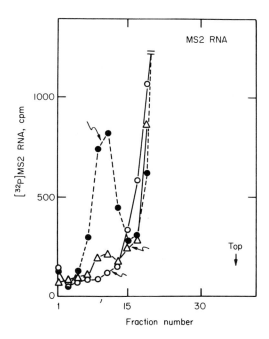

FIG. 8. The T4 mRNA-specific IF3-B4 subfraction (Table I) does not promote MS2 RNA binding to ribosomes. Binding reaction and glycerol gradient sedimentation were carried out as detailed in the text.

cistron initiation site, while it stimulates ribosome binding to the synthetase cistron initiation site. Yoshida and Rudland[5] have described an IF3 subfraction which selectively stimulates ribosome binding to the maturation protein cistron initiation site of phage R17. The T4 mRNA specific IF3 subfraction, IF3-B4 of Table I, does not bind MS2 RNA to ribosomes (Fig. 8; see Lee-Huang and Ochoa[4]).

The results obtained by the measure of ribosomal binding site are in agreement with the study of the products of MS2 RNA translation[6] (Table I). IF3-B2 (of Table I) stimulates predominantly the synthesis of coat protein. IF3-B1, the subfraction containing the IF3-i complex (see above) promotes in contrast a much higher synthesis of synthetase than coat protein. Accordingly, the addition of factor i to IF3 decreases coat protein synthesis while increasing synthetase production.[7]

The methods described here should therefore be very useful for the study and the isolation of these cistron and messenger RNA-specific translation control factors.

[8] Purification and Properties of Interference Factor i: a Cistron-Specific Translation Control Factor and a Subunit of Phage RNA Replicase

By YORAM GRONER and MICHEL REVEL

Interference factors[1-5] are characterized by their mRNA-specific effects on IF3 activity, in initiating the translation of various mRNA's. The assay purification and properties of interference factor i[2,3] (i-α) will be described here in detail.[6]

[1] M. Revel and Y. Groner, this volume [7].

[2] Y. Groner, Y. Pollack, H. Berissi, and M. Revel, *FEBS* (*Fed. Eur. Biochem. Soc.*) *Lett.* **21**, 223 (1972).

[3] Y. Groner, Y. Pollack, H. Berissi, and M. Revel, *Nature* (*London*) *New Biol.* **239**, 16 (1972).

[4] M. Revel, Y. Pollack, Y. Groner, R. Scheps, H. Inouye, H. Berissi, and H. Zeller, *Ribosomes: Structure, Function and Biogenesis, FEBS Symp.* **27**, 261 (1972).

[5] S. Lee-Huang and S. Ochoa, *Biochem. Biophys. Res. Commun.* **49**, 371 (1972).

[6] In addition to factor i, two other interference factors have been detected. Factor i-γ inhibits the translation of MS2 RNA and early T4 mRNA more than late T4 mRNA. Factor i-β inhibits late T4 mRNA more than MS2 RNA. In combination with IF3 these proteins give a series of initiation factors with very different template specificities.

Factor i blocks initiation at the MS2 coat protein cistron, but stimulates it at the MS2 synthetase cistron, on certain T4 mRNA cistrons and on T7 mRNA. Moreover, this protein becomes after RNA phage (Qβ, f2) infection the largest subunit of the RNA replicase.[7]

Assay of Interference Factor i

The assay is based on the fact that this protein inhibits translation of native MS2 RNA, but not T4 mRNA. The inhibition of MS2 RNA translation (Fig. 1) is due to the fact[3] that factor i blocks ribosome binding and initiator fMet-tRNA binding at the coat protein initiation site, which is the only site exposed on native MS2 RNA.[8] When the MS2 RNA is first unfolded by mild formaldehyde treatment,[8] initiation can take place at the three cistrons independently, and since factor i blocks only coat protein synthesis, this template is actively translated in the presence of factor i. The assay of factor i is carried out by measuring the differential effect on at least two different mRNA templates.

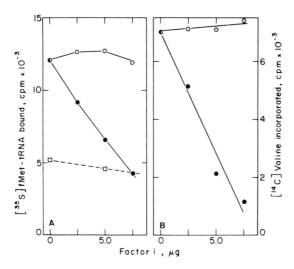

FIG. 1. Assay of interference factor i. (A) Inhibition of MS2 RNA-dependent fMet-tRNA binding to ribosomes as described by Y. Groner, Y. Pollack, H. Berissi, and M. Revel [*Nature (London) New Biol.* **239**, 16 (1972)]. (B) Inhibition of MS2 RNA translation as described in the text. ⊙—⊙, T4 mRNA; ●—●, MS2 RNA; □- - -□, no RNA. (From Groner *et al.*[3] *loc. cit.*)

[7] Y. Groner, R. Scheps, R. Kamen, D. Kolakofsky, and M. Revel, *Nature (London) New Biol.* **239**, 19 (1972).

[8] H. Lodish, *J. Mol. Biol.* **56**, 689 (1970).

Materials

TEMPLATES: MS2 RNA prepared as described by Gesteland and Spahr.[9] T4 mRNA (extracted from *Escherichia coli* 20 minutes after T4 infection of 30°) prepared according to Salser *et al.*[10]

RIBOSOMES AND FACTORS: Ribosomes were washed twice with $2 M$ NH_4Cl, 30 mM Tris·HCl, pH 7.5, 10 mM $MgCl_2$, 14 mM β-mercaptoethanol,[11] initiation factors IF1 and IF2 purified by standard procedures,[11,12] IF3 purified according to Revel and Groner[1]; High speed supernatant,[13] all from *E. coli* MRE 600. *E. coli* B tRNA from Schwarz BioResearch Inc.

Procedure. The assay mixture contained in 0.065 ml, 50 mM Tris·HCl, pH 7.5, 50 mM NH_4Cl, 13 mM $MgCl_2$, 2 mM ATP, 1 mM GTP, 5 mM phosphoenolpyruvate, 2 μg of pyruvate kinase, 0.1 mM CTP and UTP, 0.15 mM 19 amino acids (minus valine), 6 mM dithiothreitol, 0.5 μg of formyltetrahydrofolate, 1.5 nmoles of [^{14}C]valine (50 μCi/ μmole), 100 μg of high salt-washed ribosomes, 5 μl of high-speed supernatant, 5 μg of IF1, 15 μg of IF2, 3 μg of IF3, 25 μg of tRNA, 20 μg of native MS2 RNA or 100 μg of T4 mRNA. Preparations containing factor i are added in parallel to reactions containing MS2 RNA and to reactions containing T4 mRNA. After 30 minutes of incubation at 37°, the incorporation of valine into hot trichloroacetic acid-insoluble material is measured by the filter paper disk method.[11] As shown in Fig. 1 addition of factor i selectively blocks translation of MS2 RNA but not of T4 mRNA.

One unit of interference factor i activity is defined as the amount producing inhibition of 1 pmole of valine incorporation with native MS2 RNA template in the standard assay (standard assay gives 120–160 pmoles of valine incorporated with IF3 in the absence of any interference factor).

Other Assays of Factor i

The translation effect of factor i can be seen directly on initiation by measuring fMet-tRNA binding to ribosomes in the presence of MS2 or T4 mRNA[3] (Fig. 1A).

[9] R. Gesteland and P. F. Spahr, *Biochem. Biophys. Res. Commun.* **41**, 1267 (1970).
[10] W. Salser, R. Gesteland, and M. A. Bolle, *Nature* (*London*) **215**, 588 (1967).
[11] M. Revel, H. Greenshpan, and M. Herzberg, this series, Vol. 20, p. 261.
[12] J. W. B. Hershey, E. Remold-O'Donnell, D. Kolakofsky, K. F. Dewey, and R. E. Thach, this series, Vol. 20, p. 235.
[13] High Speed supernatant, free of initiation factors may be prepared without DNase as described[11] or after DNase treatment as in step 1. In this case prolonged centrifugation of S30 (80,000 g for 16 hours or 150,000 g for 2.5 hours) is used and the upper two thirds of the supernatant are dialyzed against buffer I.

Factor i can be assayed for its activity in the synthesis of Qβ RNA minus strand, using Qβ replicase devoid of i, host factor, and Qβ RNA as described by Kamen *et al.*[14]

Purification Procedure

Source. About 90% of factor i is found in the proteins which are washed off *E. coli* ribosomes by 1 M NH$_4$Cl.

Materials

Buffer I: 10 mM Tris·HCl, pH 7.5, 10 mM MgCl$_2$; 60 mM NH$_4$Cl; 7 mM β-mercaptoethanol, 5% glycerol (v/v)

Buffer II: 1 M NH$_4$Cl, 30 mM Tris·HCl, pH 7.5, 0.2 mM MgCl$_2$; 14 mM β-mercaptoethanol, 5% glycerol

Buffer III: 20 mM potassium phosphate buffer, pH 7.2, 0.2 mM MgCl$_2$, 7 mM β-mercaptoethanol, 5% glycerol

Buffer IV: 20 mM potassium phosphate buffer, pH 7.2, 0.1 M NH$_4$Cl, 0.2 mM MgCl$_2$, 7 mM β-mercaptoethanol, 5% glycerol

Buffer V: 20 mM Tris·HCl, pH 7.5, 0.2 mM MgCl$_2$, 0.2 M NH$_4$Cl, 7 mM β-mercaptoethanol, 5% glycerol

Buffer VI: 20 mM Tris·HCl, pH 7.5, 0.2 mM MgCl$_2$, 10 mM NH$_4$Cl, 7 mM β-mercaptoethanol, 3% glycerol

Buffer VII: 20 mM potassium phosphate buffer, pH 7.2, 0.2 mM MgCl$_2$, 10 mM NH$_4$Cl, 0.2 mM dithiothreitol

DEAE-cellulose, Serva, 0.65 meq/g

DEAE-Sephadex A-50, Pharmacia

Phosphocellulose, Whatman P-11, 7.4 meq/g

Hydroxyapatite, Bio-Rad

DNase, Worthington

Bacteria: *E. coli* MRE 600 is grown in a 300-liter fermentor with 4% potassium phosphate, pH 7, 0.8% glucose and 0.6% yeast extract (Difco) at 37° to mid-log phase. The culture is progressively withdrawn, cooled, and centrifuged; the cells are washed in buffer I and stored at −20°.

Step 1. Preparation of Ribosome Wash Proteins. Cells, 2 kg, are suspended in 2 liters of buffer I with 1 μg of DNase per milliliter and homogenized twice through a continuous-flow French press (Manton-Gaulin homogenizer, Everett, Massachusetts) at 7000 psi (temperature kept below 10°), in batches of 700 ml each. From S30,[11] ribosomes are sedimented in a Spinco R 30 rotor at 80,000 g for 16 hours, and resus-

[14] R. Kamen, M. Kondo, W. Römer, and C. Weissman, *Eur. J. Biochem.* **31**, 44 (1972).

pended overnight with stirring in 900 ml of buffer II at 4°, and centrifuged under the same conditions. To the supernatant, ammonium sulfate (51.6 g/100 ml) is added together with 5 ml of 1 M Tris·HCl, pH 7.8. After 20 minutes in the cold, the precipitate is collected by centrifugation at 18,000 g for 10 minutes, dissolved in buffer III, and dialyzed overnight against several changes of buffer III.

Step 2. DEAE-Cellulose Chromatography. The ribosomal wash proteins (16 g), clarified by low speed centrifugation, are applied at a concentration of 20 mg/ml to a 6.6 × 42 cm column of DEAE-cellulose, equilibrated with buffer III. The column is washed with the same buffer, and fractions of 20 ml are collected at a flow rate of 100 ml per hour. Under these conditions, more than 80% of the protein is adsorbed. A linear gradient of potassium phosphate from 20 mM to 0.3 M in buffer III is established; factor i is eluted at about 0.22 M, as shown in Fig. 2. At this stage assay of factor i activity on protein synthesis may be somewhat difficult. It is important to assay the fractions both with MS2 and T4 mRNA and to screen the fractions for a differential effect on the two templates.[15] Figure 2 indicates also the elution pattern of initiation factor IF3 and IF2 activities which may help to localize factor i. Alternatively,

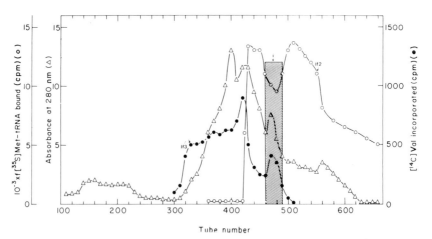

Tube number

Fɪɢ. 2. Chromatography of ribosome wash proteins on DEAE-cellulose. For details see text. Factor i is eluted in the hatched region. IF2 activity (○—○) for fMet-tRNA binding to ribosomes, and IF3 activity (●—●) for T4 mRNA-dependent valine incorporation measured according to M. Revel, H. Greenshpan, and M. Herzberg, this series, Vol. 20, p. 261. From Y. Groner, Y. Pollack, H. Berissi, and M. Revel, *FEBS (Fed. Eur. Biochem. Soc.) Lett.* **21,** 223 (1972).

[15] Stimulation of T4 mRNA translation and only slight inhibition of MS2 RNA translation may be seen at this stage.

it is possible to use a stepwise elution with 0.25 M potassium phosphate in buffer III. The proteins freed of nucleic acids or other contaminants elute as a broad peak, and factor i is found mainly in the first part of the peak.

The DEAE-cellulose fractions are precipitated with 80% saturation ammonium sulfate (pH 7.8). After 1 hour, the precipitate is centrifuged at 18,000 g for 10 minutes, dissolved in buffer IV, and dialyzed against several changes of the same buffer.

Step 3. Phosphocellulose Chromatography. The dialyzed proteins from step 2 (5.8 g) are applied at a concentration of 15 mg/ml to a 2.75 × 27 cm column of phosphocellulose equilibrated with buffer IV. Fractions of 10 ml are collected at flow rate of 34 ml/hour. The column is washed with buffer IV and factor i is found in the effluent. This step separates factor i from IF3 and some IF3-i complex which are adsorbed strongly on phosphocellulose.[1] IF2 is also retained on the column. The non-adsorbed proteins fractions are pooled (1.25 g in 350 ml).

Step 4. Chromatography on Hydroxyapatite. The proteins of step 3 in buffer IV are applied at a concentration of 5 mg/ml to a 2.8 × 38 cm column of hydroxyapatite equilibrated in buffer IV. The column is washed with the same buffer and proteins are eluted stepwise at 0.05, 0.08, 0.10, 0.13, and 0.15 M potassium phosphate (340 ml are used for each step). The bulk of factor i activity (80%) is eluted at 0.13 M phosphate and 20% at 0.15 M phosphate. The fraction eluted at 0.13 M is concentrated by vacuum dialysis and dialyzed against buffer V.

Step 5. DEAE-Sephadex Chromatography. Fractions from step 4 (150 mg protein) are applied at a concentration of 1 mg/ml to a 2.2 × 25 cm column of DEAE-Sephadex A-50 equilibrated with buffer V. Half of the material is found in the column effluent (0.2 M NH$_4$Cl), but factor i is retained on the column, from which it is eluted by a 600-ml gradient of ammonium chloride 0.2 to 0.4 M. Fraction of 10 ml are collected at a flow rate of 20 ml per hour. Factor i elutes as a sharp symmetrical peak at 0.25 M ammonium chloride. Fractions having the highest specific activity only are concentrated by vacuum dialysis (2.5 mg/ml) and dialyzed against buffer VI.

Step 6. Zone Centrifugation on Glycerol Gradients. About 0.1 ml of the fraction from step 5, is layered on a 4.9 ml linear 5–20% (v/v) glycerol gradient in buffer VII and centrifuged 16 hours at 49,000 rpm in Spinco SW 50.1 rotor at 4° (alternatively, 0.3–0.5 ml portions of step 5 may be layered on an 11-ml gradient and centrifuged 42 hours at 41,000 rpm in a Spinco SW 41 Ti rotor at 4°). Fractions of 0.2 ml are collected by puncturing the bottom of the tube and factor i activity is assayed.

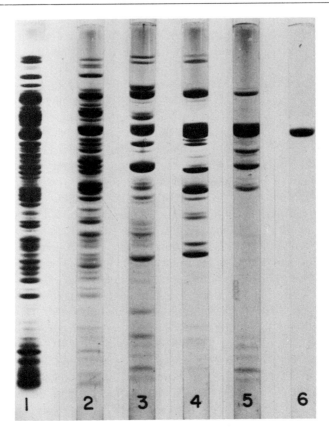

FIG. 3. Electrophoresis on SDS-polyacrylamide gels [U. K. Laemmli, *Nature* (*London*) **227**, 680 (1970)] of the proteins obtained at the different purification steps described in the text and in the table. The gels were modified to contain a gradient of polyacrylamide from 10 to 15%.

Factor i sediments at 5.5–6 S (about one-third of the gradient under the above conditions). At this stage the protein appears homogeneous upon electrophoresis on polyacrylamide gels at pH 4.5 in 8 M urea or in SDS as shown in Fig. 3. From the electrophoretic mobility in SDS–polyacrylamide gels, a molecular weight of 74,000 ± 3500 is determined by comparison with known markers.

The table summarizes the purification steps. The final preparation is purified a 100 fold over the ribosome wash proteins.[16] Factor i can

---

[16] The overall purification of factor i is actually much higher, since the ribosome associated proteins represent about one-sixth to one-eighth of the total proteins in the crude extract.

PURIFICATION OF INTERFERENCE FACTOR i

Purification stage	Protein (mg)	Total activity (units $\times 10^{-3}$ [b])	Specific activity (units/μg protein)
1. Ribosome wash proteins[a]	16,000	11,840	0.74
2. DEAE-cellulose	5,800	8,700	1.5
3. Phosphocellulose	1,250	5,750	4.6
4. Hydroxyapatite	150	3,900	26.0
5. DEAE-Sephadex	43	1,360	31.6
6. Glycerol gradient	3.5	266	76

[a] Purification from 2 kg of *Escherichia coli* MRE 600.

[b] A unit is defined as inhibition of 1 pmole in the MS2 RNA-dependent incorporation of [^{14}C]valine in the standard assay.

be stored in buffer VII in liquid air for at least one year, without loss of activity. It can be heated to 55° for 10 min without loss of activity.

Preparation of Antiserum Against Factor i

Antiserum was prepared in rabbits by three injections (intradermal at multiple sites) of 0.2 mg pure factor i (step 6) each in complete Freund's adjuvant,[17] over a period of 3 weeks. About 6 days after the last injection, bleeding was carried out by puncture of the ear vein. After clotting, serum was obtained by centrifugation for 10 minutes at 10,000 g and stored at $-20°$. To 100 ml of serum were added 40 ml of saturated ammonium sulfate (pH 8.2) at 4°. The precipitate, washed with 40% saturated ammonium sulfate, was dissolved in 0.1 M Tris·HCl, pH 8.2, 0.2 M NaCl, and dialyzed against the same buffer. The crude γ-globulin were further purified on a column of Sephadex G-150 (superfine) (2.8 $\times$ 95 cm) in the same buffer. The protein peak corresponding to 7 S was pooled, concentrated to 10 mg/ml by vacuum dialysis and stored at $-20°$. These antibodies can be used for a very convenient immunoassay of i (Fig. 4).

Identity of Factor i with Subunit of Qβ Replicase

The above antiserum gave a single band against i antigen (Fig. 4a) and was used to demonstrate the immunological identity of factor i with the largest host subunit of Qβ replicase.[7] Figure 4b shows the Ouchterlony's gel double diffusion precipitation reaction[18] of anti-i serum with a preparation of the two large subunits (I + II) and the two small

[17] A. J. Crowle, "Immunodiffusion." Academic Press, New York, 1961.
[18] O. Ouchterlony, *Acta Pathol. Microbiol. Scand.* **25**, 186 (1948).

a

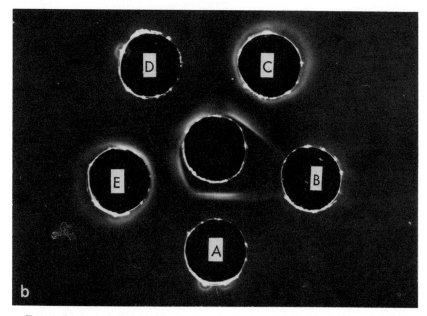

b

FIG. 4. Immunological cross-reaction between anti-i and $Q\beta$ replicase subunit I. (a) Immunoelectrophoresis of crude factor i with antiserum prepared against purified (step 6) factor i shows the purity of the antiserum used (anode at right; pH 8.8). (b) Ouchterlony's gel double diffusion precipitation reaction [O. Ouchterlony, *Acta Pathol. Microbiol. Scand.* **25**, 186 (1948)], carried out on agar plates with wells 3 mm in diameter and 5 mm apart on commercial agar-coated slides for immunodiffusion. A preparation of the two large subunits derived from $Q\beta$ replicase [R. Kamen, *Nature* (*London*) **227**, 680 (1970)] was put in well A. A preparation of the two small subunits of $Q\beta$ replicase was put in well B. Ribosome wash proteins from uninfected *Escherichia coli* were put in wells C and E, and purified factor i in well D. The center well contains anti-i antibodies.

subunits (III + IV) of $Q\beta$ replicase.[19] Reaction is visible only with the mixture of I + II, and the band fuses with that obtained with crude factor i (step 1) or pure factor i (step 6). Since subunit II is coded by the bacteriophage genome,[19] this shows the immunological identity of

[19] R. Kamen, *Nature* (*London*) **228**, 527 (1970).

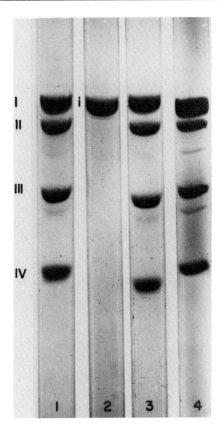

Fig. 5. Electrophoretic pattern of factor i and Qβ replicase subunits. SDS–10%
polyacrylamide gels [U. K. Laemmli, *Nature* (*London*) **227**, 680 (1970)] were
used to analyze: (1) 11 μg of Qβ replicase [R. Kamen, *Nature* (*London*) **228**, 527
(1970)]; (2) 3 μg of factor i (step 6); (3) 11 μg Qβ replicase and 1 μg of factor i;
(4) 11 μg of Qβ replicase and 2 μg of factor i. The molecular weights of factor i
and the subunits of Qβ replicase were calculated with the aid of appropriate
markers: i = 74,000; II = 65,000; III = 47,000; IV = 35,000. Factor i is identical
to the largest subunit of the Qβ replicase.

factor i with subunit I of Qβ replicase. Indeed, on SDS polyacrylamide
stacking gels,[20] factor i from uninfected cells and the large subunit of the
RNA replicase are undistinguishable (Fig. 5).

The function of i in the phage RNA synthesis can be assayed by
complementation of a form of Qβ replicase missing subunit I.[14] This form
of the enzyme is found in small amounts in Qβ-infected *E. coli* extracts.
It is active for poly(C)-dependent poly(G) synthesis and for Qβ RNA

[20] U. K. Laemmli, *Nature* (*London*) **227**, 680 (1970).

minus strand replication. It is inactive for Qβ RNA plus strand replication unless factor i is added (in addition to August's host factor[21]). This is a specific and convenient assay of factor i activity.

Properties of Factor i

The different functions of factor i in phage RNA *and* protein synthesis are best understood as resulting from the properties of this protein to recognize specific sites on the RNA template. The nature of the site recognized on MS2 (or Qβ) RNA is unclear. Qβ replicase binds strongly to a site overlapping the coat protein initiations site, and this binding blocks initiation of coat protein synthesis.[22] Binding of replicase to this site, however, is not a function of subunit i by itself, since the complete enzyme can be shown to bind and block coat protein initiation with a much higher affinity and selectivity for Qβ RNA than free factor i.[4] Factor i binds directly to RNA[4]; it also readily makes a complex with IF3 and with the ribosome.[3] It does not block IF3 dissociation activity.[4] When added to the ribosomal initiation complex, factor i increases ribosome binding to the synthetase cistron initiation site,[3] and stimulates translation of this cistron on unfolded MS2 RNA. (Experimental procedures establishing these properties of factor i are detailed elsewhere.[1]) The interaction with MS2 RNA (or Qβ) appear therefore more complex than might be expected; it is possible that interaction takes place simultaneously at several sites located far apart on the phage RNA molecule. The effect of factor i is reversible: the inhibition of native MS2 RNA translation by factor i is overcome by adding an excess of IF3 or of MS2 RNA.[4] Factor i is not a specific "repressorlike" protein for MS2 (or Qβ) RNA, since it exerts its action also on other templates. T7 mRNA translation is stimulated by factor i, whereas on T4 mRNA factor i stimulates the translation of certain cistrons while inhibiting that of others.[4] It does not inhibit poly(AUG)-dependent fMet-tRNA binding. IF3-interference factor i, therefore, appears to be involved in the control of natural mRNA translation.

[21] L. Eoyang and J. T. August, this series, Vol. 12B, p. 530.
[22] H. Weber, M. Billeter, S. Kahane, C. Weissmann, J. Hindley, and A. Porter, *Nature (London) New Biol.* **237**, 166 (1972).

[9] The Initial Steps in Protein Synthesis: Effects by Antibiotics

By JUAN MODOLELL

The initial steps of polypeptide synthesis in *Escherichia coli* proceed with (a) the formation of a fMet-tRNA·30 S subunit·mRNA complex; (b) the joining to this complex of a 50 S subunit to form a 70 S initiation complex with the fMet-tRNA located in the ribosomal donor site (P-site); (c) the binding, specified by the mRNA, of a molecule of aminoacyl-tRNA to the ribosomal acceptor site (A site); and (d) the formation of the first peptide bond (reviewed by Lucas-Lenard and Lipmann[1]). The following step, translocation of the newly formed fMet-aminoacyl-tRNA from the A to the P site, together with the movement of the mRNA and the ejection of the deacylated tRNA, will not be dealt with in this chapter. At least three factors (IF1, IF2, and IF3) are implicated in initiation (steps a and b), and two factors (EF Ts and EF Tu) are required for aminoacyl-tRNA binding (step c). In addition, initiation and aminoacyl-tRNA binding require the hydrolysis of GTP.[1] Here, I described procedures to assay steps a–d, using washed ribosomes or ribosomal subunits, either natural or synthetic messengers, the physiological initiator fMet-tRNA, crude initiation factors, and purified elongation factor EF T (a mixture of EF Ts and EF Tu). The aim is to provide relatively physiological and simple tests to investigate the interference of antibiotics with these steps of polypeptide synthesis.

Materials

MRE 600 *Escherichia coli* cells: half or late log phase; grown in rich medium. Whenever possible use fresh unfrozen cells for the preparation of ribosomes and crude initiation factors.

70 S ribosomes: prepared by alumina grinding of the cells[2] and washed once with 10 mM Tris·HCl pH 7.8, 10 mM Mg acetate, 60 mM NH$_4$Cl, 6 mM 2-mercaptoethanol, and three times with 50 mM Tris·HCl pH 7.8, 40 mM Mg acetate, 1 M NH$_4$Cl, 1 mM EDTA, 10 mM 2-mercaptoethanol, according to the procedure of Ohta *et al.*[3] The purified ribosomes can be conveniently stored unfrozen at −20° for several months in 50% glycerol

[1] J. Lucas-Lenard and F. Lipmann, *Annu. Rev. Biochem.* **40**, 409 (1971).
[2] M. W. Nirenberg, this series, Vol. 6 [3].
[3] T. Ohta, S. Sarkar, and R. E. Thach, *Proc. Nat. Acad. Sci. U.S.* **58**, 1638 (1967).

containing 10 mM Tris·HCl pH 7.8, 10 mM Mg acetate, 250 mM NH$_4$Cl, and 2 mM dithiothreitol.

Isolated ribosomal subunits: prepared as described by Staehelin and Maglott,[4] or alternative procedures, from the high salt-washed ribosomes.

Crude initiation factors (IF): prepared as described by Hershey et al.[5] Dialyzed before use against 50 mM Tris·HCl, 250 mM NH$_4$Cl, 1 mM dithiothreitol, 10% glycerol. Dialyzed solutions can be stored at 0° for 3 days to 1 week.

[³H]fMet-tRNA: prepared from E. coli B tRNA and [³H]methionine, specific activity 10 Ci/mmole (Amersham), according to the procedure of Hershey and Thach.[6] High levels of formylation can be attained in the absence of an exogenous formyl donor using, as the source of the enzymes, an undialyzed S100 extract prepared from fresh unfrozen cells.

[¹⁴C]Ala-tRNA: prepared as [³H]fMet-tRNA but replacing [³H] methionine by [¹⁴C]alanine, specific activity 0.15 Ci/mmole (Amersham). Use an extensively dialyzed S100 fraction.

Elongation factor T (EF T): a mixture of EF Ts and EF Tu, prepared as described by Parmeggiani et al.[7] or alternative procedures.[8,9] Dialyzed before use against 50 mM Tris·HCl pH 7.8, 2 mM dithiothreitol in 50% glycerol. Dialyzed solutions should be stored unfrozen at −20°. They remain active for over one month.

Phage R17-RNA: prepared as described by Eoyang and August[10] or alternative procedures[11,12]

AUG codon, and random poly(A,U,G), from Miles

GTP, puromycin, dithiothreitol, and alanine, from Sigma

Formylmethionine, from Cyclo

Formylmethionylalanine, synthesized by formylation[13] of methionylalanine obtained from Sigma

[4] T. Staehelin and D. R. Maglott, this series, Vol. 20 [47].
[5] J. W. B. Hershey, E. Remold-O'Donnell, D. Kolakofsky, K. F. Dewey, and R. E. Thach, this series, Vol. 20 [26].
[6] J. W. B. Hershey and R. E. Thach, Proc. Nat. Acad. Sci. U.S. 57, 759 (1967).
[7] A. Parmeggiani, C. Singer, and E. M. Gottschalk, this series, Vol. 20 [30].
[8] J. Gordon, J. Lucas-Lenard, and F. Lipmann, this series, Vol. 20 [29].
[9] J. Ravel and R. L. Shorey, this series, Vol. 20 [32].
[10] L. Eoyang and J. T. August, this series, Vol. 12B [137].
[11] N. R. Pace, I. Haruna, and S. Spiegelman, this series, Vol. 12B [138].
[12] D. Nathans, this series, Vol. 12B [160].
[13] J. C. Sheehan and D. M. Yang, J. Amer. Chem. Soc. 80, 1154 (1958).

Millipore HAWP 2400 or Sartorius SM 11306 filters, from Milli-
pore Corp. or Sartorius-Membranfilter GmbH, respectively
Toluene scintillation fluid, prepared by diluting 5 g of butyl-PBD
(Ciba) in 1 liter of toluene

Assay A. Formation of 70 S Initiation Complex

Principle. To check whether an antibiotic interferes with any step
of initiation, it is convenient to test for its effect on the formation of
the 70 S initiation complex using either natural (R17-RNA) or synthetic
messenger—AUG or poly(A,U,G). The reaction proceeds with the formation of a fMet-tRNA·30 S subunit·messenger complex, which, after
joining a 50 S subunit, yields the 70 S complex. It requires initiation factors
(IF) and GTP. Although mixtures of 30 S and 50 S subunits can be
used for the assay, it is simpler to employ 70 S ribosomes, which, under
the conditions used (low Mg^{2+} concentration (5 mM) and presence of
dissociation factor (IF3) in the IF preparation[14]), dissociate and provide
the required subunits. The analytical method is based on the filtration
procedure of Nirenberg and Leder[15a]: ribosomes and bound, labeled fMet-
tRNA are retained by nitrocellulose filters, while free fMet-tRNA passes
through. Retained radioactivity is measured by liquid scintillation counting.

Procedure. Each reaction mixture contains in a final volume of 30 μl:
100 mM NH$_4$Cl, 50 mM Tris·HCl pH 7.8, 5 mM Mg acetate, either
6 mM 2-mercaptoethanol or 1 mM dithiothreitol, 0.4 mM GTP, between
0.3 and 1 mg/ml IF, 1 mg/ml R17-RNA or 0.2 mg/ml of either AUG
or poly(A,U,G), 6 pmole [^{3}H]fMet-tRNA, 1 A_{260} unit of 70 S ribosomes
and varying concentrations of the antibiotic tested. The mixtures, without
[^{3}H]fMet-tRNA, are made up at 0° and ribosomes are added last. The
reaction is started by adding the [^{3}H]fMet-tRNA and the mixtures are
immediately placed at the desired temperature: 34° with R17-RNA, 25°
or 30° with AUG or poly(A,U,G). Maximal binding of [^{3}H]fMet-tRNA
is usually attained after incubation for either 10 minutes with R17-RNA
or 3–5 minutes with AUG or poly(A,U,G). Longer incubations may result in decreased binding, specially with the synthetic messengers and the
higher temperatures. The reaction is stopped by dilution of the mixtures
with 2 ml of ice-cold 100 mM NH$_4$Cl, 10 mM Tris·HCl pH 7.8, 10 mM
Mg acetate. The solutions are immediately filtered through Millipore,
Sartorius, or equivalent filters, the filters are washed with three 2-ml portions of the same buffer, and are dried and counted in toluene scintillation
fluid. Control reactions without messenger are run in each experiment,

[14] B. D. Davis, *Nature* (*London*) 231, 153 (1971).
[15] P. Leder, this series, Vol. 12B [156].
[15a] M. W. Nirenberg and P. Leder, *Science* 145, 1399 (1964).

and their values are subtracted from those of the reactions with messenger.

Discussion. Messenger stimulations vary from 3- to 10-fold or higher. Amounts of bound [^{3}H]fMet-tRNA normally found are: 0.6–2 and 1–4 pmoles per A_{260} unit of ribosomes with natural and synthetic messenger, respectively. The concentrations specified for IF, messenger, and [^{3}H] fMet-tRNA are approximate guides; thus, preliminary assays should be carried out with varying concentrations of these components to determine the conditions required for optimal binding.

When this assay indicates that an antibiotic does not interfere with initiation, the completion of the 70 S complex formed in its presence should be checked by assaying the ability of the bound [^{3}H]fMet-tRNA to react with puromycin[16] (this drug acts only on 70 S particles carrying fMet-tRNA, or another suitable substrate, bound to the ribosomal P site). Puromycin (0.2 mM) is added to the reaction mixture after maximal binding of [^{3}H]fMet-tRNA has taken place, the incubation is continued for 1 or more minutes, and bound radioactivity is determined. Completion of the complex is indicated by a release of the bound radioactivity that usually reaches 80% within 1 minute (compare with controls without antibiotic).

When an antibiotic is found to inhibit initiation, the effect can be due to interference with the formation or the stability (or both) of the 70 S complex. To help discriminate between these possibilities, the assays B and C can be carried out. Interference with the stability of the 70 S complex can also be determined with the present assay if the antibiotic is added to the reaction mixture after the complex is formed. The incubation is then continued for 5 or 10 more minutes before analyzing for bound [^{3}H]fMet-tRNA.

Assay B. Formation of 30 S Initiation Complex

Principle. 30 S subunits in the presence of IF and GTP form a complex with mRNA and fMet-tRNA. Owing to the relative instability of the complex the assays are carried out at 10 mM Mg^{2+}.

Procedure. Since 30 S subunits are inactivated to a varying degree during isolation and storage, they should be activated by heating[17] immediately before use. This is done by incubating them at 50° for 2 minutes in a buffer containing 100 mM NH$_4$Cl, 50 mM Tris·HCl pH 7.8, 30 mM Mg acetate, and 6 mM 2-mercaptoethanol.[18] Each reaction mixture con-

---

[16] This assay is not valid for those antibiotics which directly interfere with peptide bond formation.

[17] M. Grunberg-Manago, B. F. C. Clark, M. Revel, P. S. Rudland, and J. Dondon, *J. Mol. Biol.* **40**, 33 (1969).

[18] J. W. B. Hershey, K. F. Dewey, and R. E. Thach, *Nature (London)* **222**, 944 (1969).

tains in a final volume of 30 μl the same amounts of the components described in Assay A, except that Mg acetate is raised to 10 mM and 70 S ribosomes are replaced by 0.75 A_{260} unit of activated 30 S subunits. Incubation for 5 minutes at 25° or 30° is followed by filtration analysis (see Assay A). Controls without messenger should be included and their values be subtracted.

Discussion. AUG and poly(A,U,G) stimulate binding between 3- and 10-fold, but R17-RNA, in our hands, consistently has low stimulatory activity in this assay. Usually, approximately 5% or less of the 30 S subunits in the reaction mixture form initiation complexes. Thus, to prevent formation of 70 S complexes, 30 S preparations with few contaminating 50 S subunits should be used. The presence of 70 S complexes can be detected by the reactivity of the bound fMet-tRNA with puromycin (Assay A, Discussion) or by zonal centrifugation. In some cases, it may be advisable to eliminate contaminating 50 S subunits from IF preparations by DEAE-cellulose chromatography as described by Ohta et al.[3]

Assay C. Joining of the 50 S Subunit to the 30 S Initiation
Complex; Interference with the Stability
of the 70 S Complex

Principle. Addition of 50 S subunits to 30 S initiation complexes results in rapid formation of 70 S complexes. The formation of the 70 S complexes is measured by the reactivity of the bound fMet-tRNA with puromycin. Destabilization of 70 S complexes by an antibiotic is manifested as a 50 S subunit-dependent release of bound fMet-tRNA.

Procedure. 30 S complexes are prepared in a scaled-up reaction mixture as described in Assay B. After 5 or 10 minutes of incubation at 34°, 30-μl portions are either assayed for bound [³H]fMet-tRNA, as described in Assay A, or added to tubes with or without 1.5 A_{260} units of 50 S subunits and 9 nmoles of puromycin (in the four possible combinations) contained in 15 μl of 100 mM NH$_4$Cl, 50 mM Tris·HCl pH 7.8, 10 mM Mg acetate, and either 6 mM 2-mercaptoethanol or 1 mM dithiothreitol. A duplicate set of tubes contain, in addition, the antibiotic assayed. The incubation is continued for 2 or more minutes and bound [³H]fMet-tRNA is determined. A control experiment without messenger should also be performed.

Discussion. 50 S subunits may not only join preformed 30 S complexes, but also stimulate the formation of new complexes. This undesirable effect (it may partly mask the puromycin release of [³H]formylmethionine or the destabilization of the 70 S complex) is minimized using poly(A,U,G) instead of AUG and [³H]fMet-tRNA in only a slight excess. Under these conditions, and in the presence of 50 S subunits, puromycin reacts with approximately 70% of the bound [³H]fMet-tRNA within 2 minutes. Inhibition of this reaction by an antibiotic indicates interference with either the joining of the 50 S subunit or the normal sitting of the fMet-tRNA in the P site of the 70 S complex. Zonal centrifugation can be used to discriminate between these two possibilities. Destabilization of the newly formed 70 S complex is manifested as a 50 S subunit-dependent, puromycin-independent release of [³H]fMet-tRNA. Streptomycin, an antibiotic that induces this effect,[19] releases 50% of the bound [³H]fMet-tRNA in approximately 5 minutes at 34°.

Assay D. Binding of Aminoacyl-tRNA to 70 S Initiation Complex

Principle. With R17-RNA as messenger most ribosomes initiate at the coat protein cistron,[20] which codes for alanine after formylmethionine.[21] Thus, elongation factor EF T plus GTP-dependent binding of Ala-tRNA to a preformed R17-RNA·70 S ribosome·fMet-tRNA complex provides a relatively physiological assay for the binding of aminoacyl-tRNA to the initiation complex.[1] The reaction is carried out in two stages. In the first, the 70 S initiation complex is formed, and in the second, Ala-tRNA is bound to it in the presence of EF T, GTP, and the antibiotic whose effect is to be determined. Ribosome bound [³H]fMet-tRNA and [¹⁴C]Ala-tRNA are measured by filtration analysis.

Procedure. The 70 S initiation complex is formed with R17-RNA in a scaled-up mixture as described in Assay A. After incubation at 34° for 10 minutes, the mixture is chilled and 30-μl portions are either analyzed for bound [³H]fMet-tRNA or mixed with 15 μl of 100 mM NH$_4$Cl, 50 mM Tris·HCl pH 7.8, 5 mM Mg acetate, 1 mM dithiothreitol containing 8 pmoles [¹⁴C]Ala-tRNA, 4 μg EF T and varying amounts of the antibiotic assayed. The mixtures are incubated at 18° for 10 minutes, diluted with buffer, and analyzed for [³H]fMet-tRNA and [¹⁴C]Ala-tRNA bound to ribosomes as described in Assay A. A control experiment without R17-RNA is performed in parallel and its values subtracted.

Discussion. The preparations of [³H]fMet-tRNA and [¹⁴C]Ala-tRNA should be free of other acylated tRNA's, especially of Ala-tRNA and fMet-tRNA, respectively. Commercial stripped tRNA charged in the presence of only the labeled amino acid is usually satisfactory. If necessary, however, the [³H]fMet-tRNA can be freed of all unformylated aminoacyl-tRNA's by the method of Schofield and Zamecnik.[22] The binding of [¹⁴C] Ala-tRNA, under the conditions specified, should be strongly dependent

[19] J. Modolell and B. D. Davis, *Proc. Nat. Acad. Sci. U.S.* **67**, 1148 (1970).
[20] H. F. Lodish, *Nature (London)* **220**, 345 (1968).
[21] J. A. Steitz, *Nature (London)* **224**, 957 (1969).
[22] P. Schofield and P. C. Zamecnik, *Biochim. Biophys. Acta* **155**, 410 (1968).

on the presence of EF T, [³H]fMet-tRNA and R17-RNA and should approximate to a 1:1 stoichiometry with the amount of bound [³H]fMet-tRNA (after correcting for the control without messenger).[23] These properties, together with the analysis of the labeled product of the reaction (see Assay E), prove that most [¹⁴C]Ala-tRNA binds to initiation complexes and that extensive elongation does not take place in the system. If necessary, the dependence of the binding of [¹⁴C]Ala-tRNA on messenger and on EF T can be enhanced by conducting the binding at 0° and using incubations as short as 3 minutes.[24] Under these conditions, however, it may be necessary to preincubate the [¹⁴C]Ala-tRNA with EF T and GTP for 1 minute at 30° (in the ionic conditions of the assay) to help forming the GTP·EF Tu·[¹⁴C]Ala-tRNA complex.[1] Sometimes, for unclear reasons, the stoichiometry with bound [³H]fMet-tRNA deviates either above or below unity. In many cases replacing the IF preparation by a freshly dialyzed one corrects the anomalous result. This assay has been useful to help clarifying the mode of action of several antibiotics.[23,24]

Assay E. Formation of the First Peptide Bond

Principle. Ala-tRNA, bound to the fMet-tRNA·70 S ribosome·R17-RNA complex accepts the transfer of formylmethionine and fMet-Ala-tRNA is synthesized. The reaction is catalyzed by peptidyltransferase, an integral part of the 50 S subunit.[25] The ribosomes are isolated in nitrocellulose membranes, the labeled material is released from tRNA by hydrolysis with KOH, and the formylmethionylalanine is isolated by paper electrophoresis. The analytical method is a modification of those described by Ghosh and Korana[26] and Ono *et al.*[27]

Procedure. Formation of the 70 S initiation complex and binding of [¹⁴C]Ala-tRNA, in the presence and in the absence of the antibiotic tested for its effect on the formation of the peptide bond, are carried out as described in Assay D. The mixtures are chilled and portions of 150 μl are diluted and filtered through 2 nitrocellulose membranes (30-μl portions are analyzed for ribosome bound [³H]fMet-tRNA and [¹⁴C]Ala-tRNA as in Assay D). After thorough washing (Assay A), the filters are placed in vials and are extracted twice (with shaking for several minutes)

[23] J. Modolell, B. Cabrer, A. Parmeggiani, and D. Vazquez, *Proc. Nat. Acad. Sci. U.S.* **68**, 1796 (1971).

[24] B. Cabrer, D. Vazquez, and J. Modolell, *Proc. Nat. Acad. Sci. U.S.* **69**, 733 (1972).

[25] R. E. Monro, B. E. H. Maden, and R. R. Traut, *in* "Genetic Elements" (D. Shugar, ed.), p. 179. Academic Press, New York, 1967.

[26] H. P. Ghosh and H. G. Korana, *Proc. Nat. Acad. Sci. U.S.* **58**, 2455 (1967).

[27] Y. Ono, A. Skoultchi, J. Waterson, and P. Lengyel, *Nature (London)* **222**, 645 (1969).

with 1 ml of 1 M NH$_4$OH containing 5 μl of 10 mM solutions of each of the carriers formylmethionine, formylmethionylalanine, and alanine. The two eluates are placed in a small 10-ml flask and are evaporated to dryness in a rotary evaporator. The residue is suspended in 0.3 ml of 0.5 M KOH and incubated (capped) at 37° for 45 minutes. The solution is transferred to a conical centrifuge tube, and is neutralized with 20% HClO$_4$ (approximately 35 μl). The KClO$_4$ precipitated is removed by centrifugation, and the supernatant is transferred to a fresh conical tube and evaporated to dryness. The residue is carefully mixed with 25 μl of H$_2$O, and centrifuged; 15 μl of the supernatant are spotted on Whatman No. 1 paper. Standards of formylmethionine, formylmethionylalanine, and alanine (40 nmoles) can be included in different spots. Electrophoresis is performed at 80 V/cm for 50 minutes in H$_2$O, 189; pyridine, 1; acetic acid, 10 (by volume, pH 3.5). After the strips are dry, the standards can be visualized with ninhydrin[28] (alanine) or the platinic iodide reagent[29] (methionine-containing spots). The strips containing the unknowns are cut, and the radioactivity is determined by liquid scintillation. Approximate mobilities under the specified conditions are: formylmethionine 15.4 cm and formylmethionylalanine 7.5 cm, toward the anode; methionine and alanine less than 3 cm and methionylalanine 8.7 cm, toward the cathode.

Discussion. Approximately 90% of the ^{14}C radioactivity retained by the filters is recovered in the electropherogram, after correcting for the quenching of the filter paper and the size of the spotted aliquot. The quenching correction can be determined by comparing the counts obtained with identical amounts of [^{14}C]alanine spotted on nitrocellulose membranes and on filter paper that has been presoaked in electrophoresis buffer. In a typical experiment, approximately 70% of the recovered ^{14}C radioactivity moves as formylmethionylalanine. The rest is free alanine. Methionylalanine is not normally found. [3]H radioactivity is more difficult to quantify because of the strong quenching caused by the filter paper. Specific inhibition of peptide bond formation by an antibiotic is indicated by a decreased yield of formylmethionylalanine together with an increased recovery of alanine.

[28] W. Stepka, this series, Vol. 3 [79].
[29] G. Zweig and J. R. Whitaker, "Paper Chromatography and Electrophoresis," Vol. 1, p. 97. Academic Press, New York, 1967.

[10] Preparation and Assay of Ribosome Dissociation Factors from *Escherichia coli* and Rabbit Reticulocytes and the Assay for Free Ribosomes

By MICHAEL GOTTLIEB, NICOLETTE H. LUBSEN, and BERNARD D. DAVIS

The products of polysome runoff are free 70 S ribosomes.[1,2] However, physiological translation of mRNA requires initiation via ribosomal subunits. A ribosomal dissociation factor (DF) has been found that mediates the conversion of free ribosomes to subunits. This factor is identical to initiation factor IF3,[3,12] which is also involved in the determination of messenger specificity.[4] We shall describe here a procedure for assaying the dissociation activity of this factor from *E. coli,* as well as the preparation and assay procedure for an analogous factor from rabbit reticulocytes.[5] A similar preparation from rabbit reticulocytes has recently been described as a novel factor,[6] but its activity in promoting dissociation and reinitiation probably depends on the same component.

As is required for its role in the protein synthesis cycle, DF dissociates free ribosomes but not ribosomes complexed with mRNA and peptidyl-tRNA.[7] Hence methods are presented for distinguishing free from complexed ribosomes, on the basis of their stability to dissociation during sucrose gradient centrifugation under altered ionic conditions.[8,9]

E. coli Dissociation Factor

Preparation of IF3 and Free Ribosomes from *E. coli*

Principle. IF3 is found associated specifically with native 30 S subunits, from which it may be removed by washing with a high salt solution.[10] Since isolation of 30 S subunits is troublesome, it is more convenient to use a wash of the total ribosomal pellet.

---

[1] R. W. Kohler, E. Z. Ron, and B. D. Davis, *J. Mol. Biol.* **36,** 71 (1968).
[2] A. R. Subramanian and B. D. Davis, *J. Mol. Biol.* **74,** 45 (1973).
[3] A. R. Subramanian and B. D. Davis, *Nature (London)* **228,** 1273 (1970).
[4] S. Lee-Huang and S. Ochoa, *Nature (London)* **234,** 236 (1971).
[5] N. H. Lubsen and B. D. Davis, *Proc. Nat. Acad. Sci. U.S.* **69,** 353 (1972).
[6] R. Kaempfer and J. Kaufman, *Proc. Nat. Acad. Sci. U.S.* **69,** 3317 (1972).
[7] A. R. Subramanian, B. D. Davis, and R. J. Beller, *Cold Spring Harbor Symp. Quant. Biol.* **34,** 223 (1969).
[8] R. J. Beller and N. H. Lubsen, *Biochemistry* **11,** 3271 (1972).
[9] R. J. Beller and B. D. Davis, *J. Mol. Biol.* **55,** 477 (1971).
[10] A. R. Subramanian, E. Z. Ron, and B. D. Davis, *Proc. Nat. Acad. Sci. U.S.* **61,** 761 (1968).

Free runoff ribosomes, suitable for DF assay, will accumulate in cells exposed to conditions that allow ribosomes to terminate polypeptide synthesis but impair initiation,[1] e.g., treatment with antibiotics (actinomycin D, puromycin, etc.), starvation for carbon source or a required amino acid, or slow cooling (15°, 15 minutes) before harvesting. Ribosomes washed with 1 M NH$_4$Cl (which are a by-product of the IF3 preparation) are also suitable, for they lack the stabilizing ligands and have the same response to IF3 as runoff ribosomes.[3]

Materials

E. coli MRE 600 (RNase I$^-$ strain)

TKD: 10 mM Tris·HCl, pH 7.8; 50 mM KCl; 1 mM dithiothreitol (DTT)

TKM$_n$D: TKD buffer containing n mM Mg acetate

TKM$_{10}$DG$_{10}$: TKM$_{10}$D buffer containing 10% glycerol (v/v)

NH$_4$Cl, 4 M

Procedure. Ribosomes and 1 M NH$_4$Cl wash are prepared essentially as described by Iwasaki et al.[11] Ten milliliters of an S-30 (A_{260}/ml, ca. 300) was obtained from 7 g of alumina-ground cells. Ribosomes are pelleted for 2 hours in a Spinco 50 Ti rotor at 45,000 rpm (R_{av} 133,000 g), or in an IEC A321 rotor at 60,000 rpm, and resuspended in 6 ml TKM$_5$D. One-third volume (2 ml) of 4 M NH$_4$Cl is added, and the suspension is allowed to stand at 0° for at least 2 hours, but generally overnight. The suspension is recentrifuged for 3 hours at 45,000 rpm in a 50 Ti rotor. The upper four-fifths of the supernatant is carefully removed, the remainder is discarded, and the pellet is saved (see below). The protein from the supernatant is precipitated by addition of solid (NH$_4$)$_2$SO$_4$ to 70% saturation (0.47 g/ml). The precipitate is collected by centrifugation for 15 minutes at 30,000 g, dissolved in 0.5 ml of TKD (final protein concentration ca. 20 mg/ml), and dialyzed against 250 ml of the same buffer for 12 hours. The further purification of IF3 has been described elsewhere.[12,13]

The ribosomal pellet obtained from the above NH$_4$Cl washing is resuspended in 1 ml of TKM$_{10}$DG$_{10}$ and the suspension is clarified for 10 minutes at 15,000 g. The concentration is then adjusted to 500 A_{260} units/ml by appropriate dilution with TKM$_{10}$DG$_{10}$ and stored in small aliquots at $-70°$. With certain strains of E. coli (e.g., strain Q13) ribo-

[11] K. Iwasaki, S. Sabol, A. J. Wahba, and S. Ochoa, Arch. Biochem. Biophys. 125, 542 (1968).

[12] S. Sabol, M. A. G. Sillero, K. Iwasaki, and S. Ochoa, Nature (London) 228, 1269 (1970).

[13] J. S. Dubnoff and U. Maitra, this series, Vol. 20, p. 248.

somes prepared in this manner are extensively dissociated and are therefore unsuitable for the DF assay.

DF Assay for IF3 from E. coli

Principle. Incubation of IF3 with free ribosomes under the proper ionic conditions results in net dissociation of the 70 S ribosome into 50 S and 30 S subunits, which may be readily monitored by sucrose gradient centrifugation as in the assay described here. Light scattering[14,15] and electrophoretic[16] techniques have also been employed to estimate the proportion of ribosomes and subunits in a mixture.

Materials

Sucrose gradients, 10–30%, in TKM_5: 10 mM Tris·HCl, pH 7.8; 50 mM KCl; 5 mM Mg acetate
Ribosomes
IF3

Procedure. Reaction mixtures (0.1 ml) contain TKM_3D; 0.5 A_{260} unit (30 μg) ribosomes; and varying amounts of IF3 (either crude or purified). The reaction mixtures are incubated at 37° for 5 minutes and

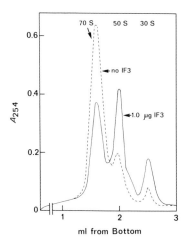

FIG. 1. Partial dissociation of 30 μg of free ribosomes of *Escherichia coli* by 1.0 μg of IF3 in TKM_3D. Adapted from A. R. Subramanian and B. D. Davis [*Nature (London)* **228**, 1273 (1970)].

[11] R. S. Zitomer and J. G. Flaks, *J. Mol. Biol.* **71**, 263 (1972).
[15] L. A. Page, S. W. Englander, and M. V. Simpson, *Biochemistry* **6**, 968 (1967).
[16] J. Talens, F. Kalousek, and L. Bosch, *FEBS (Fed. Eur. Biochem. Soc.) Lett.* **12**, 4 (1970).

then chilled in ice, and samples are layered on 4.8 ml 10–30% sucrose gradients in TKM_5. The gradients are run for 70 minutes at 4° at 45,000 rpm (R_{av} 189,000 g) in a Spinco SW 50.1 rotor. Sedimentation patterns can be determined on a gradient analyzer (e.g., ISCO, Gilford) (Fig. 1). The amount of dissociation is readily quantitated by cutting out and weighing a tracing of the peaks. Under these conditions 2 μg of IF3, or 100 μg of crude IF protein, will dissociate approximately 90% of the 30 μg of ribosomes.

Properties

Substrate Specificity. Complexed ribosomes bearing mRNA and peptidyl-tRNA (including puromycin-releasable fMet-tRNA) are insensitive to IF3 action.[7]

Kinetics and Temperature. Under the assay conditions employed the dissociation reaction is essentially complete within 2 minutes. At 0°, however, the extent of dissociation is low even after prolonged incubation.[10]

Magnesium. Mg^{2+} strongly antagonizes the extent of dissociation, and an increase in Mg^{2+} can readily reverse dissociation even at 0°. At 2 mM Mg^{2+} a 5-fold molar excess of IF3 is required for complete dissociation, and at 5 mM a 50-fold molar excess.[3]

Mechanism. The reaction of IF3 with ribosomes apparently involves an equilibrium, whose constant is dependent upon Mg^{2+}.[17] However, since there is evidence for a spontaneous equilibrium between free ribosomes and subunits, it is not yet possible to say whether IF3 attacks a 70 S ribosome or reacts with 30 S subunits, thus shifting the equilibrium:

$$\text{IF3} + 30\,\text{S} \cdot 50\,\text{S} \rightleftharpoons \text{IF3} \cdot 30\,\text{S} + 50\,\text{S} \tag{1}$$

$$30\,\text{S} \cdot 50\,\text{S} \rightleftharpoons 50\,\text{S} + 30\,\text{S} \xrightarrow{\text{IF3}} \text{IF3} \cdot 30\,\text{S} \tag{2}$$

Experiments with radioactive IF3 have indicated that only 1 mole of IF3 can be bound per mole of 30 S subunits, and none to 70 S ribosomes or 50 S subunits.[18]

Rabbit Reticulocyte Dissociation Factor

Preparation of Dissociation Factor (DF)

Principle. The dissociation factor is found associated with the native subunits,[5] from which it can be removed by treatment with 0.5 M KCl according to the method of Gilbert and Anderson.[19] As with *E. coli* DF,

---

[17] R. J. Beller and B. D. Davis, *Biochemistry* in press.
[18] S. Sabol and S. Ochoa, *Nature* (*London*) **234**, 233 (1971).
[19] J. M. Gilbert and W. F. Anderson, this series, Vol. 20, p. 542.

it is more convenient to treat the total ribosomal pellet. The high-salt wash is then partly purified in order to remove contaminants that interfere with the DF assay. The DF of rabbit reticulocytes appears to be distinct from initiation factors M1, M2A, M2B, and M3.[19a]

Materials

Standard sucrose 0.25 M sucrose, 0.1 mM EDTA (pH 7.0 with NaOH), 2 mM DTT
KCl, 4 M
DEAE-cellulose (Serva SH)
TKE: 10 mM Tris·HCl, pH 7.4; 100 mM KCl; 0.1 mM EDTA; 2 mM DTT; TK$_2$E:TKE with 200 mM KCl; TK$_4$E:TKE with 400 mM KCl
Reticulocyte lysate (prepared according to Gilbert and Anderson[19])
Ammonium sulfate (Schwarz-Mann, enzyme grade, RNase-free)

Procedure. The lysate is spun for 2.5 hours at 60,000 rpm in the IEC A321 rotor to ensure complete pelleting of the native subunits. The ribosomal pellet is resuspended in standard sucrose to 300 A_{260} units/ml (usually 1.0 ml of standard sucrose is needed per 20 ml of the original lysate volume), then 4 M KCl is added to a final concentration of 0.5 M. After standing for 30 minutes on ice the ribosomes are pelleted by centrifugation for 2.5 hours at 60,000 rpm. The supernatant (high-salt wash) is carefully aspirated and the ribosomes are discarded.

High-salt wash, 5 ml, diluted 5-fold with buffer (10 mM Tris·HCl pH 7.4, 0.1 mM EDTA, 2 mM DTT), is adsorbed on a DEAE-cellulose column (25 ml) equilibrated with TKE. The column is washed with TKE until all the hemoglobin is eluted, then further protein is removed by washing the column with 50 ml TK$_2$E, and finally the DF is eluted with TK$_4$E. The protein eluted with TK$_4$E is precipitated by addition of solid ammonium sulfate to 70% saturation. If the protein concentration is too low (less than 1 mg/ml) bovine serum albumin may be added to ensure complete precipitation (the use of Diaflo or vacuum dialysis to concentrate the fraction will reduce the yield obtained). The precipitate is resuspended in 0.2 ml TKE and dialyzed against 250 ml TKE for 3 hours. The DF preparation is rather unstable and is stored frozen in small quantities at $-70°$; 100 μg of such a preparation dissociates about 0.1 A_{260} units of ribosomes.

Alternatively, DF can be similarly prepared from native subunits isolated from a sucrose gradient. This wash can be assayed directly and has a 5 times higher specific activity.

[19a] W. C. Merrick, N. H. Lubsen, and W. F. Anderson, *Proc. Nat. Acad. Sci.* **70**, 2220 (1973).

Preparation of Ribosomes

Principle. The ribosomes used to assay DF must be free of ligands such as mRNA and peptidyl-tRNA, since DF will dissociate only free ribosomes. Ribosomes prepared from cells incubated in medium containing NaF or puromycin to induce runoff were rather insensitive to DF and dissociated only to a limited extent.[5] The most suitable substrate was ribosomes freshly prepared by the puromycin-high salt treatment described by Blobel.[20]

Materials

Reticulocyte lysate
Standard sucrose as above
Puromycin, 10 mM
High salt buffer (100 mM Tris·HCl, pH 7.4; 1 M KCl; 6 mM MgCl$_2$; 2 mM DTT)
MgCl$_2$, 0.5 mM; DTT, 2 mM

Procedure. Ribosomes are pelleted from the lysate for 1 hour at 60,000 rpm in the A 321 rotor in an IEC B 60 centrifuge, resuspended in standard sucrose to 300 A_{260} units/ml, and stored frozen in small aliquots at $-70°$ or at $0°$ (good for 3 weeks). Just before use, an aliquot is diluted to 75 A_{260} units/ml with standard sucrose. To 40 μl of this solution 10 μl of 10 mM puromycin is added, followed by 50 μl of high-salt buffer. This mixture is incubated for 10 minutes at $0°$ and then 10 minutes at $37°$; after chilling 400 μl of 0.5 mM MgCl$_2$, 2 mM DTT is added; 25 μl of this ribosome preparation is then directly added to the DF assay reaction.

DF Assay

Materials

Sucrose gradients, 15–30%, in 10 mM Tris·HCl, pH 7.4; 100 mM KCl; 3 mM MgCl$_2$
Ribosomes
DF

Procedure. The protein sample to be tested is mixed with 0.15 A_{260} units of the ribosome preparation in a total volume of 100–200 μl. The final ionic conditions should be: 10 mM Tris·HCl pH 7.4, 100 mM KCl, 1 mM MgCl$_2$, and 2 mM DTT. Since the DF reaction is very sensitive to the Mg^{2+} concentration, special care must be taken to ensure the

[20] G. Blobel, *Proc. Nat. Acad. Sci. U.S.* **68**, 832 (1971).

proper concentration. The reaction mixtures are then incubated for 10 minutes at 37°, chilled, layered over cold sucrose gradients, and centrifuged and analyzed as described above for *E. coli* DF.

Properties of DF

Substrate Specificity. Reticulocyte DF will dissociate free but not complexed reticulocyte ribosomes. It will also dissociate free *E. coli* ribosomes.

Kinetics. Under the conditions described above the DF reaction is complete within 1 minute. It should be noted that RNase and proteases can also dissociate ribosomes; these can be distinguished from DF by the fact that these reactions usually show typical enzyme kinetics rather than ceasing rapidly.

Mg^{2+} *Dependence.* The DF reaction is inhibited, and also is reversible at 37° (but not at 0°) by high Mg^{2+} concentrations (such as 5 mM).

Temperature Dependence. No reaction occurs at 0°.

Estimation of Free Ribosomes by High Salt Sucrose Gradient Centrifugation

Principle. For some experiments it may be desirable to measure the amount of free ribosomes present in an unwashed ribosome preparation (e.g., to determine suitability for use in DF assays). Such ribosomes can be routinely tested for the presence of bound peptidyl-tRNA and mRNA (which would prevent dissociation by IF3) by determining their stability to sedimentation through a sucrose gradient containing an elevated KCl concentration[8] or containing NaCl[9] instead of KCl. Complexed ribosomes with long polypeptides are stabilized by their ligands and will not dissociate under either of these conditions, whereas the free 70 S ribosomes do. Ribosomes carrying very short nascent peptides are intermediate in their stability in the Na^+ gradient but are stable in the high K^+ gradient used.[8]

The speed and length of the centrifugation, as well as the ribosome concentrations and ionic conditions, influence the patterns obtained, since dissociation does occur in the gradient as a result of hydrostatic pressure. This dissociation can be prevented by fixation of the ribosomes with glutaraldehyde.[21] For a full treatment of these phenomena see Infante and Baierlein (this volume [33]).

These conditions are suitable for analyzing *E. coli* ribosomes. For analyzing eukaryotic ribosomes gradients containing 50 mM Tris·HCl pH 7.4, 0.5 M KCl, 3 mM $MgCl_2$ have been used.[20]

---

[21] A. R. Subramanian, *Biochemistry* **11**, 2710 (1972).

Materials

> Sucrose gradients, 10–30%, containing 10 mM Tris·HCl pH 7.6, 5 mM Mg acetate, and 60 mM NaCl or 175 mM KCl
>
> TKM$_5$D$_2$: 10 mM Tris·HCl pH 7.6, 50 mM KCl, 5 mM Mg acetate, 2 mM DTT

Procedure. Samples, 0.1–0.2 ml, containing 0.1–0.3 A_{260} unit of ribosomes (usually in TKM$_5$D$_2$ buffer), are layered on 4.8-ml sucrose gradients. Use of lower NaCl or KCl, or higher Mg^{2+}, concentrations than indicated above will lead to incomplete dissociation, i.e., the subunits will not be clearly resolved and the apparent sedimentation constant will be higher. Gradients are centrifuged, analyzed, and quantitated as described above for the *E. coli* DF assay.

[11] Protein Chain Initiation in Wheat Embryo[1]

By ABRAHAM MARCUS, SAMARENDRA N. SEAL, and DONALD P. WEEKS

Wheat embryo extracts contain four soluble factors that are required for the incorporation of radioactive amino acids into protein in response to plant viral mRNA.[2] Two of these factors function in initiation[3] and two in elongation.[4] The following sections describe the resolution of the factors from wheat embryo supernatant and some of the characteristics of the initiation reactions.

Preparation of Ribosomes and Supernatant (S100)

Viable wheat embryos[5] (1.1 g) are ground with a cooled mortar and pestle in 11 ml of 1 mM MgAc$_2$, 2 mM CaCl$_2$, 90 mM KCl, 10 mM KHCO$_3$. The embryos are initially ground in 3 ml, with increments of 3 ml and 5 ml added subsequently. After centrifugation for 10 minutes at 23,000 g, 0.01 volume of both 0.1 M MgAc$_2$ and 1 M Tris·acetate (pH 7.6) is added to the supernatant, and the suspension is recentrifuged

[1] This work was supported by U.S. Public Health Service Grant GM-15122 and N.S.F. Grant GB-23041 as well as by N.I.H. Grants CA-06927 and RR-05539 awarded to this Institute and by an appropriation from the Commonwealth of Pennsylvania.
[2] S. N. Seal, J. D. Bewley, and A. Marcus, *J. Biol. Chem.* **247**, 2592 (1972).
[3] A. Marcus, J. D. Bewley, and D. P. Weeks, *Science* **167**, 1735 (1970).
[4] A. B. Legocki and A. Marcus, *J. Biol. Chem.* **245**, 2814 (1970).
[5] Preparation of the embryos is described by A. Marcus, D. Efron, and D. P. Weeks, this volume [68].

for 10 minutes at 23,000 g and then for 60 minutes at 150,000 g. The top three-fourths of the supernatant fluid is taken as S100 and stored at $-15°$.

The ribosomal pellets are suspended in 5 ml of Medium A (2 mM Tris·acetate, pH 7.4, 20 mM KCl, 1 mM MgAc$_2$, 3 mM 2-mercapto-ethanol) per initial 1.1 g of embryo and recentrifuged for 45 minutes at 150,000 g. Pellets equivalent to 2.2 g of embryo are then suspended in 2 ml of medium A, layered over 7 ml of the same medium containing 25% sucrose, and centrifuged for 135 minutes at 150,000 g. The pellet obtained is suspended in 2 ml of medium A containing 20% glycerol and clarified by centrifugation for 10 minutes at 17,000 g. Such ribosomes are referred to as $K_{20}M_1$ ribosomes. In an alternative procedure, the ribosome pellets obtained after the first wash are suspended in 2 ml of 2 mM Tris·acetate, pH 7.4, 100 mM KCl, 2 mM MgAc$_2$, 3 mM 2-mercaptoethanol, kept in ice for 30 minutes, then layered over 7 ml of 25% sucrose containing the same solution. The remainder of the procedure is identical to that used for the $K_{20}M_1$ ribosomes. This latter preparation is referred to as $K_{100}M_2$ ribosomes. The yield from 2.2 g of embryos is 28 and 22 mg of ribosomal RNA (20 A_{260} = 1 mg/ml) for $K_{20}M_1$ and $K_{100}M_2$ ribosomes, respectively.

Resolution of the Soluble Factors

All operations are carried out at 0–4° with the column chromatography done under gravitational pressure at a flow rate of approximately 10–12 ml per hour.

1. Separation of Components C and D. Seven milliliters of S100 (23 mg of protein per ml) are chromatographed on a DEAE-cellulose (Whatman DE23) column, 5 × 0.7 cm, equilibrated with medium II (1 mM Tris·acetate, pH 7.0, 2 mM MgAc$_2$, 4 mM mercaptoethanol) containing 0.1 M KCl. Elution of the column is continued with medium II, 0.1 M KCl. A void volume of 2 ml is discarded; 7 ml are collected as fraction C, and a subsequent 5 ml of wash fraction are discarded. The eluent is changed to medium II, 0.3 M KCl; 1 ml is discarded and 3.8 ml are collected as fraction D. Both of these fractions are relatively stable when kept at $-20°$ and are used as reagents (0.06 ml of fraction C and 0.05 ml of fraction D) in the ATA initiation factor assay during subsequent steps of purification (see below).

2. Purification of Factor C. Fraction C, 5.5 ml (10.7 mg of protein per milliliter) is dialyzed for 45 minutes against medium III (1 mM Tris·acetate, pH 7.6, 1 mM MgAc$_2$, 4 mM mercaptoethanol, 0.1 mM EDTA) and immediately chromatographed on a DEAE-cellulose column, 9 × 0.9 cm, equilibrated with medium III. Elution of the column is continued with medium III, 30 mM KCl, and a total volume of 10 ml is discarded. The eluent is changed to medium III, 0.15 M KCl; 2 ml are discarded, and 6 ml are collected. The preparation of factor C at this point [designated as C (Step 2)] is relatively stable at $-20°$ and is free of the other three soluble factors. Further purification can be achieved by a phosphocellulose procedure,[2] but this step results in considerably increased instability.

3. *Purification of Factor D.* Fraction D, 3.8 ml, (10.2 mg of protein per milliliter) is dialyzed for 60 minutes against 1 mM Tris·acetate pH 7.6, 1 mM MgAc$_2$, 4 mM mercaptoethanol, 0.15 M KCl, and chromatographed to a DEAE-cellulose column, 5 $\times$ 0.7 cm, equilibrated with the same solution. Elution of the column is continued with the same solution, and a total volume of 19 ml is discarded. The eluent is changed to medium II, 0.3 M KCl; 1 ml is discarded and 3 ml are collected as D (Step 2).

4. *Factor T1(E).* The 12 ml (10 ml and 2 ml) normally discarded in the purification of factor C are collected, and 3.8 g of solid $(NH_4)_2SO_4$ are added (45% saturation). The pellet obtained after centrifugation is discarded and an additional 0.85 g of $(NH_4)_2SO_4$ (55% saturation) are added. After overnight storage at 4°, the suspension is centrifuged and the pellet is dissolved in 0.7 ml of medium II, 50 mM KCl; dialyzed for 2.5 hours against 500 ml of 1 mM Tris·acetate, pH 7.3, 50 mM KCl, 4 mM mercaptoethanol with a change of the dialysis solution after 1 hour; and diluted with an equal volume of medium II, 50 mM KCl, 60% glycerol.

5. *Factor T2(E).* The first 15 ml (of the 19 ml) normally discarded in the purification of factor D are collected, and 3.71 g of solid $(NH_4)_2SO_4$ are added (35% saturation). The pellet obtained after centrifugation is discarded and an additional 1.59 g $(NH_4)_2SO_4$ are added (50% saturation). After overnight storage at 4°, the suspension is centrifuged and the pellet is processed as described for T1(E).

The amino acid incorporation assays used for the various factors are sensitive to monovalent cation concentration.[6] Consequently, the factor solutions are routinely dialyzed[7] for 1.5 to 2 hours against 1 mM Tris· acetate, pH 7.3, 50 mM KCl, 4 mM mercaptoethanol just prior to use and their K$^+$ concentrations are taken to be 50 mM. With T1(E), and T2(E), dialysis is carried out in the usual manner prior to use, in addition to the dialysis already performed as the final preparatory step. These twice-dialyzed solutions [T1(E) and T2(E)] are stable for several months at $-20°$.

---

[6] J. D. Bewley and A. Marcus, *Phytochemistry* 9, 1031 (1970).
[7] See this volume [68], footnote 17.

Preparation of Radioactive Aminoacyl tRNA's

Isolation of tRNA. The procedure is based on the method of Zubay.[8] Wheat germ, 50 g, is added to 250 ml of 0.1 M Tris·acetate pH 8, 10 mM MgAc$_2$, 5 mM EDTA pH 7, 50 mM NaCl and 10 mM 2-mercaptoethanol, in a prechilled Waring Blendor and allowed to settle to the bottom of the glass jar. The germ is then homogenized in two 30-second blendings separated by a 30-second interval. Blending speed is regulated by a rheostat set at 40–50 V. The homogenate is centrifuged at 2° for 10 minutes at 23,000 g. The supernatant is decanted through two layers of cheesecloth into a prechilled graduated glass cylinder. An equal volume of water-saturated phenol is added and the mixture is stirred for 15 minutes in the cold. Following centrifugation at 23,000 g for 10 minutes, the upper RNA-containing phase is collected by aspiration into a cooled flask, care being taken not to disturb the thick interface. The water phase is reextracted by stirring for 5 minutes at 2° with 0.5 vol of water-saturated phenol. After centrifugation and aspiration, the water phase is further extracted with one-third volume of phenol–EDTA[9] for 5 minutes. The mixture is centrifuged and the water phase is brought to 50 mM KCl with 2 M KCl, and 2 volume of cold ethanol are added. After overnight precipitation, the RNA is collected by centrifugation at 23,000 g for 10 minutes. The centrifuge tubes are drained and wiped to remove all traces of ethanol. The highly viscous pellets are pooled into 30 ml of cold 0.3 M NaAc pH 7, broken into smaller particles with a glass rod, and finally homogenized in a loose-fitting Dounce homogenizer. The centrifuge tubes and homogenizer are washed with an additional 60 ml of NaAc and the combined suspension is stirred for 10 minutes at 2°. The volume is measured, and 0.54 volume of isopropanol is added dropwise in the cold with vigorous stirring. The stirring is continued and the temperature is carefully raised to 20°. The mixture is then allowed to stand for 5 minutes before centrifugation at room temperature for 6 minutes at 5,000 g. To the supernatant, 0.44 volume (relative to the original volume) of isopropanol is added with stirring. The suspension is stirred for an additional hour at 2° and centrifuged for 10 minutes at 23,000 g. The centrifuge tubes are drained, and the pellets are dissolved in 8 ml of H$_2$O and dialyzed for 5 hours against H$_2$O at 2°. The yield is approximately 60 mg of tRNA per 50 g of wheat germ.

[8-^{14}C]Aminoacyl tRNA's. A charging incubation,[10] of total volume

---

[8] G. Zubay, *J. Mol. Biol.* 4, 347 (1962).

[9] See this volume [68], footnote 16.

[10] An appropriate mixture of several of the reagents; a procedure for preparing TMV-RNA; and the processing of the TCA-insoluble radioactivity are described in this volume [68]. ATA is aurintricarboxylic acid (Aluminon, Aldrich Chem. Co.).

2–4 ml, contains per milliliter: 1.25 μCi of each of [^{14}C]-labeled leucine, serine, lysine, glutamic acid, threonine, valine, proline, and phenylalanine (specific activities 100–300 mCi/mmole), 6.25 nmoles of each of the remaining 12 unlabeled amino acids,[11] 10 mM MgAc$_2$ 83 mM Tris· acetate (pH 8), 70 mM KCl, 1.3 mM ATP, 10 mM creatine phosphate, 53 μg creatine phosphokinase, 3 mM dithiothreitol, 1 mg of tRNA and 0.06 ml of dialyzed S100. [The S100 is freshly prepared from wheat embryo as described above and dialyzed,[7] just prior to use, against 500 ml of 2 mM Tris·acetate (pH 7.6), 5 mM 2-mercaptoethanol, 50 mM KCl for 2 hours at 2°.]

After 30 minutes at 20°, 2.5 μmoles of each of eight ^{12}C-labeled amino acids (corresponding to the ^{14}C-labeled amino acids) are added and the whole solution is applied to a 4.5 × 0.7 cm DEAE-cellulose column prewashed with 50 mM KCl, 50 mM KAc, pH 5. Elution of the column is continued with 50 mM KCl, 50 mM KAc, pH 5, and a total volume equal to 2 ml plus the sample input volume is discarded. The column is then washed with 5 ml of 0.25 M KCl, 50 mM KAc, pH 5, and the eluent is changed to 1 M KCl, 50 mM KAc, pH 5; 0.8 ml is discarded, and the ensuing 5 ml are collected. The latter solution, contains 90–95% of the tRNA present in the charging reaction and is precipitated by adding 2.5 volume of cold absolute ethanol. After overnight storage at −20°, the precipitate is collected by centrifugation (10 minutes at 23,000 g); washed twice with 10 ml of 70% ethanol, 50 mM KAc, pH 5; dissolved in H$_2$O and dialyzed for 3 hours at 2° against H$_2$O. A typical preparation contains 1.5 × 10^6 cpm per milligram of tRNA and is essentially nucleotide free.

^{14}C-Methionyl tRNA is prepared by a similar procedure with 2.5 μCi of ^{14}C-methionine added per milliliter of charging incubation. The two species of tRNA^{Met}—tRNA$_i^{Met}$ (the initiating species) and tRNA$_m^{Met}$ (the internal species)—are resolved from bulk tRNA by chromatography on benzoylated DEAE-cellulose.[12,13]

Assays

ATA Assay for Initiation Factor Activity.[10] A preliminary incubation containing in a volume of 0.35 ml, 10 μmoles of Tris·acetate, pH 8.1, 0.4 μmole of ATP, 3.2 μmoles of disodium creatine phosphate, 16 μg of creatine phosphokinase, 0.01 μmole of GTP, 10 μg of TMV-RNA, 0.9

[11] A. Marcus and J. Feeley, Proc. Nat. Acad Sci. U.S. 50, 1075 (1964).
[12] J. P. Leis and E. B. Keller, Biochem. Biophys. Res. Commun. 40, 416 (1970).
[13] H. Tarrago, O. Monasterio, and J. E. Allende, Biochem. Biophys. Res. Commun. 41, 765 (1970).

μmole of dithiothreitol, ribosomes ($K_{20}M_1$, 280 μg of RNA), 0.84 μmole of MgAc$_2$, 17.5 μmoles of KCl, 400 μmoles each of eight ^{12}C-amino acids (corresponding to the ^{14}C-amino acids, see below) and soluble factors, is carried out for 6 minutes at 30°. Thereafter, 1.2 nmoles of ATA (final concentration 30 μM), 0.6 μmole of MgAc$_2$, 60 pmoles of eight ^{14}C-aminoacyl-tRNA's (1 pmole = 450 cpm), and 0.03 ml of S100 (undialyzed) are added to a final volume of 0.41 ml; the incubation is continued for 9 minutes at 30°, and the radioactive material insoluble in hot trichloroacetic acid is determined. In the standard assay, components C and D (step 1) are used as reagents; 0.06 ml of C (0.64 mg of protein) with fractions to be tested for factor D activity and 0.05 ml of D (0.51 mg of protein) with fractions to be tested for factor C activity. In an alternative assay, particularly useful after factors C and D have been resolved free of elongation factors, the [8-^{14}C]aminoacyl-tRNA's are added to the preincubation mixture, and this preincubation is carried out for 8 minutes at 20°.

Met-tRNA Binding to Ribosomes. The reaction mixture in a total volume of 0.34 ml contains: 1.1 mM methionine, 30 mM Tris·acetate, pH 8, 1.1 mM ATP, 60 μM GTP, 10 μg of TMV-RNA, 2.6 mM dithiothreitol, 1.3 mM MgAc$_2$, 51 mM KCl, ribosomes ($K_{20}M_1$ or $K_{100}M_2$, 280 μg of RNA), 38 pmoles of unresolved [^{14}C]Met-tRNA or 34 pmoles of [^{14}C]Met-tRNA$_i$, or [^{14}C]Met-tRNA$_m$ (1 pmole = 330 cpm) and soluble factors as indicated. After incubation for 10 minutes at 20°, 4 ml of ice-cold diluting buffer (10 mM Tris·acetate, pH 7.6, 70 mM KCl, 5 mM MgAc$_2$) are added and the solution is filtered under gentle suction through nitrocellulose filters (previously washed with diluting buffer). The filter is then washed twice with ice-cold diluting buffer, dried, and counted.

Function of the Factors. The specific reactivity of factors C and D both in the ATA assay and in the Met-tRNA binding assay establishes that these two components are initiation factors. For TMV-RNA dependent amino acid incorporation the other two soluble components, T1(E) and T2(E), are required in addition to factors C and D. Direct assay for catalysis of poly(U)-dependent polymerization of phenylalanine establishes that T1(E) and T2(E) are elongation factors.[2] The individual function of each factor is further determined in a study of di- and tripeptide synthesis with STNV-RNA[14] (formation of Met-Ala and Met-Ala-Lys). Dipeptide formation requires in addition to the two initiation factors, factor T1(E). Tripeptide formation is dependent on the further addition of factor T2(E).

---

[14] S. N. Seal and A. Marcus, in preparation. See also J. Clark, Jr., and W. M. Klein, this volume, section [69].

ACTIVITY OF AMINO ACID INCORPORATION RELATIVE TO ADDITION OF
RIBOSOMES AND FACTORS C AND D[a]

Addition		^{14}C-Labeled amino acid incorporated (pmoles)	
Ribosomes	Factors C and D	TMV-RNA	STNV-RNA
$K_{20}M_1$	Step 1	15.4	10.2
$K_{20}M_1$	Step 2	13.1	4.1
$K_{100}M_2$	Step 1	14.3	5.8
$K_{100}M_2$	Step 2	7.5	2.6

[a] The ATA initiation factor assay was used.

The relative activity of the system for amino acid incorporation decreases markedly when either factor C is resolved (step 2) or when the ribosomes are washed in 0.1 M KCl ($K_{100}M_2$ ribosomes) (see the table). Attempts to restore activity by adding back either the discarded DEAE fraction (from C, step 2) or the ribosomal wash, have been unsuccessful.[15] When tested in the Met-tRNA binding assay, the resolved components (factor C, step 2, and $K_{100}M_2$ ribosomes) are as active as the initial system. A likely possibility is that protein components that can stabilize the growing ribosome·mRNA·peptidyl-tRNA complex (and thereby facilitate the growth of the peptide chain and amino acid incorporation) are removed during the purification.

Puromycin Reactivity. Met-tRNA bound to $K_{20}M_1$ ribosomes reacts with puromycin in a 5-minute incubation at 0–2°.[16] Factor T1(E) is not required for this reaction. If the binding reaction is carried out with $K_{100}M_2$ ribosomes, the amount of Met-tRNA bound is unchanged. However, the puromycin reaction now requires incubation at 30°. At 0°, the bound Met-tRNA is unreactive with puromycin. A protein factor that restores the 0° puromycin reaction can be purified from the ribosomal

[15] A protein fraction conferring puromycin reactivity has been obtained from the $K_{100}M_2$ ribosomal wash (see below). However, this fraction is completely without effect either on the complete amino acid incorporating system or on any of the partial reactions.

[16] S. N. Seal and A. Marcus, *Biochem. Biophys. Res. Commun.* **46**, 1895 (1972). The assay utilizes the Met-tRNA binding reaction, with the tubes transferred to ice after the binding incubation. Puromycin (1 mM) is then added to a final volume of 0.40 ml and the bound Met-tRNA is determined after 5 minutes. Puromycin reactivity is ascertained from the decrease in ribosomally bound Met-tRNA (relative to a nonpuromycin-containing control). Alternatively, the peptidyl puromycin is determined directly by extraction into ethyl acetate [P. Leder and H. Burstzyn, *Biochem. Biophys. Res. Commun.* **25**, 233 (1966)].

wash.[17] The factor has no effect, however, either on dipeptide synthesis or on the overall incorporation of amino acid into protein. The only recognizable function is therefore, the facilitation of the puromycin reaction.

[17] The puromycin factor is prepared from wheat germ ribosomes. Commercial wheat germ (30 g) is first allowed to soak for 4 minutes in 240 ml of 1 mM MgAc$_2$– 2 mM CaCl$_2$–90 mM KCl–10 mM KHCO$_3$ in a precooled Waring Blendor. The germ is then blended at low speed for 50 seconds (five blendings, each of 10-second duration). After filtration through two layers of cheesecloth, the suspension is centrifuged for 10 minutes at 23,000 g. The supernatant is again filtered through cheesecloth and 0.01 volume of 0.1 M MgAc$_2$ and 0.025 volume of 1 M Tris·acetate (pH 7.6) are added. The suspension is recentrifuged for 10 minutes at 23,000 g and again filtered through cheesecloth. The filtrate is then centrifuged for 2.5 hours at 78,000 g in a Spinco S 30 rotor and the pellets are suspended in 4.5 ml per tube of Medium A (see preparation of ribosomes). The contents of two tubes are combined and centrifuged for 45 minutes at 150,000 g in a Ti 50 Spinco rotor. The pellets obtained are suspended in 4 ml per tube of medium A containing 20% glycerol, clarified by centrifuging for 10 minutes at 17,000 g, and stored at −20°. The yield is 8 mg of ribosomal RNA per milliliter. For preparing the wash factor, 5 ml of the ribosome suspension are first diluted with 5 ml of Medium A, and the ribosomes are repelleted by centrifuging for 60 minutes at 150,000 g. The pellet is suspended in 9 ml of 20 mM Tris·acetate, pH 7.7, 100 mM KCl, 2 mM MgAc$_2$, 3 mM 2-mercaptoethanol and kept for 30 minutes in ice. The suspension is then centrifuged for 50 minutes at 150,00 g. The supernatant is removed and 0.1 N acetic acid is added to bring the pH to 4.5. After 10 minutes, the insoluble material is removed by centrifuging for 10 minutes at 24,000 g, and the supernatant is readjusted to pH 7.4 with 0.1 M KHCO$_3$; 6 μl of this factor added together with puromycin (in the 0° incubation) restores full puromycin reactivity to Met-tRNA bound to K$_{100}$M$_2$ ribosomes. The activity of this factor is destroyed by treatment either with 5% TCA, 0.2 N KOH (30 minutes at 30°), or 5 minutes at 50°. When stored at −20°C the activity is stable indefinitely.

[12] Initiation of Globin Synthesis: Assays

By RONALD G. CRYSTAL, NORTON A. ELSON, and W. FRENCH ANDERSON

The assays available to study the initiation process during globin synthesis have been divided into two groups: (1) those utilizing artificial messenger RNA (mRNA) templates, e.g., polyuridylic acid [poly(U)] and ApUpG; and (2) those utilizing a natural globin mRNA template, either endogenous (present on the ribosomes as isolated) or exogenous (isolated and purified; then translated on ribosomes devoid of endogenous mRNA). The assays described are the ones which have proven to be most useful in our laboratory either to purify the initiation factors or to investigate the mechanism of the initiation process. Although several of these assays are specific for globin initiation (e.g., synthesis of the naturally occurring di- and tripeptides), they may be adapted to investigate the initiation of any eukaryotic protein whose N-terminal sequence is known.

The following abbreviations will be used herein and in Chapters [13], [14], and [17]: Met$_F$, methionine donated from methionyl-tRNA$_F$; IF-M$_1$, initiation factor M$_1$; IF-M$_{2A}$, initiation factor M$_{2A}$; IF-M$_{2B}$, initiation factor M$_{2B}$; IF-M$_{2(A+B)}$, initiation factors M$_{2A}$ + M$_{2B}$ combined; IF-M$_3$, initiation factor M$_3$; EF-1, elongation factor 1 (T1); EF-2, elongation factor 2 (T2).

I. Materials

A. General

The following are used for the preparation of materials or in the assays: L-[^{14}C]leucine, L-[^{14}C]phenylalanine, L-[^{3}H]phenylalanine, L-[^{3}H]-methionine and L-[^{35}S]methionine, all of high specific activity; ATP, GTP, phosphoenolpyruvate, dithiothreitol, pyruvate kinase, ninhydrin, EDTA, and sodium heparin from Calbiochem; Tris(hydroxymethyl)aminomethane, sucrose (ultrapure, RNase free) and chromatography standards methionine-valine, methionine-valine-leucine and methionine-valine-histidine from Schwarz-Mann; L-[^{12}C]amino acids and puromycin dihydrochloride from Nutritional Biochemicals; chromatography paper (No. 3 MM) and diethylaminoethylcellulose 23 (DEAE-23) from Whatman; pancreatic RNase (protease free), phenylhydrazine, vitamin B$_{12}$, and folic acid from Sigma Chemical Company; polyuridylic acid [poly(U)] and ApUpG from Miles; cellulose nitrate filters (type HA, 0.45 μm pore size, 25 mm diameter) from Millipore Corporation; and pentobarbital from Veterinary Laboratories, Inc.

Scintillator solutions used are: Liquifluor-toluene (Liquifluor from New England Nuclear Corporation; 40 ml of Liquifluor, 1000 ml toluene), used as the standard scintillator; Bray's scintillator[1] used to dissolve Millipore filters and ethyl acetate; and Noll's scintillator[2] for sucrose gradients.

B. Preparation of Lysate

Solution 1: 0.01% (w/v) vitamin B$_{12}$, 0.1% (w/v) folic acid, in 0.9% NaCl at pH 7; store at −20°, away from light

[1] G. A. Bray, *Anal. Biochem.* **1**, 279 (1960).
[2] H. Noll, *in* "Techniques in Protein Biosynthesis" (P. N. Campbell and J. R. Sargent, eds.), Vol. 2, p. 101. Academic Press, New York, 1969.

Solution 2: 2.5% (w/v) phenylhydrazine in H_2O at pH 7; store at $-20°$ away from light

Solution 3: 140 mM NaCl, 5 mM $MgCl_2$, 50 mM KCl

Solution 4: 2 mM $MgCl_2$, 0.1 mM EDTA, pH 7, 1 mM dithiothreitol

All the materials used for the initiation assay are obtained from the reticulocytes of immature New Zealand white rabbits, 7–8 weeks old, weighing from 2.5 to 4.5 pounds. A high reticulocyte count (75–90%) is produced in the rabbits by treating them with phenylhydrazine by the following schedule:

Day 1: solution 1, 1 ml, intramuscularly, and solution 2, 0.25 ml/kg, subcutaneously

Days 2–6: solution 2, 0.25 ml/kg, subcutaneously

Day 7: no injections

Wear gloves when handling phenylhydrazine since a contact dermatitis is common. If the rabbit develops dark, watery stools, temporarily discontinue the phenylhydrazine, as this is often a preterminal toxic sign. On day 8, the rabbits are given 30 mg of pentobarbital intraperitoneally a few minutes prior to bleeding. Blood is taken by intracardiac puncture with a No. 14 Huber-point needle (Becton-Dickinson) on a 60-ml plastic syringe and transferred to 300-ml centrifuge bottles [for a GSA rotor (Sorvall)]. Approximately 60 ml of blood can be obtained from each rabbit by an experienced operator. Add 2–3 drops of heparin (500 units/ml H_2O) to each syringe and 0.5 ml of heparin to the centrifuge bottles to prevent clotting. The blood is kept on ice until all the rabbits are bled. All subsequent procedures are at 2–4° unless otherwise noted. The red cells are collected by centrifugation in the GSA rotor (8,000 g, 10 minutes). The plasma is discarded, and the red cells in each bottle are washed with 150–200 ml of solution 3 by gentle swirling. Pellet the cells (8,000 g, 10 minutes) and discard the supernatant. The red cells are washed again and separated (8,000 g, 10 minutes). The washed cells are then lysed (four volumes of solution 4 added to one volume of washed cells) and swirled gently for 1–2 minutes. The cellular debris is then pelleted (16,000 g, 20 minutes). As little as two volumes of solution 4 can be used without significant loss of material (sometimes useful if large quantities of lysate are being prepared on a single day). After centrifugation, the supernatant (lysate) is very carefully decanted into a beaker (or drawn off with an inverted 25-ml pipette) with care not to include any cell debris.

The lysate can be stored for more than 2 years without significant loss of activity in liquid nitrogen (vapor, $-196°$). It is convenient to use 50-ml plastic tubes (Falcon) for storage in liquid nitrogen.

The lysate from 10 rabbits is sufficient to prepare quantities of tRNA, supernatant fraction, 0.5 M KCl wash fraction, globin mRNA, and ribosomes for one investigator for several months.

C. Preparation of Lysate Ribosomes and Supernatant Fraction

Solution 5: 0.25 M sucrose, 1 mM dithiothreitol, 0.1 mM EDTA, pH 7

Use lysate either fresh or frozen. Any frozen material removed from liquid nitrogen is thawed as quickly as possible, and all subsequent procedures are done at 2–4°. When making preparations involving 10 or more rabbits, it is convenient to use a large-capacity rotor, such as the Beckman 35, to pellet the lysate ribosomes (143,000 g, 3 hours).

After centrifugation, decant the supernatant fraction into a beaker. If this fraction is to be used for the preparation of tRNA, it should be stored in large plastic bottles at −20° or colder. If it is to be used to make "supernatant fraction" for the globin synthesis assay, repeat the centrifugation (143,000 g, 3 hours) to clear any remaining ribosomes and slowly take the upper 75% of this supernatant (called "supernatant fraction") and store in liquid nitrogen, where it remains active for more than two years.

The pellet of lysate ribosomes is brought up in solution 5 after rinsing the surface of the pellet 3–4 times with this solution. The lysate ribosomes are then brought up into solution 5 (3–4 ml per rabbit) by first breaking the pellet with a glass stirring rod and pouring the broken pellets into a 40-ml conical centrifuge tube. They are then solubilized by intermittently spinning the conical tube in ice. When the pieces of pellet become very small, the solubilization process can be encouraged by drawing the solution up into a small Pasteur pipette a number of times. If a large amount of lysate ribosomes are being prepared (more than 10 rabbits) the broken pellets of lysate ribosomes can be poured into a beaker and very slowly stirred with a magnetic stirring bar (at 2–4°). Once the lysate ribosomes are in solution, an optical density reading (260 nm) is taken. A convenient concentration is 150 to 250 A_{260} units/ml. The usual yield from 10 rabbits is approximately 8000 A_{260} units of lysate ribosomes (higher in the warmer months).

The lysate ribosomes may be stored in liquid nitrogen for more than two years without loss of activity.

D. Preparation of 0.5 M KCl Wash Fraction and Salt-Washed Ribosomes

Solution 6: 0.25 M sucrose, 1 mM dithiothreitol, 1 mM MgCl$_2$, 10 mM KCl, 0.1 mM EDTA, pH 7

Solution 7: 20 mM Tris·HCl, pH 7.5, 0.1 mM EDTA, pH 7.5, 1 mM dithiothreitol, 350 mM KCl

Lysate ribosomes from approximately 10 rabbits are made to 200 A_{260} units/ml with solution 5. While stirring (either with a magnetic stirring bar or by twirling in a conical centrifuge tube on ice), one-seventh of the volume of the lysate ribosomes is added in the form of 4 M KCl, to make the final concentration of KCl approximately 0.5 M. The ribosomes are stirred slowly for 10 minutes and then pelleted (361,000 g, 2 hours). The resulting supernatant is called the "0.5 M KCl ribosomal wash fraction"; it is the step 1 initiation factors described in this volume [13, 14]. The upper 90% of the ribosomal wash fraction is removed and placed into a small conical centrifuge tube on ice. The remaining supernatant is discarded, and the pellet ("salt-washed ribosomes") is brought up in solution 5 to a final concentration of 100–600 A_{260} units/ml (depending upon what is convenient for subsequent assays). The salt-washed ribosomes can also be brought up in solution 6; they go into solution more slowly with this buffer, but activity (per A_{260} unit) seems to be better. In either case the pellet of salt-washed ribosomes is broken up after rinsing the surface of the pellet 3–4 times with solution 5 or 6. The small pieces are then poured into a 12-ml conical centrifuge tube on ice, and the ribosomes are solubilized as described for the lysate ribosomes. Usually this process takes 1 hour. Absorbance at 260 nm is determined and the salt-washed ribosomes and the ribosomal wash fraction are stored in liquid nitrogen in individual aliquots (1–1.5 ml); they are stable for more than two years without loss in activity.

The salt-washed ribosomes have the following properties: (1) When examined on a sucrose gradient (100 mM KCl, 5 mM MgCl$_2$), they are a heterogeneous population of small subunits, large subunits, monosomes (the major form), and polysomes. Isolated monosomes have very little endogenous mRNA (see Section I, F), but the other ribosome forms contain significant quantities of endogenous α and β globin mRNA. (2) They are significantly devoid of initiation factors IF-M$_1$, IF-M$_{2A}$, IF-M$_{2B}$, and IF-M$_3$ and elongation factor EF-1 and partially devoid of elongation factor EF-2. (3) They possess most of the aminoacyl tRNA synthetases. (4) They have significant GTPase activity (ability to hydrolyze the γ-phosphate from GTP).

The 0.5 M KCl ribosomal wash fraction (step 1 initiation factors) has the following properties: (1) It contains significant quantities of all the initiation factors. (2) It contains significant quantities of aminoacyl tRNA synthetases, EF-1, EF-2, and Mg^{2+}, and small amounts of globin mRNA and tRNA.

The Mg^{2+} in the wash fraction (removed from the lysate ribosomes)

can be easily removed by dialysis against solution 7 for 8–12 hours. If the Mg^{2+} is not removed, it is difficult to characterize any assay with increasing amounts of wash fraction, since doing this adds more Mg^{2+}. With "dialyzed ribosomal wash fraction" the Mg^{2+} optimum for the initiation assays is 4–5 mM.

The tRNA and mRNA can be removed by a batch procedure with DEAE-23. The DEAE-23 is cycled according to the manufacturer and then washed with solution 7 until the buffer exposed to the resin is at pH 7.5. In a 30-ml centrifuge tube, place 10 ml of packed, washed DEAE-23 and 10 ml of the dialyzed wash fraction. Mix with a glass stirring rod and intermittently spin on ice for 10 minutes. Pellet the DEAE-cellulose (12,000 g, 10 minutes); carefully remove the wash fraction and centrifuge again (12,000 g, 10 minutes) to remove remaining traces of DEAE-cellulose. The resulting solution is called "DEAE-treated-dialyzed-ribosomal-wash fraction." It is useful to concentrate this wash fraction by either ultrafiltration with a UM-10 membrane (Amicon Corporation) or by ammonium sulfate precipitation (the latter is more rapid). Add 42.1 g of $(NH_4)_2SO_4$ to 100 ml of the DEAE-treated-dialyzed-ribosomal-wash fraction to make a 68% solution. Pellet the precipitate (16,000 g, 20 minutes) and bring the pellet up in solution 7. Dialyze the precipitated material against solution 7 over 8–12 hours with several changes to remove the $(NH_4)_2SO_4$. If the pellet from 30 ml of the DEAE-treated-dialyzed-ribosomal-wash fraction is solubilized in 5 ml of solution 7, a 6-fold concentration is achieved. The 68% ammonium sulfate precipitation step also removes most of the hemoglobin.

This ammonium sulfate precipitated-DEAE-treated-dialyzed-ribosomal-wash fraction (called "concentrated ribosomal wash fraction" in all subsequent descriptions in this chapter) has the following properties: (1) It can be used to supply the initiation factors for any of the assays described below except for those involving cellulose nitrate filter binding; it is not useful for the latter because it still contains non-initiation factor proteins which bind methionyl-tRNA$_F$ to cellulose nitrate filters (often independent of Mg^{2+} or ribosomes); (2) because of the concentration step with $(NH_4)_2SO_4$, 5–15 μl (50–150 μg protein) of the concentrated ribosomal wash fraction is usually sufficient for any of the assays (50–100 μl) to supply saturating initiation factors; (3) it is devoid of tRNA, mRNA and more than 90% of hemoglobin, but still contains elongation factors and synthetases, and (4) it can be stored in liquid nitrogen for more than two years without loss of activity.

The procedures described for preparation of the ribosomal wash fraction can vary. For example, a better yield of initiation factors can be obtained by exposing the lysate ribosomes (at a concentration of 100 A_{260}

units/ml) to 0.5 M KCl for 30 minutes prior to centrifugation. This is the procedure used to obtain step 1 initiation factors during purification (see next chapter).

E. Preparation of Ribosomes for Artificial Templates Assays

To prepare ribosomes for initiation assays using artificial templates, the lysate ribosomes are exposed to 0.5 M KCl for 2–4 hours prior to centrifugation. This does two things: (1) the dependencies on the initiation factors are somewhat better than when the 10-minute salt wash is used, and (2) the dependency on exogenous artificial templates is better. The disadvantage of this technique is that the overall activity (per A_{260} unit of ribosomes) is decreased, probably because the ribosomes are being exposed to a solution with high salt but no magnesium for an extended period of time and the large ribosomal subunit loses 7 S RNP.[3]

F. Preparation of Ribosomes for Dipeptide Assays with Exogenous mRNA

Solution 8: 20 mM Tris·HCl, pH 7.5, 100 mM KCl, 5 mM MgCl$_2$

For the synthesis of the initial dipeptide analog (Met$_F$-puromycin) or the natural dipeptide (Met$_F$-valine) on exogenous globin mRNA we have found that the best dependency on exogenous mRNA has been with ribosomes prepared as follows: expose lysate ribosomes to 0.5 M KCl as described above (Section I, E) and solubilize in solution 5. The monosomes in this preparation contain very little endogenous mRNA and can be separated from the rest of the ribosome population by means of a 14 Ti Beckman zonal rotor. A linear gradient (15–30% sucrose) in solution 8 is made and 2000–3000 A_{260} units of salt-washed ribosomes are layered on the gradient in a volume of 25 ml (dilute with solution 8 if necessary). An overlay of 25 ml of solution 8 (without sucrose) is added, and the gradient is centrifuged for 4 hours at 45,000 rpm. The gradient is displaced with 45% sucrose and the optical density is followed with a UV detector at 254 nm. The large monosome peak is located and the center portion pooled. These monosomes are pelleted by centrifugation (361,000 g, 3 hours) and the pellet solubilized in solution 6 at a concentration of 50–100 A_{260} units/ml and stored in liquid nitrogen. These monosomes still contain a small amount of endogenous mRNA but when used in lower concentrations (0.05–0.2 A_{260} units/assay) the activity from endogenous mRNA is very low.

---

[3] G. Blobel, *Proc. Nat. Acad. Sci. U.S.* **68**, 1881 (1971).

G. Preparation of mRNA Fragment Ribosomes

Solution 9: 40 mM Tris·HCl, pH 7.5, 0.02 μg/ml pancreatic RNase
Solution 10: 25% sucrose, 500 mM KCl, 50 mM Tris·HCl, pH 7.5,
2 mM MgCl$_2$, 1 mM dithiothreitol
Solution 11: 25% sucrose, 20 mM Tris·HCl, pH 7.5, 100 mM KCl,
5 mM MgCl$_2$, 1 mM dithiothreitol

Although salt-washed rabbit reticulocyte ribosomes,[4] human (β-thalassemia) ribosomes,[5] and liver ribosomes[4,6] have all been used in this laboratory to translate exogenous globin mRNA, the ribosomes which show the most activity (per A_{260} unit), and the best dependencies (on mRNA and initiation factors) are those that have been exposed to low levels of pancreatic RNAse.[7,8] To prepare these ribosomes (called "mRNA fragment ribosomes"), lysate ribosomes are diluted with solution 5 to 150 A_{260} units/ml. Add to this an equal volume of solution 9 and incubate at 37° for 10 minutes. Layer over 3 ml of solution 10 in 10-ml centrifuge tubes, and pellet the ribosomes (368,000 g, 2 hours). Discard the supernatant and then wash the tubes carefully 4 times with solution 6. Solubilize the ribosomes in solution 6, layer the ribosomes over 3 ml of solution 11 in 10-ml centrifuge tubes, and pellet the ribosomes (368,000 g, 2 hours). Discard the supernatant, and wash the tubes carefully 4 times with solution 6. Solubilize in solution 6 to a concentration of 50 A_{260} units/ml and store in liquid nitrogen (where they are stable for more than one year).

The properties of the mRNA fragment ribosomes are as follows[8]: (1) They have essentially no intact endogenous globin mRNA; the dependency on exogenous mRNA to translate complete globin chains is almost complete, (2) 8–10% of these mRNA fragment ribosomes contain a fragment of α or β globin mRNA defined as the initiation region. (3) The dependencies on initiation factors for all initiation assays are excellent. (4) There is very little residual RNase on these ribosomes, and we have had no apparent degradation in the globin synthesis assay dependent on exogenous globin mRNA. (5) Although these ribosomes are intact as 40, 60, and 80 S particles, when the RNA contained within them is sized by sucrose gradient fractionation, it is apparent that the ribosomal RNA has been

[4] P. M. Prichard, D. J. Picciano, D. G. Laycock, and W. F. Anderson, *Proc. Nat. Acad. Sci. U.S.* **68**, 2752 (1971).

[5] A. W. Nienhuis, D. G. Laycock, and W. F. Anderson, *Nature (London) New Biol.* **231**, 205 (1971).

[6] D. J. Picciano, P. M. Prichard, W. C. Merrick, D. A. Shafritz, H. Graf, R. G. Crystal, and W. F. Anderson, *J. Biol. Chem.* **248**, 204 (1973).

[7] A. W. Nienhuis and W. F. Anderson, *J. Clin. Invest.* **50**, 2458 (1971).

[8] R. G. Crystal, A. W. Nienhuis, N. A. Elson, and W. F. Anderson, *J. Biol. Chem.* **247**, 5357 (1972).

"nicked" in several places so that when stripped of the proteins, the ribosomal RNA is no longer intact 18 S and 28 S RNA.[8]

H. Preparation of Other Material Used in Initiation Assays

1. tRNA. Rabbit reticulocyte tRNA is isolated from the supernatant fraction as described by Gilbert and Anderson.[9]

2. $tRNA_F^{Met}$ and $tRNA_M^{Met}$. These tRNA species are separated from unfractionated rabbit reticulocyte or rabbit liver tRNA by reversed-phase chromatography (RPC V) as described by Weiss *et al.*[10]; the materials for this column are now available from Miles. We have isolated three major $tRNA^{Val}$ peaks from this column; the middle peak (the largest), cochromatographs with $tRNA_F^{Met}$; this can pose a problem with valyl-tRNA dependency in the Met_F-valine initial dipeptide assay described below (Section II, B, 2).

3. Acylated tRNA. This is prepared as described by Gilbert and Anderson.[9] The synthetase preparation used is extensively dialyzed to eliminate amino acids.

4. Formyl-[³H]methionyl-$tRNA_F$. This is prepared as described by Caskey *et al.*[11] using *E. coli* transformylase as enzyme and calcium leucovorin and [³H]methionyl-$tRNA_F$ (rabbit reticulocyte) as substrates.[12]

II. Initiation Assays

The initiation assays have been divided into two general categories: (1) those using artificial templates and (2) those using natural templates (either endogenous or exogenous globin mRNA). In all these assays, the following considerations hold:

1. All assays are done with ribosomes rate limiting and all other components in saturating amounts. The quantities of initiation and elongation factors used in each assay varies with the assay, and the purity of the factor and *must* be characterized for each factor *and* each assay.

2. The values in our laboratory are listed for the $MgCl_2$ optimum, KCl optimum, etc. The $[Mg^{2+}]$ is most important; in these assays the optimum usually is between 4 and 5 mM. It can be critical in certain assays, i.e., globin synthesis, where the $MgCl_2$ optimum is very sharp.

3. The concentrated ribosomal wash fraction can substitute for the initiation factors in all assays except the binding assays utilizing cellulose nitrate filters (see Sections II, A, 2; II, B, 1).

---

[8] J. M. Gilbert and W. F. Anderson, *J. Biol. Chem.* **245**, 2342 (1970).

[10] J. F. Weiss, R. L. Pearson, and A. D. Kelmers, *Biochemistry* **7**, 3749 (1968).

[11] C. T. Caskey, B. Redfield, and H. Weissbach, *Arch. Biochem. Biophys.* **120**, 119 (1967).

[12] D. A. Shafritz and W. F. Anderson, *Nature (London)* **227**, 918 (1970).

A. Assays Using Artificial Templates

1. Polyphenylalanine Synthesis on Poly(U) at Low [Mg^{2+}]

This assay examines the "initiation" of polyphenylalanine synthesis on the artificial template, poly(U), under conditions optimal for initiation (i.e., low magnesium concentration) with phenylalanine-tRNA as a substrate. The use of acylated tRNA eliminates the acylation step and the need for phenylalanine tRNA synthetase. At high Mg^{++} concentration (10 mM) the synthesis of polyphenylalanine on the poly(U) template occurs without a requirement for initiation factors. At low Mg^{2+} concentration, however, three of the initiation factors, IF-M$_1$, IF-M$_{2A}$, and IF-M$_{2B}$ (in addition to the elongation factors, EF-1 and EF-2) are required. Because the Mg^{2+} optimum is lower (i.e., shifted down) in the presence of initiation factors, this assay is also called the "poly(U) Mg^{2+} shift assay." [13] The requirements for this assay are similar to the requirements for initiation factor IF-1 and IF-2 in prokaryotes.[14] In the mammalian system, however, it is not necessary to have N-acetylphenylalanine present to initiate the polyphenylalanine chains at low [Mg^{2+}].

The assay is usually run in 50-μl reaction mixtures for 2–3 minutes at 37°. Conditions include: 20 mM Tris·HCl, pH 7.5, 1.0 mM GTP, 2 mM phosphoenolpyruvate, 0.2 IU pyruvate kinase, 1 mM dithiothreitol, 0.2 A_{260} unit of 2–4 hours 0.5 M KCl-washed ribosomes (see Section I, E), 10 pmoles of [^{14}C] or [^{3}H]phenylalanyl-tRNA, 0.5 A_{260} unit of poly(U), IF-M$_1$, IF-M$_{2A}$, IF-M$_{2B}$, EF-1, and EF-2 (all in saturating amounts), and MgCl$_2$ (optimum varies from 4 to 7 mM). After incubation, the reaction mixtures are stopped by the addition of 10% trichloroacetic acid, and the reaction mixtures are heated at 90° for 20 minutes to digest the labeled acylated tRNA. The reaction mixtures are cooled to 0° for 10 minutes and filtered on cellulose nitrate filters with 5% TCA. The filters can be counted directly in Bray's scintillator or dried under an infrared lamp and counted in Liquifluor-toluene.

In the absence of initiation factors IF-M$_1$, IF-M$_{2A}$, and IF-M$_{2B}$, the Mg^{2+} optimum is usually 10 mM. In the presence of any two of the three initiation factors required for this assay, the magnesium optimum remains at 10 mM. However, in the presence of all three initiation factors, the magnesium optimum shifts downward to anywhere from 4 to 7 mM, and, in addition, the total activity of the assay increases.

Since EF-1 + EF-2 alone will contribute a significant amount of polyphenylalanine incorporation on the poly(U) template at 5–7 mM Mg^{2+},

---

[13] D. A. Shafritz and W. F. Anderson, J. Biol. Chem. **245**, 5553 (1970).
[14] J. Lucas-Lenard and F. Lipmann, Proc. Nat. Acad. Sci. U.S. **57**, 1050 (1967).

when assaying for the initiation factors (i.e., during purification) the assay is usually done at 3–5 mM Mg^{2+}. Under these conditions the total activity of the assay is less, but the absolute dependencies on the initiation factors are better. For example, this assay can be used easily to assay for IF-M$_1$ in the presence of saturating amounts of IF-M$_{2A}$, IF-M$_{2B}$, EF-1, and EF-2 at 3–5 mM Mg^{2+}. Similarly, IF-M$_{2A}$ can be assayed in the presence of saturating amounts of other 4 factors. The assay is rapid, convenient and very useful for the purification of IF-M$_1$, IF-M$_{2A}$, or IF-M$_{2B}$ by column chromatography (see this volume [13], [14], [17]). In addition, we use this assay at 10 mM Mg^{2+} (in the absence of initiation factors) to assay for EF-1 and EF-2.[13]

Other important aspects of this assay include the following: (1) The elongation factors EF-1 and EF-2 can be contaminated with the initiation factors (see Table III), and this can diminish dependency on the initiation factors at low Mg^{2+}. (2) Incorporation is greater when *E. coli* phenylalanine-tRNA (General Biochemicals, *E. coli* B) is used as a substrate apparently because it is deacylated less rapidly than mammalian phenylalanyl-tRNA under assay conditions.

2. Initiator tRNA Binding to ApUpG

The general concept of this assay is to examine the binding of the initiator tRNA, methionyl-tRNA$_F$, to ribosomes as directed by the codon ApUpG. This is a model reaction for an early step in mammalian initiation.[12] The ribosomes (Section I, E) used in this assay are the same as the ribosomes used for the polyphenylalanine assay described above. Better dependency on both initiation factors and ApUpG can be found if the ribosomes are washed with 0.5 M KCl a second time (but with 3 mM Mg^{++}). However, loss of activity does occur with rewashing and one has to balance the relative loss of activity to the amount gained in dependency on template and factors. This assay can also be carried out on isolated small ribosomal subunits.[15]

The assays are incubated from 2 to 2.5 minutes at 23° in 50 μl. The conditions of the assay are as follows: 20 mM Tris·HCl, pH 7.5, 1 mM dithiothreitol, 0.5 mM GTP, 4.5–5 mM MgCl$_2$, 100 mM KCl, 0.8–1.2 A_{260} units of ribosomes (Section I, E), 0.3–0.5 A_{260} unit of ApUpG, IF-M$_1$, IF-M$_{2A}$, and IF-M$_{2B}$ in saturating amounts, 10 pmoles of [^{3}H] or [^{35}S]-methionyl-tRNA$_F$. The technique of this assay, similar to the Nirenberg-Leder binding assay,[16] is crucial for success. The reactants are added in the following order: H$_2$O, Tris, dithiothreitol, GTP, MgCl$_2$, KCl, ApUpG,

[15] D. A. Shafritz, D. G. Laycock, R. G. Crystal, and W. F. Anderson, *Proc. Nat. Acad. Sci. U.S.* **68**, 2246 (1971).
[16] M. Nirenberg and P. Leder, *Science* **145**, 1399 (1964).

ribosomes, initiation factors and methionyl-tRNA$_F$. Since the methionyl-tRNA$_F$ starts the reaction (which proceeds at 2°, albeit at a slower rate), it is necessary to immediately process the reaction mixture once methionyl-tRNA$_F$ is added. Gently vortex for 1–2 seconds and incubate in a water-bath at 23°. After incubation, dilute the reaction with 3–5 ml of cold solution 8. After dilution, immediately filter on a cellulose nitrate filter with gentle suction and wash the filter with cold solution 8. Do not allow air to be sucked through the filter. The filter is removed and placed in a counting vial. After completion of the assay, Bray's scintillator is used to dissolve the cellulose nitrate filter, and the picomoles of methionyl-tRNA$_F$ binding are determined.

Several features of this assay are notable:

1. Steps 1–3 IF-M$_1$ and IF-M$_{2(A+B)}$ often have contaminating proteins that are not initiation factors but will (a) bind methionyl-tRNA$_F$ to ribosomes independent of GTP; (b) bind methionyl-tRNA$_F$ to cellulose nitrate filters independent of ribosomes; or (c) bind [^{3}H]GTP to cellulose nitrate filters independent of ribosomes. For these reasons it is preferable to use more purified factors (step 4 or better). If this is not possible, it is always necessary to determine dependencies on ribosomes, GTP, and magnesium to avoid being confused by noninitiation proteins in the initiation factor preparations which apparently bind the initiator tRNA.

2. If a magnesium curve is being done, the wash (solution 8) should have the same magnesium concentration as the reaction mixture.

3. The binding of methionyl-tRNA$_F$ to ApUpG on the ribosome appears to be a rather loose complex in that it has not been possible to detect this complex by sucrose gradients.

4. Formyl-methionyl-tRNA$_F$ can substitute for methionyl-tRNA$_F$ in this assay[12,17]; when this is done, there is a loss of the GTP and M$_{2A}$ requirements.

5. This assay is most useful for following the purification of IF-M$_1$, IF-M$_{2A}$, or IF-M$_{2B}$. Whereas the initiation factor requirements for binding of the initiator tRNA to ApUpG are the same as for binding to natural globin mRNA (Section II, B, 1, a), the GTP requirements differ.[18]

6. Methionyl-tRNA$_M$ will not substitute for methionyl-tRNA$_F$. However it will bind to ApUpG (or to globin mRNA) on ribosomes at 10 mM MgCl$_2$ in the presence of EF-1 and GTP.

3. Met$_F$-Puromycin Synthesis on ApUpG

Since methionyl-tRNA$_F$ is the initiator tRNA for α and β globin chains of rabbit reticulocytes, a model system for studying the formation of the

[17] D. A. Shafritz, P. M. Prichard, J. M. Gilbert, W. C. Merrick, and W. F. Anderson, *Proc. Nat. Acad. Sci. U.S.* 69, 983 (1972).

[18] R. G. Crystal and W. F. Anderson, *Proc. Nat. Acad. Sci. U.S.* 69, 706 (1972).

initial N-terminal dipeptide is the synthesis of the dipeptide analog, Met_F-puromycin on the template ApUpG.[19] This is similar to the formyl-methionyl-puromycin assay in prokaryotes.[20] The general concept of the assay is to synthesize the dipeptide Met_F-puromycin and then to extract it into ethyl acetate.

The ribosomes (Section I, E) used for this assay are the same as those used for the polyphenylalanine synthesis assay (Section II, A, 1).

This assay is usually done in an incubation of 50 μl at 30° for 10 minutes. Conditions include: 20 mM Tris·HCl, pH 7.5, 150 mM KCl, 4–5 mM $MgCl_2$ 0.5 mM GTP, 1 mM dithiothreitol, 1 A_{260} unit of ribosomes, 0.5 mM puromycin dihydrochloride, 0.5 A_{260} unit of ApUpG, IF-M_1, IF-M_{2A}, IF-M_{2B}, and EF-1 (in saturating amounts), and 10 pmoles of [³H]- or [³⁵S]methionyl-tRNA$_F$. After incubation, add 0.9 ml of 0.1 M KPO_4 (pH 8), puromycin peptides are extracted into 3 ml of ethyl acetate. The reaction mixture is vortexed (for 30 seconds) and centrifuged in a benchtop centrifuge for 1 minute. The top 2 ml are taken and counted in Bray's scintillator. A counting correction factor must be taken into account, since only two-thirds of the ethyl acetate is being counted, and ethyl acetate changes the efficiency of the Bray's scintillator.

This assay was useful to examine the requirements for formation of the first peptide bond in mammalian protein synthesis. However, certain characteristics of the assay must be taken into account: (1) Whereas for complete globin synthesis there is an additional requirement for IF-M_3, there is no requirement for this factor in this assay since the artificial template ApUpG is used. When the assay is done utilizing natural templates (Section II, B, 3) IF-M_3 is required. (2) The requirement for elongation factor EF-1 in this assay is not understood. In prokaryotic systems there is no requirement for an elongation factor in the formation of the first peptide bond when puromycin is the aminoacyl tRNA analog.[21] There is no evidence that EF-1 will recognize puromycin nor that the action of EF-1 in this assay is to remove the methionine-puromycin product once it is formed.[18] (3) The synthesis of Met_F-puromycin on the artificial template ApUpG is partially inhibited by the antibiotic fusidic acid.[15] This known inhibitor of translocation does not inhibit the binding of the initiator tRNA on natural mRNA templates, nor will it inhibit synthesis of the initial dipeptide on natural mRNA templates.[18]

For all these reasons we are cautious in utilizing this assay to understand the mechanism of globin initiation. However, it can be useful in

[19] D. A. Shafritz, D. G. Laycock, and W. F. Anderson, *Proc. Nat. Acad. Sci. U.S.* **68**, 496 (1971).

[20] P. Leder and H. Bursztyn, *Biochem. Biophys. Res. Commun.* **25**, 233 (1966).

[21] M. Revel, M. Herzberg, and H. Greenshpan, *Cold Spring Harbor Symp. Quant. Biol.* **34**, 261 (1969).

assaying for the initiation factors during purification, and in particular IF-M$_{2A}$ and IF-M$_{2B}$ since the requirements for these factors are absolute.[19]

There are several other dipeptide analogs that may be synthesized in the mammalian system including formyl-methionyl-puromycin on ApUpG with formyl-methionine donated from tRNA$_F^{Met}$, phenylalanyl-puromycin on poly(U), or methionyl-puromycin with the methionine donated from methionine tRNA$_M$ [the latter assay requires a template poly(A,U,G)]. For the details of these assays the reader is referred to Shafritz et al.[19] The usefulness of the latter group of assays is in having secondary assays available to examine individual initiation or elongation factors.

4. GTP Hydrolysis

Initiation factor IF-M$_{2A}$ has an ability to hydrolyze the gamma phosphate from GTP (γ^{32}) in the presence of ribosomes or 40 S ribosomal subunits.[15] This assay does not require an artificial template; however, it does necessitate using ribosomes that have been extensively washed in 0.5 M KCl (in order to remove ribosomal endogenous GTPase activity); for that reason it is included in the group of artificial template assays. Its particular usefulness comes from the knowledge that of the four initiation factors only IF-M$_{2A}$ has a ribosome dependent GTPase activity,[15] and therefore this assay can be used to purify M$_{2A}$ [usually in conjunction with the polyphenylalanine synthesis (Section II, A, 1) and Met$_F$ binding assays (Sections II, A, 2; II, B, 1, a)].

The GTP used (ICN Corporation) is labeled with ^{32}P in the gamma position. The isotope is diluted with unlabeled GTP to a concentration of approximately 10^{-4} M so that 5 μl in a 50 μl reaction mixture gives approximately 400–500 pmoles GTP.

The assay is usually done in a 50 μl reaction mixture at 37° for 10 minutes. The components of the reaction are: 20 mM Tris·HCl, pH 7.5, 1 mM dithiothreitol, 5 mM MgCl$_2$, 100 mM KCl, 1 A_{260} unit of ribosomes exposed to 0.5 M KCl for at least 2–4 hours (Section I, E), and 400–500 pmoles of [γ^{32}P]GTP. After incubation, the reaction mixture is cooled in ice, 0.5 ml of 0.02 M silicotungstate (in 20 mM H$_2$SO$_4$) is added, followed by 1.2 ml KPO$_4$ (1 mM, pH 7) and then 0.5 ml of 5% ammonium molybdate (in 4 N H$_2$SO$_4$). Vortex and add 2 ml of isobutanol:benzene (1:1). Vortex for 15 seconds twice (with caps on tubes). Centrifuge in a benchtop centrifuge for 2 minutes; take 1 ml of the upper layer and determine radioactivity in Liquifluor-toluene with 0.6% NCS (Amersham-Searle). The reaction mixture should turn yellow on addition of the ammonium molybdate. Some preparations of ammonium molybdate are inconsistent and do not react normally in forming a phosphomolybdate complex.

The major aspects of this assay are as follows[15,17]: (1) It separates

IF-M$_{2A}$ from IF-M$_{2B}$ (the latter does not have GTPase activity, nor do IF-M$_1$ or IF-M$_3$). (2) The reaction is "uncoupled" in the sense that it has not been possible to show dependencies on either ApUpG template, other initiation factors, or methionyl-tRNA$_F$. (3) The GTPase activity of IF-M$_{2A}$ can be demonstrated on the isolated small ribosomal subunit.[15] This differs from the prokaryotic factor, IF-2, in that the latter is dependent on the addition of both ribosomal subunits for its action.[22] The GTPase activity of IF-M$_{2A}$ is not stimulated by the addition of the large subunits.[15]

B. Assays with Natural Templates

1. Initiator tRNA Binding

> Solution 12: 0.1 mM potassium cacodylate, pH 5.5, 5 mM MgCl$_2$, 100 mM KCl

This assay examines the binding of the initiator tRNA to natural globin mRNA on the ribosome. It may be done in two ways: (a) analogous to binding of methionyl-tRNA$_F$ to the artificial codon ApUpG as assayed by the cellulose nitrate filters,[18,23] and (b) by sucrose gradient fractionation.[8,18]

a. Cellulose Nitrate Filter Method. This method is a modification of the assay described in Section II, A, 2 (above). The major difference in the assays is that the salt-washed ribosomes used here have significant quantities of endogenous α and β globin mRNA (Section I, D) and that the codon ApUpG is omitted. This assay is usually done at 23°, for 2–3 minutes in a 50-μl reaction mixture. The conditions of the assay are as follows: 20 mM Tris·HCl, pH 7.5, 0.5 mM GTP, 1 mM dithiothreitol, 4–5 mM MgCl$_2$, 100 mM KCl, 0.6–1.0 A_{260} units of salt-washed ribosomes (see Section I, D), IF-M$_1$, IF-M$_{2A}$, IF-M$_{2B}$ (in saturating amounts), and 10 pmoles of [^{3}H]- or [^{35}S]methionyl-tRNA$_F$. The order of reactants and techniques of the assay are identical to the artificial codon assay.

Important aspects of this assay are as follows: (1) The same caution concerning non-initiation factor proteins binding methionyl-tRNA$_F$ to cellulose nitrate filters applies here. For this reason the assay works best with step 4 IF-M$_1$, IF-M$_{2A}$, and IF-M$_{2B}$. (2) The salt-washed ribosomes are a heterogeneous mixture of ribosome forms (Section I, D); although it is clear that incubation at 2 minutes at 23° binds methionyl-tRNA$_F$ to the small ribosomal subunit only[18] (Section II, B, 1, b), we have not utilized the cellulose nitrate filter binding assay with isolated ribosomal subunits

[22] S. S. Thach and R. E. Thach, *Nature* (*London*) *New Biol.* **229**, 219 (1971).
[23] R. G. Crystal, D. A. Shafritz, P. M. Prichard, and W. F. Anderson, *Proc. Nat. Acad. Sci. U.S.* **68**, 1810 (1971).

with endogenous mRNA attached. (3) GDPCP will not substitute for GTP, and fusidic acid partially inhibits this assay.[18] The latter seems to be an artifact, since fusidic acid will not inhibit the synthesis of the N-terminal dipeptide on endogenous globin mRNA, nor will it inhibit the binding of methionyl-tRNA$_F$ to endogenous globin mRNA on the small ribosomal subunit when displayed on sucrose gradients (Section II, B, 1, b).

b. *Sucrose Gradient Method.* The binding of methionyl-tRNA$_F$ to endogenous messenger RNA on the 40 S subunit as done by sucrose gradient is identical to that with cellulose nitrate filter (Section II, B, 1, a) but after incubation the reaction is diluted with 250 μl of solution 12. This is then layered on a 15–30% linear sucrose gradient made in the same buffer. Separation of the small subunits, large subunits, monosomes, and polysomes under these conditions is easily done in a Beckman 39, 41, or 56 swinging-bucket rotor. The conditions for centrifugation are from 2–4 hours at rated speeds depending upon which rotor is used. Following centrifugation, the gradient is pumped out from below with 45% sucrose and fractions are collected. The fractions can be analyzed by (1) direct counting in Noll's scintillator; (2) binding to cellulose nitrate filters with solution 8 described in the binding assay (Section II, A, 2); and (3) precipitating the ribosomes (with their attached methionyl-tRNA$_F$) with 10% trichloroacetic acid at 0° for 10 minutes followed by filtration with 5% trichloroacetic acid on cellulose nitrate filters. (We have had the most success with this latter technique.)

The low pH of the buffer used in the gradient has been advantageous because methionyl-tRNA$_F$ is unstable and the rate of deacylation of methionine is slower at a lower pH.

c. *Initiator tRNA Binding, General.* Important aspects are as follows:

1. Although more cumbersome to use, the sucrose gradient technique has been useful for clearing up some of the discrepancies in the cellulose nitrate filter binding assay.[18] For example, initial studies with the latter binding assay suggested that the binding of methionyl-tRNA$_F$ to endogenous globin mRNA on ribosomes could be inhibited by fusidic acid. This is not true, as discussed above.

2. In addition, the sucrose gradient data clearly show that methionyl-tRNA$_F$ will bind to the small ribosomal subunit under the conditions of the assay. If the assay is carried on for longer periods of time or at higher temperature, the ^{35}S counts of the [^{35}S]methionyl-tRNA$_F$ move into the monosome region or into the polysome region.

3. Methionyl-tRNA$_M$ will not bind to natural globin mRNA template at 4–5 mM Mg^{2+} under the direction of IF-M$_1$, IF-M$_{2A}$, and IF-M$_{2B}$. It will bind to natural template at 10 mM Mg^{2+} in the presence of EF-1.[23]

4. IF-M$_1$, IF-M$_{2A}$, and IF-M$_{2B}$ are necessary and sufficient to bind

methionyl-tRNA$_F$ to endogenous template on the small subunit. IF-M$_3$ does not influence this interaction. GTP is strictly required, and GDPCP will not substitute.[18]

2. Initial Dipeptide (Met$_F$-Valine) Synthesis

Since methionyl-tRNA$_F$ is the initiator tRNA for rabbit α and β globin chains, and since valine is the N-terminal amino acid found on completed rabbit globin chains, the initial dipeptide formed during α and β chain synthesis should be methionyl-valine, with methionine donated from methionyl-tRNA$_F$. This assay can be done with (a) endogenous[18,23] or (b) exogenous mRNA[18] and can be adapted to examine the synthesis of any initial dipeptide in eukaryotic systems.

a. On Endogenous Globin mRNA. In a reaction mixture of 50 μl, the conditions are: 20 mM Tris·HCl, pH 7.5, 1 mM dithiothreitol, 1 mM GTP, 100 mM KCl, 4–5 mM MgCl$_2$, IF-M$_1$, IF-M$_{2A}$, IF-M$_{2B}$, IF-M$_3$, EF-1 (in saturating amounts), 0.6–1 A_{260} unit of salt-washed ribosomes (Section I, D), 10 pmoles of [^{12}C]valyl-tRNA, and 10 pmoles of [^{35}S]methionyl-tRNA$_F$. The addition of these components is in the same order as the binding reaction (Section II, A, 2); valyl-tRNA is added just before the methionyl-tRNA$_F$. Following addition of the [^{35}S]methionyl-tRNA$_F$, the reaction mixture is incubated for 5–10 minutes at 37°. The reaction is then placed on ice and 2 μl of 1 N NaOH is added; then 2–4 μl of 50 mM standard peptide marker methionine-valine is added. The reaction is incubated for 15 minutes at 37° to hydrolyze peptidyl-tRNA bonds and remove the dipeptide product (Met-Val) from the ribosome.

After hydrolysis the dipeptides must be identified by electrophoretic and chromatographic techniques. Since there is a large amount of protein in these assays the best first step for separation is by electrophoresis. Two buffers have been particularly useful[18,23,24]: (1) pH 3.5 (5% acetic acid, 0.5% pyridine/volume in water); (2) pH 8.0 (10 mM barbital, 15 mM sodium barbital, 30 mM triethylamine, neutralized with acetic acid).

Electrophoresis is performed on Whatman 3 MM paper in a Gilson liquid-cooled high voltage electrophorator (Model D) with standard Gilson frame and 57 × 46 cm sheets of Whatman paper. A line is drawn 4 inches from one end of the paper, and the reaction mixtures are spotted along this line in 1-inch segments with 0.5 inch between reaction mixtures. With care and a small pipette, this can be done reasonably rapidly with spreading kept to a minimum. The paper is then placed on the rack and moistened with buffer with a paintbrush excluding the area of the reaction mixtures.

[24] N. A. Elson, R. G. Crystal, and W. F. Anderson, Anal. Biochem. 54, 161 (1973).

The latter is best moistened with a bulb-type aerosol. It is vital that the reaction mixture area be completely moistened, otherwise severe "streaking" occurs. The paper is then placed in the tank with the reaction mixture side nearest the positive pole for pH 3.5 separations, and nearest the negative pole for the pH 8.0 separations. The length of time for electrophoresis must be determined by pilot runs (usually 90–240 minutes). After electrophoresis the paper is dried in a hot air oven at 70° for 25 minutes. It is stained by dipping into 0.3% (w/v) ninhydrin in acetone and dried (70°, 5 minutes). With barbital buffers the staining solution is made 10% with acetic acid immediately before use. The location of the dipeptide standards are noted, and each lane of the paper is cut into 1-cm strips corresponding to each reaction mixture. Each strip is placed in a counting vial with 10 ml of Liquifluor-toluene. This gives an efficiency of 55–60% for ^{35}S. Using the techniques described here one individual can easily do 12 reaction mixtures in one day.

b. *On Exogenous Globin mRNA.* The assay is done in an identical manner except that 0.05–0.2 A_{260} unit monosomes (Section I, F) and 0.1–0.5 A_{260} unit globin mRNA are substituted for salt-washed ribosomes. We have noted that when larger quantities of monosomes are used (>0.2 A_{260} unit/assay), the presence of small amounts of endogenous mRNA can be detected.

c. *Initial Dipeptide, General.* The initial dipeptide assay is used for two purposes: (1) examination of the mechanisms of initiation and (2) identification of the initiator tRNA and the initial dipeptide.

For the former, the assay is usually incubated for 5 minutes to ensure that the ribosomes are rate limiting. It is crucial that factors are relatively pure, particularly the IF-M$_3$ and EF-1; therefore, purification of these factors at least through phosphocellulose chromatography (step 4 in this volume [14]) is necessary. For the latter, the assay is usually run to extent (10–15 minutes).

There are additional considerations in identifying the initial dipeptide: (1) The system must be tRNA dependent; if unfractionated tRNA is used (with only one amino acid acylated), the system must be free of aminoacyl tRNA synthetases, amino acids, and ATP to avoid contamination with other aminoacyl tRNA's. (2) Electrophoretic and chromatographic systems must be used which separate N-terminal methionine-containing oligopeptides. For example, at pH 3.5, methionine-valine and methionine-leucine cochromatograph, as do methionine-proline, methionine-methionine, and methionine-tryptophan. Separation techniques for many of these oligopeptides are being published separately.[24] (3) If the concentrated ribosomal wash fraction is substituted for the initiation factors, the assay can go beyond the initial dipeptide since the wash fraction often contains EF-2.

EF-2 can be inhibited with fusidic acid ($0.4–1.0$ mM), which prevents translocation and subsequent tripeptide synthesis.[18]

3. Initial Dipeptide Analog (Met$_f$-Puromycin) Synthesis

The use of puromycin as an aminoacyl tRNA analog obviates the need for cumbersome oligopeptide separation techniques. This assay differs from Met$_F$-puromycin synthesis on the artificial codon ApUpG in that IF-M$_3$ is required.[18]

a. On Endogenous mRNA. This assay is similar to the artificial template assay (Section II, A, 3) except that $0.2–0.4$ A_{260} unit of salt-washed ribosomes (containing endogenous mRNA, Section I, D) are substituted for the 2–4-hour washed ribosomes and ApUpG. The incubation conditions are identical except that the KCl optimum is 120 mM KCl. The initiation factor requirements are: IF-M$_1$, IF-M$_{2A}$, IF-M$_{2B}$, IF-M$_3$, and EF-1.

b. On Exogenous mRNA. Identical to the assay on endogenous globin mRNA except that $0.05–0.2$ A_{260} unit monosomes (Section I, F) and $0.1–0.5$ A_{260} unit mRNA are substituted for the salt-washed ribosomes.

c. Initial Dipeptide Analog, General. This assay is simple to do and is rapid; its major drawback is the requirement for EF-1 [same as when an artificial template is used (Section II, A, 3)]. The reason for an EF-1 requirement is not clear.

Other important aspects of this assay are: (1) The requirements for Met$_F$-puromycin on natural template include all the initiation factors; hence when initially characterizing the assay it is easiest to use concentrated ribosomal wash fraction (Section I, D) as a source of factors. The factors can then be substituted for the concentrated ribosomal wash fraction. (2) It is important to use IF-M$_3$ that is not contaminated by EF-1 (see this volume [14]), since both are required for the assay. (3) Because of its convenience, this assay can be readily used to assay IF-M$_3$. (4) Methionyl-tRNA$_M$ will not substitute for methionyl-tRNA$_F$.

4. Initial Tripeptide Synthesis

In the stepwise synthesis of rabbit globin, following initiation, the initial dipeptide (Met$_F$-valine) becomes the initial tripeptides (Met$_F$-valine-leucine, α-chain; and Met$_F$-valine-histidine, β-chain).[18] This assay is identical to the initial dipeptide assay (Section II, B, 2) except 10 pmoles [^{12}C]leucyl-tRNA, 10 pmoles [^{12}C]histidyl-tRNA and EF-2 must be added to the reaction mixture. The assay is usually run for 5–20 minutes at 37°. Separation of the initial tripeptides for rabbit α and β globin can be done by electrophoresis in one step at pH 8.0. For separation of other tripeptides see the paper by Elson et al.[24]

Important aspects of the tripeptide assay include the following: (1)

While dipeptide synthesis is not inhibited by fusidic acid, tripeptide synthesis is, suggesting a translocation step prior to tripeptide synthesis.[18] (2) Unless the reactants are free of aminoacyl tRNA synthetases, amino acids and ATP, lack of tRNA dependence can be a problem. (3) Unless initiation factor requirements are being studied the concentrated ribosomal wash fraction (Section I, D) can be substituted for the factors. (4) This assay (and the dipeptide assay) can be used to examine the initiation of any eukaryotic protein, using appropriate aminoacyl tRNA's and locating the di- and tripeptides as described by Elson et al.[24] (5) Methionyl-tRNA$_M$ will not donate methionine to form an initial tripeptide.

5. Complete Globin Chains

The synthesis of intact α and β globin chains, involving the entire translational mechanisms of the cell, is, of course, the most important "initiation assay," since de novo synthesis of globin chains as directed by globin mRNA must involve the entire initiation sequence of events. The major uses of this assay in our laboratory are: (1) to test components of the cell-free system, i.e., concentrated ribosomal wash fraction, supernatant fraction, etc.,[9] (2) to examine the initiation factor requirements for globin initiation,[4,18,25] (3) to aid in the purification of IF-M$_3$,[4] and (4) to translate various homologous and heterologous mRNA's.[7,8,26] For (1)–(3), ribosomes with endogenous mRNA are used; for (4) ribosomes devoid of mRNA are used.

a. On Endogenous mRNA. The reaction is usually done in 100 μl and incubated at 37° for 20–30 minutes. Conditions include: 20 mM Tris·HCl, pH 7.5, 4.5 mM MgCl$_2$, 80–90 mM KCl, 1 mM dithiothreitol, 1 mM ATP, 0.2 mM GTP, 2 mM phosphoenolpyruvate, 0.1 IU pyruvate kinase, 40 μM [^{14}C]leucine, 40 μM 19 amino acids mixture, minus leucine, 0.1–0.2 A_{260} unit of rabbit reticulocyte tRNA, 150 μg of concentrated ribosomal wash fraction (Section I, D), 100 μg of supernatant fraction (Section I, C), and 0.1–0.2 A_{260} unit of salt-washed ribosomes (Section I, D). The reaction is started by the addition of ribosomes. After incubation, 2 ml of 10% trichloroacetic acid is added and the reactions are incubated for 20 minutes at 90° to destroy labeled aminoacyl tRNA. Then the tubes are put on ice for 10 minutes, and then filtered on cellulose nitrate filters with 5% trichloroacetic acid. The filters are dried and counted in Liquifluor-toluene.

When initiation factors (IF-M$_1$, IF-M$_{2A}$, IF-M$_{2B}$, IF-M$_3$) are substituted for the concentrated ribosomal wash fraction, they have to be in-

[25] P. M. Prichard, J. M. Gilbert, D. A. Shafritz, and W. F. Anderson, *Nature* (*London*) **226**, 511 (1970).

[26] A. W. Nienhuis and W. F. Anderson, *Proc. Nat. Acad. Sci. U.S.* **69**, 2184 (1972).

dividually characterized so that they are each saturating. The simplest way to begin is to use step 3 factors (in this case IF-M$_{2A}$ and IF-M$_{2B}$ are combined). From that point, substitution of more purified factors may be carried out. Several considerations are important with this assay:

1. The less pure initiation factors (step 3) may have cross contamination of other initiation factors (see Table III). In addition, step 3 IF-M$_1$ has a number of aminoacyl tRNA synthetases in it, and IF-M$_3$ may contain EF-1 and EF-2.

2. The supernatant fraction as used in this assay donates the elongation factors EF-1 and EF-2 and other soluble enzymes including the aminoacyl tRNA synthetases. It is important to note, however, that the supernatant fraction can contain significant quantities of initiation factors, particularly if large amounts of supernatant fraction are used. Usually 2–4 µl of the supernatant fraction (approximately 100 µg protein) is satisfactory for a 100-µl reaction mixture. This amount usually contains insignificant quantities of initiation factors. It is possible to purify all the initiation factors from the supernatant fraction if large amounts of the supernatant fraction are used and concentrated.

3. [^{14}C]Leucine is usually used as the label rather than other amino acids, since rabbit α and β chains have more leucine (35 residues out of 287 residues in both chains) than any other amino acid. [^{14}C]Valine (29 residues total) is also frequently used.

4. To identify the product of this assay, it is convenient to scale up the reaction mixture 2–5-fold and use the carboxymethyl-cellulose method of Dintzis[27] to separate rabbit α and β chains.

5. To use this assay to follow the purification of IF-M$_3$, substitute all four initiation factors for the concentrated ribosomal wash fraction until the best dependencies are determined. Column fractions during purification are then substituted for IF-M$_3$ in the assay.

6. Techniques to make this assay completely dependent on tRNA have been published in a prior volume.[28]

b. On Exogenous mRNA. The interaction of globin mRNA and rabbit reticulocyte ribosomes is very tenacious. We have examined several of the techniques available for preparing ribosomes free of mRNA (including runoff ribosomes, extensive salt washing, and use of various inhibitors) but have found that with sensitive assays (either the initiation assays described above or globin synthesis) there is always some residual globin mRNA on a significant percentage of the ribosomes. However, if lysate ribosomes are exposed to low levels of pancreatic RNase, the endogenous globin

[27] H. M. Dintzis, *Proc. Nat. Acad. Sci. U.S.* **47**, 247 (1961).
[28] J. M. Gilbert and W. F. Anderson, see this series, Vol. 20, p. 542.

mRNA is destroyed, leaving ribosomes with fragments of mRNA (Section I, G). With these ribosomes, dependency on the exogenous globin mRNA for synthesis of intact globin chains is almost absolute.[7,8]

Globin synthesis on exogenous globin mRNA using mRNA fragment ribosomes has the following conditions: 20 mM Tris·HCl, pH 7.5, 1 mM dithiothreitol, 4.5 mM MgCl$_2$, 1 mM ATP, 0.2 mM GTP, 3 mM phosphoenol pyruvate, 80–90 mM KCl, 0.1 IU pyruvate kinase, 40 μM [^{14}C]leucine, 40 μM 19 amino acids minus leucine, 0.1–0.2 A_{260} unit of rabbit reticulocyte tRNA, 150–200 μg of concentrated ribosomal wash fraction (Section I, D), 100 μg supernatant fraction (Section I, C), 0.1–0.2 A_{260} units of mRNA fragment ribosomes (Section I, G) and 0.01–0.5 A_{260} unit of globin mRNA. The assay is usually run for 30 minutes (it is linear up to 45 minutes) at 37° in 100 μl. After incubation the reaction mixtures are processed in the same manner as for globin synthesis on endogenous mRNA.

Important aspects of this assay include the following: (1) This assay has been most useful for examining the translation of several heterologous globin mRNA's including rabbit α and β, human, α, β^A, β^S, and γ, sheep α, β^A, β^B, and β^C, and goat α and several β's.[7,26] Its major advantage is a deficiency in endogenous mRNA. Since ribosomal RNA can nonspecifically stimulate endogenous mRNA, it cannot be determined (when using an homologous system) whether the protein that is being synthesized is being translated on "stimulated" endogenous mRNA or from translation of exogenously added mRNA. The assay is very sensitive and will detect 0.005 A_{260} unit of purified globin mRNA. To identify the product synthesized, [^{3}H]leucine (available in high specific activity) can substitute for [^{14}C]leucine and the reaction mixture can be scaled up 2–10-fold. Product identification of human, sheep, and goat globins are usually performed by carboxymethyl cellulose chromatography by the method of Clegg et al.[7,26,29] (2) The ratio of α to β chains synthesized in the cell free system is partially dependent on the amount of total mRNA added (less than saturating, saturating, or more than saturating). As more mRNA is added, the α/β ratio decreases. This has been documented for both rabbit and human mRNA.[30] (3) It is important to treat the 0.5 M KCl ribosomal wash fraction (used as the source of initiation factors) with DEAE-cellulose as described (Section I, D) in order to remove rabbit mRNA.

III. Summary

Several additional assays have been developed in our laboratory for the study of the globin initiation process, but the assays described here have

[29] J. B. Clegg, M. A. Naughton, D. J. Weatherall, *J. Mol. Biol.* **19**, 91 (1966).
[30] A. W. Nienhuis, P. H. Canfield, and W. F. Anderson, *J. Clin. Invest.* **52**, 1735 (1973).

been most useful (see Table I). The division into two major groups (those using artificial templates and those using natural mRNA templates) has been made because the former are easier to use (and are useful therefore for purification of the initiation factors), while the latter require more care but are more valuable in understanding the mechanism of the initiation process.

Although each assay has many uses, we have found that certain assays are more valuable for a specific purpose (e.g., purification of IF-M$_1$) or for understanding a specific mechanism (e.g., GTP requirements in initiator tRNA binding). Table II lists the assays available and ranks their usefulness.

The use of an assay for a given purpose depends on the purity of the factors available. Table III lists the initiation factors at each step of purification (described in this volume [13] and [14] for reticulocyte factors) and lists the relative amounts of contamination with other factors.

The assays described have yielded certain "rules" concerning mammalian initiation.

1. The divalent cation concentration (Mg^{2+}) is always low (4–5 mM) in the presence of the appropriate factors.

2. The monovalent cation (K$^+$) concentration is not as critical, but is usually 80–100 mM.

3. Methionyl-tRNA$_F$ is the initiator tRNA; it must be present to initiate new globin chains.

4. The factors (IF-M$_1$, IF-M$_{2A}$, IF-M$_{2B}$) are necessary and sufficient to bind the initiator tRNA to the small subunit.

5. IF-M$_3$ is necessary before the first peptide bond can be formed and is required to translate natural mRNA.

6. There is probably at least 1 GTP hydrolysis event during the initiation process.

7. Artificial mRNA templates have different factor and lower GTP requirements than natural mRNA templates; use of the latter are necessary, therefore, in order to fully understand the mechanism of globin initiation.

TABLE I

Factor and Energy Requirements for the Initiation Assays[a]

Part A. Factor Requirements for Globin Initiation

Artificial templates

	Binding of Met-tRNA$_F$ (ApUpG)	Binding of fMet-tRNA$_F$ (ApUpG)	Met$_F$-puromycin synthesis (ApUpG)	fMet$_F$-puromycin synthesis (ApUpG)	Polyphenylalanine synthesis, poly(U)
IF-M$_1$	+	+	+	+	+
IF-M$_{2A}$	+	−	+	+	+
IF-M$_{2B}$	+	+	+	+	+
IF-M$_3$	−	−	−	−	−
EF-1	−	−	+	+	+
EF-2	−	−	−	−	+

Natural globin mRNA

	Binding of Met-tRNA$_F$	Met$_F$-puromycin synthesis	Initial dipeptide synthesis	Initial tripeptide synthesis	Globin synthesis
IF-M$_1$	+	+	+	+	+
IF-M$_{2A}$	+	+	+	+	+
IF-M$_{2B}$	+	+	+	+	+
IF-M$_3$	−	+	+	+	+
EF-1	−	+	+	+	+
EF-2	−	−	−	+[b]	+

Part B. Energy Requirements for Globin Initiation[c]

Artificial templates

	Binding of Met-tRNA_F (ApUpG)	Binding of fMet-tRNA_F (ApUpG)	Met_F-puromycin synthesis (ApUpG)	fMet_F-puromycin synthesis (ApUpG)	Polyphenylalanine synthesis, poly(U)
GTP required	Yes	No	Yes	Yes	Yes
GDP substitutes for GTP	No	—	No	No	No
GDPCD substitutes for GTP	Yes	—	No	No	No
Fusidic acid inhibits	No	—[e]	Partial	Partial	Yes

Natural globin mRNA

	Binding of Met-tRNA_F	Met_F-puromycin synthesis	Initial dipeptide, endogenous mRNA	Initial dipeptide, exogenous mRNA	Initial tripeptide synthesis	Globin synthesis
GTP required	Yes	Yes	Yes	Yes	Yes	Yes
GDP substitutes for GTP	No	No	No	No	No	No
GDPCP substitutes for GTP	No	No	No	No	No	No
Fusidic acid inhibits	No[d]	Partial	No	—[e]	Yes	Yes

[a] Initiation and elongation factor requirements for each assay; +, factor is required; −, factor is not required.

[b] The requirement for EF-2 during synthesis of the initial tripeptides of α and β globin is inferred from the inhibition of tripeptide synthesis by fusidic acid, a known inhibitor of EF-2 function.

[c] Energy requirements for each assay.

[d] Demonstrated by sucrose gradient method (Section II, B, 1, b), not by cellulose nitrate filter method (Section II, B, 1, a).

[e] Not tested.

TABLE II

Specific Uses of Initiation Assays[a]

Assay	Purification				Mechanism studies					
	IF-M$_1$	IF-M$_{2A}$	IF-M$_{2B}$	IF-M$_3$	IF-M$_1$	IF-M$_{2A}$	IF-M$_{2B}$	IF-M$_3$	Energy requirements	Initiator tRNA
Artificial template										
Polyphenylalanine on poly(U)	+	+	+	−	−	−	−	−	−	−
Met$_F$ on ApUpG	+	+	+	−	±	±	±	−	−	+
fMet$_F$ on ApUpG	±	−	±	−	−	±	±	−	±	−
Met$_F$-puro on ApUpG	−	−	−	−	±	±	±	−	−	±
fMet$_F$-puro on ApUpG	−	−	−	−	−	±	−	−	±	−
GTPase	−	±	−	−	−	+	−	−	+	−
Natural template										
Met$_F$-binding										
Cellulose nitrate	±	±	±	−	+	+	+	−	−	±
Sucrose gradient	−	−	−	−	−	−	−	−	+	+
Met$_F$-puro										
Endogenous	−	±	±	+	±	±	±	+	±	±
Exogenous	−	−	−	−	±	±	±	±	±	±
Dipeptide										
Endogenous	−	−	−	−	+	+	+	+	+	+
Exogenous	−	−	−	−	±	±	±	±	±	±
Tripeptide	−	±	−	−	−	−	−	−	+	−
Globin synthesis	+	+	+	+	±	±	±	±	−	±

[a] The specific uses of the initiation assays in our laboratory: +, assay of choice; ±, alternate choice, not as useful; −, not useful either because no information is generated or because assay is too cumbersome to use for this purpose. If more than one assay ranks the same, it is listed as such. Assays rated − are useful for a number of reasons (see text), but are not in the first-line group. For example, Met$_F$-valine synthesis could be used to assay for IF-M$_3$, but there are other assays (globin synthesis or Met$_F$-puromycin on endogenous mRNA) that are much simpler to use for this purpose.

TABLE III

CONTAMINATION OF FACTORS DURING PURIFICATION[a]

Purification	IF-M$_1$	IF-M$_{2A}$	IF-M$_{2B}$	IF-M$_3$	EF-1	EF-2	R[b]	Synthetases	mRNA	tRNA	Mg^{++}
Step 1. 0.5 M KCl ribosomal wash fraction	+	+	+	+	+	+	+	+	+	+	+
a. After dialysis	++	++	++	++	++	++	++	++	+	+	−
b. After DEAE-cellulose batch	++	++	++	++	++	++	++	++	−	−	−
Step 2. (NH$_4$)$_2$SO$_4$	++	++	±	++	++	++	++	++	−	−	−
Step 3											
IF-M$_1$(DEAE)	+	−	−	±	±	±	NT	+	−	−	−
IF-M$_{2A}$(DEAE)	−	++	++	±	−	−	NT	±	−	−	−
IF-M$_{2B}$(DEAE)	−	++	++	±	−	−	NT	±	−	−	−
IF-M$_3$(DEAE)	±	±	±	+	+	+	NT	±	−	−	−
Step 4											
IF-M$_1$(G-200)	+	−	−	−	−	±	NT	±	−	−	−
IF-M$_{2A}$(G-200)	−	+	−	±	−	−	NT	−	−	−	−
IF-M$_{2B}$(G-200)	−	−	+	−	−	−	NT	−	−	−	−
IF-M$_3$(P-cell)	−	−	−	+	−	−	NT	−	−	−	−

[a] Cross contamination of initiation factors during purification steps (see this volume [13] and [14]). +, Factor present in significant amount; ±, factor can be detected if this step factor used, dependencies are not as good; −, insignificant amounts; NT, not tested. After Step 4 there is very little cross contamination. The dialysis and DEAE-cellulose steps shown under Step 1 are described in Section I, D. The (NH$_4$)$_2$SO$_4$ procedure (Step 2) used in purification of the factors and all subsequent purification steps are described in this volume [13] and [14].

[b] Release factor.

[13] Preparation of Protein Synthesis Initiation Factors IF-M$_1$, IF-M$_{2A}$, and IF-M$_{2B}$ from Rabbit Reticulocytes

By WILLIAM MERRICK, HERMANN GRAF, and W. FRENCH ANDERSON

The previous paper[1] has presented the preparation of rabbit reticulocyte 0.5 M KCl ribosomal wash fraction and the various assays which can be utilized to characterize initiation factors IF-M$_1$, IF-M$_{2A}$, IF-M$_{2B}$, and IF-M$_3$. The purpose of this and the following[2] report is to describe the separation and purification of these reticulocyte initiation factors from the 0.5 M KCl ribosomal wash fraction.

Reagents

Neutralized (NH$_4$)$_2$SO$_4$; a 50:1 mixture of (NH$_4$)$_2$SO$_4$ and (NH$_4$)$_2$CO$_3$ (w/w)

Tris·HCl, 1 M, pH 7.5 (20°)

Tris·HCl, 1 M, pH 7.9 (20°)

N-2-hydroxyethylpiperazine-N'-2-ethanesulfonic acid (HEPES)–KOH, 1 M, pH 6.2 (20°)

Neutralized EDTA, 0.1 M: a 0.1 M solution of EDTA which has been adjusted to pH 7.0 by the addition of 1.0 M KOH

Potassium acetate, 1 M, pH 5.0 (20°)

Sephadex G-200 (Pharmacia Fine Chemical, Inc.)

Fibrous DEAE-cellulose (Whatman, DE-23)

Microgranular DEAE-cellulose (Whatman, DE-32)

Microgranular CM-cellulose (Whatman, CM-32)

Phosphocellulose (Whatman, P-11)

Dithiothreitol (Calbiochem)

Buffers

0.10 M KCl Buffer A: a solution that contains per liter 20 ml of 1 M Tris·HCl, pH 7.5 (20°), 0.154 g of dithiothreitol (1 mM), 1 ml of 0.1 M neutralized EDTA, and 7.46 g of KCl

0.40 M KCl Buffer A: a solution that contains per liter 20 ml of 1 M Tris·HCl, pH 7.5 (20°), 0.154 g of dithiothreitol, 1 ml of 0.1 M neutralized EDTA, and 29.8 g of KCl

0.45 M KCl Buffer A: a solution that contains per liter 20 ml of 1 M Tris·HCl, pH 7.5 (20°), 0.154 g of dithiothreitol, 1 ml of 0.1 M neutralized EDTA, and 33.6 g of KCl

[1] R. G. Crystal, N. A. Elson, and W. F. Anderson, this volume [12].
[2] P. M. Prichard and W. F. Anderson, this volume [14].

Buffer B: a solution that contains per liter 20 ml of 1 M HEPES–KOH, pH 6.2 (20°), 0.154 g of dithiothreitol (1 mM), and 1 ml of 0.1 M neutralized EDTA

50 mM KCl Buffer B: a solution that contains per liter 20 ml of 1 M HEPES–KOH, pH 6.2 (20°), 0.154 g of dithiothreitol, 1 ml of 0.1 M neutralized EDTA, and 3.73 g of KCl

0.30 M KCl Buffer B: a solution that contains per liter 20 ml of 1 M HEPES–KOH, pH 6.2 (20°), 0.154 g of dithiothreitol, 1 ml of 0.1 M neutralized EDTA, and 22.4 g of KCl

0.10 M KCl Buffer C: a solution that contains per liter 25 ml of 1.0 M Tris·HCl, pH 7.9 (20°), 0.154 g of dithiothreitol (1 mM), 1 ml of 0.1 M neutralized EDTA, and 7.46 g of KCl

0.60 M KCl Buffer C: a solution that contains per liter 25 ml of 1.0 M Tris·HCl, pH 7.9 (20°), 0.154 g of dithiothreitol, 1 ml of 0.1 M neutralized EDTA, and 44.7 g of KCl

Assays

The poly(U)-directed polyphenylalanine synthesis assay (assay II, A, 1[1]) is used to locate the elution position of IF-M$_1$, IF-M$_{2A}$, and IF-M$_{2B}$ because of its convenience. For the purification of IF-M$_3$ the globin synthesis assay (assay II, B, 5a[1]) is employed. All the assays are performed in the presence of saturating levels of complementary factors; e.g., IF-M$_1$ activity is assayed in the presence of saturating levels of IF-M$_{2A}$, IF-M$_{2B}$, EF-1, and EF-2.

Purification Procedure

Step 1. 0.5 M KCl Ribosomal Wash. The 0.5 M KCl ribosomal wash fraction which contains all the initiation factors is prepared as described in the preceding paper.[1] All the steps involved in the purification of the initiation factors are done at 2–4° unless stated otherwise.

Step 2. 70% (NH$_4$)$_2$SO$_4$ Concentration. The 0.5 M KCl ribosomal wash fraction (approximately 700 ml at 8 mg of protein per milliliter) is made 70% of saturation in (NH$_4$)$_2$SO$_4$ by the gradual addition of 44.8 g of neutralized (NH$_4$)$_2$SO$_4$ per 100 ml of solution over a 30-minute period. After all the (NH$_4$)$_2$SO$_4$ has dissolved, the solution is allowed to stir an additional 30 minutes. The precipitated protein is collected by centrifugation at 16,000 g for 20 minutes. The pellets containing the initiation factors are dissolved in 0.10 M KCl buffer A, pooled, and dialyzed against 6 liters of 0.10 M KCl buffer A for 12–16 hours. The dialyzed ammonium sulfate fractioned ribosomal wash fraction is clarified by centrifugation at 16,000 g for 15 minutes.

This ammonium sulfate fractionation step yields a preparation of initiation factors which is approximately 7-fold more concentrated than the initial 0.5 M KCl ribosomal wash fraction and contains only 10% of the initial hemoglobin present in the ribosomal wash fraction.

Step 3. DEAE-Cellulose Chromatography. Fibrous DEAE-cellulose is precycled with 0.5 N HCl and 0.5 N NaOH and then defined several times by stirring the DEAE-cellulose in water, allowing DEAE-cellulose to settle and decanting the slow settling fines, as described by the manufacturer. The fibrous DEAE-cellulose is poured into a 1.5 × 90 cm column, packed under slight air pressure (1–2 psi) and equilibrated with sufficient 0.10 M KCl buffer A such that the column effluent has the same pH and conductivity as the equilibrating buffer.

The ammonium sulfate fractionated ribosomal wash fraction (approximately 100 ml at 20 mg protein per milliliter) is applied to the equilibrated fibrous DEAE-cellulose column (1.5 × 87 cm) at a flow rate of 1.0 ml/minute. After all the sample has been applied to the column, the column is washed with approximately 200 ml of 0.10 M buffer A until no additional UV absorbing (A_{280}) material is eluted from the column. IF-M$_1$ elutes with the protein not adsorbed to DEAE-cellulose at 0.1 M KCl.[3]

IF-M$_3$ and a mixture of IF-M$_{2A}$ and IF-M$_{2B}$ (IF-M$_{2(A+B)}$) are eluted from the DEAE-cellulose column with a linear salt gradient (350 ml of 0.1 M KCl buffer A × 350 ml of 0.45 M KCl buffer A) at a flow rate of 1.0 ml/minute; 5-ml fractions are collected. IF-M$_3$ activity is located by the globin synthesis assay while the mixture of IF-M$_{2A}$ and IF-M$_{2B}$ is located by the poly(U)-directed polyphenylalanine synthesis assay. IF-M$_3$ activity peaks at approximately 0.18 M KCl[4] and IF-M$_{2(A+B)}$ activity peaks at approximately 0.27 M KCl.[3] The resolution of IF-M$_3$ and IF-M$_{2(A+B)}$ activities is incomplete. Therefore, approximately the first 80% of the IF-M$_3$ activity is pooled, and the last 80% of IF-M$_{2(A+B)}$ activity is pooled. Under these conditions there is only minimal cross-contamination of IF-M$_3$ and IF-M$_{2(A+B)}$, although a substantial loss of recovery results.

While further purification of IF-M$_1$, IF-M$_{2A}$, IF-M$_{2B}$, and IF-M$_3$ will be given as a continuous flow, it is possible to stop the purification at this or any following step. If this is done, the various factors are concentrated either by ultrafiltration or $(NH_4)_2SO_4$ precipitation followed by dialysis against 0.1 M KCl buffer A. The percent of saturation of $(NH_4)_2SO_4$ routinely used to concentrate the initiation factors is 80% for IF-M$_1$, IF-M$_3$, and IF-M$_{2A}$ and 95% for IF-M$_{2(A+B)}$ and IF-M$_{2B}$. Ultrafiltration of these initiation factors is accomplished using a UM-10

[3] D. A. Shafritz and W. F. Anderson, *J. Biol. Chem.* **245**, 5553 (1970).
[4] D. J. Picciano, P. M. Prichard, W. C. Merrick, D. A. Shafritz, H. Graf, R. G. Crystal, and W. F. Anderson, *J. Biol. Chem.* **248**, 204 (1973).

membrane (Amicon Corp.). The concentrated preparations are clarified by centrifugation (500 g for 10 minutes) and stored in liquid nitrogen in small aliquots for assay, or large aliquots in preparation for further purification.

Further Purification of IF-M$_1$

Step 3A. 35–60% (NH$_4$)$_2$SO$_4$ Fractionation. The IF-M$_1$ obtained as the protein not adsorbed to DEAE-cellulose at 0.10 M KCl (approximately 150 ml at 7 mg of protein per milliliter) is made 35% of saturation in (NH$_4$)$_2$SO$_4$ by the addition of 19.9 g of neutralized (NH$_4$)$_2$SO$_4$ per 100 ml of solution (over a 30-minute period). After all the (NH$_4$)$_2$SO$_4$ has dissolved, the solution is allowed to continue stirring for an additional 30 minutes. The solution is centrifuged at 16,000 g for 20 minutes. The supernatant is recovered, and the pellets are discarded. The supernatant is made 60% of saturation in (NH$_4$)$_2$SO$_4$ by the gradual addition of 15.6 g of neutralized (NH$_4$)$_2$SO$_4$ per 100 ml of solution (over a 30-minute period). After all the (NH$_4$)$_2$SO$_4$ has dissolved, the solution is allowed to continue stirring for an additional 30 minutes. The precipitated IF-M$_1$ is collected by centrifugation at 16,000 g for 20 minutes. The supernatant is discarded, and the pellets containing IF-M$_1$ are dissolved in a minimal volume of 0.10 M KCl buffer A (5–10 ml) and dialyzed against 4 liters of 0.10 M KCl buffer A. After dialysis, the solution containing IF-M$_1$ is clarified by centrifugation (500 g for 10 minutes).

Step 4. Sephadex G-200 Chromatography. Sephadex G-200 is allowed to swell in 0.10 M KCl buffer A (10 ml of buffer per gram dry weight of Sephadex G-200). A slurry consisting of 2 parts 0.10 M KCl buffer A and 1 part settled, swollen Sephadex G-200 is poured into a 2.5 × 100 cm column equipped with an R25 reservoir (Pharmacia Fine Chemical, Inc.). The column is allowed to pack to a height of 90 cm, maintaining a 15 cm head pressure.

The 35–60% ammonium sulfate fractionated IF-M$_1$ (approximately 10 ml at 40 mg protein per milliliter) is applied to the Sephadex G-200 column (2.5 × 90 cm). IF-M$_1$ is eluted from the column with 0.10 M KCl buffer A at a flow rate of approximately 15 ml/hour; 5-ml fractions are collected. IF-M$_1$ activity is located by assay of 20-μl samples of the column fractions in poly(U)-directed polyphenylalanine synthesis. IF-M$_1$ elutes between 1.5 and 2 void volumes (about 55% of the bed volume).[5]

Step 5. CM-Cellulose Chromatography. CM-cellulose (Whatman,

---

[5] P. M. Prichard, J. M. Gilbert, D. A. Shafritz, and W. F. Anderson, *Nature (London)* **226**, 511 (1970).

CM-32) is precycled with 0.5 N NaOH and 0.5 N HCl as described by the manufacturer. The precycled resin is then defined several times in 0.10 M HEPES–KOH, pH 6.2 (20°). The defining is done in 0.10 M HEPES–KOH, pH 6.2 since this facilitates pH equilibration of the CM-cellulose. The CM-cellulose is poured into a 0.9 × 30 cm column equipped with an R9 reservoir (Pharmacia Fine Chemical, Inc.). The column is equilibrated with sufficient 50 mM KCl buffer B such that the conductivity and pH of the effluent are the same as the equilibrating buffer.

The Sephadex G-200 column fractions which contain IF-M$_1$ activity are pooled. As the pooled IF-M$_1$ solution (approximately 125 ml at 1 mg protein per milliliter) is stirring, an equal volume of chilled buffer B is added at a very slow rate. Next, the pH of the diluted solution is adjusted to 6.2 by the dropwise addition of 1 M potassium acetate, pH 5.0. These manipulations result in the solution being pH 6.2 and approximately 50 mM in KCl, the same pH and salt concentration as the equilibrating buffer for the CM-cellulose column (50 mM KCl buffer B). The diluted sample is applied to the CM-cellulose column (0.9 × 28 cm) at a flow rate of 12 ml per hour. The column is washed with 40 ml of 50 mM KCl buffer C and IF-M$_1$ is eluted with a linear salt gradient from 50 mM to 0.30 M KCl (300 ml of 50 mM KCl buffer B × 300 ml 0.30 M KCl buffer B) at a flow rate of 12 ml per hour; 5-ml fractions are collected. IF-M$_1$ activity elutes at approximately 0.14 M KCl.

Step 6. Phosphocellulose Chromatography. Phosphocellulose (Whatman, P-11) is precycled and equilibrated with buffer in the same manner as CM-cellulose (step 5) except that the phosphocellulose is defined in 0.1 M Tris·HCl, pH 7.9 (20°). A column of phosphocellulose is poured and packed as described above (step 5). The phosphocellulose column is equilibrated with 0.10 M KCl buffer C.

CM-cellulose column fractions which contain IF-M$_1$ activity are pooled (approximately 70 ml at 0.25 mg protein per milliliter) and mixed with 1/20 volume of 1.0 M Tris·HCl, pH 7.9, to raise the pH of the pooled fractions from 6.2 to about 7.7. This solution is applied to the equilibrated phosphocellulose column (0.9 × 28 cm) at a flow rate of 12 ml per hour. The column is washed with 40 ml of 0.10 M KCl buffer C and IF-M$_1$ is eluted with a linear salt gradient from 0.10 to 0.60 M KCl (200 ml of 0.10 M KCl buffer C × 200 ml of 0.60 M KCl buffer C) at a flow rate of 12 ml per hour; 5-ml fractions are collected. IF-M$_1$ activity elutes at approximately 0.45 M KCl. Tubes containing IF-M$_1$ activity are pooled, and the solution made 80% of saturation in neutralized $(NH_4)_2SO_4$. The 80% $(NH_4)_2SO_4$ solution is allowed to stand until a hazy precipitate forms (12–16 hours). The precipitate is collected by centrifugation at 27,000 g for 30 minutes. The super-

TABLE I
PURIFICATION OF IF-M$_1$ FROM RABBIT RETICULOCYTES

Fraction	Volume (ml)	Protein conc. (mg/ml)	Total protein (mg)	Specific activity[a] (units/mg)	Purification (fold)	Total units	Yield (%)
1. 0.5 M KCl ribosomal wash	755	8.06	6,083[c]	9.2[b]	1	55,960	100
2. 70% (NH$_4$)$_2$SO$_4$ concentration	112	18.4	2,057	20[b]	2.2	41,140	74
3. DEAE-cellulose	154	6.75	1,040	30	3.3	31,200	56
3A. 35–60% (NH$_4$)$_2$SO$_4$	10	42.0	420	58	6.3	24,360	44
4. Sephadex G-200	124	1.04	129	161	17.5	20,770	37
5. CM-cellulose	72	0.26	18.6	760	82.6	14,140	25
6. Phosphocellulose	2.0	1.12	2.24	4427	481	9,916	18

[a] One unit of activity is defined as the IF-M$_1$-dependent incorporation of 1 pmole of [^{14}C]phenylalanine into hot trichloroacetic acid-insoluble protein in the presence of saturating levels of complementary factors (EF-1, EF-2, IF-M$_{2A}$, IF-M$_{2B}$) under standard poly(U)-directed polyphenylalanine synthesis assay conditions.

[b] Since these preparations are usually dilute and contain the other initiation factors, these values may vary somewhat from one preparation to another.

[c] Preparation is from 150 rabbits.

natant is discarded, and the small pellets are dissolved in a minimal amount of 0.10 M KCl buffer A (1–3 ml). The suspended protein is dialyzed against 2 liters of 0.10 M KCl buffer A for 12–16 hours. The dialyzed preparation is stored in small aliquots in liquid nitrogen at a concentration of 1–2 mg of protein per milliliter.

Purification Table for IF-M$_1$. The values given in the purification table for IF-M$_1$ (Table I) represent average values that have been obtained several times in the purification of IF-M$_1$. In most instances, the IF-M$_1$ preparations used to assay for specific activity at each step were concentrated since the dilute solutions resulting from chromatography made activity determinations and protein measurements difficult.

Further Purification of IF-M$_{2A}$

Step 4. Sephadex G-200 Chromatography. The mixture of IF-M$_{2A}$ and IF-M$_{2B}$ obtained from the DEAE-cellulose chromatography of the 0.5 M KCl ribosomal wash fraction is concentrated to about 10 ml using a UM-10 membrane at 20–30 psi (Amicon Corp.). The concentrate is clarified by centrifugation at 500 g for 10 minutes. The mixture of IF-M$_{2A}$ and IF-M$_{2B}$ (approximately 10 ml at 20 mg protein per milliliter) is then applied to a 2.5 × 90 cm column of Sephadex G-200 which has been prepared and equilibrated with 0.10 M KCl buffer A as described above. IF-M$_{2A}$ and IF-M$_{2B}$ are eluted with 0.10 M KCl buffer A at a flow rate of 15 ml per hour; 5-ml fractions are collected. IF-M$_{2A}$ elutes just after the void volume, and IF-M$_{2B}$ elutes after 2 void volumes.[6] Tubes containing IF-M$_{2A}$ activity are pooled, and the solution is concentrated by either ultrafiltration or $(NH_4)_2SO_4$ precipitation (80%) followed by dialysis against 0.10 M KCl buffer A. The concentrated IF-M$_{2A}$ is clarified by centrifugation (500 g, 10 minutes) and stored in small aliquots in liquid nitrogen at a concentration of 5–8 mg of protein per milliliter.

Purification Table for IF-M$_{2A}$. The purification of IF-M$_{2A}$ as described above gives a final preparation which is approximately 21-fold purified with a rather poor yield of IF-M$_{2A}$ activity (28%) (Table II). This loss in activity results in part from the initial DEAE-cellulose column chromatography (step 3). There is also a considerable loss of activity with Sephadex G-200 chromatography which separates IF-M$_{2A}$ from IF-M$_{2B}$. Preliminary experiments which tested the stability of dilute solutions of step 4 IF-M$_{2A}$ indicate that a substantial amount of the IF-M$_{2A}$ activity in dilute solutions (0.2 mg/ml) is lost whereas more concentrated preparations (10 mg/ml) lost only 10–20% of their initial activity under the same conditions (18 hours at 4°). To circumvent the use of

TABLE II

PURIFICATION OF IF-M$_{2A}$ FROM RABBIT RETICULOCYTES

Fraction	Volume	Protein conc. (mg/ml)	Total protein (mg)	Specific activity[a] (units/mg)	Purification (fold)	Total units	Yield (%)
1. 0.50 M KCl ribosomal wash	755	8.06	6,083	14[b]	1	85,160	100
2. 70% $(NH_4)_2SO_4$ concentration	122	18.4	2,057	38[b]	2.7	78,170	92
3. DEAE-cellulose	10	20.1	201	215	15	43,220	51
4. Sephadex G-200	10	8.0	80	295	21	23,600	28

[a] One unit of activity is defined as the IF-M$_{2A}$-dependent incorporation of 1 pmole of [^{14}C]phenylalanine into hot trichloroacetic acid-insoluble protein in the presence of saturating levels of complementary factors (EF-1, EF-2, IF-M$_1$, IF-M$_{2B}$) under standard poly(U)-directed polyphenylalanine synthesis assay conditions.

[b] Since these preparations are usually dilute and contain the other initiation factors, these values may vary somewhat from one preparation to another.

[6] D. A. Shafritz, P. M. Prichard, J. M. Gilbert, W. C. Merrick, and W. F. Anderson, *Proc. Nat. Acad. Sci. U.S.* 69, 983 (1972).

Sephadex G-200 chromatography, studies are in progress using batch elution techniques with hydroxyapatite to separate IF-M$_{2A}$ and IF-M$_{2B}$. Pilot runs indicate that IF-M$_{2B}$ elutes at less than 0.3 M KPO$_4$, pH 7.2, while IF-M$_{2A}$ elutes at a higher salt concentration. Additional study is required to determine whether this procedure will give a higher yield of IF-M$_{2A}$ activity which is free from IF-M$_{2B}$ activity.

Further Purification of IF-M$_{2B}$

Step 4. *Sephadex G-200 Chromatography.* The separation of IF-M$_{2A}$ and IF-M$_{2B}$ by Sephadex G-200 column chromatography is described under Purification of IF-M$_{2A}$—Step 4.

Step 5. *DEAE-Cellulose Chromatography.* The fractions containing IF-M$_{2B}$ which have been separated from IF-M$_{2A}$ by Sephadex G-200 column chromatography (step 4) (approximately 130 ml at 0.2 mg protein per milliliter) are pooled and applied to a microgranular DEAE-cellulose column (0.9 × 28 cm) prepared and equilibrated with 0.10 M KCl buffer A as described above (step 3). The column is washed with 40 ml of 0.10 M KCl buffer A. IF-M$_{2B}$ is eluted from the column with a linear salt gradient from 0.10 to 0.40 M KCl (300 ml 0.10 M KCl buffer A × 300 ml of 0.40 M KCl buffer A) at a flow rate of 12 ml/ hour; 5-ml fractions are collected. IF-M$_{2B}$ activity elutes at approximately 0.15 M KCl. Fractions containing IF-M$_{2B}$ activity are pooled and

TABLE III

PURIFICATION OF IF-M$_{2B}$ FROM RABBIT RETICULOCYTES

Fraction	Volume (ml)	Protein conc. (mg/ml)	Total protein (mg)	Specific activity[a] (units/ mg)	Purification (fold)	Total units	Yield (%)
1. 0.5 M KCl ribosomal wash	755	8.06	6083	25[b]	1	152,075	100
2. 70% (NH$_4$)$_2$SO$_4$ concentration	112	18.4	2057	—	—	—	—
3. DEAE-cellulose	10	20.1	201	323	13	64,923	43
4. Sephadex G-200	130	0.18	24	1038	42	24,912	16
5. DEAE-cellulose	2.2	2.78	6.0	2005	80	12,030	8

[a] One unit of activity is defined as the IF-M$_{2B}$-dependent incorporation of 1 pmole of [^{14}C]phenylalanine into hot trichloroacetic acid-insoluble protein in the presence of saturating levels of complementary factors (EF-1, EF-2, IF-M$_1$, IF-M$_{2A}$) under standard poly(U)-directed polyphenylalanine synthesis assay conditions.

[b] Since this preparation is usually dilute and contains the other initiation factors, this value may vary somewhat from one preparation to another.

concentrated by ultrafiltration (UM-10 membrane, Amicon Corp.). The concentrated IF-M$_{2B}$ is clarified by centrifugation (500 g for 10 minutes) and stored in small aliquots in liquid nitrogen at a concentration of 1–3 mg of protein per milliliter.

Purification Table for IF-M$_{2B}$. The purification table for IF-M$_{2B}$ (Table III) indicates that a fair purification has been accomplished (approximately 80-fold); however, the yield is poor (8%). Preliminary physical studies including Sephadex G-75 chromatography,[4] SDS gel electrophoresis, and sucrose density gradient centrifugation[4] indicate that IF-M$_{2B}$ has a molecular weight of 15,000–20,000 and that most of the contaminants of the step 5 IF-M$_{2B}$ preparation are of a higher molecular weight. The poor yield may reflect preferential loss of the small molecular weight proteins in attempts to concentrate dilute solutions.

Purification of IF-M$_3$. Further purification of IF-M$_3$ obtained by DEAE-cellulose chromatography of the 0.50 M KCl ribosomal wash fraction (step 3) will be given in the following paper.[2]

[14] Preparation of Rabbit Reticulocyte Initiation Factor IF-M$_3$

By PHILIP M. PRICHARD and W. FRENCH ANDERSON

Poly(U)-directed polyphenylalanine synthesis on rabbit reticulocyte ribosomes at low Mg^{2+} concentration requires initiation factors IF-M$_1$, IF-M$_{2A}$, and IF-M$_{2B}$[1]; however, translation of globin mRNA requires IF-M$_3$ in addition to the above factors.[2,3] The exact function(s) of IF-M$_3$ remains unclear. IF-M$_3$ is required for the formation of the initial dipeptide (methionyl-valine) in hemoglobin synthesis[4] as well as globin mRNA-directed methionyl-puromycin synthesis.[5] IF-M$_3$ may also function in the recognizing and binding of natural mRNA to the ribosome.[6]

[1] D. A. Shafritz, P. M. Prichard, J. M. Gilbert, W. C. Merrick, and W. F. Anderson, *Proc. Nat. Acad. Sci. U.S.* 69, 983–987 (1972).

[2] P. M. Prichard, J. M. Gilbert, D. A. Shafritz, and W. F. Anderson, *Nature (London)* 226, 511–514 (1970).

[3] P. M. Prichard, D. J. Picciano, D. G. Laycock, and W. F. Anderson, *Proc. Nat. Acad. Sci. U.S.* 68, 2752–2756 (1971).

[4] R. G. Crystal, D. A. Shafritz, P. M. Prichard, and W. F. Anderson, *Proc. Nat. Acad. Sci. U.S.* 68, 1810–1814 (1971).

[5] R. G. Crystal and W. F. Anderson, *Proc. Nat. Acad. Sci. U.S.* 69, 706–711 (1972).

[6] S. M. Heywood, *Proc. Nat. Acad. Sci. U.S.* 67, 1782–1789 (1970).

Reagents

Tris·HCl, 1.0 M, pH 7.5 (20°)
Tris·HCl, 1.0 M, pH 7.9 (20°)
KH_2PO_4, 1.0 M (monobasic)
K_2HPO_4, 1.0 M (dibasic)
Potassium phosphate, 1.0 M, pH 7.15 (20°). This solution is made by mixing the 1.0 M mono- and dibasic potassium phosphates together so that the resultant pH is 7.15.
Neutralized EDTA, 0.10 M; a 0.10 M solution of EDTA which has been adjusted to pH 7.0 by the addition of 1.0 M KOH
DEAE-cellulose (Whatman, DE-23)
Phosphocellulose (Whatman, P-11)
Hydroxyapatite (Clarkson Chemical Co., Inc., Hypatite C)
Dithiothreitol (Calbiochem)

Buffers

0.20 M KCl Buffer A: a solution which contains per liter 20 ml of 1.0 M Tris·HCl, pH 7.5 (20°), 0.154 g of dithiothreitol (1 mM), 1.0 ml of 0.10 M neutralized EDTA and 14.9 g of KCl
0.10 M KCl Buffer D: a solution which contains per liter 50 ml of 1.0 M Tris·HCl, pH 7.9 (20°), 0.154 g of dithiothreitol, 1.0 ml of 0.10 M neutralized EDTA, and 7.46 g of KCl
0.50 M KCl Buffer D: a solution which contains per liter 50 ml of 1.0 M Tris·HCl, pH 7.9 (20°), 0.154 g of dithiothreitol, 1.0 ml of 0.10 M neutralized EDTA, and 37.3 g of KCl
50 mM KPO_4 Buffer E: a solution which contains per liter 50 ml of 1.0 M potassium phosphate, pH 7.15 (20°), 0.154 g of dithiothreitol, and 1.0 ml of 0.10 M neutralized EDTA
0.10 M KPO_4 Buffer E: a solution which contains per liter 100 ml of 1.0 M potassium phosphate, pH 7.15 (20°), 0.154 g of dithiothreitol, and 1.0 ml of 0.10 M neutralized EDTA
0.20 M KPO_4 Buffer E: a solution which contains per liter 200 ml of 1.0 M potassium phosphate pH 7.15 (20°), 0.154 g of dithiothreitol, and 1.0 ml of 0.10 M neutralized EDTA

Purification Procedure

Step 1. 0.5 M KCl Ribosomal Wash. The preparation of the 0.5 M KCl ribosomal wash fraction is described in the accompanying article by Crystal *et al.*[7]

Step 2. 70% (NH₄)₂SO₄ Concentration. The 70% $(NH_4)_2SO_4$ con-

[7] R. G. Crystal, N. A. Elson, and W. F. Anderson, this volume [12].

centration of the 0.5 M KCl ribosomal wash fraction is described in the preceding paper by Merrick *et al.* (Step 2).[8]

Step 3. Chromatography on DEAE-Cellulose. IF-M$_3$ is initially separated from the other initiation factors by DEAE-cellulose (Whatman, DE-23) chromatography at pH 7.5 (see accompanying article[8]). Small aliquots of the column fractions (10–20 μl) are assayed for IF-M$_3$ activity in the globin synthesis assay (assay II, B, 5, a^r). Column fractions are assayed for the presence of IF-M$_{2A}$ and IF-M$_{2B}$ in the poly(U)-directed polyphenylalanine synthesis assay (assay II, A, 1^r) in order that the appropriate fractions with IF-M$_3$ activity can be pooled with minimal contamination by IF-M$_{2A}$ and/or IF-M$_{2B}$. IF-M$_3$ elutes at approximately 0.18 M KCl. This and all other procedures are carried out at 2–4° unless stated otherwise.

After chromatography the protein is concentrated by ammonium sulfate fractionation. Solid ammonium sulfate is added over 5 minutes with stirring to the pooled IF-M$_3$ fractions to a final concentration of 30% (16.4 g/100 ml) and the suspension is allowed to stir slowly for an additional 30 minutes. This is followed by centrifugation at 16,000 g for 20 minutes. The supernatant fraction is carefully removed, and the precipitate is discarded. Solid ammonium sulfate is added to the 30% $(NH_4)_2SO_4$ supernatant over 5 minutes to a final concentration of 60% (18.1 g/100 ml) and allowed to stir slowly for an additional 30 minutes. The suspension is again centrifuged at 16,000 g for 20 minutes. The supernatant fraction is carefully removed and discarded. The precipitate is allowed to drain for 5 minutes.

PROCEDURE 1. This preparation can be stored and used as crude IF-M$_3$ at this stage by solubilizing the pellets in 0.20 M KCl buffer A followed by dialysis overnight against this buffer. Any precipitate that accumulates during dialysis in this or other steps is removed by centrifugation at 500 g for 5 minutes. Concentration of the protein by ammonium sulfate precipitation and subsequent dialysis has been found to be more satisfactory than concentration by ultrafiltration. IF-M$_3$ is stored in aliquots in liquid nitrogen at a concentration of 10 mg/ml. At this stage of purification IF-M$_3$ is very stable to storage in liquid nitrogen and remains active for at least one year.

PROCEDURE 2. The preparation of IF-M$_3$ can be prepared for further purification on phosphocellulose by solubilizing the pellets (obtained from ammonium sulfate fractionation as described above) in 0.10 M KCl buffer D followed by dialysis overnight against this buffer; or material stored in liquid nitrogen can be thawed, dialyzed against 0.10 M KCl buffer D, and used for phosphocellulose chromatography. For subsequent

---

[8] W. C. Merrick, H. Graf, and W. F. Anderson, this volume [13].

phosphocellulose chromatography, concentration of the pooled column fraction from DEAE-cellulose with ammonium sulfate followed by dialysis is preferable to diluting the pooled material with buffer to obtain the appropriate KCl concentration. A large amount of precipitation with concurrent loss of activity occurs even with a 2-fold dilution of the IF-M_3 solution.

Step 4. Chromatography on Phosphocellulose. The protein fraction from step 3 (procedure 2) is applied to a 1.5 × 30 cm column of phosphocellulose (Whatman, P-11) which has been equilibrated thoroughly with 0.10 M KCl buffer D. After the nonabsorbed protein has eluted from the column, IF-M_3 is eluted with a linear KCl gradient (200 ml of 0.10 M buffer D × 200 ml of 0.50 M KCl buffer D) at a flow rate of 0.25 ml/minute; fractions of 5 ml each are collected. Small aliquots of column fractions (10–20 μl) are assayed for IF-M_3 activity in the globin synthesis assay. IF-M_3 elutes at approximately 0.30 M KCl. Appropriate fractions with IF-M_3 activity are pooled. EF-1 elutes at approximately 0.40 M KCl and EF-2 at approximately 0.20 M KCl, thus satisfactorily separating IF-M_3 from the two elongation factors.

PROCEDURE 1. The IF-M_3 prepared in this manner is suitable for assays since it is free of the other initiation factors. IF-M_3 is concentrated by ammonium sulfate precipitation (30–60%) in the same manner as described above and is dialyzed against 0.20 M buffer A. Aliquots are stored in liquid nitrogen at a concentration of approximately 2 mg/ml and are stable for at least 6 months.

PROCEDURE 2. The pooled IF-M_3 fractions from the phosphocellulose column (without subsequent ammonium sulfate fractionation or dialysis) are used directly for the next step.

Step 5. Chromatography on Hydroxyapatite. Hydroxyapatite (Hypatite C, capacity = 82 mg/g, Clarkson Chemical Co., Inc.) is packed by gravity into a 1.5 × 20 cm column and equilibrated with about 20 column volumes of 50 mM KPO$_4$ buffer E. The protein solution from step 4 (procedure 2) is applied to the hydroxyapatite column. Elution is performed stepwise with (1) 0.10 M KPO$_4$ buffer E followed by (2) 0.20 M KPO$_4$ buffer E. The protein eluted by the 0.10 M KPO$_4$ step is discarded; protein eluted by the 0.20 M KPO$_4$ step is collected and pooled. The column effluent cannot be assayed directly as high phosphate concentrations inhibit globin synthesis. The 0.10–0.20 M phosphate buffer effluent is concentrated by 30–60% ammonium sulfate precipitation in the same manner as described above and dialyzed overnight against 0.20 M KCl buffer A. This preparation is stored in liquid nitrogen in small aliquots at a concentration of approximately 1 mg/ml and is stable for at least several months. IF-M_3 is relatively stable to repeated freeze-thawing.

A summary of the purification procedure is presented in the table.

PURIFICATION OF IF-M₃ FROM RABBIT RETICULOCYTES

Fraction	Volume (ml)	Protein conc. (mg/ml)	Total protein (mg)	Specific activity[a] (units/mg)	Purification	Total units	Yield (%)
1. 0.50 M KCl ribosomal wash	755	8.1	6116	19[b]	1	116,200	100
2. 70% $(NH_4)_2SO_4$ concentration	122	18.4	2057	—	—	—	—
3. DEAE-cellulose	208	2.2	458	179	9.4	81,980	71
4. Phosphocellulose	21.1	2.5	52.8	580	31	30,620	26
5. Hydroxyapatite	14.0	1.35	18.9	830	44	15,690	14

[a] A unit of activity is defined as the IF-M₃-dependent incorporation of 50 pmoles of [¹⁴C]valine into hot trichloroacetic acid-precipitable polypeptide in endogenous mRNA directed globin synthesis under standard assay conditions.

[b] Since this preparation is usually dilute and contains the other initiation factors, this value may vary somewhat from one preparation to another.

Discussion

The purification of IF-M$_3$ has presented some problems, and several observations remain to be explained. IF-M$_3$ obtained by DEAE-cellulose chromatography stimulates globin synthesis in the presence of IF-M$_1$, IF-M$_{2A}$, IF-M$_{2B}$, and supernatant protein to a level comparable to that of the ribosomal wash fraction (using the same concentration of salt-washed ribosomes). However, subsequent chromatography on phosphocellulose results in an IF-M$_3$ preparation which, although free of the other initiation factors and the elongation factors, will stimulate globin synthesis to a level only 50–60% that of the DEAE-cellulose prepared IF-M$_3$. Preliminary data indicate that a component may be split from the IF-M$_3$ by phosphocellulose chromatography.

Other purification methods such as Sephadex chromatography and sucrose gradients have not been particularly useful. IF-M$_3$ elutes at the void volume of Sephadex G-200 indicating a large molecular weight.[2] However, a significant loss in total activity is observed. A marked loss of total activity is also observed during purification of IF-M$_3$ by sucrose gradient centrifugation.

[15] Purification and Properties of the Peptide Chain Initiation Factors from Rabbit Reticulocytes

By CHARLES L. WOODLEY, YANG C. CHEN, and NABA K. GUPTA

Miller and Schweet[1] first demonstrated that the protein fraction (I fraction) obtained by washing crude reticulocyte ribosomes with 0.5 M KCl was necessary for *in vitro* hemoglobin synthesis directed by salt-washed or preincubated reticulocyte ribosomes. Shafritz *et al.*[2,3] separated two factors (M$_1$ and M$_2$) from the 0.5 M KCl ribosomal wash by column chromatographic procedures and demonstrated the requirement of these two factors for poly[r(U)]-directed polyphenylalanine synthesis at low Mg^{2+}. One of these factors (M$_1$) also stimulated the binding of fMet-tRNA$_f^{Met}$ to reticulocyte ribosomes in response to the AUG triplet codon, whereas both factors (M$_1$ and M$_2$) were necessary for binding of Met-tRNA$_f^{Met}$ to reticulocyte ribosomes in response to the same codon.[4]

[1] R. L. Miller and R. Schweet, *Arch. Biochem. Biophys.* **125**, 632 (1968).
[2] D. A. Shafritz, P. M. Prichard, J. M. Gilbert, and W. F. Anderson, *Biochem. Biophys. Res. Commun.* **38**, 721 (1970).
[3] D. A. Shafritz and W. F. Anderson, *J. Biol. Chem.* **245**, 5553 (1970).
[4] D. A. Shafritz and W. F. Anderson, *Nature* (*London*) **227**, 918 (1970).

These authors found that another factor (M_3), also present in the 0.5 M KCl ribosomal wash, was necessary in addition to M_1 and M_2 for *in vitro* translation of endogenous hemoglobin messenger.[5] Protein factors which catalyze the binding of Met-tRNA$_f^{Met}$ to the ribosomes in response to the AUG codon, have been reported to be present in the cell supernatant of rat liver[6,7] and brine shrimp embryo.[8] However, this binding of Met-tRNA$_f^{Met}$ to ribosomes in response to these factors does not require GTP.

In our peptide chain initiation studies, we use preincubated reticulocyte ribosomes, poly[r(U-G)] or poly[r(A-U-G)] messenger and I fraction.[9,10] Preincubated reticulocyte ribosome preparations retain most of the enzymatic activities (including aminoacyl tRNA synthetases and peptide chain elongation factors) necessary for polypeptide synthesis, and actively catalyze the incorporation of [^{14}C]phenylalanine in response to poly[r(U)]. Addition of I fraction was not necessary for poly[r(U)]-directed polyphenylalanine synthesis. However, these ribosomal preparations were inactive in catalyzing the transfer of methionine from Met-tRNA$_f^{Met}$ into the terminal positions of the polypeptides synthesized in response to poly[r(U-G)] and poly[r(A-U-G)] messengers, and the addition of I fraction was necessary for this transfer reaction. DEAE-cellulose chromatography of the I fraction separated at least three factors that stimulated poly[r(U-G)] and poly[r(A-U-G)]-directed methionine transfer from Met-tRNA$_f^{Met}$. All three factors were necessary to obtain maximum transfer of methionine in response to poly[r(U-G)] messenger. IF1 binds specifically Met-tRNA$_f^{Met}$ in the presence of GTP.[11] The Met-tRNA$_f^{Met}$:IF1:GTP complex is quantitatively retained on Millipore filters, and is assayed accordingly. The kinetic analysis of the transfer reactions suggests that Met-tRNA$_f^{Met} \cdot$IF1$\cdot$GTP complex formation is an essential step in the transfer of methionine from Met-tRNA$_f^{Met}$ into the terminal positions of the polypeptides synthesized in response to poly[r(U-G)] and poly[r(A-U-G)] messengers.

[5] P. M. Prichard, J. M. Gilbert, D. A. Shafritz, and W. F. Anderson, *Nature (London)* **226**, 511 (1970).

[6] E. Gasior and K. Moldave, *J. Mol. Biol.* **66**, 391 (1972).

[7] D. P. Leader and I. G. Wool, *Biochim. Biophys. Acta* **262**, 360 (1972).

[8] M. Zasloff and S. Ochoa, *Proc. Nat. Acad. Sci. U.S.* **68**, 3059 (1971).

[9] N. K. Gupta, N. K. Chatterjee, C. L. Woodley, and K. K. Bose, *J. Biol. Chem.* **246**, 7460 (1971).

[10] C. L. Woodley, Y. C. Chen, K. K. Bose, and N. K. Gupta, *Biochem. Biophys. Res. Commun.* **46**, 839 (1972).

[11] Y. C. Chen, C. L. Woodley, K. K. Bose, and N. K. Gupta, *Biochem. Biophys. Res. Commun.* **48**, 1 (1972).

Materials

The double-stranded deoxyribopolymers of defined sequences, poly-[d(T-G)]·poly[d(C-A)] and poly[d(A-T-C)]·poly[d(G-A-T)], samples of these double-stranded deoxyribopolymers were kindly provided by Dr. H. G. Khorana, Massachusetts Institute of Technology. These samples were replicated with *Escherichia coli* DNA polymerase before use by the procedure of Wells *et al.*[12,13] These newly replicated DNA-like polymers were characterized by nearest-neighbor frequency analysis.[12,13]

E. coli B, grown to ½ log phase in minimal media was purchased from Grain Processing Co., Muscatine, Iowa.

RNA polymerase was prepared from *E. coli* B according to Chamberlin and Berg.[14]

L-[^{35}S]Methionine (10,000–25,000 mCi/mmole) was purchased from Amersham-Searle and [^{14}C]valine (230 mCi/mmole) was obtained from New England Nuclear.

General Procedures

Preparation of Reticulocyte 100,000 g Cell Supernatant (S100) and Ribosomal Pellet. Rabbit reticulocyte ribosomes and 100,000 *g* supernatant were prepared by a procedure similar to Allen and Schweet.[15] A preparation scheme is given in Fig. 1. Reticulocytes were obtained from albino male rabbits (5–6 lb) which had been made anemic by 4 consecutive daily injections of 1 ml of 2.5% freshly prepared phenylhydrazine hydrochloride (adjusted to pH 7.0). No injection was given on day 5, and the rabbits were bled by cardiac puncture on day 6. The blood, approximately 600 ml from 8 rabbits, was collected in four 250-ml polyethylene centrifuge bottles containing 25 ml of heparin (1 mg/ml) in medium A (0.14 *M* NaCl, 50 m*M* KCl, and 5 m*M* MgCl$_2$). All subsequent operations were carried out at 0° to 4° unless otherwise mentioned. The cells were centrifuged for 10 minutes at 5000 rpm in a Sorvall GSA rotor. The supernatant was carefully decanted. The packed cells were gently suspended in 500 ml of medium A, and the suspension was centrifuged. This operation was repeated. Then, 500 ml of lysing medium (2 m*M* MgCl$_2$, 1 m*M* dithiothreitol, 0.1 m*M* EDTA, pH 7.0) was added to the packed cells, and the cells were lysed by vigorously

[12] R. D. Wells, E. Ohtsuka, and H. G. Khorana, *J. Mol. Biol.* **14,** 221 (1965).
[13] R. D. Wells, T. M. Jacob, S. A. Narang, and H. G. Khorana, *J. Mol. Biol.* **27,** 237 (1967).
[14] M. Chamberlin and P. Berg, *Proc. Nat. Acad. Sci. U.S.* **48,** 81 (1962).
[15] E. H. Allen and R. S. Schweet, *J. Biol. Chem.* **237,** 760 (1962).

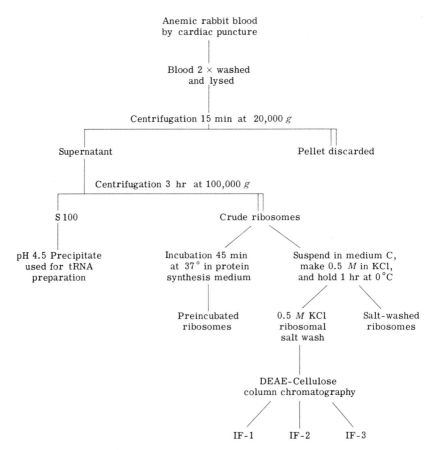

FIG. 1. Preparation scheme for reticulocyte ribosomes and initiation factors.

shaking for 10 minutes. The suspension was centrifuged at 20,000 g for 15 minutes. The 20,000 g supernatant was then centrifuged at 100,000 g for 3 hours in a Spinco ultracentrifuge Model L2-65B using a type 30 rotor.

Ribosomal Salt Wash. The 100,000 g ribosomal pellet, prepared as above, was suspended in sufficient medium C (0.25 M sucrose, 1 mM dithiothreitol, 0.1 mM EDTA, pH 7.0) to give a solution containing 300–400 A_{260} units/ml. To this ribosome suspension, 4 M KCl was slowly added, with gentle mixing, to give a final KCl concentration of 0.5 M. The suspension was held, with gentle stirring, in an ice bath for 1 hour, at which time it was centrifuged for 90 minutes at 60,000 rpm in a Spinco 65 rotor. The supernatant was dialyzed overnight against buffer D (5 mM Tris·HCl, pH 7.5, 0.1 M KCl, 1 mM dithiothreitol, 50 μM

EDTA). The ribosomal pellet was suspended in sufficient medium C to give 200–250 A_{260} units/ml.

The ribosomal salt wash preparation was usually stored in ice. Detectable loss (more than 20%) in activity was observed after 24-hour storage in ice. A similar loss in activity was also observed when the ribosomal salt wash was stored in liquid nitrogen.

Preincubated Ribosomes. In the preparation of preincubated reticulocyte ribosomes, the washed, packed cells (as prepared above from original 600 ml of blood) was lysed in 480 ml of lysing buffer and shaken for 10 minutes. The suspension was then thoroughly mixed with 120 ml of a solution containing 1.5 M sucrose and 0.15 M KCl, and was centrifuged at 20,000 g for 15 minutes. As before, the 20,000 g supernatant was further centrifuged at 100,000 g for 3 hours. The pellet obtained was gently suspended using a Potter-Elvehjem homogenizer in 30 ml of a solution containing all the components necessary for polypeptide synthesis, as described later, plus 2.5 μmoles of each of the 20 unlabeled amino acids. The solution was incubated at 37° for 45 minutes and was then diluted to 400 ml with medium B (0.25 M sucrose, 175 mM KHCO$_3$ and 2 mM MgCl$_2$). The ribosome suspension was centrifuged at 100,000 g for 3 hours. The ribosomal pellet was again suspended in 400 ml of medium B and centrifuged for 3 hours at 100,000 g. The pellet was suspended in a small volume of 0.25 M sucrose and was centrifuged at 10,000 g for 10 minutes to remove insoluble material. The final solution was adjusted to 130–150 A_{260} units/ml with 0.25 M sucrose and was stored in liquid nitrogen.

Reticulocyte tRNA. Rabbit reticulocyte tRNA was prepared from the 100,000 g supernatant of the reticulocyte cell-free lysate. The supernatant was adjusted to pH 5.0 with cold 1 M acetic acid and the solution was stirred for 15 minutes. The resulting suspension was centrifuged for 15 minutes at 10,000 rpm in a Sorvall GSA rotor. The precipitate was dissolved in 0.1 M Tris·HCl, pH 7.5; approximately 120 ml buffer was used per liter of S100. The subsequent purification steps were the same as described for rabbit liver tRNA in this series[15a]. The yield of crude reticulocyte tRNA from 1 liter of S100 was 1000–1500 A_{260} units. The tRNA$_f^{Met}$ activity in the crude tRNA preparation varied between 25 and 40 pmoles per A_{260} unit.

Precharged Reticulocyte [^{35}S]Met-tRNA$_f^{Met}$. E. coli synthetase was used to preferentially charge tRNA$_f^{Met}$ in crude reticulocyte tRNA.[15a] The incubation conditions, scaled up to a volume of 5 ml, were those described in the previous chapter. Approximately 80 A_{260} units of crude

---

[15a] K. K. Bose, N. K. Chatterjee, and N. K. Gupta, this series, Vol. 29 [43].

reticulocyte tRNA and 15 nmoles of L-[^{35}S]methionine were used in 5 ml of reaction mixture. The reaction mixture was incubated for 20 minutes at 37°, after which the reaction was stopped by the addition of 1 ml of cold 1 M sodium acetate buffer (pH 5). The reaction mixture was extracted twice with 5 ml of fresh phenol. The clear aqueous layer was dialyzed against 0.5 M NaCl–0.05 M sodium acetate (pH 5.0) (2 liters) for 6 hours and then against 20 mM sodium acetate (pH 5.0) (2 liters) for 6 hours.

Assays

Two types of assays were used to characterize the peptide chain initiation factors: (A) Amino acid incorporation into hot trichloroacetic acid-insoluble polypeptide products in response to poly[r(U-G)] and poly-[r(A-U-G)] messengers. (B) Binding of [^{35}S]Met-tRNA$_f^{Met}$ to Millipore filters in the presence of GTP and initiation factor(s).

A. Amino Acid Incorporation Assays

1. Poly[r(U-G)]-Directed Methionine Transfer from [^{35}S]Met-tRNA$_f^{Met}$. A two-stage procedure was used. In stage I, poly[r(U-G)] was synthesized by transcription of double-stranded DNA like polymer poly[d(T-G)]·poly[d(C-A)] with E. coli RNA polymerase in the presence of rUTP and rGTP.[16] In stage II, the poly[r(U-G)] messenger, containing repeating dinucleotide sequence, uridylyl guanylate, is translated into polypeptides in the presence of appropriate components required for protein synthesis. Stage I contained, per milliliter: 40 μmoles of Tris· HCl (pH 7.8), 4 μmoles of MgCl$_2$, 1 μmole of MnCl$_2$, 12 μmoles of 2-mercaptoethanol, 0.33 μmole of rUTP, 0.33 μmole of rGTP, 0.15 A_{260} units of poly[d(T-G)]·poly[d(C-A)] and approximately 600 units of E. coli RNA polymerase. After incubation at 37° for 30 minutes, the stage I reaction was cooled in ice. A 0.015-ml aliquot of the stage I reaction mixture containing poly[r(U-G)] messenger (about 0.9 nmole of each base residue) was added directly to the components of stage II (total volume 0.075 ml), so that the reaction mixture now contained, per milliliter: 36 μmole of Tris·HCl (pH 7.5), 55 μmoles of KCl, 3.5 μmoles of MgCl$_2$, 0.2 μmoles of MnCl$_2$, 20 μmoles of 2-mercaptoethanol, 0.8 μmoles of ATP, 0.2 μmoles of GTP, 3 μmoles of phosphoenolpyruvate, 0.3 IU of phosphoenolpyruvate kinase, 0.15 μmoles of each of unlabeled valine, cysteine, and methionine, 0.25 mg crude reticulocyte tRNA containing precharged [^{35}S]Met-tRNA$_f^{Met}$ (approximately 70 pmoles, 22,000 cpm/pmole), 10 A_{260} units of preincubated ribosomes and either crude

---

[16] D. S. Jones, S. Nishimura, and H. G. Khorana, *J. Mol. Biol.* **16**, 454 (1966).

ribosomal salt wash (0.5–3 mg) or purified chain initiation factors (0.05–0.15 mg). In addition, RNA polymerase, the DNA template and excess triphosphates introduced from stage I were present. After incubation at 37°, a 0.05-ml aliquot of the incubation mixture was spotted on a filter paper disk (Whatman No. 3, 2.3 cm diameter) and was assayed for hot trichloroacetic acid-insoluble radioactivity. The filter paper disks, immersed in 5% trichloroacetic acid, were heated to 90° for 20 minutes, then washed twice in cold 5% trichloroacetic acid, once in ethanol:ether (1:1), and once in ether. The paper disks were then dried and counted for radioactivity in toluene containing 4 g per liter Omnifluor (New England Nuclear).

2. *Poly[r(U-G)]-Directed [^{14}C]Valine Incorporation.* The assay system is similar to that described above for poly[r(U-G)]-directed methionine transfer from [^{35}S]Met-tRNA$_f^{Met}$ except that the incubation mixture contained 0.15 μmoles each of unlabeled methionine and cysteine and 6 nmoles of [^{14}C]valine (350 cpm/pmole); no precharged Met-tRNA$_f^{Met}$ was added.

3. *Poly[r(A-U-G)]-Directed [^{35}S]Methionine Transfer from [^{35}S]Met-tRNA$_f^{Met}$.* Reaction conditions were similar to these described above for poly[r(U-G)]-directed methionine transfer from [^{35}S]Met-tRNA$_f^{Met}$. In stage I, poly[r(A-U-G)] messenger, containing repeating trinucleotide sequence, adenylyl uridylyl guanylate, was synthesized by transcription of the double-stranded DNA-like polymer poly[d(A-T-C)·poly[d(G-A-T)] using *E. coli* RNA polymerase and rATP, rUTP, and rGTP.[17] The composition of the reaction mixture in stage II was the same as described above for poly[r(U-G)]-directed methionine transfer, except that the reaction mixture now contained poly[r(A-U-G)] messenger and no unlabeled valine and cysteine were added.

B. [^{35}S]Met-tRNA$_f^{Met}$ Binding Assay

The standard incubation mixture contained in a total volume of 0.075 ml: 20 mM of Tris·HCl, pH 7.5; 100 mM of KCl, 2 mM dithiothreitol, 0.2 mM GTP, 5 pmoles of [^{35}S]Met-tRNA$_f^{Met}$ (approximately 100,000 cpm), and 5–15 μl of either crude ribosomal salt wash or DEAE-cellulose purified fractions. The reaction mixture was incubated at 37° for 5 minutes, at which time the reaction was terminated by addition of 3 ml of cold washing buffer (20 mM Tris·HCl (pH 7.5), 100 mM potassium chloride, and 2 mM dithiothreitol). The solution was then filtered under suction through a Millipore filter (0.45 μ pore diameter). The filter was

[17] A. R. Morgan, R. D. Wells, and H. G. Khorana, *Proc. Nat. Acad. Sci. U.S.* **56**, 1899 (1966).

washed 3 times with 3 ml of cold washing buffer. The filters were dried
and counted for radioactivity in toluene as described above.

Characteristics of the Initiation Assay Procedures

Preincubated reticulocyte ribosomes actively catalyzed poly[r(U)]-
directed polyphenylalanine synthesis and the addition of the ribosomal
salt with fraction was not necessary for such incorporation.[9] However,

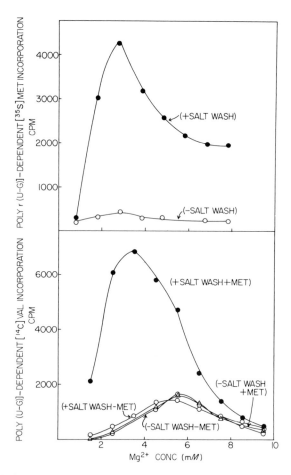

FIG. 2. Poly[r(U-G)]-directed amino acid incorporation as catalyzed by preincubated ribosomes at different Mg²⁺ ion concentrations. *Upper panel:* [³⁵S]Methionine
transfer from [³⁵S]Met-tRNA$_f^{Met}$ in the presence and in the absence of ribosomal
salt wash. *Lower panel:* [¹⁴C]Valine incorporation and the effects of addition of
ribosomal salt wash and unlabeled methionine. Reaction conditions were the same
as described in the text. The reaction mixtures were incubated at 37° for 15 minutes
and a 0.05-ml aliquot of the reaction was used for amino acid incorporation assay.

the same preincubated ribosome preparation was inactive in poly[r(U-G)]-directed methionine transfer from Met-tRNA$_f^{Met}$ into polypeptides and the addition of the ribosomal salt was necessary for this transfer reaction. The Mg^{2+} optimum for this transfer reaction was 3–4 mM (Fig. 2, upper curve). Under these conditions, poly[r(U-G)] messenger presumably directed the synthesis of a Met-(Cys-Val)$_n$ polypeptide. Preincubated ribosomes, in the absence of added ribosomal salt wash, catalyzed poly[r(U-G)]-directed [^{14}C]valine incorporation in the presence of crude tRNA and unlabeled cysteine (Fig. 2, lower curve). The Mg^{2+} optimum for [^{14}C]valine incorporation was 6–7 mM. However, in the presence of both unlabeled methionine and ribosomal salt wash, the Mg^{2+} optimum for [^{14}C]valine incorporation was shifted to 3–4 mM. This system, therefore, represents polypeptide synthesis at low Mg^{2+} (3–4 mM) which is dependent on the addition of ribosomal salt wash and the N-terminal methionine incorporation in response to the initiation codon GUG.

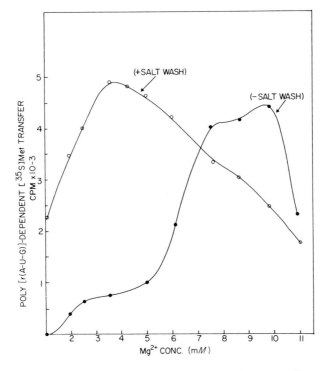

FIG. 3. Poly[r(A-U-G)] directed methionine transfer from [^{35}S]Met-tRNA$_f^{Met}$ as catalyzed by preincubated ribosomes at different Mg^{2+} ion concentrations and in the presence and the absence of ribosomal salt wash. See text for details.

The peptide chain initiation assay with poly[r(A-U-G)] messenger was performed using crude reticulocyte tRNA and unlabeled methionine so as to provide Met-tRNA$_m^{Met}$ for insertion of internal methionine into polymethionine product. However, as reported previously, in the absence of peptide chain initiation factors, Met-tRNA$_f^{Met}$ also transferred methionine into the internal positions of polymethionine products in response to poly[r(A-U-G)] messenger. The Mg^{2+} optimum for this transfer reaction was 7–8 mM (Fig. 3). At low Mg^{2+} (2–4 mM), and in the presence of excess crude tRNA, the transfer of methionine from Met-tRNA$_f^{Met}$ was very low and was increased significantly on addition of ribosomal salt wash. Under these conditions, the transferred methionine from Met-tRNA$_f^{Met}$ was mostly at the N-terminal position. This assay, therefore, represents polypeptide synthesis at low Mg^{2+}, dependent on the N-terminal methionine incorporation in response to the initiator codon, AUG.

Fractionation of the Ribosomal Salt Wash Factors by DEAE-Cellulose Chromatography

Approximately 12 ml of the 0.5 M KCl ribosomal salt wash, obtained from about 600 ml of whole rabbit blood, was dialyzed overnight against buffer D (5 mM Tris·HCl, pH 7.5, 0.1 M KCl, 1 mM dithiothreitol, 50 μM EDTA) and was applied to a microgranular DEAE-cellulose column (1.5 × 15 cm) which had been previously equilibrated with buffer D. The protein fraction which was not adsorbed to DEAE-cellulose was collected as a single fraction and labeled "fraction 1." The column was then thoroughly washed with buffer D, and was then eluted with 160 ml of a linear KCl gradient (0.1 M–0.35 M KCl) in buffer D at the rate of 40 ml per hour. Approximately 3.3-ml fractions were collected. The fractions were then dialyzed against buffer D containing 0.1 M KCl. The fractions were then concentrated to approximately 1 ml by dialysis against 20% polyethylene glycol in buffer D. The concentrated fractions were again dialyzed against buffer D for 6 hours. The fractions were then stored in ice until use.

Figure 4 shows a typical DEAE-cellulose chromatographic profile of crude ribosomal salt wash. The upper panel shows the A_{280} profile, measured using ISCO Model UA-2 ultraviolet analyzer and flow cell. A 0.015-ml aliquot of the dialyzed and concentrated fractions were assayed for their abilities to catalyze the transfer of methionine from [^{35}S]Met-tRNA$_f^{Met}$ in response to poly[r(U-G)] and poly[r(A-U-G)] messengers. Such profiles are shown in the lower panel. The DEAE-cellulose fractions were also assayed for their abilities to bind [^{35}S]Met-tRNA$_f^{Met}$ to Millipore filters in the presence of GTP. This activity eluted similarly to IF1 (Fig. 4, lower panel, ○- -○), although significant binding activity was also noted in the IF2 region.

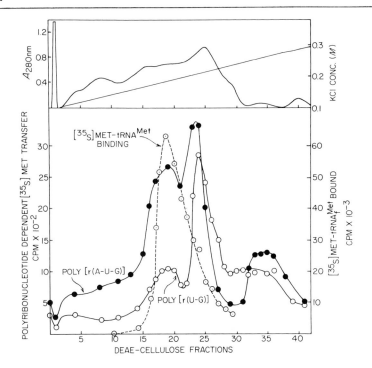

FIG. 4. DEAE-cellulose column chromatography of the 0.5 M KCl wash of crude reticulocyte ribosomes. See text for details.

Characteristics of Purified Peptide Chain
Initiation Factors

DEAE-cellulose chromatography of the ribosomal salt wash separated at least three protein peaks that stimulated poly[r(U-G)] and poly-[r(A-U-G)]-directed methionine transfer. These protein peaks eluted at 0.17 M, 0.21 M, and 0.24 M KCl, respectively, and have been designated as IF1, IF2, and IF3. The relative efficiencies of these fractions for the methionine transfer reactions in response to the two polyribonucleotide messengers were significantly different. Thus IF1 catalyzed the methionine transfer reaction approximately 3 times more efficiently in response to poly[r(A-U-G)] messenger than in response to poly[r(U-G)] messenger, although the extent of methionine transfer reactions catalyzed by IF2 in response to these two messengers was comparable (Fig. 4). The significance of the differences in the recognition of the two initiator condons, AUG and GUG is not apparent, at present.

Although the three factors independently stimulated the methionine transfer reactions, all three factors were necessary to obtain maximum transfer of methionine in response to poly[r(U-G)] messenger, and none of these factors were required for poly[r(U)]-directed polyphenylalanine synthesis.[10]

Met-tRNA$_f^{Met}$ binding factor eluted similarly to IF1 on DEAE-cellulose chromatography[11] (Fig. 4, ○- -○). This factor bound Met-tRNA$_f^{Met}$ in the presence of GTP. The complex formed was specific for reticulocyte Met-tRNA$_f^{Met}$. Other amino acyl tRNA's tested such as Met-tRNA$_m^{Met}$ (liver), Phe-tRNA^{Phe} (Retic., *E. coli*) and Val-tRNA^{Val} (Retic.) did not form a similar complex. With Met-tRNA$_f^{Met}$, complex formation did not require Mg^{2+} ion or AUG codon. It is not clear, at present, how this complex interacts with ribosomes and mRNA. However, the kinetic analysis of the transfer reactions suggests that Met-tRNA$_f^{Met}$ · IF1·GTP complex formation is an essential step in the transfer of methionine from Met-tRNA$_f^{Met}$ into the terminal positions of the polypeptides synthesized in response to poly[r(U-G)] and poly[r(A-U-G)] messengers.

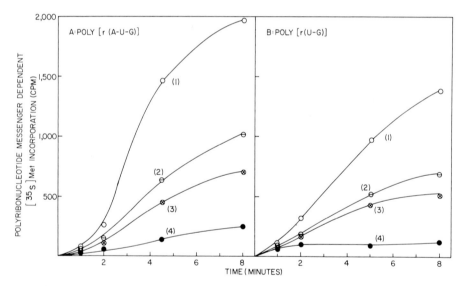

FIG. 5. Kinetics of methionine transfer from [^{35}S]Met-tRNA$_f^{Met}$·IF1·GTP complex in response to (A) poly [r(A-U-G)] and (B) poly[r(U-G)] messengers. A two-stage procedure was used. In stage I, [^{35}S]Met-tRNA$_f^{Met}$·IF1·GTP complex was formed as described in the text. A mixture of IF1 (6 μg), IF2 (10 μg), and IF3 (4 μg) was used as source of peptide chain initiation factors. The incubation mixture, in a total volume of 0.15 ml, was then mixed with 0.03 mg of crude reticulocyte tRNA containing 8 pmoles precharged [^{12}C]Met-tRNA$_f^{Met}$. To the reaction mixture were then added components for protein synthesis containing either poly[r(A-U-G)] or poly[r(U-G)] messenger. The composition in the second stage was the same as described in the text. The total reaction volume was 0.45 ml. At the indicated times, 0.1 ml aliquots were assayed for amino acid incorporation.

The results of a typical kinetic experiment are shown in Fig. 5. The experiment was carried out in two stages. In stage I, the complex with Met-tRNA$_f^{Met}$ was formed in the presence of IF1 and GTP. In stage II, the reaction mixture was diluted with 1.5 × [^{12}C]Met-tRNA$_f^{Met}$ and the transfer reaction was started by the addition of other protein-synthesizing components. Curve 1; the transfer activity in the complete system; curve 2, GTP was omitted in stage I; curve 3, peptide chain initiation factors were added in stage II; and curve 4, peptide chain initiation factors were omitted in both stage I and stage II. It is clear that in response to both poly[r(U-G)] and poly[r(A-U-G)] messengers, the Met-tRNA$_f^{Met}$·IF1· GTP complex transfer methionine more efficiently (curve 1) than free Met-tRNA$_f^{Met}$ (curve 3). The complex formed in the absence of GTP (curve 2) was inactive in the transfer reactions.

[16] Purification and Assays for the Presence and Biological Functions of Two Eukaryotic Initiation Factors from Rabbit Reticulocytes That Are Required for Initiation Complex Formation with Rabbit Reticulocyte Ribosomes[1-5]

By LINDA M. CASHION, GERALDINE L. DETTMAN, and WENDELL M. STANLEY, JR.

Initiation of protein synthesis in the cytoplasm of eukaryotes involves the formation of an initiation complex which includes an 80 S ribosome, the initiator tRNA (Met-tRNA$_i^{Met}$),[6] and an initiation region in mRNA

[1] G. L. Dettman and W. M. Stanley, Jr., *Biochim. Biophys. Acta* **287**, 124 (1972).

[2] G. L. Dettman and W. M. Stanley, Jr., *Biochim. Biophys. Acta* **299**, 142 (1973).

[3] L. M. Cashion and W. M. Stanley, Jr., *Biochim. Biophys. Acta,* in press.

[4] L. M. Cashion, P. M. Neal, and W. M. Stanley, Jr., manuscript in preparation.

[5] Partially supported by Grant E-533 from the American Cancer Society, by grants from the Cancer Research Coordinating Committee of the University of California, Irvine, California 92664, and by a grant from The Jane Coffin Childs Fund for Medical Research.

[6] Abbreviations: Met-tRNA$_i^{Met}$, methionyl-tRNA which can be enzymatically formylated (except for Met-tRNA$_f^{Met}$ from plants) by *Escherichia coli* enzymes and which functions as the initiator tRNA in the cytoplasm of eukaryotes; N-formyl-Met-tRNA$_i^{Met}$, Met-tRNA$_i^{Met}$ whose amino group of the methionyl-tRNA residue has accepted a formyl-group; Met-tRNA$_m^{Met}$, methionyl-tRNA which cannot be enzymatically formylated and which donates methionine internally into nascent polypeptide chains in the cytoplasm of eukaryotes; N-formyl-Met-tRNA$_f^{Met}$, methionyl-tRNA which has accepted a formyl group and which functions as the initiator tRNA in prokaryotes; Phe-tRNA^{Phe}, phenylalanyl-tRNA; DEAE-cellulose,

which contains the AUG initiator codon.[7] Two eukaryotic initiation factors (IF-I and IF-II) from rabbit reticulocytes are required for the proper assembly of an initiation complex with rabbit reticulocyte ribosomes.[4]

This chapter describes the purification of IF-I and of IF-II from 0.5 M KCl extracts of rabbit reticulocyte polysomes and several assays for the biological activities of each initiation factor. One function of IF-I during initiation complex formation is the binding of [35S]Met-tRNA$_i^{Met}$ to a 40 S ribosomal subunit in a GTP-dependent but AUG-, AUG(U)$_{25}$-, and mRNA-independent reaction.[3] IF-I also forms a ternary complex with [35S]Met-tRNA$_i^{Met}$ and GTP[1,2]; this complex is retained by Millipore filters. This retention property forms the basis of a convenient and routine assay for the presence of IF-I activity.

IF-II mediates the AUG, AUG(U)$_{25}$-, or globin mRNP-dependent transfer of [35S]Met-tRNA$_i^{Met}$ from a 40 S ribosomal subunit to an initiation complex containing an 80 S ribosome.[4] This transfer of the initiator tRNA can be followed by zone velocity sedimentation of the complex through preformed sucrose density gradients. [35S]Met-tRNA^{Met} in the initiation complex formed in the presence of IF-I and IF-II reacts with puromycin, resulting in the release of the [35S]methionine moiety as [35S]methionyl-puromycin.[4]

In 0.5 M KCl extracts of polysomes, the IF-I and IF-II activities are found together as a complex of molecular weight 350,000.[4] However, the IF-I and IF-II activities can be separated by appropriate step elutions from a column of DEAE-cellulose. In the presence of 1 mg of bovine serum albumin per milliliter, the molecular weights of the separated initiation factors are 30,000 for IF-I and 54,000 for IF-II.[4] In the absence of bovine serum albumin, the separated IF-I and IF-II activities each independently aggregate to molecular weights of approximately 200,000. The purifications of IF-I and of IF-II described later in this chapter take advantage of the natural complex formed between the two initiation factors.

---

diethylaminoethyl cellulose; A_{260} unit, that amount of material, which, when dissolved in 1 ml, has an absorbance of 1 at 260 nm in a 1-cm path length cuvette; AUG, ApUpGp; AUG(U)$_{25}$, ApUpG(pU)$_x$, where $x = 25 \pm 5$; mRNP, globin mRNA's with associated proteins that are released from rabbit reticulocyte polysomes by treatment with EDTA; initiation complex, a complex containing mRNA, an 80 S ribosome, the initiator tRNA, and initiation factors; IF-I, IF-2, IF-3, EF-1, and EF-2, in the uniform nomenclature proposed for initiation and translocation factors at a symposium held at Fogarty International Center, the National Institutes of Health, prokaryotic initiation factor 1 (F1, A, FI) was designated IF-1, initiation factor 2 (F2, C, FIII) IF-2, initiation factor 3 (F3, B, FII) IF-3, and eukaryotic transferase I was designated elongation factor 1 (EF-1) and transferase II is elongation factor 2 (EF-2); P site, peptidyl-tRNA binding site on the ribosome; A site, aminoacyl-tRNA binding site on the ribosome.

[7] J. Lucas-Lenard and F. Lipmann, *Annu. Rev. Biochem.* **40**, 409 (1971).

Buffers and Miscellaneous Reagents

The pH of all potassium phosphate buffers is adjusted to pH 7.0 at 20°.

Buffer A: 0.15 M KCl, 25 mM potassium phosphate

Buffer B: 25 mM KCl, 5 mM MgCl$_2$, 1 mM 2-mercaptoethanol, 50 mM Tris·HCl, pH 7.8 at 20°

Buffer C: 0.15 M KCl, 1 mM 2-mercaptoethanol, 25 mM potassium phosphate

Buffer D: 0.15 M KCl, 1 mM 2-mercaptoethanol, 10% (v/v) glycerol, 25 mM potassium phosphate

Buffer E: 18 mM MgCl$_2$, 2 mM 2-mercaptoethanol, 0.18 M Tris· HCl, pH 7.4 at 20°

Buffer F: 0.1 M sodium phosphate, pH 8.0 at 20°

Buffer G: 0.5 M KCl, 1 mM 2-mercaptoethanol, 50 mM Tris·HCl, pH 7.8 at 20°

Buffer H: 50 mM KCl, 1 mM 2-mercaptoethanol, 10% (v/v) glycerol, 25 mM potassium phosphate

Buffer I: 0.125 M KCl, 1 mM 2-mercaptoethanol, 10% (v/v) glycerol, 25 mM potassium phosphate

Buffer J: 0.25 M KCl, 1 mM 2-mercaptoethanol, 10% (v/v) glycerol, 25 mM potassium phosphate

Buffer K: 75 mM KCl, 5 mM MgCl$_2$, 1 mM 2-mercaptoethanol, 50 mM Tris·HCl, pH 7.4 at 20°

L-[^{35}S]Methionine prepared by the method of Graham and Stanley[8] and diluted to a specific activity of 10,000 Ci/mole

Rabbit liver tRNA prepared by the method of Rogg *et al.*[9] and aminoacylated by the method of Stanley[10,11]

Rabbit liver [^{35}S]Met-tRNA$_i^{Met}$, *N*-formyl-[^{35}S]Met-tRNA$_i^{Met}$, and [^{35}S]Met-tRNA$_m^{Met}$ prepared by the methods of Stanley[10,11]

GTP, trisodium salt (Sigma Chemical Co., St. Louis, Missouri 63118)

[^{3}H]GTP, tetralithium salt, specific activity 1380 Ci/mole (Schwarz/ Mann, Division of Becton, Dickinson and Co., Van Nuys, California 91401)

5′-Guanylylmethylenediphosphonate (GDPCP) (Miles Laboratories, Inc., Kankakee, Illinois 60901)

Puromycin, dihydrochloride salt (Nutritional Biochemicals Corp., Cleveland, Ohio 44128)

[8] R. Graham and W. M. Stanley, Jr., *Anal. Biochem.* **47**, 505 (1972).

[9] H. Rogg, W. Wehrli, and M. Staehelin, *Biochim. Biophys. Acta* **195**, 13 (1969).

[10] W. M. Stanley, Jr., *Anal. Biochem.* **48**, 202 (1972).

[11] W. M. Stanley, Jr., this series, Vol. 29 [44].

Pactamycin (The Upjohn Co., Kalamazoo, Michigan 49001)

Edeine, kindly given by Dr. Kurylo-Borowska, The Rockefeller University, New York, New York 10021

Millipore filters, HAWP, 0.45 μ average pore diameter, 2.7 cm diameter, 1.6 cm diameter filtering surface, premoistened with buffer A before use (Millipore Corp., Bedford, Massachusetts 01730)

Omnifluor (New England Nuclear, Boston, Massachusetts 02118), 4 grams per liter of toluene

DEAE-cellulose, Whatman DE-52, microgranular, preswollen, 1 milliequivalent per dry gram (Reeve Angel, Clifton, New Jersey 07014)

Sepharose 6B, 40-210 nm particle size (Pharmacia Fine Chemicals, Inc., Piscataway, New Jersey 08854)

Sephadex G-100, 40–120 μm particle size (Pharmacia Fine Chemicals, Inc.)

Chelex 100, 100–200 mesh (Bio-Rad Laboratories, Richmond, California 94804)

Aurintricarboxylate (ATA), ammonium salt (Mallinckrodt Chemical Works, St. Louis, Missouri 63160)

AUG prepared by the method of Last et al.[12] and dissolved in water

AUG(U)$_{25}$ prepared by the method of Stanley et al.[13] The oligoribonucleotide had a chain of 25 $\pm$ 5 and was dissolved in water.

Rabbit globin messenger ribonucleoproteins (mRNP's) prepared by the method of Burny et al.[14] The mRNP's were recovered by centrifugation at 2° for 20 hours in a Spinco type 40 rotor at 40,000 rpm (144,800 g[15]) in an L2-65B ultracentrifuge (Beckman Instruments, Inc., Spinco Division, Palo Alto, California 94304) and dissolved in buffer B.

Rabbit globin mRNA's prepared by phenol deproteinization of the rabbit globin mRNP's, precipitated with 2 volumes of ethanol, and dissolved in buffer B.

Rabbit reticulocyte polysomes, and ribosomal subunits and ribosomes preincubated in vitro with puromycin under conditions of protein synthesis and then reisolated prepared by the method of

[12] J. A. Last, W. M. Stanley, Jr., M. Salas, M. B. Hille, A. J. Wahba, and S. Ochoa, Proc. Nat. Acad. Sci. U.S. 57, 1062 (1967).

[13] W. M. Stanley, Jr., M. A. Smith, M. B. Hille, and J. A. Last, Cold Spring Harbor Symp. Quant. Biol. 31, 99 (1966).

[14] A. Burny, G. Huez, G. Marbaix, and H. Chantrenne, Biochim. Biophys. Acta 190, 228 (1969).

[15] All g forces are the maximum values for the rpm and the rotor used.

Brown et al.[16] These ribosomal subunits and ribosomes were dissolved in buffer B.

In vitro amino acid incorporation system derived from rabbit reticulocytes prepared by the method of Brown et al.[16]

IF-I and IF-II Units, Specific Activities, and Routine Assays

Definition of a Unit of IF-I Activity and of the Specific Activity of IF-I. A unit of activity is defined as that amount of material which, when incubated with GTP and $[^{35}S]$Met-tRNA$_i^{Met}$ as specified in the assay and processed as described below under Routine IF-I Assay, retains 1 pmole of $[^{35}S]$Met-tRNA$_i^{Met}$ on a Millipore filter.

The specific activity of a preparation of IF-I is the number of IF-I units per milligram of protein.[17]

Routine IF-I Assay (Assay of IF-I by Millipore Filter Retention). An aliquot (up to 45 μl) of a solution which may contain IF-I activity is added to 1.5 pmoles of $[^{35}S]$Met-tRNA$_i^{Met}$ and 70 nmoles of GTP in 50 μl of buffer C. After incubation at 25° for 15 minutes, 3 ml of cold buffer A are added. The solution is filtered in the cold through a Millipore filter (premoistened with cold buffer A) at a flow rate of 2 ml per minute. The filter is washed twice with 3-ml portions of cold buffer A, dried under an infrared lamp, and counted in the Omnifluor solution in a liquid scintillation spectrometer.

Definition of a Unit of IF-II Activity and of the Specific Activity of IF-II. A unit of IF-II activity is defined as that amount of material which, when incubated with IF-I, $[^{35}S]$Met-tRNA$_i^{Met}$, GTP, ribosomes, AUG, and puromycin as specified in the assay and processed as described below under Routine IF-II Assay, releases 1 pmole of $[^{35}S]$methionine as $[^{35}S]$methionyl-puromycin.

The specific activity of a preparation of IF-II is the number of IF-II units per milligram of protein.

Routine IF-II Assay (Assay of IF-II by Puromycin-Induced Release of $[^{35}S]$Methionine). The ternary complex of IF-I·$[^{35}S]$Met-tRNA$_i^{Met}$·GTP is generated in situ by incubation at 25° for 15 minutes of 20 μg of IF-I (purified through step 3 in Table I), 70 nmoles of GTP, and 1.5 pmoles of $[^{35}S]$Met-tRNA$_i^{Met}$ in 35 μl of buffer C. At 0°, to this mixture are added an aliquot, made up to 25 μl in buffer D, of a solution which may contain IF-II activity, 0.8 nmole of AUG (5 μl in water), 1.5 A_{260} units of rabbit reticulocyte ribosomes (10 μl in buffer B), and 25 μl of buffer E. The ini-

[16] G. E. Brown, A. J. Kolb, and W. M. Stanley, Jr., this volume [37].
[17] Protein concentration is determined by the method of O. H. Lowry, N. J. Rosebrough, A. L. Farr, and R. J. Randall, J. Biol. Chem. 193, 265 (1951) using bovine serum albumin, fraction V (Sigma Chemical Co.) as the standard.

TABLE I
PURIFICATION OF IF-I (ROUTINE IF-I ASSAY)

Procedure	Total units	Total protein (mg)	Specific activity	Yield (%)	Purification (fold)
1. Crude 0.5 M polysomal wash	174	46.4	3.7	100	—
2. First Sepharose 6B	174	11.9	14.6	100	3.9
3. DEAE-cellulose	36	1.7	20.9	20	5.6
4. Second Sepharose 6B	12	0.7	17.1	7	4.4

tiation complex is allowed to form for 10 minutes at 25°, then chilled to 0° and mixed with 0.4 nmole of puromycin in 5 μl of water. The reaction mixture is warmed to 25°, and after 30 minutes it is diluted with 1 ml of buffer F and 1.5 ml of ethyl acetate are added.[18] The two phases are mixed vigorously at room temperature with a vortex mixer for 30 seconds and then separated by centrifugation for 5 minutes at room temperature at top speed (3000 rpm; 1470 g) in a No. 221 rotor in an International desk top clinical centrifuge (Model CL, International Equipment Co., Needham Heights, Massachusetts 02194). One milliliter of the ethyl acetate (upper) phase is removed, mixed with 3 ml of Bray's solution,[19] and counted in a liquid scintillation spectrometer. The cpm are normalized to reflect the actual formation of [35S]methionylpuromycin by multiplying the observed cpm by 1.5 (incomplete recovery of the ethyl acetate phase) and then by 2 (the efficiency of extraction of methionylpuromycin into the ethyl acetate phase under these conditions is 50%[18]).

Purification Methods

Rabbit reticulocyte 0.5 M KCl polysomal extracts are used as the initial cellular fraction for the purification of both IF-I and IF-II. IF-I has been partially purified from several other cells and tissues: chicken reticulocytes, chicken liver, chick embryonic leg muscle, and rabbit liver.[3] The behavior of IF-I from all these sources is identical to that of rabbit reticulocyte IF-I during the purification which follows.

Preparation of the 0.5 M KCl Extract from Rabbit Reticulocyte Polysomes. Rabbit reticulocyte polysome pellets[16] are dissolved at 0° in buffer G at a concentration of 500 A_{260} units per milliliter. After 1 hour, the solution is centrifuged at 2° for 1.5 hours in a Spinco type 65 rotor at 58,000 rpm (293,300 g) in an L2-65B ultracentrifuge. The upper three-fourths of the supernatant is removed with a syringe and is used for the further

[18] P. Leder and H. Bursztyn, *Biochem. Biophys. Res. Commun.* **25**, 233 (1966).
[19] G. A. Bray, *Anal. Biochem.* **1**, 279 (1960).

purification of the IF-I and IF-II activities. The extract at this point contains 30 mg of protein derived per 1000 A_{260} units of polysomes.

First Gel Filtration through Sepharose 6B. The 0.5 M KCl polysomal extract (1.75 ml containing 46 mg of protein) is applied at 2° to a 0.9-cm by 35-cm column of Sepharose 6B preequilibrated with buffer H. The column is developed with buffer H at a flow rate of 6 ml per hour, and fractions are collected every 5 minutes. Each fraction is assayed for IF-I activity by the *Routine IF-I Assay,* and fractions with an IF-I specific activity greater than 2 are pooled. Seven ml of IF-I- and IF-II-containing solution are recovered. The elution position corresponds to a substance with a molecular weight of 350,000.

Step Elutions from DEAE-Cellulose (Separation of IF-I and IF-II). The IF-I- and IF-II-containing solution obtained from the Sepharose 6B gel filtration described in the preceding section is applied at 2° to a 0.5-cm by 2.5-cm column of DEAE-cellulose preequilibrated with buffer H. After application of the sample at a flow rate of less than 5 ml per hour, the column is washed with 2 ml of buffer H. (During the operation of this column, all buffers are passed through at 2° at a flow rate of 5 ml per hour.) More than 65% of the protein is not adsorbed and is washed from the column and is discarded. All the IF-I and IF-II activities are retained. IF-I is eluted from the column with buffer I. Fractions of 0.3 ml are collected, and IF-I is located by using the *Routine IF-I Assay.* Three fractions with an IF-I specific activity greater than 4 are pooled. Immediately after the elution of IF-I with buffer I, the column is washed with 2 ml of buffer D; this wash is discarded. IF-II is eluted from the column with buffer J; again 0.3-ml fractions are collected. Three fractions containing the IF-II activity, as determined by the *Routine IF-II Assay,* are pooled.

Second Gel Filtrations through Sepharose 6B. The two solutions, one containing IF-I and the other IF-II, obtained by sequential step elutions from DEAE-cellulose, are applied individually at 2° to 0.9-cm by 35-cm columns of Sepharose 6B preequilibrated with buffer D. The columns are developed at 2° with buffer D at a flow rate of 6 ml per hour and fractions are collected every 10 minutes. IF-I activity is located by the *Routine IF-I Assay,* and IF-II activity by the *Routine IF-II Assay.* Fractions containing the IF-I and the IF-II activities are pooled separately, 1 mg of bovine serum albumin per milliliter of eluate is added, and the two solutions containing the separated initiation factors are stored at −70°. The elution positions of both IF-I and IF-II correspond to substances with molecular weights of approximately 200,000.

General Considerations on the Purifications of IF-I and IF-II. The purification methods described take advantage of an unusual property of these two initiation factors; i.e., the fact that initially they are found together as a high molecular weight complex. Thus, IF-I and IF-II are co-purified as a complex of molecular weight 350,000 by gel filtration through Sepharose 6B. The two initiation factor activities then are separated by appropriate sequential step elutions from a column of DEAE-cellulose. Most high molecular weight contaminants in the separate IF-I and IF-II preparations are removed by repeating the gel filtration step.

A critical factor in the stability of IF-I activity is the protein concentration of the solution. Preparations of IF-I which contain less than 1 mg of protein per milliliter lose more than half of their activity after a day at 0°. If bovine serum albumin is added to a final concentration of 1 mg/ml, IF-I disaggregates to a molecular weight of 30,000 as judged by gel filtration, remains active following multiple freeze-thaw cycles, and can be stored at $-70°$ without significant loss of activity for several months. The addition of 10% (v/v) glycerol to the buffers used during purification increases the final yield of IF-I to an appreciable extent. Although glycerol at this concentration will interfere to some degree with the *Routine IF-I Assay,* solutions containing IF-I and 10% (v/v) glycerol almost always are diluted 5- to 10-fold during the assay; under these conditions the glycerol no longer interferes. Protein concentrations and column sizes specified under Purification Methods were developed for maximal yields of the IF-I initiation factor. However, we have not been successful in preventing rather large losses of IF-I activity (up to 80% in the case of dilute solutions) during the chromatography step on columns of DEAE-cellulose. Despite these losses of IF-I activity, DEAE-cellulose chromatography is the most effective method for the clean separation of the IF-I and the IF-II activities in the 350,000 molecular weight complex of the two initiation factors. Using the precautions developed to maintain the stability of IF-I, losses of IF-II activity are negligible.

The purification of IF-I from eukaryotic sources other than rabbit reticulocytes follows the same procedures described in the preceding sections. However, all the IF-I preparations resulting from these purifications have lower specific activities than does rabbit reticulocyte IF-I at equivalent stages of purification. These lower specific activities are probably due to lower protein concentrations in the initial 0.5 M KCl extracts of the polysomes. These extracts should be concentrated prior to subjecting them to purification since, as previously mentioned, dilute IF-I-containing solutions lose activity during the DEAE-cellulose chromatography step.

Assays and Biological Functions of IF-I and IF-II

Assay of the IF-I·[^{35}S]Met-tRNA$_i^{Met}$·GTP Ternary Complex by Millipore Filter Retention. This assay for the presence of IF-I activity is based upon the formation in solution of a very stable ternary complex between

IF-I, Met-tRNA$_i^{Met}$, and GTP. This complex is retained by Millipore filters under the conditions given previously under *Routine IF-I Assay.* Using [^{35}S]Met-tRNA$_i^{Met}$ to follow the formation of this ternary complex and the subsequent retention of the radioactive label on the filter, the *Routine IF-I Assay* is quick, specific, and suitable for the detection of IF-I activity in 0.5 M KCl extracts of polysomes and in all solutions during the purification of IF-I.

The ionic conditions of the Millipore filter retention assay have definite optima. The KCl concentration optimum is broad and level between 0.1 M and 0.15 M. Higher concentrations of KCl progressively inhibit activity; approximately 50% of the IF-I activity is not observed by this assay in the presence of 0.5 M KCl. The magnesium ion concentration optimum for the formation of the ternary complex is extremely low. Using the commercial GTP and [^{35}S]Met-tRNA$_i^{Met}$ described under *Buffers and Miscellaneous Reagents,* no added magnesium ion is required to detect maximal IF-I activity. However, after all reagents used in the assay are passed through small columns of Chelex 100, a definite magnesium ion concentration optimum between 75 μM and 100 μM can be established. Approximately two-thirds of this requirement is met by magnesium ions in the commercial GTP and the remaining one-third is associated with the tRNA preparation. IF-I is inactivated by prior exposure to a variety of sulfhydryl reagents. The presence of the optimal concentration of 2-mercaptoethanol (1 mM) increases IF-I activity 2-fold with respect to the activity measured in the presence of 0.1 mM 2-mercaptoethanol.

The concentration of [^{35}S]Met-tRNA$_i^{Met}$ specified in the *Routine IF-I Assay* is below saturation. In order to determine the maximal specific activity of IF-I by this assay, the [^{35}S]Met-tRNA$_i^{Met}$ concentration must be increased as follows: at a GTP concentration of 1.4 mM, 1 pmole of [^{35}S]Met-tRNA$_i^{Met}$ per 10^{-5} units of IF-I (determined by the *Routine IF-I Assay*) must be present during the formation of the ternary complex. For this type of assay, intended to determine the maximal specific activity of IF-I, and IF-I preparation may require dilution up to 500-fold in buffer D containing 1 mg of bovine serum albumin per milliliter. Because such large amounts of tRNA are required, these assay conditions should be used only when necessary. It should be noted that, although below the saturation concentration, the quantity of [^{35}S]Met-tRNA$_i^{Met}$ specified in the *Routine IF-I Assay* (1.5 pmoles in 50 μl) yields results directly proportional to the concentration of IF-I over a broad range throughout all stages of IF-I purification. The *Routine IF-I Assay* thus is quick, convenient, suitable, and conserves reagents.

The molar ratio of Met-tRNA$_i^{Met}$ to GTP bound to IF-I can be determined by using [^{3}H]GTP and [^{35}S]Met-tRNA$_i^{Met}$. In the early steps of IF-I purification, the binding of [^{3}H]GTP to IF-I, and its subsequent retention on Millipore filters, cannot be used to measure IF-I activity. This is due to the fact that 0.5 M KCl extracts of rabbit reticulocyte polysomes contain GTP-binding proteins which are retained on Millipore filters, but which do not represent IF-I activity. The binding of [^{3}H]GTP in the complex of IF-I·Met-tRNA$_i^{Met}$·GTP retained on Millipore filters is a quantitative assay for IF-I activity only after IF-I has been extensively purified. The ternary complex, once formed in solution and then retained on Millipore filters, is extremely stable, and will not release or exchange the [^{35}S]Met-tRNA$_i^{Met}$ if unlabeled methionine and unlabeled Met-tRNA$_i^{Met}$ are subsequently passed through the filter. The Millipore filter assay of IF-I by the retention of the ternary complex on the filter reveal little about the biological functions of IF-I, except that they are GTP-dependent, that the interaction of IF-I with aminoacyl-tRNA is specific for Met-tRNA$_i^{Met}$, and that, with highly purified preparations of IF-I, the molar ratio of Met-tRNA$_i^{Met}$ to GTP bound in the complex can be shown to be 1.[1]

Assay of the IF-I·[^{35}S]Met-tRNA$_i^{Met}$·GTP Ternary Complex by Gel Filtration Through Sephadex G-100. If the radioactive species in the ternary complex is [^{35}S]Met-tRNA$_i^{Met}$, the appearance of radioactivity in the excluded volume of a Sephadex G-100 column can be used to quantitatively follow the activity of IF-I. As stated in the preceding section, the abundance of GTP-binding proteins in 0.5 M KCl extracts of polysomes precludes the use of [^{3}H]GTP as the radioactive compound in this assay. However, as in the case of the Millipore filter retention assay, the appearance of [^{35}S]Met-tRNA$_i^{Met}$ and of [^{3}H]GTP in the excluded volume of a Sephadex G-100 column can be used to demonstrate the 1-to-1 molar ratio of these compounds bound to highly purified preparations of IF-I.[1] The most useful aspect of this assay is that it yields, in the fractions corresponding to the excluded volume of a Sephadex G-100 column, a preparation of the ternary complex which contains [^{35}S]Met-tRNA$_i^{Met}$ and either [^{3}H]GTP, γ-[^{32}P]-GTP, or nonradioactive GTP. This ternary complex is separated completely from the following unbound compounds: radioactive aminoacyl-tRNA's, other tRNA's (both aminoacylated and nonacylated), low molecular weight reagents used in the formation of the ternary complex, and some of the proteins which may contaminate the IF-I preparation. The isolated ternary complex is suitable for a variety of experimental uses.[1,2]

Assay of IF-I and of IF-II by [^{35}S]Met-tRNA$_i^{Met}$ Binding to Ribosomal Subunits and to Ribosomes. Since the IF-I·[^{35}S]Met-tRNA$_i^{Met}$·GTP complex itself is retained by Millipore filters, the binding of [^{35}S]Met-tRNA$_i^{Met}$ to ribosomal subunits and to ribosomes is analyzed by zone velocity sedimentation through preformed sucrose density gradients. By use of this technique the sequential events that finally result in the formation of an initiation complex can be demonstrated (see Table IV and Fig. 1). Although this assay is an excellent method for the characterization of the functions of the two initiation factors during the formation of an initiation complex, the assay is not convenient or practical for use during the purification of the initiation factors. On the other hand, it is the method of choice for the identification of the sites of action of various inhibitors of the initiation of protein biosynthesis.

Binding reactions are prepared as described under the *Routine IF-II Assay,* and contain either 11 μg of IF-I or 11 μg of IF-I *plus* 10 μg of IF-II (both factors purified through step 3 of Table I). After the second incubation at 25°, and before the addition of puromycin, the entire incubation mixture is layered at 2° over an exponential sucrose density gradient[20] (10% (w/v) to 25% (w/v) sucrose) in buffer K. The sucrose density gradient tube is centrifuged at 2° in a Spinco type SW 40 Ti rotor for 6 hours at 32,000 rpm (204,000 g). The contents of the tube are displaced upward by pumping 50% (w/v) sucrose under the sucrose density gradient. The displaced solution is led through a 1-cm path length flow cell mounted in a spectrophotometer set at 260 nm, and the absorbance profile is recorded on a strip chart recorder. The solution then is led from the flow cell to a fraction collector where 0.56-ml fractions are collected directly into scintillation vials. A total of 25 fractions is obtained; 5 ml of Bray's solution are added to each vial, and then the vials are counted in a liquid scintillation spectrometer.

This assay has been used to demonstrate the biological functions of the two initiation factors (IF-I and IF-II) during the assembly of an initiation complex.[4] IF-I in the presence of GTP promotes the AUG-, AUG(U)$_{25}$-, mRNA-, and mRNP-independent binding of [^{35}S]Met-tRNA$_i^{Met}$ to a 40 S ribosomal subunit. IF-II then mediates the transfer of the [^{35}S]Met-tRNA$_i^{Met}$ from a 40 S ribosomal subunit to an initiation complex containing an 80 S ribosome. This transfer is dependent upon the presence of a template such as AUG, AUG(U)$_{25}$, or the rabbit globin mRNP's. One use of the methods described in this assay is for the preparation of an initiation complex completely separated from the following: 40 S and 60 S ribosomal subunits, free [^{35}S]Met-tRNA$_i^{Met}$ and other aminoacyl-tRNA's and tRNA's, mRNA's, mRNP's, and many extraneous proteins. This isolated initiation complex

---

[20] The exponential sucrose density gradient is formed at 2° by pumping fluid from a constant-volume mixer flask which initially contains a volume of 10% (w/v) sucrose in buffer K equal to one-half the volume to be delivered to the sucrose density gradient tube. The mixer volume flask is air-tight, and its constant volume is maintained by the introduction of 28% (w/v) sucrose in buffer K from a reservoir flask at the same rate at which the pump removes fluid from the constant volume mixer flask.

(containing a radioactive methionine moiety) then can be used in an *in vitro* amino acid incorporation system to investigate the role of the initiation complex in the initiation of new polypeptide chains.

Assay of IF-II by the IF-II-Dependent Completion of an Initiation Complex as Measured by the Puromycin-Induced Release of [^{35}S]Methionine. This assay for IF-II is based upon the reaction of puromycin with the [^{35}S]methionine residue of the [^{35}S]Met-tRNA$_i^{Met}$ bound in an initiation complex and the subsequent release of [^{35}S]methionyl-puromycin. This release of [^{35}S]methionyl-puromycin is a quantitative assay for IF-II because, when the concentration of IF-I is held constant, the quantity of [^{35}S]methionyl-puromycin formed is linear with respect to the concentration of IF-II. Also, the amount of [^{35}S]methionyl-puromycin released can be accurately determined.[18] Obviously, the success of this assay for IF-II is dependent upon the availability of a supply of active IF-I completely separated from IF-II activity. The fact that the [^{35}S]methionine moiety is released in the presence of puromycin from an initiation complex formed under the direction of IF-I and IF-II in the presence of AUG, AUG(U)$_{25}$, or rabbit globin mRNP's indicates that the initiator tRNA is bound to the peptidyl site (P site) on an 80 S ribosome and that such an initiation complex is functionally active and ready to commence protein synthesis. Thus, this assay is a specific and convenient method for quantitatively following IF-II activity during all stages of its purification and also for detecting the completion of the assembly of a functional initiation complex.

Results and Discussion

The purification methods described in this chapter yield highly purified products and use a minimal number of steps. The high degrees of purity were achieved by first isolating IF-I and IF-II as a high molecular weight complex, then separating the complex into this two constituent initiation factor activities, and finally removing smaller and larger molecular weight contaminants from the separated initiation factors.

The finding that IF-I and IF-II exist naturally as a complex of molecular weight 350,000[4] is interesting. Following separation of IF-I and IF-II by step elutions from DEAE-cellulose, the individual factor activities both possess molecular weights of approximately 200,000 as determined by gel filtration through a calibrated column of Sepharose 6B. However, if the individual factor activities are passed through an identical column of Sepharose 6B, preequilibrated and developed with the same buffer, but containing 1 mg of bovine serum albumin per milliliter, the elution positions of IF-I and IF-II correspond to molecular weights of 30,000 and 54,000, respectively.[4] The significance of these complexes (the 350,000 molecular weight form that contains both IF-I and IF-II, and the two

200,000 molecular weight forms that contain the individual initiation factors) and of the much lower molecular weight forms of IF-I is unknown at this time. By the results of all of the biological assays available, it appears that both IF-I and IF-II are able to function properly, both under conditions where they exist as the lower molecular weight proteins and also under conditions where they exist as high molecular weight complexes, either formed by interaction with themselves or by interaction with each other.

It may be of some interest that an interaction between two prokaryotic initiation factors has been detected in extracts of *Escherichia coli*.[21] In this case, the complex is composed of one molecule each of IF-2 and IF-3. This IF-2·IF-3 complex is able to bind the prokaryotic initiator tRNA N-formyl-Met-tRNA$_f^{Met}$ (*E. coli*) and GTP at temperatures between 23° and 30°. This binding is sufficiently stable to be analyzed by retention of the complex on Selectron membranes and also by gel filtration of the complex through Sephadex G-100 at 4°. The authors suggested that this complex is formed on the 70 S ribosome (*E. coli*), rather than free in solution, and that the isolated quaternary complex (IF-2·IF-3·N-formyl-Met-tRNA$_f^{Met}$·GTP) is not a precursor during the initiation of prokaryotic protein biosynthesis. In view of these observations, the significance and function of the initiation factor complex detected in extracts of *E. coli* also is unclear at this time.

Because the molecular weight of IF-I which is active *in vivo* is uncertain, only tentative values for the purity of IF-I in an IF-I preparation can be calculated. The percent purity of an initiation factor is calculated by multiplying the specific activity (as defined in this chapter) of the preparation by the molecular weight of the initiation factor and then by 10^{-7} in order to normalize the units. The highest rabbit reticulocyte IF-I specific activity measured in several independent preparations is 2000. If the molecular weight of the active species is 200,000, the purity of IF-I in these preparations is 40%. However, if the molecular weight of the active form of IF-I is 30,000, or if each 30,000 molecular weight unit binds a molecule of Met-tRNA$_i^{Met}$, then the purity of the best preparations is only 6%.

Once the two initiation factors have been purified, their roles in initiation complex formation can be studied in well defined systems. Such studies require prior purification of the factors because it is necessary to remove inhibitors which are found in the 0.5 M KCl polysomal extracts. The removal of such inhibitors is indicated by an increase in the total number of units of an initiation factor during purification (see Tables II and III). In the presence of IF-I, purified through step 3 of Table I, Met-tRNA$_i^{Met}$

[21] Y. Groner and M. Revel, *Eur. J. Biochem.* **22**, 144 (1971).

TABLE II
PURIFICATION OF IF-II (ROUTINE IF-II ASSAY)

Procedure	Total units	Total protein (mg)	Specific activity	Yield (%)	Purification (fold)
1. Crude 0.5 M polysomal wash	10.0	46.4	0.2	100	—
2. First Sepharose 6B	20.3	11.9	1.7	203	8.5
3. DEAE-cellulose	5.2	1.1	4.7	52	23.5
4. Second Sepharose 6B	5.4	0.6	9.0	54	45.0

is bound only to the 40 S ribosomal subunit (Fig. 1A). This binding is not stimulated by the initiator triplet AUG, by the oligoribonucleotide AUG(U)$_{25}$, by rabbit globin mRNA's, or by rabbit globin mRNP's. The IF-I-mediated binding of Met-tRNA$_i^{Met}$ causes a significant increase (4 S) in the sedimentation coefficient of the smaller ribosomal subunit. This increase probably is due, at least in part, to a change in the conformation (a contraction of the structure) of the 40 S ribosomal subunit, since even the addition of the 200,000 molecular weight form of IF-I *plus* the Met-tRNA$_i^{Met}$ (molecular weight = 25,000) could not increase the sedimentation coefficient to such an extent.

In the presence of IF-II and either AUG, AUG(U)$_{25}$, or rabbit globin mRNP's, Met-tRNA$_i^{Met}$ is transferred from the 40 S ribosomal subunit to an initiation complex containing an 80 S ribosome (Fig. 1C–E). GTP is the only nucleotide added in order to observe the action of either IF-I or IF-II. The template requirements of the IF-II-mediated transfer reaction are interesting. The triplet AUG or an oligoribonucleotide containing an easily accessible AUG codon and lacking significant secondary structure can function during this reaction as the template. However, rabbit globin mRNA's (which contain a relatively high degree of secondary structure) are ineffective in supporting initiation complex formation (Fig. 1F) unless the specific mRNA-associated proteins remain bound to the mRNA's (Fig.

TABLE III
PURIFICATION OF IF-I (MAXIMAL SPECIFIC ACTIVITY DETERMINATIONS)

Procedure	Total units	Total protein (mg)	Specific activity	Yield (%)	Purification (fold)
1. Crude 0.5 M polysomal wash	2772	46.4	105	100	—
2. First Sepharose 6B	6676	11.9	561	241	5.4
3. DEAE-cellulose	923	1.7	543	33	5.2
4. Second Sepharose 6B	1454	0.7	2077	52	19.8

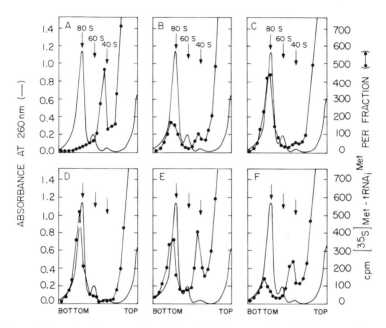

FIG. 1. Steps in the assembly of the initiation complex. Binding reactions were prepared as described in *Assay of IF-I and IF-II by* [³⁵S]*Met-tRNA*ᵢᴹᵉᵗ *Binding to Ribosomal Subunits and to Ribosomes.* As indicated below, 825 pmoles AUG, 680 pmoles AUG(U)₂₅, 0.16 A_{260} unit rabbit globin mRNP's, or 0.25 A_{260} unit rabbit globin mRNA's were added to the second incubation. (A) IF-I; (B) IF-I + IF-II; (C) IF-I + IF-II + AUG; (D) IF-I + IF-II + AUG(U)₂₅; (E) IF-I + IF-II + mRNP's; (F) IF-I + IF-II+ mRNA's.

1E). Two distinct proteins, of molecular weights 68,000 and 130,000, remain bound to the mRNA's when they are dissociated from polysomes with EDTA.[22] Considering the nature of the template requirements of the IF-I- and IF-II-mediated formation of an initiation complex, one or both of these proteins probably establishes the proper reading frame on the rabbit globin mRNA's by selecting, and making accessible to the ribosome, the correct AUG codon to serve as the initiator codon. Lacking these proteins, the rabbit globin mRNA's are incapable of functioning as the template required during the IF-II-mediated transfer of the initiator tRNA.

The methionine moiety of Met-tRNAᵢᴹᵉᵗ in an initiation complex, formed by the action of IF-I and IF-II, readily reacts with puromycin to release methionylpuromycin. This reaction has the same requirements as

[22] B. Lebleu, G. Marbaix, G. Huez, J. Temmerman, A. Burny, and H. Chantrenne, *Eur. J. Biochem.* 19, 264 (1971).

TABLE IV

QUANTITATIVE COMPARISONS OF THE COMPLETION OF AN INITIATION COMPLEX
BY TWO INDEPENDENT ASSAYS[a]

Additions or omissions[b]	Relative amount of [35S]Met-tRNA$_i^{Met}$ bound per ribosomal equivalent[c]			Relative amount of [35S]methionyl-puromycin formed[d]
	40 S	80 S	Σ(40 S + 80 S)	
1. IF-I	47	0	47	14
2. IF-II	1	13	14	0
3. IF-I + IF-II	4	29	33	35
4. IF-I + IF-II + AUG	2	92	94	100
5. IF-I + IF-II + AUG(U)$_{25}$	0	100	100	86
6. IF-I + IF-II + mRNP	27	73	100	81
7. IF-I + IF-II + mRNA	29	22	47	41
8. IF-I + GDPCP − GTP	50	0	50	ND[e]
9. IF-I + GDPCP − GTP + IF-II + AUG	73	12	85	ND
10. IF-I + IF-II + AUG + edeine	56	4	60	14
11. IF-I + IF-II + AUG + pactamycin	7	100	107	100
12. IF-I + IF-II + AUG + ATA	24	17	41	21
13. IF-I + IF-II + AUG + NaF	8	0	8	28

[a] The assays used to quantitate completion of initiation complex formation were *Assay of IF-I and of IF-II by* [35S]*Met-tRNA*$_i^{Met}$ *Binding to Ribosomal Subunits and to Ribosomes* and *Routine IF-II Assay*.

[b] In lines 8 and 9, GDPCP was substituted for GTP in the initial incubation where the ternary complex is formed. The inhibitors (lines 10–13) were added to the second incubation prior to binding of the initiator tRNA to ribosomes.

[c] The relative amounts of [35S]Met-tRNA$_i^{Met}$ bound were determined by summing the peaks of radioactivity over the 40 S ribosomal subunit and the 80 S ribosome from the sucrose density gradient. A relative activity value of 100 = 0.1 pmoles [35S]Met-tRNA$_i^{Met}$ bound to the 40 S ribosomal subunit or the 80 S ribosome under the conditions of the assay.

[d] A relative activity value of 100 = 0.1 pmole [35S]methionylpuromycin formed under the conditions of the assay.

[e] ND means that the experiment was not done.

those of initiation complex formation itself; i.e., the presence of IF-I, IF-II, and AUG, AUG(U)$_{25}$, or rabbit globin mRNP's (Table IV). This assay reveals some information about the biological activity of such an initiation complex. The fact that the methionine residue reacts with puromycin indicates that the Met-tRNA$_i^{Met}$ is bound in the P site on the 80 S ribosome. Thus, in the absence of EF-2, the initiating methionine residue is already in position to form the first peptide bond with the next amino acid residue esterified to the tRNA species to be bound, according to the base sequence of the mRNA, to the aminoacyl site (A site) on the 80 S ribosome as a preformed complex with EF-1.

Initiation complex assembly has a definite requirement for GTP. 5′-Guanylylmethylenediphosphonate (GDPCP), an enzymatically nonhydrolyzable analog of GTP, can substitute for GTP during the IF-I-mediated binding of Met-tRNA$_i^{Met}$ to a 40 S ribosomal subunit, but GDPCP cannot substitute for GTP during the IF-II-mediated transfer of Met-tRNA$_i^{Met}$ to the 80 S ribosome in the initiation complex (Table IV). The fact that GTP must be in an enzymatically hydrolyzable form during this reaction suggests that the hydrolysis of the bound GTP is an integral part of the completion of the function of IF-I and/or participates in the activity of IF-II.

The complex of IF-I·[^{35}S]Met-tRNA$_i^{Met}$·GTP can be isolated by gel filtration through a column of Sephadex G-100. This ternary complex (which was contaminated with a high molecular form of IF-II) then can be used in a highly purified and very active *in vitro* amino acid incorporation system derived from rabbit reticulocytes.[16] When this is done in the presence of large excesses of free unlabeled Met-tRNA$_i^{Met}$ and methionine, the [^{35}S]methionine is transferred rapidly and efficiently into peptide linkage using AUG(U)$_{25}$ as the template.[1,2] This proves that the methionine residue contained in the ternary complex with IF-I and GTP is a preferred precursor during the initiation of new polypeptide chains.

Using purified IF-I and IF-II, it is possible to demonstrate that various inhibitors of initiation of eukaryotic protein biosynthesis inhibit initiation complex formaton at different stages of assembly. Aurintricarboxylate (ATA), at a concentration of 60 μM, completely blocks the formation of the IF-I·Met-tRNA$_i^{Met}$·GTP complex,[2] but once the ternary complex has formed, ATA at concentrations as high as 0.4 mM does not affect the subsequent binding of the Met-tRNA$_i^{Met}$ in the ternary complex to a 40 S ribosomal subunit (Table IV). Edeine (50 μM) has no effect upon the formation of the IF-I·Met-tRNA$_i^{Met}$·GTP complex and no effect upon the IF-I-mediated binding of the initiator tRNA to a 40 S ribosomal subunit; however, edeine does prevent the IF-II-mediated transfer of the initiator tRNA from a 40 S ribosomal subunit to an initiation complex containing an 80 S ribosome. Pactamycin (0.1 mM), considered to be a potent inhibitor of eukaryotic protein biosynthesis initiation, has no effect upon the formation of the IF-I ternary complex,[2] no effect upon the IF-I-mediated binding of the initiator tRNA to a 40 S ribosomal subunit, and no effect upon the IF-II-mediated completion of an initiation complex whether the formation of the complex is determined by zone velocity sedimentation through preformed sucrose density gradients or is assayed by the puromycin-induced release of methionine. At the present, we have no explanation for these observations. NaF $(0.1\ M)$ does not prevent assembly of the IF-I ternary complex,[2] but it severely inhibits the IF-I-mediated binding of Met-tRNA$_i^{Met}$ to a 40 S ribosomal subunit and completely blocks the transfer of the initiator tRNA to an initiation complex.

In view of the intense and continuing interest in the specificity of eukaryotic protein synthesis initiation factors, studies have been carried out concerning the ability of the rabbit reticulocyte IF-I initiation factor to form a stable ternary complex at 25° in solution with eukaryotic Met-tRNA$_i^{Met}$ and GTP. Ternary complex formation requires GTP; GDPCP can replace GTP during ternary complex formation with 50% of the normal efficiency[1] and can replace GTP with equal efficiency during the IF-I-mediated binding of Met-tRNA$_i^{Met}$ to 40 S ribosomal subunits (Table IV). However, only GTP can allow the further assembly of an initiation complex. GDP, ATP, CTP, and UTP cannot substitute for GTP at any stage of initiation complex formation.

Of all of the rabbit liver aminoacyl-tRNA's tested (Met-tRNA$_i^{Met}$ Phe-tRNA^{Phe}, etc.), only Met-tRNA$_i^{Met}$ interacts with IF-I.[1] Using Met-tRNA$_i^{Met}$ preparations from two different tissues (reticulocytes and liver) and three species (rabbit, chicken, and rat), there is little or no tissue or species specificity in the formation of the rabbit reticulocyte IF-I ternary complex.[3] Interestingly, N-formyl-Met-tRNA$_i^{Met}$ (which is not the natural cytoplasmic initiator tRNA in eukaryotes) from rabbit liver, chicken liver, and rat liver can replace the Met-tRNA$_i^{Met}$ from the corresponding species with only one-tenth the efficiency of the nonformylated initiator tRNA during the formation of the ternary complex with rabbit reticulocyte IF-I.[3] It has been reported that Met-tRNA$_i^{Met}$ ($E.\ coli$) and N-formyl-Met-tRNA$_i^{Met}$ ($E.\ coli$) can interact with rabbit reticulocyte IF-I to form a complex which is retained by Millipore filters; however, this interaction is not GTP-dependent.[23]

The specificity of various IF-I preparations in the formation of the ternary complex has also been investigated. Two species (rabbit and chicken) and three functionally different tissues (reticulocytes, liver, and embryonic leg muscle) were used as sources for the IF-I preparations. Again, little or no tissue or species specificity during ternary complex formation with rabbit liver Met-tRNA$_i^{Met}$ and GTP was detected.[3] More importantly, no differences in physical or biological properties were found among the various IF-I preparations. Because IF-I from all the sources examined has a molecular weight of 350,000 in the 0.5 M KCl extracts of the polysomes,[3] the probability is very high that the IF-I initiation factor

[23] N. K. Gupta, C. L. Woodley, Y. C. Chen, and K. K. Bose, *J. Biol. Chem.* **248**, 4500 (1973).

activity is contained in a high molecular weight complex which also contains a second initiation factor corresponding to the rabbit reticulocyte IF-II initiation factor. Finally, the studies investigating the specificity of IF-I strongly indicate that at least one (IF-I) of the two initiation factors isolated from rabbit reticulocytes is a ubiquitous and obligatory factor during the initiation of protein biosynthesis in the cytoplasm of eukaryotes.

[17] Preparation of Protein Synthesis Initiation Factors from Rabbit Liver

By Dante Picciano and W. French Anderson

The procedures for the isolation and purification of protein synthesis initiation factors from enucleated mammalian cells, rabbit reticulocytes, are presented elsewhere in this volume.[1,2] The corresponding initiation factors from a nucleated mammalian source, rabbit liver, have recently been identified.[3] This report describes the methods employed in our laboratory for the isolation and partial purification of the liver initiation factors IF-M_1, IF-M_{2A}, IF-M_{2B}, and IF-M_3.

Reagents

> Sodium pentobarbital, 65 mg/ml (A. J. Buck & Son)
> Dithiothreitol (Calbiochem)
> KCl, 4.0 M
> Tris·HCl, 1.0 M, pH 7.5 (2°)
> Neutralized EDTA, 0.10 M; a 0.10 M solution of EDTA which has been adjusted to pH 7.0 by the addition of 1.0 M KOH
> MgCl$_2$, 1.0 M
> DEAE-cellulose (Whatman, DE-23)
> CM-cellulose (Whatman, CM-32)
> Sephadex G-200 (Pharmacia Fine Chemicals, Inc.)

Buffers

> Buffer A: a solution which contains per liter 50 ml of 1.0 M Tris· HCl, pH 7.5 (2°), 10 ml of 1.0 M MgCl$_2$, 5 ml of 0.10 M

[1] W. C. Merrick, H. Graf, and W. F. Anderson, this volume [13].
[2] P. M. Prichard and W. F. Anderson, this volume [14].
[3] D. J. Picciano, P. M. Prichard, W. C. Merrick, D. A. Shafritz, H. Graf, R. G. Crystal, and W. F. Anderson, *J. Biol. Chem.* 248, 204 (1973).

neutralized EDTA, 86.0 g of sucrose (0.25 M), 0.154 g of dithiothreitol (1 mM), and 2.61 g of KCl (35 mM)

Buffer B: a solution which contains per liter 50 ml of 1.0 M Tris·HCl, pH 7.5 (2°), 0.1 ml of 1.0 M MgCl$_2$, 5 ml of 0.10 M neutralized EDTA, 86.0 g of sucrose (0.25 M), 0.154 g of dithiothreitol (1 mM), and 2.61 g of KCl (35 mM)

0.10 M KCl Buffer C: a solution which contains per liter 20 ml of 1.0 M Tris·HCl, pH 7.5 (2°), 1 ml of 1.0 M MgCl$_2$, 1 ml of 0.10 M neutralized EDTA, 0.154 g of dithiothreitol (1 mM), and 7.46 g of KCl

0.20 M KCl Buffer C: a solution which contains per liter 20 ml of 1.0 M Tris·HCl, pH 7.5 (2°), 1 ml of 1.0 M MgCl$_2$, 1 ml of 0.10 M neutralized EDTA, 0.154 g of dithiothreitol, and 14.9 g of KCl

0.50 M KCl Buffer C: a solution which contains per liter 20 ml of 1.0 M Tris·HCl, pH 7.5 (2°), 1 ml of 1.0 M MgCl$_2$, 1 ml of 0.10 M neutralized EDTA, 0.154 g of dithiothreitol, and 37.3 g of KCl

Preparation of Liver Microsomal Wash Fraction

New Zealand white rabbits weighing between 1 and 2 kg (approximately 8 weeks of age) are used in all studies. Untreated rabbits which have been fasted for 18–24 hours are anesthesized by an intraperitoneal injection of sodium pentobarbital at a dosage of 1 ml per animal. The rabbits are exsanguinated by direct cardiac puncture employing a 50-ml syringe (18-gauge Huber-point needle). The animal's liver and gallbladder are removed, and the gallbladder is excised. The livers are rinsed in cold tap water and immersed in ice cold buffer A. All the following operations are carried out between 2° and 4°.

The livers are weighed and minced; 200 g of tissue are placed into a Waring Blendor which is connected to a voltage rheostat. Buffer A is added (1 ml of buffer A per gram of tissue), and the tissue is homogenized for 90 seconds with the rheostat adjusted to 45 V. The resulting homogenate is poured into 500-ml plastic centrifuge bottles and centrifuged in a Sorvall GS-3 rotor for 30 minutes at 2°. Initially the centrifuge is operated at 1500 g for 5 minutes, the speed is increased to 6000 g for an additional 5 minutes, and the speed is increased a final time to 14,000 g for the remaining 20 minutes. The resulting "postmitochondrial supernatant fraction" is covered with a lipid layer which is carefully removed and discarded. The upper four-fifths of the remaining supernatant is carefully removed and pooled. The pellets are discarded. The lipid-free postmitochondrial supernatant fraction is placed into 75-ml polycarbonate tubes and centrifuged at 95,500 g for 150 minutes at 2°. The resulting supernatant is discarded, and the free surface of the remaining pellet is carefully and quickly rinsed with 5 ml of buffer B. The rinse solution is decanted and discarded. Again 5 ml of buffer B is added, and the tube is shaken gently until the microsomal pellet has been freed from the wall of the tube. Occasionally, a small lower pellet (glycogen) remains attached to the tube; this lower pellet is discarded. The freed microsomal pellets are pooled and suspended in buffer B. Enough buffer B is employed (approximately 25–30 ml per tube) to give an absorbance per milliliter at 260 nm of 150–250. The resulting suspension is gently mixed with a magnetic stirring bar until the microsomes are in solution (approximately 2 hours) or slowly forced through the small orifice of a 50-ml syringe and then gently mixed until the microsomes are suspended (approximately 1 hour). After suspension of the microsomes, enough 4 M KCl is added over a 2-minute period (with stirring) until a final KCl concentration of 0.5 M is achieved (1 ml of 4 M KCl for each 7 ml of microsomal suspension). The resulting suspension is stirred for 1 hour. The solution is placed into 75 ml polycarbonate tubes and centrifuged at 95,500 g for 150 minutes at 2°. If the resulting supernatant fraction contains an upper lipid layer, the layer is carefully removed and discarded. The upper four-fifths of the remaining supernatant, "microsomal salt wash fraction," is carefully removed and pooled. This preparation is stored in liquid nitrogen at a concentration of 6–10 mg of protein per milliliter and is stable for at least 6 months.

Isolation of Crude Liver Initiation Factors

For all steps in the isolation and purification of IF-M$_1$, IF-M$_{2(A+B)}$, IF-M$_{2A}$, and IF-M$_{2B}$, the activities at each stage are tested in poly(U)-directed polyphenylalanine synthesis (assay II, A, 1[4]). IF-M$_3$ activity is tested by endogenous mRNA dependent globin synthesis (assay II, B, 5, a[4]).

Crude IF-M$_1$, IF-M$_{2(A+B)}$, and IF-M$_3$ are obtained from the 0.5 M KCl wash fraction of rabbit liver microsomes by DEAE-cellulose chromatography. Protein from 400 ml of crude 0.5 M KCl microsomal wash fraction (9 mg of protein per milliliter; 3600 mg total) is precipitated by the addition of solid ammonium sulfate–ammonium carbonate (50:1, w/w) to 70% saturation at 0°. The protein is sedimented by centrifugation at 16,000 g for 20 minutes at 2°. The supernatant is discarded, and the pellets are dissolved in 150 ml of 0.10 M KCl buffer C, dialyzed for 15 hours against 4 liters of the same buffer, and applied to a 1.5 × 30 cm column of DEAE-cellulose (Whatman DE-23) which has been equili-

---

[4] R. G. Crystal, N. A. Elson, and W. F. Anderson, this volume [12].

brated with the same buffer. The column is washed with 0.10 M KCl buffer C until protein no longer elutes. All material not retained on the column is pooled, fractionated by ammonium sulfate precipitation (40–60% saturation, 0°) and dialyzed against 0.10 M KCl buffer C. This crude IF-M$_1$ preparation is frozen and stored in liquid nitrogen at a concentration of 35 mg of protein per milliliter.

IF-M$_3$ and IF-M$_{2(A+B)}$ are eluted from the column with a linear KCl gradient (100 ml of 0.10 M KCl buffer C × 100 ml of 0.50 M KCl buffer C), and fractions of 5 ml each are collected every 10 minutes. The fractions displaying IF-M$_3$ activity (0.15–0.22 M KCl) in the globin synthesis assay (assay II, B, 5, a[4]) are pooled and further purified by a 30–60% ammonium sulfate fractionation. The protein pellet is dissolved in 2 ml of 0.20 M KCl buffer C and dialyzed against 4 liters of the same buffer for 12 hours. IF-M$_3$ is stored in small aliquots at a concentration of 14 mg of protein per milliliter in liquid nitrogen. The elution profile of the liver activities on DEAE-cellulose is seen in Fig. 1.

The fractions displaying IF-M$_{2(A+B)}$ activity (0.225–0.40 M KCl) in the poly(U)-directed polyphenylalanine synthesis assay at low Mg^{2+} concentration (assay II, A, 1[4]) are pooled and concentrated by the addition of ammonium sulfate to 80% saturation at 0°. The protein is sedimented by centrifugation at 16,000 g for 20 minutes at 2°. The supernatant is dis-

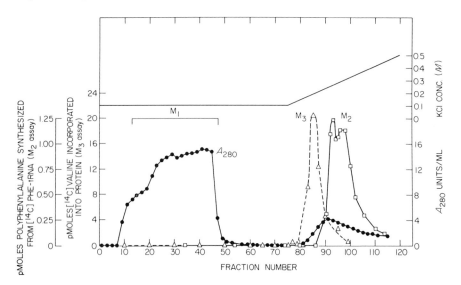

Fig. 1. The separation of liver IF-M$_1$, IF-M$_3$, and IF-M$_{2(A+B)}$ by DEAE-cellulose chromatography of liver microsomal wash fraction [see D. J. Picciano, P. M. Prichard, W. C. Merrick, D. A. Shafritz, H. Graf, R. G. Crystal, and W. F. Anderson, *J. Biol. Chem.* **248**, 204 (1973)].

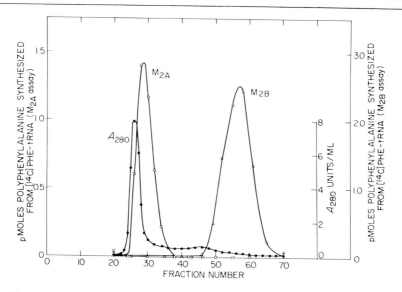

FIG. 2. Separation of IF-M$_{2A}$ and IF-M$_{2B}$ by Sephadex G-200 chromatography of crude liver IF-M$_{2(A+B)}$ [see D. J. Picciano, P. M. Prichard, W. C. Merrick, D. A. Shafritz, H. Graf, R. G. Crystal, and W. F. Anderson, *J. Biol. Chem.* **248**, 204 (1973)].

carded, and the pellet is dissolved in 0.10 M KCl buffer C and concentrated to 20 mg of protein per milliliter by ultrafiltration with a UM-10 membrane (Amicon Corp.). IF-M$_{2(A+B)}$ is further fractionated by Sephadex G-200 chromatography into two components, IF-M$_{2A}$ and IF-M$_{2B}$. The concentrated IF-M$_{2(A+B)}$ (2.1 ml; 42 mg total protein) is placed on a calibrated column (1.5 × 86 cm) of Sephadex G-200 which has been equilibrated with 0.10 M KCl buffer C. Protein is eluted with the same buffer at a flow rate of 10 ml per hour collecting 2-ml fractions. IF-M$_{2A}$ activity appears just after the void volume while IF-M$_{2B}$ activity is found as an included fraction as shown in Fig. 2. Fractions showing IF-M$_{2A}$ or IF-M$_{2B}$ activity are pooled separately, concentrated by ultrafiltration with a UM-10 membrane, and stored in aliquots in liquid nitrogen (IF-M$_{2A}$, 7.8 mg of protein per milliliter; IF-M$_{2B}$, 2.5 mg of protein per milliliter).

Purification of Liver Initiation Factors

Crude IF-M$_1$, isolated as the material not adsorbed to DEAE-cellulose at 0.10 M KCl and then precipitated by a 40–60% ammonium sulfate fractionation at 0°, is further purified by Sephadex G-200 chromatography and CM-cellulose chromatography. Both chromatographic procedures are performed in a manner identical to the corresponding chro-

TABLE I

PURIFICATION OF LIVER IF-M$_1$

Procedure	Volume (ml)	Protein (mg/ml)	Total protein (mg)	Specific activity[a] (units/mg)	Purification (fold)	Total units	Recovery of activity
1. Microsomal wash fraction	280	7.9	2210[b]	—	—	—	—
2. DEAE-cellulose eluate	260	6.8	1768	6.4	1	11,320	100
3. Ammonium sulfate precipitation (40–60%)	17.6	35	616	—	—	—	—
4. Sephadex G-200 chromatography	108	2.2	238	28.8	4.5	6,854	61
5. CM-cellulose chromatography	3.4	3.7	12.6	226	35	2,848	25

[a] One unit of activity is defined as the IF-M$_1$-dependent incorporation of 1 pmole of [^{14}C]phenylalanine into hot trichloroacetic acid insoluble protein in the presence of saturating levels of complementary factors (EF-1, EF-2, IF-M$_{2A}$, IF-M$_{2B}$) under standard poly(U)-directed polyphenylalanine synthesis assay conditions.

[b] Preparation is from 1 kg liver tissue (20–24 livers).

TABLE II

PURIFICATION OF LIVER IF-M_{2A}

Procedure	Volume (ml)	Protein (mg/ml)	Total protein (mg)	Specific activity[a] (units/mg)	Purification (fold)	Total units	Recovery of activity
1. Microsomal wash fraction	830	8.0	6640	29.4	1	195,200	100
2. DEAE-cellulose batch chromatography	240	3.2	771	113	3.8	87,120	45
3. Sephadex G-200 chromatography	64	5.5	350	185	6.3	64,750	33
4. DEAE-cellulose gradient chromatography	8	11.7	94	404	13.7	37,980	19

[a] One unit of activity is defined as the IF-M_{2A} dependent incorporation of 1 pmole of [^{14}C]phenylalanine into hot trichloroacetic acid-insoluble protein in the presence of saturating levels of complementary factors (EF-1, EF-2, IF-M_1, IF-M_{2B}) under standard poly(U)-directed polyphenylalanine synthesis assay conditions.

TABLE III

PURIFICATION OF LIVER IF-M$_{2B}$

Procedure	Volume (ml)	Protein (mg/ml)	Total protein (mg)	Specific activity[a] (units/mg)	Purification (fold)	Total units	Recovery of activity
1. Microsomal wash fraction	830	8.0	6640	—	—	—	—
2. DEAE-cellulose batch chromatography	240	3.2	771	20.2	1	15,570	100
3. Sephadex G-200 chromatography	135	1.0	135	75	3.7	10,125	65
4. DEAE-cellulose gradient chromatography	8.2	4.2	34.3	230	11.4	7,890	51

[a] One unit of activity is defined as the IF-M$_{2B}$-dependent incorporation of 1 pmole of [^{14}C]phenylalanine into hot trichloroacetic acid-insoluble protein in the presence of saturating levels of complementary factors (EF-1, EF-2, IF-M, IF-M$_{2A}$) under standard poly(U)-directed polyphenylalanine synthesis assay conditions.

TABLE IV

PURIFICATION OF LIVER IF-M₃

Procedure	Volume (ml)	Protein (mg/ml)	Total protein (mg)	Specific activity[a] (units/mg)	Purification (fold)	Total units	Recovery of activity
1. Microsomal wash fraction	395	7.7	3042	0[b]	—	0[b]	—
2. DEAE–cellulose chromatography	45.6	1.7	77.5	3.1	1	240	100
3. Ammonium sulfate precipitation (30–60%)	2	14.1	28.2	7.1	2.3	200	83

[a] One unit of activity is defined as the IF-M₃ dependent incorporation of 50 pmole of [¹⁴C]valine into hot trichloroacetic acid-precipitable polypeptide in endogenous mRNA directed globin synthesis under standard assay conditions.

[b] No IF-M₃ activity could be detected in the microsomal wash fraction.

matography of reticulocyte IF-M₁.[1] Table I summarizes the purification procedures of the fractions obtained at the various stages of liver IF-M₁ purification from 2.2 g of microsomal wash protein.

Liver IF-M$_{2A}$ and IF-M$_{2B}$, isolated from Sephadex G-200, are further purified by DEAE-cellulose (DE-23) chromatography. The peak of liver IF-M$_{2A}$ activity elutes at $0.30\,M$ KCl, and the peak of IF-M$_{2B}$ activity elutes at $0.24\,M$ KCl. Tables II and III summarize the purification procedures for liver IF-M$_{2A}$ and IF-M$_{2B}$, respectively, from 6.6 g of microsomal wash protein.

Crude liver IF-M₃, obtained from DEAE-cellulose, is further purified by a 30–60% ammonium sulfate fractionation. Preliminary attempts to purify IF-M₃ using phosphocellulose chromatography result in a very broad elution pattern as well as considerable loss of activity. However, the salt concentration at which liver IF-M₃ elutes is similar to that for reticulocyte IF-M₃ (approximately $0.30\,M$ KCl).[2] Table IV summarizes the purification for the liver IF-M₃ from 3.0 g of microsomal wash protein.

[18] Isolation, Purification, and Assay of an Initiation Factor from Rat Liver Cytosol That Promotes Binding of Aminoacyl-tRNA to 40 S Ribosomal Subunits

By David P. Leader and Ira G. Wool

The mechanism for the initiation of protein synthesis in eukaryotes appears to be similar to that in prokaryotes. Thus the process requires the initiator codon-AUG, the specific initiator tRNA-tRNA$_F^{Met}$, the small (40 S) ribosomal subunit, and three factors—IF-M₁, IF-M₂, and IF-M₃—which have been extracted from rabbit reticulocyte ribosomes.[1-9] Initiation of protein synthesis in eukaryotes does differ from that in prokaryotes as

[1] A. E. Smith and K. A. Marcker, Nature (London) 226, 607 (1970).
[2] J. C. Brown and A. E. Smith, Nature (London) 226, 610 (1970).
[3] R. Jackson and T. Hunter, Nature (London) 227, 672 (1970).
[4] D. T. Wiggle and G. H. Dixon, Nature (London) 227, 676 (1970).
[5] D. B. Wilson and H. M. Dintzis, Proc. Nat. Acad. Sci. U.S. 66, 1282 (1970).
[6] S. M. Heywood, Nature (London) 225, 696 (1970).
[7] S. M. Heywood and W. C. Thompson, Biochem. Biophys. Res. Commun. 43, 470 (1971).
[8] P. M. Prichard, J. M. Gilbert, D. A. Shafritz, and W. F. Anderson, Nature (London) 226, 511 (1970).
[9] R. G. Crystal and W. F. Anderson, Proc. Nat. Acad. Sci. U.S. 69, 706 (1972).

the initiator Met-tRNA$_F^{Met}$ is not formylated.[1] Despite this, initiation factor IF-M$_1$[10,11] (and preparations presumed to contain the factor[12-16]) will recognize and catalyze, *in vitro,* the binding of fMet-tRNA$_F^{Met}$ (as well as Phe-tRNA and *N*-acetyl-Phe-tRNA) to 40 S ribosomal subunits. The binding of nonformylated Met-tRNA$_F^{Met}$ requires in addition, IF-M$_2$ and GTP.[9,17,18]

We describe here the purification, from rat liver cytosol, of a factor which promotes the binding of Phe-tRNA to 40 S ribosomal subunits.[12] The purified factor also directs the binding of fMet-tRNA and *N*-acetyl-Phe-tRNA to 40 S subunits[13] and appears to be similar to the reticulocyte initiation factor IF-M$_1$, and analogous to the bacterial factor IF-2.

Reagents

Solid (NH$_4$)$_2$SO$_4$

Polyuridylic acid (Miles Laboratories, Inc.)

[³H]Phe-tRNA-*Escherichia coli* B tRNA aminoacylated[19,20] with [³H]phenylalanine and 19 nonradioactive amino acids

Sephadex G-25 (medium or coarse)

DEAE-cellulose: Whatman DE-52 resin

Hydroxyapatite (Clarkson Chemical Co. and Bio-Rad)

Acetic acid, 1 *M*

KOH, 1 *M*

Medium A: tris(hydroxymethyl)aminomethane (Tris)·HCl, pH 7.6, 50 m*M*; KCl, 80 m*M*; MgCl$_2$, 12.5 m*M*; sucrose, 250 m*M*

Medium B: Tris·HCl, 10 m*M*, pH 7.5; KCl, 100 m*M*; dithiothreitol, 1 m*M*; EDTA, 0.1 m*M*

Potassium phosphate buffers, pH 6.8, containing β-mercaptoethanol, 1 m*M*, and potassium phosphate at concentrations of 0.15 *M*, 0.175 *M*, 0.25 *M*, 0.35 *M* and 0.5 *M*

[10] D. A. Shafritz and W. F. Anderson, *Nature* (*London*) **227**, 918 (1970).
[11] D. A. Shafritz and W. F. Anderson, *J. Biol. Chem.* **245**, 5553 (1970).
[12] D. P. Leader, I. G. Wool, and J. J. Castles, *Proc. Nat. Acad. Sci. U.S.* **67**, 523 (1970).
[13] D. P. Leader and I. G. Wool, *Biochim. Biophys. Acta* **262**, 360 (1972).
[14] E. Gasior, P. Rao, and K. Moldave, *Biochim. Biophys. Acta* **254**, 331 (1971).
[15] E. Gasior and K. Moldave, *J. Mol. Biol.* **66**, 391 (1972).
[16] M. Zasloff and S. Ocho, *Proc. Nat. Acad. Sci. U.S.* **68**, 3059 (1971).
[17] D. A. Shafritz, D. G. Laycock, R. G. Crystal, and W. F. Anderson, *Proc. Nat. Acad. Sci. U.S.* **68**, 2246 (1971).
[18] D. A. Shafritz, P. M. Prichard, J. M. Gilbert, W. C. Merrick, and W. F. Anderson, *Proc. Nat. Acad. Sci. U.S.* **69**, 983 (1972).
[19] G. von Ehrenstein and F. Lipmann, *Proc. Nat. Acad. Sci. U.S.* **47**, 941 (1961).
[20] I. G. Wool and P. Cavicchi, *Biochemistry* **6**, 1231 (1967).

Tris·HCl buffer, 50 mM, pH 8.0, containing β-mercaptoethanol, 1 mM

Tris·HCl buffer, 10 mM, pH 7.5, containing β-mercaptoethanol, 1 mM

Tris·HCl buffer, 10 mM, pH 7.5, containing KCl, 80 mM, and MgCl$_2$, 5 mM

Toluene/PPO/POPOP scintillation fluid: 4 g of 2,5-diphenyloxazole and 0.2 g of 1,4-bis-(5-phenyloxazol-2-yl)benzene per liter of toluene

Purification Procedure

Step 1. Preparation of Cytosol Fraction. All the procedures are carried out at 0–4°. Male rats (Sprague-Dawley 100–150 g) are decapitated, the livers are quickly excised and placed in medium A (50 ml for about 50 g of liver), and the volume displaced by the tissue noted. The liver is minced with scissors and washed 3 times with medium A to remove as much blood as possible. The total volume is adjusted with medium A to 4 times the volume of the liver and the tissue is homogenized with 10 strokes of a Potter-Elvehjem homogenizer. The homogenate is centrifuged at 20,000 g for 20 minutes, the precipitate discarded and the post-mitochondrial supernatant removed and centrifuged again. The combined supernatant is filtered through glass wool and centrifuged at 100,000 g for 2 hours. The pellet consists of crude microsomes which are used for the preparation of ribosomes.[20] The postmicrosomal supernatant is the cytosol fraction containing the initiation factor.

Whereas the conditions for the preparation of the cytosol fraction are generally not critical and may vary between tissues, the ionic strength of the extraction buffer (medium A) may possibly be important. Although we have not conducted systematic studies on this point, it is conceivable that if the ionic strength were too low, the proportion of the factor associated with the microsomes might increase.

Step 2. Precipitation at pH 5.2. The pH of the cytosol fraction is slowly reduced to 5.2 by dropwise addition of 1 M acetic acid. A precipitate forms (containing ribosomes, tRNA, and aminoacyl-tRNA synthetases), and this is removed by centrifugation for 30 minutes at 10,000 g. The pH of the supernatant is adjusted to 7.0 with 1 M KOH and centrifuged again to remove any more insoluble material. The neutralized supernatant is then passed through a column of Sephadex G-25 (medium or coarse), equilibrated with Tris·HCl buffer pH 8.0 (50 mM) containing β-mercaptoethanol (1 mM). The eluate (which also contains elongation factors EF-1 and EF-2) is referred to as "G-25 fraction" and is stored frozen at $-20°$.

Step 3. Hydroxyapatite "Batch" Chromatography. The bulk of elongation factor EF-2, and much of the total protein, is now removed by hydroxyapatite chromatography.[21] The "G-25 fraction" (4 g of protein in 400 ml) is mixed with 160 ml of hydroxyapatite (Clarkson Chemical Co.) suspended in Tris·HCl buffer pH 8.0 (50 mM), containing β-mercaptoethanol (1 mM). The mixture is stirred for 1 hour, and the hydroxyapatite is sedimented (5 minutes at 1500 g). The sediment is washed 6 times with 250 ml of 0.175 M potassium phosphate, pH 6.8, containing β-mercaptoethanol (1 mM), and the washes are discarded (or used to prepare elongation factor EF-2).[21] The hydroxyapatite is then extracted 3 times with 250 ml of 0.125 M potassium phosphate buffer and the protein (about 200 mg) in the pooled extracts precipitated by adding $(NH_4)_2SO_4$ to 70% saturation. Unfortunately, the $(NH_4)_2SO_4$ precipitation cannot be used to augment the purification as the initiation factor precipitates over a wide range of salt concentrations.

Step 4. DEAE-Cellulose Chromatography. The bulk of elongation factor EF-1 is removed by DEAE-cellulose chromatography. The protein precipitated by $(NH_4)_2SO_4$ is redissolved in medium B and traces of contaminating $(NH_4)_2SO_4$ removed by gel filtration (through Sephadex G-25) or dialysis (against medium B). This redissolved precipitate can be stored in liquid nitrogen but is generally applied directly to a 1.5 × 35 cm column containing washed Whatman DE-52 resin, from which the fines have been removed and which has been equilibrated previously with medium B. The protein is eluted with medium B at a flow rate of 30 ml per hour. The protein containing fractions (about 75 mg) are pooled, concentrated to about 6 mg/ml by vacuum dialysis and stored in liquid nitrogen. A single step rather than a gradient elution is performed because the initiation factor tends to lose activity on DEAE-cellulose.

Step 5. Hydroxyapatite Column Chromatography. The concentrated eluates (150 mg protein) from chromatography of two samples on DEAE-cellulose are first passed through a Sephadex G-25 column, equilibrated with 0.15 M potassium phosphate buffer pH 6.8, containing 1 mM β-mercaptoethanol, and then applied to a 1.5 × 25 cm column of hydroxyapatite (Bio-Rad) equilibrated with the same buffer. Initially protein is eluted from the column with 0.15 M phosphate buffer at a flow rate of 30 ml per hour. Elution is continued with a linear gradient from 0.15 M to 0.35 M phosphate (250 ml) and finally with 0.5 M phosphate (40 ml). Fractions of approximately 5 ml are collected. The initiation factor is found in those fractions comprising the protein peak C eluting at about

[21] M. Schneir and K. Moldave, *Biochim. Biophys. Acta* **166**, 58 (1968).

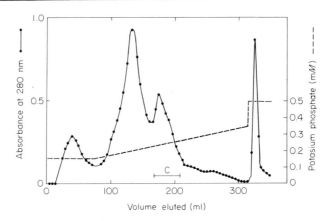

FIG. 1. Hydroxyapatite column chromatography (step 5). Protein (100 mg) eluting from DEAE-cellulose at 0.1 *M* KCl (step 4) was fractionated on a hydroxy-apatite column as described in the text [D. P. Leader and I. G. Wool, *Biochim. Biophys. Acta* **262**, 360 (1972)].

0.235 *M* phosphate (Fig. 1).[13] The peak C fractions are pooled, concentrated by vacuum dialysis, and passed through a Sephadex G-25 column, equilibrated with 10 m*M* Tris·HCl, pH 7.5 (containing 1 m*M* β-mercaptoethanol) and stored in liquid nitrogen. The factor retains activity for several months in liquid nitrogen. Hydroxyapatite chromatography removes the last traces of elongation factors EF-1 and EF-2.[13]

Assay

It is most convenient to assay the factor using the reaction in which it was originally detected—the polyuridylic acid-directed binding of (*E. coli*) [³H]Phe-tRNA to rat liver 40 S ribosomal subunits.[12] *E. coli* [³H]Phe-tRNA is prepared by standard procedures.[19,20] The method of preparing rat liver ribosomal subunits is described in a previous volume of this series.[22] The binding reaction mixture (0.1 ml) contains Tris· HCl, pH 7.5 (1 μmole), KCl (8 μmole), MgCl₂ (0.5 μmole), β-mercapto-ethanol (1 μmole), [³H]Phe-tRNA (40 μg), polyuridylic acid (10 μg), 40 S ribosomal subunits (3.6 μmoles) as well as factor containing fractions. The reaction is started by the addition of ribosomal subunits and incubation is for 15 minutes at 30°. The reaction is then stopped by addition of 5 ml of ice cold Tris·HCl buffer (10 m*M*), pH 7.5, containing KCl (80 m*M*), and MgCl₂ (5 m*M*), and filtered through nitro-cellulose filters (Millipore filter, HAWP, 25 mm in diameter, 0.45 μm pore size). The filter is washed three times with 5-ml portions of ice cold

---

[22] T. E. Martin, I. G. Wool, and J. J. Castles, this series, Vol. 20, p. 417 (1971).

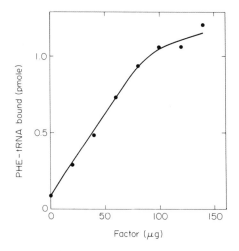

FIG. 2. Assay of binding of Phe-tRNA to 40 S ribosomal subunits promoted by purified factor. Various amounts of factor (fraction C of Fig. 1) were incubated in the binding reaction mixture described in the text [D. P. Leader and I. G. Wool, *Biochim. Biophys. Acta* **262**, 360 (1972)].

buffer,[23] dried at 120° for 30 minutes, placed in a vial with 10 ml of toluene/PPO/POPOP scintillation fluid and the radioactivity measured by scintillation spectrometry. The activity of different amounts of purified factor in this assay is shown in Fig. 2.

The best conditions for the assay are those which maximize factor-dependent binding and minimize nonenzymatic binding (which must always be measured). It is for this latter reason that it is crucial to use a magnesium chloride concentration of 5 mM. At higher concentrations, not only will nonenzymatic binding occur but, in cruder enzyme fractions, there will be binding directed by any contaminating elongation factor EF-1.[12,13] It is also important to keep the concentration of Tris · HCl low (approximately 10 mM) as this inhibits the binding reaction; variations of pH between 7 and 8 do not affect binding.[13]

It is also possible to assay the factor by its ability to catalyze the binding of fMet-tRNA to 40 S subunits in the presence of the codon, AUG.[13] However, polyuridylic acid is a cheaper and more efficient template than AUG. Anderson and co-workers have assayed reticulocyte IF-M$_1$ by its ability to complement IF-M$_2$ in polyphenylalanine synthesis[11] or Met-tRNA binding.[9,10,17,18] The binding assay described above has the advantage that it does not require other factor preparations. It has proved useful in studies of the subcellular location of eukaryotic initiation factors.[24]

[23] M. Nirenberg and P. Leder, *Science* **145**, 1399 (1964).
[24] D. P. Leader, H. Klein-Bremhaar, I. G. Wool, and A. Fox, *Biochem. Biophys. Res. Commun.* **46**, 215 (1972).

Properties of the Factor

The rat liver cytosol initiation factor has a molecular weight of about 93,000[13] and appears to contain a sulfhydryl group(s) which is required in the reduced form for activity.[12] The substrate for the factor is the 40 S subunit rather than the 80 S ribosome for addition of 60 S subunits inhibits rather than enhances the factor catalyzed binding of Phe-tRNA, N-acetyl-Phe-tRNA or fMet-tRNA to 40 S subunits.[13,15,16] Moreover, after the binding of Phe-tRNA, the 40 S subunit becomes able to reassociate with the 60 S subunit.[25]

Remarks

The main difficulty in purifying the factor is its tendency to lose activity. Thus, the overall increase in specific activity was only 5-fold. As we have been unable to devise a means of stabilizing the factor, we have been obliged to perform the purification steps in as expeditious a manner as possible, endeavoring also to maintain a high protein concentration.

Rat liver IF-M$_1$ is present in the cytosol fraction, whereas the same or analogous factors in prokaryote and reticulocytes[8] are associated with the ribosome. Since rat liver IF-M$_2$ is bound to the ribosome,[24] the separation of IF-M$_1$ from IF-M$_2$ is greatly facilitated. The subcellular fraction in which eukaryotic initiation factors are found varies and does not necessarily reflect their location *in vivo*. Thus, IF-M$_1$ from embryos of the brine shrimp, *Artemia salina* is also found in the cytosol fraction,[16] as are all three initiation factors from Krebs II ascites cells.[24]

[25] R. E. H. Wettenhall, D. P. Leader, and I. G. Wool, *Biochem. Biophys. Res. Commun.* 43, 994 (1971).

[19] Assay of Reassociation of Eukaryotic Ribosomal Subunits Catalyzed by Initiation Factors

By RICHARD E. H. WETTENHALL and IRA G. WOOL

The initiation of protein synthesis requires the formation of an initiation complex consisting of: the small ribosomal subunit; mRNA containing the initiating codon, AUG; Met-tRNA$_F$, which is the initiator aminoacyl-tRNA and which recognizes the codon AUG; GTP; and at least three protein initiation factors.[1-10] The complex associates with the

[1] C. Gutherie and M. Nomura, *Nature (London)* 219, 232 (1968).
[2] A. E. Smith and K. A. Marcker, *Nature (London)* 226, 607 (1970).

large subunit to form an 80 S (or 70 S) initiation complex. The process is formally analogous in prokaryotic and eukaryotic cells. The only difference is that the initiator aminoacyl-tRNA, Met-tRNA$_F$, is not formylated in eukaryotes.[2,3]

Analysis of the details of the biochemistry of protein synthesis, and isolation, purification, and characterization of the several factors which catalyze individual steps in the process, are dependent on assays of partial reactions.[11] The association of the small (40 S) ribosomal subunit initiation complex with the large (60 S) subunit to form an 80 S ribosomal monomer is one such partial reaction. We shall describe an assay designed to study the details of the mechanism of the reaction.[12,13]

Spontaneous reassociation of ribosomal subparticles (independent of the formation of an initiation complex) can occur in vitro[14,15]: reassociation is favored by high concentrations of magnesium,[16-19] by addition of polyamines,[17,20,21] and by the presence of some nonionic agents, such as ethanol and dioxane.[22] High concentrations of potassium and other monovalent cations, on the other hand, tend to depress spontaneous formation

[3] J. C. Brown and A. E. Smith, Nature (London) 226, 610 (1970).

[4] R. Jackson and T. Hunter, Nature (London) 227, 672 (1970).

[5] D. Housman, M. Jacobs-Lorena, U. L. RajBhandary, and H. F. Lodish, Nature (London) 227, 913 (1970).

[6] D. T. Wigle and G. H. Dixon, Nature (London) 227, 676 (1970).

[7] P. M. Prichard, J. M. Gilbert, D. A. Shafritz, and W. F. Anderson, Nature (London) 226, 511 (1970).

[8] S. M. Heywood, Nature (London) 225, 696 (1970).

[9] J. Ilan and J. Ilan, Develop. Biol. 25, 280 (1971).

[10] I. B. Pragnell, G. Marbaix, H. R. V. Arnstein, and B. Lebleu, FEBS (Fed. Eur. Biochem. Soc.) Lett. 14, 289 (1971).

[11] P. Lengyel and D. Soll, Bacteriol. Rev. 33, 264 (1969).

[12] R. E. H. Wettenhall, D. P. Leader, and I. G. Wool, Biochem. Biophys. Res. Commun. 43, 994 (1971).

[13] R. E. H. Wettenhall and I. G. Wool, J. Biol. Chem. 247, 7201 (1972).

[14] T. E. Martin, I. G. Wool, and J. J. Castles, this series, Vol. 20, p. 417 (1971).

[15] A. K. Falvey and T. Staehelin, J. Mol. Biol. 53, 1 (1970).

[16] A. Tissières, J. D. Watson, D. Schlessinger, and B. R. Hollingworth, J. Mol. Biol. 1, 221 (1959).

[17] S. Pestka, J. Biol. Chem. 241, 367 (1966).

[18] A. S. Spirin, B. Sabo, and V. A. Kovalenko, FEBS (Fed. Eur. Biochem. Soc.) Lett. 15, 197 (1971).

[19] O. P. van Diggelen, H. L. Heinsius, F. Kalousek, and L. Bosch, J. Mol. Biol. 55, 277 (1971).

[20] N. Silman, M. Artman, and H. Engelberg, Biochim. Biophys. Acta 103, 231 (1965).

[21] S. J. S. Hardy and G. Turnock, Nature (London) New Biol. 229, 17 (1971).

[22] A. S. Spirin and E. B. Lishnevskaya, FEBS (Fed. Eur. Biochem. Soc.) Lett. 14, 114 (1971).

of 40 S–60 S couples.[14,15] Finally, the association of peptidyl-tRNA with ribosomal subparticles can markedly influence *in vitro* reassociation.[14,23]

The reassociation of ribosomal subunits to form 80 S monomers, in the assay we shall describe, is carried out in circumstances where spontaneous formation of 40 S–60 S couples is kept to a minimum. That requires careful selection of ionic conditions and use of ribosomal subunits free of peptidyl-tRNA.[24] The reaction is catalyzed by a rat liver cytosol fraction (G-25 fraction)[25,26] that contains the initiation factor IF-M₁,[27] and also by purified IF-M₁.[13] Reassociation requires a template [we have generally used poly(U)], either aminoacyl-tRNA or deacylated-tRNA, and is partially dependent on the presence of GTP.

Reagents and Solutions

Escherichia coli B tRNA (Schwarz/Mann)

Puromycin dihydrochloride (Nutritional Biochemicals Corporation)

[³H]Phenylalanine (New England Nuclear Corporation)

[³H]Phe-tRNA: *Escherichia coli* B tRNA aminoacylated[28] with [³H]: phenylalanine and 19 nonradioactive amino acids

Medium A: Tris(hydroxymethyl)aminomethane (Tris)·HCl, 50 mM, pH 7.6; KCl, 80 mM; $MgCl_2$, 12.5 mM; β-mercaptoethanol-MSH, 10 mM

Medium B: Tris·HCl, 10 mM, pH 7.6; KCl, 120 mM; $MgCl_2$, 3.5 mM; MSH, 10 mM

Medium C: Tris·HCl, 50 mM, pH 7.6; KCl, 880 mM; $MgCl_2$, 12.5 mM; MSH, 10 mM.

Preparation of Ribosomal Subunits

We generally prepare liver ribosomes from male Sprague-Dawley rats of about 100–120 g[14]; however, any of the standard procedures for preparation of relatively clean particles (from any cell type) will do. The ribosomes are suspended in medium C and incubated for 15 minutes at 37° with 0.1 mM puromycin—to remove nascent peptide and dissociate the ribosomes into subparticles.[14,24] The subunits prepared in this way are free of nascent peptide but may contain residual tRNA.[29] After incu-

[23] G. R. Lawford, *Biochem. Biophys. Res. Commun.* **37**, 143 (1969).
[24] W. S. Stirewalt, J. J. Castles, and I. G. Wool, *Biochemistry* **10**, 1594 (1971).
[25] D. P. Leader, I. G. Wool, and J. J. Castles, *Proc. Nat. Acad. Sci. U.S.* **67**, 523 (1970).
[26] D. P. Leader and I. G. Wool, *Biochim. Biophys. Acta* **262**, 360 (1972).
[27] D. A. Shafritz and W. F. Anderson, *Nature (London)* **227**, 918 (1970).
[28] G. von Ehrenstein and F. Lipmann, *Proc. Nat. Acad. Sci. U.S.* **47**, 941 (1961).
[29] R. E. H. Wettenhall and I. G. Wool, unpublished results.

bation, the ribosomal suspension is layered on a 10 to 30% linear sucrose gradient and centrifuged for 4 hours at 28° in a Spinco SW 27 rotor at 27,000 rpm. The gradients are displaced with 50% sucrose using an Instrument Specialities Co., Inc. (ISCO) Model D density gradient fractionator, and the effluent is analyzed at 254 nm with an ISCO model UA-2 UV analyzer. Subunit fractions are collected.[24] The ribosomal material in the "40 S fraction" contains only 40 S subunits. However, the material in the "60 S fraction" is contaminated with variable amounts of 40 S subunits unless very narrow cuts are taken, in which case the contamination with 40 S subunits is not likely to be greater than 5%. If pure subunits are required, the 60 S fraction must be resolved further by incubating again for 15 minutes at 37° in medium C containing 0.1 mM puromycin and recentrifuging as described above.[30] In any case, the subunit fractions are dialyzed against medium A and concentrated by ultrafiltration or ethanol precipitation.[31] The concentrated subunits (20–50 OD_{260} units/ml) are stored (on ice) in medium A; they remain active in the reassociation reaction for at least 1 week.

The nature and the purity of ribosomal fractions isolated from preparative gradients can be determined on analytical gradients.[14] The fractions are suspended, generally in medium A but other buffers can be used, and incubated for 5 to 15 minutes at 30° (to reduce aggregation of the particles). The suspension is layered on a 10 to 30% sucrose gradient in medium A and centrifuged at 49,000 rpm for 50 minutes in a Spinco SW 50.1 rotor at 28°. Centrifugation at 28° minimizes ribosomal aggregation. The purity of the fractions can be assessed also from analysis of their constituent RNA.[14]

Assay of Reassociation of Ribosomal Subunits

The Effect of Magnesium and Potassium Ions. A satisfactory assay of reassociation of ribosomal subunits requires concentrations of cations that minimize spontaneous formation of 40–60 S couples. The extent to which 40 S and 60 S subunits will combine is directly proportional to the magnesium concentration (Fig. 1) and inversely proportional to the potassium concentration (Fig. 2). Thus, if the concentration of magnesium is 5 mM there is no appreciable formation of 80 S monomers unless the potassium concentration is 100 mM or less; if the concentration of potassium is 80 mM, subunits do not reassociate unless the magnesium concentration exceeds 2.5 mM.

There are two phenomena which are worth noting because they

[30] C. Sherton and I. G. Wool, *J. Biol. Chem.* **247**, 4460 (1972).
[31] M. S. Kaulenas, *Anal. Biochem.* **41**, 126 (1971).

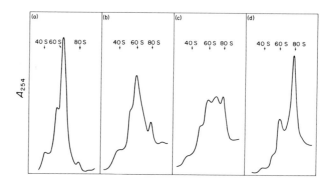

FIG. 1. The effect of the concentration of magnesium on the reassociation of ribosomal subunits. Ribosomal subunits (40 S, 7.2 μg of RNA; 60 S, 18.0 μg of RNA) were incubated for 5 minutes at 30° in 0.4 ml of buffer containing: KCl (80 mM); Tris·HCl (10 mM), pH 7.6; β-mercaptoethanol (10 mM); and varying concentrations of $MgCl_2$. After incubation, 0.2 ml of the reaction mixture was analyzed by centrifugation on 10–30% linear sucrose gradients made in the same buffer as the sample. N. B. There is a tendency for the 40 S subunits to dimerize and sediment at 55 S. (a) 1 mM Mg^{2+}; (b) 2.5 mM Mg^{2+}; (c) 3.5 mM Mg^{2+}; (d) 5 mM Mg^{2+}.

complicate the interpretation of sucrose density gradient profiles. First, a broad peak appears in the 65–75 S region of the gradients when the concentrations of ions are intermediate between those which do and those which do not favor association (cf. Figs. 1c and 2b). The peak probably does not contain a unique ribosomal species,[32] but rather a zone of 40 S and 60 S subunits which were formed by pressure-induced breakdown of weakly associated 40–60 S couples.[33–35] Second, the concentration of ions affects the aggregation of ribosomal subparticles. As the magnesium concentration is raised, the 60 S subunits dimerize and may even aggregate sufficiently to be sedimented to the bottom of the gradient.[14,29] If the concentration of potassium is less than 150 mM the small subunit may dimerize and sediment at 55 S, whereas, at concentrations greater than 150 mM it sediments at 40 S.[14,29] High concentrations (150 mM) of potassium do not prevent dimerization of the 60 S subunits.[29]

Conditions for the Formation of an Initiation Complex. A second requirement for an assay for reassociation of ribosomal subunits is that

[32] M. H. Schreier and H. Noll, *Proc. Nat. Acad. Sci. U.S.* **68**, 805 (1971).

[33] A. A. Infante and M. Krauss, *Biochim. Biophys. Acta* **246**, 81 (1971).

[34] A. S. Spirin, *FEBS (Fed. Eur. Biochem. Soc.) Lett.* **14**, 349 (1971).

[35] O. P. van Diggelen, H. Oostrom, and L. Bosch, *FEBS (Fed. Eur. Biochem. Soc.) Lett.* **19**, 115 (1971).

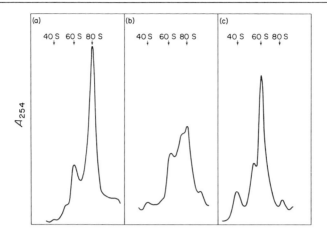

FIG. 2. The effect of the concentration of potassium on the reassociation of ribosomal subunits. Ribosomal subunits (40 S, 7.2 μg of RNA; 60 S, 18.0 μg of RNA) were incubated for 5 minutes at 30° in 0.4 ml of buffer containing: Tris·HCl (10 mM), pH 7.6; MgCl₂ (5 mM); β-mercaptoethanol (10 mM); and varying concentrations of KCl. After incubation, 0.2 ml of the reaction mixture was analyzed by centrifugation on 10 to 30% sucrose gradients made in the same buffer as the sample. N. B. The tendency for 40 S subunits to dimerize and sediment at 55 S decreases as the potassium concentration is raised. The 55 S peak seen in (c) may be due to altered 60 S particles, since it was observed when 60 S subunits were analyzed separately in the same circumstances. (a) 80 mM K⁺; (b) 100 mM K⁺; (c) 160 mM K⁺.

the concentration of ions be favorable for the formation of an initiation complex containing the 40 S ribosomal subunit, template [poly(U)], Phe-tRNA, and perhaps the initiation factor IF-M₁. The formation of that complex can be deduced from the poly(U)-dependent binding of [³H]Phe-tRNA to 40 S subunits catalyzed by IF-M₁.[25–26] The details of the assay are given in another paper in this volume [18]. There is appreciable formation of the initiation complex, but little spontaneous association of subunits, when the concentration of magnesium is 3.5 mM and potassium is 120 mM.[12] Thus, those concentrations of magnesium (3.5 mM) and potassium (120 mM) are suitable for assay of initiation factor catalyzed reassociation of ribosomal subunits.

Preparation of the Factor Promoting Reassociation of Ribosomal Subunits. The factor that catalyzes reassociation of ribosomal subunits is prepared in the cold (0 to 4°) from the 100,000 g cytosol fraction of rat liver.[25] Ribosomes, tRNA and aminoacyl-tRNA synthetases are precipitated by adjusting the pH to 5.2 with 1 M acetic acid. The precipitate is removed by centrifugation (30 minutes at 10,000 g). The pH of the supernatant is adjusted to 7.0 with 1 M KOH, and passed through a column of Sephadex G-25 equilibrated with 50 mM Tris·HCl, pH 7.6 and 1 mM MSH. The void volume eluate is referred to as "G-25 fraction."

The purified 40 S binding factor (IF-M$_1$) is obtained from the G-25 fraction by chromatography on DEAE-cellulose and hydroxyapatite[26]; the details of the procedure are given in this volume [18].

Assay of Reassociation Catalyzed by Initiation Factor. Ribosomal subunits (40 S, 10.7 μg of RNA; 60 S, 26.8 μg of RNA) are incubated for 15 minutes at 30° in 0.4 ml of medium B containing: 160 μg of Phe-tRNA; 40 μg of poly(U); 0.1 μmole of GTP; and 1.1 mg of G-25 fraction protein. The reassociation reaction is terminated by cooling the sample on ice and 0.3 ml of the mixture is layered on a 10 to 30% linear sucrose gradient in medium B (but without MSH), and centrifuged at 4° either for 90 minutes in a Spinco SW 65 rotor at 60,000 rpm, or for 105 minutes in a Spinco SW 50.1 rotor at 49,000 rpm. The delay between the end of the incubation and start of the centrifugation is 10 to 15 minutes. The distribution of ribosomal material in the gradient is determined with an ISCO density gradient fractionator and UV analyzer.

It is helpful, especially for studies of the kinetics of the reassociation reaction, to preincubate the 60 S subunits for 5 minutes at 30° in medium B containing all the components of the reassociation mixture before adding 40 S subunits. Preincubation decreases the tendency of the 60 S subunits to aggregate.

If one wishes to follow the binding of Phe-tRNA, or the synthesis of polyphenylalanine, [³H]Phe-tRNA is added to the reaction mixture. Fractions (0.25 ml) are collected from the gradients in glass vials and the volume adjusted to 1 ml with water; 10 ml of Triton-toluene scintillation fluid[36] are added and the radioactivity is measured in a liquid scintillation counter.

There is considerable formation of 80 S ribosomes when ribosomal subunits are incubated in the complete reaction mixture (Fig. 3b). Reassociation is dependent on the presence of a template [poly(U)] and Phe-tRNA (Figs. 3d and 3e). In the absence of G-25 fraction (Fig. 3c), or if that fraction is heated to 55°, there is less reassociation than in the complete system. Omission of GTP reduces reassociation by 40%, just as it reduces the G-25 fraction catalyzed binding of Phe-tRNA to the 40 S subunit.[25] If the reaction is carried out with [³H]Phe-tRNA, the radioactivity cosedimenting with 80 S ribosomes (Fig. 3b) is insoluble in hot trichloroacetic acid, indicating that synthesis of polyphenylalanine has taken place. (The G-25 fraction contains EF-1 and EF-2.[25]) However, polypeptide synthesis is not required for reassociation, since G-25 fraction-

---

[36] M. S. Patterson and R. C. Greene, *Anal. Chem.* **37**, 854 (1965).

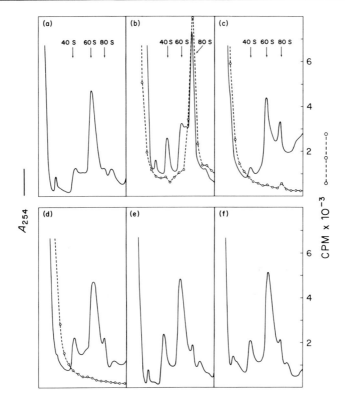

FIG. 3. Reassociation of ribosomal subunits catalyzed by G-25 fraction [R. E. H. Wettenhall, D. P. Leader, and I. G. Wool, *Biochem. Biophys. Res. Commun.* **43**, 994 (1971)]. Ribosomal subunits (40 S, 10.7 μg of RNA; 60 S, 26.8 μg of RNA) were incubated for 15 minutes at 30° in 0.4 ml of medium B containing: 160 μg of [³H]Phe-tRNA (120,000 cpm); 40 μg of poly(U); 0.2 μmole of GTP; 1.1 mg of G-25 fraction protein. In some experiments one or another component of the mixture was omitted. After incubation, samples (0.3 ml) were analyzed by centrifugation on 10–30% linear sucrose gradients; in some cases fractions were collected and the radioactivity was assayed. (a) Buffer; (b) complete; (c) −G-25 fraction; (d) −poly(U); (e) −Phe-tRNA; (f) G-25 only.

dependent reassociation is not affected by puromycin at concentrations sufficient to prevent the synthesis of protein.[12]

Purified IF-M₁, prepared from G-25 fraction,[26] will catalyze reassociation (Fig. 4), whereas purified preparations of EF-1 will not. Thus reassociation is catalyzed by initiation, rather than elongation, factors.

There is a need at times for a method of quantitating the extent of reassociation. Although no method is entirely satisfactory, there are two procedures that can be used. The extent of the reassociation of ribosomal

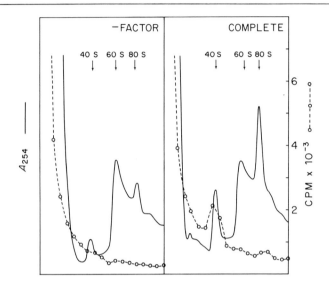

FIG. 4. Reassociation of ribosomal subunits catalyzed by purified IF-M₁ [R. E. H. Wettenhall, D. P. Leader, and I. G. Wool, *Biochem. Biophys. Res. Commun.* **43**, 994 (1971)]. The conditions of incubation and analysis were the same as in Fig. 3 except that 0.18 mg of purified IF-M₁ was used instead of the G-25 fraction.

subunits can be estimated either by cutting out and weighing the 80 S peak of the optical density tracing or by measuring the height of the peak. For the latter method it can be assumed that the area is proportional to the height of the peak; the assumption is reasonable and the method reliable only so long as the peaks are symmetrical and individual ribosomal particles are well separated (as is generally the case for the 40 S and 80 S peaks). Thus, appreciably the same results are obtained if the area of the 80 S peak is used (by cutting out the peak and weighing it) or if the height of the peak is measured. With either method the estimate of reassociated 80 S ribosomes should be corrected by subtracting the amount of 80 S material formed in control experiments (i.e., subunits incubated in buffer alone). It should be understood that neither method is exact, but they are sufficiently accurate and reproducible to allow comparisons.

Some Characteristics of the Reassociation Reaction

The assay can be used to determine the time course of the reassociation of ribosomal subunits. The reaction is stopped by chilling on ice after varying periods, and the extent of reassociation is determined. The time required for the analysis is, of course, long compared with that allowed for the reassociation reaction. For that reason we use the term kinetics to describe the time course of reassociation with some reservation. Nonetheless, reassociation does increase with time of incubation at 30° but not at 0°.[13] Reassociation at 30°, catalyzed by the G-25 fraction, is biphasic, suggesting that the process is composed of two separate reactions (Fig. 5). The first phase of the reaction, lasting 5 minutes, has a rapid initial component followed by a decline in rate; the second is characterized by prolonged linear increase in reassociation (5 to 15 minutes), with a final plateau (at 20 minutes). A small amount of reassociation occurs at 0° (but far less than at 30°). However, at 0° there is no increase in reassociation between 7 minutes (which is the shortest period for which we can make a measurement) and 20 minutes.

The G-25 fraction-catalyzed reassociation of ribosomal subunits requires tRNA, but the tRNA need not be aminoacylated.[12,13] However, the kinetics of reassociation with the two are different. With deacylated-

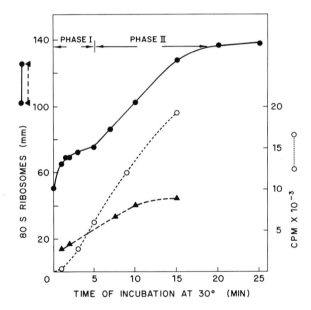

FIG. 5. Time course of the reassociation of ribosomal subunits [R. E. H. Wettenhall and I. G. Wool, *J. Biol. Chem.* **247**, 7201 (1972)]. In medium containing [³H]Phe-tRNA and all the components of the reassociation reaction, 60 S subunits were preincubated for 5 minutes at 30° before 40 S subunits were added; incubation was then continued for varying periods of time. Reassociation was assayed in the presence of G-25 fraction (●—●), and in its absence (▲- - -▲). The radioactivity insoluble in hot trichloroacetic acid in some of the assays with G-25 fraction is also given (○- - -○).

tRNA reassociation is linear for at least 15 minutes, whereas with aminoacyl-tRNA (preparations of aminoacyl-tRNA invariably contain deacylated-tRNA) the reaction is biphasic.

From analysis of the kinetics and from experiments with cycloheximide[13]—an antibiotic which specifically inhibits the binding of tRNA[Phe] to 80 S ribosomes[37]—it is evident that there are two pathways for initiation-factor-catalyzed formation of 40–60 S couples. The first pathway has faster kinetics, requires aminoacyl-tRNA, is insensitive to cycloheximide, and leads to the formation of 80 S ribosomes active in the synthesis of protein. The second pathway has slower kinetics, uses deacylated-tRNA, is sensitive to inhibition by cycloheximide, and the 80 S ribosomes formed do not synthesize protein. Factor catalyzed reassociation appears to be dependent on the prior formation of an initiation complex on the 40 S subunit. Thus, reassociation is inhibited by 10^{-5} M aurintricarboxylic acid (ATA), which interferes with the binding of poly(U) to ribosomes.[38] More to the point, if 40 S subunits are preincubated in the reassociation mixture (so as to form an initiation complex) and then ATA and 60 S subunits are added, 80 S ribosomes are formed. That is not the case when ATA is present during preincubation.

Remarks

We assume that *in vivo* initiation proceeds in the following way: In the presence of mRNA, met-tRNA$_F$ binds to the 40 S subunit in a reaction catalyzed by initiation factors. The addition of a 60 S subunit to form a stable active 80 S initiation complex then occurs rapidly. In our *in vitro* assay a similar sequence of events is promoted by the initiation factor IF-M$_1$ (or by preparations containing the factor) except that the template is synthetic [poly(U)] and the aminoacyl-tRNA is Phe-tRNA. Apparently initiation factors will select other tRNA's when the template does not contain an appropriately framed AUG codon. For example, two of the three initiation factors required for hemoglobin synthesis (IF-M$_1$ and IF-M$_2$)[7,27] are also needed for the initiation of polyphenylalanine synthesis at low concentrations of magnesium.[27]

In the cell reassociation with deacylated-tRNA is not likely to occur, because the amount of deacylated-tRNA is probably not very great, and, perhaps, because the initiation factors select for Met-tRNA$_F$. Unfortunately, we cannot exactly duplicate cellular conditions in the *in vitro* assay. In particular, it is difficult to prepare aminoacyl-tRNA free of contamina-

[37] T. G. Obrig, W. J. Culp, W. L. McKeehan, and B. Hardesty, *J. Biol. Chem.* **246**, 174 (1971).

[38] M. L. Stewart, A. P. Grollman, and M. Huang, *Proc. Nat. Acad. Sci. U.S.* **68**, 97 (1971).

tion with deacylated-tRNA[39]; moreover, aminoacyl-tRNA is deacylated at a finite rate during the assay. Thus, the deacylated-tRNA reaction is always likely to complicate the *in vitro* assay. For these reasons, it is important to understand the nature and the limitations of the reassociation assay.

[39] R. D. Mosteller, W. J. Culp, and B. Hardesty, *J. Biol. Chem.* **243**, 6343 (1968).

[20] Purification of Eukaryotic Initiation Factor 1 (EIF1) from *Artemia salina* Embryos

By MICHAEL ZASLOFF and SEVERO OCHOA

Eukaryotic initiation factor 1 (EIF1) is a protein factor isolated from high speed supernatants of encysted embryos of the brine shrimp *Artemia salina* which, like the bacterial polypeptide chain initiation factor IF2, promotes the AUG-dependent binding of fMet-tRNA$_f$ or the poly(U)-dependent binding of *N*-acetyl-Phe-tRNA to the small ribosomal subunit.[1-4] The presence of a similar factor in rat liver supernatants has been reported by others.[5,6] The rat liver and *A. salina* supernatant factors and ribosomal subunits are fully interchangeable suggesting that the two factors are probably identical.[2] An analogous factor, active with *A. salina* 40 S ribosomal subunits was demonstrated in the postribosomal supernatant of mouse fibroblasts.[2] Moreover, partially purified preparations of the mammalian initiation factor M1 from 0.5 M KCl ribosomal washes of rabbit reticulocytes[7] or calf brain[8] ribosomes can replace the *A. salina* factor for fMet-tRNA$_f$ binding to *A. salina* 40 S ribosomal subunits.[2,4] *A. salina* EIF1 can also replace M1 in poly(U) translation by reticulocyte ribosomes at low Mg^{2+} concentration.[9]

[1] M. Zasloff and S. Ochoa, *Proc. Nat. Acad. Sci. U.S.* **68**, 3059 (1971).
[2] M. Zasloff and S. Ochoa, *Proc. Nat. Acad. Sci. U.S.* **69**, 1796 (1972).
[3] R. P. McCroskey, M. Zasloff, and S. Ochoa, *Proc. Nat. Acad. Sci. U.S.* **69**, 2457 (1972).
[4] M. Zasloff and S. Ochoa, *J. Mol. Biol.* **73**, 65 (1973).
[5] D. P. Leader and I. B. Wool, *Biochim. Biophys. Acta* **262**, 360 (1972).
[6] E. Gasior and K. Moldave, *J. Mol. Biol.* **66**, 391 (1972).
[7] D. A. Shafritz, D. G. Laycock, R. G. Crystal, and W. F. Anderson, *Proc. Nat. Acad. Sci. U.S.* **68**, 2246 (1971).
[8] S. S. Kerwar, C. Spears, and H. Weissbach, *Biochem. Biophys. Res. Commun.* **41**, 78 (1970).
[9] D. J. Picciano, P. M. Prichard, W. C. Merrick, D. A. Shafritz, H. Graf, R. G. Crystal, and W. F. Anderson, *J. Biol. Chem.* **248**, 204 (1973).

Assay

The routine assay of EIF1 is based on its requirement for either the AUG-directed binding of fMet-tRNA$_f$ or the poly(U)-directed binding of N-acetyl-Phe-tRNA to *Artemia salina* 40 S ribosomal subunits.[1-4]

Assay 1

Principle. The EIF1-dependent binding of f[^{14}C]Met-tRNA$_f$ to *A. salina* 40 S ribosomal subunits, with AUG as messenger, is measured by the Millipore filter assay of Nirenberg and Leder.[10]

Reagents. The reaction mixtures, in a total volume of 0.06 ml, have the following composition:

> Tris·HCl buffer, 90 mM, pH 7.5 (25°)
> KCl, 150 mM
> Magnesium acetate, 4 mM
> Dithiothreitol (DTT), 1.5 mM
> AUG (Miles Laboratories), 0.05 A_{260} units
> *A. salina* 40 S ribosomal subunits, 0.3 A_{260} units; EIF1, 0–2 units
> f[^{14}C]Met-tRNA (*E. coli* W, 400 cpm/pmole),[1] 20 pmoles

Procedure. The above components are mixed at 0° in the order listed. The reaction is started by the addition of f[^{14}C]Met-tRNA. After incubation for 20 minutes at 0° the reaction is stopped by the addition of 3 ml of an ice-cold buffer containing 50 mM Tris·HCl, pH 7.75 (25°), 60 mM KCl, and 5 mM magnesium acetate, and the mixture is immediately filtered through a Millipore membrane (HAWP, 25 mm, 0.45 μ pore size). The filters are washed with 12 ml of the diluting buffer and dried under an infrared lamp. The retained radioactivity is measured in 10 ml of Omnifluor (New England Nuclear Corp.) in a Packard Tri-Carb liquid scintillation spectrometer.

Assay 2

Principle. The EIF1-dependent binding of N-acetyl[^{14}C]Phe-tRNA to *A. salina* 40 S ribosomal subunits with poly(U) as messenger is measured by the Millipore filter assay.

Reagents. The reaction mixtures, in a total of 0.06 ml, have the following composition:

> Tris·HCl buffer, pH 7.5 (25°), 90 mM
> KCl, 150 mM
> Magnesium acetate, 7 mM

[10] M. W. Nirenberg and P. Leder, *Science* 145, 1399 (1964).

DTT, 1.5 mM
Poly(U) (Miles Laboratories), 1.0 A_{260} unit
A. salina 40 S ribosomal subunits, 0.3 A_{260} unit; EIF1, 0–2 units
N-acetyl[^{14}C]Phe-tRNA (*E. coli* W, 680 cpm/pmole),[3] 12 pmoles

Procedure. The above components, except N-acetyl-Phe-tRNA, are mixed at 0° in the order listed. The reaction mixture is brought to 22°, and the reaction is started by the addition of N-acetyl-Phe-tRNA. After incubation for 20 minutes the reaction is stopped by the addition of 3 ml of an ice-cold buffer containing 50 mM Tris·HCl, pH 7.75 (25°), 60 mM KCl, and 10 mM magnesium acetate, then filtered through Millipore membranes, washed with 12 ml of the diluting buffer, and finally dried and counted as described for Assay 1.

One unit of EIF1 activity is defined as the amount of protein that promotes the net binding (blanks without factor being subtracted) of 1 pmole of either fMet-tRNA$_f$ or N-acetyl-Phe-tRNA under the conditions of the respective assay. Specific activity is expressed as units per milligram of protein. Protein is determined by the method of Lowry *et al.*[11] EIF1 assays at approximately the same specific activity in both reaction systems, and either assay can be used reliably in the purification of EIF1. Under the conditions described, the rate of aminoacyl-tRNA binding is linear up to about 2.5 units/assay.

Preparation of A. salina Ribosomal Subunits

Materials

> *A. salina* cysts ("Brine Shrimp Eggs," a product of Longlife Aquarium Products, Harrison, New Jersey, purchased from the Aquarium Stock Co., Inc., 31 Warren St., New York, N.Y.)
> Buffer A: Tris·HCl, 35 mM, pH 7.4 (25°); KCl, 70 mM; magnesium acetate, 9 mM; EDTA, 0.1 mM; 2-mercaptoethanol, 10 mM; sucrose, 250 mM
> Buffer B: Tris·HCl, 50 mM, pH 7.8 (25°); KCl, 700 mM; magnesium acetate, 11 mM; 2-mercaptoethanol, 20 mM
> Buffer C: Tris·HCl, 50 mM, pH 7.8 (25°); KCl, 200 mM; magnesium acetate, 10 mM; EDTA, 0.1 mM; 2-mercaptoethanol, 20 mM; sucrose, 250 mM

Procedure. All operations are conducted at 0–4°. Thirty grams (dry weight) of *A. salina* cysts are suspended in 100 ml of ice-cold 1%

[11] O. H. Lowry, N. J. Rosebrough, A. L. Farr, and R. J. Randall, *J. Biol. Chem.* **193**, 265 (1951).

NaClO, stirred for 5 minutes, and then diluted into 1 liter of ice-cold distilled water. On standing, the cysts rapidly settle, leaving in the supernatant damaged organisms and other debris. The supernatant is decanted and the sediment is washed 10 times, each time in about 400 ml of water to ensure complete removal of the hypochlorite. The pretreatment of the cysts with hypochlorite sterilizes the preparation and, in addition, facilitates the grinding step by removing the granular part of the shell[12] without damaging the embryo. The cysts are washed finally in 200 ml of buffer A, sedimented, resuspended in about 50 ml of buffer A, and then disrupted by grinding in a mortar. The thick paste is filtered through a single layer of cheese cloth which has been prewetted with buffer A, and the cloth filter is wrung to ensure complete drainage of the crude extract. One hundred milliliters of crude extract is generally obtained. The suspension is centrifuged for 15 minutes at 17,000 rpm in the Sorvall SS34 rotor to remove debris, nuclei, and mitochondria. The supernatant is carefully pipetted from above the pellet, filtered through a pad of glass wool, and then centrifuged for 30 minutes at 30,000 rpm in the Spinco No. 30 rotor to free the preparation from membranous debris and glycogen, which would otherwise sediment with the ribosomes. The upper three-fourths of the supernatant is collected and centrifuged for 130 minutes at 50,000 rpm in the Spinco Ti 60 rotor. The upper two-thirds of the high speed supernatant is collected and stored at $-20°$. This supernatant can be used as the starting material for the purification, on a limited scale, of EIF1 and the chain elongation factors EF1 and EF2.[1] The ribosomal pellet, which consists exclusively of the 80 S species,[1] is covered by a thin, clear, orange-tinted layer of membranous material, which is readily separated from the preparation by storing the pellets in ice for about 2 hours, after which time this material loosens from the surface of the dense, clear, and colorless ribosomal pellet. The pellets are rinsed gently with buffer B and then suspended in this buffer to a concentration of 400 A_{260} units/ml. Generally, about 1600 A_{260} units of 80 S ribosomes are obtained. One milliliter of this suspension is immediately layered on each of three 50-ml 15 to 30% (w/v) sucrose gradients in buffer B and centrifuged for 14 hours at 24,000 rpm in the Spinco SW 25.2 rotor. Under these conditions, the 80 S ribosomes dissociate completely into 40 S and 60 S subunits.[1]

Fractions comprising the faster 40% of the 60 S peak and the slower sedimenting 70% of the 40 S peak are pooled and immediately diluted with one volume of a buffer containing 30 mM magnesium acetate and 20 mM 2-mercaptoethanol. The subunits are concentrated to a clear

[12] J. E. Morris and B. A. Afzelius, *J. Ultrastruc. Res.* **20**, 244 (1967).

pellet by centrifugation for 10 hours at 40,000 rpm in the Spinco Ti 60 rotor. It is important to note that, if the centrifugation is carried out at higher speeds, the resulting pellets yield, rather than a faintly opalescent solution, a milky, turbid suspension of virtually inactive ribosomes. The supernatants are discarded and the clear, colorless pellets are rinsed gently with buffer C. The 40 S and 60 S subunits are suspended in buffer C to a concentration of 200 and 400 A_{260} units/ml, respectively, and stored at $-20°$ after adding one volume of glycerol. The subunits retain full activity in either the EIF1-dependent assays or in poly(U) translation for several months. As assayed by poly(U) translation, the 40 S subunits show less than 2–5%, and the 60 S subunits less than 5%, contamination with the other species. When an equimolar mixture of subunits is suspended in buffer A they reassociate extensively to 80 S ribosomes.

Purification Procedure

Materials

A. salina cysts

Phosphocellulose (Whatman P-11), 7.4 meq/g

Sephadex G-200 (Pharmacia), 40–120 μ

Carboxymethylcellulose (Whatman CM-52), 1.0 meq/g

Hydroxyapatite (BioRad HTP)

Buffer A: 60 mM KCl; 30 mM Tris·HCl, pH 7.75 (25°); 9 mM magnesium acetate; 0.1 mM EDTA; 10 mM 2-mercaptoethanol; 10% glycerol (w/v)

Buffer B: 100 mM KCl; 50 mM Tris·HCl, pH 7.75 (25°); 0.1 mM EDTA; 10 mM 2-mercaptoethanol; 10% glycerol (w/v)

Buffer C: 150 mM KCl; 50 mM Tris·HCl, pH 7.75 (25°); 0.1 mM EDTA; 10 mM 2-mercaptoethanol; 10% glycerol (w/v)

Buffer D: 280 mM KCl; 50 mM Tris·HCl, pH 7.75 (25°); 0.1 mM EDTA; 10 mM 2-mercaptoethanol; 10% glycerol (w/v)

Buffer E: 200 mM KCl; 50 mM Tris·HCl, pH 7.2 (4°); 0.1 mM EDTA; 10 mM 2-mercaptoethanol; 10% glycerol (w/v)

Buffer F: 10 mM KCl; 30 mM Tris·HCl, pH 7.2 (4°); 0.1 mM EDTA; 10 mM 2-mercaptoethanol; 10% glycerol (w/v)

Buffer G: 100 mM KCl; 30 mM Tris·HCl, pH 7.5 (25°); 0.1 mM EDTA; 10 mM 2-mercaptoethanol; 10% glycerol (w/v)

Buffer H: 100 mM potassium phosphate, pH 7.15; 0.1 mM EDTA; 1 mM DTT; 10% glycerol (w/v)

Buffer I: 170 mM potassium phosphate, pH 7.15; 0.1 mM EDTA; 1 mM DTT; 10% glycerol (w/v)

Buffers are prepared by dilution of stock salt solutions which have been filtered through Millipore membranes (0.45 μm pore size).

Step 1. Preparation of Homogenate. All operations are conducted at 0°–4°. Three kilograms of brine shrimp eggs were separately processed in 1-kg batches through the ammonium sulfate fractionation step. Commercial preparations of brine shrimp cysts generally contain large quantities of sea sand and other dense debris. Before the cysts can be disintegrated in a French press, the sand must be removed. This is readily accomplished by suspending 1 kg of cysts in 6 liters of saturated NaCl. The cysts, of lower density than the salt solution, rise slowly to the surface and are decanted, leaving behind the dense debris which has settled. This procedure is repeated until all sand has been removed. The suspension is then passed through a sieve (No. 80 U.S. Standard Series), and the cake is resuspended in about 4 liters of glass distilled water. The suspension is stirred gently and the cysts are allowed to settle, at which time the supernatant is poured off and a second portion of water is added. The washing is repeated until the preparation is free of salt. The cysts are then resuspended in about 1 liter of buffer A, allowed to settle, and finally resuspended in a total volume of 3200 ml of this buffer. The preparation is then disrupted by one passage through a French press at 6000–9000 psi. The chocolate-brown homogenate is centrifuged in the Sorvall GSA rotor at 10,000 rpm for 90 minutes. The supernatant, a milky-orange in color, is decanted and filtered through a layer of glass wool.

Step 2. Acid Fractionation. To the crude homogenate (2400 ml) 1.0 N acetic acid is added dropwise with stirring to bring the pH of the suspension to 5.3. The preparation is stirred for about 5 minutes at which time it is cleared of precipitated material by centrifugation in the Sorvall GSA rotor at 10,000 rpm for 1 hour. The supernatant is decanted from the heavy orange precipitate and filtered through a layer of glass wool. The pH of the preparation is then adjusted to 6.8 by dropwise addition of 1.0 N KOH. As an alternative to acid fractionation, the crude homogenate can be centrifuged at 100,000 g for about 3 hours. The factor activity partitions primarily in the postribosomal supernatant and is of a specific activity comparable to that of the preparation obtained by the acid step.

Step 3. Ammonium Sulfate Fractionation. Finely powdered ammonium sulfate (668 g) is added to the solution from the previous step (2050 ml) to 55% saturation at 0°. The suspension is stirred for at least 1 hour and then allowed to settle overnight. The precipitate is collected by centrifugation in the Sorvall GSA rotor, at 10,000 rpm for 1 hour, and discarded. The supernatant (about 2000 ml) is brought to 78% saturation by addition of 354 g of ammonium sulfate and stirred for about 2 hours;

the precipitate is harvested by low speed centrifugation. The precipitate is resuspended in about 100 ml of buffer B to a concentration of about 80 mg/ml. The brown turbid solution contains large amounts of high molecular weight glycogen, which is removed by centrifugation in the 60 Ti Spinco rotor at 50,000 rpm for 90 minutes. The clear, deep-brown supernatant is carefully pipetted off the colorless, viscous layer at the bottom of the tube and stored in ice. The three 1-kg batches are generally worked up over three 2-day intervals, and the undialyzed ammonium sulfate cuts are stored in ice during the interim periods. After the third batch of cysts has been processed through the ammonium sulfate step, the ammonium sulfate fractions are pooled and dialyzed against several changes in buffer B (up to a volume of 15 liters), until the conductivity of the solution equals that of the buffer.

Step 4. Phosphocellulose Chromatography. Phosphocellulose was washed with 25% NaCl, until the washes were colorless, and then with distilled water to remove the salt, packed into a column (8 × 24 cm), and equilibrated in buffer B. The solution from the preceding step (490 ml, 46 mg protein/ml) is applied to the column followed by washing with one column volume of buffer B, and then with buffer C, until the A_{280} of the effluent falls to about 0.03 unit above that of the buffer. This removes large amounts of inactive protein. The activity is eluted from the column with buffer D, fractions of 10 ml being collected. The factor emerges slightly behind the main protein peak. Fractions of highest specific activity are pooled (110 ml) and the protein is concentrated by precipitation with ammonium sulfate (66 g) to 90% saturation. The suspension is stirred overnight and the precipitate is dissolved in about 20 ml of buffer E. This solution, containing about 18 mg/ml of protein, is dialyzed against 10 volumes of buffer E. The preparation can be frozen at this point and stored at −20° for at least 1 month without appreciable loss of activity.

Step 5. Gel Filtration on Sephadex G-200. The solution from the preceding step is applied to a column (2.5 × 90 cm) of Sephadex G-200. This column is developed with buffer E at a flow rate of 0.2 ml/min. Fractions of 3 ml are collected. The peak of factor activity elutes at approximately 1.5 × the void volume of the column. Fractions of highest specific activity are pooled and concentrated by ultrafiltration in an Amicon cell, fitted with a PM 30 membrane, to a volume of about 15 ml. This solution contains 16 mg/ml of protein.

Although the increase in specific activity achieved on G-200 is rarely greater than 2-fold, this step is essential in that it effectively separates EIF1 from substantial amounts of EF1 and other high molecular weight proteins which will otherwise contaminate EIF1 in the subsequent steps.

Step 6. Carboxymethyl Cellulose Chromatography. The protein solution from the previous step is dialyzed with several changes against a total of 2 liters of buffer F, equilibration being carefully ascertained by conductivity measurements. The solution is then applied to a column (2.5 × 40 cm) of carboxymethyl cellulose previously equilibrated in buffer F. The column is washed free of nonadsorbing protein with buffer F until the A_{280} of the effluent is negligible. The column is then developed with a linear salt gradient generated from 800 ml of buffer F and 800 ml of buffer G. The flow rate is maintained at 0.6 ml/minute and 10-ml fractions are collected. The factor elutes as a symmetrical peak centered approximately half way through the gradient. In dilute solutions near neutral pH the factor appears to be more stable at KCl concentrations in the neighborhood of 100 mM. Thus, 1 ml of a buffer containing 1 M KCl, 50 mM Tris·HCl, pH 7.2 (4°), 0.2 mM EDTA, and 5 mM β-mercaptoethanol is included in each collecting tube. Fractions of highest specific activity are pooled and the protein concentrated by ultrafiltration to a volume of 12.5 ml at a concentration of 0.97 mg/ml.

Step 7. Hydroxyapatite Chromatography. An aliquot of the solution of the preceding step (5 mg), dialyzed against 2 liters of buffer H overnight, is applied to a column (0.6 × 17 cm) of hydroxyapatite previously equilibrated in buffer H. The column is washed with this buffer until the A_{280} of the effluent is negligible. It is then developed with buffer I at a flow rate of 0.15 ml per minute. Elution is continued until all the protein

PURIFICATION OF SUPERNATANT CHAIN INITIATION FACTOR (EIF1)
FROM *Artemia salina* EMBRYOS

Step	Volume (ml)	Protein (mg)	Units[b] (× 10⁻⁴)	Specific activity[c]	Yield (%)
1. Crude extract[a]	7500	220,000	—	—	—
2. pH 5.3 supernatant	5730	103,510	104	10	100
3. (NH₄)₂SO₄ fractionation	490	22,700	60	26	58
4. Phosphocellulose chromatography	21	372	26	703	25
5. Gel filtration on Sephadex G-200	15	242	18	730	17
6. CM-cellulose chromatography	13	12	4.3	3,610	4
7. Hydroxyapatite chromatography[d]	2.4	0.5	1.6	31,300	1.5

[a] From 3 kg (dry weight) of *A. salina* cysts.

[b] One unit = 1 pmole of fMet-tRNA (*Escherichia coli*) bound under standard assay conditions (20 minutes at 0°).

[c] Units per milligram of protein.

[d] Based on 12 mg of step 6 factor.

has emerged, 1.2-ml fractions being collected. Faster moving contaminants elute at the buffer front while EIF1 activity elutes as a broad peak of low protein concentration after approximately three column volumes (15 ml) have emerged. The fractions of highest specific activity are pooled and concentrated, from 40 ml to 5 ml by ultrafiltration and then further to 1 ml by dialysis against a concentrated solution of Carbowax 6000 (Union Carbide) in buffer B. The protein solution is made 50% (w/v) in glycerol and stored at $-20°$. A summary of the purification procedure is given in the table.

Properties

As a 50% glycerol solution in buffer B, stored at $-20°$, step 7 EIF1 shows no loss of activity over a period of 6 months, even at protein concentrations as low as 50 $\mu g/ml$. Stock solutions of step 7 factor are generally diluted immediately prior to assay to a working concentration of about 15 $\mu g/ml$ (500 units/ml) in a solution of the composition of buffer B and containing 0.25 mg/ml bovine serum albumin (Armour).

The procedure described yields a preparation of EIF1 purified over 3000-fold over the crude extract. On SDS gel electrophoresis step 7 factor gives rise to a single major band representing about 70–80% of the staining material and several minor components. The preparation is free of EF1 and EF2.

On SDS gel electrophoresis the major band displays the mobility of a polypeptide of molecular weight about 74,000, while under nondenaturing conditions it elutes from Sephadex G-200 at a volume consistent with that of a globular protein of molecular weight about 145,000. We thus infer that *A. salina* EIF1 is composed of two subunits of equivalent mass.[4]

EIF1 requires a free 40 S ribosomal subunit for expression of its activity.[1-4] Thus, it will catalyze the template directed binding of fMet-tRNA$_f$ and N-acetyl-Phe-tRNA solely to the free 40 S ribosomal subunit, but not to either the 60 S or 80 S species. The N-acylaminoacyl-tRNA is bound to the 40 S subunit on a site defined as the initiation site since the subsequent addition of the 60 S subunit and puromycin results in the direct conversion of the bound N-acylaminoacyl-tRNA to the dipeptide analog, N-acylaminoacylpuromycin.[3] Both the binding reaction and the synthesis of the puromycin derivatives occur in the absence of GTP or other ribonucleoside triphosphates.[3] EIF1 will also promote the template-directed binding of Met-tRNA$_f$ and Phe-tRNA to the 40 S ribosomal subunit; however, the bound aminoacyl-tRNA's are found to be less reactive with puromycin than their corresponding N-blocked derivatives.[3] Equivalent results have been obtained with tRNA species from either *E. coli* or *A. salina*.[3]

Under the conditions of the standard binding assays, EIF1 has been shown to function catalytically.[4] The reaction is inhibited by low concentrations of edeine and aurintricarboxylic acid, inhibitors of polypeptide chain initiation in both eukaryotic and prokaryotic systems.[1,2] EIF1 is sensitive to SH binding reagents such as N-ethylmaleimide and p-chloromercuribenzoate.[1,2] The prokaryotic and eukaryotic initiation factors IF2 and EIF1 are not interchangeable[2] even though they catalyze identical reactions in their respective systems.

[21] Assay and Isolation of a 40 S Ribosomal Subunit · Met-tRNA$_f^{Met}$ Binding Factor from Rabbit Reticulocytes

By J. M. CIMADEVILLA and B. HARDESTY

A protein factor from rabbit reticulocytes promotes codon-directed binding of various forms of tRNA^{Phe} and tRNA$_f^{Met}$ to 40 S ribosomal subunits.[1] The factor with poly(U) or ApUpG is active with either the N-acetylaminoacyl-tRNA, aminoacyl-tRNA or the deacylated forms of these species of tRNA. It appears to have little or no activity for Met-tRNA$_M^{Met}$ or other species of tRNA tested. Factor-dependent binding of tRNA to the 40 S ribosomal subunit is severely inhibited by the 60 S subunit. This factor-dependent binding also is inhibited by aurintricarboxylic acid and edeine, both of which inhibit peptide initiation on eukaryotic ribosomes. GTP or other ribonucleoside triphosphates have no effect on binding under any conditions tested. Met-tRNA$_f^{Met}$ bound to 40 S subunits with the binding factor is sensitive to hydrolysis by Met-tRNA hydrolase considered elsewhere in this volume.

The tRNA binding factor prepared as described below is virtually free of EF-1, EF-2, Phe-tRNA synthetase, and Met-tRNA synthetase. It appears to be a basic protein of molecular weight about 85,000 that is sensitive to inactivation by sulfhydryl reactive reagents and heat.

The factor considered here appears to be similar or identical to tRNA binding factors from brine shrimp,[2] rat liver,[3,4] ascites cells,[5] wheat,[6] and

[1] J. M. Cimadevilla and B. Hardesty, unpublished observations.
[2] M. Zasloff and S. Ochoa, Proc. Nat. Acad. Sci. U.S. 68, 3059 (1971).
[3] E. Gasior and K. Moldave, J. Mol. Biol. 66, 391 (1972).
[4] D. Leader and I. Wool, Biochim. Biophys. Acta 262, 360 (1972).
[5] D. Leader, H. Klein-Brehaar, I. Wool, and A. Fox, Biochem. Biophys. Res. Commun. 46, 215 (1972).
[6] A. Marcus, J. Biol. Chem. 245, 962 (1970).

yeast.[7] It has some of the characteristics of initiation factor M$_1$ from rabbit reticulocytes described by Anderson and co-workers.[8]

Materials and Solutions

Preparation of Reticulocytes, Unfractionated tRNA, and Unwashed Ribosomes. A detailed description of the preparation of these materials has been given previously.[9] Unfractionated tRNA is isolated from rabbit liver with phenol. Ribosomes used for preparation of ribosomal subunits were from reticulocytes isolated from rabbits treated with phenylhydrazine. The 40–70 ammonium sulfate enzyme fraction used for charging tRNA is prepared as described below.

40 S Ribosomal Subunits. Ribosomal subunits are prepared from ribosomes isolated from reticulocytes that have been previously incubated with NaF. The procedure used is a modification of the method of Falvey and Staehelin.[10] It is presented in another chapter in this volume.[11]

Phe-tRNA. Phe-tRNA is formed from unfractionated rabbit liver tRNA and reisolated with phenol using the procedures previously described.[9] For the formation of Phe-tRNA, each milliliter of the reaction mixture contains 100 mM Tris·HCl, pH 7.5; 20 mM β-mercaptoethanol; 10 mM MgCl$_2$; 2.0 mM ATP; 16 μM phenylalanine ([^{14}C]phenylalanine at 100 Ci/mole); 1.0 mg of unfractionated rabbit liver tRNA; and 1.0 mg of protein from the 40–70 S ammonium sulfate enzyme fraction.

B.F. Buffer. The binding factor buffer solution (B.F. buffer) contains 10 mM Tris·HCl, pH 7.5; 5.0 mM β-mercaptoethanol; 1.0 mM dithioerythritol, and 0.1 mM EDTA.

Assay of Met-tRNA$_f^{Met}$ Binding Factor Activity

Principle and Characteristics. The basis for the assay system is the increase in binding of tRNA to 40 S ribosomal subunits observed in the presence of the factor. Binding is measured by the nitrocellulose filter procedure of Nirenberg and Leder.[12] N-Acetyl-Phe-tRNA, Phe-tRNA, deacylated tRNA^{Phe} or the similar forms of tRNA$_f^{Met}$ have been used for assay of the factor. Binding of tRNA to the 40 S ribosomal subunit is dependent upon an appropriate form of mRNA, as indicated for Phe-tRNA with poly(U) in Table I. Phe-tRNA is recommended for routine proce-

[7] A. Toraño, A. Sandoval, C. SanJosé, and C. Heredia, *FEBS (Fed. Eur. Biochem. Soc.) Lett.* **22**, 11 (1972).

[8] D. Shafritz, D. Laycock, R. Crystal, and W. Anderson, *Proc. Nat. Acad. Sci. U.S.* **68**, 2246 (1971).

[9] B. Hardesty, W. McKeehan, and W. Culp, this series, Vol. 20, p. 316 (1971).

[10] A. Falvey and T. Staehelin, *J. Mol. Biol.* **53**, 1 (1970).

[11] J. Morrisey and B. Hardesty, this series, Vol. 29 [60].

[12] M. W. Nirenberg and P. Leder, *Science* **145**, 1399 (1964).

TABLE I

Characteristics of the Standard Assay System

| | tRNA Binding (pmoles) | | |
| | Factor addition | | |
Additions	−	+	Factor activity
Complete[a]	1.20	4.85	3.65
− poly(U)	0.08	0.04	0
− 40 S subunits	0.02	0.01	0
+ 60 S subunits	0.55	1.45	0.90
4 mM Mg^{2+}	0.20	0.55	0.35
+ GTP	1.20	4.80	3.60
N-Acetyl-Phe-tRNA[b]	1.30	4.56	3.26

[a] The complete system is the standard assay system with 0.36 μg of protein from the hydroxylapatite fraction prepared as described below.

[b] In place of Phe-tRNA.

dures because of its ease of preparation, availability and cost of poly(U), and stability in the binding assay system. Under some conditions, Met-tRNA hydrolase[11,13] will interfere with assay using Met-tRNA.

In all cases binding factor activity is calculated as the increase in tRNA binding observed in the absence and presence of the factor. Typical results are presented in Table I. Factor-dependent binding of tRNA is nearly linear with added factor in the standard assay system to at least 6 pmoles of total Phe-tRNA bound. The binding factor is adsorbed to glass. Dilute solutions of the binding factor should be handled in polyethylene or similar plastic apparatus. Also, adsorption of the factor to glass is decreased or prevented by higher concentrations of KCl. Dilution of the purified factor for assay may be made in Pyrex tubes with solutions containing 0.5 M KCl.

Assay Procedure. Each assay reaction mixture contains in a total volume of 0.25 ml: 20 mM Tris·HCl (pH 7.5), 100 mM KCl, 8 mM MgCl$_2$, 10 mM β-mercaptoethanol, 20 μg of poly(U), 60 pmoles of Phe-tRNA (about 100 μg of tRNA charged generally with [^{14}C]phenylalanine, 100 Ci/mole), 90 μg of 40 S ribosomal subunits. Binding factor activity is added, generally in an amount sufficient to give 4–6 pmoles total of Phe-tRNA bound. A pair of similar reaction mixtures, one containing binding factor, is prepared for each condition to be tested. Reaction mixtures are incubated for 4 minutes at 37°. About 7 ml of cold solution containing 20 mM Tris·HCl (pH 7.5), 100 mM KCl, and 8 mM MgCl$_2$ are added

[13] J. Morrisey and B. Hardesty, *Arch. Biochem. Biophys.* 152, 385 (1972).

and the mixture is quickly filtered through modified cellulose filter (0.45 μm pore size, type HAWG, Millipore Corp., Bedford, Massachusetts) as described by Nirenberg and Leder.[12] The filters are washed twice more with about 7 ml of the same cold solution and dried; bound ^{14}C is determined by a liquid scintillation procedure. Binding factor activity is expressed normally as picomoles of tRNA bound per milligram of protein.

Isolation Procedure

The Met-tRNA$_f^{Met}$ binding factor is isolated from the lysate of rabbit reticulocytes by five fractionation steps following centrifugation to remove ribosomes from the reticulocyte lysate. First the factor is precipitated with ammonium sulfate between the levels of 40% and 70% saturation. The ammonium sulfate fraction is then chromatographed on DEAE-cellulose. The fractions containing the binding factor are pooled and the factor is chromatographed first on cellulose phosphate, then on pulverized glass, and finally on hydroxylapatite. All steps are carried out at about 0° unless otherwise indicated.

High Speed Supernatant. The procedure for preparation and lysis of rabbit reticulocytes is described in detail elsewhere.[9] Ribosomes are collected from the lysate by centrifugation for 1.5 hours at 36,000 rpm in a Model 42 rotor (Spinco Division, Beckman Instruments, Inc., Palo Alto, California). The supernatant fluid is used as the high speed supernatant.

Ammonium Sulfate. The high speed supernatant fluid is adjusted to pH 6.5 by the careful addition of 1.0 M acetic acid. Solid ammonium sulfate is added slowly with stirring to give 40% saturation (22.6 g/100 ml). After 30 minutes of gentle stirring, the resulting precipitate is removed from suspension by centrifugation at 15,000 g for 20 minutes. The binding factor is precipitated by addition of ammonium sulfate to give 70% saturation (18.2 g/100 ml of supernatant solution from the preceding step). The resulting precipitate is collected by centrifugation (15,000 g for 20 minutes) and resuspended in a volume of buffer solution equal to 1% of the original volume of the high speed supernatant. The buffer solution contains 10 mM Tris·HCl (pH 7.5); 0.1 mM β-mercaptoethanol; and 0.2 mM EDTA. The slurry of precipitate in buffer solution is then dialyzed against three changes of 2 liters each of the same buffer solution for a total of 14–18 hours. The resulting fraction contains protein at a concentration of about 50 mg/ml and can be stored at −90° with little or no loss of binding factor activity.

DEAE-Cellulose. DEAE-cellulose, about 150 g of coarse mesh (Sigma Chemical Co., St. Louis, Missouri) is washed by suspension, gravitational settling, and decantation three times in a total of about 8 liters of a solution containing 0.1 M Tris·HCl, pH 7.5, and 1 mM EDTA and then three times in deionized water. Finally the DEAE-cellulose is suspended in 3 liters of B.F. buffer modified by the addition of KCl to give a final concentration of 40 mM. Excess solution is decanted, and the slurry is poured into a 5 × 60 cm glass column to give a bed height to about 55 cm. At least 4 liters of the same modified B.F. buffer is percolated through the column over about a 72-hour period.

Approximately 6.0 g of protein in the dialyzed ammonium sulfate fraction is applied to the column. Nonabsorbed protein is removed from the column with 2.0 liters of B.F. buffer modified to contain 40 mM KCl. About 55% of the total protein and 11% of the binding factor activity initially applied to the column is removed by this washing procedure.

Binding factor activity is eluted from the column with 3 liters of B.F. buffer containing 0.1 M KCl. Fractions with an adsorbance of 280 nm light greater than 0.20 are pooled to give a combined volume of about 750 ml and then concentrated by ultrafiltration (UM10 filter in a Model 402 pressure cell, Amicon Corp., Lexington, Massachusetts) to a final volume of about 50 ml. The binding factor activity and protein recovered after concentration are given in Table II. This concentrated fraction is applied immediately to the phosphocellulose column described below. The distribution of binding factor activity eluted from the DEAE-cellulose column can be checked with 0.10-ml aliquots of the fractions collected. No significant loss of binding factor activity occurs during concentration. The ratio of absorption at 280 nm to absorption at 260 nm is generally 1.0 for the 40–70 ammonium sulfate fraction applied to the DEAE-cellulose column and about 1.5 for the pooled active fractions after concentration. This appears to reflect removal of most but not all of the nucleic acid present in the 40–70 ammonium sulfate fraction.

The elution profile of binding factor activity from DEAE-cellulose

TABLE II

PURIFICATION OF THE INITIATOR tRNA BINDING FACTOR

Fraction	Total volume (ml)	Total protein (mg)	Total units[a] (×10⁻³)	Specific activity (units/ mg)	Yield (%)
High speed supernatant	2,000	69,000	278	4	100
40–70% Ammonium sulfate	120	6,000	120	20	43.2
DEAE-cellulose	80	520	83	160	29.9
Cellulose phosphate	5.4	6.25	31.2	5,000	11.2
Pulverized glass	3.1	2.06	22.4	10,800	8.1
Hydroxylapatite	2.0	1.54	17.0	11,000	6.1

[a] Picomoles of Phe-tRNA bound as determined in the standard assay system.

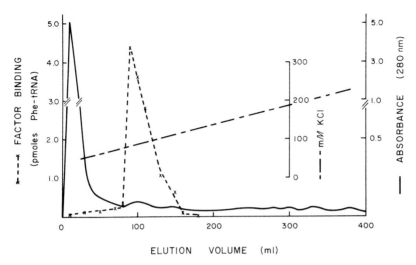

FIG. 1. Chromatography of the Met-tRNA$_f^{Met}$ binding factor on DEAE-cellulose.

under conditions similar to those used for the preparative procedure described above are shown in Fig. 1. For this figure a 1.5 × 30 cm bed of DEAE-cellulose prepared as described above was loaded with 100 mg of 40–70 ammonium sulfate protein and then washed with 25 ml of B.F. buffer containing 50 mM KCl. The column was then eluted with a linear 50 mM to 350 mM KCl gradient in B.F. buffer.

Cellulose Phosphate. Cellulose phosphate, about 25 g of medium mesh (exchange capacity of 0.87 mg/gm, Sigma Chemical Co.) is washed by suspending and decanting three times in a total of 4 liters of a solution containing 0.1 mM Tris·HCl, pH 7.5, and 1 mM EDTA and then three times in deionized water. The cellulose phosphate is then suspended in 1 liter of B.F. buffer modified to contain 0.1 M KCl. Excess fluid is decanted, and cellulose phosphate is poured into a 2.5 × 30 cm glass column to give a bed height of about 20 cm. Then, about 2.0 liters of the same B.F. buffer containing 0.1 M KCl is passed through the column.

About 500 mg of protein in the DEAE-cellulose fractionation are applied to the column. Protein that is not adsorbed to the phosphocellulose is removed by percolating 0.5 liter of B.F. buffer containing 0.1 M KCl through the column. About 85% of the protein applied to the column is removed by this procedure; however, binding factor activity is below detectable levels in these fractions. The column is then eluted with a linear 400-ml gradient of 0.10 M to 0.80 M KCl in B.F. buffer. Binding factor activity is eluted as a single peak at a KCl concentration of about 350 mM as shown in Fig. 2. Activity is determined in 0.01-ml aliquots of 6-ml

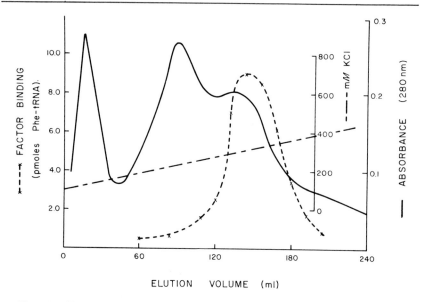

FIG. 2. Chromatography of the Met-tRNA$_f^{Met}$ binding factor on cellulose phosphate.

fractions. The fractions containing binding factor activity are pooled (combined volume is generally about 60 ml) and concentrated by ultrafiltration (UM10 filter in a Diaflo pressure cell) to a volume of about 5 ml. This concentrated solution is dialyzed overnight against 2.0 liters of B.F. buffer containing 0.1 M KCl.

Pulverized Glass. Pyrex test tubes are broken with a hammer after wrapping in multiple layers of heavy wrapping paper. Then, the glass is pulverized with a mortar and pestle. Larger pieces of pulverized glass are removed by sifting the pulverized glass through a double layer of cheesecloth. The sifted glass is allowed to stand at room temperature for about 12 hours in concentrated nitric acid and then washed ten times in deionized water. Very fine particles of glass are removed by decanting the water from the glass before the fine particles can settle. A 2 × 40 cm column is packed to a bed height of 15 cm with this pulverized glass. B.F. buffer containing 0.10 M KCl, about 1.5 liters, is percolated through the column for 72 hours.

About 6 mg of protein in the concentrated and dialyzed fraction from the preceding cellulose phosphate fractionation step is applied to the column. Nonadsorbed protein is washed from the column with 200 ml of B.F. buffer containing 0.10 M KCl. About 4 mg of protein are removed from the column by this procedure. The binding factor is eluted from the column with 200 ml of B.F. buffer containing 0.5 M KCl. Fractions having an absorption at 280 nm of 0.04 or greater are pooled (total combined volume of about 15 ml), and immediately concentrated to 3 ml (protein concentration of about 0.8 mg/ml) by ultrafiltration. Preparations to be used at this stage of purity are stored at $-90°$ without dialysis. These preparations of binding factor are estimated to be about 70% to 80% pure but contain significant amounts of EF-II estimated[14] to account for 5% or more of the total protein. This EF-II can be separated from binding factor activity by chromatography on hydroxylapatite as described below. Pooled fractions from the glass column to be chromatographed on hydroxylapatite are concentrated and used immediately without dialysis.

Hydroxylapatite. Hydroxylapatite (Hypatite C, Clarkson Chemical Co., Inc., Williamsport, Pennsylvania), about 15 g wet weight, is suspended in 500 ml of solution containing 0.15 M potassium phosphate, pH 7.5, and 5 mM β-mercaptoethanol. Excess solution is decanted, and the hydroxylapatite is poured gently into a 0.9 × 15 cm glass column to give a bed height of 5 cm. About 500 ml of the 0.15 M phosphate solution is percolated through the column over a 72-hour period.

About 2 ml of protein from the preceding fractionation step on ground glass is applied to the column. Nonadsorbed protein is washed from the column with 75 ml of the same 0.15 M phosphate solution used in preparation of the column. EF-II is removed from the column by this procedure. Binding factor activity is eluted from the column with 100 ml of solution containing 0.35 M potassium phosphate, pH 7.5, plus 5 mM β-mercaptoethanol. Fractions collected preferably in polyethylene tubes with absorption at 280 nm above 0.02 are pooled to give a combined volume of about 10 ml, concentrated by ultrafiltration, and then dialyzed against B.F. buffer modified to contain 0.50 M KCl and 50% glycerol. The final volume and protein concentration of the dialyzed samples are given in Table II.

Comments

The Met-tRNA$_f^{Met}$ binding factor is sensitive to inactivation by sulfhydryl reactive reagents. Care should be taken throughout the procedure to protect the binding factor from heavy metal ions and to maintain it in solutions containing a sulfhydryl compound. No loss of binding factor activity has been detected that can be attributed to inactivation by thiol reagents, as has been reported for bacterial IF-2.[15] The factor is avidly bound

---

[14] B. Hardesty and W. McKeehan, this series, Vol. 20, p. 330 (1971).
[15] Y. B. Choe, R. Mazumder, and S. Ochoa, *Proc. Nat. Acad. Sci. U.S.* **62**, 1181 (1969).

Fig. 3. SDS polyacrylamide gel electrophoresis of binding factor fractions. Pooled fractions containing the Met-tRNA$_f^{Met}$ binding factor from cellulose phosphate (A: 54 μg of protein) and from hydroxylapatite (B: 45 μg of protein) were electrophoresed in the presence of sodium dodecyl sulfate as described by K. Weber and M. Osborn [*J. Biol. Chem.* **244**, 4406 (1969)].

to glass surfaces, a property utilized in one of the chromatographic steps outlined above. An apparently similar phenomena has been described for bacterial IF-2.[16] Serious losses may occur due to absorption on glass vessels. This appears to be particularly serious when dilute solutions of relatively pure material is involved. Polyethylene vessels appear to be a satisfactory substitute for glass.

An indication of the composition and purity of the fractions from cellulose phosphate and hydroxylapatite is given in Fig. 3, for which these fractions were electrophoresed in polyacrylamide SDS gel columns. The major band in the hydroxylapatite fraction, apparently the binding factor, contains more than 70% of the total protein in the fraction as judged by scans of the stained gels with 550 nm light.

The maximum specific activity obtained for the reticulocyte binding factor has been near 11,000 pmoles Phe-tRNA bound per milligram of protein as determined in the standard assay system. A maximum specific activity of 11,800 would be expected for pure binding factor of a molecular weight of 85,000 in one Phe-tRNA were bound for each molecule of

---

[16] D. Kolakofsky, K. Dewey, and R. Thatch, *Nature* (*London*) **223**, 694 (1969).

factor. Ochoa and his co-workers have obtained specific activities of up to 38,000 pmoles of Met-tRNA bound per milligram of the binding factor from *Artemia salina*.[17] The molecular weight of this factor is estimated to be 145,000 thus indicating that each molecule of factor is involved in binding several molecules of fMet-tRNA under the conditions used. The basis for this important difference in the two systems is not clear.

[17] M. Zasloff and S. Ochoa, *J. Mol. Biol.* **73**, 65 (1973).

Section II

Elongation Factors in Protein Synthesis

[22] Elongation Factor Tu and the Aminoacyl-tRNA·EFTu·GTP Complex

By DAVID LEE MILLER and HERBERT WEISSBACH

The bacterial elongation factors EFTu and EFTs promote the binding of AA-tRNA[1] to ribosomes via an AA-tRNA·EFTu·GTP intermediate.[2] When this ternary complex interacts with ribosomes in the presence of messenger-RNA, the AA-tRNA is transferred to the ribosome and GTP is hydrolyzed with the formation of EFTu·GDP and P_i. The EFTu·GDP complex dissociates very slowly. The function of EFTs is to catalyze the exchange of the tightly bound GDP with free GTP to form EFTu·GTP, which can interact with another molecule of AA-tRNA, thus allowing EFTu to function catalytically in the binding cycle.[3,4] These reactions are summarized in the following equations:

$$EFTu·GDP + EFTs \rightleftharpoons EFTu·EFTs + GDP \tag{1}$$

$$EFTu·EFTs + GTP \rightleftharpoons EFTu·GTP + EFTs \tag{2}$$

$$AA\text{-}tRNA + EFTu·GTP \rightleftharpoons AA\text{-}tRNA·EFTu·GTP \tag{3}$$

$$AA\text{-}tRNA·EFTu·GTP + ribosome·mRNA \rightarrow$$
$$AA\text{-}tRNA·ribosome·mRNA + EFTu·GDP + P_i \tag{4}$$

Factor EFTu interacts with a number of diverse substances, among which are GDP, GTP, ppGPP, AA-tRNA (but not tRNA), EFTs, and possibly components of the ribosome. In addition, EFTu and EFTs have recently been identified as two of the three host-donated components of bacteriophage Qβ RNA polymerase.[5] The EFTu·EFTs complex is called EFT.

Chapters in Volume 20 of this series contain descriptions of the identification and purification of the factors and the reaction of the ternary

---

[1] The following abbreviations are used: AA-tRNA, aminoacyl-tRNA; Phe-tRNA, phenylalanyl-tRNA; PEP, phosphoenolpyruvic acid; DTT, 1,4-dithiothreitol; ppGpp, guanosine 5'-pyrophosphate, 3'-pyrophosphate; GMP·PCP, guanylyl methylenediphosphonate; GMP·PNP, guanylyl imidodiphosphonate; BAL, 2,3-dimercaptopropanol.

[2] J. Lucas-Lenard and F. Lipmann, *Annu. Rev. Biochem.* 40, 409 (1971).

[3] H. Weissbach, D. L. Miller, and J. Hachmann, *Arch. Biochem. Biophys.* 137, 262 (1970).

[4] D. L. Miller and H. Weissbach, *Biochem. Biophys. Res. Commun.* 38, 1016 (1970).

[5] T. Blumenthal, T. A. Landers, and K. Weber, *Proc. Nat. Acad. Sci. U.S.* 69, 1313 (1972).

complex with the ribosomes.[6-8] The purposes of this article are to describe convenient assays for the factors, and to give details on the interaction of the purified factors with AA-tRNA and ribosomes. A procedure to obtain a crystalline preparation of EFTu is also presented.

Assays for EFTu, EFTs, and AA-tRNA·EFTu·GTP Formation

Assay for EFTu

The earlier assays for EFTu and EFTs were based upon their being required for phenylalanine polymerization in the presence of ribosomes, poly(U), and EFG.[6] EFTu can be more conveniently assayed by measuring the extent of binding of ^{3}H-GDP to EFTu, by adsorbing the EFTu·^{3}H-GDP complex on to a cellulose nitrate filter.[9] The major protein present in *Escherichia coli* extracts, which tightly binds GDP and is retained on cellulose nitrate filters, is EFTu. Since the dissociation constant of EFTu·GDP is about 3 nM, amounts of EFTu in the picomole range can be measured.

Materials

^{3}H-GDP, 1–5 Ci/mmole, 400–2000 cpm/pmole
Cellulose nitrate filters, 25 mm diameter (Millipore Corp. type HA or Schleicher and Schuell type C-5)
Binding buffer [250 mM Tris·HCl, pH 7.4, 50 mM MgCl$_2$, 250 mM NH$_4$Cl, and 25 mM dithiothreitol (DTT)], 5 times working concentration
Wash buffer: 10 mM Tris·HCl, pH 7.4, 10 mM MgCl$_2$, and 10 mM NH$_4$Cl
EFTs, 30 units/μl (the definition of a unit of EFTs is described in the next section)
Dioxane-based scintillation fluid[10]

Method. An aliquot containing 5–50 pmoles of EFTu is added to a 10 × 75 mm test tube in a total volume of 200 μl, containing 40 μl of binding buffer, 2.5 μM ^{3}H-GDP, and 10–30 units of EFTs. The reaction mixture is allowed to equilibrate for 5 minutes at 37°, then is diluted with 2 ml of wash buffer, filtered, and washed three times with 3-ml portions

---

[6] J. Gordon, J. Lucas-Lenard, and F. Lipmann, this series, Vol. 20, p. 281.
[7] A. Parmeggiani, C. Singer, and E. M. Gottschalk, this series, Vol. 20, p. 291.
[8] J. M. Ravel and R. L. Shorey, this series, Vol. 20, p. 306.
[9] R. Ertel, B. Redfield, N. Brot, and H. Weissbach, *Arch. Biochem. Biophys.* **128**, 331 (1968).
[10] G. A. Bray, *Anal. Biochem.* **1**, 279 (1960).

of the same buffer. The filters are dissolved in 10 ml of scintillation fluid, and the radioactivity is measured. A control containing no EFTu is also run.

Calculations. EFTu as normally isolated contains an equivalent of tightly bound GDP that will dilute the ^{3}H-GDP and lower the specific activity. The following equation is used to correct for the dilution if more than 50 pmoles of EFTu·GDP are added to the reaction mixture.

$$\text{Tu (pmoles)} = \frac{C_b}{(Q - C_b)/\text{GDP}}$$

where C_b = cpm ^{3}H-GDP bound to the filter; Q = initial specific activity of ^{3}H-GDP in cpm/pmole; and GDP = pmoles ^{3}H-GDP added, usually 500 pmoles.

Comments. The exchange reaction between ^{3}H-GDP and EFTu·GDP proceeds slowly in the absence of EFTs, especially at 0°; however, impure EFTu contains enough EFTs to allow equilibration in 5 minutes at 37°. If there is doubt about the completeness of the exchange, a 10-minute reaction can be run for comparison or more EFTs can be added (60 units). The complex adsorbed to the filter withstands extensive washing with buffer; however, EFTu is readily inactivated at pH levels below 6 and at temperatures above 50°. In the absence of Mg^{2+}, GDP dissociates leaving free EFTu, which is unstable.

Solutions of ^{3}H-GDP sometimes contain an unidentified radioactive impurity which adheres to the filters. This impurity can be removed by a small-scale chromatographic separation on DEAE-cellulose. The impurity is washed off with water, and the ^{3}H-GDP can be eluted with a dilute salt solution. In addition, the scintillation fluid itself sometimes gives a chemiluminescent reaction independent of the presence of radioactivity, which can be quenched by adding 50 μl of 5% aqueous trichloroacetic acid.

Assay for EFTs

The assay for EFTs is based upon its ability to catalyze the exchange between free GDP and EFTu·GDP, as Eq. (1) indicates. The rate of exchange is measured over a period of 5 minutes at 0°. At this temperature there is little exchange of EFTu·GDP with exogenous nucleotides in the absence of EFTs.

Method. An aliquot containing EFTs (less than 20 units) is added to a tube containing 35 pmoles of EFTu·GDP (free of EFTs) and 40 μl of binding buffer in a total volume of 200 μl. The reaction is started by adding 500 pmoles of ^{3}H-GDP, and the reaction is allowed to proceed for 5 minutes at 0°. The mixture is diluted with 2 ml of ice-cold wash buffer and is rapidly filtered and washed as previously described. Another mixture containing no EFTs is treated similarly and is used as a blank for the determination.

Definition of the Unit. One unit of EFTs is that amount which will catalyze the exchange of 1 pmole of EFTu·GDP with ^{3}H-GDP in 5 minutes under the assay conditions.

Comments. The rate of exchange is constant only during the first half of the reaction, after which it slows down. In general, the amount of EFTs added should exchange less than 20 pmoles of GDP with EFTu·GDP in 5 minutes. A kinetic analysis has not yet been performed on the reaction; nevertheless, the assay quantitatively determines EFTs activity under the conditions used.

The binding of ^{3}H-GDP to free EFTu is very rapid,[3] and one may be misled to think that EFTs is contaminating the preparation. This rapid reaction is eliminated by adding a small amount of unlabeled GDP to the assay mixture before adding ^{3}H-GDP. Low pH (below 7) and the absence of Mg^{2+} both dissociate EFTu·GDP and accelerate the uncatalyzed exchange reaction, thus mimicking EFTs.

Assay for AA-tRNA·EFTu·GTP Formation (Ternary Complex)

Reaction (3) in the introductory section can be assayed readily by a variation of the filter assay for EFTu·GDP previously described. Ravel *et al.*[11] and Gordon[12] reported that AA-tRNA prevents EFTu·GDP from binding to cellulose nitrate filters, and showed that this effect is due to the formation of the ternary complex, AA-tRNA·EFTu·GTP. The following procedure assays ternary complex formation by measuring the decrease in EFTu·GTP bound to the filter in the presence of Phe-tRNA. The method uses phosphoenolpyruvate (PEP) and pyruvate kinase to convert the GDP present in EFTu·GDP (and contaminating the GTP) to GTP, because GDP binds to EFTu much more strongly than GTP, and inhibits the formation of EFTu·GTP.[3,13]

Materials

PEP, trisodium salt
Pyruvate kinase, crystalline, 150 IU/mg in 2 M (NH$_4$)$_2$SO$_4$
Mixture of [^{3}H]GTP and [γ-^{32}P]GTP (sp. act. 400–1000 cpm/pmole)
EFTu·GDP
Phe-tRNA from yeast or *E. coli*

[11] J. M. Ravel, R. L. Shorey, and W. Shive, *Biochem. Biophys. Res. Commun.* **29**, 68 (1967).
[12] J. Gordon, *Proc. Nat. Acad. Sci. U.S.* **59**, 179 (1968).
[13] D. Cooper and J. Gordon, *Biochemistry* **8**, 4289 (1969).

Method. The reaction mixtures contained EFTu·GDP (60 pmoles), PEP (0.3 μmole), pyruvate kinase (1–2 IU), [γ-^{32}P-^{3}H]GTP (500 pmoles), and 10 μl of binding buffer in a total volume of 50 μl. The reaction mixtures (run in duplicate) are incubated for 10 minutes at 37°. One of the mixtures is diluted to 2 ml and filtered; the filter is washed as earlier described. The amount of ^{32}P and ^{3}H and the ratio of ^{32}P:^{3}H are determined. A low value indicates incomplete conversion of GDP to GTP, suggesting the presence of a GTPase activity or a failure in the phosphorylation system. A ratio of ^{32}P:^{3}H of 0.7–1.0, is satisfactory for the purposes of the assay. The second reaction mixture is placed in ice and cooled to 0°, and Phe-tRNA up to 30 pmoles is added. After 3 minutes, the reaction mixture is filtered and the filter is washed as before. The difference between the amounts of EFTu·GTP retained on the filter with and without Phe-tRNA is used to calculate the amount of ternary complex formed. In the presence of limiting amounts of Phe-tRNA it has been possible to obtain excellent stoichiometry between Phe-tRNA added and complex formation (Fig. 1).[14,15]

Comments. The dissociation constant of the ternary complex has not been accurately determined; however, it appears to be less than 10 n*M*.[16] Therefore, AA-tRNA at concentrations above 0.1 μ*M* reacts quantitatively

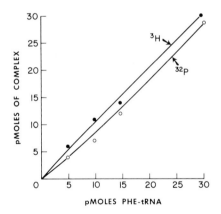

FIG. 1. Formation of the ternary complex as a function of Phe-tRNA concentration. The details of the reaction are described under Assay for AA-tRNA·EFTu·GTP Formation. From H. Weissbach, B. Redfield, and N. Brot, *Arch. Biochem. Biophys.* **145**, 676 (1971).

[14] H. Weissbach, B. Redfield, and N. Brot, *Arch. Biochem. Biophys.* **145**, 676 (1971).
[15] C. M. Chen and J. Ofengand, *Biochem. Biophys. Res. Commun.* **41**, 190 (1970).
[16] D. L. Miller, M. Cashel, and H. Weissbach, *Arch. Biochem. Biophys.* **154**, 675 (1973.)

with EFTu·GTP. Although most of our studies have used Phe-tRNA, EFTu·GTP has been shown to react with several other AA-tRNA's including those of valine, leucine, arginine,[17] glycine, alanine,[18] lysine,[11] serine,[19] and phenyllactyl-tRNA$_{phe}$.[20] EFTu·GTP also reacts strongly with yeast Phe-tRNA.[15] No interaction of EFTu·GTP has been detected with the following substances: deacylated tRNA (60 μM),[12] N-acetyl Phe-tRNA (0.2 μM),[11] N-formyl methionyl-tRNA met$_f$, or methionyl tRNA met$_f$.[19] Also EFTu·GDP cannot replace EFTu·GTP in the reaction with AA-tRNA.[12]

Generally, we have been satisfied with 80–90% conversion of EFTu· GDP to EFTu·GTP. In the subsequent reaction with AA-tRNA, the ^{32}P and ^{3}H activities bound to the filter should drop to the same extent; any excess ^{3}H activity left on the filter represents unconverted EFTu·GDP.

The sites on EFTu for AA-tRNA binding are different from those for GTP binding, and it is possible to prepare EFTu that is active in guanosine nucleotide binding, but will not react with AA-tRNA. Specifically, EFTu· GDP treated with −SH reagents retains its ability to interact with EFTs and guanosine nucleotides, but will not bind Phe-tRNA.[21] The inactivation of this −SH group may occur during purification, but has not been observed to any great extent in our procedures.

Purification and Isolation of EFTu·GDP

For studies of the structure and reactivity of EFTu·GTP and the ternary complex, it is necessary to have large amounts of pure protein. We have found that the following simple procedure gives relatively high yields of pure EFTu·GDP.[22] It is derived from the method of Lucas-Lenard and Lipmann.[6]

Materials

    Mid-log E. coli B cells, 2 lb
    Guanosine diphosphate
    Dithiothreitol
    DNase
    β-Mercaptoethanol

[17] J. Ofengand and C. M. Chen, J. Biol. Chem. 247, 2049 (1972).
[18] S. Fairfield and S. Pestka. Unpublished experiments.
[19] Y. Ono, A. Skoultchi, A. Klein, and P. Lengyel, Nature (London) 220, 1304 (1968).
[20] S. Fahnestock, H. Weissbach, and A. Rich, Biochim. Biophys. Acta 262, 62 (1972).
[21] D. L. Miller, J. Hachmann, and H. Weissbach, Arch. Biochem. Biophys. 144, 115 (1971).
[22] D. L. Miller and H. Weissbach, Arch. Biochem. Biophys. 141, 26 (1970).

Method

All operations were performed at 0–4°.

Extraction. Two pounds of *E. coli* B cells suspended in 1500 ml of a buffer solution containing 50 mM Tris (pH 7.8), 10 mM MgCl$_2$, 10 mM mercaptoethanol, and 5 mg DNase was broken in a Manton-Gaulin mill. The mixture was centrifuged (30 minutes at 20,000 g) to remove the cell debris, and the supernatant liquid was recentrifuged for 2 hours at 58,000 rpm in a Beckman Model L2-65B preparative ultracentrifuge using a 60 Ti rotor. The sediment was discarded, and the supernatant fraction was retained (HSSN).

Ammonium Sulfate Fraction. The supernatant solution (1200 ml) was brought to 0.30 saturation by the addition of 200 g of ammonium sulfate. The suspension was centrifuged, and the sediment was discarded. The pH of the supernatant fraction was adjusted to an apparent value of 7.8 at 0° (glass electrode) by the careful addition of 7 N ammonia. Ammonium sulfate was added to 0.7 saturation, and the precipitate, which contained the transfer factors, was removed by centrifugation. The precipitate was dissolved in 200 ml of buffer I (50 mM Tris, pH 7.8; 10 mM MgCl$_2$; 1 mM DTT), and the solution was dialyzed vs. 8 liters of buffer I. The dialyzed solution was centrifuged to remove any insoluble material, and the supernatant was saved [(NH$_4$)$_2$SO$_4$ fraction].

Chromatography on DEAE-Sephadex A-50. The (NH$_4$)$_2$SO$_4$ fraction (350 ml) was diluted about 2-fold to obtain a protein concentration of about 30 mg/ml, and KCl was added to a concentration of 0.1 M. At this stage GDP was added to a final concentration of 1 mM, because it improved the separation of EFTs and EFTu and increased the yield of EFTu. One-half of the solution (350 ml) was applied to a 90 × 2.5-cm column of DEAE-Sephadex which had been equilibrated in buffer I containing 0.10 M KCl. The column was washed at a rate of 1 ml/min with 500 ml of this buffer to which 0.2 mM GDP had been added. The elution was continued with GDP-free buffer until the EFTs activity in the effluent was low (1000 units/ml), which required about 2 liters of buffer. The fractions containing EFTs were reserved for the further purification of this protein.[23] At this point the KCl concentration was raised to 0.35 M, and the EFTu was eluted from the column (1st DEAE-Sephadex). This procedure was repeated with the remainder of the ammonium sulfate fraction. Transfer factor EFG also was eluted with EFTu, but the two proteins could be separated in the following chromatography.

Gradient Elution from DEAE-Sephadex A-50. The fractions contain-

[23] J. Hachmann, D. L. Miller, and H. Weissbach, *Arch. Biochem. Biophys.* **147**, 457 (1971).

ing EFTu (and also EFG) were concentrated by ultrafiltration (Diaflo, Amicon Inst.) or by precipitation with ammonium sulfate (0.6 saturation). The solution (about 50 ml) was dialyzed vs. buffer I, the protein concentration was adjusted to 30 mg/ml, and the Cl⁻ concentration was adjusted to 0.15 M with KCl. The solution was applied to an 80 × 2.5-cm column of DEAE-Sephadex A-50, equilibrated in buffer I containing 0.10 M KCl. The column was eluted with a linear gradient starting with 0.8 liter of 0.10 M KCl in buffer I in the mixing chamber and 0.8 liter of 0.35 M KCl in buffer I in the reservoir. Fractions of 16 ml were collected, and fractions 65–75, which contained the EFTu activity, were pooled (2nd DEAE-Sephadex).

Sephadex G-100 Chromatography. The EFTu fraction was concentrated by ultrafiltration (Diaflo) or with Aquacide (CalBiochem) to a volume of 10 ml. One-third of the concentrated EFTu fraction was applied to a 95 × 2.5-cm Sephadex G-100 column equilibrated with buffer I. The column was eluted with buffer I at a rate of 0.4 ml/min, and 5-ml fractions were collected. The EFTu emerged in fractions 50–56. The procedure was repeated twice more on the remainder of the DEAE-Sephadex preparation, and the EFTu containing fractions from the three separations were pooled (G-100 Sephadex).

Ammonium Sulfate Extraction. A number of enzymes have been purified and crystallized by extraction with ammonium sulfate.[24] This technique also was found to be effective in purifying EFTu·GDP. The EFTu-containing fractions from the Sephadex G-100 chromatography were concentrated to 25 ml by Diaflo ultrafiltration and were made 5 mM in DTT and 50 μM in GDP. Ten grams of ammonium sulfate was slowly added; the mixture was allowed to stand for 15 minutes, then centrifuged. The precipitate was extracted with 10-ml portions of progressively more dilute ammonium sulfate solutions in buffer I containing 5 mM DTT and 50 μM GDP as follows: Three extractions with 0.50 saturation followed by two extractions with 0.45 saturation and 0.40 saturation, and finally four extractions with 0.35 ammonium sulfate saturation. During each extraction the precipitate was thoroughly mixed for 15 minutes, and the mixture was then centrifuged. Eighty percent of the EFTu activity was found in the soluble extracts prepared with 0.35 saturated ammonium sulfate solution. The first fraction extracted with 0.35 saturated ammonium sulfate was reserved for crystallization. The other extracts were dialyzed vs. buffer I to remove ammonium sulfate and GDP and stored in liquid N₂.

Crystallization. It was found that thin needlelike crystals of EFTu·GDP formed in the 0.35 saturation ammonium sulfate extracts when they were

[24] T. P. King, *Biochemistry* 11, 367 (1972).

kept at 0° for 2–5 days. Although the yield was low, the crystals grew to 2 mm in length. Slow raising of the ammonium sulfate concentration to 0.40 saturation resulted in a high yield of much smaller crystals. The solubility of EFTu·GDP is strongly affected by Mg^{2+} and Cl^-; the protein will crystallize from a buffer composed of 1 mM $MgSO_4$, 20 mM Tris·SO_4, pH 8, and 0.2 M ammonium sulfate, if Cl^- is removed.

Comments

A summary of the purification procedure is shown in the table. Until the G-100 Sephadex chromatography there is enough EFTs contaminating the EFTu so that no extra EFTs need be added to the assay mixtures. The most difficult part of the procedure is the separation of EFTs from EFTu, because they bind together tightly. As indicated in Eq. (1), GDP effects this separation and, therefore, is added to the dialyzed protein before the initial chromatography; however, enough cellular GDP remains associated with the protein after dialysis to allow a partial separation of EFTu and EFTs without added GDP. EFTu·GDP purified through the Sephadex G-100 step exchanges with ³H-GDP in 5 minutes at 0° to the extent of only 5–10%; therefore, at this state the EFTu is suitable for use in EFTs assays.

Purity of the Preparation. The preparation of EFTu appears homogeneous by disc gel electrophoresis and ultracentrifugation. It is judged to contain less than 0.1% EFTs, less than 0.1% EFG, and it has a slight, probably endogenous, GTPase activity. The protein isolated by this procedure contains about one equivalent of tightly bound zinc. The metal ion may not be functionally significant, since it can be removed by dialysis vs. BAL without impairing the protein's activity in the phenylalanine polymerization assay.

Properties of EFTu. As the EFTu·GDP complex, the protein is stable for months at 4° as a crystalline suspension in 2 M ammonium sulfate, in

PURIFICATION OF EFTu·GDP[a]

Fraction	Total activity (pmoles $\times$ 10^{-6})	Specific activity (pmoles/mg)
HSSN	3.0	170
$(NH_4)_2SO_4$ fraction	3.5	330
1st DEAE Sephadex	2.8	2,200
2nd DEAE Sephadex	2.1	6,500
Sephadex G-100	1.8	14,000
Extraction	1.3	23,000
Crystallization	—	23,000

[a] From D. L. Miller and H. Weissbach, *Arch. Biochem. Biophys.* **141,** 26 (1970).

buffer I. The protein will not tolerate ordinary freezing and thawing, but it is unaffected by repeated freezing in liquid nitrogen. Dilute solutions of the protein are best stored in a liquid nitrogen freezer. At pH levels below 7, EFTu is irreversibly denatured. Apparently, free EFTu is the species which is unstable to acid, because the protein can be stabilized at lower pH by excess GDP. EFTu·GDP is rapidly denatured by heating at 55° and by tryptic digestion and is sensitive to sulfhydryl reagents. The molecular weight of EFTu is about 42,000 and the protein binds one equivalent of GDP or GTP. The dissociation constants of EFTu·GDP and EFTu· GTP are 3 nM and 0.3 μM, respectively, in the presence of 10 mM Mg^{2+}. The dissociation constant of EFTu·GDP in the absence of Mg^{2+} is 1 μM.[13] EFTu will bind 2′-deoxy-GDP nearly as strongly as it binds GDP. The protein binds GMP·PNP nearly as strongly as it binds GTP[25]; however, it binds GMP·PCP relatively weakly. EFTu does not interact to a measurable extent with the following nucleotides: CDP, UDP, ADP, dADP, ATP, GMP.

Preparation of GDP-Free EFTu, EFTu·GMP·PCP, and EFTu·GTP

Preparation of EFTu

GDP can be removed from small quantities of EFTu by prolonged dialysis against a Mg^{2+}-free buffer.[3] Removal of 75% of the GDP from 2 mg of EFTu·GDP requires 3 days of dialysis vs. 4 liters of 50 mM Tris· HCl (pH 8), 1 mM EDTA, 1 mM DTT, the buffer being changed daily. The extent of removal of GDP can be best assayed by first labeling the EFTu with ^{3}H-GDP.

This method is unsatisfactory for preparing several milligrams of GDP-free EFTu, because longer times are required for complete removal of the nucleotide and losses due to denaturation of the unstable protein occur. To eliminate this problem, the following procedure was developed in which GDP is displaced by a high concentration of the weakly bound GTP analog, GMP·PCP.

Preparation of EFTu·GMP·PCP

Although the methylene analog of GTP, GMP·PCP,[26] binds weakly to EFTu (K_{diss} = 0.1 mM in 10 mM $MgCl_2$), it simulates the action of

[25] F. Eckstein, M. Kettler, and A. Parmeggiani, *Biochem. Biophys. Res. Commun.* **45**, 1151 (1972).

[26] GMP·PCP is available from Miles Laboratories or can be readily synthesized according to T. C. Myers, K. Nakamura, and J. W. Flesher, *J. Amer. Chem. Soc.* **85**, 3292 (1963).

GTP by forming a ternary complex that will bind to ribosomes.[27] To prepare EFTu·GMP·PCP, 10 mg of Mg^{2+}-free EFTu·GDP is passed through a 1.5 × 85 cm column of Bio-Gel P-4 equilibrated with 1 mM GMP·PCP in 50 mM Tris·HCl (pH 8), 1 mM EDTA, 1 mM DTT, at a flow rate of 1 ml/min. About 85% of the GDP can be removed by this method.

The EFTu·GMP·PCP complex can be precipitated in 2 M ammonium sulfate and stored in liquid nitrogen indefinitely. From it free EFTu can be easily prepared by passage through a 1 × 15 cm column of Bio-Gel P-4.

Preparation of EFTu·GTP

Small-Scale Preparation. Small amounts (10 nmoles or less) of EFTu·GTP can be conveniently prepared by scaling up the assay procedure for the ternary complex. EFTu·GDP (3 nmoles), [^{3}H, γ-^{32}P]GTP (15 nmoles), PEP (2 μmoles), and pyruvate kinase (2 IU) are incubated 10 minutes at 37° with 75 μl binding buffer in a total volume of 350 μl. When 90% of the EFTu·GDP has been converted to EFTu·GTP as determined on an aliquot of the reaction mixture using the filter assay, the excess GTP and PEP are removed by passing the mixture through a 0.9 cm × 12 cm column of Sephadex G-25 or Biogel P-4 equilibrated with 50 mM Tris·HCl (pH 7.4)–10 mM $MgCl_2$. The EFTu·GTP prepared in this manner is contaminated by small amounts of free GTP and pyruvate kinase (4% of the total protein).

Large-Scale Preparation. Large amounts of EFTu·GTP can be prepared from the easily dissociable EFTu·GMP·PCP simply by adding an equivalent amount of GTP to EFTu·GMP·PCP in a Mg^{2+}-containing buffer. One should be sure that the GTP used is relatively free of GDP (<2%) since GDP contaminating the GTP will bind to EFTu preferentially. EFTu·GTP hydrolyzes slowly so it cannot be kept for more than 8 hours without forming a substantial quantity of EFTu·GDP.

Preparation of the AA-tRNA·EFTu·GTP Complex (Ternary Complex)

The preparation of the ternary complex is based on the filter characteristics of the ternary complex described above in the section on assay of the ternary complex. The complex is routinely prepared, labeled with ^{3}H, ^{32}P and ^{14}C: [^{14}C]Phe-tRNA·EFTu·[γ-^{32}P-^{3}H]GTP.

Procedure

The incubations, conditions, and components are as described for the formation of the ternary complex (see assay section) except that the reac-

---

[27] R. L. Shorey, J. M. Ravel, and W. Shive, *Arch. Biochem. Biophys.* **146**, 110 (1971).

tion components are scaled up 20-fold. Thus the incubations contain 1200 pmoles EFTu·GDP in a total reaction volume of 1 ml. After the preliminary incubation at 37° to form EFTu·GTP, the incubations are placed in ice and 500–600 pmoles of [^{14}C]Phe-tRNA (60–80% pure from *E. coli*) are added. After incubation at 0° for 3 minutes, the mixture is filtered through a stack of 2 nitrocellulose filters and the filters are washed with 0.5 ml of a buffer containing 10 mM Tris·HCl pH 7.4, 10 mM MgCl$_2$ and 10 mM NH$_4$Cl. The filtrate and wash are combined and dithiothreitol is added to the filtrate to a final concentration of 1 mM. Ten microliters of filtrate contain between 3 and 4 pmoles of complex. The specific activities (cpm/pmole) for the three isotopes used in the preparations of the complex are about ^{3}H = 550; ^{14}C = 600, and ^{32}P = 500.

Comments

The filtrate prepared as described above can be used directly as a source of ternary complex in studies on the interaction of the complex with ribosomes (see next section). Under the conditions used, unreacted EFTu· GTP has been removed by the filter and virtually all the Phe-tRNA added is incorporated into the complex. Thus the filtrate contains ternary complex, unreacted [γ-^{32}P-^{3}H]GTP plus trace amounts of unreacted Phe-tRNA. The filtrate also contains pyruvate kinase, PEP, and pyruvate. These components have not been found to interfere with studies on the reactivity of the ternary complex. The complex in the filtrate is stable for 8 hours and occasionally has been kept for 24 and 48 hours. Usually appreciable decomposition is seen when the complex is kept overnight. The amount of complex in the filtrate when first prepared can be determined by performing, along with the large incubation, a small-scale incubation (see assay section on ternary complex) and measuring the amount of [γ-^{32}P-^{3}H]GTP retained on a nitrocellulose filter in the presence and in the absence of Phe-tRNA.

Large amounts of the ternary complex can be formed by the reaction of AA-tRNA with GTP (free of GDP) and EFTu·GMP·PCP. To 40 nmoles of EFTu·GMP·PCP in 1 ml of buffer solution (50 mM Tris·HCl, pH 7.4, 10 mM MgCl$_2$) are added 60 nmoles of GTP and 40 nmoles of Phe-tRNA in 0.4 ml acetate buffer (1 mM NaOAC, pH 5.4, 10 mM MgCl$_2$). The formation of the ternary complex occurs immediately at 0° and is relatively stable for several hours. This procedure eliminates contamination by pyruvate kinase and PEP; however, the product contains some unreacted EFTu and the displaced GMP·PCP.

Interaction of the Ternary Complex with Ribosomes

The ternary complex, prepared containing ^{3}H, ^{14}C, and ^{32}P as described in the preceding section can be used for ribosome binding studies.

$$[^{14}C]Phe\text{-}tRNA\cdot EFTu\cdot[\lambda\text{-}^{32}P\text{-}^{3}H]GTP \xrightarrow{\text{Rib-poly(U)}}$$
(not retained on filter)

Rib-poly(U)-[^{14}C]Phe-tRNA $\Big\}$ Both products
+   retained on
EFTu·[^{3}H]GDP $\Big\}$ filter
+
^{32}P$_i$ $\Big\}$ Not retained on filter

In the presence of ribosomes and poly(U), [^{14}C]Phe-tRNA is transferred to the ribosomes, and EFTu·[^{3}H]GDP and ^{32}P$_i$ are formed. The reaction, therefore, results in GTP hydrolysis, which can be used as an assay, although a more convenient filter assay has been developed. As seen in the above reaction, the ternary complex or substrate of the reaction is not retained on a nitrocellulose filter whereas two of the products of the reaction—[^{14}C]Phe-tRNA bound to ribosomes and EFTu·[^{3}H]GDP—are retained by the filter.

Materials

Ternary complex, ([^{14}C]Phe-tRNA·EFTu·[^{32}P-^{3}H]GTP), prepared as described in the preceding section. The specific activities of all three labels should be greater than 500 cpm/pmole.

E. coli ribosomes, NH$_4$Cl washed, and prepared as described elsewhere[28]

Poly(U), 1 mg/ml

Binding mix: 0.25 M Tris·HCl pH 7.4, 0.4 M NH$_4$Cl, 0.4 M KCl, and 50 mM MgCl$_2$

Method

The incubations are performed in ice and contain, in a total volume of 50 μl, 10 μl of binding mix, 5 μg of poly(U), 3–6 pmoles of ternary complex, and 1–3 A_{260} ribosomes. After 2 minutes at 0° the incubations are diluted with 3 ml of a cold wash buffer containing 50 mM Tris·HCl, pH 7.4, 12 mM MgCl$_2$, and 0.16 M NH$_4$Cl, and rapidly filtered through a nitrocellulose filter. The filter is washed with 6 ml of the wash buffer and assayed for radioactivity.

Comments

In the absence of poly(U) no retention of labels is seen on the filter whereas with poly(U) a ^{3}H:^{14}C ratio of about 1 is obtained. No ^{32}P$_i$ is seen on the filter in either case. It has been observed in kinetic experi-

[28] N. Brot, E. Yamasaki, B. Redfield, and H. Weissbach, *Biochem. Biophys. Res. Commun.* **40**, 698 (1970).

ments, with some ribosome preparations, that at early time points the $^3H:^{14}C$ ratio is >1.[14] When only 30 S subunits are used, there is no hydrolysis of the GTP in the ternary complex and the intact complex interacts with the 30 S particle resulting in the retention of all three labels.[14]

[23] Elongation Factor T-Dependent GTP Hydrolysis: Dissociation from Aminoacyl-tRNA Binding

By J. P. G. BALLESTA

The elongation factor EF T (Tu + Ts) is required in the mechanism of polypeptide chain elongation for the formation of a ternary complex, aminoacyl-tRNA·EF Tu·GTP, which subsequently binds to the ribosome.[1] In the binding process the GTP molecule is hydrolyzed and a binary complex, EF Tu·GDP, is released. EF Ts displaces the GDP molecule from this complex regenerating the original EF T (Tu + Ts) that can initiate a new cycle.

In the presence of methanol, we have observed that the EF T-dependent GTP hydrolysis, strongly stimulated by the alcohol, is uncoupled from the aminoacyl-tRNA binding to the ribosomes, which is not affected.[2] This ribosome- and EF T-dependent uncoupled GTPase has some interesting characteristics and properties that will be described in this paper.

Reagents

Ribosomes prepared from *Escherichia coli* strain MRE 600 or D-10 cells in the late phase of exponential growth[3] and repeatedly washed in 20 mM Tris·HCl (pH 7.8) buffer containing 1 M ammonium chloride, 40 mM Mg(acetate)$_2$, 2 mM EDTA and 10 mM β-mercaptoethanol, until no endogenous GTPase activity was detected. They were dialyzed against 20 mM Tris·HCl (pH 7.8) buffer containing 150 mM ammonium chloride, 20 mM Mg(acetate)$_2$, 4 mM dithiothreitol and 50% glycerol and stored at 20° at 40–45 mg/ml. [γ-^{32}P]GTP was prepared following the procedure of Glynn and Chappell.[4] The final concentration of radioactive GTP was 0.05 mM and was maintained at $-20°$.

[1] J. Lucas-Lenard and F. Lipmann, *Annu. Rev. Biochem.* **40**, 409 (1971).
[2] J. P. G. Ballesta and D. Vazquez, *Proc. Nat. Acad. Sci. U.S.* **69**, 3058 (1972).
[3] J. Modolell, *in* "Methods in Molecular Biology" (J. A. Last and A. I. Laskin, eds.), Vol. 1, p. 1. Dekker, New York, 1971.
[4] I. M. Glynn and J. B. Chappell, *Biochem. J.* **90**, 147 (1964).

[^{14}C]Phenylalanyl-tRNA$_{Phe}$ prepared from tRNA$_{Phe}$, which was charged with radioactive phenylalanine (902 cpm/pmole) using a extensively dialyzed S100 fraction. The tRNA$_{Phe}$ was a preparation from the Oak Ridge National Laboratory.[5] A solution of [^{14}C]Phe-tRNA 4.5 μM was stored.

Five times concentrated standard buffer (5 × SB) containing 0.1 M Tris·HCl (pH 7.8), 0.5 M ammonium chloride, 25 mM magnesium acetate, and 5 mM dithiothreitol

Polyuridylic acid, 1 mg/ml solution in distilled water, stored at $-20°$

Methanol, analytical grade

Elongation factor EF T, prepared from the supernatant of alumina ground *E. coli* B cells following the method of Parmeggiani, Singer, and Gottschalk,[6] which results in electrophoretically pure EF T. In each of the purification steps, the peaks of EF T activity tested by standard methods, such as polyphenylalanine synthesis system or EF T-dependent retention of GTP on nitrocellulose filters[7] coincide with those of the methanol stimulated reaction which we will describe in this contribution. The results obtained in the last step of the purification procedure are shown in Fig. 1. The factor was dialyzed against 50 mM Tris·HCl (pH 7.8), 2 mM dithiothreitol in 50% glycerol, and stored at $-20°$.

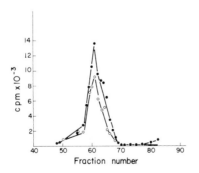

FIG. 1. Purification of EF T by polyacrylamide gel electrophoresis. Fraction aliquots from the last step on EF T purification procedure (A. Parmeggiani, C. Singer, and E. M. Gottschalk, this series, Vol. 20, p. 291) were tested for GTP retention to nitrocellulose filters (●) and for methanol-dependent GTP hydrolysis (○).

[5] A. D. Kelmers, H. O. Weeren, J. F. Weiss, R. L. Pearson, M. P. Stulberg, and G. D. Novelli, this series, Vol. 20, p. 9.

[6] A. Parmeggiani, C. Singer, and E. M. Gottschalk, this series, Vol. 20, p. 291.

[7] J. E. Allende and H. Weissbach, *Biochem. Biophys. Res. Commun.* **28**, 32 (1967).

Alternatively EF T was stored at 0° in 2.66 M $(NH_4)_2SO_4$; in this case the ammonium concentration in the concentrated standard buffer was 0.1 M. In both cases the concentration of the EF T solution stored was 2–2.5 mg/ml.

Assay Procedure

The total reaction mixture of 0.05 ml contains: 5 × SB 0.01 ml, ribosomes 0.001 ml, polyuridylic acid 0.005 ml, [^{14}C]Phe-tRNA 0.005 ml, EF T 0.001 ml, methanol 0.01 ml, water 0.015 ml. When some of the components were omitted, they were replaced by an equal volume of distilled water except when EF T in 2.66 M $(NH_4)_2SO_4$ was used; this was replaced by ammonium sulfate solution of the same molarity. Reactions were started by the addition of 0.003 ml of [γ-^{32}P]GTP. Incubation can be carried out at either 0° for 10 minutes or 30° for 5 minutes. At the end of the incubation period 0.02-ml aliquots are taken and diluted in 2 ml of ice-cooled standard buffer for measuring binding of Phe-tRNA by filtering through nitrocellulose filters.[8] 0.1 ml of 0.7 M perchloric acid and 0.005 ml of 50 mM KH_2PO_4 are added to the remaining 0.03 ml of the incubation mixture to determine hydrolysis of GTP by estimation of the extraction of the split ^{32}P with isopropyl acetate.[9]

Omitting some of the components in this reaction mixture the minimum requirements for the GTP hydrolysis have been studied, and the results obtained are presented in the table. As shown in this table, only ribosomes, EF T and methanol are required for GTP hydrolysis. Consequently, in a simplified system with only these components, the GTPase activity is uncoupled from aminoacyl-tRNA binding.

EFFECT OF ALCOHOL ON EF T-DEPENDENT ACTIVITIES

Reaction mixture[a]	Hydrolysis of GTP		Binding of Phe-tRNA	
	−Methanol (pmoles)	+Methanol (pmoles)	−Methanol (pmoles)	+Methanol (pmoles)
Complete	16.5	44.9	2.17	2.51
−Polyuridylic acid	10.5	47.9	0.11	0.25
−Phe-tRNA	10.6	45.5	—	—
−EF T	10.8	9.5	0.86	0.88

[a] Reaction mixtures, as described in the text, were incubated 5 minutes at 30°.

[8] M. W. Nirenberg and P. Leder, *Science* **145**, 1399 (1964).
[9] B. E. Wahler and A. Wollenberger, *Biochem. Z.* **329**, 508 (1958).

Methanol-Dependent Uncoupled GTPase Activity. Characteristics
of the Reaction

The characteristics of the uncoupled GTPase activity have been
studied in the simplified system in the absence of polyuridylic acid and
Phe-tRNA.

There is a slight dependence of the extent of this uncoupled GTP
hydrolysis on the period of time that the ribosomes have been in contact
with the alcohol. The action of the methanol in the ribosomes must be
very rapid, however, because 2 minutes of incubation at 0° is enough
to permit subsequently maximum hydrolysis.

Maximum GTP hydrolysis takes place at 20% (v/v) methanol concentration. Higher concentration of alcohol start being inhibitory of the
EF T-dependent activities.

At 5 mM Mg^{2+} concentration a maximum of the methanol dependent
GTPase hydrolysis, is found in the absence of polyuridylic acid and Phe-
tRNA, whereas in the presence of these reagents the optimum concentration of this Mg^{2+} ion is 20 mM.

The uncoupled GTPase is not affected, or even enhanced by some
inhibitors of the coupled activity, such as thiostrepton and siomycin,
which appears to block mainly aminoacyl-tRNA interaction with the
ribosome.[2,10] On the other hand, the uncoupled GTPase activity is blocked
by the prior formation of the complex EF Tu·GMPPCH$_2$P·ribosome
probably preventing binding of EF T to the pertinent site on the ribosome.

[10] J. Modolell, B. Cabrer, A. Parmeggiani, and D. Vazquez, *Proc. Nat. Acad. Sci.
U.S.* **68**, 1796 (1971).

[24] The Binding of *Escherichia coli* Elongation Factor G
to the Ribosome

By JAMES W. BODLEY, HERBERT WEISSBACH, and NATHAN BROT

Principle. Elongation factor G (EF G) is one of the soluble factors
involved in the process of peptide chain elongation. It is required for
translocation to occur, i.e., the movement of the ribosome relative to the
mRNA. GTP is required for EF G to function in the overall reaction,
and GTP is hydrolyzed to GDP and P$_i$ (for a recent review see Lucas-
Lenard and Lipmann[1]). In addition, EF G in the presence of ribosomes,
but in the absence of other components of protein synthesis, catalyzes

[1] J. Lucas-Lenard and F. Lipmann, *Annu. Rev. Biochem.* **40**, 409 (1971).

the hydrolysis of GTP.[2] Both the EF G-dependent translocation and the "uncoupled" GTP hydrolysis are inhibited by the steroid antibiotic fusidic acid.[3] Recent studies have shown that EF G binds to the ribosome in a reaction dependent on either GTP or GDP to form a labile ribosome·EF G·GDP complex. The inhibition by fusidic acid of uncoupled GTP hydrolysis appears to result from the stabilization of the ribosome·EF G·GDP complex[4-6] through the binding of the antibiotic.[7] The resulting ribosome·EF G·GDP·fusidic acid complex dissociates relatively slowly and hence inhibits the catalytic action of EF G in the hydrolytic reaction. As mentioned above, the binding of EF G to ribosomes requires either GDP or GTP, and in the latter case, complex formation is associated with immediate and irreversible cleavage of GTP and release of P_i:

Ribosome + EF G + GDP (GTP) + fusidic acid $\rightleftharpoons$
Ribosome·EF G·GDP·fusidic acid (+P_i)

The ribosome complex is sufficiently stable so that the EF G-dependent binding to the ribosome of either radiolabeled GDP or fusidic acid (or 24,25-dihydrofusidic acid) can be detected by a variety of means. The most convenient of these is by filtration through nitrocellulose filters (Millipore Corp.), a technique widely used in the quantitation of ribosome complexes. The method described here is sufficiently rapid, specific, and sensitive so that it may be used as a routine assay for EF G.

Materials

Reaction Buffer: 0.1 M Tris·HCl (pH 7.4), 0.1 M NH$_4$Cl, 0.1 M magnesium acetate, 10 mM dithiothreitol

Wash Buffer: 10 mM Tris·HCl (pH 7.4), 10 mM NH$_4$Cl, 10 mM magnesium acetate, 10 mM fusidic acid (sodium salt)

EF G, 100–2000 units/ml.[8] See references cited in footnotes 6 and 8 for purification procedures.

Ribosomes, 750 A_{260}/ml, either 70 S or 50 S subunits washed at

---

[2] Y. Nishizuka and F. Lipmann, *Arch. Biochem. Biophys.* **116**, 344 (1966).

[3] N. Tanaka, T. Kinoshita, and H. Masukawa, *Biochem. Biophys. Res. Commun.* **30**, 278 (1968).

[4] J. W. Bodley, F. J. Zieve, L. Lin, and S. T. Zieve, *J. Biol. Chem.* **245**, 5656 (1970).

[5] J. W. Bodley, F. J. Zieve, and L. Lin, *J. Biol. Chem.* **245**, 5662 (1970).

[6] N. Brot, C. Spears, and H. Weissbach, *Arch. Biochem. Biophys.* **143**, 286 (1971).

[7] A. Ikura, T. Kinoshita, and N. Tanaka, *Biochem. Biophys. Res. Commun.* **41**, 1545 (1970).

[8] One unit of EF G is defined as that amount of protein which, in the presence of excess ribosomes, causes the retention of 1 pmole of GDP to Millipore filters: J. H. Highland, L. Lin, and J. W. Bodley, *Biochemistry* **10**, 4404 (1971).

least three times with buffer containing either 0.5 or 1.0 M
NH$_4$Cl. See Brot *et al.*[6] for isolation of ribosomes.
Nitrocellulose filters, Type HAWP 0.45 μm, 25 mm (Millipore
Corp.)
For nucleotide binding:
 ^{3}H-GDP or ^{3}H-GTP, 0.01 mM (1–5 Ci/mmole) (New England
 Nuclear Corp.)
 Fusidic acid, 30 mM (sodium salt)
For [^{3}H]24,25-dihydrofusidic acid binding
 [^{3}H]24,25-Dihydrofusidic acid (sodium salt), prepared from
 fusidic acid by reductive tritiation by the method of Godtfred-
 sen and Vangedal,[9] 1 μM, at least 10 Ci/mmole
GDP, 1 mM

Procedure. The incubation, final volume 50 μl, contains the following components which are added at 0°: 25 μl water and 5 μl each of
reaction buffer, EF G, ribosomes, and either 1 mM GDP:GTP (for
^{3}H-24,25-dihydrofusidic acid binding[10]) or 30 mM fusidic acid (for
^{3}H-GDP binding). The reaction is initiated by the addition of 5 μl of
the appropriate radiolabeled ligand. Generally the binding reaction is
conducted at 0° for 5 minutes, but the ribosome·EF G·GDP·fusidic acid
complex forms rapidly and is quite stable so that, when large numbers
of routine assays are involved, reaction times may be varied up to 1 hour
or more.

The reaction is terminated by the addition of approximately 3 ml
of ice-cold wash buffer, and the diluted reaction is immediately filtered
through a Millipore filter which has been previously rinsed with wash
buffer. The reaction tube is rinsed once with wash buffer, and then two
more 3-ml aliquots are passed through the filter. Because the complex
dissociates during dilution and filtration, it is important to perform these
operations rapidly and uniformly. Best results are obtained by dispensing
the wash buffer from a 500-ml wash bottle. Under most circumstances
repetitive filtrations can be performed at intervals of 30 to 60 seconds.
Following the last wash the filter is immediately removed from the filter
housing, dried, and assayed for radioactivity in a toluene-based fluor or
the wet filter can be dissolved directly in the scintillation fluid described
by Bray.[11]

---

[9] W. O. Godtfredsen and S. Vangedal, *Tetrahedron* **18**, 1029 (1962).
[10] The details of the binding of [^{3}H]24,25-dihydrofusidic acid binding to the ribosome·
EF G·GDP complex will be reported elsewhere: N. Richman, G. R. Willie, and
J. W. Bodley, *Fed. Proc., Fed. Amer. Soc. Exp. Biol.* **31**, 898 (1972); G. R.
Willie and J. W. Bodley (in preparation).
[11] G. A. Bray, *Anal. Biochem.* **1**, 279 (1960).

Comments. In order to apply the present assay technique to the study of the interaction between the ribosome and EF G, the ribosome must first be washed free of bound EF G by repeated exposure to "high salt." When adequately treated in this way, 250 μg (3.75 A_{260}) of ribosomes in the absence of exogenous EF G bind negligible amounts of either [³H]GDP or [³H]24,25-dihydrofusidic acid.

The binding of GDP to the ribosomes is linear over a relatively wide range of EF G concentrations. While the maximum amount of GDP binding varies with different ribosome preparations, binding over the linear range is quite constant. The binding of [³H]24,25-dihydrofusidic acid to the ribosome is hyperbolically related to EF G concentration, presumably because of the higher dissociation constant.

It is now apparent that translocation factors from eukaryote sources can bind GTP or GDP in the absence of ribosomes.[12-15] *E. coli* EF G, however, does not cause the retention of either GDP, GTP or 24,25-dihydrofusidic acid in the absence of ribosomes. However, in order to obtain this result EF G must be free of EF Tu, which does bind GDP and GTP; therefore, [³H]GDP binding to ribosomes does not provide a good estimate of EF G in crude extracts which contain EF Tu. Similarly, although [³H]24,25-dihydrofusidic acid appears to bind only to the ribosome·EF G·GDP complex, for reasons that are not entirely clear this assay too requires at least partial purification of EF G.

[12] S. Raeburn, R. S. Goor, J. A. Schneider, and L. S. Maxwell, *Proc. Nat. Acad. Sci. U.S.* **61**, 1428 (1968).
[13] J. W. Bodley, L. Lin, M. L. Salas, and M. Tao, *FEBS (Fed. Eur. Biochem. Soc.) Lett.* **11**, 153 (1970).
[14] L. Montanaro, S. Sperti, and A. Mattioli, *Biochim. Biophys. Acta* **238**, 493 (1971).
[15] E. Bermek and H. Matthaei, *Biochemistry* **10**, 4906 (1971).

[25] Binding of the Elongation Factors EF 1 and EF 2 to 80 S Ribosomes in a Cell-Free System from Porcine Brain of the Hypothalamic Region

By Dietmar Richter

Various assays have been used to study the function of bacterial and mammalian elongation factors associated with peptide chain elongation.[1] Recently, several laboratories have reported on the interaction of the

[1] J. Lucas-Lenard and R. Lipmann, *Annu. Rev. Biochem.* **40**, 409 (1971).

bacterial EF Tu and EF G with a specific ribosomal site, which revealed that both factors compete for the same ribosomal region.[2-5] These experiments have been extended to a mammalian cell-free system from porcine brain of the hypothalamic region, using 80 S ribosomes and elongation factors EF 1 and EF 2[6]; the bacterial equivalent to the former is EF T, and to the latter, EF G. This article reports on three assays for studying ribosomal binding of EF 1 and EF 2 from brain: (1) binding of EF 1 to ribosomes in the presence of Phe-tRNA and GTP; (2) complex formation between ribosomes and EF 2 and GTP; and (3) competition of EF 1 and EF 2 for the common ribosomal binding site.

Materials

Buffer 1: 50 mM Tris·HCl (pH 7.9), 25 mM KCl, 5 mM dithiothreitol, 5 mM MgCl$_2$, 0.25 M sucrose

Buffer 2: 50 mM Tris·HCl (pH 7.9), 5 mM MgCl$_2$, 0.5 M KCl, 5 mM dithiothreitol, 0.5 M sucrose

Buffer 3: 20 mM Tris·HCl (pH 7.9), 5 mM dithiothreitol

Buffer 4: 20 mM Tris·HCl (pH 7.9), 1 mM dithiothreitol, 5 mM MgCl$_2$, 50 mM KCl

Buffer 5: 20 mM Tris·HCl (pH 7.9), 1 mM dithiothreitol, 5 mM MgCl$_2$

To remove dust particles and to reduce the blanks of the assays described later, buffers 3, 4, and 5 are passed through Millipore filters before use. All operations are carried out at 4° unless otherwise stated.

Fusidic acid (Leo Pharmaceuticals, Copenhagen, Denmark)

Porcine hypothalamus fragments (Oscar Mayer, Madison, Wisconsin)

Millipore filter: diameter 25 mm, pore size 0.45 μm

tRNA from *Escherichia coli* (Miles Laboratories)

[14C]Phenylalanine (Schwarz/Mann), specific activity 460 mCi/mmole

[3H]Phenylalanine (Schwarz/Mann), specific activity 7 Ci/mmole

[3H]GTP (Schwarz/Mann), specific activity 9.2 Ci/mmole

[2] D. Richter, *Biochem. Biophys. Res. Commun.* **46**, 1850 (1972).
[3] N. Richman and J. W. Bodley, *Proc. Nat. Acad. Sci. U.S.* **69**, 686 (1972).
[4] B. Cabrer, D. Vazquez, and J. Modolell, *Proc. Nat. Acad. Sci. U.S.* **69**, 733 (1972).
[5] D. L. Miller, *Proc. Nat. Acad. Sci. U.S.* **69**, 752 (1972).
[6] D. Richter, *J. Biol. Chem.* **248**, 2853 (1973).

General Procedures

Preparation of 80 S Ribosomes and EF 1 and EF 2
from Hypothalamus Tissue

The method for isolation of 80 S ribosomes from brain tissue has been described by Zomzely-Neurath.[7] In brief, the following steps are involved: 20 g of freeze-dried hypothalamic fragments are dissolved in 400 ml of buffer 1, kept for 20 minutes at 4°, and then homogenized in a glass tube with a Teflon pestle. The homogenate is centrifuged for 10 minutes at 15,000 g, and the supernatant fraction is treated with deoxycholate, layered on a discontinuous sucrose gradient, and centrifuged.[7] The ribosomal pellet is dissolved in a small volume of distilled water. The yield is 50 A_{260} (absorbance at 260 nm) units/ml with a total volume of 20 ml. For further purification, the ribosomal fluid is incubated in the presence of 1 mM puromycin and 0.5 M KCl at 37° for 15 minutes.[8] Of this solution, 15 ml are layered on 6 ml of buffer 2 and centrifuged for 3 hours at 105,000 g. The ribosomal pellet is then rinsed with distilled water and stored as a pellet in liquid nitrogen until used. These ribosomes are free of peptidyl-tRNA and mRNA.

The procedure for the isolation of mammalian EF 1 and EF 2 has been described.[9-11] Homogenate from hypothalamic fragments (20 g) is prepared as described above and centrifuged at 105,000 g for 3.5 hours. The supernatant fraction is treated with 43 g of $(NH_4)_2SO_4$/100 ml of fluid, pH 7.9; the precipitate is dissolved in buffer 3 and dialyzed against 2 × 2 liters of the same buffer. The protein concentration is 80 mg/ml; the total volume is 10 ml. EF 1 and EF 2 are separated by Sephadex G-200 gel filtration.[9,10,12] Further purification of EF 1 and EF 2 can be achieved by chromatography on hydroxyapatite columns.[11] Optimal conditions for phenylalanine synthesis in this system are achieved at 6 mM Mg^{2+}; the pH of the reaction mixture should be kept between 7.9 and 8.0. Incubation is carried out at 37° for 15 minutes. The activity of the elongation factors and ribosomes can vary; a good preparation should incorporate 5–6 pmoles of phenylalanine into hot trichloroacetic acid-insoluble material per minute, per OD unit (A_{260}) of ribosomes.

[7] C. E. Zomzely-Neurath, *in* "Methods in Molecular Biology" (J. A. Last and A. I. Lasikin, eds.), p. 147. Dekker, New York, 1972.
[8] G. Blobel and D. Sabatini, *Proc. Nat. Acad. Sci. U.S.* 68, 390 (1971).
[9] B. Hardesty, W. McKeehan, and W. Culp, this series, Vol. 20, p. 317.
[10] B. Hardesty and W. McKeehan, this series, Vol. 20, p. 330.
[11] K. Moldave, W. Galasinski, and P. Rao, this series, Vol. 20, p. 337.
[12] D. Richter and F. Klink, this series, Vol. 20, p. 349.

E. coli tRNA is charged with phenylalanine in the presence of activating enzymes from *E. coli*. The specific activity of the phenylalanyl-tRNA used in these experiments ranges between 300 and 500 pmoles of phenylalanine per milligram of tRNA.

Procedures to Study the Interaction of the Elongation Factors with Ribosomes

Ribosomal Binding of EF 1. To demonstrate that EF 1 is bound to 80 S ribosomes, a method is used whereby ribosomes are incubated with EF 1, GTP, and [^{14}C]Phe-tRNA (first incubation step). Then ribosomes carrying EF 1 and [^{14}C]Phe-tRNA are recovered by gel filtration. The ribosome·EF 1·Phe-tRNA complex can also be recovered by centrifugation.[13] In a second incubation step, the recovered ribosomes are complemented with EF 2 for polyphenylalanine synthesis. A typical reaction mixture of 100 μl contains: 20 mM Tris·HCl, pH 7.9, 5 mM MgCl$_2$, 50 mM KCl, 1 mM dithiothreitol, 2 OD units (A_{260}) of ribosomes, 80 μg of poly(U), 100 μM GTP, 150 pmoles of [^{14}C]Phe-tRNA, and 100 μg of EF 1. After incubation at 4° for 30 minutes, the mixture is diluted with 100 μl of buffer 4 and filtered through a Sepharose-2B column (0.9 × 4.2 cm). The column is prepared using a 2.5-ml glass syringe and equilibrated with buffer 4. Fractions of 0.2 ml are collected; 10-μl aliquots of each fraction are used to follow the ^{14}C radioactivity. The ribosomal-bound [^{14}C]Phe-tRNA that comes off the column in the void volume is collected and combined. That these ribosomes carry not only [^{14}C]Phe-tRNA but also EF 1 can be demonstrated in a third incubation step: a reaction mixture of 50 μl of buffer 4 contains 0.2 OD units (A_{260}) of ribosomes carrying 12 pmoles of [^{14}C]Phe-tRNA and is incubated at 37° for 15 minutes with 30 μg of EF 2, 1 mM GTP, and 25 pmoles of newly added [^{3}H]Phe-tRNA. The reaction is stopped with 1 ml of 5% trichloroacetic acid and analyzed for polyphenylalanine as described.[12] The table shows that [^{14}C]Phe-tRNA previously bound to the ribosomes by EF 1 and isolated by gel filtration, together with the ribosomes, is incorporated into polyphenylalanine when EF 2 is added (line 2), whereas newly added EF 1 causes no further increase (line 3). Incorporation of [^{3}H]Phe-tRNA, newly added to the second incubation step, depends again only on EF 2 (line 2), which shows that EF 1 has been sufficiently bound to ribosomes in the first incubation step.

Ribosomal Binding of EF 2. Either GTP, GDP, or β,γ-methylene-guanosine 5′-triphosphate is required to bind EF 2 to ribosomes. The eukaryotic ribosome-EF 2-guanosine nucleotide complex is rather stable,

---

[13] F. Ibuki and K. Moldave, *J. Biol. Chem.* **243**, 44 (1968).

POLYPHENYLALANINE SYNTHESIS WITH EF 2 AND RIBOSOMES
CARRYING EF 1 AND [14C]Phe-tRNA

Incubation of ribosomes complexed with EF 1 and [14C]Phe-tRNA	Polyphenylalanine synthesized	
	[3H]Phe (pmoles)	[14C]Phe (pmoles)
Alone	0.3	0.05
With EF 2	18.5	9.8
With EF 1 + EF 2	18.9	9.7

in contrast to the bacterial one which requires the antibiotic, fusidic acid.[14,15] To optimize stabilization of the complex, fusidic acid can be added. Ribosomal-bound EF 2 is easily measured by an indirect method using radioactively labeled guanosine nucleotide as an indicator for ribosomal-bound EF 2; in the assay described here, [3H]GTP is used. The ribosome·EF 2·[3H]GTP complex can be isolated on Millipore filters. A reaction mixture of 100 μl contains: 20 mM Tris·HCl, pH 7.9, 5 mM MgCl$_2$, 50 mM KCl, 1 mM dithiothreitol, 20 mM fusidic acid (neutralized), 0.3–0.5 OD units (A_{260}) of ribosomes, 50 pmoles of [3H]GTP, and 20–40 μg of EF 2. The reaction is started by the addition of [3H]GTP, kept for 10 minutes at 4°, and stopped with 1 ml of buffer 5. The mixture is passed through a Millipore filter presoaked in buffer 5. The filter is then washed three times with 3 ml of the same buffer and counted in Bray's solution.[16] Before EF 2 is added to the incubation mixture, it is helpful to preincubate the elongation factor at 50° for 5 minutes in buffer 5; this does not inactivate EF 2 but reduces the GTP binding by EF 2 alone. In contrast to the bacterial equivalent, EF G, EF 2 binds GTP or GDP by itself. Therefore, ribosomal binding of EF 2 has to be corrected by subtracting a blank caused by EF 2 and GTP.

To demonstrate directly that EF 2 is bound to the ribosomes in a guanosine nucleotide-dependent reaction, a similar experiment can be performed as described for the ribosomal binding of EF 1. In Fig. 1 ribosomes are preincubated with EF 2, with or without GTP as outlined above, and then isolated by gel filtration on Sepharose 2B columns (see section on ribosomal binding of EF 1), with the exception that radioactivity in the eluate is followed by measuring the tritium activity of the ribosome·EF 2·[3H]GTP complex. In the experiment where [3H]

[14] J. W. Bodley, L. Lin, M. Salas, and M. Tao, FEBS (Fed. Eur. Biochem. Soc.) Lett. 11, 153 (1970).
[15] D. Richter, L. Lin, and J. W. Bodley, Arch. Biochem. Biophys. 147, 186 (1971).
[16] G. A. Bray, Anal. Biochem. 1, 279 (1960).

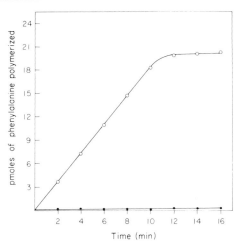

FIG. 1. Activity of the ribosome·EF 2·guanosine nucleotide complex in poly-phenylalanine synthesis. ○—○, Ribosomes preincubated with EF 2 + GTP; ●—●, ribosomes preincubated with EF 2.

GTP is omitted, the eluate is assayed for ribosomes by following the absorbancy at 260 nm. In a second incubation step at 37° for 15 minutes, 20 μg of complexed ribosomes are complemented with 20 μg of poly(U), 1 mM of GTP, 40 μg of EF 1, and 25 pmoles of [^{14}C]Phe-tRNA; the total volume of the reaction mixture is 100 μl and contains the components listed for buffer 4. Figure 1 shows that ribosomes preincubated with EF 2 and [^{3}H]GTP and reisolated by gel filtration, polymerize phenyl-alanine when the complementary factor EF 1 is added (upper curve). If GTP is omitted from the preincubation step no EF 2 is bound to the ribosomes and no polyphenylalanine is synthesized (lower curve).

Competition of EF 1 and EF 2 for a Common Ribosomal Binding Site

To demonstrate that EF 1 and EF 2 share and compete for a common ribosomal site, it is necessary to bind one enzyme to the ribosome while the other is assayed for its function. In the assay described here, first EF 2 will be bound to the ribosomes, then the function of EF 1 will be measured by determination of Phe-tRNA binding to the EF 2·ribosome complex. Ribosomal Phe-tRNA binding is assayed by a Millipore filter assay.

Reaction Mixture I. Complex formation between EF 2 and ribosomes is carried out in the presence of unlabeled GTP. A total reaction mixture of 50 μl containing 20 mM Tris·HCl (pH 7.9), 50 mM KCl, 5 mM

$MgCl_2$, 1 mM dithiothreitol, 0.1 mM GTP, 20 mM fusidic acid, 0.2 OD units (A_{260}) of 80 S ribosomes, and 20 μg of poly(U), is preincubated with 50 μg of EF 2 at 4° for 10 minutes.

Reaction Mixture II. This mixture is prepared in a separate experiment; it contains in 20 μl: 50 μg of EF 1, 0.1 M GTP, 20 mM Tris·HCl, pH 7.9, 50 mM KCl, 5 mM $MgCl_2$, 20 pmoles of [^{14}C]Phe-tRNA, and 1 mM dithiothreitol. This mixture is also kept at 4° for 10 minutes.

Reaction Mixture III. After preincubation, reaction mixtures I and II are combined, mixed, and incubated at 4° for the times indicated in Fig. 2. The reaction is stopped with 1 ml of buffer 4, and the mixture is passed through a Millipore filter presoaked in the same buffer. The filter is washed three times with 4 ml of buffer 4, dried under an infrared lamp, and counted in Liquifluor scintillation fluid. The radioactivity obtained represents the ribosomal-bound [^{14}C]Phe-tRNA retained by the Millipore filter.[17] Figure 2 shows that ribosomal-bound EF 2 inhibits EF 1-directed Phe-tRNA binding (lower curve); if EF 2 is omitted from the first incubation step, the binding site on the ribosomes is no longer blocked and therefore ribosomal binding of Phe-tRNA can take place (upper curve of Fig. 2).

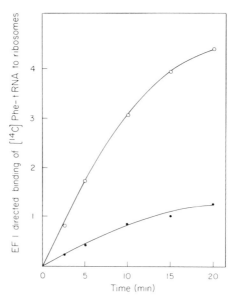

Fig. 2. Inhibition of EF 1-dependent [^{14}C]Phe-tRNA binding to ribosomes complexed with EF 2, GTP, and fusidic acid (●—●). ○—○, Ribosomes + GTP + fusidic acid.

[17] J. Ravel and R. A. L. Shorey, this series, Vol. 20, p. 306.

Comments

Ribosomes complexed with EF 2 and guanosine nucleotide are not only inactive in accepting EF 1-dependent Phe-tRNA binding, but also inactive in accepting nonenzymatically bound Phe-tRNA at the elevated Mg^{2+} concentration of 20 mM. The competition test described here can also be carried out for the reverse reaction. In this case, EF 1 is first bound to the ribosomes; the ribosomes are then assayed for EF 2-dependent GDP binding. In any event, the results of these experiments show that both enzymes have at least an overlapping region on the ribosomes, and that once bound to them, either one of the two factors inhibits the function of its counterpart.[6]

[26] Separation of Cytoplasmic and Mitochondrial Elongation Factors from Yeast

By DIETMAR RICHTER and FRITZ LIPMANN

Only relatively recently has it been firmly established that the eukaryotic organism contains two sets of protein-synthesizing apparatus: one, located in the cytoplasm, which represents the typical eukaryotic system, and another in mitochondria, which closely resembles the prokaryotic system.[1-3] Functionally and in chromatographic behavior, the elongation factors from the cytoplasm and mitochondria are quite different.[4] They can be isolated in two ways: (A) mitochondria and cytoplasm are separated by discontinuous centrifugation, and then the mitochondria are lysed mechanically or by osmotic shock to yield mitochondrial factors; (B) cells are disrupted by high shear forces which cause the breakage not only of cell walls but also of the organelles. Thus, these homogenates contain cytoplasmic as well as mitochondrial elongation factors which can be separated by chromatography.[4] Method B is rapid and convenient since it avoids isolation of the organelles; method A is essential, however, for identification.

This article describes both methods of isolation of mitochondrial and cytoplasmic elongation factors from yeast cells. A brief method for the isolation of mitochondrial ribosomes is also included. Since the isolation

[1] A. W. Linnane, J. M. Haslam, H. B. Lukins, and P. Nagley, *Annu. Rev. Microbiol.* **26**, 163 (1972).
[2] P. Borst, *Annu. Rev. Biochem.* **41**, 333 (1972).
[3] H. Küntzel, *Curr. Top. Microbiol. Immunol.* **54**, 94 (1971).
[4] D. Richter and F. Lipmann, *Biochemistry* **9**, 5065 (1970).

of the two cytoplasmic elongation factors from yeast has been published elsewhere,[5] we have put more emphasis on the separation of the two sets of elongation factors as well as on the isolation and purification of the mitochondrial elongation factors. Because of parallels between bacterial and mitochondrial elongation factors, the nomenclature for prokaryotic elongation factors is used for the latter: EF T is the aminoacyl-tRNA binding factor, and EF G is the peptidyl translocase. For the cytoplasmic elongation factors, EF 1 and EF 2 are used, as generally applied in eukaryotic protein synthesis.

It appears that mitochondrial ribosomes and elongation factors are interchangeable with their bacterial, but not with their cytoplasmic, counterparts in the eukaryotic cell.[4,6-8] The only case where a mitochondrial (or bacterial) elongation factor is compatible with eukaryotic ribosomes is EF T; however, its cytoplasmic counterpart, EF 1, does not react with mitochondrial or prokaryotic ribosomes.[4,9,10] For analysis during isolation of mitochondrial factors it is preferable to complement them with the more easily available ribosomes and elongation factors from *Escherichia coli.*

Reagents and Materials

Buffer 1: 20 mM Tris·HCl, pH 7.4, 1 mM dithiothreitol
Buffer 2: 5 mM K phosphate, pH 7.2, 1 mM dithiothreitol. All buffers are adjusted at 4°.
Hydroxylapatite (Clarkson Chemical Co., Hypatite C)
DEAE-cellulose (Bio-Rad)
[^{14}C]Phenylalanine, specific activity 460 mCi/mmole
Dithiothreitol (RSA Corporation)
Yeast tRNA (Miles Laboratories)
Triton X-100 (Rohm and Haas)

Unless otherwise stated, all operations are carried out at 4°.

Strains

Saccharomyces fragilis, ATCC No. 10022.
Saccharomyces cerevisiae, strain 18A; the mitochondrial DNA-

[5] D. Richter and F. Klink, this series, Vol. 20, p. 349.
[6] M. Grandi and H. Küntzel, *FEBS (Fed. Eur. Biochem. Soc.) Lett.* **10**, 25 (1970).
[7] A. Perani, O. Tiboni, and O. Ciferri, *J. Mol. Biol.* **55**, 107 (1971).
[8] A. H. Scragg, *FEBS (Fed. Eur. Biochem. Soc.) Lett.* **17**, 111 (1971).
[9] I. Krisko, J. Gordon, and F. Lipmann, *J. Biol. Chem.* **244**, 6117 (1969).
[10] D. Richter, *Biochem. Biophys. Res. Commun.* **38**, 864 (1970).

depleted "petite" mutants II-1-40 and III-1-7 can be obtained by treatment of the cells with ethidium bromide.[11,12]

General Procedures

Assay Method for Polyphenylalanine Synthesis

The function of the two elongation factors from yeast cytoplasm can be studied with 80 S cytoplasmic ribosomes from yeast or mammals; mitochondrial elongation factors can be combined with either mitochondrial or bacterial ribosomes.

Polyphenylalanine synthesis is measured by determination of ^{14}C radioactivity incorporated into hot trichloroacetic acid-insoluble protein. A typical assay (125 μl) contains: 50 mM Tris·HCl, pH 7.4, 100 mM NH$_4$Cl, 60 mM KCl, 10 mM Mg(CH$_3$COO)$_2$, 1 mM dithiothreitol, 2 mM GTP, 100–150 μg of ribosomes, 30 μg of poly(U), 10–20 pmoles of [^{14}C]Phe-tRNA (350 pmoles of [^{14}C]Phe per milligram of yeast tRNA), and 50–70 μg of each of the enzymes. Incubation is carried out at 37° for 10 minutes; the radioactive protein is analyzed as described.[5] Protein concentration can be estimated by the method of Lowry et al.[13] or of Warburg and Christian.[14]

Growth Conditions

The following yeast strains can be used: S. cerevisiae (wild-type 18A and two mutants[11] lacking mitochondrial DNA, II-140 and III-1-7), and S. fragilis. The yeast cells are grown in a medium containing, per liter: 5 g of yeast extract, 10 g of peptone, 6 g of (NH$_4$)$_2$HPO$_4$, 2 g of MgSO$_4$, 9 g of KCl, and 33 ml of a 60% lactate syrup. The pH is adjusted to 4.5 with concentrated HCl, and the medium is sterilized at 15 psi for 20 minutes. A 10-liter carboy is inoculated with 300 ml of an overnight culture containing 7.0 A_{450} (absorbance at 450 nm) units/ml. Cells are grown at 30° in a New Brunswick fermentor under aeration (2 liters of air per minute) with stirring (800 rpm). The culture is grown to a turbidity of 8.0 as measured from the 450-nm absorbance, and is quickly cooled by passing it through a copper cooling coil. Cells are harvested in a Sharples continuous flow centrifuge. The "petite" mutants are grown

[11] E. S. Goldring, L. J. Grossman, D. Krupnick, D. R. Cryer, and J. Marmur, J. Mol. Biol. 52, 323 (1970).
[12] P. O. Slonimski, G. Perrodin, and J. H. Croft, Biochem. Biophys. Res. Commun. 30, 232 (1968).
[13] O. H. Lowry, W. J. Rosebrough, A. L. Farr, and R. J. Randall, J. Biol. Chem. 193, 265 (1951).
[14] O. Warburg and W. Christian, Biochem. Z. 310, 384 (1941).

under similar conditions except that lactate is replaced by 1.5% glucose. The yield for all strains varies between 7 and 12 g of cells (wet weight) per liter of medium. The well-packed cells are kept overnight at 4° without loss of activity of the factors.

Isolation of the Elongation Factors

Method A

Elongation Factors from Isolated Mitochondria. For the isolation of mitochondria, spheroplasts from 1 kg of yeast cells (wet weight) are prepared[15] and gently homogenized for 15 seconds in a Waring Blendor; the mitochondrial particles are isolated and washed according to the method of Mattoon and Balcavage.[16] Washed mitochondria are suspended in buffer 1 with 10 mM Mg(CH$_3$COO)$_2$, and yield 4–5 mg of mitochondrial protein per milliliter of solution. The mitochondrial suspension is passed through a French press at 6000 psi. The homogenate is centrifuged at 105,000 g for 2 hours, the resulting S100 fraction is treated with 43 g of (NH$_4$)$_2$SO$_4$/100 ml, pH 6.9, stirred for 15 minutes, and centrifuged at 20,000 g for 15 minutes. The pellet is dissolved in 5 ml of buffer 2 and dialyzed against 2 liters of the same buffer. About 100 mg of protein are applied to a hydroxyapatite column (0.8 × 12 cm) previously equilibrated with buffer 2, and the column is washed with 100 ml of the same buffer. Mitochondrial EF G elutes at 30 mM and EF T at 70 mM phosphate buffer; all buffers contain 1 mM dithiothreitol, and the pH is adjusted to 7.2. Fractions containing either EF T or EF G are combined and concentrated using a Diaflo Model 50 ultrafiltration cell with a PM-10 membrane. Both factors can be stored for several months in liquid nitrogen. For further purification steps of the mitochondrial elongation factors see method B.

Preparation of Mitochondrial Ribosomes. Washed mitochondria from 1000 g of yeast cells (wet weight) are diluted with an equal volume of 40 mM Tris·HCl buffer, pH 7.4, and 20 mM Mg(CH$_3$COO)$_2$ and lysed by adding 1/20 of the volume of a 20% Triton X-100 solution. The ribosomes are pelleted at 105,000 g for 2 hours and redissolved in 20 mM Tris·HCl (pH 7.4) and 10 mM Mg(CH$_3$COO)$_2$. The A_{260}:A_{280} of this preparation is 1.93; the yield is 10–15 mg of ribosomal protein.

Method B

Separation of Cytoplasmic and Mitochondrial Elongation Factors from Yeast Homogenates. Yeast cells (200 g wet weight) are resuspended in

---

[15] E. A. Duell, S. Inoue, and M. F. Utter, *J. Bacteriol.* **88**, 1762 (1964).
[16] J. R. Mattoon and W. X. Balcavage, this series, Vol. 10, p. 135.

400 ml of buffer 1 and disrupted in a Manton-Gaulin mill.[17] The homogenate is centrifuged at 5000 g for 10 minutes, and the pellet is reextracted twice with 300 ml of buffer 1. The supernatant fractions are combined and further clarified by centrifugation at 18,000 g for 20 minutes. The pH of the supernatant fluid is readjusted with 1 M Tris·HCl to 7.4. After centrifugation at 78,000 g for 2 hours, two-thirds of the supernatant fractions are collected; the combined fractions are referred to as S100. The yield is 1040 ml or 9310 mg of protein.

Ammonium Sulfate Fractionation. The pH of the S100 fraction is adjusted to 6.8 with 10% acetic acid, and 70 g of ammonium sulfate/100 ml of fluid are added. The slurry is stirred for 1 hour, then centrifuged at 18,000 g for 1 hour. The supernatant fluid is decanted and the protein pellet is reextracted three times with 150 ml of 25% (step 1), three times with 22% (step 2), and finally three times with 18% (step 3) ammonium sulfate solutions (w/w). All solutions contain 1 mM dithiothreitol, and the pH is adjusted to 6.8 with NH_4OH. The supernatant fractions of each step are combined and reprecipitated with 20 g of ammonium sulfate/100 ml of fluid. The precipitates are collected by centrifugation at 18,000 g for 15 minutes, dissolved in buffer 1, and dialyzed against 3 liters of the same buffer. The yield for step 1 is 20 ml, or 935 mg of protein; for step 2, 47 ml, or 2700 mg of protein; and for step 3, 56 ml, or 1860 mg of protein. Step 1 contains no mitochondrial elongation factors, but cytoplasmic EF 1 and some cytoplasmic EF 2. The second step contains most of the mitochondrial EF T but is almost free of its complementary factor, EF G; this step also contains the bulk of cytoplasmic EF 1 and EF 2. Mitochondrial EF G is found in step 3 together with some EF T. The latter step also contains some cytoplasmic EF 2. The ammonium sulfate fractions can be stored for several months in liquid nitrogen without loss of activity.

Separation of Mitochondrial EF T and Cytoplasmic EF 1. The cytoplasmic EF 1 can be isolated from either step 1 or step 2 of the ammonium sulfate fractionation by gel filtration on Sephadex G-200 columns. The molecular weight of this EF 1 is high (220,000), and hence it can be separated rather easily from the rest of the elongation factors which are of lesser size (between 60,000 and 90,000). Figure 1 shows typical elution and activity profiles of cytoplasmic EF 1 compared with those of its mitochondrial counterpart, EF T. From the ammonium sulfate step 2, 80 mg of protein are layered on top of a Sephadex G-200 column (4.5 × 80 cm) equilibrated with buffer 1. Elution is carried out with the same buffer; 4.5 ml per fraction are collected. Aliquots (10 μl) of the eluted fractions are assayed for polyphenylalanine synthesis with an excess of ribosomes

---

[17] J. Gordon, J. Lucas-Lenard, and F. Lipmann, this series, Vol. 20, p. 281.

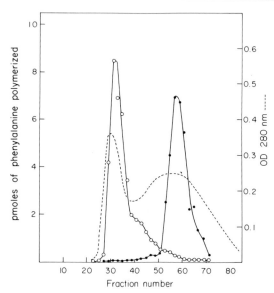

FIG. 1. Separation of mitochondrial EF T (●—●) and cytoplasmic EF 1 (○—○).

and complementary factors. The cytoplasmic EF 1 is free of cytoplasmic EF 2 and mitochondrial elongation factors. The mitochondrial EF T is contaminated by small amounts of cytoplasmic EF 2 and mitochondrial EF G. For further purification of the mitochondrial EF T, see below.

Separation of Mitochondrial EF G and Cytoplasmic EF 2. The ammonium sulfate step 3 contains most of the mitochondrial EF G as well as some cytoplasmic EF 2. The molecular weights of both factors are very close and do not separate on a Sephadex column. However, they both have different affinities for DEAE-cellulose. Figure 2 shows that mitochondrial EF G is recovered from the column at about 0.2 *M* KCl, whereas the cytoplasmic EF 2 comes off at 0.3 *M* KCl. DEAE-cellulose columns (1.2 × 15 cm) are equilibrated with buffer 1 containing 0.1 *M* KCl. From the ammonium sulfate steps 2 or 3, 50–80 mg of protein are dialyzed against the same buffer and applied to the columns, which are then washed with 100 ml of the same KCl concentration. Linear gradients from 0.1–0.4 *M* KCl in buffer 1 (50 × 50 ml) are used; 3.5 ml per fraction are collected. Aliquots (10 μl) of the eluted fractions are analyzed for polyphenylalanine activity in the presence of ribosomes and complementary factors as described above. Both cytoplasmic EF 2 and mitochondrial EF G are free of contamination by other elongation factors.

Purification of Mitochondrial EF T. The ammonium sulfate step 2,

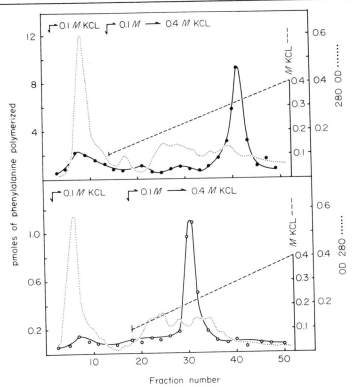

FIG. 2. Chromatography of mitochondrial EF G (O—O) and cytoplasmic EF 2 (●—●) on DEAE-cellulose columns.

which has most of the mitochondrial EF T, is used for its further purification. This includes protamine sulfate treatment, stepwise chromatography on hydroxylapatite columns, anion-exchange chromatography on DEAE-cellulose, and gel filtration on Sephadex G-200. The steps are summarized in Table I.

Protamine Sulfate Step. Neutralized protamine sulfate solution, 0.8 ml, is added per 10 ml of protein solution (step 2 of the ammonium sulfate fractionation). The solution is stirred for 20 minutes and the resulting precipitate is removed by centrifugation at 15,000 g for 15 minutes. The supernatant fluid is dialyzed against 2 liters of buffer 2. The yield is 45 ml, or 2480 mg of protein (Table I).

Hydroxylapatite Step. A hydroxylapatite column (4.5 × 9 cm) is packed and equilibrated with buffer 2. The protein fraction of the protamine sulfate step containing mitochondrial EF T is then applied to this column; the latter is washed with 500 ml each of 10 mM and 30 mM phosphate buffers, pH 7.2, both of which contain 1 mM dithiothreitol. Mitochondrial EF T is eluted with 70 mM phosphate buffer with 1 mM dithiothreitol. Fractions with mitochondrial EF T activity are combined and 43 g of $(NH_4)_2SO_4$/100 ml of solution are added. The slurry is stirred for 20 minutes and then centrifuged at 18,000 g for 20 minutes. The precipitate is dissolved in buffer 1 and dialyzed against 2 liters of the same buffer. The yield is 12 ml, or 424 mg of protein (Table I).

DEAE-Cellulose Step. A DEAE-cellulose column (1.3 × 23 cm) is equilibrated with buffer 1 containing 0.1 M KCl. The mitochondrial EF T from the hydroxylapatite step is applied to the column, which is then washed with 200 ml of buffer 1 containing 0.1 M KCl; the mitochondrial EF T is eluted with a linear gradient from 0.1 to 0.5 M KCl in buffer 1 (200 × 200 ml), and 4 ml/tube are collected. Fractions containing mitochondrial EF T are combined and concentrated in a Diaflo Model 50 ultrafiltration cell with a UM-10 membrane. The yield is 1.5 ml, or 31 mg of protein.

Sephadex G-150 Step. Mitochondrial EF T is further purified by gel filtration through a Sephadex G-150 column (1.3 × 75 cm). Buffer 1 is used to equilibrate the column and to elute mitochondrial EF T; 2.5 ml/tube are collected. The tubes with mitochondrial EF T are combined and concentrated as described for the DEAE-cellulose step. The yield is 1.5 ml, or 12 mg of protein.

TABLE I

PURIFICATION STEPS OF THE MITOCHONDRIAL EF T AND EF G
FROM *Saccharomyces cerevisiae*, STRAIN 18A[a]

	Total protein (mg)		Specific activity (nmoles/ 10 min/mg)		Total enzyme units (nmoles/ 10 min)	
	T	G	T	G	T	G
1. S100	9310		0.100		931	
2. Ammonium sulfate	2720	1860	0.187	0.250	509	465
3. Protamine sulfate	2480	1815	0.190	0.252	471	457
4. Hydroxylapatite	424	211	0.420	0.430	178	90.7
5. DEAE-cellulose	31.1	8.4	5.5	7.0	171	58.8
6. Sephadex G-150	12.1	3.2	9.2	12.7	111	40.6

[a] Activities for EF G and EF T are determined by polyphenylalanine synthesis in the presence of an excess of the complementary *Escherichia coli* factor and *E. coli* ribosomes [D. Richter and F. Lipmann, *Biochemistry* **9**, 5065 (1970)]. The activities are derived from experiments with a linear dependence on either EF G or EF T concentration.

TABLE II

PURIFICATION OF MITOCHONDRIAL EF T AND EF G FROM *Saccharomyces cerevisiae* STRAIN 18A AND THE "PETITE" MUTANTS II-1-40 AND III-1-7

| | Enzyme units (nmoles/10 min)/g of total S100 protein | | | | | |
| | 18A | | II-1-40 | | III-1-7 | |
Purification steps	T	G	T	G	T	G
1. S100	100	100	76	94	65	82
2. Ammonium sulfate	55	50	45	47	37	43
3. Protamine sulfate	51	49	39	45	32	31
4. Hydroxylapatite	19	10	12	8	—	—
5. DEAE-cellulose	18	6	10	5	—	—
6. Sephadex G-150	12	4	6	3	—	—

Further Purification of Mitochondrial EF G. Step 3 of the ammonium sulfate fractionation is used to purify mitochondrial EF G, using a similar method to that described for mitochondrial EF T, but with the following exceptions: mitochondrial EF G is eluted from the hydroxyapatite column (4.5 × 6.0 cm) with 30 mM phosphate buffer, and DEAE-chromatography (column size 1.2 × 24 cm) is carried out with a linear gradient from 0.1 to 0.4 M KCl in buffer 1 (150 × 150 ml). The results of the various purification steps of mitochondrial EF T and EF G are summarized in Table I.

Mitochondrial Elongation Factors from "Petite" Mutants

Mitochondrial elongation factors can be obtained from various yeast strains including mutants lacking mitochondrial DNA. This shows that mitochondrial EF T and EF G are coded by nuclear, not by mitochondrial, DNA.[18] Table II compares mitochondrial EF T and EF G from the wild-type strain 18A and from the two "petite" mutants II-1-40 and III-1-7. Mitochondrial elongation factors from all these strains are identical in their functional, protein chemical, and immunological properties.[18]

[18] D. Richter, *Biochemistry* **10**, 4422 (1971).

[27] Soluble Protein Factors and Ribosomal Subunits from Yeast. Interactions with Aminoacyl-tRNA

By A. Toraño, A. Sandoval, and C. F. Heredia

The elongation of the peptide chain on the ribosomes requires the participation of protein factors which are present in the soluble fraction of both prokaryotic and eukaryotic cells. Yeast has two different protein synthesizing systems: a mammalian-type one in the cytoplasm[1] and other in the mitochondria with properties similar to those from bacteria.[2,3] Two elongation factors, EF-1 and EF-2, have been found in both the cytoplasm[4,5] and the mitochondria[2,3] of yeast cells. EF-1 promotes a GTP-dependent binding of aminoacyl-tRNA to the ribosomes. EF-2 is involved in the translocation of the growing peptide chain from the acceptor to the donor site of the ribosome.

In vitro assays for these factors are usually based on their complementarity for the polymerization of amino acids from [14C]aminoacyl-tRNA by purified ribosomes. EF-1 is also assayed by its capacity to catalyze the binding of [14C]phenylalanyl-tRNA to the ribosomes·poly(U) complex in a GTP-dependent reaction. This factor can discriminate against *N*-acetylphenylalanyl-tRNA for this enzymatic binding reaction.[6] Recent experiments have shown the existence in the soluble fraction of yeast extracts of an activity, different from the two elongation factors, which promotes a GTP-dependent binding of *N*-acetylphenylalanyl-tRNA to the ribosome·poly(U) complex.[7]

This article describes the preparation of a supernatant fraction from the yeast *Saccharomyces fragilis* × *Saccharomyces dobzanskii* and the interactions of this fraction with aminoacyl and *N*-acetylaminoacyl-tRNA, the isolation from this supernatant fraction of partially purified elongation factors, and the preparation of ribosomes and active ribosomal subunits and their interactions with aminoacyl-tRNA.

[1] R. K. Bretthauer, L. Marcus, J. Chaloupka, H. O. Halvorson, and R. M. Bock, *Biochemistry* **2**, 1079 (1963).
[2] D. Richter and F. Lipmann, *Biochemistry* **9**, 5065 (1970).
[3] A. H. Scragg, H. Morimoto, V. Villa, J. Nekkorocheff, and H. O. Halvorson, *Science* **171**, 908 (1971).
[4] F. Klink and D. Richter, *Biochim. Biophys. Acta* **114**, 431 (1966).
[5] M. S. Ayuso and C. F. Heredia, *Biochim. Biophys. Acta* **145**, 199 (1967).
[6] M. S. Ayuso and C. F. Heredia, *Eur. J. Biochem.* **7**, 111 (1968).
[7] A. Toraño, A. Sandoval, C. SanJosé, and C. F. Heredia, *FEBS* (*Fed. Eur. Biochem. Soc.*) *Lett.* **22**, 11 (1972).

Reagents

Buffer A: 10 mM Tris·HCl (pH 7.5), 5 mM Mg(CH$_3$—COO)$_2$, 10 mM mercaptoethanol, 10 mM KCl

Buffer B: 0.5 M potassium phosphate (pH 6.5), 10 mM mercaptoethanol

Buffer C: 30 mM Tris·HCl (pH 7.2), 0.5 M NH$_4$Cl, 0.1 M Mg(CH$_3$—COO)$_2$, 0.25 M sucrose and 5 mM mercaptoethanol

Buffer D: 50 mM Tris·HCl (pH 7.7), 12 mM Mg(CH$_3$—COO)$_2$, 20 mM mercaptoethanol, 0.8 M KCl

Buffer E: 35 mM Tris·HCl (pH 7.7), 9 mM Mg(CH$_3$—COO)$_2$, 0.25 M sucrose, 70 mM KCl

Assay Methods

Binding of Aminoacyl-tRNA to the Ribosomes. The complete reaction mixture for the binding of aminoacyl-tRNA to the ribosomes contains in a final volume of 0.1 ml the following components: 50 mM Tris (adjusted to pH 6.5 with acetic acid), 100 mM NH$_4$(CH$_3$—COO), 50 μg of poly(U) (Boehringer), 1 mM GTP, 10 mM Mg(CH$_3$—COO)$_2$, 15–30 pmoles of [^{14}C]phenylalanyl or N-acetyl-[^{14}C]phenylalanyl-tRNA (200–300 pmoles of [^{14}C]amino acid per milligram of tRNA), 50–100 μg of soluble proteins and 0.5–2 A_{260} units of ribosomes or ribosomal subunits as indicated. The reaction is started by addition of the aminoacyl-tRNA. After incubation at 30°, usually for 20 minutes, the mixture is diluted with 3 ml of cold Tris·acetate buffer containing 100 mM NH$_4$(CH$_3$—COO) and 10 mM Mg(CH$_3$-COO)$_2$ and filtered through Millipore filters[8] (0.45 μm, 2.5 cm). The filters are washed three times with 3 ml each time of the same buffer, dried, and counted in a liquid scintillation spectrometer. Preparation of [^{14}C]phenylalanyl-tRNA is carried out by charging deacylated crude yeast tRNA (Sigma) with [^{14}C]phenylalanine (200 μCi/μmole) followed by phenol extraction and ethanol precipitation.[9] Acetylation of phenylalanyl-tRNA was performed as described elsewhere.[10]

Polymerization of Amino Acids. The polymerization of amino acids is followed by determination of the radioactivity incorporated into hot tri-chloroacetic acid-insoluble material. The reaction mixture contains, in a final volume of 0.1 ml, the following components: 50 mM Tris (adjusted to pH 6.5 with acetic acid), 100 mM NH$_4$(CH$_3$—COO), 50 μg of poly(U), 1 mM GTP, 10 mM Mg(CH$_3$—COO)$_2$, 15–30 pmoles of [^{14}C]

[8] M. Nirenberg and P. Leder, *Science* **145**, 1399 (1964).
[9] C. F. Heredia and H. O. Halvorson, *Biochemistry* **5**, 946 (1966).
[10] A. L. Haenni and F. Chapeville, *Biochim. Biophys. Acta* **114**, 135 (1966).

phenylalanyl-tRNA, 50–100 μg of "soluble protein fraction" and 0.5–2 A_{260} units of ribosomes or ribosomal subunits as indicated. Incubation is at 30° usually for 10 minutes. After this time, 0.05 ml of a solution of bovine albumin (4 mg/ml) is added as carrier followed by 3 ml of cold 5% trichloroacetic acid. The mixture is heated at 90° for 15 minutes, cooled, and filtered through glass fiber disks (Whatman GF/C, 2.5 cm). The precipitate is washed three times with 3 ml each time of cold 5% trichloroacetic acid, dried, and counted in a liquid scintillation spectrometer.

Preparative Procedures

Growth of the Yeast. The yeast hybrid *Saccharomyces fragilis* × *Saccharomyces dobzanskii* has been used for the preparation of the soluble protein factors and the ribosomes. The yeast cells are grown at 30° in a medium containing, per liter, the following components: 2 g of KH_2PO_4, 8 g of $(NH_4)_2SO_4$, 0.016 g of $MgCl_2$, 0.3 g of yeast extract (Difco), 20 g of glucose and 0.5 ml of a solution containing per liter: 16.5 g of $MgCl_2$, 2 g of $MgSO_4 \cdot 7H_2O$, 1 g of NaCl, 0.5 g of $FeSO_4 \cdot 7H_2O$, 0.5 g of $ZnSO_4 \cdot 7H_2O$, 0.5 g of $MnSO_4 \cdot 7H_2O$, 0.05 g of $CuSO_4 \cdot 5H_2O$, and 10 ml of 0.1 N H_2SO_4. The solutions of salts and glucose are sterilized separately by autoclaving at 120° and 2.5 atm for 20 minutes and mixed at the time of inoculation.

Yeast cells from an agar slant are inoculated in 25 ml of a medium containing 2% glucose and 0.3% yeast extract and incubated with vigorous agitation at 30° for 24 hours. Appropriate amounts of this inoculum are transferred to 10-liter flasks containing 7 liters of the culture medium described above and incubated at 30° with aeration. When the cultures have reached an optical density of 0.9 at 660 nm they are cooled by addition of ice. The cells are harvested by centrifugation in a continuous-flow rotor at 4° and washed three times by centrifugation with cold buffer A. The pelleted cells are placed on a porous disk for 30–60 minutes at 4°, and they are kept frozen at −20° until used.

Preparation of the Crude Extracts. Frozen yeast cells are suspended (25% w/v) in cold buffer A, and they are broken in a Ribi cell fractionator (Sorvall Model RF-1) at 20,000 psi and 5°. Alternatively the yeast cells can be broken by grinding in a cold mortar with 2–3 times its weight of sand or glass beads and extracted with 2.5 volumes of buffer A. Alumina must be avoided for breaking the cells since most of the amino acid polymerization factors are adsorbed on it. The homogenates are centrifuged in the cold, first at 8000 g for 10 minutes to eliminate unbroken cells, debris, and sand or glass beads if the second procedure is used, and then twice at 20,000 g for 20 minutes. The 20,000 g supernatant is further centrifuged at 105,000 *g* for 90 minutes. The supernatant after this centrifugation is referred to as S105 fraction. The pellet contains the crude ribosomes.

Purification of the Soluble Protein Factors. One volume of the S105 supernatant is adjusted to a protein concentration of 6.5 mg/ml with cold buffer A and then treated with 0.1 volume of a suspension of alumina C_γ (Sigma) containing 10 mg/ml, dry weight. After stirring for 10 minutes at 4°, the gel is separated by centrifugation in the cold at 5000 rpm for 10 minutes, and the supernatant is discarded. The gel is washed with cold 0.1 *M* potassium phosphate buffer pH 6.5 containing 10 m*M* mercaptoethanol. The proteins that remain adsorbed in the gel are eluted by suspending the gel in 0.5 *M* potassium phosphate buffer (buffer B) (about one-tenth of the original S105 volume). After stirring for 5 minutes at 4°, the suspension is centrifuged in the cold at 5000 rpm for 10 minutes, and the pellet was discarded. The supernatant (0.5 *M* potassium phosphate eluate) is dialyzed for 3–4 hours against cold buffer A. The resulting preparation is referred to as soluble protein fraction. The specific activity of this fraction is 15–20 times greater than that of the original S105, as determined by the poly(U)-directed polymerization of [¹⁴C]phenylalanine from [¹⁴C]phenylalanyl-tRNA using purified yeast ribosomes.[5]

A simple method to obtain preparations of EF-1 with low EF-2 activity from the 0.5 *M* potassium phosphate eluate is as follows: 1–2 ml of the 0.5 *M* potassium phosphate eluate (adjusted to 1–2 mg of protein per milliliter with buffer B) are placed in a test tube, and the tube is immersed in a water bath (60°) with agitation. When the temperature inside the tube has reached 55°, the tube is maintained at this temperature for 3–4 minutes and then rapidly cooled in an ice bath. The preparation is then dialyzed for 3–4 hours against cold buffer A. By this treatment usually about 90% of the EF-2 activity originally present in the preparation is inactivated, as measured by the loss of phenylalanine polymerization from [¹⁴C]phenylalanyl-tRNA while factor EF-1 remains active. With preparations of different protein content it is advisable to make a time course heat inactivation curve at 55° to establish the optimal conditions.

Factor EF-2 is obtained by filtration of the 0.5 *M* potassium phosphate eluate through Sephadex G-200. For this purpose the protein concentration is adjusted with buffer B to about 2 mg/ml and 3–5 ml of this preparation is passed through a Sephadex G-200 column (2 cm × 45 cm) equilibrated with buffer A. The proteins are eluted with buffer A. Fractions of 4 ml are collected at a flow rate of 0.2 ml/min, and EF-2 activity is determined in 0.1 ml aliquots of each fraction by its complementation with EF-1 for the poly(U)-directed polymerization of phenylalanine from [¹⁴C]phenylalanyl-tRNA. EF-2 is quite unstable and has to be used soon after prep-

TABLE I

BINDING OF AMINOACYL-tRNA TO PURIFIED YEAST RIBOSOMES
PROMOTED BY SUPERNATANT FACTORS. EFFECT OF
N-ETHYLMALEIMIDE (NEM)[a]

Expt. No.	Additions	Aminoacyl-tRNA bound	
		N-(Ac)-Phe-tRNA (pmoles)	Phe-tRNA (pmoles)
1	None	0.3	0.4
	Soluble protein fraction	2.3	—
	NEM-treated soluble protein fraction	0.3	4.3
2	Elongation factor 1	—	3.1
	NEM-treated factor 1	—	2.7
3	None	0.5	0.4
	Soluble protein fraction	2.5	—
	Elongation factor 1	0.4	2.5
	Elongation factor 2	0.4	—

[a] Conditions as described in the text using 2 A_{260} units of ribosomes and approximately 50 μg of each of the protein fractions. When indicated protein fractions were preincubated with N-ethylmaleimide (20 mM) at 30° for 5 minutes. From A. Toraño, A. Sandoval, C. SanJosé, and C. F. Heredia, FEBS (Fed. Eur. Biochem. Soc.) Lett. **22**, 11 (1972).

aration. A method for the isolation of elongation factors from *Saccharomyces cerevisiae* has been published.[11]

A summary of the interactions of the soluble protein factors with phenylalanyl-tRNA and N-acetylphenylalanyl-tRNA is shown in Table I. EF-1 interacts with phenylalanyl-tRNA and promotes binding of this compound to the ribosomes. N-acetylphenylalanyl-tRNA cannot be substituted for phenylalanyl-tRNA as substrate for this binding reaction. There is, however, another activity in the "soluble protein fraction" which promotes binding of N-acetylphenylalanyl-tRNA to the ribosomes. This activity, in contrast with EF-1, is sensitive to treatment with N-ethylmaleimide. These two factor-dependent binding reactions require poly(U), GTP, ammonium and magnesium ions.[6,7]

Purification of the Ribosomes. For the purification of the ribosomes two different procedures can be used. One is a modification of the method developed by Bruenning and Bock.[12] Crude ribosomes are suspended in cold buffer C to an optical density of about 300 A_{260} units/ml. Nine milliliters of this ribosomal suspension are layered over 3 ml of buffer C in which the concentration of sucrose has been increased to 2 M. After cen-

[11] D. Richter and F. Klink, this series, Vol. 20, p. 349.
[12] G. Bruenning and R. M. Bock, *Biochim. Biophys. Acta* **149**, 377 (1967).

trifugation at 105,000 g in a Spinco 40 rotor for 4 hours, three layers can be observed. An upper layer, an interphase, and a lower layer which contains the ribosomes. The two first layers are carefully removed with a syringe and discarded. The bottom layer is diluted twice with cold buffer A and passed through a Sephadex G-25 column equilibrated with the same buffer. The ribosomes are eluted with this buffer. Fractions are collected, and the optical density at 260 nm is recorded. The fractions containing the ribosomes are pooled, diluted with 2–3 volumes of buffer A, and centrifuged at 105,000 g for 2 hours, the ribosomal pellet is suspended in the appropriate amount of buffer A. The ribosomal suspension is centrifuged at 10,000 g for 10 minutes to remove denatured material and then adjusted with buffer A to the desired concentration. The ribosomes are distributed in small aliquots and kept either under liquid nitrogen or at −70°. Under these conditions they remain fully active for the polymerization of amino acids for at least 1 month. The ribosomes obtained in this way are completely dependent on the supernatant factors for the polymerization of amino acids from aminoacyl-tRNA.[6]

Purified ribosomes can also be obtained by an alternative procedure[6] in which crude ribosomes are suspended in cold buffer C to an optical density of about 300 A_{260} units/ml. The suspension is stirred overnight at 4°, and the ribosomes are diluted 1:1 with buffer A, sedimented by centrifugation at 105,000 g for 2 hours, resuspended in buffer A, and sedimented again by centrifugation. Finally, they are suspended in buffer A, centrifuged (10,000 g for 10 minutes) to remove denatured material, adjusted to the desired concentration, and kept in small aliquots as indicated above.

Preparation of Ribosomal Subunits. A procedure to obtain active ribosomal subunits from other species of *Saccharomyces* has been previously reported.[13] The method described here is a slight modification of that developed for rat liver[14] and *Artemia salina*[15] ribosomes. Crude ribosomes are suspended in a dissociation buffer (buffer D). The ribosomal suspension is adjusted to an optical density of 400 A_{260} units/ml. About 2 ml of this ribosomal suspension is layered on the top of a 50-ml linear sucrose gradient (15 to 30%) made in the same buffer and centrifuged at 2–5° in the SW 25.2 rotor of the Spinco ultracentrifuge for 12–14 hours at 22,500 rpm. After centrifugation fractions of the gradient (2 ml) are collected and their optical density at 260 nm is measured. Fractions corresponding to the 40 S and 60 S regions are pooled separately and diluted with one

[13] E. Battaner and D. Vazquez, this series, Vol. 20, p. 446.
[14] T. E. Martin and I. G. Wool, *Proc. Nat. Acad. Sci. U.S.* **60**, 569 (1968).
[15] M. Zasloff and S. Ochoa, *Proc. Nat. Acad. Sci. U.S.* **68**, 3059 (1971).

TABLE II

RECONSTITUTION OF 80 S RIBOSOMES FROM THEIR SUBUNITS[a]

Particles	Phenylalanine polymerized (pmoles)
80 S ribosomes	12.0
40 S subunits	0.5
60 S subunits	1.6
40 + 60 S subunits	10.3

[a] Conditions as described in the text using 2, 1.2, and 0.7 A_{260} units of 80 S, 60 S, and 40 S particles, respectively.

volume of 20 mM β-mercaptoethanol, 30 mM Mg(CH$_3$—COO)$_2$. The subunits are sedimented by centrifugation at 40,000 rpm and 2–5° for 14 hours in the Spinco 50 rotor. At this stage the 40 S subunits are quite pure, but the 60 S subunits are still contaminated with both 40 S subunits and 80 S monomers. For a more extensive purification, the pellets corresponding to each subunit are separately resuspended in about 1 ml of the buffer D and rerun on separate sucrose gradients as indicated above. After re-

TABLE III

BINDING OF PHE-tRNA AND N-(Ac)-PHE-tRNA TO RIBOSOMES
AND RIBOSOMAL SUBUNITS

Expt. No.[a]	Ribosomal particle	Additions	Aminoacyl-tRNA bound	
			N-(Ac)-Phe-tRNA (pmoles)	Phe-tRNA[b] (pmoles)
1	80 S	—	2.8	4.2
	40 S	—	3.1	2.5
	60 S	—	<0.1	<0.1
2	40 S	—	0.2	0.2
	40 S	Soluble protein fraction	0.4	0.5
	60 S	—	<0.1	<0.1
	60 S	Soluble protein fraction	0.2	0.3
	40 S + 60 S	—	0.7	0.8
	40 S + 60 S	Soluble protein fraction	2.3	3.6

[a] Expt. 1 was carried out at 20 mM Mg^{2+} and 1 mM spermidine, without added GTP and soluble factors, using 1.4, 1, and 0.5 A_{260} units of 80 S, 60 S, and 40 S particles, respectively. Expt. 2 was carried out at 10 mM Mg^{2+} and 1 mM spermidine using 2.3 and 1.3 A_{260} units of 60 S and 40 S particles, respectively.

[b] The soluble protein fraction used for the binding of Phe-tRNA was preincubated with N-ethylmaleimide as indicated in Table I, to inactivate EF-2 and the N-(Ac)-Phe-tRNA binding activity.

cording the optical densities of the gradients at 260 nm, the fractions corresponding to each peak are pooled, and the subunits are pelleted by centrifugation at 40,000 rpm for 14 hours at 2–5° in the Spinco 50 rotor. The pellets are suspended in buffer E and stored in 50% glycerol at −20°. Under these conditions the subunits retain full activity at least for several weeks. Cross contamination is less than 5% in the case of 40 S and about 10% in the case of the 60 S as estimated by the percentage of phenylalanine polymerized by each fraction as compared with the control (Table II).

The results in Table III summarize the interactions of the ribosomal particles with aminoacyl-tRNA. At high magnesium concentration (Expt. 1) both the 40 S subunits and the 80 S ribosomes bind phenylalanyl-tRNA or N-acetylphenyl-tRNA[16] in a reaction which requires poly(U) but not GTP or soluble protein factors. This binding reaction does not occur with the 60 S subunits alone. At lower magnesium concentrations (Expt. 2) both phenylalanyl-tRNA or N-acetylphenylalanyl-tRNA bind preferentially to the 80 S ribosomes (Table I) or to the combination of 40 S plus 60 S subunits (Table III) in response to the soluble factors described above.

Acknowledgments

We are indebted to Miss C. Moratilla for expert technical assistance and to Miss C. Estévez for the typing of the manuscript.

[16] D. Vazquez, E. Battaner, R. Neth, G. Heller, and R. E. Monro, *Cold Spring Harbor Symp. Quant. Biol.* 34, 369 (1969).

[28] The Use of Inhibitors in Studies of Protein Synthesis

By SIDNEY PESTKA

Many antibiotics and other inhibitors have been found to block protein synthesis in various steps. Although the mechanism and precise mode of action of most antibiotics has not been definitively evaluated, their use often can help elucidate not only their own mode of action, but the steps in protein biosynthesis itself. It is wise to note that antibiotics and other inhibitors cannot be used indiscriminately by blindly following procedures of other laboratories, for their use often requires specific reaction conditions. It is therefore advisable when using an antibiotic to use appropriate controls to ascertain that the inhibitions produced are indeed those which are desired. Often the concentration of an antibiotic is critical producing different effects at different concentrations.

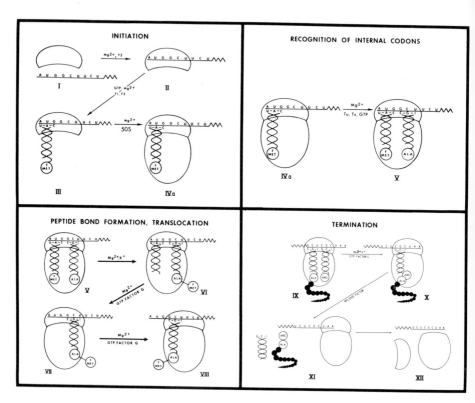

Fig. 1. Schematic summary of protein synthesis. The semilunar cap (I) represents the free 30 S subunit. Initiation of protein synthesis involves attachment of mRNA to a 30 S subunit (I) to form complex II; this process requires Mg^{2+} as well as initiation factor F3. Subsequent attachment of fMet-tRNA in response to initiation codon AUG to form complex III requires GTP and initiation factors F1 and F2. Junction of the 50 S subunit to complex III produces complex IVa; it is probable that prior to complex IVa an intermediate state IV exists where fMet-tRNA is in a nonpuromycin reactive state; the transition from IV to IVa possibly occurs on hydrolysis of GTP. Enzymatic recognition of internal codons involves factors Tu, Ts, and GTP: the Tu·GTP·Ala-tRNA complex binds to the ribosomes in response to the GCU codon to form complex V. Peptide bond formation occurs by transfer of the fMet (peptidyl) group to form fMet-Ala (VI); peptidyl transfer requires only ribosomes. Translocation involves several coordinate processes: release of deacylated tRNA$_f^{Met}$ to form state VII; one codon movement of mRNA and ribosome with respect to each other, precisely positioning the next codon UCU into position for translocation and coordinate movement of peptidyl-tRNA (fMet-Ala) from the "A" to "P" site resulting in state VIII. By repetition of the codon recognition step Ser-tRNA would enter the A-site in response to the codon UCU. Complex IX represents a peptidyl-tRNA with a polypeptide almost completed. Transpeptidation and translocation produces complex X with a completed protein still attached to tRNA and a termination codon, UAA, in the next recognition site. In response to a release factor, the completed protein is released and perhaps also tRNA after translocation (XI). Provided no further cistrons are to be translated, the ribosome may be dissociated into 30 S and 50 S subunits with release of mRNA (XII). Alternatively, mRNA may be degraded prior to this stage.

262

In this article, I shall summarize the current knowledge about the mode of action of various inhibitors on pro- and eukaryotic protein synthesis. No doubt, these views will be modified as our understanding of their mode of action increases and our knowledge of protein synthesis expands and changes. For convenience, the effect of each antibiotic on various steps of protein synthesis which can be specifically assayed will be considered. For a detailed review on inhibitors of ribosome function and protein synthesis several reviews can be consulted.[1-4] References given in these reviews may be omitted. This should avoid extensive duplication.

For the purposes of this chapter, protein synthesis will be considered in terms of the individual reactions which can be measured. These are schematically summarized in Fig. 1. The effects of inhibitors on protein synthesis will be considered in terms of these steps.

It should be noted that in general most experiments in protein synthesis are interpreted in terms of the acceptor-donor site model for protein synthesis (Fig. 2). The illustrations of Fig. 1 have been drawn also in terms of the acceptor–donor model for ribosomal function. Recent experiments on protein synthesis with antibiotics have thrown some question onto the suitability of the simple donor–acceptor model for interpreting experiments on protein synthesis. Another model for ribosomal function which appears to be consistent with most of the facts is shown in Fig. 3. This is presented so that the reader can readily follow the discussion concerning several antibiotics. In addition, the presentation of both models

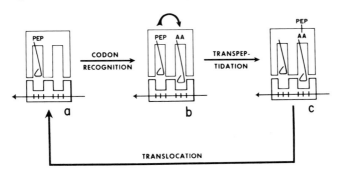

FIG. 2. Schematic illustration of the donor–acceptor site model for ribosome function following the detailed illustration of Fig. 1.

[1] S. Pestka, *Annu. Rev. Microbiol.* **25**, 487 (1971).
[2] B. Weisblum and J. Davies, *Bacteriol. Rev.* **32**, 493 (1968).
[3] D. Gottlieb and P. D. Shaw, "Antibiotics I, Mechanism of Action." Springer-Verlag, Berlin and New York, 1967.
[4] J. Corcoran, "Antibiotics II, Mechanism of Action." Springer-Verlag, Berlin and New York, in press.

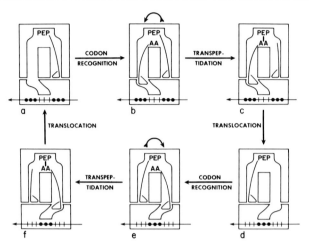

Fig. 3. The ribosome epicycle; a model of ribosome sites and ribosome function for codon recognition, transpeptidation, and translocation. Major features of the model are as follows: (a) There are two or more sites for tRNA binding on the 50 S subunit. Although the model would be essentially similar for any number of sites greater than two, the minimum number of sites on the 50 S subunit is two. (b) The two sites on the 50 S subunit are functionally similar but not identical. Transpeptidation can occur in both directions as illustrated (b,e). As a consequence of this, each site can contain aminoacyl-tRNA, peptidyl-tRNA, or deacylated tRNA. The direction of transpeptidation depends on the nature of the constituents of the sites rather than on the sites themselves. (c) There is only one site for tRNA binding on the 30 S subunit; this is the decoding site. By a conformational change (shown in the figure) or by a rotation of the subunit the site can align with either 50 S site. Because of this separate relative movement over many angstroms, the subunits should be separate particles. (d) Translocation involves movement of mRNA along the 30 S subunit, realignment of the 30 S subunit decoding site with the second 50 S site, and removal of deacylated tRNA. It is probable that removal of tRNA is the step dependent on factor G and GTP; realignment of the 30 S site with the second 50 S site may or may not be dependent on factor G. It is possible that factor Tu may have a role here. (e) In order for the active center of the peptidyltransferase to interact in identical ways with peptidyl-tRNA and aminoacyl-tRNA in the two sites, it may be necessary that the active center of the peptidyltransferase rotate with respect to the sites. However, since the C-C-A(peptidyl) end of peptidyl-tRNA does not bind firmly to ribosomes, it may be possible for the 3'-terminal portion of peptidyl-tRNA to align identically with the catalytic center of the peptidyltransferase without rotation of the catalytic center. This may be possible because of free rotation around each of the bonds of the C-C-A(peptidyl) end. Analogously, although the C-C-A(amino acid) of aminoacyl-tRNA is the fixed moiety, the free rotation around each bond of this end may place the C-C-A(amino acid) from either 50 S site in the same stereochemical position with respect to the peptidyltransferase.

TABLE I
Inhibitors of Protein Synthesis

Supernatant	30 S (40 S)	50 S (60 S)
	In Prokaryotes	
Folic acid antagonists	Aminoglycosides	Chloramphenicol
Tosylphenylalanyl	Streptomycin	Macrolides
chloromethane	Dihydrostreptomycin	Niddamycin
	Paromomycin	Carbomycin
	Neomycin	Spiramycin III
	Kanamycin	Tylosin
	Gentamycin	Leukomycin
	Bluensomycin	Erythromycin
	Spectinomycin	Chalcomycin
	Kasugamycin	Oleandomycin
	Edeine	Lankamycin
	Edeine A	Methymycin
	Edeine B	Lincomycin
		Streptogramin A group
		Ostreogrycin A
		Synergistin A (PA114A)
		Streptogramin A
		Vernamycin A
		Mikamycin A
		Streptogramin B group
		Ostreogrycin B
		Synergistin B (PA114B),
		etc.
		Viridogrisein (etamycin)
		Thiostrepton group
		Thiostrepton
		(bryamycin, thiactin)
		Siomycin A
		(sporangiomycin,
		A-59)
		Althiomycin
		Micrococcin
		Bottromycin A_2
	In Both	
Aminoalkyl adenylates	Pactamycin	Puromycin
Guanylyl-5′-methylene	Aurintricarboxylic acid	4-Aminohexose pyrimidine
diphosphonate		nucleosides
Fusidic acid		Gougerotin
		Amicetin
		Blasticidin S
		Plicacetin
		Bamicetin
		Sparsomycin

(*Continued*)

TABLE I (*Continued*)

Supernatant	30 S (40 S)	50 S (60 S)
		Tetracyclines
		Chlortetracycline
	In Eukaryotes	
Diphtheria toxin		Glutarimides
		Cycloheximide
		(Actidione)
		Acetoxycycloheximide
		Streptovitacin A
		Ipecac alkaloids
		Emetine

in this place presents an opportunity for the reader to evaluate their suitability. It is hoped as well, however, that their presentation will stimulate some individuals to test the predictions of the two models.

Inhibitors of protein synthesis can be classified in many ways. In Table I, inhibitors are classified according to their site of inhibition in protein synthesis: supernatant, small subunit (30 S, 40 S) or large subunit (50 S, 60 S). In addition, in Table I the inhibitors are classified as to whether they inhibit protein synthesis in prokaryotes, eukaryotes or both. In a few cases which will be clarified in the text, the location of the inhibitors with respect to a given sector of the table is arbitrary and represents a best estimate from the present data. Tetracycline probably inhibits both the small and large subunit functions, for evidence suggests that tetracycline may not be a specific inhibitor of the 30 S subunit as has been previously considered. Functions associated with the large subunit may also be inhibited, at least at high tetracycline concentrations.

30 S Inhibitors

Aurintricarboxylic Acid

In cell-free extracts, aurintricarboxylic acid prevents attachment of mRNA to ribosomes and subunits of both pro- and eukaryotes (Table II). Accordingly, on inhibition of protein synthesis, subunits accumulate at the expense of polysomes. At high aurintricarboxylic acid concentrations, additional effects are observed. For specific inhibition of initiation, it is particularly important to adjust concentrations so that other reactions are not inhibited, for aurintricarboxylic acid can inhibit many reactions in protein synthesis because of its ability to bind to various proteins. Studies on chain initiation in mammalian cell-free systems have indicated that

TABLE II

EFFECT OF ANTIBIOTICS ON THE STEPS OF PROTEIN SYNTHESIS[a]

Antibiotic	Initiation: mRNA binding	Initiation: tRNA^Met binding	Codon recognition: AA-tRNA binding	Codon recognition: Mis-coding	Transpeptidation: Catalytic center	Transpeptidation: C-C-A(AA) binding	Transpeptidation: C-C-A(PEP) binding	Ab-normal 30–50 S, 40–60 S couples	Translocation: tRNA release	Translocation: Ribosome subunit exchange	Translocation: mRNA ribosome movement	Translocation: GTP hydrolysis	Translocation: Termination
Aurintricarboxylic acid	●	—											—
Streptomycin	—	●	●	●	○	—		●	—	—	—	—	○
Neomycin, kanamycin	—	—		●	○	○		●	○	○	—	—	○
Spectinomycin	—	—	○		—			—	○	○	—	—	
Kasugamycin	—	●	○		—			—		○			
Pactamycin	—	●	●		●					●			●
Tetracycline	—	○	○		—	—	—	●		●	—		○
Edeine	—	●	○		—		—	○					○
Puromycin	—	—			●	●	○						○
Gougerotin, amicetin	—	—			●	●	●			○	○		○
Chloramphenicol	—	—			●	●				●	○		
Sparsomycin	—	—			●	●	●				○		
Anisomycin					●	○	○						
Niddamycin, carbomycin					○	○	○		○				
Spiramycin III, tylosin					○	○	○		○				
Erythromycin					○	○	●		○	○			
Lincomycin			○		○	○	○		—	○			
Streptogramins A		●			●								
Thiostrepton			●		●						●	●	—
Micrococcin											●	—	—
Althiomycin											●	—	—
Cycloheximide		●	—						●		●	—	—
Emetine								●			●		

[a] Effect of antibiotics on various steps have been estimated from studies described in the text. In some cases it was conjectured from studies by closely related antibiotics or from genetic evidence. In many cases no experimental data were reported, and it was felt that a reasonable guess could not be made: these represent gaps in our knowledge and appear as empty places in the table. Obviously, adequate interpretation of the table requires reference to the appropriate section in the text. Most of the column headings are self-explanatory except, perhaps, for the following: tRNA^Met binding, refers to the binding of initiator tRNA to ribosomes; in bacteria this is fMet-tRNA_f^Met, in mammalian cells, Met-tRNA_f^Met or possibly even other species of initiator tRNA. Interference with binding by an antibiotic does not imply interference with rate of formation of the complex, for it may equally be due to destabilization of the complex after its formation. Binding of C-C-A(AA) and C-C-A(PEP) refer to interaction of the aminoacyl termini of peptidyl- and AA-tRNA with the substrate binding sites of the peptidyltransferase. Release of tRNA refers to release of deacylated tRNA species after transfer of its polypeptide to AA-tRNA. —, no effect; ○, small effect; or large effect at high antibiotic concentration; or effect may not be relevant to antibiotic action *in vivo*; ●, major effect.

initiation by natural mRNA is much less sensitive to aurintricarboxylic acid than is poly(U)-directed synthesis of polyphenylalanine.[5] In contrast, Huang and Grollman[6] have found reinitiation on native globin mRNA to be sensitive to low concentrations of aurintricarboxylic acid (0.01–0.1 mM). Their results with various sources of aurintricarboxylic acid suggest that some of the reported variability may be due to differences in the components present in the commercial preparations. Ordinarily intact cells are impermeable to aurintricarboxylic acid so that experiments on whole cells are precluded.

Streptomycin

Binding of streptomycin occurs to a single site on the 30 S subunit containing a structure including protein S12. The K_{Diss} for the complex of dihydrostreptomycin at 25° with 70 S ribosomes from K12 strains of *Escherichia coli* is 94 nM.[7] Streptomycin inhibits chain elongation probably by interferring with aminoacyl-tRNA binding to ribosomes. Sometime after polypeptide synthesis has ceased, peptidyl-tRNA is slowly released from ribosomes, which themselves are released from mRNA as subunits or 70 S monomers.[8,9] After release the subunits reassociate into 70 S monomers which cannot dissociate as readily. These streptomycin monosomes contain mRNA and are 70 S ribosomes irreversibly inactivated by streptomycin; they are inactive in both cells and extracts, and may be aberrant initiation complexes.[10,11] Streptomycin also stimulates miscoding which occurs in cells and cell-free extracts. At high concentrations, streptomycin can more directly interfere with translocation and peptide bond formation. The precise causes of lethality are still uncertain, but are related to its irreversible binding to ribosomes. Perhaps, the aberrant initiation complexes and a cyclic dynamic blockade of polyribosomes[12] can explain both the results in cell-free systems and the lethality of the antibiotic.

Neomycin and Kanamycin

These two aminoglycoside antibiotics also interact with 30 S subunits, probably at several sites. They produce extensive miscoding, which is due

[5] M. B. Mathews, *FEBS* (*Fed. Eur. Biochem. Soc.*) *Lett.* **15**, 201 (1971).

[6] M. T. Huang and A. P. Grollman, *Mol. Pharmacol.* **8**, 111 (1972).

[7] F. N. Chang and J. G. Flaks, *Antimicrob. Ag. Chemother.* **2**, 294 (1972).

[8] J. Modolell and B. D. Davis, *Nature* (*London*) **224**, 345 (1969).

[9] M. Kogut and E. Prizant, *FEBS* (*Fed. Eur. Biochem. Soc.*) *Lett.* **12**, 17 (1970).

[10] L. Luzzato, D. Apirion, and D. Schlessinger, *J. Mol. Biol.* **42**, 315 (1969).

[11] J. C. LeLong, M. A. Cousin, D. Gros, M. Grunberg-Manago, and F. Gros, *Biochem. Biophys. Res. Commun.* **42**, 530 (1971).

[12] P. F. Sparling and B. D. Davis, *Antimicrobial Ag. Chemother.* **1**, 252 (1972).

to deoxystreptamine or streptamine moieties. In the case of kanamycin, and probably neomycin, the S12 protein is involved in its inhibitory action. Inactive 60 S monosomes form in the presence of neomycin. They inhibit peptide bond formation. Recently, neomycin has also been shown to inhibit peptide bond formation on eukaryotic ribosomes at high concentration.[13] These antibiotics like streptomycin are bactericidal. It is of note that neither neomycin nor kanamycin interferes significantly with binding of dihydrostreptomycin to ribosomes.[7]

Spectinomycin

Spectinomycin is bacteriostatic, not bactericidal as are some other aminoglycosides. Its effects are reversible. Protein synthesis is inhibited through interaction with a region of the 30 S subunit involving protein S5.[14] Inhibition of protein synthesis by spectinomycin requires the presence of cytidylic or guanylic acid residues in synthetic templates; it may interfere with mRNA · 30 S subunit interactions occurring during translocation. Spectinomycin stabilizes polyribosomes; and the dominance of sensitivity over resistance in heterozygotes suggests that in the presence of spectinomycin sensitive ribosomes block the movement of resistant ribosomes along the message.[12] It does not inhibit peptide bond formation, AA-tRNA binding, initiation, or termination reactions (Table II). Nor does it produce miscoding.

Kasugamycin

Kasugamycin, like spectinomycin, is bacteriostatic, and its effects are reversible. Protein synthesis is inhibited through interaction with the 30 S subunit. Kasugamycin inhibits initiation complex formation on 30 S subunits as well as on 70 S ribosomes.[15,16] Although enzymatic and non-enzymatic binding of Phe-tRNA is inhibited by kasugamycin, aminoacyl-tRNA binding is not as sensitive to the antibiotic as is fMet-tRNA binding. The antibiotic does not produce miscoding (Table II) in cell-free extracts; nor does it inhibit peptide bond synthesis.[17] Strains of *Escherichia coli* resistant to kasugamycin have been found to be deficient in methylation of 16 S ribosomal RNA.[18] Methylation of 16 S RNA in cell-free extracts

[13] J. A. Schneider and E. Maxwell, *Biochemistry* 12, 475 (1973).

[14] A. Bollen, J. Davies, M. Ozaki, and S. Mizushima, *Science* 165, 85 (1969).

[15] A. Okuyama, N. Machiyama, T. Kinoshita, and N. Tanaka, *Biochem. Biophys. Res. Commun.* 43, 196 (1971).

[16] T. L. Helser and J. E. Davies, *Bacteriol. Proc.* P82 (1971).

[17] S. Pestka, *J. Biol. Chem.* 247, 4669 (1972).

[18] T. L. Helser, J. E. Davies, and J. E. Dahlberg, *Nature (London) New Biol.* 233, 12 (1971).

followed by reconstitution into 30 S particles converts 16 S RNA from kasugamycin-resistant strains to 16 S RNA, which behaves similarly to 16 S RNA from strains sensitive to the antibiotic.[19] It was concluded that the kasugamycin A locus involved in these mutations to resistance corresponds to an RNA methylase.

Pactamycin

Pactamycin inhibits initiation in eukaryotes by interferring with appropriate functional attachment of initiator tRNA to the initiation complex (Table II).[20-25] It binds to 30 S and 40 S subunits and alters the structure of the initiation complex causing destabilization and dissociation. At higher concentrations, inhibition of translocation occurs. In intact cells and extracts, conversion of polysomes to monoribosomes occurs rapidly as proteins are completed in the presence of pactamycin. In intact reticulocytes and reticulocyte lysates, pactamycin has been shown to be an effective inhibitor of initiation at concentrations which permit completion and release of globin chains.[21,22] By following the differential incorporation of initiator fMet-tRNA$_f^{Met}$ and internal amino acids,[26] it was confirmed that pactamycin can specifically block initiation in reticulocyte lysates. In extracts of *E. coli,* although initiation reactions can be inhibited,[27] concentrations which inhibit initiation are so close to those which inhibit elongation that it has been questioned whether the antibiotic can be useful as a specific inhibitor of initiation.[23] Pactamycin can be used in studies with intact cells as well as cell-free extracts. It has been used effectively to determine the gene order of certain RNA viruses infecting animal cells.[28-30]

[19] T. L. Helser, J. E. Davies, and J. E. Dahlberg, *Nature (London) New Biol.* **235**, 6 (1972).

[20] I. H. Goldberg, M. L. Stewart, M. Ayuso, and L. Kappen, *Fed. Proc., Fed. Amer. Soc. Exp. Biol.* **32**, 1688 (1973).

[21] J. S. MacDonald and I. H. Goldberg, *Biochem. Biophys. Res. Commun.* **41**, 1 (1970).

[22] M. Stewart-Blair, I. Yanowitz, and I. H. Goldberg, *Biochemistry* **10**, 4198 (1971).

[23] M. L. Stewart and I. H. Goldberg, *Biochim. Biophys. Acta* **294**, 123 (1973).

[24] M. Ayuso and I. H. Goldberg, *Biochim. Biophys. Acta* **294**, 118 (1973).

[25] L. S. Kappen, H. Suzuki, and I. H. Goldberg, *Proc. Nat. Acad. Sci. U.S.* **70**, 22 (1973).

[26] H. F. Lodish, D. Housman, and M. Jacobsen, *Biochemistry* **10**, 2348 (1971).

[27] L. B. Cohen, A. E. Herner, and I. H. Goldberg, *Biochemistry* **8**, 1312 (1969).

[28] D. F. Summers and J. V. Maizel, *Proc. Nat. Acad. Sci. U.S.* **68**, 2852 (1971).

[29] R. Taber, D. Rekosh, and D. Baltimore, *J. Virol.* **8**, 395 (1971).

[30] B. E. Butterworth and R. R. Rueckert, *Fed. Proc., Fed. Amer. Soc. Exp. Biol.* **31**, 407 (1972).

Tetracyclines

Tetracycline inhibits binding of aminoacyl-tRNA to the ribosomal acceptor site in extracts and in intact cells.[31,32] At low tetracycline concentrations the amount of polysomes is diminished. At higher tetracycline concentrations, polysomes appear to be increased in size[33]; this may be a consequence of the ability of the tetracyclines to inhibit termination[34,35] as well as aminoacyl-tRNA binding to ribosomes. It is possible to demonstrate other effects on various steps of protein synthesis: inhibition of initiation and of peptide bond formation. Tetracycline may bind preferentially to 30 S subunits although binding to both 30 S and 50 S subunits is found.[36–38] Recent evidence suggests that high concentrations of tetracycline probably inhibit 50 S functions, such as binding of aminoacyl-oligonucleotides to ribosomes.[39] Tetracycline inhibits termination of protein synthesis in cell-free extracts as measured by release of fMet from fMet-tRNA bound to ribosomes or by release of the aminoterminal coat protein fragment of R17-RNA.[34,35] At 0.5 mM tetracycline, there is greater than 90% inhibition of termination.

Edeine

Edeine is a potent inhibitor of protein synthesis in bacterial cell-free extracts. It inhibits aminoacyl-tRNA and fMet-tRNA binding to the donor site at 10–20 mM Mg^{2+}, but to both sites at lower Mg^{2+}.[40] Thus, at high Mg^{2+}, edeine and tetracycline are complementary in that they inhibit donor and acceptor sites, respectively. Edeine inhibits binding of aminoacyl-tRNA to 30 S subunits almost completely at low and high Mg^{2+}. Tu-dependent binding of Phe-tRNA to 70 S ribosomes is inhibited only if no initiator tRNA is bound to ribosomes.[41] The antibiotic stabilizes the association of subunits in 70 S ribosomes and binds to both 30 and 50 S subunits.

[31] G. Suarez and D. Nathans, *Biochem. Biophys. Res. Commun.* **18**, 743 (1965).
[32] M. Hierowski, *Proc. Nat. Acad. Sci. U.S.* **53**, 594 (1965).
[33] E. Cundliffe, *Mol. Pharmacol.* **3**, 401 (1967).
[34] Z. Vogel, A. Zamir, and D. Elson, *Biochemistry* **8**, 5161 (1969).
[35] M. R. Capecchi and H. A. Klein, *Cold Spring Harbor Symp. Quant. Biol.* **34**, 469 (1969).
[36] L. E. Day, *J. Bacteriol.* **91**, 1917 (1966).
[37] R. E. Connamacher and H. G. Mandel, *Biochim. Biophys. Acta* **166**, 475 (1968).
[38] I. H. Maxwell, *Mol. Pharmacol.* **4**, 25 (1968).
[39] S. Pestka and R. Harris, in preparation.
[40] W. Szer and Z. Kurylo-Borowska, *Biochim. Biophys. Acta* **224**, 477 (1970).
[41] W. Szer and Z. Kurylo-Borowska, *Biochim. Biophys. Acta* **259**, 357 (1972).

50 S Inhibitors

Puromycin

Puromycin, through its resemblance to the aminoacyl-adenylyl end of aminoacyl-tRNA, inhibits protein synthesis by competing with aminoacyl-tRNA for this site on the 50 S subunit of ribosomes. In place of aminoacyl-tRNA, it accepts nascent peptides causing premature release of incomplete polypeptide chains. As a result of these actions, puromycin produces breakdown of polysomes. The K_m for puromycin depends on the particular cell-free system used. When polyribosomes are used and the incorporation of [³H]puromycin is examined, the K_m for puromycin on bacterial and rat liver polyribosomes are 4 μM and 8 μM, respectively.[17,42,43] When fMet- or acetyl-phenylalanyl-puromycin is the product being studied, the K_m for puromycin is approximately 0.2 mM.[44,45]

There are many useful analogs of puromycin which function well in cell-free systems or in intact cells. Substitution of the amino group of puromycin with an hydroxyl group permits puromycin to function and the products on the ribosomes are peptidyl esters.[44] The K_m for hydroxypuromycin is about two orders of magnitude higher than that of puromycin in comparable systems. Studies on aminoacyl-oligonucleotides, which represent the termini of aminoacyl-tRNA's, have also proved valuable. Studies with C-A-C-C-A(Phe), C-A-C-C-A(Ac-Phe), and similar derivatives have indicated that stable binding of these compounds requires a free amino group.[46] The dissociation constant for binding of C-A-C-C-A-(Phe) to ribosomes was reported to be 36 nM in the presence of 20% ethanol.[47] Nascent peptides can be transferred to C-A-C-C-A(Phe) and similar aminoacyl-oligonucleotides as to puromycin.[48] Recently the synthesis of an interesting carbocyclic analog of puromycin has been described.[49] The compound contains a cyclopentane carbon skeleton replacing the ribose ring. It competes with puromycin in peptidylpuromycin

[42] S. Pestka, *Proc. Nat. Acad. Sci. U.S.* **69**, 624 (1972).

[43] S. Pestka, R. Goorha, H. Rosenfeld, C. Neurath, and H. Hintikka, *J. Biol. Chem.* **247**, 4258 (1972).

[44] S. Fahnestock, H. Neumann, V. Shashoua, and A. Rich, *Biochemistry* **9**, 2477 (1970).

[45] S. Pestka, *Arch. Biochem. Biophys.* **136**, 80 (1970).

[46] T. Hishizawa, J. L. Lessard, and S. Pestka, *Proc. Nat. Acad. Sci. U.S.* **66**, 523 (1970).

[47] R. J. Harris and S. Pestka, *J. Biol. Chem.* **248**, 1168 (1973).

[48] J. L. Lessard and S. Pestka, *J. Biol. Chem.* **247**, 6901 (1972).

[49] S. Daluge and R. Vince, *J. Med. Chem.* **15**, 171 (1972).

synthesis on polyribosomes almost as well as puromycin, itself. The K_i for its inhibition of peptidylpuromycin synthesis is 10 μM.[50]

Although the mode of action of puromycin is one of the best understood of the antibiotics, there are some uncertainties centered around it. Is release of the nascent chain from the ribosome after reaction with puromycin dependent on translocation? Can ribosomes continue translation internally within a cistron after puromycin releases the nascent chain? Are ribosomes released from polysomes at the position where the nascent chain is released or must they travel to a terminator codon before release? Is a supernatant factor required for release of ribosomes from mRNA after the puromycin reaction? Thus, although the action of puromycin is understood, many questions still revolve around it.

4-Aminohexose Pyrimidine Nucleosides

Amicetin, gougerotin, and blasticidin S inhibit transpeptidation reactions (Table II). They can inhibit binding of C-A-C-C-A(Phe)[47] and stabilize binding of C-A-C-C-A(Ac-Leu)[51] to ribosomes from E. coli. They may also inhibit some translocation related events for polysomes are stabilized by gougerotin and amicetin.[52,53] Binding of blasticidin S to E. coli ribosomes is localized to the 50 S subunit and was not inhibited by puromycin.[54] Although in general these antibiotics are similar in action, they differ specifically in various systems. Thus, amicetin does not inhibit transpeptidation on rat liver polysomes as well as does gougerotin stabilize binding of C-A-C-C-A(Ac-Leu)[51] to ribosomes from E. coli. very effectively.[52] The inhibition of peptidyl-[³H]puromycin by these antibiotics seems to show two phases, a noncompetitive and a competitive phase.[17,55] The significance of this is discussed in the section on chloramphenicol, below.

Chloramphenicol

Chloramphenicol binds specifically to 50 S subunits.[56] It inhibits functional attachment of the aminoacyl end of aminoacyl-tRNA to the 50 S

[50] S. Pestka, R. Vince, S. Daluge, and R. J. Harris, Antimicrobial. Ag. Chemother. 4, 37 (1973).
[51] M. Celma, R. E. Monro, and D. Vazquez, FEBS (Fed. Eur. Biochem. Soc.) Lett. 6, 273 (1970).
[52] H. L. Ennis, Antimicrobial. Ag. Chemother. 1, 197 (1972).
[53] S. R. Casjens and A. J. Morris, Biochim. Biophys. Acta 108, 677 (1965).
[54] T. Kinoshita, N. Tanaka, and H. Umezawa, J. Antibiot. 23, 288 (1970).
[55] S. Pestka, H. Rosenfeld, R. J. Harris, and H. Hintikka, J. Biol. Chem. 247, 6895 (1972).
[56] D. Vazquez, Biochim. Biophys. Acta 114, 277 (1965).

subunit, thus inhibiting transpeptidation, but may also have a direct inhibitory effect on the peptidyltransferase (Table II).[17,45,57] The antibiotic decreases the rate of polysome breakdown, but ribosomes can slowly progress along mRNA in the absence of peptide bond formation.[58] Mitochondrial protein synthesis as well as bacterial protein synthesis is inhibited by chloramphenicol. The peculiarity of the inhibition of transpeptidation by chloramphenicol does not seem consistent with the acceptor–donor site model of ribosome function as currently promulgated. When peptidyl-puromycin synthesis was examined with polyribosomes, the reaction seemed to show two phases of inhibition by chloramphenicol (Fig. 4) suggesting that peptidyl-tRNA existed in two states on the ribosome with respect to sensitivity to chloramphenicol and its ability to react with puromycin.[17] Similar observations have been made with the 4-aminohexose pyrimidine nucleoside antibiotics with the use of bacterial and mammalian polyribosomes.[17,55] In the case of rat liver polyribosomes, the 4-aminohexose pyrimidine nucleosides as well as cycloheximide show two phases of inhibition of the reaction.[55] The model of ribosome function shown in Fig. 3 was presented to explain these observations.

The two modes of inhibition by chloramphenicol may mean there are

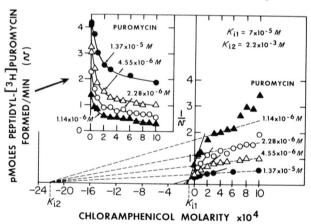

FIG. 4. Kinetics of chloramphenicol inhibition of peptidyl-[³H]puromycin formation on native polyribosomes. Chloramphenicol concentration was varied as indicated on the abscissa at four different puromycin concentrations. Polyribosomes were added last to start the reactions. The data of the left panel are plotted and calculated as a Dixon plot in the right panel. ●, $1.37 \times 10^{-5} M$ puromycin; △, $4.55 \times 10^{-6} M$ puromycin; ○, $2.28 \times 10^{-6} M$ puromycin; ▲, $1.14 \times 10^{-6} M$ puromycin.

[57] S. Pestka, Proc. Nat. Acad. Sci. U.S. 64, 709 (1969).
[58] C. Gurgo, D. Apirion, and D. Schlessinger, FEBS (Fed. Eur. Biochem. Soc.) Lett. 3, 34 (1969).

two sites for chloramphenicol action or perhaps two classes of ribosomal states amenable to chloramphenicol inhibition. Equilibrium dialysis, in fact, suggests that there are two ribosomal sites for binding of chloramphenicol to washed ribosomes.[48] In any case, the inhibition of peptidyl-puromycin synthesis on polyribosomes by chloramphenicol appears to be asymmetric with respect to peptidyl-tRNA and puromycin. About half of the peptidyl-tRNA which is available for reaction with puromycin is in a state where chloramphenicol inhibits the reaction competitively with respect to puromycin; the other half of peptidyl-tRNA reacts with puromycin in such a way that the inhibition by chloramphenicol is noncompetitive (Fig. 4).

In the context of the current donor–acceptor model (Fig. 2), no clearly discernible mechanism is evident to explain asymmetry of inhibition of peptidyl-puromycin synthesis by antibiotics. If peptidyl-tRNA is in the donor site, it must by definition react with puromycin; if it is on the acceptor site, it cannot react with puromycin. The present evidence points to the existence of peptidyl-tRNA in two distinct states, each of which can react with puromycin to form peptidyl-puromycin. This is inconsistent with the current model (Fig. 2), but it is consistent with the model illustrated in Fig. 3, where these antibiotics may interfere with one 50 S site better than the other. Other alternative explanations, however, may also be possible. Since the two models (as illustrated in Figs. 2 and 3) make different predictions in many cases, it should be possible to distinguish them experimentally.

Sparsomycin

Sparsomycin is one of the most effective inhibitors of peptide bond formation with both 70 S and 80 S ribosomes (Tables I and II). It prevents breakdown of polysomes. At low concentrations sparsomycin inhibits C-A-C-C-A(Phe) binding to ribosomes,[47,57,59] but at high concentration sparsomycin stimulates this binding,[47,59] similarly to its stimulation of C-A-C-C-A(Ac-Phe) or U-A-C-C-A(Ac-Leu) binding to ribosomes.[46,60] Sparsomycin also stimulates the formation of inert complexes between the peptidyl end of peptidyl-tRNA and ribosomes.[61] In all cases studied, sparsomycin is a competitive inhibitor of the puromycin reaction whether the synthetic model systems or native polyribosomes are used to study transpeptidation.[17,62] Although the precise molecular mechanism needs to be elucidated, it is clear that in both pro- and eukaryotes sparsomycin is

[59] M. Yukioka and S. Morisawa, *Biochim. Biophys. Acta* **254**, 304 (1971).
[60] R. E. Monro, M. L. Celma, and D. Vazquez, *Nature (London)* **222**, 356 (1969).
[61] A. E. Herner, I. H. Goldberg, and L. B. Cohen, *Biochemistry* **8**, 1335 (1969).
[62] I. H. Goldberg and K. Mitsugi, *Biochemistry* **6**, 383 (1967).

an effective inhibitor of transpeptidation. The K_i for its inhibition of peptidyl-puromycin synthesis is about 0.2 μM with bacterial polyribosomes[17] and about 1 μM with polyribosomes from rat liver.[55] Other partial reactions of protein synthesis which have been studied do not appear to be sensitive to sparsomycin except for release or termination,[34,35,63] which is generally inhibited by agents interfering with transpeptidation in model systems.

Anisomycin

Anisomycin inhibits peptide bond synthesis in eukaryotes only (Tables I and II). The antibiotic stabilizes polyribosomes.[64] Anisomycin is a competitive inhibitor of puromycin with an inhibition constant, K_i, equal to 12 μM as determined with polyribosomes from rat liver.[55] It also inhibits release of methionine from Met-tRNA[fMet] in a model system for termination with reticulocyte ribosomes.[65]

Macrolides

Macrolides bind to a common region on 50 S subunits.[66-69] It is difficult to make many generalizations about the mode of action of the macrolide antibiotics. Their influence on peptidyl-transfer is highly dependent on the substrates and assays used.[70-73] Very often they stimulate, rather than inhibit peptidyl transfer in model systems with washed ribosomes and diverse synthetic donors. In whole cells, however, it is unlikely that these macrolide antibiotics can inhibit transpeptidation on polyribosomes.[17] In fact, studies with native polyribosomes indicate that none of the macrolides can in fact inhibit peptidyl-puromycin synthesis even though model reactions in synthetic systems with washed ribosomes and synthetic donors are very strongly inhibited by some of these antibiotics. Direct effects of the macrolides on initiation through interaction with free 50 S subunits have not been excluded nor in fact included as a major mode of action.

[63] E. Scolnick, R. Tompkins, C. T. Caskey, and M. Nirenberg, *Proc. Nat. Acad. Sci. U.S.* 61, 768 (1968).
[64] A. P. Grollman, *J. Biol. Chem.* 242, 3226 (1967).
[65] A. L. Beudet and C. T. Caskey, *Proc. Nat. Acad. Sci. U.S.* 68, 619 (1971).
[66] J. M. Wilhelm and J. W. Corcoran, *Biochemistry* 6, 2578 (1967).
[67] J. C.-H. Mao and M. Putterman, *J. Mol. Biol.* 44, 347 (1969).
[68] D. Vazquez, *Life Sci.* 6, 845 (1967).
[69] B. Weisblum, C. Siddhikol, C. J. Lai, and V. Demohn, *J. Bacteriol.* 106, 835 (1971).
[70] J. C.-H. Mao and E. E. Robishaw, *Biochemistry* 11, 4864 (1972).
[71] J. C.-H. Mao and E. E. Robishaw, *Biochemistry* 10, 2054 (1971).
[72] H. Teraoka and K. Tanaka, *Biochim. Biophys. Acta* 232, 509 (1971).
[73] J. Cerna, I. Rychlik, and P. Pulkrabek, *Eur. J. Biochem.* 9, 27 (1969).

It is very likely that the specific actions of the various macrolides will differ, for in intact protoplasts and in model systems they show different effects. For example, erythromycin stabilizes polyribosomes in protoplasts whereas spiramycin produces their rapid breakdown.[74,75] Some unusual effects with tylosin, spiramycin, and erythromycin have been noted. Their inhibition of reactions appears to reach a plateau at partial inhibition.[45,47,70,76] Perhaps, this effect is an indication of ribosomal heterogeneity although other explanations may be invoked.

Bacterial mutants resistant to erythryomycin were shown to have 50 S subunits with an altered affinity for the macrolides, and an altered 50 S protein was identified in the erythromycin-resistant strains.[77,78] In addition the phenomenon of inducible resistance to the macrolides has been described. When sensitive cells of *Staphylococcus aureus* are explosed to a subinhibitory concentration of erythromycin (10^{-8} to 10^{-7} M) in an otherwise complete medium for growth, the cells become resistant to macrolides, lincosamides, and to streptogramin-type B antibiotics.[69,79] In one strain of *S. aureus*, N^6-dimethyladenine, which is not normally present in 23 S rRNA, is found after resistance has been induced by erythromycin.[80] Ribosome reconstitution experiments have shown that the antibiotic sensitivities of the reconstituted ribosomes depend on the source of the 23 S rRNA; 23 S rRNA containing N^6-dimethyladenine confers upon the reconstituted ribosomes resistance to the above antibiotics.[81]

Lincosamides

Lincomycin binds to 50 S subunits.[82] In model systems such as fMet-puromycin formation, lincomycin inhibits peptide bond synthesis. With washed ribosomes the bound lincomycin interferes with proper positioning of the acceptor (and possibly donor) substrates, namely peptidyl- and aminoacyl-tRNA, so that peptide bond synthesis is interdicted and binding of C-A-C-C-A(Phe) to ribosomes is inhibited.[47] In the intact cell,

[74] E. Cundliffe and K. McQuillen, *J. Mol. Biol.* 30, 137 (1967).
[75] E. Cundliffe, *Biochemistry* 8, 2063 (1969).
[76] M. L. Celma, R. E. Monro, and D. Vazquez, *FEBS (Fed. Eur. Biochem. Soc.) Lett.* 13, 247 (1970).
[77] K. Tanaka, H. Teraoka, M. Tamaki, E. Otaka, and S. Osawa, *Science* 162, 576 (1968).
[78] E. Otaka, H. Teraoka, M. Tamaki, K. Tanaka, and S. Osawa, *J. Mol. Biol.* 48, 499 (1970).
[79] H. Hashimoto, H. Oshima, and S. Mitsuhashi, *Jap. J. Microbiol.* 12, 321 (1968).
[80] C. J. Lai and B. Weisblum, *Proc. Nat. Acad. Sci. U.S.* 68, 856 (1971).
[81] C. J. Lai, B. Weisblum, S. R. Fahnestock, and M. Nomura, *J. Mol. Biol.* 74, 67 (1973).
[82] F. N. Chang and B. Weisblum, *Biochemistry* 6, 836 (1967).

however, lincomycin probably does not interfere with transpeptidation, for it is clear that transpeptidation on native polyribosomes is not inhibited by lincomycin.[17] In addition, the rapid breakdown of polyribosomes in the presence of lincomycin[75] indicates that lincomycin inhibits initiation or peptide bond formation only when the peptide chains are small. Like other inhibitors of fMet-puromycin synthesis, lincomycin inhibits termination as measured by release of formylmethionine from fMet-tRNA.[34]

Streptogramin Group

Streptogramin A antibiotics bind tightly to 50 S subunits. The dissociation constant for the binding of vernamycin A to ribosomes is $2 \times 10^{-8} M$.[83] The binding of the streptogramin A group is enhanced by the streptogramin B antibiotics. With washed ribosomes and synthetic donors, streptogramin A compounds inhibit transpeptidation, such as fMet- or Ac-Phe-puromycin synthesis.[45,84] The streptogramin A antibiotics also are potent inhibitors of termination as measured by fMet release from fMet-tRNA.[34] They are also strong inhibitors of C-A-C-C-A(Phe) binding to ribosomes.[47] However, they do not inhibit protein synthesis readily after protein synthesis has begun,[85,86] nor do they inhibit peptidyl-puromycin synthesis on polyribosomes.[17] They produce a rapid decrease in number of polysomes with concomitant increase in number of free subunits and ribosomes.[75] It is likely they inhibit initiation or some early events in protein synthesis. Inhibition of fMet-tRNA binding to ribosomes by vernamycin A has been reported.[87] Streptogramin B antibiotics may inhibit protein synthesis by fixing peptidyl- or aminoacyl-tRNA ends onto 50 S subunits, thus interferring with the necessary coordinated movements of these moieties on the ribosomes.

Thiostrepton Group

Thiostrepton, siomycin A, thiopeptin, and multhiomycin are similar antibiotics. Thiostrepton and siomycin A bind to 50 S ribosomal subunits.[88,89] All these antibiotics are strong inhibitors of translocation through inhibition of ribosomal functions, an integral part of the translocational

[83] H. L. Ennis, *Biochemistry* **10**, 1265 (1971).
[84] R. E. Monro and D. Vazquez, *J. Mol. Biol.* **28**, 161 (1967).
[85] H. L. Ennis, *Mol. Pharmacol.* **2**, 543 (1966).
[86] H. Yamaguchi and N. Tanaka, *J. Biochem.* (*Tokyo*) **61**, 18 (1967).
[87] H. L. Ennis and K. E. Duffy, *Biochem. Biophys. Acta* **281**, 93 (1972).
[88] B. Weisblum and V. Demohn, *J. Bacteriol.* **101**, 1073 (1970).
[89] K. Tanaka, S. Watanabe, H. Teraoka, and M. Tamaki, *Biochem. Biophys. Res. Commun.* **39**, 1189 (1970).

events.[89-94] They inhibit ribosomal elongation factor G-dependent GTP hydrolysis as well as formation of factor G·GTP complexes with ribosomes. These antibiotics do not inhibit transpeptidation.[17] They do, however, also inhibit GTPase activity of elongation factor Tu and the enzymatic binding of aminoacyl-tRNA to ribosomes.[94-100] In the intact cell it appears that inhibition of aminoacyl-tRNA binding by these antibiotics is predominant over inhibition of translocation.[99,100] The studies with this group of antibiotics is a good example of the contribution inhibitory agents can make in understanding protein synthesis. The observation that these antibiotics affecting ribosomal function inhibited both EF G and EF Tu function led to the concept that these elongation factors shared a ribosomal site. This suggestion has been confirmed by the demonstration of competition between EF Tu and EF G functions.[101-104]

Micrococcin

The antibiotics micrococcin and micrococcin P appear to be identical. The antibiotic is inhibitory chiefly to gram-positive microorganisms. Although the complete structure of micrococcin is not known, the structures of several of the component fragments have been determined with the suggestion that the antibiotic has some structural resemblances to the thiostrepton group. The antibiotic inhibits protein synthesis in cell-free extracts from *Escherichia coli*. It appears to inhibit translocation through its interaction with the ribosome rather than through the soluble elonga-

[90] S. Pestka, *Biochem. Biophys. Res. Commun.* **40**, 667 (1970).
[91] B. Weisblum and V. Demohn, *FEBS (Fed. Eur. Biochem. Soc.) Lett.* **11**, 149 (1970).
[92] S. Pestka and N. Brot, *J. Biol. Chem.* **246**, 7715 (1971).
[93] J. W. Bodley, L. Lin, and J. Highland, *Biochem. Biophys. Res. Commun.* **41**, 1406 (1970).
[94] J. Modolell, B. Cabrer, A. Parmeggiani, and D. Vazquez, *Proc. Nat. Acad. Sci. U.S.* **68**, 1796 (1971).
[95] T. Kinoshita, Y.-F. Liou, and N. Tanaka, *Biochem. Biophys. Res. Commun.* **44**, 859 (1971).
[96] H. Weissbach, B. Redfield, E. Yamasaki, R. C. Davis, S. Pestka, and N. Brot, *Arch. Biochem. Biophys.* **149**, 110 (1972).
[97] T. Tanaka, K. Sakaguchi, and H. Yonehara, *J. Biochem. (Tokyo)* **69**, 1127 (1971).
[98] S. Watanabe, *J. Mol. Biol.* **67**, 443 (1972).
[99] E. Cundliffe, *Biochem. Biophys. Res. Commun.* **44**, 912 (1971).
[100] M. Cannon and K. Burns, *FEBS (Fed. Eur. Biochem. Soc.) Lett.* **18**, 1 (1971).
[101] N. Richman and J. Bodley, *Proc. Nat. Acad. Sci. U.S.* **69**, 686 (1972).
[102] B. Cabrer, D. Vazquez, and J. Modolell, *Proc. Nat. Acad. Sci. U.S.* **69**, 733 (1972).
[103] D. L. Miller, *Proc. Nat. Acad. Sci. U.S.* **69**, 752 (1972).
[104] D. Richter, *Biochem. Biophys. Res. Commun.* **46**, 1850 (1972).

tion factor G,[92] but it does not inhibit GTPase activity associated with ribosomes and elongation factor G. It does not inhibit other steps in protein synthesis which have been measured such as enzymatic or nonenzymatic aminoacyl-tRNA binding to ribosomes, transpeptidation, and GTP hydrolysis associated with either EF G or EF Tu.[17,92] It is likely that micrococcin is inhibiting translocation through its interaction with the ribosome in such a way that GTP hydrolysis is uncoupled from the other translocational events. In this respect micrococcin differs from the thiostrepton group of antibiotics. As a consequence of its inhibition of translocation, polyribosomes are stabilized by micrococcin. Consistent with these results in cell-free extracts, genetic studies have determined that resistance and sensitivity to micrococcin map close to the genetic markers for other antibiotics which inhibit ribosomal function.[105]

Glutarimide Antibiotics

The glutarimide antibiotics interact with 60 S, but not 50 S, ribosomal subunits.[106,107] In reticulocyte lysates cyclohexamide interferes with chain elongation at concentrations where initiation is not blocked.[108] In addition, studies on native polyribosomes have indicated that cycloheximide can inhibit transpeptidation.[13,55] Whether this is a major site of action of the antibiotic, however, has not been definitively established. It has also been suggested that cycloheximide inhibits translocation.[109,110] Additionally, in cell-free extracts it has been suggested that the glutarimide antibiotics interfere with chain initiation.[110]

Ipecac Alkaloids

Emetine and similar active alkaloids are effective inhibitors of protein synthesis in eukaryotes. Emetine does not prevent transpeptidation[55] or translocation of peptidyl-tRNA from acceptor to donor sites[111]; however, it does stabilize ribosomes on polysomes,[112] and perhaps, the resultant 80 S·emetine complexes can no longer carry out aminoacyl-tRNA binding or some translocational functions. Emetine inhibits elongation rather than initiation.[108] Another alkaloid cryptopleurine appears to be an in-

---

[105] D. Dubnau, C. Goldthwaite, I. Smith, and J. Marmur, *J. Mol. Biol.* **27**, 163 (1967).

[106] S. S. Rao and A. P. Grollman, *Biochem. Biophys. Res. Commun.* **29**, 696 (1967).

[107] M. R. Siegel and H. D. Sisler, *Biochim. Biophys. Acta* **103**, 558 (1965).

[108] H. F. Lodish, *J. Biol. Chem.* **246**, 7131 (1971).

[109] B. S. Baliga, A. W. Pronezuk, and H. N. Munro, *J. Biol. Chem.* **244**, 4480 (1969).

[110] T. G. Obrig, W. J. Culp, W. L. McKeehan, and B. Hardesty, *J. Biol. Chem.* **246**, 174 (1971).

[111] B. S. Baliga, S. A. Cohen, and H. N. Munro, *FEBS* (*Fed. Eur. Biochem. Soc.*) *Lett.* **8**, 249 (1970).

[112] A. P. Grollman, *J. Biol. Chem.* **243**, 4089 (1968).

hibitor of transpepdiation, particularly in yeast cell-free systems.[55,113] Some cryptopleurine-resistant mutants are being examined in yeast.[113] Mitochondrial protein synthesis as well as cytoplasmic protein synthesis of rat liver is inhibited by emetine.[114]

Inhibitors of Supernatant Factors

Fusidic Acid

Fusidic acid inhibits translocation in both pro- and eukaryotes through inhibition of elongation factor G or T2 activity, respectively.[115–123] In the presence of fusidic acid the ribosome·EF G·GDP complex is stabilized. Thus, fusidic acid inhibits catalytic GTP hydrolysis by preventing the dissociation of the GDP once the complex has formed so that EF G cannot turn over. Fusidic acid-resistant G-factor has been reported.[124,125] Radioactive fusidic acid binds to ribosomes in the presence of EF G.[126] Because of the common ribosomal sites of interaction of EF G and EF Tu, fusidic acid secondarily inhibits aminoacyl-tRNA binding to ribosomes.[127]

Tosylphenylalanyl chloromethane

Tosylphenylalanyl chloromethane irreversibly inhibits bacterial elongation factor T.[128] Neither ribosomes nor EF G are inhibited at concentrations which inhibit EF T function.

[113] J. E. Davies, personal communication.

[114] P. S. Lietman, Mol. Pharmacol. 7, 122 (1971).

[115] N. Tanaka, T. Kinoshita, and H. Masukawa, Biochem. Biophys. Res. Commun. 30, 278 (1968).

[116] S. Pestka, Proc. Nat. Acad. Sci. U.S. 61, 726 (1968).

[117] N. Brot, R. Ertel, and H. Weissbach, Biochem. Biophys. Res. Commun. 31, 563 (1968).

[118] R. W. Erbe and P. Leder, Biochem. Biophys. Res. Commun. 31, 798 (1968).

[119] S. Pestka, J. Biol. Chem. 244, 1533 (1969).

[120] J. W. Bodley, F. J. Zieve, L. Lin, and S. T. Zieve, Biochem. Biophys. Res. Commun. 37, 437 (1969).

[121] N. Tanaka, T. Nishimura, T. Kinoshita, and H. Umezawa, J. Antibiot. 22, 181 (1969).

[122] M. Malkin and F. Lipmann, Science 164, 71 (1969).

[123] J. Waterson, M. L. Sopori, S. L. Gupta, and P. Lengyel, Biochemistry 11, 1377 (1972).

[124] N. Tanaka, G. Kawano, and T. Kinoshita, Biochem. Biophys. Res. Commun. 42, 564 (1971).

[125] A. Bernardi and P. Leder, J. Biol. Chem. 245, 4263 (1970).

[126] A. Okura, T. Kinoshita, and N. Tanaka, J. Antibiot. 24, 655 (1971).

[127] E. Cundliffe, Biochem. Biophys. Res. Commun. 46, 1794 (1972).

[128] J. Jonak, J. Sedlacek, and I. Rychlik, FEBS (Fed. Eur. Biochem. Soc.) Lett. 18, 6 (1971).

Diptheria Toxin

Diptheria toxin catalyzes a reversible reaction in which the adenosine diphosphate ribose moiety of NAD is transferred to eukaryotic elongation factor T2. This modification of EF T2 inactivates its transferase activity in protein synthesis.[129-133]

The foregoing has been a summary of the current knowledge of many useful inhibitors of protein synthesis. They can be effectively used in further developing our understanding of protein synthesis.

[129] R. J. Collier and A. M. Pappenheimer, *J. Exp. Med.* **120**, 1019 (1964).
[130] R. J. Collier, *J. Mol. Biol.* **25**, 83 (1967).
[131] T. Honjo, Y. Nishizuka, O. Hayaishi, and I. Kato, *J. Biol. Chem.* **243**, 3553 (1968).
[132] R. S. Goor and E. S. Maxwell, *J. Biol. Chem.* **245**, 616 (1970).
[133] J. Everse, D. A. Gardner, N. O. Kaplan, W. Galasinski, and K. Moldave, *J. Biol. Chem.* **245**, 899 (1970).

[29] The Elongation Steps in Protein Synthesis by Eukaryotic Ribosomes: Effects of Antibiotics

By L. Carrasco, E. Battaner, and D. Vazquez

Growth of the peptide chain in protein synthesis takes place in the elongation phase, which consists of repeated cycles of aminoacyl-tRNA binding to ribosomes, peptide bond formation, and translocation.[1] The elongation phase, according to the two-sites translocating model, starts by binding of the aminoacyl-tRNA determined by the corresponding trinucleotide of the mRNA to the ribosomal A-site. Peptide bond formation then takes place by transfer of the Met or the peptidyl moiety of the Met-tRNA$_F$ or peptidyl-tRNA bound to the P site in such a way that the —COOH group of methionine or the terminal amino acid of the peptidyl is linked to the α-NH$_2$ group of the amino acid moiety of the aminoacyl-tRNA bound to the A site. The stripped tRNA$_F$ or tRNA is then released from the P site and the peptidyl-tRNA formed and attached to the A site is then moved to the P site in a complex step known as translocation, leaving the A site free to start a new cycle of the elongation phase by taking place binding of a new aminoacyl-tRNA. A number of antibiotics and other compounds are known to block protein synthesis by eukaryotic systems, some of them by acting on the elongation phase.[2-4]

[1] J. Lucas-Lenard and F. Lipmann, *Annu. Rev. Biochem.* **40**, 409 (1971).
[2] D. Vazquez, T. Staehelin, M. L. Celma, E. Battaner, R. Fernandez-Muñoz, and

The reaction of peptide bond formation by 80 S type ribosomes takes place at a catalytic center, known as peptidyltransferase, that is an integral part of the 60 S subunit.[5] For the reaction of peptide bond formation the 3'-terminal moiety of peptidyl-tRNA (CCA-peptidyl) binds to the donor site of the peptidyltransferase center whereas the 3'-terminal moiety of aminoacyl-tRNA (A-aminoacyl) binds to the acceptor site. In the presence of methanol or ethanol it is possible to replace the normal substrates by their 3'-terminal moieties or other small analogs using the reaction assay known as "fragment reaction." Replacing peptidyl-tRNA by CACCA-(Ac[³H]Leu) and aminoacyl-tRNA by puromycin the fragment reaction takes place as follows[6]:

$$\text{CACCA-(Ac[}^3\text{H]Leu)} + \text{puromycin} \xrightarrow[\text{ribosomes}]{\text{ethanol}} \text{Ac[}^3\text{H]Leu-puromycin} + \text{CACCA} \quad (1)$$

Ethanol appears to be required in the fragment reaction assay only for the interaction of the simplified substrates to either the donor or acceptor sites of peptidyltransferase. The individual steps of substrate binding to the donor site [reaction (2) indicated below] and to the acceptor site [reaction (3) indicated below] of the peptidyltransferase center can be studied following reactions[4,7,8]:

$$\text{CACCA-(Ac[}^3\text{H]Leu)} + \text{ribosome} \underset{}{\overset{\text{ethanol}}{\rightleftharpoons}} \text{ribosome·CACCA-(Ac[}^3\text{H]Leu)} \quad (2)$$

$$\text{UACCA-([}^3\text{H]Leu)} + \text{ribosome} \underset{}{\overset{\text{ethanol}}{\rightleftharpoons}} \text{ribosome·UACCA-([}^3\text{H]Leu)} \quad (3)$$

In the system of reaction (2) the antibiotic sparsomycin by interacting with the peptidyltransferase center induces formation of a very stable complex displacing the equilibrium of the reaction totally to the right[4,9]:

$$\text{CACCA-(Ac[}^3\text{H]Leu)} + \text{ribosome} + \text{sparsomycin} \xrightarrow[\text{sparsomycin}]{\text{ethanol}}$$

$$\text{CACCA-(Ac[}^3\text{H]Leu)·ribosome·sparsomycin} \quad (4)$$

R. E. Monro, in "Inhibitors: Tools in Cell Research" (T. Bücher and H. Sies, eds.), p. 100. Springer-Verlag, Berlin, 1969.

[3] S. Pestka, Annu. Rev. Microbiol. 25, 487 (1971).

[4] E. Battaner and D. Vazquez, Biochim. Biophys. Acta 254, 316 (1971).

[5] D. Vazquez, E. Battaner, R. Neth, G. Heller, and R. E. Monro, Cold Spring Harbor Symp. Quant. Biol. 34, 369 (1969).

[6] R. E. Monro, this series, Vol. 20, p. 472.

[7] M. L. Celma, R. E. Monro, and D. Vazquez, FEBS (Fed. Eur. Biochem. Soc.) Lett. 6, 273 (1970).

[8] M. L. Celma, R. E. Monro, and D. Vazquez, FEBS (Fed. Eur. Biochem. Soc.) Lett. 13, 247 (1971).

[9] R. E. Monro, M. L. Celma, and D. Vazquez, Nature (London) 222, 356 (1969).

The above systems are very useful to study peptide bond formation and substrate interaction with the peptidyltransferase center. These reactions are well resolved since the systems do not require the presence of the 40 S ribosomal subunit and mRNA and interactions with the ribosome in the immediate vicinity of the catalytic center can be specifically investigated. These resolved systems are also suitable for testing the effect of protein synthesis inhibitors in order to know their precise target.[4]

We describe in this contribution suitable methods to study the individual steps of the elongation phase (aminoacyl-tRNA binding to the A site of the ribosomes, peptide bond formation and translocation) in protein synthesis and summarize the results obtained when specific inhibitors of protein synthesis by eukaryotic cells are used in the different systems described.

Material and Reagents

Yeast cytoplasmic ribosomes and their ribosome subunits from *Saccharomyces cerevisiae* were prepared and stored as previously described.[10]

Human tonsil ribosomes were obtained as already described.[11] These ribosome preparations are washed three times (after the centrifugation step on a sucrose discontinuous gradient) with 50 mM Tris·HCl buffer, pH 7.4, containing 0.5 M KCl, 10 mM MgCl$_2$ and 7 mM 2-mercaptoethanol. Ribosomes were then stored at $-17°$ in 50 mM Tris·HCl, pH 7.4, containing 25 mM KCl, 5 mM MgCl$_2$, 7 mM 2-mercaptoethanol, and 50% glycerol. Final ribosome concentration was approximately 20 mg/ml (13.1 absorbance units at 260 nm was 1 mg of ribosomes). No loss of activity was observed during 3 months of storage under these conditions.

Elongation factors 1 (EF 1) and 2 (EF 2) from human tonsils were separated and purified as previously described.[12,13] Purity of EF 2 after phosphocellulose column chromatography[13] is approximately 50%.[14]

[^{14}C]Phe-tRNA was prepared from *Escherichia coli* B tRNA (Sigma) that was charged with [^{14}C]phenylalanine (513 mCi/mmole) (The Radiochemical Centre, Amersham, England) using a partially purified supernatant preparation of *E. coli*.[15] [^{14}C]Phe-tRNA binding to human tonsil

[10] E. Battaner and D. Vazquez, this series, Vol. 20, p. 446.
[11] R. Neth, G. Heller, and H. Matthaei, *Hoppe-Seyler's Z. Physiol. Chem.* 349, 1514 (1968).
[12] E. Bermek and H. Matthaei, *FEBS (Fed. Eur. Biochem. Soc.) Lett.* 10, 121 (1970).
[13] W. Galasinski and K. Moldave, *J. Biol. Chem.* 244, 6527 (1969).
[14] E. Bermek and H. Matthaei, *Biochemistry* 10, 4906 (1971).
[15] T. W. Conway, *Proc. Nat. Acad. Sci. U.S.* 51, 1216 (1964).

ribosomes was measured following essentially methods already described using other mammalian systems using buffer C.[16]

CACCA-(Ac[³H]Leu) (20 Ci/mmole) and UACCA-([³H]Leu) (20 Ci/mmole) were prepared following methods already described.[6] The fragment reaction assay to test peptide bond formation has also been described.[5,6,17] Binding of CACCA-(Ac[³H]Leu) (either in the presence or in the absence of sparsomycin) and UACCA-([³H]Leu) to *S. cerevisiae* ribosomes has been carried out following essentially methods already described,[7-9] but ribosome bound radioactivity was measured by filtration through Millipore filters[4] rather than pelleting the ethanol precipitate.

Poly(U), dithiothreitol, and GTP were from Sigma; 2-mercaptoethanol from Fluka, Millipore HAWP 2400 (from Millipore Corp.), or Sartorius SM 11306 (Membranfilter GmbH) filters were used. Scintillation fluid was prepared by diluting 5 g of butyl-PBD (Ciba) in 1 liter of toluene was used to measure radioactivity retained by filtration through Millipore filters. To measure [¹⁴C]Phe-puromycin or Ac[³H]Leu-puromycin extracted by ethyl acetate a scintillation fluid was used prepared by addition of 250 ml of methoxyethanol to 750 ml of toluene in which 5 g of butyl PBD (Ciba) have been dissolved previously.

The following buffers were used in the assays described below:

Buffer A: 20 mM Tris·HCl, pH 7.4, containing 60 mM KCl and 7 mM MgCl$_2$

Buffer B: 50 mM Tris·HCl, pH 7.4, containing 60 mM KCl, 11 mM MgCl$_2$ and 8 mM 2-mercaptoethanol

Buffer C: 20 mM Tris·HCl, pH 7.4, containing 60 mM KCl and 5 mM MgCl$_2$

Sources of the inhibitors of protein synthesis used in this work were as follows:

Actinobolin (Parke Davis); amicetin, chartreusin, cycloheximide (Upjohn); adrenochrome and gougerotin (Calbiochem); anisomycin (Pfizer); aurintricarboxylic acid (ATA) (May & Baker); blasticidin S and bottromycin A$_2$ (Institute of Applied Microbiology, Tokyo, Japan); emetine (Wellcome); fusidic acid (Leo); puromycin (Serva and Nutritional Biochemicals); sparsomycin (National Cancer Institute, Bethesda, Maryland); and tenuazonic acid (Merck Sharp & Dohme). Edeine A$_1$ was a gift from Dr. Z. Kurylo-Borowska (Rockefeller University, New York). Pederine

[16] J. H. Matthaei and G. H. Schoech, *Biochem. Biophys. Res. Commun.* **27**, 638 (1967).

[17] R. Neth, R. E. Monro, G. Heller, E. Battaner, and D. Vazquez, *FEBS (Fed. Eur. Biochem. Soc.) Lett.* **6**, 198 (1970).

was given to us by Professor M. Pavan (Institute of Entomology, University of Pavia, Italy). Diphtheria toxin was a gift from Dr. E. Bermek (Max-Planck Institute for Experimental Medicine, Göttingen, Germany).

Isolation of the Complex Ribosome·[¹⁴C]Phe-tRNA

To obtain ribosome·[¹⁴C]Phe-tRNA (with [¹⁴C]Phe-tRNA enzymatically bound to the ribosomal A site) the following components are incubated for 30 minutes at 37° in a final volume of 5 ml of buffer A: poly(U), 200 μg/ml; [¹⁴C]Phe-tRNA, 50 pCi/ml; 600 μg of protein per milliliter of EF 1 preparation; GTP, 0.2 mM; and human tonsil ribosomes, 0.7 mg/ml. Complex formation is started by addition of GTP. The incubation mixture is chilled on ice at the end of the incubation period, and a 50-μl sample is taken to estimate formation of complex ribosome·[¹⁴C] Phe-tRNA by Millipore filtration.[16] Approximately 12% of the total ribosomes in the incubation mixture form the complex ribosome·[¹⁴C]Phe-tRNA. This total mixture is taken and placed over 7 ml of a 10% sucrose solution in buffer B in 12-ml tubes of the Spinco/Beckman ultracentrifuge, and centrifugation is carried out for 3 hours at 105,000 g at 4°. The pellet was carefully rinsed for three times, without removing the pellet, with buffer B and resuspended in 0.5 ml of the same buffer. Absorbancy at 260 nm and formation of the complex ribosome·[¹⁴C]-Phe-tRNA is then estimated. The complex suspension is stored at 0° and is stable for at least 2 weeks.

Translocation of [¹⁴C]Phe-tRNA from the A Site to the P Site of the Ribosome

The ribosome·[¹⁴C]Phe-tRNA complex prepared as described above was used to study translocation. For this purpose, 100-μl mixtures were used that contained in buffer B 2.4 pmoles of ribosome·[¹⁴C]Phe-tRNA complex, 70 μg of EF 2, and 4 mM GTP. The reaction was started by addition of GTP and incubation carried out for 15 minutes at 37°. (Protein synthesis inhibitors were added to the incubation mixture, to test their effect on translocation, prior to the addition of GTP, and the mixture was preincubated for 10 minutes at 37° to facilitate inhibitor binding before addition of GTP.) To stop the translocation reaction, the mixture was chilled on ice, at the end of the incubation period, and placed over 12 ml of a 10% sucrose solution in buffer B in tubes of the Spinco/Beckman ultracentrifuge and centrifuged for 3 hours at 105,000 g at 4°. The pellet was resuspended in 100 μl of buffer B at 24 pmoles/ml. The extent of the translocation reaction was estimated by the puromycin reac-

TABLE I
EFFECTS OF PROTEIN SYNTHESIS INHIBITORS IN THE ELONGATION PHASE[a]

Inhibitor	Concentration	[^{14}C]Phe-tRNA binding	Translocation	[^{14}C]Phe-puromycin formation
Adrenochrome	0.1 mM	23	90	90
Anisomycin	0.1 mM	91[f]	115	16
ATA	0.1 mM	15	76	76
Blasticidin S	0.2 mM	112	90	23
Edeine A$_1$		14[b]	72[c]	140[c]
Emetine		74[d]	88[e]	87[d]
Diptheria toxin	125 μg/ml	100	11	108
Gougerotin	0.1 mM	101	85	73
Fusidic acid	0.2 mM	95	97	90
Pederine	20 μM	91	30	97
Sparsomycin	0.1 mM	100	154	19
Tenuazonic acid	0.1 mM	92	101	50

[a] Assays were carried out using human tonsil ribosomes under the experimental conditions described above. Figures given in this table are percentage of control reaction in the absence of inhibitor. Diptheria toxin was always assayed in the presence of 10 μM NAD. The following compounds produced no effect or only marginal difference from control in the above reaction assays: Bottromycin A$_2$ (0.1 mM), chartreusin (0.2 mM), cycloheximide (0.1 mM) and amicetin (0.1 mM).
[b] 1 μM edeine A$_1$.
[c] 0.1 mM edeine A$_1$.
[d] 0.1 mM emetine.
[e] 10 μM emetine.
[f] 10 μM anisomycin.

tion with this resuspended material as described below. Some 25–30% of bound [^{14}C]Phe-tRNA was translocated.

Formation of [^{14}C]Phe-Puromycin (Puromycin Reaction)

Reaction with puromycin of the ribosome·[^{14}C]Phe-tRNA complex obtained after translocation as indicated above, was studied in 100-μl incubation mixtures containing in buffer B: 0.7 pmole of translocated [^{14}C]Phe-tRNA complex and 1 mM puromycin. Incubation was carried out for 30 minutes at 37°. (Protein synthesis inhibitors were added to the incubation mixture, to test their effect on peptide bond formation, prior to addition of puromycin and preincubated for 10 minutes at 37°.) Puromycin reaction was stop by addition of 0.5 ml of ammonium bicarbonate 1 M, pH 9, and [^{14}C]Phe-puromycin formed extracted with 1 ml of ethyl acetate.[18]

[18] P. Leder and H. Bursztyn, *Biochem. Biophys. Res. Commun.* **25**, 233 (1966).

TABLE II

EFFECTS OF PROTEIN SYNTHESIS INHIBITORS ON PEPTIDE BOND FORMATION[a,c]

Inhibitor		Ac-Leu-puromycin formation			CACCA-(Ac[3H]Leu) binding	Sparsomycin-induced CACCA-(Ac[3H]Leu) binding	UACCA-([3H]Leu) binding
		Tonsil ribosomes	Yeast				
			Ribosomes	60 S			
Actinobolin	0.1 mM	75	35	—[b]	96	—	100
	1 mM	31	—	—	—	—	—
Amicetin	1 mM	73	72	12	85	—	90
Anisomycin	0.1 mM	5	2	0	43	35	43
Blasticidin S	0.1 mM	12	5	—	89	17	—
	1 mM	—	—	—	82	—	70
Bottromycin A₂	0.2 mM	90	—	—	93	—	—
Cycloheximide	0.1 mM	—	—	—	94	40	96
	1 mM	80	84	—	78	19	—
Gougerotin	0.1 mM	—	3	0	94	40	100
	1 mM	16	—	—	—	22	—
Sparsomycin	0.1 mM	7	2	0	177	—	147
Tenuazonic acid	0.1 mM	22	100	—	—	—	—

[a] Ac-Leu-puromycin formation was studied after the fragment reaction assay. Yeast ribosomes were used for the experiments on UACCA-([3H]Leu) and CACCA-(Ac[3H]Leu) binding induced or not by sparsomycin. In the case of sparsomycin-induced binding of CACCA-(Ac[3H]Leu), all the values are corrected for the binding in the absence of sparsomycin.

[b] Dash shows that the indicated experimental conditions have not been studied.

[c] Figures given in this table are percentage of control reaction in the absence of inhibitor.

Inhibitors of the Elongation Phase of Protein Synthesis

The effect of a number of inhibitors of protein synthesis by eukaryotic systems has been tested on (a) EF 1-dependent binding of [^{14}C]Phe-tRNA to the A site of human tonsil ribosomes, (b) translocation of [^{14}C]Phe-tRNA to the P site (c) reaction of the ribosome · [^{14}C]Phe-tRNA complex after translocation with puromycin. Some of the results obtained are briefly summarized in Table I. The effect of protein synthesis inhibitors on peptide bond formation has also been studied in the fragment reaction assay using human tonsils and *S. cerevisiae* ribosomes, and the results obtained are summarized in Table II. The effect of protein synthesis inhibitors has also been tested on CACCA-(Ac[^{3}H]Leu) and UACCA-([^{3}H]Leu) binding to the donor- and the acceptor-site, respectively, of *S. cerevisiae* ribosomes (Table II); in a similar system the effect of protein synthesis inhibitors on the formation of the complex CACCA-(Ac[^{3}H]Leu) · ribosome · sparsomycin has been studied (Table II).

Adrenochrome and edeine A$_1$ were found to inhibit enzymatic binding of Phe-tRNA. Diphtheria toxin and pederine are good inhibitors of translocation (Table I).

Anisomycin, blasticidin S, sparsomycin, and tenuazonic acid were found to inhibit peptide bond formation by human tonsil ribosomes in the puromycin and the fragment reaction. Small or no inhibitory effect by actinobolin, amicetin, and gougerotin was found in the puromycin reaction, but these compounds are inhibitors in the fragment reaction assay mainly when 60 S ribosome subunits were used (Table II). It is very interesting the differential results obtained with tenuazonic acid which is an inhibitor of peptide bond formation by human tonsil ribosomes, but not by yeast ribosomes.

Concerning the specific step affected by peptide bond formation inhibitors, it was found that anisomycin blocked binding of substrates to the donor and the acceptor sites of the peptidyltransferase center. Sparsomycin, blasticidin S, cycloheximide, and gougerotin were found to block formation of the complex CACCA-(Ac[^{3}H]Leu) · ribosome · sparsomycin either by affecting the donor site of the peptidyltransferase center or blocking the sparsomycin binding site.

Section III
Termination Factors in Protein Synthesis

[30] Mammalian Release Factor; *in Vitro* Assay and Purification[1]

By C. T. CASKEY, A. L. BEAUDET, and W. P. TATE

Peptide chain termination and several of its intermediate steps can be studied in mammalian extracts. The codon directed release of peptides from ribosomal bound peptidyl-tRNA is investigated in mammalian extracts by modification of the formylmethionine (fMet) release assay previously described for bacterial extracts.[2] As shown in reaction (1), f[³H]Met-

f[³H]Met-tRNA + reticulocyte ribosomes → f[³H]Met-tRNA·ribosome

f[³H]Met-tRNA·ribosome + RF + terminator codon + GTP → f[³H]Met   (1)

tRNA can be bound to reticulocyte ribosomes at a magnesium ion concentration of 55 mM without protein factors, GTP, polynucleotides, or oligonucleotides.[3] The release of fMet from these ribosomal complexes is stimulated by one of three oligonucleotides of defined sequence (UAAA, UAGA, and UGAA) or randomly ordered polynucleotide templates containing the codons UAA, UAG, or UGA.[3,4] Thus far the triplets UAA, UAG, or UGA have not stimulated fMet release with mammalian extracts. The codon directed release of fMet is stimulated by GTP and requires a mammalian protein release factor (RF). The *in vitro* release of fMet from fMet-tRNA·ribosome intermediates is believed to be analogous to the release of peptides upon peptide chain termination in mammalian cells.

Two additional methods for investigation of mammalian peptide chain termination are also available. The first studies fMet-tRNA hydrolysis as a partial reaction of the total process.[4] The hydrolysis of ribosomal bound f[³H]Met-tRNA which occurs in reactions containing RF and 20% ethanol does not require terminator codon recognition and is not stimulated by GTP. This partial reaction of peptide chain termination, outlined below (reaction 2), differs, therefore, from reaction (1) in its specificity.

RF + f[³H]Met-tRNA·ribosome + 20% ethanol → f[³H]Met   (2)

The RF-mediated hydrolysis of f[³H]Met-tRNA on ribosomes in this re-

[1] This work was supported by Grant GM-18682-02.
[2] C. T. Caskey, R. Tompkins, E. Scolnick, T. Caryk, and M. Nirenberg, *Science* **162**, 135 (1968).
[3] J. L. Goldstein, A. L. Beaudet, and C. T. Caskey, *Proc. Nat. Acad. Sci. U.S.* **67**, 99 (1970).
[4] A. L. Beaudet and C. T. Caskey, *Proc. Nat. Acad. Sci. U.S.* **68**, 619 (1971).

action has been separated from the intermediate steps of reaction (1) which require the participation of GTP and oligonucleotide codons.

The binding of RF to ribosomes can be determined indirectly with radioactive oligonucleotides. While UA[³H]AA is used exclusively in these studies, it appears likely that UAG[³H]A and UG[³H]AA would also be effective. The ribosomal binding of RF can be examined by the quantitation of the amount of radioactive RF·UA[³H]AA·ribosome formed.[5]

$$\text{RF} + \text{ribosomes} + \text{UA[}^3\text{H]AA} + \text{GDPCP} \xrightarrow{\text{ETOH}} \text{RF·UA[}^3\text{H]AA·ribosomes} \quad (3)$$

The RF·UA[³H]AA·ribosome complex is stabilized for isolation by addition of 20% ethanol to all reactions. Following this stabilization the radioactively labeled ribosomal complex can be isolated and quantitated by centrifugation, Sephadex column chromatography, or by Millipore filtration. The recovery of the complex on Millipore filters is enhanced 2-fold by GTP and 5-fold by equivalent amounts of GDPCP. It is not known whether GTP or GDPCP is a component of the ribosomal complex.

By use of the three in vitro methods outlined above, total and partial reactions of peptide chain termination can be investigated in mammalian extracts. This report outlines the methodology of the in vitro assays and the purification and characterization of rabbit reticulocyte RF.

Preparation of fMet-tRNA·Ribosome Intermediates

Formyl methionine-tRNA, the initiator tRNA species of Escherichia coli protein biosynthesis is used as a peptidyl-tRNA analog.[6] The fMet-tRNA has high affinity for the ribosomal site and the quantitation of f[³H]Met release is simple. Both mammalian and E. coli f[³H]Met-tRNA (formylated with E. coli transformylase) bind equally well to reticuloctye ribosomes at a magnesium ion concentration of 55 mM. The f[³H]Met-tRNA·ribosome complexes are reactive with puromycin yielding 95–100% of the radioactivity as f[³H]Met-puromycin in less than 5 minutes.[3] The ribosomal binding and puromycin reactivity of mammalian and E. coli fMet-tRNA were indistinguishable. Highly purified E. coli tRNA$_f^{Met}$ (1400 pmoles of methionine acceptance per A_{260} unit) (available through NIGMS) is routinely used. The [³H]Met-tRNA$_f$ binds to reticulocyte ribosomes as well as f[³H]Met-tRNA$_f$, but the [³H]Met-tRNA·ribosome complexes do not react with puromycin to form [³H]Met-puromycin or RF to form [³H]Met without other factors.

The precise level of ribosomes and f[³H]Met-tRNA needed for

[5] W. P. Tate, A. L. Beaudet, and C. T. Caskey, in preparation.
[6] M. S. Bretscher and K. A. Marcker, Nature (London) 211, 380 (1966).

f[^{3}H]Met-tRNA·ribosome intermediate formation varies with f[^{3}H]Met-tRNA binding activity of the ribosomal preparation. Therefore, preliminary studies are necessary to determine the quantity of ribosomes needed to bind 85% of the f[^{3}H]Met-tRNA. A typical binding reaction is incubated 15 minutes at 30° and contains in 5.0 ml: 50 mM Tris·HCl, pH 7.0; 55 mM MgCl$_2$; 80 mM NH$_4$Cl; 300–500 A_{260} units of reticulocyte ribosomes and 2 nmoles $E. coli$ f[^{3}H]Met-tRNA. Aliquots of 0.01 ml are examined for quantity of f[^{3}H]Met-tRNA·ribosome intermediate formed by the technique of Nirenberg and Leder[7]; f[^{3}H]Met-tRNA precipitable by trichloroacetic acid; and free f[^{3}H]Met by ethyl acetate extraction.[8] Typical values are: 3–5 pmoles f[^{3}H]Met-tRNA·ribosome; 3–5 pmoles f[^{3}H]Met-tRNA; and 0.3–0.5 pmole f[^{3}H]Met. These values do not change with storage at −170° over several months. This 5.0 ml quantity of f[^{3}H]Met-tRNA·ribosome intermediate is adequate for 500 determinations. It is stored at −170° in aliquots of 0.3 ml.

Synthesis and Purification of Oligonucleotides

The oligonucleotide sets (UAA and UAAA; UGA and UGAA) are synthesized enzymatically by polynucleotide phosphorylase from the doublets UA or UG and ADP.[9] The UAGA is synthesized in two steps. The UAG is synthesized and isolated as previously described.[10] The UAGA is synthesized enzymatically by polynucleotide phosphorylase, UAG and ADP. The oligonucleotides are isolated from the reaction mixture by DEAE-cellulose chromatography as described by Petersen and Reeves[11] except that urea is omitted. The tri- and tetranucleotides isolated by this method have less than 2% contaminating ultraviolet absorbing material as analyzed by T2 ribonuclease digestion and base ratio analysis.[9] Synthesis and isolation of UA[^{3}H]AA is performed as described above except that reactions contained [^{3}H]ADP (20.3 Ci/mmole) and nonradioactive UA.

Codon-Directed f[^{3}H]Met Release

Mammalian RF from rabbit reticulocytes, guinea pig, and Chinese hamster liver participate in fMet release from fMet-tRNA·ribosome substrates with oligonucleotides UAGA, UAGG, UAAA, or UGAA but not with AAAA, UUUU, UAA, UAG, or UGA.[4] Furthermore, randomly

[7] M. W. Nirenberg and P. Leder, $Science$ 145, 1399 (1964).
[8] P. Leder and H. Bursztyn, $Biochem. Biophys. Res. Commun.$ 25, 233 (1966).
[9] P. Leder, M. F. Singer, and R. L. C. Brimacombe, $Biochemistry$ 4, 1561 (1965).
[10] R. Thach, in "Procedures in Nucleic Acid Research" (G. L. Cantoni and D. R. Davies, eds.), p. 520. Harper, New York, 1966.
[11] G. B. Petersen and J. M. Reeves, $Biochim. Biophys. Acta$ 129, 438 (1966).

ordered poly(U,G,A), (U,A$_3$), and (U,A$_3$,G$_{0.5}$) stimulate release of f[^{3}H]Met. A typical release reaction is incubated 15 minutes at 24° and contains in 50 μl; 3.0 pmoles of f[^{3}H]Met-tRNA·ribosome intermediate; 20 mM Tris·HCl pH 7.4; 60 mM KCl; 11 mM MgCl$_2$; 0.1 mM GTP; 0.1 A$_{260}$ unit UAAA, UAGA, or UGAA; or alternatively 0.2 A$_{260}$ units of poly(U,G,A), (U,A$_3$), (U,A$_3$,G$_{0.5}$), and partially purified mammalian RF.

Reactions are terminated by addition of 0.25 ml of 0.1 N HCl and 1.5 ml of ethyl acetate. The aqueous and ethyl acetate phases after 10 seconds of vortex mixing are separated by centrifuging in a desk-top clinical centrifuge for 5 minutes. One milliliter of ethyl acetate (upper phase) is transferred to 10 ml of Biosolv (3 liter toluene, 100 ml Beckman Biosolv, and 300 ml Beckman Fluoralloy) counting fluid, and the radioactivity is quantitated in a Beckman LS-233. Total f[^{3}H]Met release is 2.15 × (cpm-bgd) since only a 1.0-ml aliquot is counted and 70% of the released f[^{3}H]Met is extracted into ethyl acetate. We routinely report total f[^{3}H]Met released.

The rate of f[^{3}H]Met release directed by tetranucleotides is markedly stimulated (3–4-fold) by GTP, inhibited by GDPCP, and unaffected by GDP. The rate of this release directed by tetranucleotides is markedly inhibited in reactions with a potassium ion concentration above 60 mM. The rate of f[^{3}H]Met directed by polynucleotides is also stimulated by GTP (1.5–2.0-fold), but is not inhibited by potassium ions up to 150 mM. Assay of RF in column fractions (which contain variable potassium ion concentration) is most commonly determined with polynucleotides.

Ethanol-Directed f[^{3}H]Met Release

The hydrolysis of ribosomal bound fMet-tRNA with RF can occur in the absence of terminator codon recognition if reactions contain 20% ethanol, methanol, or acetone. The addition of ethanol stimulates ribosomal binding of RF and thus eliminates requirement for codon. Release of f[^{3}H]Met under these conditions is not stimulated by GTP and is partially inhibited by potassium ion concentrations above 120 mM. The f[^{3}H]Met release requires both ethanol and partially purified RF (fraction IV). Little release is observed with crude RF.

A typical reaction is incubated 20 minutes at 24° and contains in 0.05 ml: 20 mM Tris·HCl; 60 mM potassium chloride, 11 mM magnesium chloride; 3.0 pmoles of f[^{3}H]Met-tRNA·ribosome; variable quantities of partially purified RF; and 20% ethanol. The quantity of f[^{3}H]Met released is determined as described above.

The relative rates of RF dependent f[^{3}H]Met release as directed by UAAA, poly(U,A$_3$), and 20% ethanol are shown in Fig. 1.

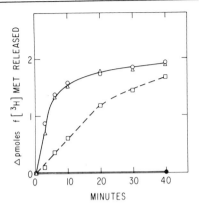

FIG. 1. Reticulocyte release factor activity. At the indicated time intervals, f[³H]Met released from a 0.03-ml portion of a 0.25-ml reaction was determined. Each tube was incubated at 24° and contained: 15.25 pmoles of f[³H]Met-tRNA$_f$ · ribosome intermediates; 65 μg of reticulocyte RF (fraction V); 0.1 mM GTP; 1.0 A_{260} unit poly(A₃,U) (△), or 0.5 A_{260} unit UAAA (○), or 20% ethanol (v/v) (□), or no template (●); and other components as described. The zero time value of 0.18 pmole was subtracted from all values.

Reticulocyte RF Binding to Ribosomes

The ribosomal binding of *E. coli* RF1 (UAA or UAG) and RF2 (UAA or UGA) to *E. coli* ribosomes has been shown to occur with the indicated codon specificity.[12] This ribosomal binding is rapidly quantitated using radioactive terminator codons. The complex RF·UA[³H]AA·ribosome (reaction 3) is quantitated by retention of the radioactively labeled UAAA on Millipore filters.[12] Two modifications of the procedure permitted its application to mammalian extracts. Mammalian RF binds to reticulocyte ribosomes with UA[³H]AA but not with UA[³H]A, and this RF binding is stabilized by GDPCP. The requirements for retention of the radioactive complex on Millipore filters is given in Table I.

A typical reaction is incubated at 4° for 15 minutes and contains in 0.05 ml; 1.8 A_{260} units of reticulocyte ribosomes; 20.0 pmoles of UA[³H]AA (10.0 Ci/mmole); 0.1 mM GDPCP; 50 mM Tris·acetate pH 7.4; 20 mM magnesium acetate; 0.1 M ammonium acetate; 20% ethanol; and purified RF (fraction IV). The reaction is terminated by addition of 0.5 ml of cold buffer containing 50 mM Tris·acetate pH 7.4; 20 mM magnesium acetate; 0.1 M ammonium acetate; and 10% ethanol. The radioactive complex is collected on a Millipore filter, washed with 10 ml of the above cold buffer, dried, and counted in 10 ml of POPOP counting fluid.

The methods employed to study the partial reactions of peptide chain

[12] E. M. Scolnick and C. T. Caskey, *Proc. Nat. Acad. Sci. U.S.* **64**, 1235 (1969).

TABLE I
REQUIREMENTS FOR UA[³H]AA BINDING WITH RETICULOCYTE
RELEASE FACTOR (RF)[a]

Condition	UA[³H]AA bound (pmoles)
Complete	2.06
−RF	0.07
−Ribosomes	0.02
−Ethanol	0.07
−GDPCP	0.40
−GDPCP, +GTP	0.85
−GDPCP, +GDP	0.43

[a] Requirements for UA[³H]AA binding with reticulocyte RF; reactions contain as indicated 1.8 A_{260} unit ribosomes, 15 µg of reticulocyte RF, 20% v/v ethanol, 0.1 mM guanine nucleotide, and 11 pmoles of UA[³H]AA.

termination described above have been useful in probing the intermediate events involved and the possible role of peptidyl transferase in this process.[3-5,13]

Purification of Rabbit Reticulocyte Release Factor, and Ribosomes

Lysate. New Zealand white rabbits (about 4–5 pounds) are injected intramuscularly with 0.1 mg of vitamin B_{12} and 1 mg of folic acid in 0.9% NaCl pH 7.0. For six successive days they are injected subcutaneously with 2.5% phenylhydrazine (0.1 ml/lb), pH 7.0.[14] The rabbit's lateral ear vein is lacerated, and 50–70 ml of blood is obtained under a vacuum of 5–7 psi. The cells from the heparinized blood are collected by centrifuging at 10,000 g for 10 minutes at 4°, washed twice by suspending in 2 volumes of buffer (0.14 M NaCl, 50 mM KCl, and 5 mM MgCl₂) and collected by repeated centrifuging. The packed cells are lysed by adding 4 volumes of hypotonic buffer (2 mM MgCl₂, 1 mM dithiothreitol, 0.1 mM EDTA, pH 7.0). The suspension is stirred for several minutes at 4° and centrifuged at 15,000 g for 30 minutes at 4°. This supernatant fraction can be stored at −170° without loss of RF activity. Typically a preparation of lysate from 20 rabbits is needed for RF isolation and purification.

Ribosomes. The rabbit reticulocyte lysate is centrifuged at 105,000 g for 3 hours. The ribosomal pellet is suspended in 0.1 of the original lysate volume in 50 mM Tris·HCl pH 7.4, 0.5 mM KCl, 30 mM β-mercaptoethanol, 2 mM MgCl₂, and 15% (v/v) glycerol by stirring 16 hours at 4°. The ribosomes are collected by centrifuging at 105,000 g for 3 hours at

[13] C. T. Caskey, A. L. Beaudet, E. M. Scolnick, and M. Rosman, *Proc. Nat. Acad. Sci. U.S.* **68**, 3163 (1971).
[14] J. M. Gilbert and W. F. Anderson, this series, Vol. 20, p. 542.

4°. The ribosomal pellet is resuspended in 3 volumes of buffer (50 mM Tris·HCl pH 7.4, 0.25 M sucrose, 2 mM MgCl$_2$, 3 mM β-mercapto-ethanol), and after a low speed centrifugation (30,000 g for 15 minutes) is stored at $-170°$C. This ribosomal preparation is used in the preparation of f[^{3}H]Met-tRNA·ribosome intermediates and is free of RF activity.

RF Purification. A preparation of rabbit reticulocyte lysate from 20 animals is used for the following purification scheme. The lysate is sub-jected to 105,000 g centrifugation for 3 hours at 4° to pellet the ribosomes. The RF activity in the supernatant of the lysate (fraction I) is partially purified by ammonium sulfate precipitation (42.4 g/100 ml) at 4° main-taining the pH at 7.5. The precipitate containing RF (fraction II) is collected by centrifuging (30,000 g, 15 minutes, at 4°) and dissolved in 50–100 ml of buffer A (0.1 M KCl, 20 mM Tris·HCl pH 7.8, 1 mM dithio-threitol, and 0.1 mM EDTA). Fraction II is dialyzed against 50 volumes of buffer A for 18 hours. Since fractions I and II hydrolyze fMet-tRNA independent of terminator codon, the RF activity determinations on these fractions are difficult and unreliable. Fraction II is applied to a column (2.5 × 50 cm) packed with DEAE-Sephadex equilibrated with buffer A. A column of this size could accommodate 40 g of protein in fraction II. After sample application, the column is washed with buffer A (2.5 column volumes) followed by 1600–2000 ml of a solution containing all the components of buffer A and a linearly increasing concentration gradient of potassium chloride (0.1 M to 0.7 M) (Fig. 2A). Fractions of 10 ml are collected at a flow rate of 30–40 ml/hr. The RF activity of column frac-tions can be determined using UAAA, UAGA, UGAA, or poly(U,G,A) (Fig. 2A). Recently the reticulocyte RF has fractionated with apparent heterogeneity on DEAE-Sephadex using 2000 ml of solution A containing a linearly increasing concentration gradient of potassium chloride (0.1 to 0.5 M). The trailing shoulder of RF (Fig. 2A) can be more widely spread by such a gradient but has not been completely separated from the major RF peak at this time. The peak RF activities elute at 0.31 and 0.34 KCl under these conditions. Since the characterization of the later eluting RF activity is not complete at this time, the remaining purification will deal with the major RF fraction eluting at 0.31 M KCl. The fractions (fraction III) containing RF activity are pooled, concentrated by pressure filtration, and dialyzed against buffer B (50 mM Tris·HCl pH 7.8, 0.1 M KCl, 1 mM DTT, and 0.1 mM EDTA).

Fraction III can be further purified by use of phosphocellulose column chromatography. Fraction III is applied to a 30 × 1.5 cm (or 30 × 2.5 cm) column packed with phosphocellulose equilibrated in buffer B. After sample application, the column is washed with 2–3 column volumes of buffer B followed by 400 ml (or 1200 ml if the larger column is used)

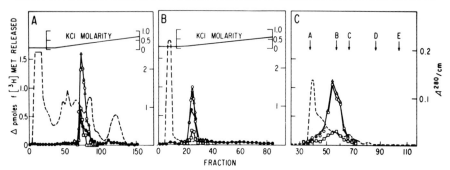

FIG. 2. Reticulocyte release factor (RF) purification.

(A) Fraction II (3.36 g of protein in 82 ml) was applied to a 43 × 2.5-cm DEAE-Sephadex column; washed with 230 ml of buffer A; and eluted with a KCl gradient (2000 ml). Fractions were 15.5 ml. All release reactions were 10 minutes, 24°, and contained: 0.2 A_{260} unit poly(A,G,U) (●); 3.05 pmoles of f[³H]Met-tRNA$_f$· ribosome intermediates; 0.1 mM GTP; 0.015 ml of each column fraction, and other components described. Activity was determined with 0.1 A_{260} unit UAAA (○), UGAA (△), or UAGA (□) under the identical conditions except that all fractions were dialyzed against buffer C prior to assay with 0.025 ml of each fraction. Values for f[³H]Met release are stimulation by oligonucleotide. The peak A_{260} value (fraction 12) is 26.5 A_{260}/cm. Note that the left-hand scale for f[³H]Met release is constant in panels A, B, and C, while the right-hand scale for A_{260}/cm is 10-fold less in panel C than in A and B. - - -, A_{280}.

(B) Fraction III (300 mg of protein in 59 ml) was applied to a 33 × 2.5-cm phosphocellulose column, washed with 100 ml of buffer B, and eluted with a linear (1200 ml) KCl gradient. Fractions, 15.5 ml, were assayed as described in (A) except that 0.015 ml, rather than 0.025 ml, of each dialyzed fraction was assayed with tetranucleotide.

(C) Fraction IV (5.84 mg of protein in 4.0 ml) was applied to a 92 × 2.5-cm Sephadex G-200 column and eluted. Fractions were 3.8 ml. Release assays were incubated 10 minutes, 24° and contained: 2.92 pmoles of f[³H]Met-tRNA$_f$·ribosome intermediates; 0.1 mM GTP; where indicated 0.1 A_{260} unit UAAA (○), UGAA (△), or UAGA (□); 0.025 ml of each fraction; and other components as indicated. The zero time value (0.24 pmole) is subtracted. Each purified protein marker was applied separately in 4.0 ml, and its position of elution was determined by absorbance at 280 nm or 416 nm (cytochrome c). Arrow A indicates dextran blue (void volume, 142.5 ml); arrow B, pyruvate kinase (217 ml, MW 237,000); arrow C, aldolase (252 ml, MW 158,000); arrow D, ovalbumin (327 ml, MW 43,000); and arrow E, cytochrome c (392 ml, MW 12,523).

of buffer B containing a linear concentration gradient of KCl (0.1–0.4 M) (Fig. 2B). The 5-ml fractions are eluted at a rate of 30 ml per hour. The RF activity is determined as described above and elutes at 0.21 M KCl (Fig. 2B). The active fractions are pooled and concentrated by pressure filtration (fraction IV).

Fraction IV can be further purified by several alternative methods.

Sephadex G-200 column chromatography has proved most convenient and reliable. Fraction IV (4.0 ml) is applied to a 92 × 2.5 cm Sephadex G-200 column equilibrated with buffer C (20 mM Tris·HCl pH 7.8, 50 mM KCl, 1 mM DTT, and 0.1 M EDTA). The elution position of RF with respect to marker proteins is given in Fig. 2C. This elution position for RF would give an estimated molecular weight of 255,000. In separate studies where the G-200 chromatography was done as upward flow in 0.3 M KCl, the molecular weight estimate was 200,000. The fractions containing RF are pooled and concentrated (fraction V) by pressure filtration. Fraction V has no detectable EF1 or EF2 activity.[4] A summary of the results of the above purification procedures is given in Table II.

Two purification procedures, hydroxylapatite and sucrose gradient sedimentation, have been found successful with reticulocyte RF. These procedures are recommended when RF of high purity is needed. Use of the above procedures and judicious pooling of only the most active fractions can achieve purification of a homogeneous fraction, as judged by acrylamide gel analysis.

A purification of RF by hydroxylapatite column chromatography has been found similar to Sephadex G-200. The RF fraction is dialyzed against buffer D (10 mM KH$_2$PO$_4$ pH 7.4, 1 mM DTT, and 0.1 mM EDTA) prior to application to a 10 × 2 cm column of hydroxylapatite equilibrated in buffer D. The column is washed successively with 4–5 column volumes of buffer D, buffer D containing 0.1 M KH$_2$PO$_4$, pH 7.4, and buffer D

TABLE II

Reticulocyte Release Factor (RF) Purification[a]

Fraction	Protein (mg)	Specific activity[b]	Yield (%)
I　S100	14,796	—	—
II　Ammonium sulfate	1,822	—	—
III　DEAE-Sephadex	294	13.9	100
IV　Phosphocellulose	7.2	198	35
V　Sephadex G-200	1.2	572	17

[a] For calculation of specific activity, the rate of f[^{3}H]Met release was determined at 15 minutes, 24°, in reactions containing limiting levels of RF, 2.92 pmoles of f[^{3}H]Met-tRNA$_f$·ribosome intermediates, 0.2 A_{260} unit of poly(A,G.U), 0.1 mM GTP, and other components as described. The percent yield is the cumulative recovery based on the activity of fraction III. Activity was calculated as the increase in f[^{3}H]Met release due to the addition of oligonucleotide. Activity determinations on fractions I and II are difficult because of the presence of a codon-independent fMet-tRNA hydrolase activity in these fractions.

[b] Picomoles of f[^{3}H]Met per milligram in 15 minutes.

containing 0.2 *M* KH$_2$PO$_4$, pH 7.4. The RF activity elutes in the final buffer.

Reticulocyte RF can be purified by sucrose gradient sedimentation. A 0.5 ml fraction of RF is sedimented through a 5–20% sucrose gradient (13 ml) in an SW 41 rotor of Beckman L265B ultracentrifuge at 39,000 rpm for 25 hours at 4°. The RF activity of 0.25-ml fractions are given

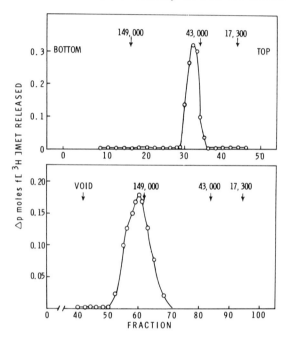

Fig. 3. (A) The major fraction of partially purified reticulocyte release factor (RF) (○—○) was sedimented through a 5 to 20% sucrose gradient (13 ml) in an SW 41 rotor of Beckman L265B ultracentrifuge at 39,000 rpm for 25 hours at 2°. Proteins, aldolase (149,000), ovalbumin (43,000), and myoglobin (17,800) were used as molecular weight markers (arrows). The sedimentation profiles have been superimposed in this diagram. Fractions were 0.25 ml. All release reactions were 15 minutes at 24° and contained: 0.1 A_{260} units of UAAA, 3.00 pmoles of f[^{3}H]met-tRNA$_f$·ribosome intermediates, 0.1 m*M* GTP, 0.010 ml of each fraction, and buffer components as described. (B) The major fraction of partially purified reticulocyte RF (4-ml sample) was chromatographed on a Sephadex G-200 column (100 × 2.5 cm) and eluted with 50 m*M* Tris·HCl, 0.3 *M* KCl, 1 m*M* EDTA, and 0.1 m*M* DTT. Fractions were 4 ml. Release assays were incubated 15 minutes at 24° and contained 0.1 A_{260} units of UAAA, 3.0 pmoles of f[^{3}H]Met-tRNA$_f$·ribosome intermediate; 0.1 m*M* GTP, 0.010 ml of each fraction, and buffer components as described. Each purified protein marker was applied separately in 4 ml, and its position of elution was determined by absorbance at 280 nm or 410 nm (myoglobin). The void volume as determined from the elution volume of dextran blue is also shown.

in Fig. 3 together with the sedimentation positions of proteins of known molecular weight. The estimated molecular weight of RF based on these studies is 54,000. This is a useful final purification step because of the strikingly different migration of RF (on the basis of molecular weight) on Sephadex G-200 and when sedimented in a sucrose gradient (Fig. 3A, 3B).

Polyacrylamide Gel Analyses of Reticulocyte RF

The composition of the highly purified major RF fraction (fraction V, sedimented through a 5–20% sucrose gradient as described—legend Fig. 3A) is examined by 0.1% sodium dodecyl sulfate-10% polyacrylamide gel electrophoresis as described by Laemmli.[15] After electrophoresis of the bromophenol dye to the end of an 11 cm gel (3 hours at 2 mA/gel), the gel was removed from the glass tube and stained with 0.2% Coomassie Brilliant Blue in 50% methanol, 7% acetic acid for 2 hours at 37°. The gels were destained with 7% acetic acid, 5% methanol over 30 minutes at 1 A in a quick gel destainer apparatus (Canalco). These studies suggest the major fraction of reticulocyte RF contains a single component of estimated molecular weight 54,000.

[15] U. K. Laemmli, *Nature (London)* **227**, 680 (1970).

[31] An Assay for Protein Chain Termination Using Peptidyl-tRNA

By JOHN R. MENNINGER[1] and CHARLINE WALKER

This assay for protein chain termination has the virtues of being performed at low magnesium ion concentrations in the absence of solvent perturbants (e.g., ethanol) with unpurified transfer RNA and a random copolymer message. It may therefore be of use in exploring the process of protein chain termination in eukaryote systems where inefficient isolation methods preclude the extensive purification of components. Since the nucleic acid moiety of the substrate for the assay is a typical transfer RNA, use of this technique gives answers complementary to those involving the unique methionine $tRNA_F$.[2] The ability to perform the measurements in the absence of, e.g., ethanol or acetone may allow one to per-

[1] Supported by funds from the U.S. Public Health Service (GM 15123).
[2] C. T. Caskey, E. Scolnick, R. Tompkins, G. Milman, and J. Goldstein, this series, Vol. 20, p. 367.

ceive the control of protein chain termination unimpeded by the distorting effects of structure perturbants.

Preparation of Substrate

The substrate for the assay is a set of oligolysyl-tRNA's synthesized by a cell-free system in response to the message poly(rA). Since the substrate is synthesized in a separate step, the incubation parameters may be modified to yield maximum product without concern for their relation to physiological conditions. Below is our current protocol for synthesizing oligolysyl-tRNA's in a cell-free extract derived from *Escherichia coli* K12.

The incubation mixture contains, per milliliter: 100 μmoles of ammonium cacodylate (pH 7.2), 100 μmoles of NH_4Cl, 15 μmoles of magnesium acetate, 5.0 μmoles of PEP, 0.63 μmole of neutralized ATP, 0.06 μmole of CTP, 0.30 μmole of GTP, 20 μg of pyruvate kinase (Boehringer), 40 nmoles of [^{14}C]lysine (specific activity 50–100 μCi/μmole), 3–6 A_{260} (absorbance at 260 nm) units of poly(rA) (Miles), 5 μmoles of β-mercaptoethanol, <50 A_{260} units of *E. coli* tRNA (General Biochemicals), 75 μl of S100 enzymes, and 40 A_{260} units of washed ribosomes. This mixture is incubated at 37° for 40 minutes, then the pH is reduced by addition of 1/10 volume of 1.0 M sodium acetate buffer (pH 4.5). During the incubation the progress of oligolysyl-tRNA formation can be estimated by assaying the cold acid-precipitable radiolysines. A satisfactory final value is 100 pmoles of lysine per A_{260} unit of tRNA.

The entire reaction mixture is next applied to a column of at least equal volume of benzoylated DEAE-cellulose. The column resin should be cycled first with buffer TRG and preequilibrated with buffer TI before use. Buffer TI contains 10 mM sodium acetate (pH 4.5), 10 mM $MgSO_4$, 1 mM $Na_2S_2O_3$, and 0.3 M NaCl. Buffer TRG contains the same components except 2.0 M NaCl and 25% (v/v) ethanol. Protein elutes from the column in buffer TI, as evidenced by the emergence of an absorbance peak at 280 nm. The nucleic acid is eluted by applying buffer TIII, which contains the same components as TI except 1.5 M NaCl and 10% (v/v) ethanol. The fractions containing high absorbance at 260 nm are pooled, and the nucleic acids are precipitated by addition of 3 volumes 95% ethanol. It is also satisfactory to remove protein and ribosomal RNA by phenol extraction and precipitation in high salt.[3]

The precipitated nucleic acids are collected by sedimentation at 45,000 g and resuspended at 0° in 10 mM potassium acetate (pH 5.0). This solution is then brought to 5 mM magnesium acetate, and any remaining poly(rA) is allowed to precipitate at 0° for 30 minutes. After removing the precipitate by sedimentation, as above, the oligolysyl-tRNA prepara-

[3] J. R. Menninger, *Biochim. Biophys. Acta* **240**, 237 (1971).

tion can be concentrated by ethanol precipitation or used directly after dialysis against an appropriate solution.

There is little, if any, fractionation of the tRNA's during these isolation procedures and although the oligo-[¹⁴C]lysyl-tRNA's are isotopically pure all the other tRNA's are also present, albeit largely unacylated. It is possible to analyze the length distribution of the oligolysines by alkaline hydrolysis of the oligolysyl-tRNA's followed by chromatography on phosphocellulose.[4] This is not necessary for the *E. coli* system since we have shown[3] that all the length classes participate in the chain termination assay to roughly equal extents.

There are four assays of the oligolysyl-tRNA preparation that are normally performed: cold acid-precipitable radioactivity, which measures tRNA-bound oligolysines of length ≥ 1 (cold trichloroacetic acid does not precipitate free oligolysines); the direct phosphocellulose paper assay, which measures free oligolysines of length ≥ 2; the total phosphocellulose paper assay, which measures total oligolysines of length ≥ 2, and the phosphocellulose paper spot assay, which measures total lysines. One preparation had the following values, respectively: 160, 3.2, 108, 199 pmoles per A_{260} unit.

Assay of Oligolysines

The general assay has been described previously.[5] We review here our current methodology and some typical values. The assay is based on the electrostatic attraction between the protonated amino groups of oligolysines and the phosphate groups of phosphocellulose paper (Whatman P-81) that exists at low pH. Complete reaction mixtures or other solutions to be assayed are spread on the paper with a micropipette, ≤ 45 μl per 3 × 4 cm piece of Whatman P-81. This paper is immediately immersed in a solution containing, per liter: 6 moles of urea, 20 ml of 1.0 N acetic acid, and 50 ml of 1.0 M sodium or potassium acetate buffer (pH 5.0). This solution should be filtered (millipore HA 0.45 μ) before use. We usually have about 1 liter of the solution in a plastic basin of 20 cm diameter into which 30 such samples can be placed without serious cross-contamination. (We try to avoid allowing papers to lie against one another for long periods of time.) A blank paper, in the same basin, carried through the various washes will have a radioactive count rate of less than 100 cpm, well below the background of control incubations in an experiment.

The papers remain for 10 minutes or longer in the urea solution at

[4] J. R. Menninger, M. C. Mulholland, and W. S. Stirewalt, *Biochim. Biophys. Acta* **217**, 496 (1970).
[5] M. S. Bretscher, H. M. Goodman, J. R. Menninger, and J. D. Smith, *J. Mol. Biol.* **14**, 634 (1965).

room temperature with occasional agitation. The solution is then aspirated from the tilted basin and is replaced with an equal volume of 0.35 M NaCl, 50 mM sodium formate (pH 3.5). This wash is repeated twice more with periods of 10 minutes of occasional gentle agitation between. This salt concentration is sufficient to remove all the free lysines and lysyl-tRNA's from the paper. Oligolysines of length ≥ 2 are largely retained, provided they are not attached to tRNA. (Washing with buffer containing 0.65 M NaCl will remove most of the free dilysines.[5]) A low-salt wash to improve counting efficiency (50 mM NaCl; 50 mM sodium formate, pH 3.5) is followed by the final wash of 5% (v/v) of 1.0 M sodium formate (pH 3.5) in absolute ethanol, which speeds drying the papers on aluminum foil at 70°.

The radioactivity is assessed in a toluene scintillator: 4.0 g of 2,5-diphenyloxazole (PPO); 0.25 g of 1,4-bis-2-(4-methyl-5-phenyloxazole) benzene (dimethyl POPOP); toluene to 1 liter. The papers may be removed from the scintillator and replaced with other samples without danger of radiocontamination although we usually count the empty bottles briefly to ensure that no fragments of the previous sample remain (count rate < 40 cpm). The efficiency of counting [^{14}C]lysine in a particulate scintillation spectrometer can easily be measured by spotting a known sample on a piece of P-81 paper that has been carried through the series of washes. We find values in the range 0.75–0.80 cpm/dpm for the Packard Model 3375.

If the reaction mixture or other solution is spread directly on the phosphocellulose paper, only free oligolysines (≥ 2) are retained after washing. To assay total oligolysines one must treat briefly with alkali to hydrolyse the oligolysyl-tRNA ester bond. The solution is brought to around 0.3 M KOH and incubated 10 minutes, 37°, before neutralizing with an equivalent amount of acetic acid. This treatment suffices to hydrolyze all the oligolysines from oligolysyl-tRNA but does not degrade oligolysines. To assay total lysines we spot a sample on a piece of P-81 paper that has been taken through the wash regime and count the radioactivity after drying.

By suitably combining the four assays described above, one can estimate the tRNA-bound oligolysines (length ≥ 2), lysyl-tRNA and the fraction of oligolysines that are free from tRNA. This last value is usually no more than 5% and represents the background due to spontaneous hydrolysis against which one measures chain termination.

Preparation of Ribosomes, Release Factors, and Poly[r(A,U)]

Ribosomes from *E. coli* may be prepared by any of the standard techniques. We have found it necessary to wash thoroughly in 1.0 M NH$_4$Cl to remove all release factor activity from *E. coli* K12 ribosomes.[3] This is in contrast to *E. coli* B ribosomes which require only 0.1 M NH$_4$Cl to strip off the release factor activity.[6] Excessive washing at these salt concentrations may tend to remove ribosomal proteins; four to six sedimentations of 90 minutes at 150,000 g from 1.0 M NH$_4$Cl is sufficient. One can substitute two sedimentations through 35% sucrose in 1.0 M NH$_4$Cl.[6] Washing in high salt also removes peptidyl-tRNA hydrolase from the ribosomes, which would otherwise cause a high message-independent release of oligolysines.[4]

We have used postribosomal supernatant, purified from the peptidyl-tRNA hydrolase, as a source of release factors.[2,7] After two sedimentations of a crude extract at 150,000 g to remove ribosomes, the supernatant is passed over Sephadex G-25 previously equilibrated with 20 mM Tris·HCl, pH 7.8; 2 mM β-mercaptoethanol. The protein peak is located by its absorbance of 280 nm light, then passed through a column of phosphocellulose (Whatman P-11) previously equilibrated with the same buffer. The protein emerging from phosphocellulose is precipitated with 0.56 g of ammonium sulfate per milliliter (final volume). This material contains the bulk of the release activity, is superior to protein eluted from DEAE-cellulose[3] and is free from peptidyl-tRNA hydrolase. Release factor activity for this assay can be separated from other proteins by chromatography over DEAE-Sephadex (Menninger and Mottur, unpublished results; see also footnotes 2 and 7). The crude material can be stored at $-70°$ for months with good retention of activity.

The message for the assay of protein chain termination is a random copolymer poly[r(A,U)]. The ratio of the two nucleotides is important in determining the ability of the RNA to stimulate chain termination. The data in Fig. 1 show the dependence of release on the ratio A/U. Commercial suppliers of random poly[r(A,U)] tend to make only polymers with A/U = 1, which is not suitable for this assay.

We have made our own copolymers with commercially available polynucleotide phosphorylase. The incubation mixture contains: 0.10 M Tris·HCl (pH 9.0), 10 mM MgCl$_2$, 0.5 mM EDTA (pH 7), 47 mM ADP, 9 mM UDP, 1 mg/ml polynucleotide phosphorylase (P-L Biochemicals, type 1, *Micrococcus lysodeikticus*), one drop toluene. The pH must be adjusted to approximately 8.5–9.0 by adding KOH, using pH paper as a guide (requires approximately 40 μmoles/ml).

The incubations were carried out at 37° for 15 hours with addition of 1/100 volume of 0.6 M MgCl$_2$, 0.6 M NH$_4$Cl at 7 hours to precipitate

[6] M. R. Capecchi, *Proc. Nat. Acad. Sci. U.S.* **58**, 1144 (1967).
[7] H. A. Klein and M. R. Capecchi, *J. Biol. Chem.* **246**, 1055 (1971).

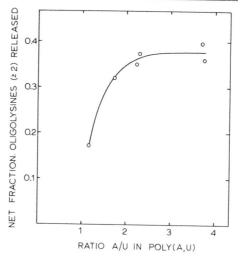

Fig. 1. Oligolysines released from oligolysyl-tRNA, stimulated by poly[r(A,U)] of different base ratios. The incubations were for 40 minutes at 37°, and components of the reaction mixture (50 μl) are described in the text. The various poly[r(A,U)] were present at an absorbance (260 nm) of 7. In control experiments the net release stimulated by poly(rA) (A_{260} = 6) was 0.11. Release in the absence of message has been subtracted from the illustrated data. The experiments were performed by George Mottur.

inorganic phosphate. The viscosity was obviously increased after 2 hours in these conditions. The RNA product was isolated by extracting the reaction mixture with 80% phenol, and thoroughly dialyzing the recovered aqueous phase against 1 mM Tris·HCl, pH 7. The absorbance (260 nm) of such a preparation is approximately 100. The course of the reaction can be estimated by ascending thin-layer chromatography in 250 μm cellulose layers (with fluorescent indicator, Analtech) in 0.1 M phosphate (pH 6.8)/ammonium sulfate/n-propanol (100/60/2). The polymer remains at the origin of the chromatogram while precursor nucleoside-diphosphates have R_f's greater than 0.37. The base ratios of the product synthesized in the reaction described were determined by the method of Katz and Comb[8] as A/U = 3.7. The RNA product base ratio is approximately linear with precursor nucleotide-diphosphate ratio; slope is 0.7.

Assay for Protein Chain Termination

The reaction volume normally used is 50 μl and contains: 0.10 M ammonium cacodylate (pH 7.2), 0.11 M NH₄Cl, 10 mM magnesium acetate, 50 μM of each of the twenty amino acids, 6 mM β-mercapto-

[8] S. Katz and D. G. Comb, *J. Biol. Chem.* **238,** 3065 (1963).

ethanol, 1/10 volume of supernatant enzymes containing release factors, 40 A_{260} units per milliliter of washed ribosomes and approximately 10^4 cpm of oligo-[^{14}C]lysyl-tRNA. These are incubated at 37° for 40 minutes and are either assayed directly for released oligolysines (45 μl on 3 × 4 cm piece of P-81 phosphocellulose paper) or receive 5 μl of 4 N KOH to hydrolyze oligolysyl-tRNA's. These latter are incubated for 10 minutes at 37° and 5 μl of 4 N acetic acid is added before a 45 μl sample is applied

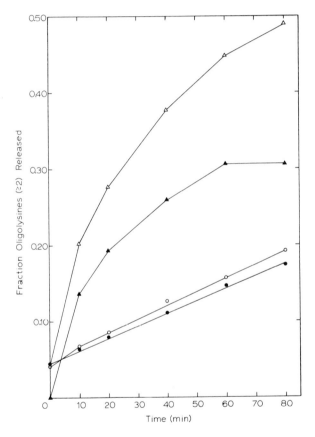

FIG. 2. Time course of oligolysine release stimulated by poly(rA) and poly [r(A,U)]. The 37° incubations contained the components described in the text and 0.3 mM GTP. Oligo-[^{14}C]lysyl-tRNA was present at an absorbance (260 nm) of 9.8 (140 pmoles of lysines (length > 2) per A_{260} unit), poly(rA) (○—○), and poly[r(A,U)] (△—△) (A/U ratio = 1.8) at an absorbance of 6.9 (260 nm). The control incubation without added mRNA is shown (●—●) as well as the poly[r(A,U)]-stimulated release (▲—▲) corrected for the average release stimulated by poly(A) and the message-less control. This figure is reproduced, with permission, from J. R. Menninger, *Biochim. Biophys. Acta* **240**, 241 (1971), Fig. 2.

to the phosphocellulose paper. The dilution of the reaction volume during hydrolysis/neutralization is taken into account in tabulating the data.

Preparations of oligolysyl-tRNA's vary. The required sensitivity of the assay can usually be obtained by adding between 10 and 20 A_{260} units per milliliter of our typical preparations (specific activity of lysine approximately 80 $\mu Ci/\mu mole$). At these concentrations the substrate is limiting for the reaction rate.

The time course of release stimulated by poly(rA) and poly[r(A,U)] are shown in Fig. 2. We have used the fraction of lysines in peptides of length ≥ 2 released from tRNA in 40 minutes in the standard assay as our measure of protein chain termination activity. Because of the indefinite peptide length of the substrate, a standard specific activity cannot easily be determined. As a control we always determine the fraction released in the presence of poly(rA) which binds the substrate oligolysyl-tRNA's to the ribosome but which contains no chain termination signals.

The ion requirements for this chain termination assay are similar to those for other ribosome-stimulated reactions: magnesium from 5 to 15 mM (optimal at 8–15 mM), potassium or ammonium from 100 to 300 mM (Tris ion does not satisfy this requirement). The spontaneous release of oligolysines increases with increasing pH and so also does the poly[r(A,U)]-stimulated termination. There is a range of pH in which the discrimination of message-stimulated release from the spontaneous background is satisfactory. We have used ammonium cacodylate as the buffer because the useful range is around pH 7.0 where Tris is a poor buffer. It seems now to be difficult to obtain cacodylic acid with which to make such buffers but the more expensive HEPES or equivalent can be used satisfactorily. (It is worth noting that 1.0 M ammonium cacodylate at pH 7.6 changes to pH 7.2 on dilution to 0.1 M.) This assay requires no energy source in the form of ATP. We have noticed a small but consistent stimulation by GTP which is not shown by the analog of GMP-PCP. This stimulation is saturated by 0.3 mM GTP.

Section IV

Ribosome Structure and Function

[32] Preparation and Characterization of Free and Membrane-Bound Polysomes

By H. BLOEMENDAL, E. L. BENEDETTI, and W. S. BONT

In most eukaryotic cells ribosomes are found either attached to membranes or lying free in the cytoplasmic ground substance. The occurrence of these two classes of ribonucleoprotein particles in several types of cells was originally reported by Palade.[1]

Moreover, it appeared that in rapidly growing cells, such as embryonic tissue, most of the ribosomes were free and probably engaged in the synthesis of protein for internal use. On the other hand, in differentiated cells where large amounts of protein are produced and exported, the particles were found mostly in close association with the endoplasmic reticulum membranes. In few instances, as in tissue culture,[2] membrane-bound ribosomes had been described in cells producing little if any protein for export.

In order to enable a correlation between morphological features and functional aspects, the need was felt to isolate separately the two classes of ribosomes. The apparent prerequisite for this aim was to avoid the use of media that would be harmful for membranes. In fact most of the available methods for the isolation of ribosomes and ribosomal clusters were based on the treatment of the postmitochondrial fraction with detergents. Progress in this field was achieved when polysomes, active both *in vivo* and *in vitro,* could be isolated without the use of any detergent.[3-6] Under this condition two well-defined fractions of ribosomes were obtained: one consisting of free ribosomal clusters, and the other of membrane-bound ribosomes.

Nomenclature

For operational purposes the following nomenclature is proposed. Free polysomes are those polysomes accumulating in a pellet after centrifugation of the postmitochondrial supernatant through a discontinuous

[1] G. E. Palade, *Microsomal Particles Protein Synth., Pap. Symp. 1st Biophys. Soc., 1958,* p. 36.
[2] B. Attardi, B. Cravioto, and G. Attardi, *J. Mol. Biol.* **44,** 47 (1969).
[3] H. Bloemendal, W. S. Bont, and E. L. Benedetti, *Biochim. Biophys. Acta* **87,** 177 (1964).
[4] T. E. Webb, G. Blobel, and V. R. Potter, *Cancer Res.* **24,** 1299 (1964).
[5] H. Bloemendal, W. S. Bont, M. de Vries, and E. L. Benedetti, *Biochem. J.* **203,** 177 (1967).
[6] G. Blobel and V. R. Potter, *J. Mol. Biol.* **26,** 279 (1967).

sucrose gradient. Bound polysomes are polysomes isolated in a fraction containing fragments of the endoplasmic reticulum and which can be freed from the membranes of the endoplasmic reticulum by the use of a detergent. Rough membranes are membranes that have ribosomes attached. Smooth membranes are membranes devoid of ribosomes.

Media

A certain flexibility in the composition of the media is permissible.

Medium A: 0.35 M sucrose, 50 mM tris(hydroxymethyl)aminomethane (Tris)·HCl, pH 7.6, 25 mM KCl, 10 mM MgCl$_2$ (or Mg acetate). The sucrose should be RNase-free (commercially available from British Drug Houses).
Medium B: Medium A without sucrose
Medium C: 50 mM Tris·HCl, pH 7.6, 5 mM MgCl$_2$, 70 mM KCl

Preparation of Rat Liver Polysomes

Large-Scale Preparation

Rats, preferably not older than 3 months, are killed by decapitation; the livers are collected and rinsed with ice-cold medium A (2.5 volumes per gram wet weight of liver). A Teflon–glass homogenizer is used. The clearance between tube (diameter 25 mm) and pestle should not exceed the range of 0.15–0.25 mm. The glass tube is cooled in ice. Five strokes at 500–1000 rpm are applied. Nuclei and mitochondria are removed by centrifugation at 15,000 g for 10 minutes in rotor 30 of a Spinco ultracentrifuge or another type of centrifuge with comparable rotor. Three-quarters of the supernatant fraction is sucked off and collected; then 15 ml are layered on top of a discontinuous sucrose gradient consisting of 10 ml of 2 M sucrose and 10 ml of 1.5 M sucrose in medium B. Centrifugation at 75,000 g_{av} in rotor 30 is for 16 hours. The separation achieved is schematically depicted in Fig. 1. A pellet at the bottom and three colored fractions in the 1.5–2 M sucrose interphase are obtained.

After the run the upper (red) layer is sucked off and either discarded or used as source of soluble enzymes for incorporation experiments. Then the wall of the tube is punctured slightly above band 1. This band, which contains predominantly smooth membranes, is collected. The same manipulation is repeated with band 3, in which the bound polysomes accumulate. Thereafter the 2 M sucrose layer is sucked off, and the part of the tube with the pellet is cut out. For electron microscopy and chemical analysis all separated fractions are freed from sucrose by washing in sucrose-free medium. For incorporation studies it is not necessary to wash the free polysomes. Instead they are immediately suspended in sucrose-

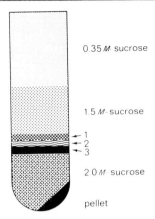

0.35 M sucrose

1.5 M sucrose

← 1
← 2
← 3

2.0 M sucrose

pellet

FIG. 1. Schematic representation of the separation obtained after centrifugation of a postmitochondrial supernatant fraction through a discontinuous sucrose gradient.

free medium and kept for 1 hour on ice. If stored at $-21°$ or lower temperature, the polysomes keep their activity for months. Freezing of the polysomes should be performed in small portions in order to prevent repeated thawing and freezing in subsequent experiments.

For incorporation studies *in vitro* with membrane-bound polysomes it is recommended to keep this fraction throughout in the 100,000 g fraction or to avoid a sucrose-concentration of the medium lower than 1 M.[7] This is achieved by diluting the band 3 fraction with medium B to a final sucrose concentration of 1.1 M. The diluted suspension is centrifuged through a cushion of 1 ml 1.25 M sucrose in rotor 50 at 150,000 g for 2 hours.

Small-Scale Preparation

When only small amounts of polysomes are wanted, the centrifugation procedure can be carried out in a small rotor, e.g., rotor No. 50 of the Spinco ultracentrifuge. The gradient consists of the following layers: 2.5 ml of 2 M sucrose in medium B, 2.5 ml of 1.5 M sucrose in medium B, and 5 ml of the postmitochondrial fraction. Centrifugation is performed at 100,000 g_{av} for 3 hours.

Preparation of Spleen Polysomes

The procedure described for the isolation of rat liver polysomes is also applicable for polysomes from many other tissues. However, modifications may be required in some cases. As an example, the isolation of

[7] W. S. Bont, J. Geels, A. Huizinga, K. Mekkelholt, and P. Emmelot, *Biochim. Biophys. Acta* **262**, 514 (1972).

polysomes from spleen is described. When this technique is used without modification for the isolation of mouse spleen polysomes, most of the polysomes are lost in the 15,000 g pellet. For this reason, the 7000 g supernatant has to be used.[8] Moreover, a satisfactory yield of intact polysomes cannot be obtained from normal spleens without the use of a potent RNase inhibitor. Rat liver supernatant contains such an inhibitor.[9-13] The inhibitor is a protein of limited specificity. It inhibits alkaline RNase from different sources, e.g., spleen RNase.[8,14]

Spleens of normal mice are homogenized in 6 ml (containing approximately 200 mg of protein) of rat liver supernatant per gram of tissue by five strokes of a motor-driven Teflon–glass homogenizer (2000 rpm). If a purified inhibitor preparation is used (DEAE-cellulose material[13]) the amount of added protein may be reduced to 6 mg.

The homogenate is centrifuged at 1250 g_{av} for 3 minutes and subsequently at 7000 g_{av} for 10 minutes in a Sorvall RC2-B preparative centrifuge. The supernatant fraction is used for the isolation of polysomes. Samples (18 ml) are layered on top of a discontinuous gradient, similar as described for rat liver polysomes, however, the middle layer is 1.0 M in sucrose. These gradients are centrifuged at 94,000 g_{av} for 20 hours at 0° in the SW 27 swinging-bucket rotor of the Spinco L-2 preparative ultracentrifuge. The pellet contains the free polysomes whereas a large amount of monomers and small polysomes are retained in the 2.0 M sucrose layer. The 1 M:2 M interphase contains smooth and rough membranes in addition to 80 S ribosomal monomers. In order to obtain a separation of bound polysomes from monomeric ribosomes an additional centrifugation step is required. For this purpose the interphase fraction from the discontinuous gradient is diluted to a sucrose concentration lower than 10% and 5 ml is applied onto a continuous sucrose gradient (31 ml, 10–34% w/v) in medium C. The gradient is layered on a sucrose cushion (3 ml, 68% sucrose w/v). Centrifugation is for 1.5 hours at 94,000 g_{av} in a Spinco SW 27 rotor. The gradient can be monitored automatically at 254 nm. A typical scanning is shown in Fig. 2. Fractions 2–5 contain smooth and rough membranes, fractions 25–32, mainly monomers.

[8] J. T. M. Burghouts, A. L. H. Stols, and H. Bloemendal, *Biochem. J.* **119**, 749 (1970).

[9] J. S. Roth, *J. Biol. Chem.* **231**, 1085 (1958).

[10] W. S. Bont, G. Rezelman, and H. Bloemendal, *Biochem. J.* **95**, 15C (1965).

[11] G. Blobel and V. R. Potter, *Proc. Nat. Acad. Sci. U.S.* **55**, 1283 (1966).

[12] H. Sugano, I. Watanabe, and K. Ogata, *J. Biochem. (Tokyo)* **61**, 778 (1967).

[13] A. A. M. Gribnau, J. G. G. Schoenmakers, and H. Bloemendal, *Arch. Biochem. Biophys.* **130**, 48 (1969).

[14] P. V. Northup, W. S. Hammond, and M. F. La Via, *Proc. Nat. Acad. Sci. U.S.* **57**, 273 (1967).

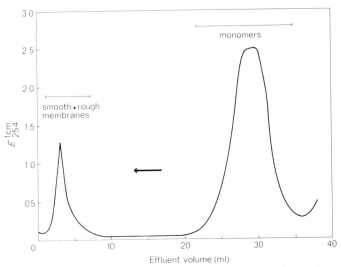

FIG. 2. Separation of the total membrane fraction from free ribosomal monomers. (The arrow indicates the direction of sedimentation.) The bound spleen polysomes concentrate on the sucrose cushion.

Electron Microscopy

For electron microscopic studies, the pellet and the layers after isolation are resuspended and repeatedly washed by centrifugation, each time at 100,000 g for 30 minutes in medium B. The pellets obtained are then fixed in 2% OsO_4 in phosphate buffer at pH 7.1 for 1 hour. The embedding is performed in Vestopal W or in Epon. Thin sections can be stained in uranyl acetate and lead hydroxide.[15]

For high-resolution studies the pellets can be resuspended and studied as a spread preparation. For that purpose the resuspended material in medium B is carefully spread at various concentrations (optimum about 50 μg RNA per milliliter) on carbon-coated grids. The preparation is stained immediately after the spreading either in an aqueous solution of uranyl acetate (1% at unadjusted pH 4.3) or in a solution of uranyl-oxalate.[16] A solution of uranyl acetate in acetone (10^{-5} M) may also be used.[17] Several drops of the staining solutions are added onto the grids. During this step the excess of material and contaminants are washed away.

[15] E. L. Benedetti, W. S. Bont, and H. Bloemendal, *Lab. Invest.* **15**, 196 (1966).

[16] J. L. Mellema, E. F. J. van Bruggen, and M. Gruber, *J. Mol. Biol.* **31**, 75 (1968).

[17] M. Beer, P. Bartl, T. Koller, and A. P. Erickson, *in* "Methods in Cancer Research," (H. Busch, ed.), Vol. 4, p. 283. Academic Press, New York, 1971.

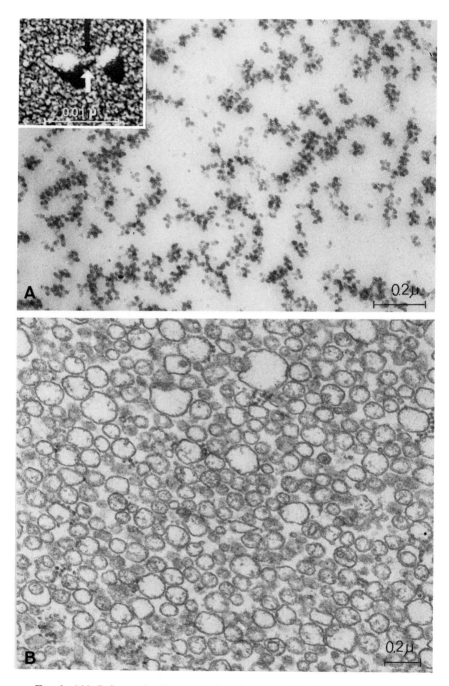

FIG. 3. (A) Polysomal pellet consisting almost exclusively of ribosomal clusters of various length, some showing helical configuration. Thin section stained with

318

The staining solution is then removed with filter paper, and the preparation is very rapidly dried.

Spread and unstained preparations can be shadowed with carbon platinum evaporation either from one direction at an angle of 25° or while rotating the specimen at 40 rpm, the angle of shadowing being 10°. Usually before shadowing the spread preparation is fixed for 5 minutes in 0.1% glutaraldehyde dissolved in medium B. For this purpose several drops of the solution are added on top of the grids. Thereafter the excess of glutaraldehyde is removed very carefully without drying the sample. The latter is finally thoroughly washed with twice distilled water and dried.

Characterization of Polysomes

The characterization of isolated free or bound polysomes can be achieved in several ways: morphological, functional, and biochemical.

Morphological Characterization

Electron microscopic observation as thin sections of the embedded pellet originating from the fraction at the bottom of the 2 M sucrose layer reveals that the preparation consists essentially of clusters of ribosomes (polysomes) (Fig. 3A). These clusters are formed by a variable number of monomers, held together by a thin strand (see inset, Fig. 3A).[18] A distribution curve of polysomes composed of various numbers of monomers can be obtained by plotting the number of monomers per polysome against the number of polysomes detected.[19]

In the isolated pellet, most of the polysomes show a helical configuration. Examination of the thin sections obtained from the pelleted layers (compare Fig. 1, bands 1, 2, 3) reveals that these fractions consist of membrane profiles studded with a variable number of ribosomes.

In the upper two layers the number of smooth membranes is preponderant (Figs. 3B and 4A) in comparison with layer 3, where most of the membrane fragments are associated with ribosomal clusters (Fig. 4B). The distribution of ribosomes at the membrane surface is irregular, and some segments of the membrane profile have a smooth outer surface. Moreover, a substantial part of the polysomal array is lying free, and only

[18] E. L. Benedetti, H. Bloemendal, and W. S. Bont, C. R. Acad. Sci. 259, 1353 (1964).
[19] E. L. Benedetti, A. Zweers, and H. Bloemendal, Biochem. J. 108, 765 (1968).

uranyl acetate and lead hydroxide. *Inset.* Two monomers bridged by a strand. Spread preparation shadowed from one direction with platinum carbon. (B) Layer 1 consisting mainly of smooth membranes.

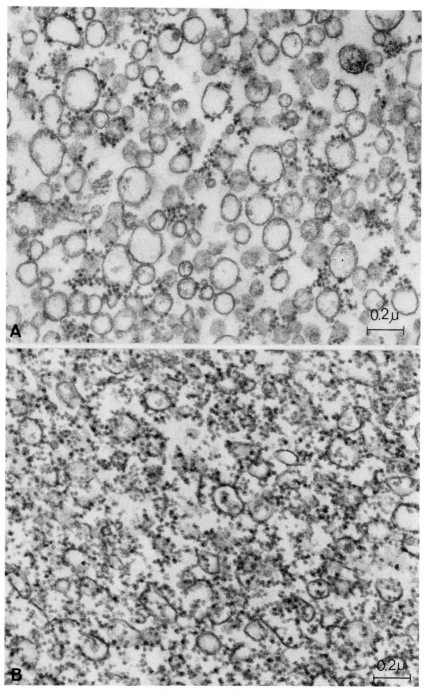

FIG. 4. (A) Layer 2 showing that only a small proportion of the membranes is studded with ribosomal array. (B) Layer 3. In this fraction the major part of the membranes is associated with ribosomal clusters.

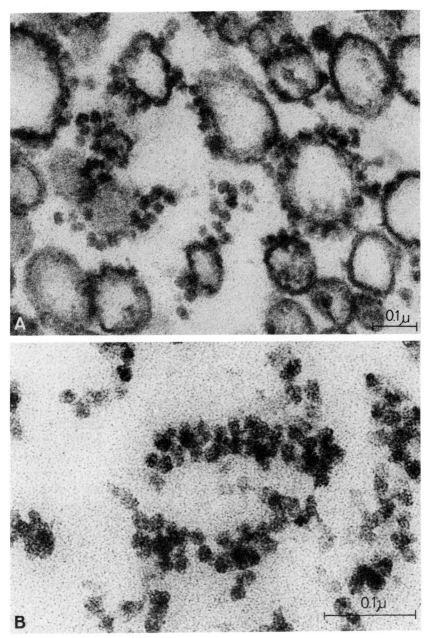

FIG. 5. (A) Layer 3. Some polysomes are attached to the membranes via one or few monomers. (B) Free polysomes having helical configuration.

one or two monomers appear to be connected with the membranes[20] (Fig. 5A). These particles like most polysomes in the pellet show a helical configuration (Fig. 5B).

Functional Characterization

It is well established that the polysomes are the site of protein biosynthesis in the intact cell. Hence it is plausible that also the biosynthetic activity *in vitro* of the isolated particles is considered as a specific property. In order to compare the two classes of polysomes, preparations of both are incubated with amino acids, cofactors, and energy under identical conditions. In such comparative studies one has to keep in mind that incorporation of amino acids by bound polysomes (band 3) is dependent on the physical state of the membranous part which in turn is affected by the manipulations necessary for the isolation. Hence, as mentioned the material derived from band 3 should be kept either in a medium with a sucrose concentration of at least 1 M or in postmicrosomal supernatant fraction. When transferred into a medium with lower sucrose concentration the incorporation of amino acids is impaired.[7]

When bound polysomes have to be freed from membranes, this can be achieved by treatment with ionic or nonionic detergents.[21,22] Since ribonuclease may be released when the endoplasmic reticulum membranes are solubilized by deoxycholate, it is necessary also for this reason to suspend band 3 in postmicrosomal supernatant prior to the addition of the detergent. An inhibitor present in the latter fractions, which counteracts the alkaline ribonuclease, protects the polysome against breakdown.[10–13,23,24]

1. Incorporation of Amino Acids in Vivo. Rats are starved for 2 days. At 30 minutes before injection with the radioactive amino acid (e.g., 8 μCi of [¹⁴C]leucine), they are fed and per rat 1.5 μmoles of the amino acid dissolved in 0.5 ml of 0.9 M sodium chloride is injected intravenously. Rats are killed, respectively, after 5, 15, 60, and 120 minutes. The different fractions are isolated as described under large-scale preparation p. 314.

The graphical representation of the experiment is shown in Fig. 6.

[20] E. L. Benedetti, W. S. Bont, and H. Bloemendal, *Nature (London)* **210**, 1156 (1966).

[21] F. O. Wettstein, T. Staehelin, and H. Noll, *Nature (London)* **197**, 430 (1963).

[22] S. Olsnes, R. Heiberg, and A. Pihl, *Biochim. Biophys. Acta* **272**, 75 (1972).

[23] A. A. M. Gribnau, J. G. G. Schoenmakers, M. van Kraaikamp, and H. Bloemendal, *Biochem. Biophys. Res. Commun.* **38**, 1064 (1970).

[24] A. A. M. Gribnau, J. G. G. Schoenmakers, M. van Kraaikamp, M. Hilak, and H. Bloemendal, *Biochim. Biophys. Acta* **224**, 55 (1970).

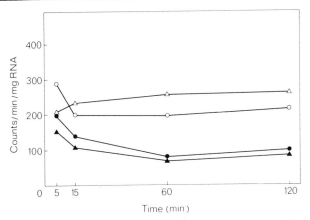

FIG. 6. Incorporation of [¹⁴C]leucine *in vivo* into the protein of free and membrane-bound polysomes. ○, pellet; △, band 3; ●, pellet after DOC treatment; ▲, band 3 after DOC treatment.

2. Incorporation of Amino Acid in Vitro. The incubation mixture contains 0.02 μmole of DL-[¹⁴C]leucine (7 mCi/mmole), 0.125 μmole of ATP, 0.063 μmole of GTP, 1.25 μmoles of phosphoenolpyruvate, 7 μg of pyruvate kinase (EC 2.7.2.40), 0.69 mg of 2-mercaptoethanol, 0.01 μmole of 20 amino acids (minus leucine), 0.05 ml of supernatant fraction (2 mg of protein), and a sample of the thrice-washed polysome preparation corresponding to 125 μg of RNA. The whole mixture is suspended in medium B; the final volume is 0.25 ml. Incubation is for 30 minutes at 37°. The reaction is stopped by addition of 2 ml of 10% perchloric acid.

After centrifugation the precipitate is resuspended in 3 ml of 5% TCA, boiled for 15 minutes at 90° (to hydrolyze aminoacyl-tRNA), and poured onto Millipore filters (diameter 25 mm, pore size 0.45 μm). After

TABLE I

In Vitro AMINO ACID INCORPORATION INTO FREE- AND BOUND POLYRIBOSOMES

Polyribosomes	Washing procedure	Radioactivity (cpm/mg RNA)	
		Complete system	Minus energy
Bound	Sucrose-free	580 (1197)[a]	130 (89)
Bound	1.1 M sucrose	1279 (1190)	100 (75)
Free	Sucrose-free	1572 (1384)	75 (68)

[a] Values in parentheses were obtained when incubation follows treatment of free and bound polysomes with deoxycholate (1% w/v final concentration).

TABLE II

RNA, PROTEIN AND PHOSPHOLIPID CONTENT IN CENTRIFUGAL FRACTIONS
OBTAINED FROM 10 g OF FRESH RAT LIVER

Fraction	RNA (mg)	Protein (mg)	RNA:protein ratio	Phospholipid (μg P)
Band 1	7.0	65.1	0.11	1064
Band 2	1.1	4.0	0.28	65
Band 3	4.8	8.4	0.57	110
Pellet	9.4	9.1	1.03	6

repeated washing with 5% TCA, the filters are placed in glass counting vials and dried at 65° for 30 minutes. The scintillation fluid used is 10 ml of a solution of 0.3% 2,5-diphenyloxazole (PPO) and 0.2% 4,4-bis(2-5-phenyl)oxazolylbenzene (POPOP) in toluene. Radioactivity is measured in a liquid scintillation spectrometer. All assays are performed in duplicate. The result of a typical experiment is given in Table I. Values in parentheses were obtained when incubation was performed after treatment of free and bound polysomes with deoxycholate (1% w/v final concentration).

Biochemical Characterization

A rather rough characteristic of polysomes is the ratio of RNA to protein content. Polysomes contain almost equal quantities of protein and RNA.

For protein estimation, the Lowry method[25] is convenient using albumin or casein as a standard. For routine analysis RNA can be determined after hydrolysis of samples with 5% v/v perchloric acid and measuring the absorption at 260 nm. At that wavelength, a value of 22 may be assumed as an equivalent of 1 mg/ml. Furthermore the phospholipid content of separated fractions is frequently given in order to show that the free polysomes are not or only slightly contaminated with membranous material.

In Table II the various estimates are depicted for a typical separation experiment.

Comments

The procedure described allows the separation of free- and membrane-bound polysomes. However, it should be stressed that the availability of this procedure will not answer in a conclusive way whether

[25] O. H. Lowry, N. J. Rosebrough, A. L. Farr, and R. J. Randall, *J. Biol. Chem.* 193, 265 (1951).

or not the two fractions isolated without detergent are structurally and functionally representative for the two distinct classes of ribosomes identified by electron microscopy in intact cells. Even if it cannot be completely ruled out that an artificial redistribution of particles may occur during the isolation our method, nevertheless, allows a direct comparative analysis in vitro of the functional properties of the ribosomes being either free or in association with membranes.

The exact nature of the binding is still unclarified. Several studies indicate that certain metabolic events affecting either the ribosomes or the membranes or both are involved in the binding.

Good evidence has been provided that the large ribosomal subunit is the part of the ribosomes which is in close contact with the membranes.[26,27]

At least in HeLa cells even within the class of membrane-bound polysomes the occurrence of two populations has been claimed.[28] One is characterized by tight binding of the ribosomal cluster directly through the 60 S subunit, whereas in the other population the attachment to the membranes is mediated by an entity sensitive to ribonuclease and EDTA (messenger RNA?).

Evidence has been provided that there are several factors which facilitate the binding of ribosomes to membranes. A protein on the membranes of the ER seems to be responsible for the attachment of the polysomes. This protein is an enzyme which catalyzes disulfide interchange and apparently is masked when polysomes are attached to the membranes.[29] The binding mechanism is claimed to be affected by the interaction with steroid hormones.

On the other hand, ribosomal protein was also considered to be involved in the binding.

A difference in electrophoretic pattern between the protein of the two classes of polysomes has been reported.[30] From this study it appeared that the large subunit of free monomers carries a protein that is absent in the membrane-bound ribosomes. This special protein may preclude attachment of ribosomes to membranes. Other observations are consistent with the possibility that the binding is dependent on the nascent polypeptide chain.[31]

[26] D. D. Sabatini, Y. Tashiro, and G. E. Palade, *J. Mol. Biol.* **19**, 503 (1966).

[27] E. Shelton and E. L. Kuff, *J. Mol. Biol.* **22**, 23 (1966).

[28] M. Rosbash and S. Penman, *J. Mol. Biol.* **59**, 227 (1971).

[29] D. W. James, B. R. Rabin, and D. J. Williams, *Nature (London)* **224**, 371 (1969).

[30] B. R. Fridlender and F. O. Wettstein, *Biochem. Biophys. Res. Commun.* **39**, 247 (1970).

[31] G. Blobel and D. D. Sabatini, *J. Cell Biol.* **45**, 130 (1970).

Webb *et al.*[4] calculated the ratio of free and bound polysomes as found after application of the detergent-free method. These authors ascertained that in normal rat liver about one-half of the population of polysomes is attached to membranes. In contrast, most of the polysomes are lying free in hepatoma. However, it has to be kept in mind that any quantita-

TABLE III

Biosynthetic Activity for a Number of Specific Proteins by Free- and Membrane-Bound Polysomes

Protein	Type of polysomes		Reference
	Bound	Free	
Albumin	+	−	a
Albumin	+	−	b
Albumin	+	−	c
Albumin	+	−	d
Catalase	+	+	b
Ferritin	−	+	c
Ferritin	−	+	e
Fibroin	−	+	f
Fibroin	+	−	g
Glycoprotein	+	−	h
Glycoprotein	+	−	i
Immunoglobulin	++[n]	+	j
Immunoglobulin	++[n]	+	k
NADPH–cytochrome c Reductase	+	+	l
Proinsulin	+		m
Transferrin	+	−	d

[a] M. Takagi, T. Tanaka, and K. Ogata, *Biochim. Biophys. Acta* **217,** 140 (1970).
[b] M. Takagi, T. Tanaka, and K. Ogata, *Biochim. Biophys. Acta* **217,** 148 (1970).
[c] S. J. Hicks, J. W. Drysdale, and H. N. Munro, *Science* **164,** 584 (1969).
[d] M. C. Ganoza and C. A. Williams, *Proc. Nat. Acad. Sci. U.S.* **63,** 1370 (1969).
[e] C. M. Redman, *J. Biol. Chem.* **244,** 4308 (1969).
[f] H. Shigematsu, H. Takeshita, and S. Onoreda, *J. Biochem. (Tokyo)* **60,** 140 (1966).
[g] J. Daillie, L. Grasset, J. C. Prudhomme, J. P. Beck, and J. P. Ebel, *FEBS (Fed. Eur. Biochem. Soc.) Lett.* **13,** 321 (1971).
[h] G. C. Priestley, M. L. Pruyn, and R. A. Malt, *Biochim. Biophys. Acta* **190,** 154 (1969).
[i] T. Hallinan, C. N. Murty, and J. H. Grant, *Arch. Biochem. Biophys.* **125,** 715 (1968).
[j] B. Lisowska-Bernstein, M. E. Lamm, and P. Vassalli, *Proc. Nat. Acad. Sci. U.S.* **66,** 425 (1970).
[k] C. J. Scherr and J. W. Uhr, *J. Immunol.* **106,** 69 (1971).
[l] G. R. Ragnotti, G. R. Lawford, and P. N. Campbell, *Biochem. J.* **112,** 139 (1969).
[m] M. A. Permutt and D. M. Kipnis, *Proc. Nat. Acad. Sci. U.S.* **69,** 505 (1972).
[n] Mainly synthesized in bound polysomes.

tive analysis of this kind has to be taken with great care. The isolation of the *total* amount of cytoplasmic polysomes from rat liver or other mammalian tissues has so far not been achieved, for the following reason. In order to minimize damaging of polysomal arrays and their attachment to membranes the homogenization is performed with a loose-fitting plunger. As a consequence, losses of large fragments of rough endoplasmic reticulum occur already after the first centrifugation step. Electron microscopy reveals that further losses occur in all subsequent centrifugation steps.[32] Hence it is doubtful that this method allows any quantitative analysis of the dynamics of the polysomes during the process of protein synthesis which involves cyclic steps of rapid exchange of monomers between free and membrane-bound polysomes and/or a cyclic dissociation and association of the ribosomal subunits.[33]

In spite of all limitations, our method has been considered as a useful tool in the study of the specific functional activity of each of the two distinct classes of polysomes. This is illustrated by a number of results listed in Table III. A critical evaluation of this type of results has been given by Campbell.[34]

Addendum

A very efficient isolation procedure for polyribosomes from tissue culture has been described by Gielkens *et al.*[35] Optimal yields, especially of the heavier classes of polysomes can be obtained at a rather high KCl concentration in the presence of nonionic detergent (e.g., nonidet p-40 from Shell Corp.). In this method the yield of polysomes is pH dependent only at low salt concentration, while at high salt concentration no influence of the pH is observed. Refreshing of growth medium several hours before harvesting the cells, increases the ratio of polysomes to 80 S monomers.

[32] H. Bloemendal, A. Zweers, and A. H. Stols, unpublished results.
[33] C. Baglioni, I. Bleiberg, and M. Zauderer, *Nature (London) New Biol.* **232**, 8 (1971).
[34] P. N. Campbell, *FEBS (Fed. Eur. Biochem. Soc.) Lett.* **7**, 1 (1970).
[35] A. L. J. Gielkens, A. J. M. Berns, and H. Bloemendal, *Eur. J. Biochem.* **22**, 478–484 (1971).

[33] Pressure-Induced Dissociation of Ribosomes[1]

By R. Baierlein and A. A. Infante

When uncomplexed ribosomes are centrifuged at high speeds, the sedimentation patterns are often anomalous. The anomalies can be traced to the dissociation of the ribosomes during sedimentation, and the dissociation, in turn, can be traced to the effect of hydrostatic pressure on the equilibrium between ribosomes and ribosomal subunits.[2,3] The hydrostatic pressure, which increases throughout the length of a sucrose gradient and can easily reach 1000 atmospheres, radically distorts the ribosome–subunit equilibrium in favor of dissociation. In consequence, either the subunits display apparent sedimentation coefficient (s) values larger than those for isolated subunits that are centrifuged separately or, if the duration of centrifugation is insufficient to resolve the subunits, a single band is observed, which appears to denote an apparently reduced sedimentation rate of the ribosome. Both of these results can lead to an incorrect judgment of the real sedimentation rates and to misinterpretation of the conformational state of the subunits and ribosomes.

An outline of this paper is in order. There is first a section on the theory of pressure-induced dissociation; then a section on preparations. Next comes a section with a half-dozen tests to determine whether a system is affected by pressure-induced dissociation. How to avoid such dissociation is discussed briefly, and the final section explains at length several significant uses. The methods presented here have been used with sea urchin ribosomes; pressure-induced dissociation of ribosomes has, however, also been reported for rat liver[4] and *Escherichia coli*[5] ribosomes, and so the methods are very likely of general value.

The Theory of Pressure-Induced Dissociation

The two basic requirements for pressure-induced dissociation are (a) that there exist a dynamic equilibrium between single ribosomes and subunits:

$$\text{ribosomes} \rightleftarrows \text{small subunit (SS)} + \text{large subunit (LS)}$$

[1] Supported by funds from the U.S. Public Health Service (HD03753-04).
[2] A. A. Infante and M. Krauss, *Biochim. Biophys. Acta* **246**, 81 (1971).
[3] A. A. Infante and R. Baierlein, *Proc. Nat. Acad. Sci. U.S.* **68**, 1780 (1971).
[4] J. Hauge, *FEBS* (*Fed. Eur. Biochem. Soc.*) *Lett.* **17**, 168 (1971).
[5] O. P. Van Diggelen, H. Ostrom, and L. Bosch, *FEBS* (*Fed. Eur. Biochem. Soc.*) *Lett.* **19**, 115 (1971).

and (b) that the molecular volume (V) of the ribosome exceed the sum of the volumes of the two subunits:

$$\Delta V \equiv V_{\text{ribosome}} - (V_{\text{LS}} + V_{\text{SS}}) > 0 \tag{1}$$

If the volume change ΔV is indeed positive, the subunits have to elbow aside solution in order to reassociate. This requires energy, in an amount equal to the work $P(x)\Delta V$ done, where $P(x)$ is the hydrostatic pressure at the position x in the centrifuge tube. Thus, an increase in pressure will increase the energy required for reassociation, and so will push the ribosome–subunits equilibrium in the direction of further dissociation. The energy here is in some ways analogous to an extra term in the energy of chemical binding, and so one may expect the law of mass action for the reacting system to have the form:

$$\frac{[\text{SS}][\text{LS}]}{\text{ribosome}} = K_0 e^{P(x)\Delta V/RT} \tag{2}$$

Here K_0 is the equilibrium constant at zero pressure, and RT is the product of the gas constant and the absolute temperature. The entire expression on the right is the equilibrium constant at position x, denoted subsequently by $K(x)$. (Because the centrifuge tube is sealed under ambient pressure, the pressure at the meniscus is 1 atmosphere, but for so modest a pressure the exponential is very close to unity. Thus, K_0 is, in practice, equivalent both to the equilibrium constant at the meniscus and to that on the laboratory bench.) Provided ΔV is itself unaffected by pressure and hence has the same value for all locations within the tube, the equation is indeed correct, as rigorous derivations by Josephs and Harrington[6] and by Ten Eyck and Kauzmann[7] demonstrate. The implication is that the equilibrium constant depends, through $P(x)$, on both rotor speed and position down the tube. The pressure is proportional to (rotor speed)[2] and increases in a roughly linear manner with distance (within a typical centrifuge tube); thus the equilibrium constant itself grows at an extremely rapid rate with both rotor speed and position.[3]

It is worth noting that a positive ΔV need not mean a literal increase in molecular volume, nor does it necessarily signify a conformational change. For example, if charged groups partially neutralize one another when the subunits associate, the reduced electrostriction of the aqueous surroundings can produce a volume change whose influence would be correctly described by Eq. (2).

The volume change that one can expect is small, a few hundred milliliters per mole in order of magnitude.[3,6] Typical determinations of buoyant

[6] R. Josephs and W. F. Harrington, *Biochemistry* 7, 2834 (1968).
[7] L. F. Ten Eyck and W. Kauzmann, *Proc. Nat. Acad. Sci. U.S.* 58, 888 (1967).

densities and of molecular weights are *not* sufficiently accurate to enable one to deduce ΔV from them. To succeed, one would need at least four-figure accuracy in the densities and weights, which is well beyond present standards.

Preparations and Materials

Cell-Free Extracts from Sea Urchin Eggs[2]

The buffers used for homogenization and sucrose gradients contained 50 mM triethanolamine and 5 mM MgCl$_2$; concentrations of KCl ranged from zero to 0.80 M. The pH is adjusted after the addition of all salts to 7.8 with HCl at 4°. Homogenization medium also contains 0.25 M sucrose. The sucrose used in the homogenization medium and in sucrose gradients was a commercial ribonuclease-free grade. Sucrose gradients of 15% (w/w) to 30% (w/w) are used in the preparation and analysis of the ribosomes.

The sea urchin eggs are resuspended in 4–8 volumes of homogenization medium and homogenized with 2 strokes of a tight-fitting Dounce homogenizer. A 15,000 g supernatant (S15) of this homogenate contains at least 90% of the total ribosomes in the eggs. When the unfertilized egg is used, virtually all the ribosomes in the S15 are free of peptidyl-tRNA. When the homogenization buffer contains KCl concentrations of 0.4 M or higher, the eggs are more difficult to homogenize, and so it is convenient to prepare the S15 in 0.10 M KCl and then adjust the KCl concentration to the desired concentration. The S15 must be used soon after preparation because freezing results in a large and variable degree of ribosome degradation. This is true over a wide range in the ionic strength of the homogenization buffer. Especially prominent in the sedimentation pattern of a frozen S15 is a peak at approximately 50 S.

Isolation of Ribosomes and Subunits

The isolation and examination of the sedimentation properties of ribosomes and subunits in sucrose gradients depend strongly upon several factors which influence the ribosome–subunit equilibrium. These will be described in detail in the next section. In general, ribosomes can be prepared from gradients containing low concentrations of KCl (50 mM–0.10 M is used in most of our studies), and subunits from gradients containing high KCl (0.5 M or greater). This is shown in Fig. 1. The subunits formed in high KCl are active in poly(U)-directed protein synthesis and can be reassociated by simply recombining peaks A and B and reducing the KCl concentration by dilution. Peak A contains exclusively 18 S RNA, and peak B contains only 28 S RNA.

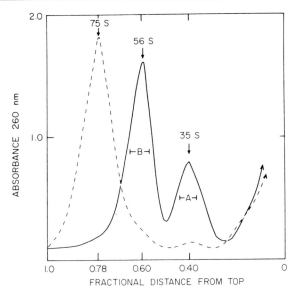

FIG. 1. Sedimentation of ribosomes of *Strongylocentrotus purpuratus* in low and high ionic strength. Unfertilized eggs of *S. purpuratus*, 0.05 ml, were homogenized in 0.4 ml of 50 mM triethanolamine–5 mM $MgCl_2$ containing either 0.10 M KCl (- - -) or 0.50 M KCl (—). A supernatant fraction (S15) was prepared and sedimented at 41,000 rpm in the SW 41 rotor at 2.5° for 5 hours through 15–30% sucrose gradients containing the same buffer as used in the homogenization.

Stability of Ribosomes and Subunits

We have found that prolonged exposure of the subunits to high KCl results in loss of their ability to reassociate.[8] Even dialysis for 12 hours to reduce the KCl concentration causes extensive inactivation. Once the KCl concentration is reduced, however, the subunits can be kept on ice separately for at least 24 hours with no loss in ability to recombine. Freezing of the isolated subunits in high or low KCl results in the immediate and complete loss of their ability to reassociate.

The ability of isolated ribosomes to dissociate in high KCl is preserved for at least 3 days if they are kept at 0° and in low ionic strength solution. However, freezing, even at −80°, immediately destroys this property of the ribosome. Clearly, at least for sea urchin material, freezing of crude homogenates, or of isolated ribosomes and subunits will not permit an equilibrium between ribosomes and subunits to be observed or studied. Also exposure to high KCl concentrations for prolonged periods should be avoided.

[8] A. A. Infante and R. Baierlein, unpublished results, 1972.

Complexed Ribosomes

When extracts are prepared from embryonic stages in which polysomes are present, a 75 S peak is found even at 0.5 M KCl. This peak is due to polysome degradation during the preparation and may be avoided by including bentonite in the homogenization buffer.[9] This complexed ribosome may be dissociated in high KCl by prior incubation with puromycin at 2 μmole/ml to remove the nascent polypeptide. The complexed ribosome is also very resistant to dissociation by high pressure and so is not included in our consideration of the subunit–ribosome equilibrium observed with free ribosomes.

Tests for Dissociation

In this section we describe some of the anomalous patterns that can result from sedimentation of free ribosomes in sucrose gradients. Simultaneously, we offer a variety of tests to determine whether pressure-induced dissociation is affecting ribosome sedimentation patterns.

The Duration of Centrifugation

The sedimentation properties of ribosomes and subunits in sucrose gradients is strongly influenced by the duration of centrifugation and by the ionic conditions. For sea urchin, (and probably most eukaryotic) ribosomes in the range from 0.15 M to 0.5 M KCl in 5 mM Mg^{2+}, either pure ribosomes or subunits can be obtained at 41,000 rpm in the Spinco SW 41 rotor simply by varying the length of centrifugation time. Figure 2 demonstrates this. At the end of a short centrifugation (2.5 hours, Fig. 2a) the ribosomal material has sedimented as though it were composed purely of 75 S particles. Longer centrifugation (3.7 hours, Fig. 2b) broadens the ribosomal zone and produces an apparent reduction in the sedimentation coefficient (s), to about 67 S. After 5 hours (Fig. 2c), two well-separated peaks, corresponding to sedimentation coefficients of 55 S and 65 S, appear. These latter peaks are, respectively, the small and large ribosomal subunits, which would sediment at 35 S and 56 S if they were centrifuged separately.[2]

Pressure-induced dissociation provides an explanation of these results. When the ribosomal material starts to sediment, it is predominantly in the form of ribosomes moving at about 75 S. As the material reaches higher pressures, the equilibrium (ribosomes $\rightleftarrows$ subunits) shifts continuously toward further dissociation. The subunits move more slowly than the ribosome, and so, as they become a larger fraction of the total material, the absorbance peak moves more slowly than 75 S. *This gives the appear-*

[9] A. A. Infante and P. N. Graves, *Biochim. Biophys. Acta* **246**, 100 (1971).

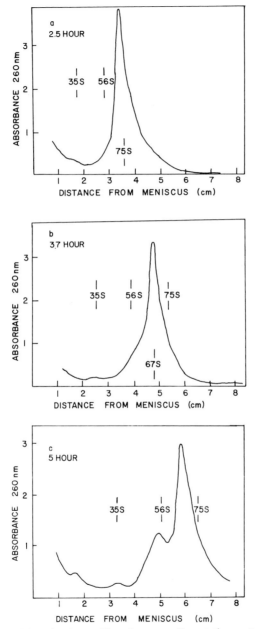

FIG. 2. Effect of length of centrifugation on sedimentation pattern of ribosomes. An S15 extract was layered onto 15–30% sucrose gradients in buffer containing 0.20 M KCl and centrifuged at 41,000 rpm for the indicated times.

ance of a reduction in the sedimentation coefficient of ribosomes. Ultimately, the difference between the sedimentation velocities of the two subunits makes itself felt, and two separated peaks emerge from one broad, slow peak. This process yields subunits with apparently *increased* rates of sedimentation.

Thus, prior to assigning sedimentation values to subunits or to ribosomes, it is important to assess the effect of varying the duration of centrifugation upon both the sedimentation profile and the apparent *s* value of the components in the gradient. For example, it is likely that a pressure-induced dissociation of ribosomes is occurring if, with increased duration of centrifugation, the apparent sedimentation rate of a ribosome band decreases.

Effect of Ribosome Concentration

Because there is an active subunit–ribosome equilibrium, the subunit: ribosome ratio depends on the concentrations: the lower the concentration of ribosomal material, the greater the subunit:ribosome ratio in any preparation at any given pressure. This behavior follows directly from the law

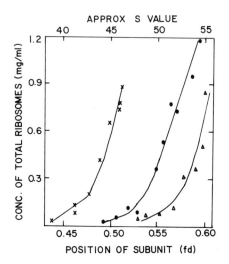

FIG. 3. The relationship between the sedimentation of the small subunit and the concentration of ribosomes. The concentration of ribosomes (both subunits) was calculated from the absorbance profiles obtained after monitoring the sucrose gradients and using 1 mg/ml = 12 A_{260} units. Of the S15 of a purified ribosome preparation, 0.4 ml was layered onto each gradient. Sedimentation was for 5 hours in the SW 41 at 41,000 rpm as described in Fig. 1. The final location of the small subunit in the gradients is given as the fractional distance (fd). ×, 0.30 M KCl; ●, 0.24 M KCl; △, 0.20 M KCl.

of mass action as expressed in Eq. (2). The more intuitive explanation is this: lowering the concentration reduces the probability that two subunits will meet and hence pushes the equilibrium in the direction of further dissociation. Variations in the ribosome concentration can strikingly change the apparent equilibrium distribution and also the apparent sedimentation rate of ribosomes and subunits.[2,3,10,11] The latter effect is shown in Fig. 3 where, in three different KCl concentrations, the s value of the small subunit appears to increase with increasing ribosome concentration. The wide range of "artifactual" s values of the small subunits shown here (from about 38 S to 56 S) is yet another pitfall that should be avoided in studying ribosomes and a rather simple way of demonstrating the strong effects of pressure and ionic strength on the sedimenting ribosome. The influence of ribosome concentration on the s value of the subunits can be calculated on the basis of an active equilibrium that is influenced by pressure and the calculated values based on the pressure theory adequately explain these results.[3]

Fixation of Ribosomes with Glutaraldehyde

The problem of pressure effects upon the sedimenting ribosome can be eliminated by fixing the ribosome preparation with glutaraldehyde prior to centrifugation. Glutaraldehyde at 4% will, within 30 seconds after addition,[8] prevent the dissociation of ribosomes by high pressure (and by high KCl). This is true whether the ribosomes have been isolated or are in a crude homogenate (S15). Figure 4 presents some results that support this; in panel a, the S15 was made in 0.3 M KCl and equal aliquots were centrifuged on parallel gradients. One aliquot was fixed prior to centrifugation, the other served as a control. The unfixed preparation becomes completely dissociated during the centrifugation whereas the fixed preparations contain virtually only the 75 S ribosome.

The reassociation of subunits is also blocked by fixation. Subunits were isolated from high KCl gradients and fixed with glutaraldehyde separately before being recombined in low KCl. The fixed subunits do not reassociate, under conditions where the unfixed subunits almost completely form the 75 S ribosome.[8]

The fixation method is also very useful in determining the equilibrium distribution of ribosomes and subunits under certain ionic conditions. For example (Fig. 4b), when ribosomes in high KCl (0.5 M) are centrifuged in a gradient containing high KCl, only subunits are observed at 41,000 rpm in the Spinco SW 41 rotor. If the same preparation is fixed prior to

[10] A. S. Spirin, *FEBS (Fed. Eur. Biochem. Soc.) Lett.* **14**, 349 (1971).
[11] R. S. Zitomer and J. Flaks, *Fed. Proc., Fed. Amer. Soc. Exp. Biol.* (Abstract) **30**, 1312 (1971).

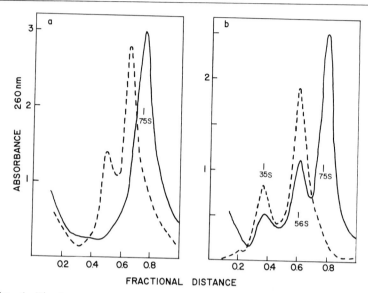

FIG. 4. Fixation prevents pressure-induced dissociation. (a) One half of an S15 in 0.3 M KCl was fixed by addition of glutaraldehyde to 4%, and the other half remained unfixed. Each portion was centrifuged in parallel gradients (0.3 M KCl) at 41,000 rpm for 5 hours, as in Fig. 1. (b) The same as in (a) except that isolated ribosomes (prepared from low salt gradients as in Fig. 1) were used. The KCl concentration was made 0.5 M prior to fixation, and the gradients contained 0.5 M KCl. —, Fixed; - - -, unfixed.

centrifugation, however, about 60% of the subunits are seen to be associated in the 75 S ribosome. Under these conditions of high KCl, the hydrostatic pressure found at the top of the sucrose gradient almost immediately dissociates the predominant 75 S ribosome species present in the equilibrium mixture (at atmospheric pressure)—unless those ribosomes have been fixed.

Overlay Method

If pressure is dissociating the ribosomes during centrifugation, then starting the ribosomes at high pressure should produce early dissociation. To test this, ribosomes may be layered onto an abbreviated sucrose gradient (50%, of normal length, for example) and then the tube filled to the top with buffer or mineral oil. The overlay immediately subjects the ribosomes to high pressure. Subunits may now form after a brief centrifugation, although such dissociation and separation may be absent when the ribosomes are layered onto the top of a full length gradient and centrifuged for the same period of time. Succinctly, one tries to compare the results of centrifugation in which only the pressure has been changed, indeed, vastly so, by the overlay.

Effect of Rotor Speed

Perhaps the most critical factor that influences the sedimentation properties of free ribosomes is the rotor speed. For any given rotor, the higher the speed, the greater the tendency for subunits to be formed. This is illustrated in Fig. 5, where ribosomes in the same homogenate (S15) were centrifuged in separate sucrose gradients at different rotor speeds. As the rotor speed was increased, the time of centrifugation was decreased such that the product (rotor speed)^2 × (elapsed time) remained constant, at least to a good approximation. Because sedimentation

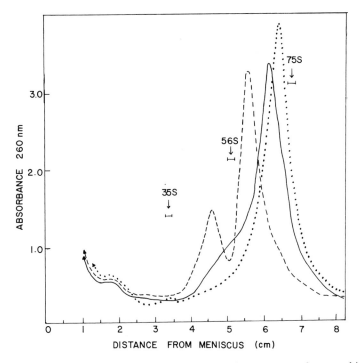

FIG. 5. How rotor speed changes the sedimentation patterns of sea urchin ribosomes. S15 extract was layered onto a 15–30% sucrose gradient in buffer containing 0.24 M KCl. Centrifugation in the SW 41 rotor was at (····) 24,000 rpm for 14 hours, (—) 30,000 rpm for 9.25 hours, (- - -) 41,000 rpm for 5 hours. The ranges of the positions attained in parallel gradients by the isolated small (35 S) and large (56 S) subunits, and by 75 S ribosomes are indicated. The ribosome range was determined in 50 mM KCl, where no dissociation is perceptible. The abscissa gives distance from the meniscus, which itself is 7 cm from the rotor axis.

velocity is proportional to (rotor speed)^2, one would, perhaps naively, expect to find the same pattern in the same location down the tube at all three speeds tested. Yet, at 24,000 rpm there is only a single peak, with a sharp trailing edge; at 30,000 rpm, there is a single, slower peak with a shoulder on the trailing edge; and at 41,000 rpm, there are two distinct peaks. The latter peaks, although at 48 S and 61 S, are in fact the small (35 S) and large (56 S) subunits.[2] Therefore, as the rotor speed increases—and, with speed, the hydrostatic pressure—there is a clear shift toward dissociation.

When studying the sedimentation properties of ribosomes, one must, therefore, ascertain the effects of varying the rotor speed. This simple expedient is useful and sufficient in determining whether pressure effects are influencing the sedimentation properties under any set of conditions. If pressure is playing a role, then with higher speeds the apparent s value of the subunits will become lower, or a single band of ribosomes may be resolved into the two subunits.

Indeed, Lengyel and his associates used different rotor speeds to examine the sedimentation properties of pretranslocation and posttranslocation complexes of bacterial ribosomes.[12] Despite previous reports[13–15] describing differences in the sedimentation properties of these ribosomes (which suggested different conformational states), these workers found that the differences are manifested only at high rotor speeds, not at low speeds. Thus, pressure evidently effects the ribosomes in the two functional states to different extents.

Numerical Estimation

The last test is a theoretical one: a numerical estimate of whether pressure-induced dissociation could be significant under current or anticipated centrifugations. To generate such an estimate, one may reason as follows.

Because the pressure increases with distance down the tube, the effect is strongest at the maximum distance that the ribosomal material reaches in the tube; call that distance x_{max}. The meniscus is located at x_0, all distances being measured from the rotor axis. In good approximation, the pressure at x_{max} is the product of the "height" $(x_{max} - x_0)$, the average density in the interval, ρ_{av}, and the average value of the centrifugal force

[12] J. Waterson, M. L. Sopori, S. L. Gupta, and P. Lengyel, *Biochemistry* **11**, 1377 (1972).

[13] S. L. Gupta, J. Waterson, M. L. Sopori, and P. Lengyel, *Biochemistry* **10**, 4410 (1971).

[14] D. M. Chuang and M. V. Simpson, *Proc. Nat. Acad. Sci. U.S.* **68**, 1474 (1971).

[15] M. H. Schreier and H. Noll, *Proc. Nat. Acad. Sci. U.S.* **68**, 805 (1971).

(per unit mass) in the interval, g_{av}. If the pressure is expressed in atmospheres, the numerical relation is

$$P(x_{max}) = (9.68 \times 10^{-4})(x_{max} - x_0)\rho_{av}g_{av} \qquad (3)$$

when the distances are measured in cm, the average density is given in gm/ml, and g_{av} is expressed as a multiple of normal gravity. The centrifugal force (per unit mass) is proportional to the distance, x, from the rotor axis to the contemplated position in the tube, and so $g(x_{max})$ is simply $(x_{max}/x_0)\ g(x_0)$, which means that g_{av} can readily be computed as

$$g_{av} = 1/2[1 + x_{max}/x_0]g(x_0) \qquad (4)$$

To facilitate further computation, Eq. (2) can be rewritten as

$$K(x_{max}) = K_0 10^{0.00529 P(x_{max})\Delta V/T} \qquad (5)$$

correct if the pressure is given in atmospheres, ΔV in milliliters per mole, and T in degrees kelvin. The value of ΔV does, of course, involve some guessing; a value of several hundred milliliters per mole is likely to be reasonable. Then one can evaluate the exponent, and see whether $K(x_{max})$ differs sufficiently from K_0 to make an appreciable difference in the degree of dissociation (at the concentration of ribosomal material in the tube).

Here is an example. Suppose the Spinco SW 50 is run at 45,000 rpm and that material sediments about 80% of the way down the tube. The distance x_0 from the rotor axis to the meniscus is 5.97 cm, and x_{max} is that number plus 80% of 4.76 cm, the full tube length, whence $x_{max} = 9.78$ cm. The value of $g(x_0)$ is 135,000 (times normal gravity) and so Eq. (4) yields

$$g_{av} = 1/2[1 + 9.78/5.97] \times 135,000 = 178,000$$

If the average density is 1.12 g/ml, then Eq. (3) states that

$$P(x_{max}) = 9.68 \times 10^{-4} \times 3.81 \times 1.12 \times 178,000$$
$$= 735 \text{ atmospheres}$$

Next, if the temperature is 4°, whence $T = 277°$ kelvin, and if one chooses $\Delta V = 250$ ml/mole as a reasonable potential value, then Eq. (5) yields

$$K(x_{max}) = K_0 10^{0.00529 \times 735 \times 250/277}$$
$$= K_0 10^{3.51}$$

The change in equilibrium constant is large, and its effect can be made more obvious if one compares the extent of dissociation at x_{max} with that at the meniscus, as follows. Suppose $K_0 = 10^{-9}$ mole/liter and that [ribosome] + [LS] = 2×10^{-7} mole/liter. That *sum* of concentrations (in moles/liter) is the relevant quantity because it does not change during the dissociation reaction; for sea urchin ribosomes, the value corresponds to a density of ribosomal material of about 0.6 mg/ml. At the meniscus the degree of dissociation is about 7%, but at x_{max} it has become 94%.

How to Avoid Pressure-Induced Dissociation

There are several techniques by which one can avoid pressure-induced dissociation (or at least greatly mitigate the effect). The less pressure, the less effect, and so the obvious technique is to avoid large pressures by using low rotor speeds. At any fixed point in the centrifuge tube, the pressure grows only as the square of the rotor speed (to excellent approximation), but the equilibrium constant depends exponentially on the pressure, and so a 2-fold increase in rotor speed, say, can produce an astonishing amount of dissociation and a radically different sedimentation profile. Keep the rotor speed low.

Next one can adjust the ionic conditions (e.g., K^+ and Mg^{2+}) so that they favor the intact ribosome. Here one is really adjusting conditions so that the first factor in Eq. (2), K_0, is small, implying that the ribosome is favored. (Low rotor speed tends to keep the second factor small, that is, close to unity.)

Even when the equilibrium constant is fixed, the extent of dissociation depends on the concentration of ribosomal material. One uses this to advantage by keeping the material concentration high.

Finally, if the ribosomes are not to be used for further study, fixation with glutaraldehyde prior to centrifugation will prevent dissociation, apparently completely. This seems to be the surest way to avoid dissociation, and one need no longer worry about adjusting ionic conditions or rotor speeds, but one should always ask whether the fixation process itself alters the ribosome/subunit distribution prior to centrifugation or affects the sedimentation characteristics of the ribosomal particles. This method has already proved useful in studying *E. coli* ribosomes.[16–18]

Uses of Pressure-Induced Dissociation

Preparation of Active Subunits

To date, the most useful means of preparing eukaryotic subunits is by exposing uncomplexed ribosomes to high concentrations of KCl.[19] Although the removal of magnesium from the media (either through

[16] A. S. Spirin, *FEBS (Fed. Eur. Biochem. Soc.) Lett.* **15**, 197 (1971).
[17] A. R. Subramanian and B. D. Davis, *Proc. Nat. Acad. Sci. U.S.* **68**, 2453 (1971).
[18] A. R. Subramanian, *Biochemistry* **11**, 2710 (1972).
[19] T. E. Martin, I. G. Wool, and J. J. Castles, this series, Vol. 20, p. 417.

dialysis or by complexing with EDTA) results in dissociation of eukaryotic ribosomes, the subunits that are formed generally appear to have abnormally low sedimentation rates, are incapable of reassociating, and are not active in poly(U)-directed protein synthesis.[8,19] Therefore, this method of producing subunits and of shifting the equilibrium must be considered a drastic and not very useful procedure for eukaryotic ribosomes.

We have found that pressure-derived subunits formed in moderate KCl solution during sedimentation at high rotor speeds are more active in supporting poly(U) translation than are the high salt-derived subunits. It seems likely that the exposure to high KCl may have some deleterious effect on the subunits that is avoided by the combination of moderate KCl concentrations and high pressure. The latter procedure may therefore provide the most gentle means of preparing large amounts of subunits for use in protein synthesis.

K_0 and ΔV from Sedimentation Patterns

Both the equilibrium constant (at one atmosphere) and ΔV can be extracted from the sedimentation patterns. Succinctly, one simulates centrifugation on a computer and then adopts those (trial) values of K_0 and ΔV that provide the best agreement between the actual and the simulated sedimentation patterns.

Here is the scheme in a bit more detail. Experimental data at different salt concentrations and different rotor speeds will provide an array of sedimentation patterns. To reproduce theoretically the essential features of the patterns, one needs only the following elements.

1. The equilibrium constant at the meniscus, K_0, is assumed to be a function of salt concentration.

2. The variation of the equilibrium constant $K(x)$ is given by Eq. (2), where ΔV is taken to be independent of both pressure and salt concentration.

3. As the particles sediment, reversible reactions change the local concentrations so that equilibrium at the local value of $K(x)$ is achieved.

4. Diffusion is negligible.

These elements can readily be incorporated into a computer program. First, the computer can let each species (subunit or ribosome) drift at its own sedimentation rate for a short time (such that the particles drift at most a few millimeters). Next, the computer assesses local concentrations and reequilibrates at the local value of the equilibrium constant. And then it reiterates. To calculate absorbance at 260 nm, one can weight the concentrations of ribosome, large subunit and small subunit by their absorbance per mole, often taken to be in the ratio of their molecular weights.

To fit the patterns with the theory, one has a number of adjustable parameters: a separate value of K_0 for each different salt condition and the single value of ΔV. If one has profiles for, say, three rotor speeds at each of several salt concentrations, then there are far more profiles than adjustable parameters. Once the computer is given trial values of the parameters, it can readily print out profiles, and judicious search can produce a satisfying fit to the data, together with an estimate of uncertainty (based on how well neighboring parameter values yield profiles that resemble the data). The end results are a value for ΔV, a set of equilibrium constants, and a test of how well the theoretical elements can reproduce the data, especially the assumption that there is a reversible reaction sufficiently dynamic to preserve equilibrium during sedimentation.

Use of Pressure Cell

One can use pressure to shift the equilibrium to a point where the subunits are an appreciable proportion of the ribosomal preparation and hence the degree of dissociation is easier to study. This can be done in a pressure cell similar in principle to the Zo-Bell & Oppenheimer model described in detail by Morita.[20] To determine the equilibrium constant at any specified pressure, one pumps up the pressure, allows the ribosomal preparation to reach equilibrium at the high pressure, and then fixes the ribosomes while they are still under pressure. The system illustrated in Fig. 6 is convenient for this. The end of a 1-ml disposable plastic syringe is closed by flaming. To the syringe is added a stainless steel ball bearing or a glass bead, the solution of ribosomes (usually 0.3 ml), and an 8% glutaraldehyde solution (usually 0.4 ml) in the same buffer as used for

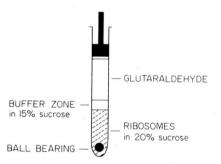

FIG. 6. Apparatus for fixing ribosomes under pressure. Details using a 1-ml plastic B-D disposable syringe are given in text.

[20] R. Y. Morita, in "Methods in Microbiology" (J. R. Norris and D. W. Ribbons, eds.), Vol. 2, p. 243. Academic Press, New York, 1970.

the ribosomes. A small "buffer zone" (0.1 ml) is used to separate the ribosomes and the glutaraldehyde. In order to stabilize the three layers, sucrose is present at 20% and 15% in the ribosome and "buffer zone" layers, respectively. A thin wire is placed in the syringe, and then the plunger is inserted. The deformation of the plunger by the wire allows the air to escape, and then the wire is removed. Up to six syringes can be placed in the pressure cell at one time. After the desired pressure is attained a valve on the cell is closed, the cell is disconnected, and the ribo-some–subunit mixture is fixed by inverting the cell and allowing the ball bearing to mix the contents of the syringe. The distribution of ribosomes and subunits can then be measured on sucrose gradients. (Other methods, such as light scattering or analysis on acrylamide gels, should also be useful here to measure the equilibrium constant.) Figure 7 presents some results using this method. In (a) the S15 solution was fixed at 1 atm pressure, and under these ionic conditions virtually only the 75 S species is observed. In (b) the preparation was fixed at 8000 psi and in (c) at

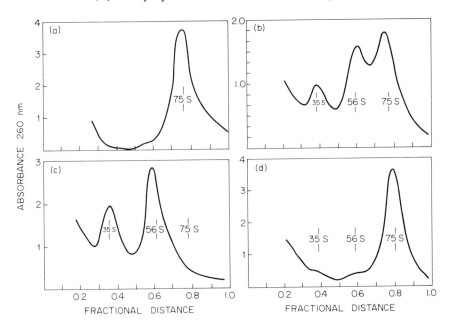

FIG. 7. Shift in the equilibrium distribution of ribosomes and subunits by pressure. The procedure used to fix the equilibrium mixture under pressure is given in the text. An S15 in 0.3 M KCl (5 mM Mg^{2+}) was divided into 4 equal parts. In (a), (b), and (c) the preparation was fixed at 15 (1 atm), 8000, and 14,000 psi, respectively. In (d) the preparation was subjected to 14,000 psi and then the pressure was returned to 1 atm before fixation. Analyses of the fixed preparations were on sucrose gradients (0.3 M KCl) centrifuged at 41,000 rpm for 5 hours as in Fig. 1.

14,000 psi. At 8000 psi, approximately equimolar amounts of ribosomes and subunits are present, corresponding to an equilibrium constant of 10^{-7} M. At 14,000 psi only subunits are observed. In (d) the S15 was subjected to 14,000 psi and the pressure was then reduced to 1 atm prior to fixing. Within 2 minutes of releasing the pressure, all the subunits present at 14,000 psi have become reassociated at 1 atm. This method might allow one to determine the rate constants in the reaction.

Determination of the equilibrium constant at various pressures K_p provides data as shown in Fig. 8. To analyze the data, one notes that Eq. (2) can be applied to a pressure cell provided only that $P(x)$ is replaced by the pressure that exists in the cell. The right-hand side then becomes K_p, and taking logarithms yields

$$\log K_p = \log K_0 + \frac{5.29 \times 10^{-3}}{T} \Delta V P \qquad (6)$$

provided the pressure is in atmospheres, ΔV in ml/mole, and T in degrees kelvin. In the graph of $\log K_p$ versus P, the slope yields the change in molar volume that occurs when the ribosome dissociates; the intercept, determined by extrapolation, yields the value of the equilibrium constant

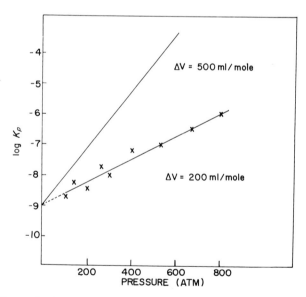

FIG. 8. Determination of K_0 and ΔV for ribosomes. $\times$, Values of K_p at various pressures were determined from graphs as in Fig. 7, for 0.3 M KCl and 5 mM Mg^{2+}. The line using a ΔV of 500 ml/mole is calculated; 500 ml/mole was the value arrived at by comparing computer-simulated gradient profiles with actual sedimentation patterns.

at zero pressure. The K_0 under these conditions is 10^{-9}, which is in excellent agreement with the value determined by fitting the sedimentation data with computer generated profiles. The ΔV change is 200 ml/mole, which differs from the value of 500 estimated by computer simulation.[3] To select parameter values by comparing simulated and actual sedimentation profiles is intrinsically difficult, and so the agreement is acceptable. To our knowledge, the pressure-cell method is the only one which is sensitive enough to permit accurate estimation of ΔV values for ribosomes.

Utility of Pressure Measurements

Subtle effects of ions (such as Mg^{2+}) and pH changes on ribosome dissociation and reassociation can best be determined if the equilibrium is easily measured. This is not so at 1 atm for sea urchin ribosomes and other eukaryotic systems in which the ribosome predominates. By using pressure to shift the equilibrium to a point where subunits and ribosomes are in approximately equimolar amounts, effects on the equilibrium can be more easily assayed. Here one would be asking questions concerning the number of ions involved in the equilibrium and how they are acting to facilitate dissociation or association. For example, does the Mg^{2+} concentration affect the equilibrium purely stoichiometrically?

One must, however, append words of caution.[21] Pumping up the pressure from 1 atm to 600, say, may itself change the pH, the ionic concentrations, and the temperature.[20] By working near 4° (the temperature at which the density of water is maximal and hence where the variation in temperature with respect to pressure is zero, at least at low pressure) one can minimize the temperature change. The order of magnitude will be only 0.5° for 1000 atm. One feels that ionic concentrations (such as Mg^{2+}) will not be seriously affected, but pH and the properties of buffers are much more difficult to assess.

Despite the precautionary injunctions, one can ascertain K_0 and ΔV, and they are useful in determining the thermodynamic constants for the ribosome equilibrium. Knowledge of the thermodynamics of subunit association should be useful in understanding both the structure of ribosomes and the forces involved in binding of subunits.

Finally, determination of the effect on K_0 and ΔV of adding ligands to the ribosome, such as mRNA, and aminoacyl-tRNA should also be amenable by this method, and such studies could yield useful information concerning ribosome function and variations in the state of the ribosome.

[21] A. M. Zimmerman, ed., "High Pressure Effects on Cellular Processes." Academic Press, New York, 1970.

[34] The Dissociation of Rat Liver Ribosomes by Formamide: Extraction of a 5 S RNA·Protein Complex

By Mary L. Petermann

The 80 S ribosomes of animal tissues do not dissociate readily to subunits, even when the bound magnesium has been decreased to a fairly low level. Beyond some critical point further loss of magnesium does lead to dissociation, but the subunits may unfold, lose proteins and 5 S RNA, and become inactive. One way to retain activity is to carry out the dissociation in the presence of 0.5–1.0 M KCl.[1] A different type of procedure, developed in this laboratory, depends on the use of an agent that weakens hydrogen bonding, such as formamide[2] or urea.[3] Formamide first dissociates rat liver ribosomes to active 40 S and 55 S subunits. At higher concentrations it converts both subunits to inactive, unfolded particles, and detaches the 5 S RNA from the large subunit in the form of a ribonucleoprotein complex (RNP).[4] The preparation of this RNP, and some of its properties, are described below.

Ribosomes. Adult male rats are given 25 mg of phenobarbital per day, in the drinking water, for 7 days before sacrifice, to reduce the amount of RNase in the liver.[3] Ribosomes are isolated and stored as described previously,[5] except that the solution is clarified by centrifuging at 20,000 g for 15 minutes just before the magnesium precipitation step.[3]

Extraction of RNP. To dissociate the ribosomes, and extract the RNP,[4] 6 ml of ribosome solution (about 80 mg or 1120 A_{260} units) are mixed with 1 ml of 0.7 M KCl and dialyzed in a medium-sized Zeineh dialyzer (Biomed Instruments, Inc., Chicago) for 2.5 hours at 5° against buffer containing 30 mM KCl, 0.1 mM MgCl$_2$, 3 mM potassium phosphate pH 7.3. The bound magnesium decreases to 0.22 equivalent per mole of RNA phosphate,[6] but the ribosomes do not dissociate. At this stage they can be mixed with 0.05 volume of 40% sucrose (Schwarz/Mann, ultra pure), frozen rapidy in a dry ice–alcohol bath, and stored at −20°.

[1] T. E. Martin and I. G. Wool, *J. Mol. Biol.* **43**, 151 (1969).
[2] M. L. Petermann, A. Pavlovec, and M. G. Hamilton, *Biochemistry* **11**, 3925 (1972).
[3] M. L. Petermann and A. Pavlovec, *Biochemistry* **10**, 2770 (1971).
[4] M. L. Petermann, M. G. Hamilton, and A. Pavlovec, *Biochemistry* **11**, 2323 (1972).
[5] M. L. Petermann, this series, Vol. 20, p. 429.
[6] M. L. Petermann and A. Pavlovec, *Biochemistry* **6**, 2950 (1967).

Formamide (Fisher, reagent grade) is adjusted with 0.2 N HCl until a sample diluted with dialysis buffer to 8.9 M (40% by volume) has a pH of 7.3. The ribosome solution is set in an ice bath, and the neutralized formamide is added slowly, with good stirring, to a final concentration of 8.9 M. After standing for 30 minutes at 0° the mixture is dialyzed in the Zeineh dialyzer for 2.5 hours against 20 mM KCl, 0.2 mM MgCl₂, 5 mM potassium phosphate pH 7.8. The removal of formamide is checked by measuring the refractive index of the solution in an Abbé refractometer. The solution is layered, in 6-ml portions, over 3 ml of 0.3 M sucrose containing the same buffer, and centrifuged for 1 hour at 150,000 g, in a Spinco 50 rotor, at 5°. The solution above the sucrose is removed in 1- or 2-ml portions, and the samples of low absorbancy at 260 nm (usually the top 5 ml) are pooled. The yield is about 60 A_{260} units. This material has been called the formamide extract.[4] Unless it is to be used immediately, or fixed, one-third volume of 40% sucrose is added, and the mixture is frozen rapidly in a dry ice–alcohol bath and stored at −20°.

Electrophoretic Patterns of Formamide Extracts. Polyacrylamide gel electrophoresis is carried out in Tris-borate buffer at pH 8.7, in a slab of 10% gel, with methylene blue staining for RNA or Coomassie Blue for protein.[4] The slab is photographed with a yellow filter and Polaroid 46-L film, and the film blackening is measured in a densitometer. A typical slab (Fig. 1B) shows a tRNA band, a 5 S RNA band, and a sharp slower band that always travels 35% as far as the tRNA.[4] Since a similar sharp band appears with the protein stain (Fig. 1C) this material repre-

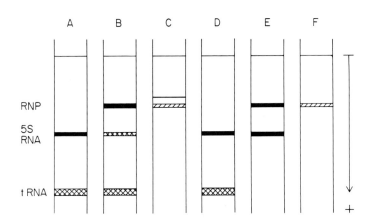

FIG. 1. Polyacrylamide gel electrophoresis of formamide extracts. A, marker RNA's; B, C, and D, extract of whole ribosomes; E and F, extract of large subunits. A, B, D, and E, RNA stain; C and F, protein stain; D, treated with sodium dodecyl sulfate before electrophoresis.

sents a complex of RNA and protein. A faint slower protein band appears in extracts made from whole ribosomes, but an extract of isolated large subunits shows only a single protein, in the RNP position (Fig. 1F); with the RNA stain it shows the RNP band and a 5 S RNA band, but no tRNA (Fig. 1E).

Stability of RNP. The RNA in the complex is readily converted to free 5 S RNA.[4] When the formamide extract is treated with sodium dodecyl sulfate before analysis, the RNP band disappears and the 5 S RNA band becomes more intense (Fig. 1D). A similar result is obtained when the RNP is treated with pronase.[7] The complex is also dissociated by a variety of treatments such as ethanol precipitation or lyophilizing, but it can be stored at $-20°$ in 10% sucrose. When samples are kept at 5° for 3 or 4 days before analysis, the composition of the solvent has a marked effect on RNP stability. At pH 7.8 a KCl concentration above 50 mM is unfavorable, and NaCl is worse. The stability is unchanged when the pH is lowered to 6.5 in cacodylate buffer, but is markedly decreased in acetate at pH 5.0. In 40 mM KCl, 1 mM EDTA at pH 7.8, 3 buffers, 5 mM potassium phosphate, 3 mM Tris, and 5 mM triethanolamine, give similar results. A series of experiments was carried out in 5 mM potassium phosphate, pH 7.8 containing different amounts of KCl and $MgCl_2$. With no KCl, and a range of 0.1 to 0.5 mM $MgCl_2$, 70% of the complex remained after 3 days. As the KCl concentration was increased from 0 to 35 mM the optimal amount of $MgCl_2$ decreased, and in 40 mM KCl the complex was most stable (81%) when all the Mg^{2+} was sequestered by EDTA. The most useful buffer has been 20 mM KCl, 0.2 mM $MgCl_2$, 5 mM potassium phosphate pH 7.8. The short-term effect of increased $MgCl_2$ was also examined. When the complex was kept in 20 mM KCl, 5 mM potassium phosphate, pH 7.8 for only 60 minutes before analysis the $MgCl_2$ concentration could be varied from 0.1 to 3 mM with no decrease in the percentage of RNP.

Isolation and Properties of RNP. To separate the RNP from the remaining ribosomal subunits and the tRNA the formamide extract is concentrated to 4 ml by further dialysis against the same buffer, in a Schleicher-Schuell membrane supported in a 500-ml vacuum flask, under reduced pressure. The outside buffer is stirred magnetically, and the inside solution is mixed by slowly bubbling nitrogen. The material is then fractionated on Sephadex G-200, in the same buffer. The subunits are eluted at the void volume. The RNP, with an elution volume corresponding to a molecular weight of about 80,000, is clearly separated from the tRNA.[4] The recovery of 5 S RNA in the RNP fraction is about 14

[7] G. Blobel, *Proc. Nat. Acad. Sci. U.S.* **68**, 1881 (1971).

A_{260} units (1 mg), 70% of the theoretical amount; free 5 S RNA would be eluted just ahead of tRNA.

After Sephadex fractionation the RNP is much less stable; half the material that has emerged from the column in the RNP fraction is converted to free 5 S RNA in only 1 day. To obtain stable material for further characterization the formamide extract has been fixed with formaldehyde before Sephadex fractionation. The elution pattern is similar.[4] The isolated RNP fraction has a sedimentation coefficient of 6.8 S, and a buoyant density in CsCl of 1.568 g/ml, corresponding to an RNA content of 48.5%.[4]

The protein of the RNP complex has a molecular weight of 41,000, as measured by polyacrylamide gel electrophoresis in sodium dodecyl sulfate,[4] and the properties of the complex suggest that it contains one protein molecule for each 5 S RNA (40,000 daltons). Thus its molecular weight, measured on Sephadex, is about 80,000, and its buoyant density agrees with an RNA content of about 50%.

Blobel[7] has extracted a similar 5 S RNP from rat liver with EDTA, and reported a molecular weight of 35,000 for its protein. Other preparations have been derived from yeast[8] and reticulocytes.[9] These complexes can also be extracted by EDTA or urea, but formamide gives a higher ratio of RNP to free 5 S RNA. Although the formamide treatment seems to extract only this RNP from the large subunit, attempts to reassemble active large subunits are limited by the instability of both the RNP and the residual 41 S particle.

[8] A. G. Mazelis and M. L. Petermann, *Biochim. Biophys. Acta,* **312,** 111 (1973).
[9] B. Lebleu, G. Marbaix, G. Huez, J. Temmerman, A. Burny, and H. Chantrenne, *Eur. J. Biochem.* 19, 264 (1971).

[35] Preparation of Ribosomal Subunits by Large-Scale Zonal Centrifugation

By PAUL S. SYPHERD and JOHN W. WIREMAN

The increase in studies on ribosome structure and function has placed a large demand on purified ribosomal subunits. The introduction of large-scale zonal rotors was an important step in solving this problem, and the more recent application of hyperbolic sucrose gradients has increased the capacity up to 2 g of applied ribosomes while maintaining good resolution. We describe in this section a modification of the procedure of Eikenberry *et al.*[1] which permits easy loading and unloading of the zonal rotor with relatively simple equipment. The procedure is described for a Beckman Ti 15 rotor and ribosomes from *Escherichia coli.*

Reagents

2000 ml TM-4 buffer (10^{-2} M Tris·HCl, pH 7.4; 10^{-4} M MgCl$_2$)
800 ml 50% (w/v) sucrose in TM-4 buffer
700 ml 7.4% (w/v) sucrose in TM-4 buffer
200 ml 45% (w/v) sucrose in TM-4 buffer
2000 ml 60% (w/v) sucrose in TM-4 buffer
Ribosomes in TM-4 buffer (7000–30,000 OD$_{260}$ units, representing 0.5–2 g of ribosomes)

Equipment

Beckman Ti 15 rotor
2-Chamber mixing device with 100-ml capacity
2-Chamber mixing device with 2000-ml capacity (see Fig. 1)
Peristaltic pump to deliver 1 liter/hour

Procedure

1. All solutions and the rotor should be cooled to 5°. Completely fill the Ti 15 rotor with TM-4 buffer and secure the top. The rotor is brought to 3000–4000 rpm with the upper bearing and seal assembly in place.

2. The sample is applied in an inverse linear gradient. Fill the mixing side of a simple gradient-mixing device with the 50 ml of the ribosome suspension and the other side with 50 ml of 5% sucrose in TM-4.

3. The sample is pumped into the rotor at the rotor edge with a peristaltic pump. The sample will layer beneath the buffer and displace buffer through the rotor center. Stop the pump just as the mixing chamber is emptied and with sample still filling the line. Avoid air in the input lines.

4. If a gradient-producing pump is not available, a simple 2-chamber gradient mixer can be used (Fig. 1). The mixing side should be filled with 700 ml of 7.4% sucrose in TM-4 buffer and sealed, and the second chamber filled with 800 ml of 50% sucrose in TM-4 buffer.

5. While mixing, connect the large mixing chamber to the tube, which now contains some sample solution, open the port connecting the 2 chambers and start the peristaltic pump. The gradient will be pumped beneath the sample and push the sample toward the center of the rotor.

[1] E. F. Eikenberry, T. A. Bickle, R. R. Traut, and C. A. Price, *Eur. J. Biochem.* **12,** 113–116 (1970).

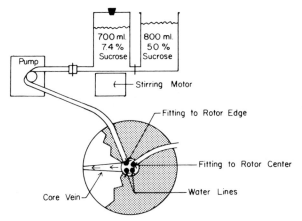

FIG. 1. Placement of apparatus for loading the zonal rotor with a 700-ml 7.4% to 34% hyperbolic gradient. The gradient mixer has a total capacity of 2 liters, and has the left side sealed after filling with sucrose. The gradient is pumped into the rotor with a peristaltic pump which delivers 1000–1500 ml per hour. Both the sample and the gradient are pumped into the rotor at the rotor edge.

Buffer will be displaced from the rotor through the center vein. The volume of liquid in the mixing side of the chamber should remain constant. Pump 700 ml of gradient into the rotor, i.e., until 100 ml of 50% of sucrose remain in the outer chamber. The loading is then completed

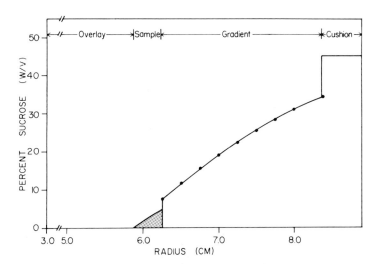

FIG. 2. The placement and shape of the 7.4% to 34% hyperbolic gradient. The sample is loaded as an inverse gradient with respect to ribosome concentration, in 0–5% sucrose concentration. The sample is denoted by the shaded area.

by pumping 200 ml of 45% sucrose through the outer vein as a cushion. The gradient pitch and position in the rotor is shown in Fig. 2. The gradient may conveniently be run for 5.5 hours at 35,000 rpm, 8 hours at 30,000 rpm, or 15 hours at 22,000 rpm for bacterial ribosomes. Shorter times will be necessary for the ribosomes of eukaryotes.

6. To recover the separated subunits, the rotor is slowed to 3000–4000 rpm and the upper bearing and seal assembly repositioned. Pump 60% sucrose in buffer into the tube leading to the rotor edge. The rotor contents will be displaced through the rotor center. The first 800 ml from the rotor may be discarded, and then 20-ml fractions collected until 60% sucrose emerges from the rotor (about 50 fractions). Although OD_{260} is the most accurate wavelength to locate the subunit peaks, in practice OD_{305} may be used ($OD_{260}/OD_{305} = 45$ for ribosomes). A typical OD_{305} profile is shown in Fig. 3.

7. The ribosomal subunits may be concentrated from the fractions in any of several ways. We have routinely recovered particles after first

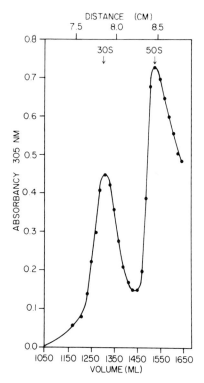

FIG. 3. Separation of *Escherichia coli* ribosomes. A total of 15,000 A_{260} units (1 g) was run for 5.5 hours at 34,000 rpm in the Beckman Ti 15 rotor.

reducing the sucrose concentration by dialysis against buffer, followed by ammonium sulfate precipitation. After dialysis, add $1\,M$ MgSO$_4$ to make the final concentration $10^{-2}\,M$. Add solid (NH$_4$)$_2$SO$_4$ in the ratio of 60 g/100 ml of ribosome suspension. Allow the precipitate to form in the cold, then centrifuge for 20 minutes at 12,000 g. Resuspend the precipitate in TM-2 buffer ($10^{-2}\,M$ Tris·HCl, pH 7.4, $10^{-2}\,M$ Mg^{2+}), and dialyze into TM-2. The magnesium concentration must be maintained at 10 mM until the (NH$_4$)$_2$SO$_4$ is removed. Concentration of pooled fractions can also be done by low pH precipitation or by pressure filtration.

Comments

In much of the earlier work in separating ribosomes on zonal rotors, a linear gradient was used. A gradient which is formed linearly will actually be concave, or exponential, as a function of sucrose concentration versus radius. The hyperbolic gradient is far superior in terms of resolution and capacity.[2] Eikenberry et al.[1] provided a useful discussion of some theoretical aspects of the hyperbolic zonal gradient, including the effects of increasing the length of the gradient and sample size. As they pointed out, the system they described is a compromise of several variables which produces the optimum separation. There are several advantages in placing the sample at the 6 cm position rather than at the rotor center. The sample at the 6 cm position occupies a narrow symmetrical band, rather than a wide asymmetrical band adjacent to the rotor core. In addition, the running time, and therefore diffusion, is reduced by the higher centrifugal force at the 6 cm position. We have investigated several different configurations of the hyperbolic gradient, with no improvement in resolution over that shown in Fig. 3 for the 7.4%–34% gradient.

The procedure described here employs a simple and inexpensive device for forming the hyperbolic gradient. The gradient can be formed by a programmed gradient pump (e.g., Beckman Model 141) by following the manufacturer's recommendations (see, also Eikenberry et al.[1]). In earlier experiments we formed the gradient first, moved it to the center of the rotor with 45% sucrose, and then layered the ribosome gradient through the inside of the rotor, followed by the 800 ml buffer overlay. This necessitated pumping solutions through the outer, then inner channels of the rotor core. We have found that the rotating rulon seal must be fitted precisely to allow pumping into the center of the core against a rotor full of sucrose solution. The method outlined here permits complete loading and unloading by pumping only through the outer veins of

[2] A. S. Berman, *Nat. Cancer Inst. Monogr.* **21**, 41 (1966).

the rotor core, and virtually eliminates sample loss. Loss of sample is a common experience when it is injected through the core center.

The ribosome separation shown in Fig. 3 is typical when the running time has been chosen correctly. Significantly longer times will result in driving the smaller subunits into the area occupied by the larger ones and reduce the purity of both. This is easily diagnosed by the loss of good separation of the peaks. It is not uncommon to find on the heavy side of the 50 S particles a sharp peak, which consists of 50 S ribosomes that have piled up at the interface between the end of the gradient and the 45% sucrose cushion.

Separations like that shown in Fig. 3 yield 30 S ribosomes of high purity (i.e., greater than 95%), as determined by sedimentation analysis, polyuridylic acid-directed polyphenylalanine synthesis, and disc gel analysis of proteins. The method of ribosome preparation frequently affects the purity of the 50 S particles. We have found that in some cases ribosomes prepared by ammonium sulfate precipitation and separated in TM-4 buffer yield 50 S preparations which appear homogeneous by sedimentation analysis, but which have appreciable 30 S backgrounds in protein-synthesizing systems. This apparently is due to the presence of 30 S aggregates, probably dimers, in the 50 S peak. The aggregate can easily be seen by analysis in acrylamide-agarose gels. This problem may be eliminated by using TMK buffer ($10^{-2} M$ Tris, $10^{-1} M$ KCl, $10^{-3} M$ Mg^{2+}) in the centrifugation procedures instead of TM-4. When separations are performed in sucrose gradients which contain salt, shorter running times are required. This is probably due to the ribosomes assuming a more compact configuration, and consequently sedimenting at a faster rate.

[36] Separation of Large Quantities of Eukaryotic Ribosomal Subunits by Zonal Ultracentrifugation

By CORINNE C. SHERTON, RALPH F. DI CAMELLI, and IRA G. WOOL

Bacterial ribosomes dissociate into subunits if the concentration of magnesium is lowered,[1] and large quantities of pure ribosomal subunits (2 g) can be separated by centrifugation in a zonal rotor.[2] However, the ribosomes of eukaryotic cells do not dissociate completely when sus-

[1] A. Tissières, D. Schlessinger, and F. Gros, *Proc. Nat. Acad. Sci. U.S.* **46**, 1450 (1960).

[2] E. F. Eikenberry, T. A. Bickle, R. R. Traut, and C. A. Price, *Eur. J. Biochem.* **12**, 113 (1970).

pended in low concentrations of magnesium—complete dissociation also requires the addition of a chelating agent.[3–5] Unfortunately, subunits prepared in that way are irreversibly altered; they do not recombine to form 80 S monomers, nor are they active in protein synthesis.[3,4,6]

Eukaryotic ribosomes can be dissociated by high concentrations of potassium (0.3–1.0 M) and the subunits will recombine to form monomers which synthesize protein in the presence of added template RNA.[3,7–21] In some cases complete dissociation requires preincubation of the particles with puromycin to remove nascent peptide and messenger RNA.[3,10,16,18,19,22]

Smaller quantities of eukaryotic ribosomal subunits (60–500 mg) have been separated in a zonal rotor.[18,21,23–28] Attempts to separate larger amounts of mammalian subunits have hitherto not been completely suc-

[3] T. E. Martin, F. S. Rolleston, R. B. Low, and I. G. Wool, *J. Mol. Biol.* 43, 135 (1969).

[4] H. Lamfrom and E. Glowacki, *J. Mol. Biol.* 5, 97 (1962).

[5] Y. Tashiro and P. Siekevitz, *J. Mol. Biol.* 11, 149 (1965).

[6] Y. Tashiro and T. Morimoto, *Biochim. Biophys. Acta* 123, 523 (1966).

[7] T. E. Martin and I. G. Wool, *Proc. Nat. Acad. Sci. U.S.* 60, 569 (1968).

[8] T. E. Martin and I. G. Wool, *J. Mol. Biol.* 43, 151 (1969).

[9] S. A. Bonanou and H. R. V. Arnstein, *FEBS (Fed. Eur. Biochem. Soc.) Lett.* 3, 348 (1969).

[10] G. R. Lawford, *Biochem. Biophys. Res. Commun.* 37, 143 (1969).

[11] H. R. V. Arnstein, *Biochem. J.* 117, 55P (1970).

[12] A. K. Falvey and T. Staehelin, *J. Mol. Biol.* 53, 1 (1970).

[13] A. M. Reboud, J. P. Reboud, C. Wittmann, and M. Arpin, *Biochim. Biophys. Acta* 213, 437 (1970).

[14] K. Terao and K. Ogata, *Biochem. Biophys. Res. Commun.* 38, 80 (1970).

[15] E. Bermek, H. Monkemeyer, and R. Berg, *Biochem. Biophys. Res. Commun.* 45, 1294 (1971).

[16] E. Busiello, M. DiGirolamo, and L. Felicetti, *Biochim. Biophys. Acta* 228, 289 (1971).

[17] M. S. Kaulenas, *Biochem. Biophys. Res. Commun.* 43, 1081 (1971).

[18] B. Mechler and B. Mach, *Eur. J. Biochem.* 21, 552 (1971).

[19] C. H. Faust, Jr. and H. Matthaei, *Biochemistry* 11, 2682 (1972).

[20] A. M. Reboud, M. Arpin, and J. P. Reboud, *Eur. J. Biochem.* 26, 347 (1972).

[21] B. A. M. van der Zeijst, A. J. Kool, and H. P. J. Bloemers, *Eur. J. Biochem.* 30, 15 (1972).

[22] G. Blobel and D. Sabatini, *Proc. Nat. Acad. Sci. U.S.* 68, 390 (1971).

[23] L. H. Kedes, R. J. Koegel, and E. L. Kuff, *J. Mol. Biol.* 22, 359 (1966).

[24] E. S. Klucis and H. J. Gould, *Science* 152, 378 (1966).

[25] S. Bonanou, R. A. Cox, B. Higginson, and K. Kanagalingam, *Biochem. J.* 110, 87 (1968).

[26] B. M. Mullock, R. Hinton, M. Dobrota, D. Froomberg, and E. Reid, *Eur. J. Biochem.* 18, 485 (1971).

[27] M. L. Petermann and A. Pavlovec, *Biochemistry* 10, 2770 (1971).

[28] B. A. van der Zeijst and H. Bult, *Eur. J. Biochem.* 28, 463 (1972).

cessful.[29] We have now adapted the use of a hyperbolic sucrose density gradient[2] in a Spinco Ti 15 zonal rotor to achieve excellent separation of up to 1.4 g of rat liver or muscle ribosomes.[30]

Media

Medium A: tris(hydroxymethyl)aminomethane (Tris)·HCl (20 mM), pH 7.8; KCl (830 mM); MgCl$_2$ (12.5 mM); 2-mercaptoethanol (MSH) (20 mM)

Medium B: Tris·HCl (20 mM), pH 7.8; KCl (500 mM); MgCl$_2$ (3 mM); MSH (20 mM)

Medium C: Tris·HCl (200 mM), pH 7.8; KCl (800 mM); MgCl$_2$ (125 mM)

Medium D: Tris·HCl (200 mM), pH 7.8; KCl (800 mM); MgCl$_2$ (30 mM)

Medium E: Tris·HCl (10 mM), pH 7.6; KCl (80 mM); MgCl$_2$ (12 mM)

Medium F: Tris·HCl (10 mM), pH 7.6; KCl (500 mM); MgCl$_2$ (5 mM)

Medium G: Tris·HCl (50 mM), pH 7.6

Reagents

Carbon; decolorizing, alkaline (Norit-A) (Fisher Scientific Co.)
Puromycin dihydrochloride (Nutritional Biochemical Corp.)
Sodium dodecyl sulfate (SDS), 20% in water (Fisher Scientific Co.)

Preparation of Ribosomes. We generally prepare muscle[3,31] and liver[3,8] ribosomes from male Sprague-Dawley rats that weigh 100–120 g. A modification[32] which we have found valuable in the preparation of skeletal muscle ribosomes is described in detail elsewhere in this volume.[33] However, any of the standard procedures for the preparation of relatively uncontaminated particles (from any cell type) will do. One should choose conditions of centrifugation likely to yield the maximum number of pure particles. Ribosome pellets can be stored at −20° for several months without loss of activity.

---

[29] T. E. Martin, I. G. Wool, and J. J. Castles, this series, Vol. 20, p. 417.

[30] C. C. Sherton and I. G. Wool, *J. Biol. Chem.* **247**, 4460 (1972).

[31] W. S. Stirewalt, J. J. Castles, and I. G. Wool, *Biochemistry* **10**, 1594 (1971).

[32] R. Zak, J. Ettinger, and D. A. Fischmann, in "Research in Muscle Development and the Muscle Spindle" (R. Pizybylski, J. Vander Meullen, M. Victor, and B. Banker, eds.), p. 163 (Excerpta Med. Found. Int. Congr. Ser. No. 240). Elsevier, Amsterdam, 1971.

[33] C. C. Sherton and I. G. Wool, this volume [49].

TABLE I
PREPARATION OF SOLUTIONS FOR ZONAL SEDIMENTATION IN MEDIUM A

Solution	Medium C (ml)	KCl (g)	2.5 M KCl (ml)	60% sucrose[a] (ml)	1 M MSH (ml)	Volume[b] (ml)
Medium A (1.0347)[c]	100	—	300	—	20	1000
7.4% Sucrose (1.0592)	100	—	300	123	20	1000
38% Sucrose (1.1803)	100	55	—	633	20	1000
45% Sucrose (1.2100)	150	84	—	1,125	30	1500

[a] The sucrose (60%, w/v in water) was treated with Norit-A as described in the text to remove material absorbing in the ultraviolet.
[b] Made to volume with water.
[c] Density, g/cm^3.

Separation of Ribosomal Subunits

Preparation of Ribosomes. Ribosome pellets are suspended in medium A (Table I) or medium B (Table II) by gentle homogenization, and aggregates are removed by centrifugation for 20 minutes at 2000 g. The final volume should be 40 ml, and should contain 13,000–15,000 A_{260} units (1.2–1.4 g) of ribosomes. The suspension can be kept in ice until it is loaded into the rotor.

Preparation of the Gradient. The sucrose used for the gradients should be freshly prepared and free of material that absorbs in the ultraviolet. Contaminants that absorb in the ultraviolet are removed from a 60% sucrose solution by heating with Norit-A (approximately 70 g per liter) and filtering twice through Whatman No. 1 filter paper; 20 mM MSH is included in the gradient to preserve the activity of the subunits. Since

TABLE II
PREPARATION OF SOLUTIONS FOR ZONAL SEDIMENTATION IN MEDIUM B

Solution	Medium D (ml)	KCl (g)	2.5 M KCl (ml)	60% sucrose[a] (ml)	1 M MSH (ml)	Volume[b] (ml)
Medium B (1.0243)[c]	100	—	168	—	20	1000
7.4% Sucrose (1.0496)	100	—	168	123	20	1000
38% Sucrose (1.1680)	100	—	168	633	20	1000
45% Sucrose (1.1938)	150	47	—	1,125	30	1500

[a] The sucrose (60%, w/v in water) was treated with Norit-A as described in the text to remove material absorbing in the ultraviolet.
[b] Made to volume with water.
[c] Density, g/cm^3.

the MSH is gradually oxidized, only the 45% sucrose solution (in which the ribosomes are never actually suspended during separation) can be frozen, stored, and reused.

A choice between medium A and medium B should be made on the basis of the behavior of the particular ribosome preparation that is to be used (see Remarks). Intact, active subunits can be prepared with either medium, but medium B gives better subunit separation with rat liver and skeletal muscle ribosome preparations and increases the yield 38% for the 40 S subunit and 56% for the 60 S subunit.

Loading the Zonal Rotor. A hyperbolic sucrose density gradient is generated in a Spinco Ti 15 zonal rotor using a Beckman Model 141 Gradient Pump with a hyperbolically shaped program cam[2] (the gears are selected so the total delivery will be 2 liters). The heavy solution line is filled with 38% sucrose. The light solution line, the three pump cylinders, and the line leading to the outer fitting of the loading head assembly (peripheral line) are equilibrated with 7.4% sucrose. The program cam is set at 0 (at which setting only 7.4% sucrose from the light solution reservoir is pumped into the rotor). It is the shape of the cam (actually its height as it rotates) which determines the proportions of the total outflow coming from the light and heavy solution lines.

The assembled rotor is placed in a specially adapted Beckman L2-65B centrifuge. With the zonal operation switch set for load and the refrigeration set at 26°, the rotor shield is put in place, and the rotor is started at 2500 rpm. When the rotor is at speed, the water-cooled loading head assembly is attached, and the gradient is generated at a speed control setting of 860 (power switch on low). Thus, the hyperbolic gradient is pumped into the rotor through the peripheral line, beginning with 7.4% and ending with 38% sucrose.

The formation of the gradient is complete when a total of 795 ml of both sucrose solutions have entered the pump, although it remains for a portion of the gradient to be transferred from the pump to the rotor. The heavy solution line is now transferred to the 45% sucrose reservoir. (The time at which the change is made can be indicated with a mark on the program cam.) While that is being done, the pump should be turned off and care taken that no air bubbles enter the line. Pumping is resumed. When the mixing of the gradient is completed (maximum height of the program cam) the program cylinder is arrested, so that only 45% sucrose enters the mixing cylinder. Pumping is continued until the rotor is full (total capacity 1665 ml), and the top of the gradient flows out the line from the center fitting of the loading head assembly (central line). The inner sector of the rotor now contains the 795 ml of hyperbolic gradient, and the outer sector the 870 ml of 45% sucrose cushion.

The peripheral line is disconnected from the pump. The continuous output and mixing cylinders and the heavy solution line are then equilibrated with medium B (the program cylinder remains arrested). A syringe containing 20 ml of 7.4% sucrose solution is attached to the end of the central line. The end of the peripheral line is placed in a beaker containing 45% sucrose solution. Bubbles that have accumulated under the outer rim of the rotor are now removed by slowly pushing them out with the 7.4% sucrose in the syringe, and then pulling 45% sucrose solution back in through the peripheral line. This operation should be repeated 2 to 3 times, until no bubbles are seen.

The suspended ribosomes which have been stored on ice are now dissociated by incubation for 15 minutes at 37° in 0.1 mM puromycin. The sample is introduced onto the hyperbolic gradient as a linear inverse gradient, produced with a two-chambered linear gradient maker. The mixing chamber contains 40 ml of 7.4% sucrose and the second chamber contains the 40 ml of ribosome suspension. The mixing chamber is connected to the central line, and the sample is slowly loaded onto the hyperbolic gradient by pulling 45% sucrose solution out of the rotor through the peripheral line, using a syringe.

The sample is overlayed with 680 ml of medium B pumped in through the central line (speed control 230 low for first 200 ml; 460 low for remainder), as 45% sucrose is removed from the peripheral line. The first 100 ml of the latter are combined with the 45% sucrose solution

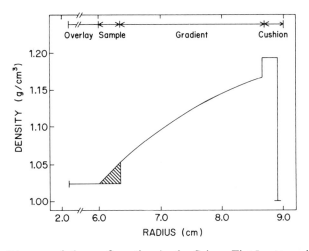

Fig. 1. Diagram of the configuration in the Spinco Ti 15 rotor prior to zonal sedimentation of ribosomal subunits. The densities are of sucrose solutions in medium B at room temperature (23°). The shaded area represents 1.35 g of ribosomes in an inverse linear sucrose gradient.

displaced during loading of the sample, and the absorption at 260 nm is determined to discover whether there has been a leak across the rotating rulon seal from the central line to the peripheral one. The loading head is then removed, and the rotor cap is attached to seal the rotor.

A continuous gradient of sucrose has been formed, starting at a radius of 6.0 cm from the rotor core, with the sample contained in the first 90 ml of linear gradient. At 6.35 cm the concentration of sucrose in the gradient is 7.36%, and at 8.66 cm it is 37.2%. Between 8.66 and 8.89 cm is a cushion of 45% sucrose solution. The relationships are shown diagrammatically in Fig. 1.

After a vacuum is formed, centrifugation is at 13,500 rpm for 17 hours at 26°. The zonal operation switch may now be placed on run, activating the centrifuge's normal safety circuits. If there has been a slight leak (2–3% of the total sample) across the central seal of the rotating rulon seal in the loading head assembly, the total running time should be increased by 10% because not all the buffer overlay was actually introduced into the rotor.

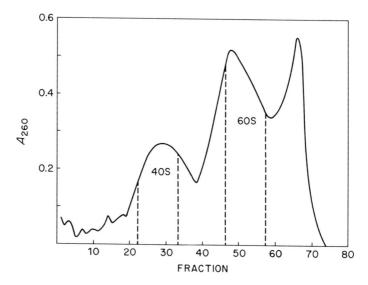

FIG. 2. Sedimentation of liver ribosomal subunits in a zonal rotor [C. C. Sherton and I. G. Wool, *J. Biol. Chem.* **247**, 4460 (1972)]. Liver ribosomal subunits (1.38 g) were separated by centrifugation for 17 hours at 13,500 rpm in a Spinco Ti 15 zonal rotor using a 7.4 to 38.0% hyperbolic sucrose density gradient [E. F. Eikenberry, T. A. Bickle, R. R. Traut, and C. A. Price, *Eur. J. Biochem.* **12**, 113 (1970)] in medium A. Fractions (10 ml) were collected, and those indicated by the interrupted lines were pooled; the other fractions were discarded; 117 mg of 40 S and 243 mg of 60 S subunits were recovered.

Unloading the Zonal Rotor. When the separation is completed the rotor speed is reduced to 2500 rpm; the operation switch is placed on load, the vacuum is released, and the rotor cap is removed. The pump is connected to the peripheral line and is equilibrated with untreated 60% sucrose (w/v in water)—a total of 2.5 liters is required for unloading. The loading head is attached to the rotor, and the contents of the rotor are displaced with 60% sucrose.

A total of 1 liter of solution is removed through the central line and discarded (the first 200 ml are at a speed control setting of 230 low; the second 200 ml at 460 low; the remainder at 895 low). Now 10-ml fractions are collected (the flow rate is 40 ml per minute) until the contents of the rotor have been completely displaced by 60% sucrose (a total of some 75–80 fractions). A 1:50 dilution of each fraction is made with a 1:1 mixture of the 7.4% and 38% sucrose solutions, and the absorption at 260 nm determined.

Precipitation of Ribosomal Subunits. The fractions containing the subunits, either 40 S or 60 S (Figs. 2 and 3), are pooled and dialyzed against

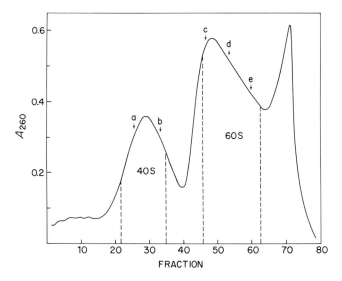

FIG. 3. Sedimentation of liver ribosomal subunits in a zonal rotor. Liver ribosomal subunits (1.35 g) were separated by centrifugation for 18.75 hours at 13,500 rpm in a Spinco Ti 15 zonal rotor using a 7.4 to 38.0% hyperbolic sucrose density gradient [E. F. Eikenberry, T. A. Bickle, R. R. Traut, and C. A. Price, *Eur. J. Biochem.* **12**, 113 (1970)] in medium B. The fractions indicated by the interrupted lines were pooled; 162 mg of 40 S and 413 mg of 60 S subunits were recovered. The analysis of fractions a–e is in Fig. 5.

8 liters of medium E for at least 36 hours with 3 to 4 changes of buffer. Dialysis is necessary because high concentrations of sucrose and potassium interfere with precipitation of ribosomes by ethanol.[34] Cold 95% ethanol (0.2 of a volume) is added to the dialyzed subunits and the suspension is kept at 0° for at least 1 hour (to ensure complete precipitation of the particles). The subunits are collected by centrifugation at 10,000 g for 15 minutes; they may be stored at $-20°$.

Purification of 60 S Subunits

The 60 S subunits isolated in this way are contaminated with small amounts of 40 S subunits and for some purposes must be resolved further.[30] Fractions containing 60 S subunits collected from centrifugation in a zonal rotor (about 740 mg) are incubated again for 15 minutes at 37° in medium A or B containing 0.1 mM puromycin. They are recentrifuged in the same way in a zonal rotor using a hyperbolic sucrose

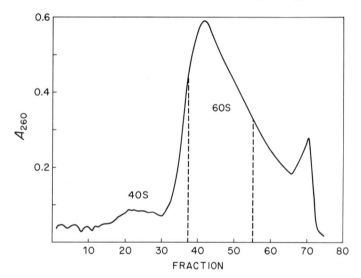

Fig. 4. Sedimentation of 60 S subunits of liver ribosomes in a zonal rotor [C. C. Sherton and I. G. Wool, *J. Biol. Chem.* **247**, 4460 (1972)]. Fractions containing 736 mg of 60 S subunits from three previous centrifugations in a zonal rotor were pooled and centrifuged again for 17 hours at 13,500 rpm in a Spinco Ti 15 zonal rotor using a 7.4 to 38.0% hyperbolic sucrose density gradient [E. F. Eikenberry, T. A. Bickle, R. R. Traut, and C. A. Price, *Eur. J. Biochem.* **12**, 113 (1970)] in medium A. The fractions indicated by the interrupted lines were pooled; 324 mg of 60 S subunits were recovered.

[34] M. S. Kaulenas, *Anal. Biochem.* **41**, 126 (1971).

gradient in either medium A or B, and fractions are collected (Fig. 4), dialyzed, and precipitated with ethanol as described above.

Analysis of the Purity of Ribosomal Subunits

The purity of the subunit fractions can be determined by zonal centrifugation (Fig. 5). Samples in medium E are layered on 5.2 ml of a 10 to 30% linear sucrose gradient in medium F. Centrifugation is at 60,000 rpm for 40 minutes at 26° in a Spinco SW 65 rotor, or at 48,000 rpm for 55 minutes at 26° in a Spinco SW 50.1 rotor. The distribution of particles is determined with an ISCO (Instrument Specialities Company) density gradient fractionator and UV analyzer.

The 40 S subunit fractions from the first centrifugation in the zonal rotor are generally free of contamination with 60 S subunits (Fig. 5,a and b). The slower sedimenting 60 S subunit fractions from the first centrifugation in the zonal rotor show some contamination with 40 S particles (Fig. 5c). The faster sedimenting 60 S subunit fractions have decreasing 40 S contamination, but contain a small amount of 90 S particles–60 S dimers (Fig. 5,d and e). The 60 S particles which have been purified by a second centrifugation in the zonal rotor are free of contamination (Fig. 6a).

Another method of determining whether subunit preparations are contaminated is analysis of their RNA.[3] Sufficient 20% SDS is added to a suspension of ribosomal particles in medium E to give a final concentration of 0.1%. The sample is incubated for 5 minutes at 37°, and then layered onto 5.2 ml of a 5 to 20% linear sucrose gradient in medium G. Centrifugation is at 4° for 2 hours at 60,000 rpm in a Spinco SW 65 rotor or for 2.75 hours at 48,000 rpm in a Spinco SW 50.1 rotor. The presence of 18 S (small subunit) and 28 S (large subunit) RNA is determined with an ISCO density gradient fractionator and UV analyzer.

The 40 S subunit fractions from the first zonal centrifugation contain only 18 S RNA (Fig. 5,f and g). The 60 S subunit fractions from the first zonal centrifugation contain predominantly 28 S RNA, but also a small amount of 18 S RNA, and usually some RNA that sediments between 18 and 28 S; the latter is probably from breakdown of 28 S RNA[3,20] (Fig. 5,h–j). The 60 S particles which have been purified by a second centrifugation in the zonal rotor are free of contamination with 18 S RNA (Fig. 6b).

Remarks

Active ribosomal subunits have been obtained from a variety of eukaryotic cells, including rat liver[8,10,13,14,20,22,27] and skeletal muscle,[3,7,8] mouse liver[12] and plasmocytoma tumor,[18,19] rabbit reticulocytes[9,11,16] and

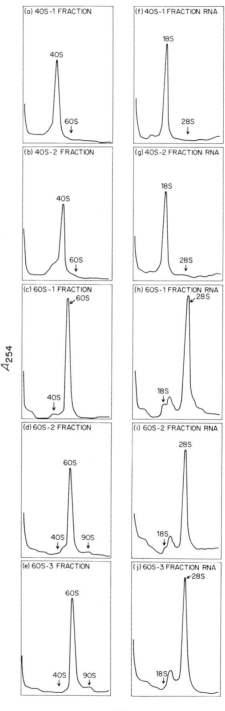

364

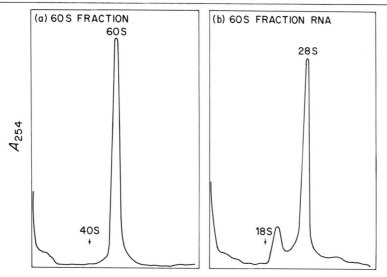

FIG. 6. Sedimentation in sucrose gradients of a purified 60 S subunit fraction and its RNA. The 60 S subunits collected from the zonal centrifugation depicted in Fig. 4 were analyzed as in Fig. 5 to determine the purity of the particles and their RNA.

skeletal muscle,[8] dog pancreas,[35] human tonsil[15] and reticulocytes,[36] HeLa cells,[37] insects,[17,34] yeast,[21] and protozoa.[8] Ribosomal subunits have been separated by unfolding in urea[27] and by ion-exchange chromatography[35]; but in all other cases, modifications of the original Martin and Wool procedure[3] of centrifugation through high concentrations of potassium were used. In selecting a method for separating large quantities of eukaryotic ribosomal subunits, it is important to consider the best means of dissociating the ribosomes, the optimal conditions for the separation of the sub-

[35] S. R. Dickman and E. Bruenger, *Biochemistry* **8**, 3295 (1969).

[36] J. E. Fuhr, C. Natta, A. Bank, and P. A. Marks, *Biochim. Biophys. Acta* **240**, 70 (1971).

[37] E. A. Zylber and S. Penman, *Biochim. Biophys. Acta* **204**, 221 (1970).

FIG. 5. Sedimentation in sucrose gradients of subunit fractions and their RNA. The purity of the subunit fractions collected from the zonal centrifugation depicted in Fig. 3 was determined (a–e) after dialyzing the samples against medium E. The samples were analyzed on 10 to 30% linear sucrose gradients in medium F, centrifugation was at 48,000 rpm for 55 minutes at 26° in a Spinco SW 50.1 rotor. The RNA of the subunit fractions was analyzed (f–j) after suspending the particles in medium G with 0.1% SDS, and incubating for 5 minutes at 37°. The samples were centrifuged in 5 to 20% linear sucrose density gradients in medium G at 48,000 rpm for 2.75 hours at 4° in a Spinco SW 50.1 rotor.

units, and the type of zonal centrifugation that will yield the largest amounts of pure subunits.

Polysomes are resistant to dissociation. Several methods have been used to reduce their numbers and hence to increase the yield of subunits. Incubation of HeLa cells at 42° greatly reduces the proportion of polysomes present[37]; incubation at 0 to 4° has the same effect on cricket nymph[17] and human reticulocyte[36] polysomes. The most efficient, and the most generally used, method of dissociating polysomes is incubation with puromycin to release the nascent peptide and probably the messenger RNA. Incubation, generally with 0.1–0.5 mM puromycin, may be in high concentrations of potassium[19,22,30,37] as described here, or at low monovalent cation concentrations in the presence of supernatant factors and an energy generating system.[3,10,15,16,18] Falvey and Staehelin,[12] prepared subunits from "runoff" ribosomes by first incubating mouse liver polysomes in the presence of all the components necessary for protein synthesis; they used that procedure because they found that subunits prepared with puromycin were less active. Nonetheless, our method does produce active subunits and has the advantage of not requiring the addition of extraneous proteins during the separation procedure.

The conditions of centrifugation are also important for the separation of active subunits in good yield. Centrifugation is at 26° because at 4° 40 S and 60 S rat liver and muscle subunits form 55 S and 90 S aggregates,[3,8] respectively. However, others have been able to isolate eukaryotic ribosomal subunits at lower temperatures (0 to 10°),[12,17–19,21,37] and mouse plasmocytoma subunits are less active when prepared at 20° rather than at 5°.[19]

One or more ribosomal proteins has sulfhydryl groups which must be kept reduced if the function of the particles is to be preserved.[15,38] Active subunits have been prepared without addition of either 2-mercaptoethanol or dithiothreitol,[11,12,17,18,37] even though a sulfhydryl reducing reagent has been shown to be an absolute requirement for the preparation of active subunits by others.[3,7,8,21]

The separation of subunits is generally in relatively high concentrations of potassium and low concentrations of magnesium ions (the use of other mono- and divalent cations has been investigated[37,39]). It is thought that potassium replaces magnesium ions which stabilize the interaction between the subunits. Thus, the ratio of potassium to magnesium ions must be high enough so that dissociation is complete.[10,18,19,22] Furthermore, the concentration of potassium ions must be high enough to re-

[38] B. S. Baliga and H. N. Munro, *Nature (London) New Biol.* 233, 257 (1971).
[39] A. M. Reboud, M. Buisson, and J. P. Reboud, *Eur. J. Biochem.* 26, 354 (1972).

move factors or supernatant proteins adventitiously bound to the surface of the subunit,[11,39,40] yet not so high as to extract ribosomal proteins.[13,18-20] On the other hand, the magnesium concentration must not be so low as to destabilize the structure of the ribosomal subunit, and thus produce irreversible unfolding and the release of the 5 S RNA from the 60 S subunit.[10,18,19] We believe that the use of 500 mM potassium and 3 mM magnesium (medium B) best fits these criteria. The ratio of potassium to magnesium concentrations in medium B is 167; high enough to produce complete dissociation of the particles and better separation of the subunit peaks than is obtained with 830 mM potassium and 12.5 mM magnesium (medium A) (cf. Figs. 2 and 3). The subunits prepared in medium B are intact (cf. Fig. 5), and do not contain proteins removed by higher concentrations of potassium.[41]

Centrifugation in a zonal rotor is the only practicable method to prepare large quantities of ribosomal subunits. Linear sucrose density gradients allow zone broadening (sectorial dilution) which may result in contamination of subunit fractions; moreover, they can only be used to separate at the most 300 mg of ribosomes.[23-27] Earlier we reported a rather unsuccessful attempt to separate 1.0 g of rat liver ribosomes using a linear gradient.[29] Equivolumetric sucrose density gradients—ones in which sample particles of like density pass through a constant volume of gradient per unit time—have been devised to eliminate the problem of sectorial dilution.[42] However, they have been successfully used only for the separation of up to 130 mg of ribosomes.[21,28] The hyperbolic sucrose density gradient developed by Eikenberry et al.—in which an initially stable zone, such as the ribsome sample in its inverse gradient, does not undergo sectorial dilution—has been used by them to separate up to 2 g of E. coli ribosomes.[2] The same gradient has also been used to separate 500 mg of mouse plasmocytoma ribosomes.[18] We have modified the concentration of ions in the hyperbolic gradient, added a sulfhydryl reducing agent, and adjusted the temperature to achieve an efficient means for the preparation of eukaryotic ribosomal subunits; the method allows for the preparation of large quantities of pure, intact subunits from rat liver or muscle ribosomes.[30]

[40] S. M. Heywood, Cold Spring Harbor Symp. Quant. Biol. 34, 799 (1969).
[41] C. C. Sherton and I. G. Wool, unpublished results (1972).
[42] M. S. Pollack and C. A. Price, Anal. Biochem. 42, 38 (1971).

[37] A General Procedure for the Preparation of Highly Active Eukaryotic Ribosomes and Ribosomal Subunits[1]

By GLENN E. BROWN, ALFRED J. KOLB, and WENDELL M. STANLEY, JR.

This article describes a highly reproducible method for preparing ribosomes and ribosomal subunits from a variety of eukaryotic cells. The number of sources used (rabbit reticulocytes, rabbit liver, chicken reticulocytes, chicken liver, and embryonic chicken leg muscle) suggests that this method can be applied to cells from many eukaryotic sources. The resultant ribosomes exhibit essentially no endogenous polypeptide synthetic ability, but retain extremely high amino acid polymerization activity when challenged by poly(U); 30 nanomoles of phenylalanine incorporated per milligram of ribosomes, or 150 phenylalanine residues polymerized per 80 S ribosome. Separated and recombined subunits also exhibit this high response to poly(U).

Processed ribosomes and their separated and reassociated subunits are capable of translating natural messenger RNA's[2] and artificially synthesized oligoribonucleotides containing an AUG initiation codon at the 5' terminus—but only in the presence of 0.5 M KCl extracts of polysomes (which contain, among other proteins, initiation factors[2a-2e]). Thus, these ribosomes and their subunits can be used in constructing highly active *in vitro* systems suitable for the assay of messenger RNA, initiation factors,[2a-2e] elongation factors, tRNA, and other components of the protein biosynthetic system which are usually associated with ribosomes prepared by other techniques.

Buffers, Materials, and Reagents

The pH of all buffers is adjusted with HCl and measured at 20°.

Buffer A: 2.5% phenylhydrazine (v/v) (Aldrich Chemical Co., Inc., Milwaukee, Wisconsin 53233), 1 mM glutathione (Schwarz/

[1] This study was aided partially by Grant E-533 from the American Cancer Society, Inc., by grants from the Cancer Research Coordinating Committee of the University of California, and by grants from the School of Biological Sciences, University of California, Irvine, California 92664.

[2] Qβ RNA, tobacco mosaic virus RNA, yeast messenger RNA's, and rabbit globin messenger RNA's.

[2a] G. L. Dettman and W. M. Stanley, Jr., *Biochim. Biophys. Acta* **287**, 124 (1972).

[2b] G. L. Dettman and W. M. Stanley, Jr., *Biochim. Biophys. Acta* **299**, 142 (1972).

[2c] L. M. Cashion and W. M. Stanley, Jr., *Biochim. Biophys. Acta*, **324**, 410 (1973).

[2d] L. M. Cashion, G. L. Dettman, and W. M. Stanley, Jr., this volume [16].

[2e] L. M. Cashion, P. M. Neal, and W. M. Stanley, Jr., *Proc. Nat. Acad. Sci. U.S.*, in press.

Mann, Division of Becton, Dickinson and Co., Orangeburg, New
York 10962), pH 7.0

Buffer B: 0.13 M NaCl, 5 mM KCl, 7.5 mM MgCl$_2$

Buffer C: 2 mM MgCl$_2$, 1 mM 2-mercaptoethanol

Buffer D: 0.1 M KCl, 7.5 mM MgCl$_2$, 1 mM 2-mercaptoethanol,
0.1 mM EDTA, 50 mM Tris·HCl, pH 7.8

Buffer E: 1.5 M sucrose, 50 mM KCl, 2 mM MgCl$_2$, 1 mM 2-mer-
captoethanol, 50 mM Tris·HCl, pH 7.8

Buffer F: 25 mM KCl, 5 mM MgCl$_2$, 50 mM Tris·HCl, pH 7.8

Buffer G: 0.25 M KCl, 10 mM MgCl$_2$, 10 mM Tris·HCl, pH 7.4

Buffer H: 0.5 M KCl, 1mM 2-mercaptoethanol, 50 mM Tris·HCl,
pH 7.8

Buffer I: 0.15 M KCl, 1mM MgCl$_2$, 50 mM Tris·HCl, pH 7.8

Buffer J: 0.3 M KCl, 1 mM MgCl$_2$, 50 mM Tris·HCl, pH 7.8

Heparin: sodium salt (Sigma Chemical Co., St. Louis, Missouri
63178, grade 1)

Nembutal: Nembutal Sodium (sodium pentobarbital injection, 50
mg/ml, Abbott Laboratories, North Chicago, Illinois 60064)

Tissue grinder: 0.006 inch to 0.009 inch clearance between the
serrated Teflon pestle and the glass receptacle (A. H. Thomas
Co., Philadelphia, Pennsylvania 19105; size C, catalog No.
3431-E25)

Collodion bags: S and S Collodion Bags No. 100 for Protein Concentration (Schleicher and Schuell, Inc., Keene, New Hampshire
03431)

Puromycin: dihydrochloride salt (Nutritional Biochemicals Corp.,
Cleveland, Ohio 44128)

ATP: disodium salt (Sigma Chemical Co.)

GTP: trisodium salt (Sigma Chemical Co.)

Creatine phosphate: dipotassium salt, 3½ hydrate (Calbiochem,
Los Angeles, California 90054)

Creatine phosphokinase: from rabbit muscle, substantially salt-free
powder, EC 2.7.3.2 (Sigma Chemical Co.)

Energy mix: prepared at room temperature by adding together
3.12 ml of 1 M Tris·HCl (pH 7.8), 0.76 ml of 1 M KCl, 60
mg of ATP, 15 mg of GTP, 0.35 g of creatine phosphate, 2 mg
of creatine phosphokinase, 70 μl of 2-mercaptoethanol, and adjusting the final volume to 5 ml with H$_2$O. The final pH should
be about 7.6 when measured at 20°. Aliquots are frozen in a
dry ice–ethanol bath and stored at −70°.

Transfer RNA: prepared by the method of Rogg et al.[3] from in-

---

[3] H. Rogg, W. Wehrli, and M. Staehelin, *Biochim. Biophys. Acta* **159**, 13 (1969).

dividually frozen livers (Pel-Freez Biologicals, Inc., Rogers, Arkansas 72756), freed of gall bladders and bile, obtained from 24-hour-fasted female chickens or female New Zealand white rabbits. The various species of rabbit liver or chicken liver aminoacyl-tRNA's are prepared as described by Stanley.[4,5]

L-[^{35}S]Methionine: specific activity between 15,000 Ci/mole and 40,000 Ci/mole, prepared by the method of Graham and Stanley[6]

L-[^{14}C]Phenylalanine: uniformly labeled, specific activity 455 Ci/ mole (Schwarz/Mann, Division of Becton, Dickinson and Co., Van Nuys, California 91401)

L-[^{14}C]Valine: uniformly labeled, specific activity 195 Ci/mole (ICN Chemical and Radioisotope Division, a division of International Chemical and Nuclear Corp., Irvine, California 92664)

Poly(U): potassium salt, lyophilized, $s_{20,w} = 11$ (P-L Biochemicals, Inc., Milwaukee, Wisconsin 53205)

AUG(U)$_x$: ApUpG(pU)$_x$, $x = 25 \pm 5$, prepared by the method of Stanley et al.[7]

Globin mRNA's: prepared as described by Evans and Lingrel[8] from rabbit reticulocytes

DEAE-cellulose: Whatman DE-52, microgranular, preswollen, 1 meq per dry gram (Reeve Angel, Clifton, New Jersey 07014)

Miscellaneous Procedures

Protein Concentrations. These were measured by the method of Lowry et al.[9] using bovine serum albumin (fraction V, Sigma Chemical Co.) as the standard.

Ribosome and Ribosomal Subunit Concentrations. These were measured by the absorbance at 260 nm. One milligram of ribosomes or of ribosomal subunits dissolved in 1 ml has an absorbance at 260 nm of 12. One milligram of ribosomes or of ribosomal subunits is equivalent to 12 A_{260} units.[10]

Transfer RNA Concentrations. These were measured by the absorb-

[4] W. M. Stanley, Jr., *Anal. Biochem.* **48**, 202 (1972).

[5] W. M. Stanley, Jr., this series, Vol. 29 [44].

[6] R. Graham and W. M. Stanley, Jr., *Anal. Biochem.* **47**, 505 (1972).

[7] W. M. Stanley, Jr., M. A. Smith, M. B. Hille, and J. A. Last, *Cold Spring Harbor Symp. Quant. Biol.* **31**, 99 (1966).

[8] M. J. Evans and J. B. Lingrel, *Biochemistry* **8**, 3000 (1969).

[9] O. H. Lowry, N. J. Rosebrough, A. L. Farr, and R. J. Randall, *J. Biol. Chem.* **193**, 265 (1951).

[10] One A_{260} unit is that amount of material, which, when dissolved in 1 ml, has an absorbance of 1 at 260 nm in a 1-cm path length cuvette.

ance at 260 nm. One milligram of tRNA (potassium salt) dissolved in 1 ml has an absorbance at 260 nm of 24.

Messenger RNA Concentrations. Poly(U): measured by the absorbance at 260 nm. One milligram of poly(U) (potassium salt) dissolved in 1 ml has an absorbance at 260 nm of 27. AUG(U)$_x$: measured by its absorbance at 260 nm. One milligram of AUG(U)$_x$ (potassium salt) dissolved in 1 ml has an absorbance at 260 nm of 27. Globin mRNA's, yeast mRNA's, tobacco mosaic virus RNA, Qβ RNA; measured by their absorbance at 260 nm. One milligram of these mRNA's (potassium salt) dissolved in 1 ml has an absorbance at 260 nm of 24.

Preparation of Supernatant Fractions and Crude Polysomes

Rabbit Reticulocytes. Crude polysomes are prepared essentially by the procedure of Allen and Schweet.[11] Female New Zealand white rabbits (Mission Laboratories Supply, Inc., Rosemead, California 91770) weighing 3.5–4 lb are injected subcutaneously with buffer A: 1 ml each on days 1 and 2, 0.8 ml on day 3, and 0.6 ml on day 4. On the first day each rabbit also receives a 1-ml intramuscular injection of folic acid and vitamin B$_{12}$.[12] The rabbits are sacrificed on day 6 by injection of 0.9 ml of Nembutal into an ear vein and subsequent removal of as much blood as possible (approximately 65 ml per rabbit) by cardiac puncture. Withdrawal syringes (100 ml) and needles (15 gauge by 3.5 inches) are prerinsed with 0.5% (w/v) heparin to prevent clotting. The blood is chilled to 0° immediately.

The blood is centrifuged at 2° for 1 minute at 10,000 rpm (12,100 g[13]) in an SS-34 rotor in a Sorvall RC2-B centrifuge (Ivan Sorvall, Inc., Norwalk, Connecticut 06856); the cells are resuspended in 3 cell volumes of buffer B, and recentrifuged. This rinse is repeated once. Two cell volumes of buffer C are added at 0° to lyse the cells osmotically. After 1 minute, 0.12 cell volume of 1 M KCl is added and the suspension is centrifuged at 2° for 10 minutes at 10,000 rpm in an SS-34 rotor. The supernatant is centrifuged at 2° for 3 hours at 60,000 rpm (361,000 g) in a Beckman Spinco 60 Ti rotor (Beckman Instruments, Inc., Spinco Division, Palo Alto, California 94304).[14] The middle third of the supernatant is carefully removed with a syringe and is saved for later use as

---

[11] E. H. Allen and R. S. Schweet, *J. Biol. Chem.* **237**, 760 (1962).

[12] J. M. Gilbert and W. F. Anderson, this series, Vol. 20, p. 542.

[13] All g forces given in this article are the maximum values for the rpm and rotors used.

[14] This long centrifugation is necessary to completely free the supernatant of ribosomal subunits. Unpublished data, G. E. Brown and W. M. Stanley, Jr., 1971.

the supernatant fraction. It contains between 60 and 80 mg of protein per milliliter. The residual fluid around the polysome pellets is aspirated and discarded. Each pellet contains between 700 and 1000 A_{260} units of polysomes (600–800 A_{260} units per rabbit). The polysome pellets and aliquots of the supernatant are frozen in a dry ice–ethanol bath and stored at $-70°$.

Rabbit Liver. Crude polysomes are prepared either from fresh rabbit livers or from individually frozen livers (Pel-Freez Biologicals, Inc.), freed of gall bladders and bile, obtained from 24-hour-fasted female New Zealand white rabbits. Frozen livers are thawed and then either the fresh livers or the thawed material are homogenized at $2°$ in buffer D (2 ml of buffer per gram of liver) by 10 up-and-down strokes with a motor-driven serrated Teflon pestle at 1100 rpm in a tissue grinder. The homogenate is centrifuged at $2°$ for 10 minutes at 10,000 rpm in an SS-34 rotor. The supernatant is carefully decanted, layered over 4 ml of cold buffer E, and centrifuged at $2°$ for 4 hours at 60,000 rpm in a Beckman Spinco 60 Ti rotor. The pellets are frozen in a dry ice–ethanol bath and stored at $-70°$. Each pellet contains approximately 250 A_{260} units of polysomes (750 to 1000 A_{260} units per rabbit).

Chicken Reticulocytes. Polysomes and the supernatant fraction are prepared from 1.5–3 lb adult White Leghorn chickens (strain Shaver Starcross; Demler Farms, Anaheim, California 92805). Reticulocytosis is induced by 5 daily subcutaneous injections of 0.3 ml of buffer A per chicken. The chickens are sacrificed on the sixth day by an intravenous injection of 0.7 ml of Nembutal per chicken and subsequent removal of as much blood as possible (approximately 50 ml per chicken) by cardiac puncture. The rest of the procedure is identical to that used for rabbit reticulocytes with the following exceptions. After the addition of 1 cell volume of buffer C, the cells receive 5 up-and-down strokes in a tissue grinder at $2°$ followed by the addition of 0.5 cell volume of 0.15 M KCl. The homogenate is centrifuged for 30 minutes at $2°$ at 11,000 rpm (14,500 g) in an SS-34 rotor. The supernatant is layered over 8 ml of buffer F containing 1.5 M sucrose at $2°$ and is centrifuged at $2°$ for 5 hours at 60,000 rpm in a Beckman Spinco 60 Ti rotor. Each pellet contains 700–900 A_{260} units of polysomes (500–600 A_{260} units per chicken). The supernatant fraction contains approximately 130 mg of protein per milliliter.

Chicken Liver. Polysomes are prepared from fresh livers, freed of gall bladders and bile, obtained from 24-hour-fasted adult White Leghorn chickens (strain Shaver Starcross). The procedure is identical to that used for rabbit liver polysomes. Each pellet contains approximately 300 A_{260} units of polysomes (400–500 A_{260} units per chicken).

Embryonic Chicken Leg Muscle. The preparation of polysomes is based on the procedure of Heywood et al.[15] Leg muscles are removed from 14-day-old chicken embryos (strain Shaver Starcross) and homogenized at 2° in buffer G (equal amounts, w/v) by 10 up-and-down strokes in a tissue grinder. The homogenate is centrifuged at 2° for 30 minutes at 11,000 rpm in an SS-34 rotor. The supernatant is layered over 8 ml of buffer F containing 1.5 M sucrose and centrifuged at 2° for 5 hours at 60,000 rpm in a Beckman Spinco 60 Ti rotor. The pellets are frozen in a dry ice–ethanol bath and stored at −70°. Each pellet contains approximately 600 A_{260} units of polysomes (25–30 A_{260} units per embryo).

Preparation of Crude Initiation Factors

A preparation of crude initiation factors is obtained by modifying the procedure of Miller and Schweet.[16] The crude polysome pellets are dissolved in buffer H at a concentration of 425 A_{260} units/ml. The solution is stirred at 0° for 1 hour and then is centrifuged in a Beckman Spinco 65 rotor for 2 hours at 2° at 50,000 rpm (218,000 g). The upper four-fifths of the supernatant is recovered by syringe and is used as the crude initiation factor preparation. It contains between 23 and 26 mg of protein per milliliter. It is frozen in aliquots in a dry ice–ethanol bath and stored at −70°.

Experiments involving natural mRNA's use a crude initiation factor preparation prepared essentially in the same manner except that the crude polysome pellets are dissolved at a concentration of 130 A_{260} units/ml. After centrifugation, the upper four-fifths of the supernatant is vacuum concentrated at 0° in collodion bags surrounded by the extraction buffer. The final solution contains about 60 mg of protein per milliliter.[17a] We found subsequently that there are no qualitative differences between the two initiation factor preparations.

The ribosomes in the pellets, after either procedure, are relatively inactive in *in vitro* amino acid incorporation reactions, whether or not they are supplemented by the crude washes, and the ribosomes therefore are discarded.

Preincubation with Puromycin

The preincubation with puromycin is based on the procedure of Lawford.[17] The incubation contains, in a volume of 10 ml: 300 A_{260} units of

---

[15] S. M. Heywood, R. M. Dowben, and A. Rich, *Biochemistry* 7, 3289 (1968).

[16] R. L. Miller and R. Schweet, *Arch. Biochem. Biophys.* 125, 632 (1968).

[17] G. R. Lawford, *Biochem. Biophys. Res. Commun.* 37, 143 (1969).

[17a] Approximately 75% of the protein in the two factor preparations represents non-polysomal proteins sedimented with crude polysomes during the 3 hour preparative centrifugation of the lysate.

crude polysomes in 1 ml of buffer F, 4 ml of the homologous reticulocyte supernatant fraction, 0.8 ml of 0.1 M $MgCl_2$, 0.8 ml of a solution of 8.75 $\times$ 10^{-4} M puromycin dihydrochloride, and 0.8 ml of the energy mix.

The preincubation with puromycin is at 37° for 45 minutes, after which 1.5 g of sucrose is added and dissolved. The solution is chilled to 0° for 10 minutes, filtered through 5 cm of packed glass wool in a disposable Pasteur pipette, centrifuged at 2° for 10 minutes at 10,000 rpm in an SS-34 rotor, and then passed ascendingly through a 2.5 × 50 cm column of Sephadex G-100 at 2°. Immediately after the introduction of the sample into the bottom of the column, a more dense solution of 10 ml of 20% (w/v) sucrose is used to raise the sample into the gel. The column buffer is buffer I and the 20% (w/v) sucrose is dissolved in buffer I. The column then is developed ascendingly. Routinely, a 2-meter hydrostatic head results in an elution time of approximately 3 hours for the ribosomal subunits. Fractions are collected every 10 minutes, and the tubes containing the ribosomal subunits are pooled. Fifty microliters of 1 M $MgCl_2$ are added to each 10 ml of the subunit solution (equivalent to the contents of 1 Oak Ridge-type tube for the Beckman Spinco 40 rotor). The reassociated ribosomes are pelleted by centrifugation at 2° for 2 hours at 35,000 rpm (110,900 g) in a Beckman Spinco 40 rotor. The pellets are rinsed with buffer I, frozen in a dry ice–ethanol bath and stored at −70°. Each pellet contains between 50 and 65 A_{260} units of ribosomes.

If ribosomal subunits are desired, only the peak fractions of the Sephadex G-100 column effluent containing the ribosomal subunits are pooled (12–15 A_{260} units per milliliter in a volume of 7–9 ml). One milliliter is layered over each sucrose density gradient in a Beckman Spinco SW 27 tube (1 × 3.5 inches). Six exponential sucrose density gradients are formed simultaneously with a 6-channel pump by pumping from a constant volume gradient mixer containing 105 ml of 15% (w/v) sucrose in buffer I. The reservoir contains excess 30% (w/v) sucrose in buffer I. The sucrose density gradients are centrifuged in a Beckman Spinco SW 27 rotor for 12 hours at 27,000 rpm (131,000 g) at 2°. The contents of each sucrose density gradient tube are displaced upward by pumping 50% (w/v) sucrose under the sucrose density gradient. The contents of the tube are led through a 1 cm path length flow cell mounted in a double-beam spectrophotometer set at 260 nm and the absorbance profile is recorded automatically on a strip-chart recorder. The ribosomal subunits are collected in tubes in an ice bath and pelleted by centrifugation through a 0.2-ml layer of buffer I containing 1 M sucrose at 2° for 4 hours at 50,000 rpm in a Beckman Spinco 65 rotor. The pellets are immediately aspirated as dry[18] as possible (the 40 S ribosomal subunits

---

[18] Excessive amounts of sucrose inhibit the *in vitro* incorporation of amino acids into hot acid-insoluble material. Unpublished data, G. E. Brown, 1971.

tend to go quickly and spontaneously back into solution). Each 40 S ribosomal subunit pellet contains between 2.5 and 6 A_{260} units; each 60 S ribosomal subunit pellet contains between 6 and 10 A_{260} units. The pellets are dissolved in buffer F at the desired concentration and are used immediately.

Amino Acid Incorporations

Amino acid incorporation reaction mixtures contain in a volume of 50 μl the following: 1 A_{260} unit of ribosomes that have been preincubated with puromycin and reisolated, 20 μl of the rabbit reticulocyte supernatant fraction, 4 μl of 0.14 M MgCl$_2$, 4 μl of energy mix, 5 μl of the crude rabbit reticulocyte initiation factors or 5 μl of buffer H, and (1) in the case of incorporations using poly(U), 50 μg of poly(U), 1 A_{260} unit of unfractionated rabbit liver or chicken liver tRNA, 4 nmoles uniformly labeled L-[^{14}C]phenylalanine (diluted with unlabeled phenylalanine to a specific activity of 10 Ci/mole), 10 nmoles each of 19 unlabeled amino acids (no phenylalanine), or (2) in the case of incorporations using AUG(U)$_x$, 0.165 A_{260} units AUG(U)$_x$, 1 A_{260} unit of rabbit liver L-[^{35}S]Met-tRNA$_i^{Met}$ (specific activity 6000 Ci/mole), 1 A_{260} unit of unfractionated rabbit liver tRNA aminoacylated with 19 unlabeled amino acids (no methionine), or (3) in the case of incorporations using globin mRNA's, 0.04 A_{260} units of globin mRNA's, 1 A_{260} unit of unfractionated rabbit liver tRNA aminoacylated with L-[^{14}C]valine (specific activity 195 Ci/mole) and with 19 unlabeled amino acids (no valine), 12.5 nmoles each of 20 unlabeled amino acids, and 5 μl of 50 mM MgCl$_2$ instead of 0.14 M MgCl$_2$. For some of the incorporations involving globin mRNA's, the ribosomes used in the incorporation reactions are dissolved in buffer F (200 A_{260} units/ml) and incubated at 37° for 30 minutes before use (as indicated in the legend to Table VII). Incorporation tubes are incubated at 37° for 1 hour or as indicated in the appropriate figure legends. If isolated subunits are being examined instead of the reassociated ribosomes, each 50-μl reaction mixture contains 0.1 A_{260} units 40 S ribosomal subunits and/or 0.2 A_{260} units 60 S ribosomal subunits or (to serve as a control) 0.3 A_{260} units unseparated subunits kept at 0° during the subunit separations and then processed identically to the subunits during the concentration step.

The extent of amino acid incorporation is evaluated by adding to each tube 0.15 ml of 1 M KOH, heating at 90° for 5 seconds, adding 4 ml of 5% trichloroacetic acid, heating at 90° for 10 minutes, cooling at 0° for 10 minutes, centrifuging at room temperature in a No. 221 rotor in an International desk-top clinical centrifuge (Model CL, International Equipment Co., Needham Heights, Massachusetts 02194) at top speed (3000 rpm; 1470 g) for 5 minutes, and aspirating and discarding the supernatant. This procedure is repeated once. In lieu of the second centrifugation, the contents of each tube are filtered at 5 ml per minute through Whatman type GF 83 glass fiber disks (2.7 cm diameter; 1.6 cm diameter filtration surface). The disks are washed twice with 5 ml of cold 5% trichloroacetic acid, dried, and counted in a liquid scintillation spectrometer or in a gas flow counter.

Results

The absorbance profiles of preparative sucrose density gradients containing rabbit reticulocyte ribosomal subunits (Fig. 1) show that the polysomes and ribosomes are completely converted to subunits following the preincubation with puromycin and gel filtration through Sephadex G-100. Figure 1B is identical to Fig. 1A except that the KCl concentration in the sucrose density gradient was 0.3 M (buffer J) instead of 0.15 M (buffer I). A very slight shoulder is sometimes seen on the light side of the 60 S ribosomal subunit peak.[19,20]

Ribosomal subunits prepared on sucrose density gradients containing 0.3 M KCl migrate slightly, but consistently, slower than those prepared on sucrose density gradients containing 0.15 M KCl. This indicates either a loosening of the ribosomal subunit structure and/or a loss of ribosomal material. The ion concentrations employed in the experiment diagrammed in Fig. 1A are just high enough to accomplish the complete conversion of the 80 S ribosomes to subunits, since lowering either the Tris·HCl or the KCl concentration results in incomplete conversion.[21] Although the separation of the ribosomal subunits appears better in sucrose density gradients containing 0.3 M KCl (and the yield of subunits can therefore be higher), the amino acid incorporation potential of the recovered ribosomal subunits is greatly reduced.

Figure 2 shows the absorbance profiles of the RNA's from rabbit reticulocyte ribosomal subunit pellets which have been dissolved in 5 mM Tris·HCl, pH 7.8, 0.5% (w/v) sodium dodecyl sulfate, incubated at 37° for 5 minutes, and layered over 15% (w/v) to 28% (w/v) exponential sucrose density gradients prepared in 5 mM Tris·HCl, pH 7.8. An RNA absorbance pattern from unseparated subunits is shown in Fig. 2A for comparison. The absorbance profile in Fig. 2B indicates complete purity of the 40 S ribosomal subunit. The 60 S ribosomal subunit RNA (Fig. 2C) is slightly contaminated with RNA derived from the 55 S dimer of the 40 S

---

[19] This shoulder is due to dimers of the 40 S ribosomal subunits which sediment at 55 S. The amount of the 55 S dimer increases with the age of the ribosomal subunits. Therefore, separation of the ribosomal subunits on sucrose density gradients should be carried out immediately after the elution of the subunits from Sephadex G-100.

[20] Unpublished data, W. M. Stanley, Jr., 1971.

[21] Unpublished data, G. E. Brown, 1971.

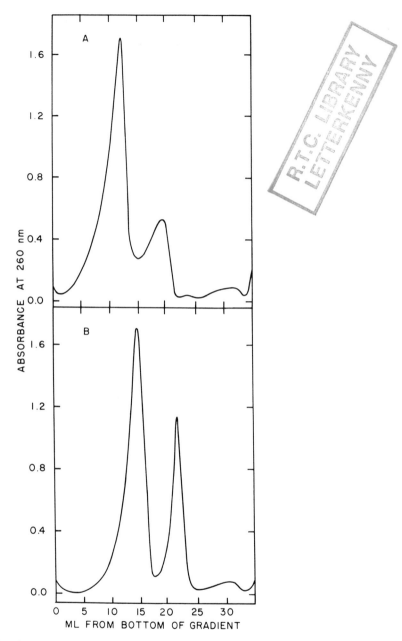

FIG. 1. Absorbance profiles of preparative sucrose density gradients of rabbit reticulocyte ribosomal subunits. (A) The sucrose density gradient contains buffer I. (B) The sucrose density gradient contains buffer J. 40 S subunits were collected from the beginning of the 40 S peak to an absorbance at 260 nm of 0.4 on the heavy side of the peak. 60 S subunits were collected from an absorbance at 260 nm of 1.4 on the light side of the 60 S peak to an absorbance at 260 nm of 1.2 on the heavy side of the peak.

377

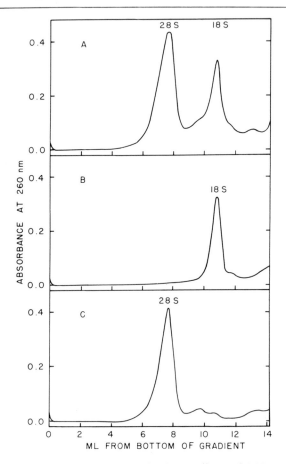

Fig. 2. Absorbance profiles of sucrose density gradients of RNA extracted from isolated rabbit reticulocyte ribosomal subunits. (A) 1.5 A_{260} units of RNA from unseparated subunits. (B) 0.5 A_{260} unit of RNA from isolated 40 S subunits. (C) 1.0 A_{260} unit of RNA from isolated 60 S subunits. These sucrose density gradients were centrifuged at 2° for 6 hours at 36,000 rpm (230,000 g) in a Beckman Spinco SW 40 Ti rotor.

ribosomal subunit. These dimers sediment close to the region that was selected in the original preparative sucrose density gradient to represent 60 S ribosomal subunit material. This contamination of the 60 S ribosomal subunits has been confirmed by electrophoresis of the RNA from the separated ribosomal subunit peaks through gels of polyacrylamide.[21] Absorbance scans at 260 nm of these gels also indicate the complete absence of tRNA, aminoacyl-tRNA, and peptidyl-tRNA from both subunits, and a normal amount of 5 S ribosomal RNA in the 60 S subunits: i.e., a 1:1 molar ratio with the 28 S RNA of the 60 S ribosomal subunit.

TABLE I

PHENYLALANINE POLYMERIZATION REQUIREMENTS OF RABBIT RETICULOCYTE
RIBOSOMES THAT HAVE BEEN PREINCUBATED WITH
PUROMYCIN AND REISOLATED[a]

Ingredients	Phenylalanine residues incorporated per 80 S ribosome per hour
Complete system	85
− Poly(U)	0
− Ribosomes	0
− Supernatant	0
− tRNA	7.5

[a] Incorporation conditions are described under Amino Acid Incorporations. Each incorporation is performed in 50 μl that contain 40 μg of poly(U), 50 μg of rabbit liver tRNA, and 1 A_{260} unit of reassociated ribosomes that have been preincubated with puromycin and reisolated. Crude initiation factors are not included. The values in the table represent the averages of duplicate assay tubes; values from tubes incubated at 0° have been subtracted (equivalent to 2.5 phenylalanine residues per 80 S ribosome per hour).

TABLE II

POLYMERIZATION ACTIVITY OF RIBOSOMES FROM VARIOUS SOURCES THAT
HAVE BEEN PREINCUBATED WITH PUROMYCIN AND REISOLATED[a]

	Phenylalanine residues incorporated per 80 S ribosome per hour	
	Source of supernatant fraction	
Source of ribosomes	Rabbit reticulocyte	Chicken reticulocyte
Rabbit reticulocytes	63	33
Rabbit liver	40	23
Chicken reticulocytes	54	35
Chicken liver	36	24
Embryonic chicken leg muscle	42	19

[a] Incorporation conditions are described under Amino Acid Incorporations. Each incorporation is performed in 50 μl that contain 40 μg poly(U), 1 A_{260} unit of ribosomes, and 50 μg tRNA (rabbit liver tRNA in the case of rabbit ribosomes and chicken liver tRNA in the case of chicken ribosomes). Crude initiation factors are not included. Chicken reticulocyte supernatant is used at 15 μl per 50-μl incorporation since greater amounts are inhibitory. Values from incorporations not including poly(U) in no case exceeded the backgrounds given below. The values in the table represent the averages of duplicate assay tubes; values from tubes incubated at 0° have been subtracted (equivalent to 0.5 to 0.8 phenylalanine residues per ribosome per hour).

Table I summarizes the activity and some of the requirements of the *in vitro* amino acid incorporation system using reassociated rabbit reticulocyte ribosomes that have been preincubated with puromycin and re-isolated. More than 30 nmoles of phenylalanine (equivalent to 150 residues per 80 S ribosome) are incorporated per milligram of ribosomes if the incorporation is allowed to continue for 2 hours at 37°. Although the system already exhibits at least a 10-fold stimulation upon the addition of exogenous rabbit liver tRNA, the system can be made completely dependent upon added tRNA by passing the supernatant fraction through a column of DEAE-cellulose at a KCl concentration of 0.3 *M*. However, this is not done routinely, since a significant fraction of the phenylalanine polymerization activity is lost by so doing.

Table II compares the polymerization activities of ribosomes from 5 sources that have been preincubated with puromycin and reisolated. All the ribosomes show good activities when programmed with poly(U) in the presence of the rabbit reticulocyte supernatant fraction (ranging from 36 to 63 phenylalanine residues incorporated per ribosome

TABLE III

PolY(U)-Dependent Elongation Ability and Functional Purity of
Rabbit Reticulocyte Ribosomal Subunits Isolated on Sucrose
Density Gradients Containing Either 0.15 *M* or 0.3 *M* KCl[a]

Type of ribosomal subunits used	Phenylalanine residues incorporated per 40 S or 60 S subunit per hour
0.15 *M* KCl[b]	
40 S	0
60 S	2
40 S + 60 S	15
0.3 *M* KCl[b]	
40 S	0
60 S	1
40 S + 60 S	2
Control[c]	50

[a] Incorporation conditions are described under Amino Acid Incorporations. Initiation factors are not added. The values in the table represent the averages of duplicate assay tubes from which values from tubes not containing poly(U) have been subtracted: backgrounds range from 2.12 to 2.66 phenylalanine residues per 40 S or 60 S subunit per hour.

[b] Indicates the KCl concentration in the sucrose density gradients used to isolate the subunits. Buffer I contains 0.15 *M* KCl; buffer J contains 0.3 *M* KCl.

[c] Control ribosomes represent an aliquot of the ribosomal subunit solution, used to prepare the subunits, which was kept at 0° in buffer I during the subunit separation and then processed identically to the isolated ribosomal subunits during the concentration step.

per hour at 37°) and no endogenous activities. The chicken reticulocyte supernatant fraction is less active for any of the types of ribosomes than is the corresponding rabbit fraction (Table II). Hence, all the other data reported in this article were obtained using the supernatant fraction from rabbit reticulocytes.

Table III summarizes the results of poly(U)-stimulated incorporations of phenylalanine into hot acid-insoluble material using separated and isolated rabbit reticulocyte ribosomal subunits. These results demonstrate the validity of the earlier statement that ribosomal subunits separated on sucrose density gradients containing 0.15 M KCl have considerably higher activity than those separated on sucrose density gradients containing 0.3 M KCl. Table IV indicates that although the separated and reassociated ribosomal subunits exhibit less activity than the unseparated ribosomal subunits, held as a control at 0° for a time equivalent to the time required for subunit isolation, this loss of activity can be largely restored by the addition of the crude initiation factor preparation.[22]

TABLE IV

EFFECT OF THE 0.5 M KCl WASH OF CRUDE RABBIT RETICULOCYTE POLYSOMES
ON POLY(U)-DEPENDENT INCORPORATION BY ISOLATED AND RECOMBINED
RABBIT RETICULOCYTE RIBOSOMAL SUBUNITS[a]

Type of ribosomal subunits used	Phenylalanine residues incorporated per 40 S or 60 S subunit per hour	
	+0.5 M KCl wash	−0.5 M KCl wash
0.15 M KCl[b]		
40 S + 60 S	36	21
0.3 M KCl[b]		
40 S + 60 S	11	3
Control[c]	51	50

[a] Incorporation conditions are described under Amino Acid Incorporations. The values in the table represent the averages of duplicate assay tubes from which values from tubes not containing poly(U) have been subtracted: backgrounds range from 2.25 to 2.42 phenylalanine residues per 40 S or 60 S subunit per hour.

[b] Indicates the KCl concentration in the sucrose density gradients used to isolate the subunits. Buffer I contains 0.15 M KCl; buffer J contains 0.3 M KCl.

[c] Control ribosomes are defined in the legend to Table III.

[22] The crude initiation factor preparation contains over 30 proteins of both ribosomal and nonribosomal origin as indicated by electrophoresis through polyacrylamide gels containing sodium dodecyl sulfate (G. L. Dettman, thesis, University of California, Irvine, 1972). Not only are initiation factors present,[2a-2e] but a large number of ribosome-associated proteins of unknown origin and function also are present. It therefore is not unexpected that ribosomal subunits which have exhibited a

TABLE V

METHIONINE RESIDUES INCORPORATED FROM RABBIT LIVER L-[^{35}S]Met-tRNA$_i^{Met}$
PER HUNDRED 80 S RABBIT RETICULOCYTE RIBOSOMES PER HOUR[a]

Additions	Methionine residues incorporated per hour
None	0.1
0.5 M KCl wash[b]	0.5
AUG(U)$_x$	1.2
0.5 M KCl wash[b] + AUG(U)$_x$	13.0

[a] Incorporation conditions are described under Amino Acid Incorporations. The values in the table represent the averages of duplicate assay tubes; values from tubes incubated at 0° have been subtracted (equivalent to 0.04 methionine residues per hundred 80 S ribosomes per hour).

[b] From rabbit reticulocyte polysomes.

Table V shows the activities and factor requirements of the *in vitro* amino acid incorporation system when challenged with AUG(U)$_x$ using reassociated rabbit reticulocyte ribosomes that have been preincubated with puromycin and reisolated. The system is markedly dependent upon the addition of both AUG(U)$_x$ and the 0.5 M KCl polysomal extract. This indicates that processed ribosomes are severely depleted of messenger RNA activity and initiation factors, but can respond to their addition. [2a-2e] Table VI lists data from the same experiment using separated and recombined ribosomal subunits. The fact that [^{35}S]methionine is incorporated from the rabbit liver initiator tRNA, Met-tRNA$_i^{Met}$, in response to the initiation codon AUG at the 5' terminus of a synthetic oligoribonucleotide indicates that a process of polypeptide chain initiation analogous to *in vivo* protein biosynthesis initiation is being observed. [2a-2e,23] The values given in Tables V and VI represent relative rather than maximal activities since the activities of the ribosomes and their separated and reassociated subunits,

---

partial loss of polypeptide chain elongation activity upon storage at 0°, or exposure to nonoptimal ionic conditions, can have their polymerization capacity partially restored by the addition of all the proteins contained in the crude initiation factor preparation.

[23] *In vitro* amino acid incorporations utilizing AGU(U)$_{\bar{n}}$, GUU(U)$_{\bar{n}}$, poly(A), and poly(U) as messenger RNA's are not stimulated by the addition of the crude initiation factor preparations. Met-tRNA$_m^{Met}$ will not donate methionine into hot-acid-precipitable material when the *in vitro* amino acid incorporation systems contain AUG(U)$_x$ and the crude initiation factor preparation, but will do so when the polymer is random poly(A,U,G). Unpublished data, G. E. Brown and W. M. Stanley, Jr., 1971.

TABLE VI

METHIONINE RESIDUES INCORPORATED FROM RABBIT LIVER L-[^{35}S]Met-tRNA$_i^{Met}$
PER HUNDRED 40 S OR 60 S RABBIT RETICULOCYTE RIBOSOMAL SUBUNITS[a]

| | Methionine residues incorporated per 100 40 S or 60 S ribosomal subunits per hr | | | | |
| | Type of ribosomal subunits used | | | | |
Additions	40 S	60 S	40 S + 60 S	Control[b]	None[c]
None	0.12	0.12	0.10	0.26	0.07
0.5 M KCl wash[d]	0.16	0.16	0.31	0.76	0.08
AUG(U)$_x$	0.10	0.16	0.28	1.24	0.06
0.5 M KCl wash[d] + AUG(U)$_x$	0.11	0.42	3.90	10.20	0.02

[a] Incorporation conditions are described under Amino Acid Incorporations. The values in the table represent the averages of triplicate assay tubes; values from tubes incubated at 0° have been subtracted (equivalent to 0.21 methionine residues per hundred 40 S or 60 S ribosomal subunits per hour). Ribosomal subunits were isolated on sucrose density gradients containing buffer I.

[b] Control ribosomes are defined in the legend to Table III.

[c] Values corresponding to no ribosomes were calculated by assuming that 0.3 A_{260} unit control ribosomes was added per 50 μl incorporation.

[d] From rabbit reticulocyte polysomes.

when challenged by AUG(U)$_x$ and the crude initiation factor preparation, can be increased manyfold by concentration and purification of the initiation factors in the crude initiation factor preparation.[2a-2e,21]

Table VII shows the response of the reassociated ribosomal subunits to the natural messenger RNA's for rabbit globins. This translation is completely dependent upon the addition of the crude initiation factor preparation. Much of the activity in the incorporation mixtures containing the factor preparation, but lacking the added messenger RNA's, is due to two sources: (1) globin messenger ribonucleoproteins remaining in the supernatant fraction (Olsen et al.,[24] Schapira et al.,[25] Jacobs-Lorena and Baglioni,[26] and Bonanou-Tzedaki et al.[27]); (2) a small amount of messenger RNA activity that remains with the ribosomes after the preincubation of the polysomes with puromycin, purification of the ribosomal subunits

[24] G. D. Olsen, P. Gaskin, and D. Kabat, Biochim. Biophys. Acta **272**, 297 (1972).

[25] G. Schapira, L. Reibel, and F. Cuault, Biochemie **54**, 465 (1972).

[26] M. Jacobs-Lorena and C. Baglioni, Proc. Nat. Acad. Sci. U.S. **69**, 1425 (1972).

[27] S. A. Bonanou-Tzedaki, I. B. Pragnell, and H. R. V. Arnstein, FEBS (Fed. Eur. Biochem. Soc.) Lett. **26**, 77 (1972).

TABLE VII
EFFECT OF THE 0.5 M KCl WASH OF CRUDE RABBIT RETICULOCYTE
POLYSOMES ON GLOBIN mRNA-DEPENDENT INCORPORATION BY
RIBOSOMES THAT HAVE BEEN PREINCUBATED WITH
PUROMYCIN AND REISOLATED[a]

| Messenger | Picomoles valine incorporated[b] per mg of 80 S ribosomes per $\frac{1}{2}$ hour | |
	+0.5 M KCl wash	−0.5 M KCl wash
	Rabbit reticulocyte ribosomes[c]	
Rabbit globin mRNA	115.4 (102.0)[d]	2.3 (1.0)[d]
None	49.6 (10.8)	2.1 (0.5)
	Rabbit liver ribosomes[c]	
Rabbit globin mRNA	118.0	0.6
None	7.1	0.6
	Chicken reticulocyte ribosomes[c]	
Rabbit globin mRNA	131.0	1.2
None	7.8	0.7

[a] Incorporation conditions are described under Amino Acid Incorporations. Rabbit liver tRNA and rabbit reticulocyte supernatant fraction were used throughout. The incorporations were at 26° for 30 minutes. The values in the table represent the averages of duplicate assay tubes; only the counter background has been subtracted (equivalent to 5.1 picomoles valine per milligram of ribosomes per $\frac{1}{2}$ hour).

[b] From rabbit liver tRNA aminoacylated with L-[^{14}C]valine (specific activity 195 Ci per mole) and the other 19 unlabeled amino acids. The products of these *in vitro* amino acid incorporations have been identified as complete α and β chains of rabbit globin by chromatography on carboxymethyl cellulose [H. M. Dintzis, *Proc. Nat. Acad. Sci. U.S.* **47**, 247 (1961)].

[c] Source of ribosomes used in the amino acid incorporations listed in the 2 lines below.

[d] Values in parentheses represent data from an identical experiment except that the ribosomes which have been preincubated with puromycin and reisolated have received a second preincubation (in buffer F) at 37° for 30 minutes just prior to their use in the amino acid incorporation system.

by gel filtration, reassociation to 80 S ribosomes by raising the magnesium ion concentration, and recovery by centrifugation. The messenger RNA activity that remains with these ribosomes can be largely inactivated by dissolving the ribosomes in buffer F and incubating them at 37° for 30 minutes before adding them to the amino acid incorporation mixture (data in parentheses; Table VII). The overall polymerization capacity of the ribosomes is diminished only about 10% in the case of rabbit and chicken reticulocyte ribosomes (20% in the case of rabbit liver ribosomes) by this additional treatment. Messenger RNA activity is not present in the crude initiation factor preparations since similar amino acid *in vitro* incorporation results have been obtained with initiaton factor preparations which have been passed through columns of DEAE-cellullose under conditions where globin messenger RNA's and messenger ribonucleoproteins would have been retained. The reassociated rabbit reticulocyte ribosomal subunits also translate Qβ RNA, yeast messenger RNA's, and tobacco mosaic virus RNA, but only in the presence of a crude initiation factor preparation.[21]

Discussion and Conclusions

It is well known that eukaryotic ribosomes actively engaged in protein biosynthesis are extremely resistant to dissociation into subunits mainly due to the peptidyl-tRNA moieties which they bear. In the past, eukaryotic ribosomal subunits have been obtained by treatments such as exposure to metal ion chelators[28,29] or high concentrations of monovalent salts[30-32] which inevitably result in ribosomal subunits of greatly impaired physical[31-33] and functional[30-34] integrity. Incubation of polysomes in the presence of sodium fluoride has been used to prevent reassociation of ribosomal subunits prior to reinitiation,[35,36] but unfortunately this treatment does not allow dissociation of the messenger RNA from the ribosomes upon termination of the polypeptide chain.[37-39] Attempts to promote polypeptide chain completion during a preincubation under conditions of *in vitro* protein synthesis leave the ribosomes still attached to messenger RNA and/or messenger RNA fragments[40] and to peptidyl-tRNA.[41] Preincubation with puromycin under conditions of *in vitro* protein synthesis, although removing the nascent peptide chain and thus destabilizing the 80 S ribosome, is

[28] M. L. Petermann and A. Pavlovec, *Biochemistry* 6, 2950 (1967).
[29] J. Stahl, G. R. Lawford, B. Williams, and P. N. Campbell, *Biochem. J.* 109, 155 (1968).
[30] T. E. Martin, I. G. Wool, and J. J. Castles, this series, Vol. 20, p. 417.
[31] C. Godin, J. Kruh, and J. C. Dreyfus, *Biochim. Biophys. Acta* 182, 175 (1969).
[32] A. M. Reboud, J. P. Reboud, M. Arpin, and C. Wittmann, *Biochim. Biophys. Acta* 232, 171 (1971).
[33] J. C. S. Clegg and H. R. V. Arnstein, *Eur. J. Biochem.* 13, 149 (1970).
[34] E. Gravela, *Biochem. J.* 121, 145 (1971).
[35] R. D. Nolan and H. R. V. Arnstein, *Eur. J. Biochem.* 10, 96 (1969).
[36] W. Hoerz and K. S. McCarthy, *Proc. Nat. Acad. Sci. U.S.* 63, 1206 (1969).
[37] B. Colombo, C. Vesco, and C. Baglioni, *Proc. Nat. Acad. Sci. U.S.* 61, 651 (1969).
[38] C. Vesco and B. Colombo, *J. Mol. Biol.* 47, 335 (1970).
[39] M. Terada, S. Metafora, J. Banks, L. W. Dow, A. Bank, and P. A. Marks, *Biochem. Biophys. Res. Commun.* 47, 766 (1972).
[40] E. A. Zylber and S. Penman, *Biochim. Biophys. Acta* 204, 221 (1970).
[41] G. Blobel and D. D. Sabatini, *J. Cell Biol.* 45, 130 (1970).

not in itself sufficient to allow dissociation of tRNA and messenger RNA bound to the ribosomes.[42,43]

We have combined the destabilizing effect of removal of the nascent peptide chain (by preincubation of the polysomes under conditions of protein synthesis in the presence of puromycin) with passage of the ribosomes through a Sephadex G-100 column equilibrated with a buffer containing a low concentration of magnesium ion, but of approximately physiological ionic strength. This gel filtration step achieves a rapid equilibration of the ribosomes with this buffer, dissociates the 80 S ribosomes into 40 S and 60 S subunits, and separates the subunits from tRNA, messenger RNA and messenger RNA fragments, puromycin and puromycin derivatives, and essentially all the proteins and other components present in the preincubation mixture. We feel that it is the gel filtration step that is essential for the preparation of ribosomes extremely responsive to the addition of messenger RNA, tRNA, elongation factors, and initiation factors.[2a-2e] However, there does remain with the processed ribosomes a low level of messenger RNA activity that can be activated and translated at low magnesium ion concentrations in the presence of a crude initiation factor preparation in the *in vitro* amino acid incorporation mixture. This activatable messenger RNA activity can be reduced severalfold by further incubation of the processed ribosomes. This second preincubation does not significantly affect the amino acid polymerization potential of the ribosomes, and it does allow the ribosomes to give at least a 10-fold response to exogenous rabbit globin messenger RNA's in the presence of a crude initiation factor preparation.

This general procedure of preincubating polysomes *in vitro* under conditions of protein synthesis in the presence of puromycin, and then subjecting them to gel filtration through Sephadex G-100 under the proper ionic conditions, has been used successfully in our laboratory with polysomes from rabbit reticulocytes, rabbit liver, chicken reticulocytes, chicken liver, and embryonic chicken leg muscle. The preincubated and reassociated ribosomes are responsive to rabbit globin messenger RNA's, yeast messenger RNA's, $Q\beta$ RNA, and tobacco mosaic virus RNA. As in the case of the synthetic oligoribonucleotide with the AUG initiation codon at the 5' terminus, these natural messenger RNA's are translated only in the presence of a crude initiation factor preparation. We have purified and characterized the biological functions of two factors in the crude initiation factor preparations from rabbit reticulocyte polysomes.[2a-2e]

[42] I. B. Pragnell and H. R. V. Arnstein, *FEBS* (*Fed. Eur. Biochem. Soc.*) *Lett.* **9**, 331 (1970).
[43] G. Blobel and D. Sabatini, *Proc. Nat. Acad. Sci. U.S.* **68**, 390 (1971).

In this article, we have presented a detailed procedure for the preparation of eukaryotic ribosomes and ribosomal subunits. This procedure has been successful with polysomes from all cells and tissues with which we have dealt. Thus, these methods should be effective with ribosomes obtained from other eukaryotic sources. Processed ribosomes are severely depleted of initiation factors[2a-2e] and endogenous messenger RNA activity, but will participate in translation with high efficiency when supplemented with these substances. Thus, the procedures which we have described are useful for preparing *in vitro* amino acid incorporation systems which can be used as sensitive assays for messenger RNA, tRNA, elongation factors, initiation factors,[2a-2e] and other components of the protein biosynthetic apparatus which are usually associated with ribosomes prepared by other techniques.

Acknowledgments

We thank Ms. Marcia Witte for her excellent technical assistance and Demler Farms for the donation of adult Shaver Starcross Leghorn chickens.

[38] The Characterization of Ribosomal Subunits of Eukaryotes by Ultracentrifugal Techniques. Criteria of Purity

By MARY G. HAMILTON

The problem of establishing homogeneity for a complicated structure like a ribosome is not easily solved. Perhaps such a concept should be irrelevant. The subunits of eukaryotic cytoplasmic ribosomes, in their "active" state, seem to behave more like supramolecular complexes than simple ribonucleoproteins.[1] Moreover, if heterogeneity exists among the population of small subunits of *Escherichia coli* ribosomes,[2] it probably exists for eukaryotic ribosomes as well.

Despite these reservations it is important to try to characterize the subunits as fully as possible. Fortunately, techniques are now available for the complete characterization of ribosomal subunits.[3] One can separate subunits by zonal centrifugation,[4] although pressure-induced dissociation

---

[1] M. G. Hamilton, A. Pavlovec, and M. L. Petermann, *Biochemistry* **10**, 3424 (1971).

[2] C. G. Kurland, *Annu. Rev. Biochem.* **41**, 377 (1972).

[3] M. G. Hamilton and M. E. Ruth, *Biochemistry* **8**, 851 (1969).

[4] P. S. Sypherd and J. Wireman, this volume [35]; C. C. Sherton *et al.*, this volume [36].

PHYSICAL AND CHEMICAL PROPERTIES OF ACTIVE SUBUNITS OF RAT LIVER RIBOSOMES AND THE SUBUNITS OBTAINED BY EDTA TREATMENT[a,b]

Subunit	Sedimentation coefficient, $s_{20,w}$ (S)	Molecular weight ($\times 10^{-6}$)	Buoyant density in CsCl, θ (g/ml)	Protein content	
				From $1/\theta^c$ (%)	From chemical analysis (%)
Active small	40.9 (0.3, 4)	1.50 (0.03, 11)	1.551 (0.002, 10)	55	54.5 (0.67, 3)
Active large	59.1 (0.5, 3)	3.00 (0.08, 5)	1.614 (0.002, 12)	43	44.7[d]
EDTA small	28.6[e]	1.18 (0.04, 9)	1.593 (0.002, 9)	47	49.5 (2.06, 3)
EDTA large	49.9[e]	3.12 (0.02, 5)	1.602 (0.002, 8)	45	44.7 (0.97, 3)

[a] This table is taken from M. G. Hamilton, A. Pavlovec, and M. L. Petermann [Biochemistry 10, 3424 (1971)]. The data on the EDTA subunits are from M. G. Hamilton and M. E. Ruth, Biochemistry 8, 851 (1969)].

[b] All measurements except the chemical analyses were made on formaldehyde-fixed samples. The standard error of the mean and the number of determinations are given in parentheses.

[c] From a calibration line given by M. G. Hamilton (this series, Vol. 20, p. 512).

[d] Single determination; the value has been corrected for the presence of 15% of small subunits.

[e] $s_{20,w}^0$, from a study of sedimentation coefficient as a function of concentration on unfixed samples in 30 mM KCl–0.2 mM MgCl2–1 mM potassium phosphate, pH 7.0.

may sometimes complicate the separation.[4a] The subunits can be fixed with formaldehyde to prevent degradation. This is useful in general but essential for the buoyant density measurements which can be used for an estimate of protein content as well as purity. The ultraviolet absorption optical system of the analytical ultracentrifuge permits measurements at very low concentrations. With only a few A_{260} units one can measure sedimentation coefficients and buoyant densities, and if the solutions are monodisperse, the molecular weight can be measured directly by equilibrium centrifugation, one of the few absolute methods. The table reproduces the results of such measurements on the subunits of rat liver ribosomes. Tashiro and Yphantis[5] and Hill et al.[6] have also used analytical ultracentrifugation to measure molecular weights of ribosomes and subunits.

Criteria of Purity: Sedimentation Analysis and Buoyant
 Density Measurements

Although monodispersity of size is measurable by sedimentation analysis,[7] with ribosomal subunits a single sedimenting boundary may represent more than one species. Ribosomal subunits can both open up to sediment more slowly or dimerize to sediment faster. By some natural perverseness the opened large forms of rat liver ribosomes[8] sediment at almost exactly the same rate as active small subunits and, *vice versa*, dimerized active small ones sediment at almost exactly the same rate as large ones.[9] One simple test for such cross-contamination is to examine the sample after EDTA treatment.[10] This converts the active forms to inactive forms of lower sedimentation (s) rates but has no effect on the "opened" forms. Examples of these cases are illustrated in Figs. 1 and 2 for fractions taken from zonal separations of ribosomes dissociated by the urea method of Petermann and Pavlovec.[11] In Fig. 1 small subunit dimers cosedimenting with the large subunits (upper left) are revealed by EDTA treatment as a 30 S boundary (lower left). In Fig. 2, an opened 41 S form of the large subunit cosedimenting with the small subunits (upper left) is unchanged by EDTA treatment (lower left), but the active 40 S small subunits now sediment at 28 S.

Because the buoyant density in CsCl depends on the protein content

[4a] R. Baierlein and A. A. Infante, this volume [33].
[5] Y. Tashiro and D. A. Yphantis, *J. Mol. Biol.* 11, 174 (1965).
[6] W. E. Hill, G. P. Rossetti, and K. E. Van Holde, *J. Mol. Biol.* 44, 263 (1969).
[7] R. Trautman, S. P. Spragg, and H. B. Halsall, *Anal. Biochem.* 28, 396 (1969).
[8] M. L. Petermann, A. Pavlovec, and M. G. Hamilton, *Biochemistry* 11, 3925 (1972).
[9] M. G. Hamilton, unpublished observations, 1971.
[10] M. L. Petermann, this series, Vol. 20, p. 429.
[11] M. L. Petermann and A. Pavlovec, *Biochemistry* 10, 2770 (1971).

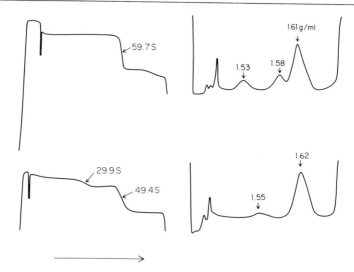

Fig. 1. Large-subunit fraction from a zonal separation of rat liver ribosomes dissociated by treatment with 2 M urea. Comparison by sedimentation velocity (left side) and isodensity equilibrium centrifugation (right side) of samples fixed with HCHO before (upper row) and after (lower row) EDTA treatment. The patterns were obtained by densitometry of UV films.

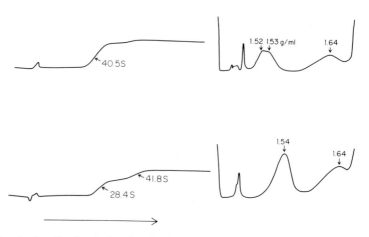

Fig. 2. Small subunit fraction from a zonal separation of rat liver ribosomes dissociated by urea. Upper patterns, control. Lower patterns, after EDTA treatment. See legend of Fig. 1. The patterns on the left were obtained by the photoelectric scanner, and the right-hand ones were obtained by densitometry of UV films.

of the subunit and the active forms differ in protein content (see the table), another criterion of homogeneity is available. This analysis reveals cross-contamination of subunits even more dramatically than does the sedimentation pattern. Figure 1 shows a large subunit fraction examined in this way. While the sedimentation pattern shows mainly a single boundary (upper left), the isodensity pattern (upper right) shows three bands which correspond to small subunits, large subunits, and an intermediate band, which may represent residual ribosomes. Again, EDTA treatment (lower right) corroborates the existence of a mixture. The small subunit does not lose protein after EDTA treatment unless it is physically separated from the reaction mixture.[9] When fixed immediately after the addition of EDTA, it retains its low buoyant density. This is a convenience in this assay since it remains distinguishable from the large subunit, as is shown in Figs. 1 and 2, lower right panels.

Technical Details

The procedures for the main variations of analytical ultracentrifugation have been fully described in this series[12] and elsewhere.[13] Chervenka[14] gives a simplified and practical guide. The calculations, with statistical treatment of the data, have been programmed for a desk-top computer.[15] We use Dr. Trautman's programs for the Olivetti Underwood Programma 101 Desk-Top Computer. Velocity sedimentation, in a density gradient with suitable markers of known s rates, and isodensity equilibrium centrifugation can be performed in preparative ultracentrifuges. Molecular weight measurements by classical equilibrium centrifugation, however, require an analytical ultracentrifuge.

A minimum of 2 A_{260} units, at a concentration of 0.6–0.8 A_{260} units/ml, is required for the measurements. Naturally, one would like more material for replicate analyses.

Preparation of Sample. Fixation with formaldehyde, which is necessary for the buoyant density measurement, is also useful in preserving the structure for the lengthy measurements. Buffers such as triethanolamine-HCl (TEA-HCl), which lack primary amino groups, should be used rather than Tris·HCl. Fractions from the sucrose gradient separation are dialyzed overnight against 2 M formaldehyde, 10 mM TEA·HCl, pH 7.4, 0.2 mM

[12] H. K. Schachman, this series, Vol. 4, p. 32; J. Vinograd, Vol. 6, p. 854.
[13] R. Trautman and M. G. Hamilton, *in* "Principles and Techniques in Plant Virology" (C. I. Kado and H. O. Agrawal, eds.), p. 491. Van Nostrand-Reinhold, Princeton, New Jersey, 1972.
[14] C. H. Chervenka, "A Manual of Methods for the Analytical Ultracentrifuge." Spinco Division of Beckman Instruments, Inc., Palo Alto, California, 1969.
[15] R. Trautman, *Ann. N.Y. Acad. Sci.* **164**, 52 (1969).

$MgCl_2$, and for 1 or 2 days against at least two lots of 0.1 M HCHO, 30 mM KCl, 10 mM TEA·HCl, pH 7.4, 0.2 mM $MgCl_2$ (FTKM), a suitable solvent for all the centrifugation studies. If necessary, the sample can be concentrated to 0.6–0.8 A_{260} units/ml by vacuum dialysis in a Schleicher-Schuell collodion bag.

EDTA Test for Cross-Contamination of Zonal Fractions. An excess of EDTA, 0.1 or 0.2 volumes of 0.1 M Na_2 EDTA (depending on the $MgCl_2$ content of the separation buffer), which has been adjusted to pH 7 with NaOH, is added to a 2 A_{260} unit aliquot. After a few minutes the sample is dialyzed overnight against buffered 2 M HCHO without $MgCl_2$, but switched to FTKM for the final dialysis.

Sedimentation Velocity. These runs should be performed first because the samples can be recovered after thorough mixing of the cell contents. The ultraviolet absorption optical system should be used preferably with the photoelectric scanner optics, but the film system can also be used. Multicell runs save time and improve accuracy by eliminating the temperature variable when comparisons of s rates are required. The FTKM buffer has a density of 1.002_3 g/ml and a relative viscosity of 1.024_4. For this solvent and a partial specific volume, $\bar{v}$, of 0.63, the "salt correction" for calculating[12,14] sedimentation in water, after the temperature correction to 20° has been made, is 1.03.

Isodensity Equilibrium Centrifugation. These measurements have been described.[16] We now add 0.1 volume of 0.1 M NaH_2PO_4, 0.1 M $Na_2H PO_4$ to ensure a constant pH near 7.5.[17] Also, the higher speeds available in a titanium rotor, 48,000 or 52,000 rpm, are advantageous in increasing the gradients and thus displaying more species. When small subunits are present, an initial density of 1.58 g/ml is useful.

Classical Equilibrium Centrifugation. The requirements for monodispersity and lack of cross-contamination must be stressed. Since the high speed method[18] must be used (the present-day centrifuges are reputed to be unreliable at the lowest available speeds), complications due to the presence of heterogeneity should be avoided. The principles and practice of molecular weight measurements have been concisely described by Van Holde.[19] The photoelectric scanner optics must be used for UV measurements.

The runs are made in the cold in the An-J rotor in multichannel cells[18] equipped with sapphire windows. No oil base is used. The solvent side

[16] M. G. Hamilton, this series, Vol. 20, p. 512.
[17] H. A. Wood, *Virology* **43**, 511 (1971).
[18] D. A. Yphantis, *Biochemistry* **3**, 297 (1964).
[19] K. E. Van Holde, "Sedimentation Equilibrium." Fractions 1, p. 1. Spinco Division of Beckman Instruments, Inc., Palo Alto, California, 1969.

holds 0.13 ml of dialyzate and the sample side 0.10 ml of a solution of about 0.6 A_{260} units/ml. Nine samples can be run at one time, and conveniently over the weekend; this is sufficient time to reach "practical" equilibrium, i.e., no change in the pattern with time. Figure 3 provides a guide for the selection of speed for the equilibrium run. This has been calculated from an equation given by Yphantis[18] for the desired concentration distribution. Reestablishing equilibrium at several higher speeds, e.g., 20% and 40% higher, improves precision by providing additional data. This usually takes an additional 24 hours for each new speed, and by clearing the meniscus area also provides baselines that can be used for the lower speeds. Establishment of the correct baseline is a problem with the high speed method. The solution columns can be scanned at 265 nm and 280 nm at the slowest scan speed to magnify the x dimension. x is converted to radius, r, in the usual way.[12,14] The y dimension is converted to absorbancy, c, by means of the calibration "steps" which accompany each scan.[14] Since the absorbance at 280 nm is about half that at 260 nm, the 280 nm readings provide an optical dilution for the region near the base of the solution column. The data should be plotted on graph paper as log c vs. r^2 and inspected for linearity. Frequently, points near the top of the solution column show a distinct curvature. This may indicate a poor choice of baseline, although heterogeneity can give a positive curvature and nonideality a negative one to the line. Readings less than 0.1 A unit probably should be discarded. Although the deficiencies of the log c vs. r^2 plot

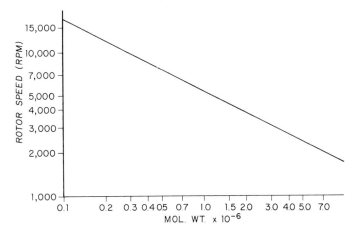

FIG. 3. Graph for selection of speed for high speed equilibrium centrifugation of a ribonucleoprotein with $\bar{v} = 0.63$, in a buffer with a density of 1.002 g/ml in a 3 mm column at 5°. Adapted from K. E. Van Holde, "Sedimentation Equilibrium." Fractions 1, p. 1. Spinco Division, Beckman Instruments, Inc., Palo Alto, California, 1969.

as an index of homogeneity have been noted,[20] it is the simplest method to apply to scanner data. In the equation

$$M_w = \frac{4.606RT}{(1 - \bar{v}\rho)\omega^2} \frac{d(\log c)}{d(r^2)}$$

M_w is the weight-average molecular weight; R, the gas constant (8.313×10^7 ergs deg^{-1} mol^{-1}); T, the absolute temperature; ρ, the solution density (the solutions are so dilute that the solvent density of 1.002 g/ml may be used); $\bar{v}$, the partial specific volume[21]; ω, the angular velocity (2π rpm/60); and $[d(\log c)]/d(r^2)$ is the slope of the line obtained from the plot of log c vs. r^2. We have found[22] that replicate measurements are essential for precision with the UV optical system (see also the table).

When more material is available, the molecular weight measurements should be made with the interference optical system. This requires about 0.1 ml of a solution containing 4–6 A_{260} units/ml. Plate reading is laborious (unless automated)[22a], but more detailed and precise calculations of the molecular weight distribution can be made.[23] Double exposures with schlieren optics,[24] after masking of the interference image, provide additional data (dc/dr) for a plot of

$$\log \left[\frac{1}{\bar{r}} \frac{dc}{dr} \right] \text{ vs. } r^2,$$

the slope of which is proportional to M_z, the Z-average molecular weight in the equation

$$M_z = \frac{4.606RT}{(1 - \bar{v}\rho)\omega^2} \frac{d \log \left[\frac{1}{r} \frac{dc}{dr} \right]}{d(r^2)}$$

Identity of M_w and M_z is an indication of homogeneity.

Comments on Other Methods

Obviously, with the chemical composition from the buoyant density[16] and the size of the RNA molecule, one can estimate the molecular weight of a ribosomal subunit. The molecular weight of RNA can be measured

[20] H. Fujita and J. W. Williams, *J. Phys. Chem.* **70**, 309 (1966).
[21] The partial specific volume can be estimated from the composition of the particle. In the expression, $\bar{v} = \bar{v}_{RNA}$ (%RNA/100) + $\bar{v}_{protein}$ (%protein/100), $\bar{v}_{RNA}$ is 0.53 (measured pycnometrically for NaRNA) and $\bar{v}_{protein}$ is 0.74 (calculated from the amino acid composition). For rat liver ribosomal subunits this value coincides with the reciprocal of the buoyant density in CsCl, but for other ribosomes it may not.
[22] M. G. Hamilton, *Biochim. Biophys. Acta* **134**, 473 (1967).
[22a] Photographic labeling of fringes as suggested by S. J. Edelstein and G. H. Ellis [*Analytical Biochemistry* **43**, 89 (1971)] helps in plate alignment on the microcomparator.
[23] D. C. Teller, T. A. Horbett, E. G. Richards, and H. K. Schachman, *Ann. N.Y. Acad. Sci.* **164**, 66 (1969).
[24] C. H. Chervenka, *Anal. Chem.* **38**, 356 (1966).

by sedimentation equilibrium,[22] but this requires purified RNA. A simple treatment with the detergent sodium dodecyl sulfate releases the RNA for examination by electrophoretic analysis in polyacrylamide gels or sedimentation analysis. Both methods can reveal the cross-contamination of large and small subunits discussed above, but obtaining the molecular weight by these methods may be a problem. The gel technique depends on calibration with molecules of known size and conformation and may be in serious error.[25] Estimating molecular weight from sedimentation velocity studies is also dependent on calibration curves or assumptions about shape and hydration. Sedimentation in formaldehyde is not the answer for eukaryotic ribosomal RNA's,[9,26] although it seems to work for bacterial ones and viral RNA's.[26,27] A new electrophoretic technique using a nonaqueous solvent may prove useful.[28]

[25] I. B. Dawid and J. W. Chase, *J. Mol. Biol.* **63**, 217 (1972).
[26] M. L. Fenwick, *Biochem. J.* **107**, 851 (1968).
[27] H. Boedtker, *J. Mol. Biol.* **35**, 61 (1968).
[28] D. Z. Staynov, J. C. Pinder, and W. B. Gratzer, *Nature (London)* **235**, 108 (1972).

[39]   Hydrogen-Tritium Exchange Studies on Ribosomes[1]

By Melvin V. Simpson

The method of hydrogen-tritium exchange (HX) has been used in the study of some of the conformational properties of ribosomes.[2-8] The method is very suitable for detecting *changes* in ribosomal conformation, particularly those which occur while the ribosome is functioning.[4-7] Reviews on the theoretical aspects as well as the experimental aspects of HX have appeared[9,10] as well as a detailed description and discussion by

[1] Much of the research upon which this article is based was supported by Grants Nos. GB 5597 and GB 8375 from the National Science Foundation.
[2] L. A. Page, S. W. Englander, and M. V. Simpson, *Biochemistry* **6**, 968 (1967).
[3] M. I. Sherman and M. V. Simpson, *Cold Spring Harbor Symp. Quant. Biol.* **34**, 220 (1969).
[4] M. I. Sherman, D. Chuang, and M. V. Simpson, *Cold Spring Harbor Symp. Quant. Biol.* **34**, 109 (1969).
[5] M. I. Sherman and M. V. Simpson, *Proc. Nat. Acad. Sci. U.S.* **64**, 1388 (1969).
[6] D. Chuang, H. A. Silberstein, and M. V. Simpson, *Arch. Biochem. Biophys.* **144**, 778 (1971).
[7] D. Chuang and M. V. Simpson, *Proc. Nat. Acad. Sci. U.S.* **68**, 1474 (1971).
[8] M. I. Sherman, *Eur. J. Biochem.* **25**, 291 (1972).
[9] A. Hvidt and S. O. Nielson, *Advan. Protein Chem.* **21**, 287 (1966).
[10] S. W. Englander, *in* "Poly-α-Amino Acids: Protein Models for Conformational Studies" (G. Fasman, ed.), p. 339. Dekker, New York, 1967.

Englander[11] of general procedural details. We refrain from repeating much of the material in the latter publication. Rather, we describe here modifications of the technique found to be most suitable for the study of ribosomes along with pertinent discussion. It is therefore imperative that the discussion by Englander[11] be carefully consulted as a companion article, particularly with respect to the effects of column size, flow rates, Sephadex bead sizes, and for discussions of loading the columns, of obtaining very early time points, of the use of the rapid dialysis apparatus, and of storage of the high specific activity THO. Much valuable discussion and data on ribosomal HX may be found in the thesis by Sherman[12] in much greater detail than could appear here.

While some of our early studies dealt with mammalian ribosomes[2] all our recent work has been with preparations from *Escherichia coli,* strain MRE 600, and it is the technique involving these ribosomes that is described here.

The HX properties of ribosomes are most conveniently studied by first labeling with tritium (T) the exchangeable hydrogens of the ribosome. This is accomplished by a *preincubation* step in which the ribosomes are equilibrated or near-equilibrated with THO solvent for a number of days. The free THO is then removed by gel filtration or by a special rapid dialysis procedure. One can then assess the effect of various conditions on the HX behavior of the ribosome by an *incubation* step under the desired test conditions (e.g., addition of GTP, addition of streptomycin, alteration of Mg^{2+} concentration), samples being removed from the incubation mixture at appropriate intervals. Each removed sample is again subjected to either of the above fractionation procedures to remove the THO produced by exchange-out, and the ribosomes are then counted for the determination of their specific activity.

This description represents the outline of the procedures in principle. In practice, and for special purposes, shortcuts are often taken or combinations of the gel filtration and rapid dialysis procedures are used. These will be described later.

Discussion of Methods

General Considerations

Preincubation Step and Stability of the Ribosome. Inasmuch as the preincubation period during which ribosomes are equilibrated with THO

---

[11] S. W. Englander, this series, Vol. 12B, p. 379.
[12] M. I. Sherman, Thesis, State University of New York at Stony Brook, 1969. University Microfilms, Inc., 300 North Zeeb Rd., Ann Arbor, Michigan 48106.

is fairly long (3–4 days), the physical and functional stability of the ribosomes to standing should first be ensured. This is especially vital where ribosomal subunits are concerned since these tend to be less stable than the undissociated monomer. If the stability of the ribosomes and other experimental conditions permit, the time for equilibration can be shortened by raising the temperature. If the stability of the ribosomes to standing is not great, the labeling time can be shortened by carrying out the preincubation only partly to equilibrium. If this is combined with elevated temperatures, the preincubation period can be made quite short. It should be appreciated, however, that the farther away the reaction is from equilibrium, the less will be the labeling of the slowly exchanging hydrogens. If these are the hydrogens which happen to be primarily involved in the conformational change being studied, the detection of this change will become more difficult.

It is interesting to note that at the point at which equilibrium appears to be reached (3 to 4 days at 2°), a great many of the exchangeable hydrogens are still unlabeled with tritium and show no sign of becoming labeled even after an additional week of preincubation. These hydrogens become labeled fairly rapidly, however, if the preincubation with THO is carried out with either or both ribosomal subunits rather than with the 70 S monomer.

Purity of the Ribosomal Preparation. In many types of studies, purity of the ribosomes (e.g., freedom from ribosomal subunits) is relatively unimportant since the experimental design is often geared to the detection of the functional activity of only the active ribosome. In HX studies, nonfunctional impurities which become labeled during the preincubation step can raise the background appreciably, making alterations in ribosomal HX rate more difficult to detect. In particular, contaminating ribosomal subunits and tRNA affect the HX rate in greater proportion than the extent of their presence since their HX rate is faster than that of ribosomes. Steps should therefore be taken to remove those species of contaminants which could raise the HX background. It might be pointed out that small molecules, as well as macromolecules possessing either little three-dimensional structure or an unstable three-dimensional structure, will probably not cause such background interference since their rate of exchange will be so high as to be complete before the first point on the ribosomal HX curve is taken.

Preparation of Ribosomes. The method selected to prepare ribosomes will depend upon the particular experimental aims of the investigator. In some of our studies,[3–5,8] we have found Kurland's[13] technique to be

[13] C. G. Kurland, *J. Mol. Biol.* 18, 90 (1966).

especially effective in yielding stable ribosomes. However, contamination by subunits and by 100 S dimers has been found in these preparations. The subunits may be removed by differential ultracentrifugation, and the dimers may be removed by dialysis against buffer containing 10^{-3} M Mg^{2+}, then against buffer containing 10^{-2} M Mg^{2+}. In other experiments,[6,7] we have used NH_4Cl-washed ribosomes and have increased the proportion of active monomers by the addition of poly(U) and the isolation of the polysome peak.

For HX studies on ribosomes in a complex milieu (e.g., one containing tRNA or mRNA), where determination of the ribosome content of a sample by optical methods is difficult, it has been found convenient to use ribosomes prelabeled with ^{14}C, obtained by growing E. coli in minimal medium supplemented with, per milliliter, 10 mg of thymidine and 5 mg of [^{14}C]uracil (5 μCi/mg). After harvesting and washing, the cells could be frozen. The isolated ribosomes should have a specific activity of between 10,000 and 20,000 cpm/mg. Their specific activity should be determined at the double label settings to be used for the run itself, as this value is needed when calculating H/ribosome.

Incubation Temperature. The temperature of the *preincubation* step was discussed earlier. When possible, *incubation* temperatures are kept close to $0°$ (we have found it convenient to use 2–4°) as the ribosomal exchange rate can increase rapidly with temperature,[2] although this should be checked for each type of ribosomal preparation. Such an elevated exchange rate would mean that many rapidly exchanging hydrogens would have already exchanged out before the first sample could be taken, and so not be measured. On the other hand, many ribosomal functions occur poorly or not at all at $0°$, e.g., translocation. If these functions are to be studied by HX, it is necessary to experimentally seek an effective compromise temperature. For example, temperatures in the range of 12–15° have been found to be practicable for studies on the action of streptomycin[4,5] and on translocation.[6,7]

Selection of Technique for Measuring Ribosomal
    Hydrogen Exchange

Several methods have been used to remove unbound THO from the equilibrated ribosomal solution after preincubation. These methods are also subsequently used to free the ribosomal incubation mixture of the THO resulting from tritium exchange-out during the incubation run. Each method has its advantages and its limitations, and the selection of the method depends upon the particular experimental objectives. A discussion of the various techniques follows:

Gel Filtration—Single-Column Method. The gel filtration method is useful when very early time points (i.e., a few minutes or less) are required, and for this purpose single-column runs are performed. In such a run, the removal of the unbound THO after preincubation, the subsequent incubation, and the removal of the free THO produced by the tritium exchange-out during the incubation, all take place on the gel filtration column. Measurable exchange-out begins when the ribosomes have moved down the column far enough to be separated from the retarded high specific activity THO.

There are a number of disadvantages to the single column technique. First, the method is tedious because each time point on the HX curve (and there might easily be a dozen or more points) requires a separate column. The procedure becomes even more cumbersome when more runs than one are being done simultaneously. Second, the radioactivity background tends to be high because small amounts of the unbound THO, which possesses a very high specific activity, are not completely separated from the ribosome peak. However, it is possible to sacrifice yield and use only the leading edge of the peak. Third, and most important, it is difficult to study the effect of added substances. For example, in studying ribosomal translocation, the effect of added GTP might be difficult to assess because GTP would be retarded on the gel and therefore be separated from the ribosomes. The difficulties increase if a number of substances must be added. Because most of our ribosomal studies have been concerned with the effect of added substances and varied conditions on the ribosomal HX pattern, we have not used the single-column technique very much, and it will not be described here in detail.

Gel Filtration—Two-Column Method. The last two objections in the preceding paragraph can be overcome by the introduction of a second column: After preincubation of the ribosomes with THO and removal of the unbound THO by gel filtration, the ribosome sample is then incubated under the appropriate experimental conditions (e.g., with G factor and GTP, with spermine, with an antibiotic, etc.). At appropriate intervals, samples are withdrawn and passed through a fresh gel filtration column, a separate column being used for each withdrawn sample. This method removes restrictions on what can be added to the incubation mixture and also reduces the background. However, the tedium of the procedure is not relieved and only one run can conveniently be done at a time by a single person.

If it is desired to obtain earlier time points than this method is usually capable of yielding (although not approaching the very early times of the single column technique), the column runs can be speeded up by switching the gel to the coarse grade of Sephadex G-25. This, however, requires a higher starting concentration of ribosomes since this grade of Sephadex gives less resolution and the trailing edge of the ribosomal peak usually is discarded because of contamination with unremoved THO.

The Rapid Dialysis Method. The rapid dialysis procedure is a far less tedious method of removing THO. After preincubation of ribosomes with THO, the incubation mixture is transferred to a sac on the rapid dialysis rack. Samples are then removed from the sac at desired intervals and analyzed for ribosomal specific activity. The substance whose influence on the ribosome is being tested can be added directly to the dialysis sac at any time or, alternatively, to the external dialyzing buffer. Since an entire HX curve can be obtained by sampling a single sac and manipulations are simple, it is not difficult to do a number of runs simultaneously.

The rapid dialysis method suffers from a number of disadvantages. Because large amounts of unbound high specific activity THO are present at the start of the dialysis, this must first be removed, before exchanged-out THO can be measured. This removal occurs during the first 40 minutes or so of the dialysis when the removal is complete. Hence, no valid exchange-out data can be obtained during this 40-minute period. Moreover, because of the very high specific activity of the THO, extreme care must be taken that no residue of THO remains sequestered in the folded, clamped, part of the sac. Despite these disadvantages, this technique has proved extremely useful for ribosome work and a considerable amount of data could be collected without undue expenditure of energy.

The rapidity of the dialysis depends on the ratio of the surface area of the sac to the "thickness" of the layer of solution inside the sac. Thus the volume of inner solution must therefore be kept limited, 1 ml being preferred; in any case no more than 2 ml should be used.

The Combined Gel Filtration–Rapid Dialysis Method. The combined gel filtration–rapid dialysis method[14] is free from most of the shortcomings of the two individual methods yet for the most part retaining the advantages. Here, the equilibrated mixture after preincubation is first subjected to gel filtration with the removal of unbound THO. The ribosome fraction from the column eluate is then transferred to the rapid dialysis apparatus. Thus, only one column per run is required and moderately early (about 10–15 minutes) points can be obtained. Because of the dilution resulting from the initial gel chromatography step, and since the maximum volume which can be subjected to rapid dialysis is about 1–2 ml per sac, it is desirable to start with a somewhat higher initial concentration of ribosomes (3–4 mg/ml). For experiments in which multiple time points are required in order to construct an HX curve, we have found this combined method to be the most useful, and it is recommended.

[14] S. W. Englander and R. Staley, *J. Mol. Biol.* **45**, 277 (1969).

The Single Time-Point Method. Finally, with the use of the single time-point technique,[12] relatively large numbers of samples can be measured. This procedure is similar to the combined gel filtration-dialysis method except that once in the dialysis sacs, all samples are incubated simultaneously for a single given period of time and only a single time point is obtained for each sample under study. The incubation is started by simultaneously lowering the dialysis racks, all of which are suspended from a single horizontal bar, into the external dialysis solution. Incubation is stopped by raising the bar, thus lifting all racks out of the dialyzate. The technique is particularly useful for serial type measurements other than a time series, e.g., relating the effect of the concentration of a substance to its effect on the rate of HX of the ribosome.

Experimental Details

Safety and Contamination Considerations. The vial containing the stock THO is stored in a small airtight jar which is kept in a sealed desiccator (silicone stopcock grease should be used) in the deep freeze. It is opened only in the fume hood.

All transfers of the high specific activity stock THO solutions are done in the fume hood. Ribosome preincubations are also done in the hood in an ice bath. Once diluted, further handling of the THO is done in a cold box similar in shape to a beer cooler. This box should be vented to the outside of the building preferably through the fume hood system.

Handling of all solutions and all glassware should be done with disposable plastic gloves. High specific activity solutions or solutions containing more than nominal amounts of radioactivity are discarded in the appropriate radioactive waste receptacle.

Preincubation of Ribosomes. Ribosomes at a concentration of 1–4 mg/ml (depending upon which procedure is selected to follow the preincubation) are incubated with THO at 2–4° for a period of 3 days, at which point equilibrium is reached or nearly reached. We have routinely used the following reaction mixture for our early studies: 50 mM cacodylate pH 7.3, 50 mM KCl and 10 mM Mg acetate. The same buffer was used both for the preincubation and for the incubation run itself, although switching buffers is permissible. In later studies, we have used much more complex media for the preincubation (as well as for the incubation), particularly in studies of HX during ribosome function, e.g., translocation, and this has caused no difficulties. Whenever possible, however, we have tried to avoid using Tris buffer since Tris tends to catalyze HX, probably by amine catalysis.[2]

The amount of high specific activity THO (1 Ci/g, biological grade) used in the preincubation mixture is such that the final specific activity of the medium is 10 mCi/ml. Any transfers of this solution are done in the fume hood using disposable plastic gloves which are also used for all further manipulations of solutions containing tritium. For further details of handling and storing this high specific activity water, see Englander.[11]

After the preincubation period, three 0.025-ml samples of the medium are pipetted into 500-ml volumetric flasks and brought to volume with water. One-milliliter samples of the resultant solutions are removed to determine in triplicate the specific activity of the tritiated solution; these samples should contain not more than 10^5 cpm.

The Gel Filtration Technique—Two-Column Procedure. After pre-incubation of the ribosome solution (ca. 1.5 mg/ml ribosomes) with THO, solid sucrose is added to make a final concentration of 1–2%. The solution is then layered below the buffer head, just above the surface of a calibrated Sephadex (G-25 fine bead) column. (For details of setting up and dealing with these columns, see Englander.[11]) Elution is started immediately using a solution that is unlabeled, but otherwise identical to the preincubation medium. When the ribosomes are eluted, a sample of the eluate is taken for specific activity determination and the remaining volume of eluate is pipetted into a test tube or flask containing the appropriate reaction mixture. Several parallel flasks can be run if two laboratory workers are available. The flask is then incubated for the desired length of time under appropriate conditions. At each desired time interval, a sample is removed and is chromatographed on a Sephadex column under conditions identical to those used in the column described earlier. One can also use an eluent identical to the medium in the control flask or possibly some other eluent which has been established as not affecting the ribosomal HX pattern. It should be clear from this description that each time point requires a separate Sephadex column.

After the elution, a sample of the eluate is taken for determination of ribosomal specific activity. For this purpose, if ^{14}C-labeled ribosomes were used in the experiment, a 1-ml sample of the eluate is pipetted directly into a scintillation vial for double label counting. If ^{14}C-labeled ribosomes were not used, a 1.2-ml sample is taken for spectrophotometry, and 1 ml of this is then transferred to the scintillation vial. The scintillation solution described by Bray[15] or some other suitable mixture miscible with water may be used. Eluate fractions preceding the ribosome peak are used as blanks both for the radioactivity and the spectrophotometric analyses.

It is often helpful to have an incubation flask serve as its own control. Thus, one incubates a flask containing a control reaction mixture and

[15] G. A. Bray, *Anal. Biochem.* 1, 279 (1960).

takes several time points. The substance whose effect is under study is then added and sampling at intervals is continued. Any effect of the added substance on HX will be seen as a change in slope of the HX curve. In our experience, this change is usually sharp and immediate.

Column sizes may be varied (see Englander[11]). In studies in this laboratory, columns 8 cm in height and 3 cm in diameter have been used for one-column runs while 6×3 cm columns are routinely used for two-column runs.

The Rapid Dialysis Technique. The rapid dialysis technique and the apparatus specifically designed for it have been described in detail by Englander and Crowe.[16,17] Before use, the dialysis tubing is boiled for 20 minutes in a solution of 0.2% disodium EDTA–0.6% sodium carbonate; after this, the dialysis tubing is stretched.[11] The tubing is attached to the dialysis rack and hung from the lip of a 500-ml graduated cylinder. After sampling for specific activity, 1–2 ml (2 ml maximum) of the preincubated tritiated ribosomal solution (3 mg ribosomes/ml is adequate for ten time point determinations) is pipetted into the sac, and the tubing is fastened between the upper jaws of the rack. The time of the run begins when the tubing and the rack are submerged in the external solution in the cylinder. After about 2 minutes of spinning, the rack is moved briskly up and down in the cylinder to distribute the solution evenly throughout the dialysis sac. After 10 minutes, the rack is transferred to a cylinder containing fresh buffer. The rack is again transferred to fresh solutions at 20 and 30 minutes. The highly radioactive dialyzate in the first cylinder should be poured into a liquid radioactive waste receptacle.

At approximately 40 minutes, the background should be low enough to allow readings to be taken. Prior to removal of a sample, the dialysis rack is partly drawn out of the dialyzate and hung from the top of the cylinder. The upper jaws are opened, and by squeezing the tubing upward, any fluid trapped in the sac between the jaws is expelled, as this somewhat sequestered part of the sample may occasionally have been prevented from exchanging completely. With a precooled pipette, 0.1–0.3 ml of dialyzate (i.e., the outer solution) is drawn up and transferred directly to a liquid scintillation vial for determination of the background value for that time point. The time of withdrawal of the sample should be noted. Buffer is then added to that vial to make a total volume of 1 ml. The pipette is then carefully inserted about halfway down into the sac so as to avoid contamination of the sample with dialysis sac "dust"

---

[16] S. W. Englander and D. Crowe, *Anal. Biochem.* **12**, 579 (1965).
[17] The dialysis racks are commercially available from Hoefer Scientific Instruments, 520 Bryant Street, San Francisco, California 94107; Cat. No. MD 102.

at the bottom (which would interfere with spectrophotometric assays) and a volume is removed equivalent to the dialyzate volume taken. If ribosome concentration is to be measured by ^{14}C content, then the removed sample is pipetted directly into a scintillation vial, buffer is added to make a total of 1 ml, and the sample is counted. If ^{14}C-labeled ribosomes are not used, the removed sample is transferred to a small conical centrifuge tube, and 1.2 ml of buffer are added. After centrifugation for removal of dialysis sac "dust," the absorbancy of the sample is measured at 260 nm. If, on occasion, the sample is not perfectly clear (as judged by its absorbance at 321 nm), then a light-scattering correction should be made.[11] After the spectrophotometric assay, 1 ml of the sample is pipetted into a scintillation vial for assay of ^{3}H content.

Immediately after withdrawal of the sample from the sac for the above assays, the sac is closed, submerged completely, and allowed to spin once more until time for the next sample. Each time a sample is taken, an equal volume of dialyzate is also taken for radioactivity background. In this way, a series of points is obtained for construction of an HX curve. The outer solution should be changed about once every hour.

Addition of substances to the reaction mixture is performed by either adding the substance directly to the sac at a convenient point, e.g., immediately after removal of a sample, or by adding it to the outer solution and permitting it to dialyze in. Expensive or rare substances, or those which are nondialyzable, would of course be added directly to the sac. However, if studies such as those on the effect of pH, ions, type of buffer are being carried out, it is faster and more convenient simply to shift the dialysis sac to a different external solution. The lag involved because of dialysis time is very short, not exceeding several minutes for the usual ions in biological work.

The Combined Gel Filtration–Rapid Dialysis Method. The initial column run in this method is identical with that in the Gel Filtration Technique—Two Column Procedure, with three exceptions if earlier time points are desired (see earlier discussion of the Combined Method): First, Sephadex G-25 *coarse* grade beads are used; second, the initial ribosome concentration is raised to 3–4 mg/ml; third, only the ribosomal material constituting the *leading* half of the ribosomal peak is used, the contaminated trailing half being discarded.

After the initial column run, a 1–2 ml ribosome sample from the pooled column eluate is pipetted into a dialysis sac as described previously for the Rapid Dialysis Technique. Since a single individual can conveniently do three dialysis runs simultaneously, 1–2 ml samples would be pipetted into each of three sacs. It may be necessary to change the outer solution once, 10 minutes after transfer of the eluted ribosomal solution to the dialysis sacs. Sampling is then carried out as described for the Rapid Dialysis Technique. However, using the Combined Method, the first sample can be obtained at least as early as 18 minutes after the beginning of the run as opposed to 40–50 minutes for the Rapid Dialysis method.

Single Time-Point Technique. For this technique,[12] a 1.5 mg/ml ribosome solution is used and this is sufficient to run 5 samples. After the preincubation, the sample is passed through a Sephadex G-25, coarse grade, column as described in the preceding section. Fractions from the leading edge of the eluted ribosomal peak are pooled to give a final volume of almost 5 ml. Samples of 0.9 ml volume are then transferred to small test tubes which already contain the basic incubation mixture plus the particular added substances whose effects are under study. The number of samples which can be taken is not limited by handling problems, but rather by the number of dialysis racks available and by the dimensions of the whole setup. Five samples is not an unreasonable number but any number over six would require starting with more ribosomes. These samples should include one "standard" ribosome sample for normalization with other runs. The final volume in the test tubes should be such that a sample of 1–2 ml can be removed for addition to the dialysis sac.

The dialysis apparatus is set up prior to any of the above manipulations. It consists of two ring stands (or an appropriate "Flexaframe") arranged so that a horizontal bar may easily be moved up and down. A number of dialysis racks are suspended from the bar by monel, polyethylene, or Teflon hooks. Beneath each rack stands a 500-ml graduated cylinder containing external dialyzing solution, and beneath each cylinder is a magnetic stirrer.

When the samples have been transferred and all the sacs closed (approximately 15 minutes), the bar is lowered, simultaneously immersing all the racks in the dialysis solution. The run then begins. The racks are then detached from the bar and spun for the desired time.

One minute prior to the end of the incubation, the racks are reattached to the bar. When the bar is raised, the incubation is completed. Immediately upon raising the bar, the racks are quickly shaken to remove solution clinging to the outside of the sac, preventing further dialysis. Samples are removed from the sacs for analysis as described in previous sections, and the run ends. Samples are also removed from the dialyzate for radioactivity background. Control experiments have shown that, with this technique, HX values for identical samples of untreated ribosomes give a standard deviation of 2 to 3%.[12] However, attempts to extend this technique by removing only part of the sample and continuing the incubation for more time points have, for reasons not yet understood, produced unreliable values.

Calculations

The numerical determination of H/ribosome gives the number of unexchanged hydrogens on the ribosome at the given time. The equation used for obtaining these values is as follows[18]:

$$H/\text{ribosome} = (C/D)/C_0 \times 111 \times E_p;$$

where C is the difference in the cpm/ml of tritium between a sample and an equivalent volume of its dialyzate (or eluent immediately preceding the ribosomal peak when the gel filtration method is used); D is the optical density of the sample at 260 nm. If ^{14}C-labeled ribosomes are used, the value of ^{14}C/OD$_{260\ nm}$ is predetermined and OD values are calculated from the ^{14}C content of each sample; 111 is the atom concentration of hydrogen in H_2O; E_p is the "molar" absorbancy of the ribosome solution at 260 nm, i.e., the product of $E_{1\ cm}^{1\%}$ and the molecular weight. For bacterial ribosomes, the value varies between 4.0 and 4.2×10^7 cm^{-1}, depending upon the method of preparation. A value of 5.56×10^7 cm^{-1} has been used for rat liver ribosomes.[2]

[18] S. W. Englander, *Biochemistry* **2**, 798 (1963).

[40] The Inactivation and Reactivation of *Escherichia coli* Ribosomes

By ADA ZAMIR, RUTH MISKIN, ZVI VOGEL, and DAVID ELSON

When *E. coli* ribosomes or either of their subunits are assayed under certain conditions for any of a number of biological activities, they can be shown to exist in one of several different states: active, inactive, or partially active. The state is determined by the past treatment of the ribosome. These states are reversibly interconvertible and ribosomes can be brought from one to another by relatively mild treatments. Among the factors that influence these interconversions *in vitro* are temperature, the ionic environment, and interactions between subunits or between the ribosome and certain other macromolecules that participate in protein synthesis. 70 S, 50 S, and 30 S ribosomes all undergo these interconversions, although the characteristics of the process may vary somewhat in each case. Table I lists those ribosomal activities that we have tested so far

TABLE I

RIBOSOMAL ACTIVITIES WHICH ARE REVERSIBLY INACTIVATED AND REACTIVATED

> 30 S subunit activities
>> Nonenzymatic binding of aminoacyl-tRNA[a,b]
>> Enzymatic binding of aminoacyl-tRNA[b,c]
>> Formation of the initiation complex[b]
>> Binding of dihydrostreptomycin[d,e]
>> Association with the 50 S subunit[b,f]
>
> 50 S subunit activities
>> Peptidyltransferase activity[g]
>> Ability to catalyze the termination reaction[h]
>> Binding of chloramphenicol, erythromycin, and sparsomycin[i]
>> Association with the 30 S subunit[b,f]

[a] A. Zamir, R. Miskin, and D. Elson, *J. Mol. Biol.* **60**, 347 (1971).
[b] R. Miskin, Ph.D. Thesis, Weizmann Institute of Science, 1972, and unpublished results.
[c] Y. Kaufmann and A. Zamir, *J. Mol. Biol.* **69**, 357 (1972).
[d] Z. Vogel, T. Vogel, A. Zamir, and D. Elson, *J. Mol. Biol.* **54**, 379 (1970).
[e] R. Miskin and A. Zamir, *Nature (London) New Biol.* **238**, 78 (1972).
[f] R. Miskin and A. Zamir, in preparation.
[g] R. Miskin, A. Zamir, and D. Elson, *J. Mol. Biol.* **54**, 355 (1970).
[h] Z. Vogel, A. Zamir, and D. Elson, *Biochemistry* **8**, 5161 (1969).
[i] Z. Vogel, T. Vogel, A. Zamir, and D. Elson, *J. Mol. Biol.* **60**, 339 (1971).

and found to be affected; there are probably others. Each of these activities is abolished under certain conditions and restored under others, some, but not all, with exactly parallel kinetics.

In general, activity is lost when ribosomes are depleted of certain cations (see below). If the depleted ions are restored in the cold the ribosomes remain largely inactive or regain activity only very slowly. However, if the solution is heated, the ribosomes rapidly become active and remain so when subsequently chilled. This shows that a change, presumably conformational, has taken place in the ribosome itself.

These reversible changes in activity seen *in vitro* may have implications regarding the mechanism of ribosome function *in vivo*.[1-3] However, the primary purpose of this article is to describe the phenomenon to workers in the field, to enumerate the conditions that induce changes in activity, to point out that failure to take them into account may lead to error in interpreting the results of *in vitro* studies of ribosomal function, and to suggest how such errors may be avoided.

[1] A. Zamir, R. Miskin, and D. Elson, *J. Mol. Biol.* **60**, 347 (1971).
[2] R. Miskin, A. Zamir, and D. Elson, *J. Mol. Biol.* **54**, 355 (1970).
[3] Y. Kaufmann, R. Miskin, and A. Zamir, *FEBS (Fed. Eur. Biochem. Soc.) Lett.* **22**, 315 (1972).

Certain methods often employed to prepare ribosomes and separate their subunits cause inactivation. On the other hand, most assays are done under conditions that cause activity to reappear, at widely varying rates determined by the extant conditions. If initially inactive ribosomes are used to study a particular ribosomal reaction under assay conditions that cause activation of the ribosomes, the requirements, kinetics and other characteristics of the activation will be superimposed on those of the reaction under study and may even override and obscure them (see footnotes 4 and 5 for examples). The true characteristics of the reaction may be discernible only if concurrent activation is eliminated. This may be done by pretreating the ribosomes to bring them to full activity before assaying them for the reaction under study; and, in fact, there are many reports that the preheating of ribosomes in an assay medium has enhanced one or another activity (or caused it to appear) when the ribosomes were subsequently assayed at the same temperature or in the cold (e.g., see footnotes 6–13).

In what follows we shall describe the effect of various conditions and factors on the inactivation and reactivation of ribosomes, and describe typical procedures for producing active and inactive ribosomes and subunits. Full experimental details will not be given; they can be found in the original publications.[1-5,14-18]

The major factors affecting the conversion of ribosomes from one state to another are monovalent cations, Mg^{2+} ions and temperature. Considerable effects are also exerted by polynucleotides, bound fMet-tRNA, streptomycin, short-chain aliphatic alcohols, sulfhydryl compounds, and the interaction of one subunit with the other subunit. Active ribosomes or subunits generally remain active at $0°$ under suitable ionic conditions. Similarly,

[4] Y. Kaufmann and A. Zamir, J. Mol. Biol. 69, 357 (1972).

[5] Z. Vogel, T. Vogel, A. Zamir, and D. Elson, J. Mol. Biol. 54, 379 (1970).

[6] M. Lubin, Fed. Proc., Fed. Amer. Soc. Exp. Biol. 23, 994 (1964).

[7] S. Pestka and M. Nirenberg, J. Mol. Biol. 21, 145 (1966).

[8] H. Kaji and Y. Tanaka, J. Mol. Biol. 32, 221 (1968).

[9] T. Nakamoto and E. Hamel, Proc. Nat. Acad. Sci. U.S. 59, 238 (1968).

[10] C. S. McLaughlin, J. Dondon, M. Grunberg-Manago, A. M. Michelson, and G. Saunders, J. Mol. Biol. 32, 521 (1968).

[11] B. E. H. Maden, R. R. Traut, and R. E. Monro, J. Mol. Biol. 35, 333 (1968).

[12] H. Teraoka, J. Mol. Biol. 48, 511 (1970).

[13] A. Kikuchi and R. Monier, FEBS (Fed. Eur. Biochem. Soc.) Lett. 11, 157 (1970).

[14] R. Miskin, Ph.D. Thesis, Weizmann Institute of Science, 1972, and unpublished results.

[15] R. Miskin and A. Zamir, Nature (London) New Biol. 238, 78 (1972).

[16] R. Miskin and A. Zamir, in preparation.

[17] Z. Vogel, A. Zamir, and D. Elson, Biochemistry 8, 5161 (1969).

[18] Z. Vogel, T. Vogel, A. Zamir, and D. Elson, J. Mol. Biol. 60, 339 (1971).

inactive ribosomes remain wholly or largely inactive, even in an activating medium, if the temperature is kept close to 0°; they become active if heated, at a rate that is greatly accelerated with rising temperature. Therefore, in order to minimize interconversion during the assay of activity, assays must be carried out at 0°. The standard tests for activity that we have used in studying the interconversions were: for 30 S subunits, the nonenzymatic binding of Phe-tRNA in the presence of poly(U), assayed by membrane filtration[1,19]; for 50 S subunits, peptidyltransferase activity, assayed by a modification of the "fragment" or "alcohol" reaction.[2,20]

Inactivation

Effect of Monovalent Cations.[1,2] The state of activity of both 30 S and 50 S subunits depends on specific monovalent cations. When active subunits dissolved in a buffer containing NH₄Cl are dialyzed so as to remove the NH₄Cl or to replace the NH₄⁺ ions with an equal concentration of another monovalent cation, activity is reduced or lost (Table II). The effect is clearly a specific ion effect. NH₄⁺ is most effective in maintaining the active state or conformation; Rb⁺, K⁺, and Cs⁺ are less effective; and Na⁺ and Li⁺ are ineffective. The effect is absolute in the case of the 50 S subunits. However, the 30 S subunits show a sizable activity in the absence

TABLE II
EFFECT OF MONOVALENT CATIONS ON MAINTENANCE OF RIBOSOMAL ACTIVITY[a]

Salt in dialysis medium	Relative activity (%)	
	50 S subunits[b] (peptidyltransferase)	30 S subunits[c] (Phe-tRNA binding)
NH₄Cl	100	100
RbCl	90	41
KCl	55	61
CsCl	22	43
NaCl	0	13
LiCl	0	9
None	0	24

[a] For each subunit, identical aliquots of a ribosome solution in a buffer containing NH₄Cl were dialyzed at 4° for 20 hours against otherwise identical buffers containing equal concentrations of the salts listed above, or no salt. Activity was assayed in the cold under standard conditions.
[b] R. Miskin, A. Zamir, and D. Elson, *J. Mol. Biol.* **54**, 355 (1970).
[c] A. Zamir, R. Miskin, and D. Elson, *J. Mol. Biol.* **60**, 347 (1971).

[19] M. Nirenberg and P. Leder, *Science* **145**, 1399 (1964).
[20] R. E. Monro and K. A. Marcker, *J. Mol. Biol.* **25**, 347 (1967).

of salt. This is due, at least in part, to a slow reactivation during the assay, which can occur with 30 S subunits inactivated in this way (discussed below). Na$^+$ and Li$^+$ ions lower this residual activity, probably by replacing remaining NH$_4^+$ ions.

Effect of Mg^{2+} Ions.[1,2] In a solution containing 0.1 M NH$_4$Cl the lowering of the Mg^{2+} concentration has a drastic effect on the aminoacyl-tRNA binding activity of isolated and fully reactivated 30 S subunits. Activity is slightly but noticeably reduced at 4 mM Mg^{2+}, nearly abolished at 1 mM, and fully abolished at 0.5 mM. 70 S ribosomes retain full binding activity down to 2 mM Mg^{2+}. At lower concentrations there is complete dissociation into subunits and the binding activity drops in parallel with that of the isolated 30 S subunit. Interaction with the 50 S subunit evidently

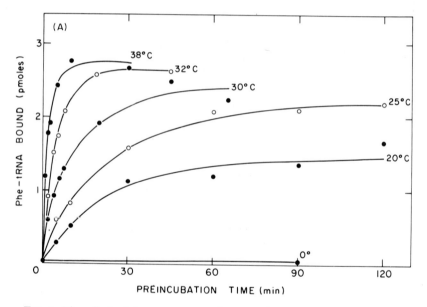

Fig. 1. The effect of temperature on the rate of reactivation. Previously inactivated subunits were preincubated in an activating medium at the specified temperature. Aliquots were transferred directly to cold assay medium and assayed in the cold for Phe-tRNA binding (30 S), or were chilled and then assayed in the cold for peptidyltransferase (50 S). Note different time scales.

(A) 30 S subunits inactivated at low Mg^{2+}; reactivated in 50 mM Tris·HCl (pH 7.2), 20 mM Mg acetate, 200 mM NH$_4$Cl, 2 mM dithiothreitol [A. Zamir, R. Miskin, and D. Elson, *J. Mol. Biol.* **60**, 347 (1971)].

(B) 30 S subunits inactivated by NH$_4^+$ depletion; reactivation as in (A) [A. Zamir, R. Miskin, and D. Elson, *J. Mol. Biol.* **60**, 347 (1971)].

(C) 50 S subunits inactivated by NH$_4^+$ depletion; reactivated in 10 mM Tris·HCl (pH 7.4), 5 mM Mg acetate, 50 mM NH$_4$Cl [R. Miskin, A. Zamir, and D. Elson, *J. Mol. Biol.* **54**, 355 (1970)].

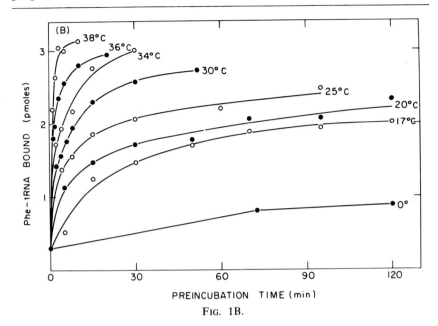

FIG. 1B.

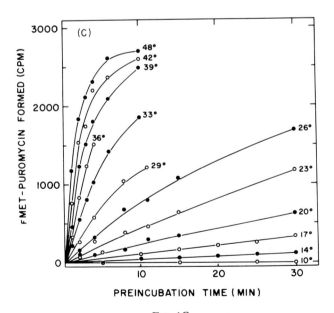

FIG. 1C.

stabilizes the active conformation of the 30 S subunit. The isolated 50 S subunit is more stable to these changes and retains full peptidyltransferase activity down to at least 1 mM Mg^{2+}; it has not been tested at lower Mg^{2+} concentrations. Therefore, the use of a low Mg^{2+} concentration to dissociate 70 S ribosomes into subunits is likely to produce inactive 30 S subunits.

Summary. The 30 S subunit can be inactivated in two different ways: (a) by depleting it of essential monovalent cations (NH$_4^+$, K$^+$, Rb$^+$, Cs$^+$) at high Mg^{2+} concentrations, and (b) by reducing the Mg^{2+} concentration to about 1 mM, even in the presence of appropriate monovalent cations. The two inactive 30 S forms differ in certain characteristics. The 50 S subunit is inactivated by treatment (a) but not by (b).

Reactivation

Ribosomes inactivated as described above can be returned to full activity by restoring the needed cations. The process is strongly temperature dependent.

Effect of Temperature.[1,2] The effect of temperature on the rate of reactivation of inactivated 30 S and 50 S subunits is shown in Fig. 1. During the time studied, 50 S subunits were not reactivated at low temperature, and the same was true of 30 S subunits inactivated at low Mg^{2+}. However, 30 S subunits inactivated by NH$_4^+$ depletion underwent a slow reactivation even at 0°. The rate of reactivation increased with rising temperature. In the presence of NH$_4^+$ ions, 50 subunits appear to reach the same final level of activity at all activating temperatures after a sufficiently long incubation (not apparent in figure). This has not always been the case with other activating cations. With 30 S subunits, the final level rises with rising temperature, particularly below 30°.

If 50 S subunits are chilled to 0° before reactivation is complete, they remain partially reactivated; i.e., the population of subunits retains the level of peptidyltransferase activity reached when heating was stopped.[17] This makes it possible to collect samples during the course of activation or inactivation, store them at 0°, and assay them subsequently. 30 S subunits are much more labile in the sense that the level of activity is easily altered by changes in ions, ion concentration, temperature, etc. As a rule, we have tried to assay 30 S subunits immediately after reactivation.

In all cases studied, the reactivation of isolated subunits followed first-order kinetics, showing that each ribosome acts as an independent unit during the process. Rate constants varied widely under different temperature and salt conditions, but, in general, 30 S reactivation is slower than 50 S reactivation. The Arrhenius activation energy is similar for the two subunits, about 27 kcal/mole in the temperature range 30–40°.

Effect of Monovalent Cations.[1,2] Figure 2 shows the influence of different monovalent cations on the heat reactivation of subunits. In general, the effects parallel those seen in the preservation of activity (compare Table II). NH_4^+, K^+, Rb^+, and Cs^+ ions reactivate both subunits. Li^+ is completely ineffective. Na^+ is ineffective with 50 S subunits, but causes partial reactivation of the two inactive 30 S forms. Under otherwise identical conditions, the rate of reactivation rises with rising concentration of the activating salt.

Our experience suggests that there is a maximum level of activity which a given population of subunits can attain if incubated long enough under a variety of conditions. The 50 S subunit appears to be able to reach this level under most, but not all, conditions. Cs^+, in particular, has often given submaximal reactivation, and in one or two instances this has also been observed with K^+ at low concentration (e.g., Fig. 2D). Results with the 30 S subunit have been much more variable. In general, NH_4^+ appears usually to be the most effective activating monovalent cation for both subunits with respect to both the rate and the final level of reactivation.

Effect of Mg^{2+} Ions.[1,2] Under otherwise identical conditions the rate of reactivation of the 30 S subunit increases with rising Mg^{2+} concentration. The 50 S subunit shows an optimum Mg^{2+} concentration which is different for different NH_4^+ concentrations (see below).

Interrelationship between Mg^{2+} and Monovalent Cations.[1,2] It is clear that the effects of Mg^{2+} and monovalent cations cannot be treated separately but are closely interdependent.[1,2,21,22] This is most evident with the 50 S subunit, where not only the absolute concentration of each ion affects the rate of reactivation, but also the ratio of one to the other. For each NH_4^+ concentration examined there was an optimum Mg^{2+} concentration which increased with rising NH_4^+ concentration, suggesting a balance between two opposing effects.[2] This type of relationship was not seen with the 30 S subunit.[1]

Effect of pH.[2] The dependence of reactivation rate on pH was determined only for the 50 S subunit. The rate rose to a maximum at about pH 7 and remained constant up to pH 8.6, the highest pH tested.

Additional Factors. A number of additional factors have been found to affect the rate of reactivation. Most of these observations apply to or were made on only one of the subunits, and they are listed according to the subunit affected.

---

[21] P. Spitnik-Elson, personal communication.
[22] A. Ghysen, A. Bollen, and A. Herzog, *Eur. J. Biochem.* **13**, 132 (1970).

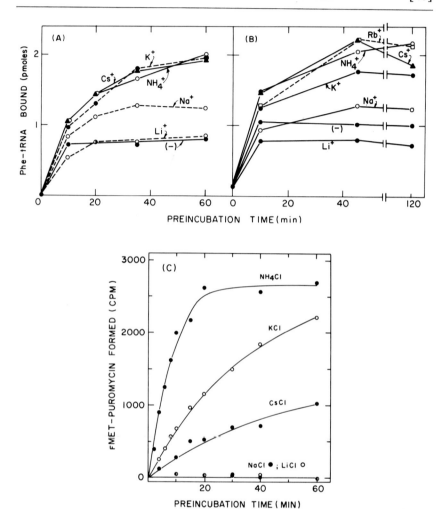

Fig. 2. The effect of different monovalent cations on reactivation. Previously inactivated subunits were preincubated and assayed as in Fig. 1.

(A) 30 S subunits inactivated by NH_4^+ depletion at low Mg^{2+}; reactivated at 36° in 50 mM Tris·HCl (pH 7.2), 4 mM Mg acetate, 100 mM salt (chloride) [A. Zamir, R. Miskin, and D. Elson, *J. Mol. Biol.* **60**, 347 (1971)].

(B) 30 S subunits inactivated by NH_4^+ depletion alone; reactivated as in (A) [A. Zamir, R. Miskin, and D. Elson, *J. Mol. Biol.* **60**, 347 (1971)].

(C) 50 S subunits inactivated by NH_4^+ depletion; reactivated at 26° in 15 mM Tris·HCl (pH 7.4), 5 mM Mg acetate, 100 mM salt [R. Miskin, A. Zamir, and D. Elson, *J. Mol. Biol.* **54**, 355 (1970)].

(D) 50 S subunits inactivated by NH_4^+ depletion; reactivated at 40° in 20 mM Tris·HCl (pH 7.4), 1 mM Mg acetate, 50 mM salt [R. Miskin, A. Zamir, and D. Elson, *J. Mol. Biol.* **54**, 355 (1970)].

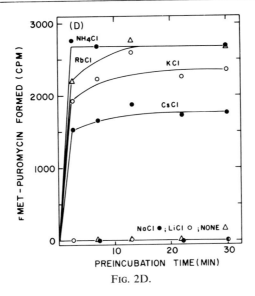

FIG. 2D.

Additional Factors—50 S Subunits

Aliphatic Alcohols.[2] The rate of restoration of peptidyltransferase activity to inactivated 50 S subunits is enhanced by the inclusion of aliphatic alcohols in the reactivation medium. Alcohol does not abolish the requirement for specific cations and heat but, in their presence, markedly accelerates the reactivation process. For example, under suboptimal activating conditions (17°, 50 mM NH$_4$Cl, 5 mM Mg acetate) 20% methanol (v/v) increased the first-order rate constant 9-fold, with only a slight effect on the Arrhenius activation energy. At equal volume fractions, methanol, ethanol, *n*-propanol, and *n*-butanol were equally effective, giving maximum enhancement at a volume fraction of 0.15. (Precaution: traces of aldehydes in the alcohol inactivate the ribosomes.) Alcohol is also essential for the expression of peptidyltransferase activity in the fragment reaction assay[20]; however, its effect on reactivation rate is unrelated to this.

Bound Substrate.[2] When inactivated by NH$_4^+$ depletion, 70 S ribosomes maintain their associated state and exhibit about 50% activity in binding fMet-tRNA in the cold in the presence of the trinucleotide messenger AUG and initiation factors (see below). These ribosomes are incapable of peptide bond formation (formation of fMet-puromycin) unless heat reactivated. When reactivated under otherwise identical conditions, ribosomes carrying bound fMet-tRNA regained peptidyltransferase activity considerably faster than did ribosomes to which the substrate was not attached, even though it was present in the reactivation medium.

Sulfhydryl Compounds.[14] Sulfhydryl reducing compounds do not affect peptidyl transferase activity when present during the assay, and treatment of ribosomes with sulfhydryl blocking reagents was not found to affect this activity.[23] Nevertheless, we have observed that there is a significant loss of activity when 50 S subunits are stored at 0° in the presence of sulfhydryl reducing compounds such as β-mercaptoethanol, dithiothreitol, cysteine, or thioglycolate. When compared with otherwise identical controls stored in parallel without the sulfhydryl compound, active 50 S subunits lost activity and inactive subunits lost the ability to be reactivated (Fig. 3). The loss was greater in the case of inactive subunits. It appears, therefore, that the active configuration of the 50 S subunit or the transition to it requires at least one disulfide bond and that reduction is more readily accomplished when the subunit is in the inactive conformation.

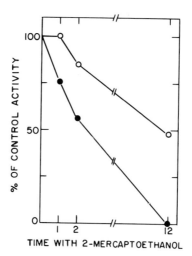

TIME WITH 2-MERCAPTOETHANOL

Fig. 3. The effect of mercaptoethanol on the activity of 50 S subunits during storage at 0°. 50 S subunits (10 mg/ml) were stored at 0°. Active subunits were stored in 20 mM Tris·HCl (pH 7.4), 5 mM Mg acetate, 100 mM NH₄Cl, 20 mM β-mercaptoethanol, and the active control in the same medium minus mercaptoethanol. At the times shown, 100-μg samples were assayed for peptidyltransferase, with 2 mM mercaptoethanol present in all cases. Inactive subunits and the inactive control (no mercaptoethanol) were stored in the same media minus NH₄Cl; 100-μg samples were reactivated for 15 minutes at 40° in 20 mM Tris·HCl (pH 7.4), 10 mM Mg acetate, 100 mM NH₄Cl, 4 mM mercaptoethanol, chilled, and assayed. Each point represents the activity of a sample as percent of the corresponding control activity, assayed at the same time. The control activities for 1, 2, and 12 days were (cpm): active (○): 2959, 3286, 2744; inactive (●): 3072, 3117, 1664 (R. Miskin, Ph.D. Thesis, Weizmann Institute of Science, 1972, and unpublished results).

[23] R. E. Monro, *J. Mol. Biol.* **26**, 147 (1967).

For practical purposes, it is best to store 50 S subunits in the active form without a sulfhydryl reducing compound.

Interaction with the Opposite Subunit. (See section with same title under 30 S subunits, below.)

Additional Factors—30 S Subunits

Polynucleotides.[5] When present in the preincubation medium, poly(U) markedly increases the rate of reactivation of inactive 30 S subunits for the subsequent nonenzymatic binding of Phe-tRNA. This effect is due to an interaction between the polynucleotide and the inactive subunit, to which poly(U) binds readily. When another reactivable activity was examined—streptomycin binding (see below)—poly(U) again showed a large effect, increasing the reactivation rate manyfold. Other synthetic polyribonucleotides showed much smaller enhancing effects, as did tRNA and ribosomal RNA; natural messenger RNA was not tested. Poly(C) caused enhancement at low temperatures and inhibition at higher temperatures. However, more data are required, particularly on the effect of changing conditions and on the binding of polynucleotides to inactive subunits, before it can be decided whether polynucleotides in general have a significant effect on reactivation rate, or whether poly(U) is a special case.

Sulfhydryl Compounds.[1] When inactive 30 S subunits are stored for a number of days at 0° in the absence of a sulfhydryl compound such as β-mercaptoethanol or dithiothreitol, there is a gradual decline in their ability to be reactivated. Part of this decline, presumably due to air oxidation, is prevented if the storage medium contains a sulfhydryl compound, or is reversed if the compound is added to the reactivation medium. It appears, therefore, that the active configuration of the 30 S subunit requires at least one reduced sulfhydryl group, either for its formation, its activity, or both. 30 S subunits in the active form are less sensitive to air oxidation and do not lose activity during similar periods of storage without a sulfhydryl compound. It seems best to store 30 S subunits in the active form and, in contrast to 50 S subunits, the presence of a sulfhydryl compound may provide an additional measure of security. In any case, however, the 30 S subunit is relatively labile and should be prewarmed before assay in order to ensure maximal activity.

Interaction with the Opposite Subunit.[14] As mentioned above, an active 30 S subunit is stabilized against inactivation at low Mg^{2+} concentration when it is associated with a 50 S subunit in the 70 S ribosome. Reactivation is also influenced by association with the opposite subunit, as shown by a number of observations made with isolated 30 S and 50 S subunits and mixtures of the two. The effect was much more pronounced when the 30 S and 50 S subunits were produced together by dissociating 70 S ribosomes than when isolated 30 S and 50 S subunits were mixed.

Most of these studies dealt with the reactivation of the ability of initially inactive 30 S subunits to bind Phe-tRNA nonenzymatically, and may be summarized as follows. When inactive 30 S subunits are activated in the presence of active 50 S subunits under certain conditions, Phe-tRNA binding activity begins to appear only after an initial lag period. The lag becomes longer as ribosome concentration is raised, temperature is lowered, or the ratio Mg^{2+}/NH^{4+} is increased—i.e., under conditions that favor the increasing association of subunits. Lag periods as long as 2 hours have been observed. If the conditions are changed in the opposite direction so as to favor dissociation, the lag decreases and eventually disappears. Such a lag has never been seen in the absence of 50 S subunits, and would appear to be caused by interaction between the two subunits.

This conclusion appeared paradoxical at first, since inactivation severely impairs the ability of the 30 S subunit to interact with the 50 S subunit (see below). The matter was therefore investigated further and can be understood in the following way. The reactivation of the 30 S subunit is not a one step process, but proceeds through one or more intermediate stages, so that some activities appear before others (see following section for an example). There are indications that the 30 S subunit regains the ability to associate with the 50 S subunit early in the reactivation process, before Phe-tRNA binding ability appears. Apparently the 50 S subunit stabilizes this metastable intermediate 30 S configuration and hinders the additional conformational changes that are required to restore Phe-tRNA binding ability, thus causing the observed lag.

The effect of 30 S subunits on the reactivation of 50 S subunits has not been studied systematically, but there are indications of an analogous situation, that is, that the restoration of peptidyltransferase activity is slower in the presence of 30 S subunits than in their absence.

Streptomycin.[15] At concentrations close to the ribosome concentration, streptomycin (and its derivative dihydrostreptomycin) exerts a drastic effect on the interconversions between ribosomes active and inactive in binding Phe-tRNA. The drug severely inhibits both the activation of inactive ribosomes and the inactivation of active ribosomes isolated from streptomycin-sensitive (but not streptomycin-resistant) bacteria. Although seen with isolated 30 S subunits, the effect is much more pronounced in the presence of 50 S subunits and is most pronounced with unfractionated 70 S ribosomes.

It appears from this that streptomycin interacts with both active and inactive forms of the ribosome and has the effect of stabilizing whichever form it attaches to. Since fully inactivated 30 S subunits do not bind streptomycin[5] (Table I), it was not apparent, at first, how the drug could affect the inactive ribosome. However, kinetic analyses with 70 S ribosomes showed that during the course of reactivation the ability to bind the antibiotic is fully restored before the ability to bind Phe-tRNA begins to appear. The experimental observations indicate that streptomycin binds at this stage and hinders further conformational changes required to restore Phe-tRNA binding ability.

A streptomycin binding intermediate was not detected with isolated 30 S subunits, and both drug binding and Phe-tRNA binding activities appeared to be restored simultaneously. Nevertheless, the existence of such an intermediate must be assumed to explain the effect of streptomycin on the reactivation of isolated 30 S subunits. Presumably this intermediate form is too labile to be detected by the procedure employed. The interaction with the 50 S subunit (which also occurs early in reactivation—see preceding section) thus appears to stabilize the intermediate.

As noted, inactive 70 S ribosomes regain their ability to bind streptomycin much more rapidly than their ability to bind Phe-tRNA when incubated at elevated temperatures. This is also true at 0°, where considerable drug-binding activity may appear within an hour in an activating medium such as the assay medium; Phe-tRNA binding activity does not appear in this time. (This is true only for 70 S ribosomes; isolated 30 S subunits are not appreciably reactivated for either activity under these conditions.) Therefore, in relevant experiments inactive ribosomes should not be put into an activating medium sooner than necessary.

Association–Dissociation Behavior.[14,24] The state of activity of the ribosomal subunits affects their physical interactions. Isolated inactive 30 S subunits of both types (NH_4^+ depletion or low Mg^{2+}) have a pronounced tendency to form aggregates, mostly dimers sedimenting at about 50 S. Dimerization is favored by a high Mg^{2+}/NH_4^+ ratio; at 10 mM Mg^{2+} it can be reduced but not abolished by raising the NH_4^+ concentration to 160 mM. (The sedimentation analysis is performed in the cold to avoid reactivation.) Active 30 S subunits have a much smaller tendency to dimerize. However, at high Mg^{2+} concentrations they may, while not showing a distinct dimer peak, sediment in a broad peak with an elevated sedimentation constant (e.g., 30–38 S) which may, perhaps, indicate a dynamic equilibrium between aggregated and nonaggregated forms.

The tendency of 50 S subunits to undergo self aggregation is much lower; here, too, inactive subunits dimerize somewhat more readily than

[24] R. Miskin, unpublished results.

inactive ones. Under certain conditions (see below) inactive 70 S ribosomes can be formed which do not dissociate into subunits. These form dimers under certain conditions where active 70 S ribosomes do not dimerize.

Inactive 30 S subunits show a markedly reduced affinity for the 50 S subunit. When inactivated at low Mg^{2+} they undergo virtually no association to 70 S ribosomes. When inactivated by NH_4^+ depletion the ability to associate is impaired but not lost, and it is possible to prepare inactive 70 S ribosomes by NH_4^+ depletion at a Mg^{2+} concentration high enough to prevent full dissociation. Kikuchi and Monier have also observed that inactivated 30 S subunits combine less readily with 50 S subunits.[13] 50 S ribosomes are less affected, but here, too, inactivation lowers the affinity for

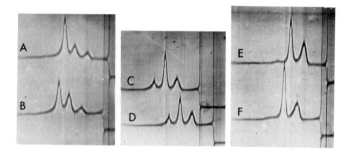

Fig. 4. Association behavior of active and inactive subunits.

Pattern	50 S subunit	30 S subunit	$s_{20,w}$ (S)
A	Active	Active	30, 44, 58
B	Inactive	Active	29, 44, 65
C	Active	Inactive (NH_4^+ depletion)	31, 48, 60
D	Inactive	Inactive (NH_4^+ depletion)	28, 48, 62
E	Active	Inactive (low Mg^{2+})	27, 46
F	Inactive	Inactive (low Mg^{2+})	28, 46

Isolated 50 S subunits were inactivated by NH_4^+ depletion and isolated 30 S subunits either by NH_4^+ depletion or at low Mg^{2+}. They were reactivated to produce active subunits by a 15-minute incubation at 40° in 10 mM Tris·HCl (pH 7.2), 2 mM Mg acetate, 100 mM NH_4Cl (50 S), or 10 mM Tris·HCl, 20 mM Mg acetate, 200 mM NH_4Cl, 2 mM dithiothreitol (30 S). The sedimentation buffer was 10 mM Tris·HCl, 10 mM Mg acetate, 160 mM NH_4Cl, 2 mM dithiothreitol. 30 S subunits (2.5 mg/ml) were mixed with 50 S subunits (5.7 mg/ml) and analyzed at 50,740 rpm in a Beckman-Spinco Model E analytical ultracentrifuge at 4.9° (A, B) or 8.5° (C–F) to prevent activation during the run. (R. Miskin, Ph.D. Thesis, Weizmann Institute of Science, 1972, and unpublished results.)

the 30 S subunit, although it does not abolish it (see also footnotes 4, 25). Figure 4 illustrates these relationships under one set of salt conditions.

In summary, inactivation impairs the ability of each subunit to interact with the other, and increases its tendency toward self aggregation, including dimer formation. Activation has the opposite effect. The effect is most pronounced with the 30 S subunit.

Activity States and Ribosomal Conformation. The conversion of a ribosomal subunit from an active to an inactive state is not caused by the loss of any ribosomal component (other than inorganic salt) essential to activity. This is shown by the fact that inactive subunits can be centrifuged down, resuspended in buffer, and restored to full activity without the addition of supernatant. The difference between active and inactive subunits is therefore conformational. The presently available direct evidence for this is of three types: (a) As mentioned above, 30 S subunits undergo sulfhydryl oxidation and 50 S subunits disulfide reduction more readily in the inactive form, indicating that the groups in question are situated differently in the active and inactive forms.[1,2] (b) The active 30 S subunit sediments more rapidly than the inactive form. Parallel samples of inactivated and reactivated subunits in the same medium were analyzed simultaneously in the same centrifuge rotor at low temperature. In each of 14 experiments with 6 different subunit preparations, the active subunits sedimented faster, with a mean difference of $s_{20,w} = 2.1$.[14,26] This shows that there is either a difference in shape or hydration or else a different equilibrium between aggregated and nonaggregated forms, indicating a conformational difference in either case. No such difference has been seen with the 50 S subunit.[2] (c) Active and inactive 30 S subunits were labeled with radioactive *N*-ethylmaleimide, a reagent that reacts covalently with sulfhydryl groups, and the labeled proteins were identified. At a concentration of 5 mM the reagent reacted with seven proteins in each subunit form. Only three of these were the same in the active and inactive forms; i.e., eight 30 S sulfhydryl proteins reacted differently to the reagent in different activity states.[26,27] The 50 S subunit has not yet been examined in this way.

Ribosomal Activities Affected by Inactivation and Reactivation Treatment. In addition to the activities assayed routinely in most of the studies mentioned above, we have found several other ribosomal activities to be abolished and restored by the treatments which affect Phe-tRNA binding

[25] Z. Vogel, T. Vogel, A. Zamir, and D. Elson, *Eur. J. Biochem.* **21**, 582 (1971).
[26] P. Spitnik-Elson, A. Zamir, R. Miskin, Y. Kaufmann, Y. H. Ehrlich, I. Ginzburg, and D. Elson, *in* "RNA Viruses/Ribosomes," edited by H. Bloemendal, E. M. J. Jaspers, A. van Kammen, and R. J. Planta, North Holland Press. *FEBS (Fed. Eur. Biochem. Soc.) Symposia* **27**, 251 (1972).
[27] I. Ginzburg, R. Miskin, and A. Zamir, *J. Mol. Biol.* (1973), in press.

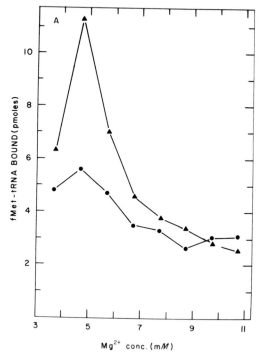

Fig. 5. Formation of the initiation complex by activated and inactivated ribosomes. The binding of fMet-tRNA in the presence of poly(AUG) and initiation factors was assayed at different Mg^{2+} concentrations essentially as described by R. Miskin, A. Zamir, and D. Elson [*J. Mol. Biol.* **54**, 355 (1970)]. (A) Samples, 100 μg, of 70 S ribosomes active (▲) as prepared and subsequently inactivated (●) by NH_4^+ depletion. (B) Samples, 30 μg, of 30 S subunits inactivated by NH_4^+ depletion and reactivated. (C) Samples, 30 μg, of 30 S subunits inactivated at low Mg^{2+} and reactivated. Inactivation and reactivation treatments are shown in Tables III and IV. B and C: ○, active; ●, inactive.

in the 30 S subunit and peptidyltransferase activity in the 50 S subunit (Table I).

50 S Activities. Peptidyltransferase was routinely assayed with a modification[2] of the "fragment reaction,"[20] which requires alcohol but does not require the 30 S subunit, messenger RNA, or initiation factors. This activity was also assayed in the absence of alcohol by the reaction of puromycin with the initiation complex (fMet-tRNA bound to 70 S ribosomes in the presence of AUG and initiation factors) with results that fully paralleled those obtained with the fragment reaction, i.e., peptide bond synthesis occurred with activated but not with inactivated ribosomes.[17]

The ability of the ribosome to catalyze the termination reaction of

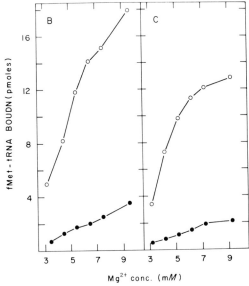

FIG. 5,B and C.

protein synthesis (the hydrolytic release of the completed polypeptide chain from ribosome-bound tRNA)[28-30] is inactivated and reactivated simultaneously with peptidyltransferase, suggesting that both activities may share a common 50 S component.[17]

Of the antibiotics that interact with the 50 S subunit and interfere with peptide bond formation, three have been examined. The reactions assayed were the binding of chloramphenicol and erythromycin and the formation of the sparsomycin-induced complex of acetyl-Leu-tRNA with the subunit (see Vogel *et al.*[18] for literature citations). All three activities disappeared and reappeared in parallel with the peptidyltransferase activity.[18] Teraoka has also reported a heat-dependent activation of erythromycin binding capacity and has attributed it, as we do, to a conformational change in the 50 S subunit.[12]

All the above 50 S functions appear to be inactivated and reactivated simultaneously, suggesting that their sites either overlap or are coupled to each other through the structure of the ribosome.

30 S Activities. In addition to the nonenzymatic binding of Phe-tRNA [poly(U)],[1] we have examined the nonenzymatic binding of Lys-tRNA

[28] M. C. Ganoza, *Cold Spring Harbor Symp. Quant. Biol.* **31,** 273 (1966).
[29] M. R. Capecchi, *Proc. Nat. Acad. Sci. U.S.* **58,** 1144 (1967).
[30] C. T. Caskey, R. Tompkins, E. Scolnick, T. Caryk, and M. Nirenberg, *Science* **162,** 135 (1968).

[poly(A) or poly(AUG)][5,14] Val-tRNA [poly(AUG)][14] and Met- and fMet-tRNA [poly(AUG)],[14] and the enzymatic binding of Phe-tRNA [poly(U)] mediated by elongation factor Tu in the presence of 50 S subunits.[14] All these activities were abolished and restored by the treatments described above. The same is true of the ability of the isolated 30 S subunit to form the initiation complex [the enzymatic binding of fMet-tRNA in the presence of poly(AUG) and initiation factors], where inactivated subunits show very little activity (Fig. 5).[14] The case is different, however, when inactive 70 S ribosomes are tested, provided that inactivation was accomplished by NH_4^+ depletion at a Mg^{2+} concentration high enough to preserve undissociated 70 S ribosomes (see above). Such inactivated ribosomes show about 50% of the activity of fully activated ribosomes at the optimum Mg^{2+} concentration and an even higher relative activity at other Mg^{2+} concentrations (Fig. 5).[14] The fMet-tRNA bound to such ribosomes cannot react with puromycin.[17] The retention of this binding activity when non-enzymatic Phe-tRNA binding ability has been lost appears to be due to association with the 50 S subunit. This association apparently preserves the activity of the 30 S subunit until the moment of complex formation, when the 70 S ribosome undergoes a transient dissociation.[31]

The binding of streptomycin and dihydrostreptomycin has been discussed above.

In contrast with our present information about 50 S reactivation kinetics, it is clear that not all of the 30 S activities are reactivated in parallel. Although the various aminoacyl-tRNA binding activities appear to be restored synchronously, at least two activities are restored much more rapidly: the binding of streptomycin and the ability to associate with the 50 S subunit (see above).

Typical Procedures for Inactivating and Reactivating E. coli ribosomes

The procedures described here are commonly employed in our laboratory with ribosomes prepared from *E. coli* MRE600 and washed with 0.5 or 1 *M* NH_4Cl, and subunits isolated from them by large-scale zonal centrifugation.[25,32] The procedures can be used with ribosomes prepared in other ways and from other bacterial strains, and the conditions can be varied with respect to time, temperature, and buffer composition. In each case it is advisable to assay the activity in question in order to ascertain whether inactivation or reactivation has reached the desired level.

Salts and other materials should be of analytical grade. Dialysis tubing

[31] M. Noll and H. Noll, *Nature (London) New Biol.* **238**, 225 (1972).
[32] P. Spitnik-Elson, *FEBS (Fed. Eur. Biochem. Soc.) Lett.* **7**, 214 (1970).

TABLE III

MEDIA FOR THE INACTIVATION OF 30 S, 50 S, AND 70 S RIBOSOMES

Treatment	Ribosome treated	Concentration (mM)			
		Mg acetate	NH$_4$Cl	Tris·HCl, pH 7.5	β-Mercapto-ethanol
30 S functions					
NH$_4$ depletion	30 S	5	—	20	6
NH$_4$ depletion	70 S	10	—	20	6
Low Mg^{2+}	30 S, 70 S	0.5	100	20	6
NH$_4$ depletion in low Mg^{2+}	30 S, 70 S	0.5	—	20	6
50 S functions					
NH$_4$ depletion	50 S, 70 S	1–10	—	20	—

should be boiled for 30 minutes in 10 mM EDTA (brought to pH 7.5 with NaOH), rinsed with doubly distilled water, and stored at 4° in 1 mM EDTA (pH 7.5); it should be rinsed thoroughly with water before use.

Inactivation.[1,2] Ribosomes or subunits in a volume not exceeding 10 ml are dialyzed at 4° for 18–20 hours against 500 ml of an inactivating medium. Ribosome concentrations ranging from 2 to 70 mg/ml have given similar results. The outside fluid is changed at least twice. Appropriate media are shown in Table III.

The 30 S subunits inactivated at low Mg^{2+}, whether in the presence or in the absence of NH$_4$Cl, have identical properties but, as noted above, differ from subunits inactivated by NH$_4^+$ depletion alone. Since isolated 30 S subunits were usually exposed to a low Mg^{2+} concentration when the 70 S ribosomes were dissociated, they can be converted to the NH$_4^+$ depleted type only by first reactivating them (see below) and then inactivating them by NH$_4$ depletion. On the other hand, NH$_4$-depleted 30 S subunits can be converted directly to the low Mg^{2+} type without reactivation.

Reactivation.[1,2] Ribosomes or subunits are incubated for 30 minutes at

TABLE IV

MEDIA FOR THE REACTIVATION OF 30 S, 50 S, AND 70 S RIBOSOMES

	Ribosome treated	Concentration (mM)			
		Mg acetate	NH$_4$Cl	Tris·HCl (pH 7.5)	Dithio-threitol
30 S functions	30 S, 70 S	20	200	20	2
50 S functions	50 S, 70 S	1–10	100	20	—

40° in an activating medium. The concentration of isolated subunits can be varied widely without affecting the rate of reactivation. At high concentrations of 70 S ribosomes or mixtures of 30 S and 50 S subunits there may be a lag period before 30 S functions begin to appear. Appropriate media are shown in Table IV.

Acknowledgment

We are grateful to Mr. Dan Haik for the preparation of ribosomes and subunits.

[41] Preparation of [³H]Anisomycin and [³H]Gougerotin: Binding to Eukaryotic Ribosomes

By Mariano Barbacid and David Vazquez

A number of antibiotics are known to block protein synthesis at the ribosome level.[1,2] Studies on the interaction of such antibiotics with the bacterial ribosome have been very useful in learning their mode of action and the nature of their binding sites on the ribosome.[3] Studies of this sort were not possible with eukaryotic ribosomes because of the lack of highly radioactive antibiotics active on ribosomes of this type. The antibiotic anisomycin is known to act on the peptidyltransferase center of eukaryotic ribosomes[4] whereas gougerotin, amicetin, blasticidin S, and fusidic acid are active on bacterial as well as eukaryotic systems.[1,2] We describe in this contribution a method for the preparation of pure [³H]-anisomycin and [³H]gougerotin of high specific activity and the assays for their binding to ribosomes to calculate the association and dissociation constants and number of available binding sites. Basically similar methods might be useful in obtaining some other radioactive antibiotics which also bind to eukaryotic ribosomes.

Principle

High specific activity and purity are required for the use of antibiotics in binding studies. Methods to obtain highly radioactive antibiotics can

[1] D. Vazquez, T. Staehelin, M. L. Celma, E. Battaner, R. Fernandez-Muñoz, and R. E. Monro, in "Inhibitors: Tools in Cell Research" (T. Bücher and H. Sies, eds.), p. 100. Springer-Verlag, Berlin, 1969.

[2] S. Pestka, Annu. Rev. Microbiol. 25, 487 (1971).

[3] R. Fernandez-Muñoz, R. E. Monro, R. Torres-Pinedo, and D. Vazquez, Eur. J. Biochem. 23, 185 (1971).

[4] D. Vazquez, E. Battaner, R. Neth, G. Heller, and R. E. Monro, Cold Spring Harbor Symp. Quant. Biol. 34, 369 (1969).

be summarized as follows: (a) Chemical synthesis using radioactive precursors, (b) biochemical synthesis by the microorganism producer of the antibiotic, adding a radioactive precursor to the growth medium, (c) chemical modification of the antibiotic introducing some radioactive atoms or groups, and (d) tritium exchange labeling of the nonradioactive antibiotic. This tritium exchange can be carried out either by the Wilzbach method (tritiation for long periods of time in tritium atmosphere under normal pressure and temperature) or the catalytic gas exposure (similar to the Wilzbach method but using a catalyzer) or the catalytic exchange in aqueous media (heating the antibiotic, in a sealed tube, with a tritiated solvent and a catalyzer). Methods (a) and (b) are hardly useful for obtaining radioactive pure compounds in the cases of antibiotics active on eukaryotic systems since synthesis of these compounds in most cases is insufficiently studied. Method (c) can be useful only in some specific cases (e.g., obtaining [^{3}H]dihydrofusidic acid from fusidic acid). For this reason we have chosen the catalytic exchange variant of method (d), which usually is less damaging to the complex antibiotic molecule and gives higher specific activities than the other variants of the same method.

Tritiation by catalytic exchange of the antibiotics amicetin, anisomycin, blasticidin S, fusidic acid, and gougerotin was carried out by the Radiochemical Centre (Amersham, England).[5] The products obtained after tritiation contain an unknown amount of ^{3}H, ^{3}H$_2$O, and an undefined number of compounds derived by breakage of the antibiotic molecule due to the pressure and temperature conditions of the tritiation method. High resolution is required for the separation of the antibiotics from some of their derived impurities which have a great similarity in their chemical structures. Methods basically used for us to purify the pure radioactive products from the tritiated preparation are (a) high voltage paper electrophoresis, (b) column chromatography, and (c) chromatography on silicic acid–glass fiber adsorbent.

Evaluation of antibiotic activity in the crude tritiated preparations is required. This was carried out by simple biological methods, such as the inhibitory activity on either polyphenylalanine synthesis, or the puromycin reaction or the fragment reaction. When the crude tritiated preparation does not show any biological activity (e.g., in the case of tritiated blasticidin S), it is discarded. In some cases the tritiated preparation has some antibiotic activity (e.g., in the case of tritiated fusidic acid), which in the procedure of purification was found to be due to components different from the nontritiated antibiotic. On the other hand, [^{3}H]anisomycin and [^{3}H]gougerotin were prepared from crude preparations ob-

[5] M. Barbacid and D. Vazquez, *Anal. Biochem.* (in press).

tained by tritium catalytic exchange of the nonradioactive antibiotics which were initially shown to have antibiotic activity. The methods developed for their purification and binding to eukaryotic ribosomes are described in this contribution.

Reagents and Equipment

Preparation of [³H]Anisomycin

Anisomycin (Pfizer): 100 mg of the antibiotic were labeled by the tritium catalytic exchange method in aqueous media at the Radiochemical Centre (Amersham, England). The crude preparation contained 115 mCi in a volume of 25 ml.

Whatman chromatography column 2.54 × 45 cm. Carboxymethyl cellulose (CM-23) from Whatman

Elution buffer: Ammonium carbonate solution ranging from 2 to 50 mM on linear gradient

Shandon high voltage electrophoresis apparatus water cooled

Whatman papers 3 MM and 17 for electrophoresis

Electrophoresis buffer: ethanolamine 50 mM carried to pH 10 with 1 M HCl

Shandon chromatography chamber 25 × 9 × 25 cm

ChromAR Sheet 500 Mallinckrodt

Solvent system: Butanol (2)–petroleum benzine boiling range 40° to 60°–2 N ammonia (25:5:4)

Elution apparatus[6]

Packard Tri-Carb liquid scintillation spectrometer

Virtis M.D. 10-145-MRBA lyophilizer

Preparation of [³H]Gougerotin

Gougerotin (Calbiochem): 25 mg of the antibiotic were labeled by the tritium exchange method in an aqueous media at the Radiochemical Centre (Amersham, England). The crude radioactive preparation contained 80 mCi in a total volume of 25 ml.

Whatman chromatography column 2.54 × 45 cm

Phosphocellulose (P-11) from Whatman

Elution buffer: A linear gradient ranging from 0.15 to 1.5 M ammonia using a formic acid–ammonium buffer pH 2.8

Shandon chromatography chamber 25 × 9 × 25 cm

Solvent system I: tert-butanol–ethyl methyl ketone–water–ammonia 25% (4:3:2:1)

Solvent system II: acetonitrile–0.1 M ammonium acetate (7:3)

---

[6] L. A. Heppel, this series, Vol. 12A [38a].

Liquid scintillation spectrometer, paper, electrophoresis and elution apparatus were used as described above for preparation of [³H]-anisomycin

Binding of Radioactive Antibiotics to Eukaryotic Ribosomes

Ribosomes from *Saccharomyces cerevisiae* were prepared as previously described⁷ with the only modification that cells were broken in a mortar with purified sea sand (1:2, w/w). The ribosomes were finally resuspended in a 50 mM ammonium–maleate buffer, pH 6.5, containing 5 mM Mg^{2+} and 5 mM 2-mercaptoethanol.

Cellulose nitrate tubes, 0.6 ml

High speed ultracentrifuge

Procedure for Purification of [³H]Anisomycin

This procedure was carried out at 0–4°, and the pH of the tritiated preparation was kept between 3.5 and 6.0 unless otherwise specifically indicated. CM-cellulose column chromatography, high voltage electrophoresis, and chromatography on ChromAR Sheet 500 were carried out sequentially for this purification.

CM-Cellulose Column Chromatography. CM-cellulose (CM-23 fines reduced from Whatman) was suspended in 15 volumes (w/v) of 0.5 N NaOH and left sedimenting for 30 minutes. The supernatant liquor was decanted off and the cellulose was washed with water following the same procedure until pH 8. This step was repeated twice with 0.5 N HCl and finally washed with water until pH 4.

The equilibration step was carried out by pouring the exchanger into a volume of ammonium carbonate buffer containing one equivalent of ammonia. The suspension was left out for 10 minutes, and the supernatant liquor was decanted off. This step was repeated four times. The exchanger was poured into a glass cylinder, 20 volumes of the 2 mM ammonium carbonate buffer added and the suspension was allowed to settle down for 20 minutes. The column was packed with this CM-cellulose to give a bed volume of 75 ml; 2 mM ammonium carbonate was passed through the column overnight.

Aliquots (10 ml) of the crude radioactive material obtained from the Radiochemical Centre in the tritiation of anisomycin were lyophilized and then dissolved in 4 ml of 2 mM ammonium carbonate and poured onto the top of the chromatography column. Then 150 ml of the same buffer was passed through the column to wash the unbound radioactivity (45%).

⁷ E. Battaner and D. Vazquez, this series, Vol. 20 [46].

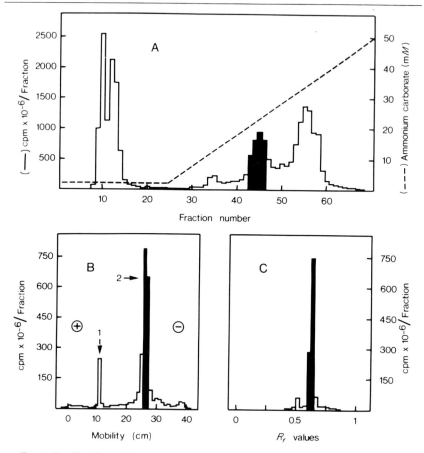

Fig. 1. Purification of [³H]anisomycin. (A) Profile of radioactivity of CM-cellulose column chromatography. Elution was carried out with a linear gradient of ammonium carbonate 2–50 mM (- - - -) at a flow rate of 100 ml/hour, and 7.5-ml fractions were collected in tubes containing 1.5 mmoles of formic acid to keep the pH under 6.5. Fractions in black bars (43–46) were taken for the next step of purification. (B) Profile of the high voltage electrophoresis using Whatman No. 17 paper; electrophoresis was carried out in 50 mM ethanolamine·HCl buffer, pH 10, for 4.5 hours at a voltage of 40 V/cm. The eluted from material at mobility 25–27 cm toward the cathode were taken for the following step of purification. (C) Chromatographic profile in ChromAR Sheet 500 (Mallinckrodt) of the material eluted on the electrophoresis. Solvent used was *sec*-butanol–petroleum benzene 40–60°–2 N ammonia (25:5:4); development of the chromatography was carried out at room temperature until the front has moved 16 cm. [³H]Anisomycin was detected by UV light and eluted with 10 mM formic acid; R_f values and total radioactivity of the eluate are shown in black.

Bound radioactivity was eluted with 320 ml of a linear gradient of ammonium carbonate ranging from 2 to 50 mM. Finally 100 ml of 250 mM ammonium carbonate was passed through the column to wash any residual radioactivity bound. Flow rate of the eluate was in all cases 100 ml/hour, and 7.5 ml fractions were collected in tubes containing 1.5 mmoles of formic acid to decrease the pH of the eluate. Samples (3 μl) of each fraction were mixed with 2 ml of scintillator fluid, Bray's 4% (w/v) Cab-O-Sil, and radioactivity was measured. The profile is well reproducible and practically is necessary only to assay the fractions of the peak eluting between 20 and 25 mM ammonium carbonate (Fig. 1A).

The assay of the fractions was carried out by high voltage paper electrophoresis using 50 mM ethanolamine·HCl buffer pH 10 at a voltage of 60 V/cm for 3.5 hours. Strips of 14 × 65 cm; 3 MM Whatman paper was used; 55 cm was the distance between the cooling plates, and the other 10 cm were required to connect with the electrodes. Aliquots 1 cm wide were cut from 1.5 cm-wide strips of the electrophoresis paper, mixed in vials with 2.5 ml of a scintillation fluid [toluene containing 5% (w/v) butyl-PBD], and radioactivity was measured.

In this analytical high voltage paper electrophoresis, the profiles obtained were compared with controls of 0.2 μmole of nonradioactive anisomycin per centimeter in width of the strip. The anisomycin spot was detected with a reagent of secondary amines,[8] which was prepared immediately before spraying by mixing equal volumes of 5% solution of sodium nitroprusside in 10% aqueous acetaldehyde and 2% sodium carbonate. Anisomycin appeared as a blue spot on a red background.

The electrophoretic profile showed mainly two peaks similarly to the preparative profile presented in Fig. 1B. Peak 1 was not identified whereas peak 2 moved electrophoretically at the same position of the control of nonradioactive anisomycin. The ratio of radioactivity in peaks 1 and 2 obviously differed with the fractions and increased with the molarity of the eluting buffer; when the ratio of radioactivity in peaks 1 and 2 was higher that 0.5, fractions were discarded. On the other hand, some of the early fractions collected at buffer concentrations 20–22 mM do not contain peak 1 but show a different peak of impurities when analyzed electrophoretically at pH 8.6 (50 mM triethanolamine·HCl buffer and 100 V/cm for 45 minutes); fractions showing a ratio of radioactivity in this peak to radioactivity in peak 2 higher than 0.5 were also discarded. Chromatography fractions with associated radioactivity and moving electrophoretically at the same position as the spot of the control anisomycin

[8] "Data for Biochemical Research" (R. M. C. Dawson, D. C. Elliot, W. H. Elliot, and K. M. Jones, eds.), p. 519. Oxford Univ. Press, London and New York, 1969.

(shown in black in Fig. 1A) were pooled (making approximately 30 ml) lyophilized and redissolved in 0.2–0.4 ml of 10 mM HCl.

Preparative High-Voltage Paper Electrophoresis. The 0.2–0.4 ml of the [³H]anisomycin solution obtained by CM-cellulose column chromatography were taken on a strip 14 × 57 cm of preparative paper Whatman No. 17 using for the contact with the electrodes a thinner paper (Whatman No. 1) to diminish the intensity through the electropheregram, allowing the use of a higher voltage.

Application of the radioactive material was carried out in the center of Whatman No. 17 paper in a band 5 cm broad as near as possible to the anodic side of the paper. Samples of 1 μmole of nonradioactive anisomycin and 0.5 μl of the radioactive material were placed on both sides at 1 cm distance. The nonradioactive anisomycin was used in order to know the position of the antibiotic spot after staining as indicated above. The 0.5 μl sample of radioactive material was used to analyze the profile of radioactivity which is shown in Fig. 1B.

Electrophoresis was carried out using a 50 mM ethanolamine·HCl buffer, pH 10, for 4.5 hours at 40 V/cm with the maximum possible cooling. The two controls on the sides of the main sample define the position of the [³H]anisomycin to be eluted which was usually taken in a strip 2–3 × 7 cm. It is convenient to take separately two strips of 1 cm of the paper before and after the theoretical position of the [³H]anisomycin in the sample in case that due to some distortion phenomenon the electrophoretic mobility of the central sample differs from the side samples.

Samples were eluted in the elution apparatus at 4° overnight with 10 mM formic acid. The 9–10 ml eluted were lyophilized and resuspended in 0.25 ml of 10 mM HCl. Quantitative estimation of radioactivity eluted in the central strip and the adjacent 1-cm strips localize exactly the position of [³H]anisomycin and the eluates which should be taken.

Chromatography on ChromAR Sheet 500 (Mallinckrodt). This is a chromatographic medium composed of approximately 70% silicic acid and 30% glass fiber available in sheets of 8 × 80 inches; it has the advantage over the standard cellulose papers of having a higher resolution, a smaller diffusion, and more reproducible results. ChromAR Sheets were used for the final purification of the [³H]anisomycin obtained by preparative high-voltage paper electrophoresis and for analysis of the final [³H]anisomycin obtained. For analytic chromatograms the ChromAR Sheets were used without prior washing, and radioactivity was estimated in 0.5-cm strips as described above under high-voltage paper electrophoresis. For preparative chromatograms, the ChromAR sheets were washed by descending elution with 1 ml of 10 mM formic acid per square centimeter, which was shown not to affect the chromatographic medium for the subsequent use in chromatography. The preparative sheets cut across (14 × 20 cm) and developed by ascending chromatography, introducing 0.5 cm of the bottom end of the sheet in the solvent, applying sample, and controls at 2 cm (similarly to high voltage paper electrophoresis) and stopping the development after the front has moved 16 cm.

sec-Butanol–petroleum benzine 40 to 60°–2 N ammonia (25:5:4) was the solvent. The development of the chromatography was carried out at room temperature (about 1.5 hours). The chromatogram was then dried, and the [³H]anisomycin was detected by UV light. This made it possible to cut correctly the part of the chromatogram containing [³H]anisomycin which was usually 0.8–1 × 7 cm (Fig. 1C shows the profile of radioactivity). Elution was carried out with 10 mM formic acid, and the eluate (4–5 ml) was immediately lyophilized and dissolved in diluted HCl to give a final solution of [³H]anisomycin at pH 5.5–6.0, which was stored at −20°. For the subsequent purpose of quantitative estimation and purity of the [³H]anisomycin finally obtained, it was necessary to elute a similar piece of a ChromAR Sheet which had been treated exactly like that containing the radioactive antibiotic; this eluate was used as a control in the spectrophotometric spectra of [³H] anisomycin to calculate its concentration and purity.

Specific Activity of [³H]Anisomycin Obtained. To know the specific activity of the [³H]anisomycin preparation, the radioactivity and concentration of the product had to be measured. Radioactivity was measured in the scintillator spectrometer the efficiency of which was calculated with a standard of tritiated water. Concentration of the antibiotic was carry out using a Cary 15 spectrophotometer to obtain the continuous UV spectrum in the range 200–320 nm (Fig. 3A). In this range there are three maxima at 224, 275, and 280 nm which coefficients of molar extinction are 10800, 1800, and 1600, respectively.[9] The specific activity of the [³H]anisomycin obtained was 285 mCi/mmole. The yield was 2.4%.

Procedure for Purification of [³H]Gougerotin

All steps of this procedure were carried out at 0–4° unless otherwise is specifically indicated. P-cellulose column chromatography and chromatography on ChromAR Sheet 500 were carried out sequentially for this purification.

P-Cellulose Column Chromatography. P-cellulose precycling, equilibration, packing, and settling was carried out similarly as described above for CM-cellulose; 7.5 ml of the crude tritiated preparation of gougerotin

---

[9] "The Merck Index," 8th ed. (P. G. Stecher, ed.), p. 86. Merck, Rahway, New Jersey, 1968.

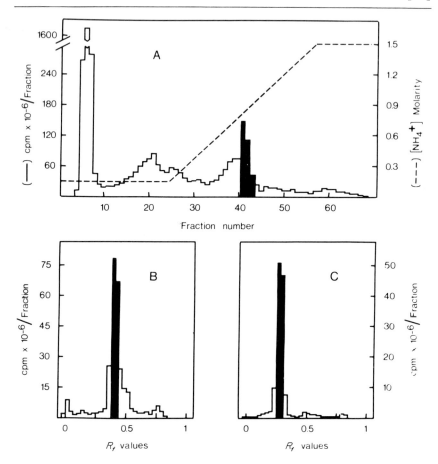

Fig. 2. Purification of [³H]gougerotin. (A) Profile of radioactivity of P-cellulose column chromatography. Elution was carried out with a formic acid–ammonia buffer, pH 2.8, containing a linear gradient of ammonium ion ranging from 0.15 to 1.5 M (----). Flow rate of the elution was 60 ml/hour and 10-ml fractions were collected. Fractions in black (41–43) were taken for the next step of purification. (B) Chromatographic profile in ChromAR Sheet 500 (Mallinckrodt) of the material eluted on the chromatography column. Solvent used was *tert*-butanol–ethyl methyl ketone–water–25% ammonia (4:3:2:1); development of the chromatography was carried out at room temperature until the front had moved 16 cm. [³H]Gougerotin was detected by UV light and eluted with distilled water; R_f values and total radioactivity of the eluate taken for the next step are shown in black. (C) Profile of radioactivity of the final step of purification carried out in ChromAR Sheet 500 similarly to the previous step but using the solvent acetonitrile–0.1 M ammonium acetate (7:3). R_f values and radioactivity of the final eluate are shown in black.

were lyophilized, redissolved in 2 ml of formic acid–ammonia buffer, pH 2.8, 0.15 M in ammonium ion and finally poured on the top of the chromatography column. Then 150 ml of the same buffer was passed through the column to wash the unbound radioactivity (65%). Bound radioactivity was eluted with 300 ml of a linear gradient of the buffer formic acid–ammonia, pH 2.8, with an ammonium ion concentration ranging from 0.15 to 1.5 M. Finally 150 ml of the same buffer containing 2.5 M ammonium was passed through the column to wash any residual bound radioactivity. Flow rate of the elution was 60 ml/hour, and 10-ml fractions were collected. The profile of radioactivity showed a peak at the middle of the gradient (2–3 fractions) preceded by a shoulder (2 fractions) (Fig. 2A). It is possible to take only the main peak, discarding the shoulder of radioactivity or alternatively to analyze the different fractions by high voltage paper electrophoresis and pool fractions with more than 50% radioactivity moving electrophoretically as control gougerotin. In this case Whatman paper No. 1 and formic acid–ammonia buffer pH 2.8, of an ammonia concentration 50 mM were used for the electrophoresis which was carried out for 45 minutes at 70 V/cm. The gougerotin control (0.5 μmole) moved 30 cm toward the cathode. In the analysis of the chromatography fractions, there were peaks of impurities moving 15 and 20 cm. Fractions containing more than 50% radioactivity as gougerotin were pooled, lyophilized, and redissolved in 0.2–0.4 ml of distilled water.

Chromatography on ChromAR Sheet 500. The method followed was similar to that described above for the purification of [³H]anisomycin, but the solvents were different. ChromAR Sheet was washed similarly, but using distilled water. Purification of [³H]gougerotin was carried out in two sequential steps. [³H]Gougerotin was directly detected in the ChromAR Sheet by UV absorption. Solvent I [*tert*-butanol–ethyl methyl ketone–water–25% ammonia (4:3:2:1)] was used for the first step (Fig. 2B), and solvent II [acetonitrile–0.1 M ammonium acetate (7:3)] for the second one (Fig. 2C). Solvent I had to be prepared immediately before use to have reproducible R_f whereas solvent II can be stored for a time. The eluate after the second chromatography was lyophilized, redissolved in water, and stored at $-20°$.

Specific Activity of the [³H]Gougerotin Obtained. Radioactivity and concentration of the product were determined to know the specific activity similarly as in the case of [³H]anisomycin described above. The absorption spectra of gougerotin depends on the pH.[10] An aqueous solution of

[10] J. J. Fox, Y. Kuwada, K. A. Watanabe, T. Ueda, and E. B. Whipple, *Antimicrob. Ag. Chemother.*, p. 518 (1964).

gougerotin shows two maxima at 266 and 234 nm, and their coefficients of molar extinction are, respectively, 9400 and 8900.[11] At an acid pH there is only one maximum at 276 nm with a coefficient of molar extinction of 13,300. The spectra of [³H]gougerotin solution in water and 0.1 N HCl were determined to confirm specific activity and purity of the compound. The specific activity of the [³H]gougerotin preparation was 110 mCi/mmole. The yield of the preparation was 1.6%.

Purity of the [³H]Antibiotic Preparations

The following methods were used in order to know the purity of the [³H]antibiotic obtained: (a) high voltage paper electrophoresis, (b) chromatography, (c) spectroscopy, and (d) biological assays.

Purity of [³H]Anisomycin. The final [³H]anisomycin preparation was 100% pure according to the following criteria: (a) High voltage paper electrophoresis carried out in 50 mM triethanolamine·HCl, pH 8.6, 100 V/cm for 45 minutes as described in the preparation procedure; (b) chromatography in *n*-butanol–water–acetic acid (60:10:3) as described in the preparation procedure; (c) UV spectroscopy in the range 200–320 nm (Fig. 3A). This is the most conclusive method of analysis of all those used. Total coincidence with the spectrum of nonradioactive anisomycin was obtained, implying 100% purity. Very small impurities would be detected by a slight raise of the minimum at 242 and mainly at 208 nm; and (d) the ³H-anisomycin obtained inhibited poly(U)-directed polyphenylalanine synthesis by *Saccharomyces cerevisiae* ribosomes. However the most convincing biological assay to test the purity of the [³H]anisomycin preparation was the study of binding to ribosomes until saturation in different steps. For this assay [³H]anisomycin and *S. cerevisiae* ribosomes were mixed in such a ratio that more than 50% of the antibiotic added was bound (30 mg ribosomes/ml and 10⁻⁶ M [³H]anisomycin); the ribosomes were pelleted by sedimentation in the ultracentrifuge.[3,12] The supernatant containing the unbound antibiotic was used for a new binding assay (30 mg ribosomes/ml); this procedure was repeated until total binding of the [³H]anisomycin. A control was centrifuged similarly in each step, containing the same amount of ribosomes as above; an amount of the [³H]anisomycin preparation containing the same radioactivity as the supernatant was used for a similar step in the saturation experiment. Identity of the percentages of binding in the controls and

---

[11] H. Umezawa, *in* "Index of Antibiotics from Actinomycetes," p. 313. Univ. of Tokyo Press, Japan, 1967.

[12] R. Fernandez-Muñoz, R. E. Monro, and D. Vazquez, this series, Vol. 20 [51].

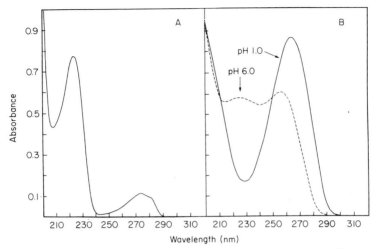

Fig. 3. UV spectra (200–320 nm) of the [³H]anisomycin and [³H]gougerotin preparations. (A) Spectrum of [³H]anisomycin, showing the three maxima at 224, 275, and 280 nm. Totally identical spectrum was found when nonradioactive anisomycin was studied. (B) Spectra of [³H]gougerotin in aqueous solution (pH 6.0) and in 0.1 N HCl (pH 1.0). Identical spectra were found for nonradioactive gougerotin under similar conditions. As a control to study the spectra of radioactive antibiotics, a ChromAR Sheet eluate was used as indicated in the text.

the different steps on saturation of binding was obtained, demonstrating a 100% purity of the preparations since any possible impurities presumably would not bind to ribosomes or would bind with different affinity than [³H]anisomycin.

Purity of [³H]Gougerotin. The final [³H]gougerotin preparation was 100% pure according to the following criteria: (a) high voltage paper electrophoresis in 50 mM formic acid–ammonia, pH 3.7, at 70 V/cm for 50 minutes. (b) Chromatography in six different solvents (solvents I and II as used in the purification procedure; solvent III, 25% ammonia–*n* butanol–*n* propanol–ethanol–water (9:8:8:2:2); solvent IV, *n*-butanol–formic acid–*n*-propanol–acetone–water–trichloroacetic acid (10:5:5:4:4:2); solvent V, methanol–isopropanol–25% ammonia–water (9:6:3:2); solvent VI, water–*n* propanol–25% ammonia (9:10:1). (c) UV spectroscopy in the range 200–320 nm (Fig. 3B) (100% purity of [³H] gougerotin was obtained; small impurities would be detected mainly by a raise of the minimum at 228 nm in an aqueous solution, pH 6.0, and also at the minimum at 240 nm in 0.1 N HCl). (d) Saturation binding of [³H]gougerotin to *S. cerevisiae* ribosomes similarly to that described above for [³H]anisomycin (25 mg ribosomes/ml and 0.75 μM [³H] gougerotin were used for these experiments).

Binding of [³H]Anisomycin and [³H]Gougerotin
    to Eukaryotic Ribosomes

We have studied binding of [³H]anisomycin and [³H]gougerotin to *Saccharomyces cerevisiae* ribosomes according to the sedimentation method previously described in binding studies of other radioactive antibiotics to bacterial ribosomes.[3,12] Binding of [³H]anisomycin was studied at an antibiotic concentration ranging from 8×10^{-8} to 10^{-5} M and at a ribosome concentration of 5 mg/ml (at antibiotic concentration below 10^{-6} M) or 10 mg/ml (at 10^{-6} or higher antibiotic concentration). A 50 mM ammonium maleate buffer, pH 6.5, was used containing 5 mM Mg^{2+} and 5 mM 2-mercaptoethanol. Binding of [³H]gougerotin was also studied in these ionic conditions, but using a ribosome concentration of 9, 15, and 21 mg/ml for an antibiotic concentration ranging from 10^{-7} to 10^{-6} M, 10^{-6} to 10^{-5} M, and 10^{-5} to 7×10^{-5} M, respectively. Scatchard plots of

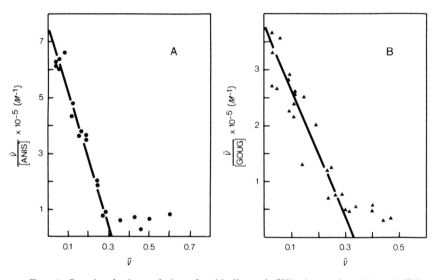

Fig. 4. Scatchard plots of data for binding of [³H]anisomycin (A) and [³H]-gougerotin (B) to *Saccharomyces cerevisiae* ribosomes. Binding was studied by the sedimentation method of ribosomes in the ultracentrifuge [R. Fernandez-Muñoz, R. E. Monro, R. Torres-Pinedo, and D. Vazquez, *Eur. J. Biochem.* **23**, 185 (1971); R. Fernandez-Muñoz, R. E. Monro, and D. Vazquez, this series, Vol. 20 [51]]. A 50 mM ammonium maleate buffer, pH 6.5, was used containing 5 mM Mg^{2+} and 5 mM 2-mercaptoethanol. Binding of [³H]anisomycin was studied at an antibiotic concentration ranging from 8×10^{-8} to 10^{-5} M and at a ribosome concentration of 5 mg/ml (at antibiotic concentration below 10^{-6} M) or 10 mg/ml (at 10^{-6} M or higher antibiotic concentration). Binding of [³H]gougerotin was studied at ribosome concentration of 9, 15, and 21 mg/ml for an antibiotic concentration ranging from 10^{-7} to 10^{-6} M, 10^{-6} to 10^{-5} M, and 10^{-5} M to 7×10^{-5} M, respectively.

the results obtained are shown in Fig. 4. From these results the association $(2.3 \times 10^6 \ M^{-1})$ and dissociation $(4.25 \times 10^{-7} \ M)$ constants were obtained for [³H]anisomycin and the association $(1.1 \times 10^6 \ M^{-1})$ and dissociation $(9.11 \times 10^{-7} \ M)$ constants were obtained for [³H]gougerotin binding to *Saccharomyces cerevisiae* ribosomes.

[42] Ribonuclease Sensitivity of Aminoacyl-tRNA: an Assay for Codon Recognition and Interaction of Aminoacyl-tRNA with 50 S Subunits

By SIDNEY PESTKA

Binding of aminoacyl-tRNA to ribosomes was shown to be an early step in protein biosynthesis.[1-3] In these early studies, ultracentrifugation was used to sediment the tRNA·ribosome·template complexes. For greater sensitivity and speed than previously reported, Nirenberg and Leder,[4] developed a convenient assay for studying these complexes. Their assay involved binding of ribosomes to a nitrocellulose membrane filter. Any tRNA or template bound to the ribosomes was consequently also bound to the filter. Both nonspecific binding in the absence of a template and a specific stimulation of binding aminoacyl-tRNA to ribosomes in the presence of template codons were observed. Triribonucleotide diphosphates as well as larger polyribonucleotides turned out to be effective templates. With all 64 possible triplet codons, their assay permitted determination of the codon assignments for each amino acid.[5-7] Furthermore, the assay has been used effectively in the study of numerous other aspects of protein biosynthesis

[1] R. Arlinghaus, G. Favelukes, and R. Schweet, *Biochem. Biophys. Res. Commun.* **11**, 92 (1963).

[2] T. Nakamoto, T. W. Conway, J. E. Allende, G. J. Spyrides, and F. Lipmann, *Cold Spring Harbor Symp. Quant. Biol.* **28**, 227 (1963).

[3] A. Kaji and H. Kaji, *Biochem. Biophys. Res. Commun.* **13**, 186 (1963).

[4] M. Nirenberg and P. Leder, *Science* **145**, 1399 (1964).

[5] M. Nirenberg, T. Caskey, R. Marshall, R. Brimacombe, D. Kellogg, B. Doctor, D. Hatfield, J. Levin, F. Rottman, S. Pestka, M. Wilcox, and F. Anderson, *Cold Spring Harbor Symp. Quant. Biol.* **31**, 11 (1966).

[6] J. H. Matthaei, H. P. Voigt, G. Heller, R. Neth, G. Schoech, H. Kübler, F. Amelunxen, G. Sander, and A. Parmeggiani, *Cold Spring Harbor Symp. Quant. Biol.* **31**, 25 (1966).

[7] H. G. Khorana, H. Büchi, H. Ghosh, N. Gupta, T. M. Jacob, H. Kössel, R. Morgan, S. A. Narang, E. Ohtsuka, and R. D. Wells, *Cold Spring Harbor Symp. Quant. Biol.* **31**, 39 (1966).

and codon recognition. For example, codon recognition and aminoacyl-tRNA binding to 30 S subunits could also be studied by their assay.[8-10]

Aminoacyl-tRNA bound to 70 S ribosomes in the presence of a template is resistant to digestion by pancreatic ribonuclease. On the basis of this observation, a convenient sensitive assay for codon recognition and aminoacyl-tRNA binding to ribosomes was devised.[11] The protection occurred as a result of the association of aminoacyl-tRNA with the 50 S subribosomal particle; thus, the assay provided a sensitive functional test involving the specific interaction of tRNA with the 50 S subunit of *Escherichia coli* ribosomes. Aminoacyl-tRNA attached to 30 S subunits was sensitive to ribonuclease digestion. The kinetics of attachment of 50 S subunits to 30 S ribosome·polyribonucleotide template·aminoacyl-tRNA complexes could be studied by the assay. Furthermore, the assay provides an additional parameter for evaluating the site of action of antibiotics and other substances interfering with protein synthesis as well as for determining the functional integrity of aminoacyl-tRNA binding to ribosomes.

Determination of Aminoacyl-tRNA Binding to Ribosomes by Ribonuclease Sensitivity

Reagents

5 × Standard Buffer: 0.25 M TrisAc, pH 7.2, 0.25 M KAc, 0.10 M MgAc$_2$

NH$_4$Cl-washed *E. coli* ribosomes,[12] stored in 1 M NH$_4$Cl, 10 mM MgCl$_2$, and 10 mM Tris·HCl, pH 7.2, at a concentration of 700–2000 A_{260} units/ml. For these assays, ribosomes were diluted in this same buffer to a concentration of 580 A_{260} units/ml.

Poly(U): 48 A_{260} units/ml. Poly(A), poly(UA), poly(UG), poly(UC), and poly(AC) were each dissolved in water to a concentration of 45 A_{260} units/ml.

UpUpU, ApApA, dissolved in water at concentrations of 50–100 A_{260} units/ml

[^{14}C]Aminoacyl-tRNA: about 500–1000 pmoles of [^{14}C]Phe-tRNA (20–40 A_{260} units) per milliliter. The [^{14}C]Phe-tRNA was prepared as reported previously.[13] Other aminoacyl-tRNA preparations were similarly made.

[8] H. Matthaei, F. Amelunxen, K. Eckert, and G. Heller, *Ber. Bunsenges. Phys. Chem.* **68**, 735 (1964).

[9] I. Suzuka, H. Kaji, and A. Kaji, *Biochem. Biophys. Res. Commun.* **21**, 187 (1965).

[10] S. Pestka and M. Nirenberg, *J. Mol. Biol.* **21**, 145 (1966).

[11] S. Pestka, *J. Biol. Chem.* **243**, 4038 (1968).

[12] S. Pestka, *J. Biol. Chem.* **243**, 2810 (1968).

[13] S. Pestka, *J. Biol. Chem.* **241**, 367 (1966).

Pancreatic ribonuclease: 2.6 μg/ml dissolved in 1 × Standard Buffer (50 mM TrisAc, pH 7.2, 50 mM KAc, and 20 mM MgAc$_2$).

For each reaction mixture in a volume of 0.05 ml were added 0.010 ml of 5 × Standard Buffer, 0.005 ml of NH$_4$Cl-washed *E. coli* ribosomes, 0.005 ml of poly(U), 0.005 or 0.010 ml of [^{14}C]aminoacyl-tRNA, and water to bring the volume to 0.050 ml. Each 0.050-ml reaction mixture contained the following components: 50 mM TrisAc, pH 7.2; 50 mM potassium acetate; 20 mM magnesium acetate; 0.10 M NH$_4$Cl (from the 1.0 M NH$_4$Cl in the ribosome solution); 2.9 A_{260} units of NH$_4$Cl-washed ribosomes; about 5 pmoles of [^{14}C]Phe-tRNA (0.2 A_{260} unit); and 0.24 A_{260} unit of poly(U). Reactions were incubated at 24° for 10 minutes.

To determine the amount of aminacyl-tRNA bound to ribosomes, at exactly 10 minutes of incubation at 24°, 0.013 μg of pancreatic ribonuclease in a volume of 0.005 ml of 1 × Standard Buffer was added to each tube. Incubation at 24° was continued for an additional 5 minutes. Reactions were stopped by adding 2 ml of cold 10% (w/v) trichloroacetic acid to each tube to precipitate [^{14}C]aminoacyl-tRNA. The contents of each tube were then filtered onto a nitrocellulose or other appropriate filter for retaining the precipitate. The tube and filter were then washed three times with 3-ml portions of 5% trichloroacetic acid. The filters were then dried under an infrared lamp and counted in a scintillation spectrometer as previously described.[14] For brevity, this assay will be cited as the "RNase-TCA" assay, for it depends on protection of [^{14}C]Phe-tRNA appropriately bound to ribosomes from digestion by RNase. Other filters such as glass fiber filters or filter paper disks could be used with appropriate precipitation techniques to detect the undigested [^{14}C]Phe-tRNA. For comparison, the assay of Nirenberg and Leder[4] was utilized; the reaction mixtures were identical with those described above except that 0.005 ml of 1 × Standard Buffer without pancreatic RNase was added to each tube. For brevity, this assay will be called the "nitrocellulose filter" assay because it depends on trapping ribosomes and any compounds bound to them on nitrocellulose filters.

Characteristics of the Assay

The degradation of free and ribosome-bound [^{14}C]Phe-tRNA by pancreatic ribonuclease is shown in Fig. 1. [^{14}C]Phe-tRNA in the presence or absence of ribosomes and poly(U) was incubated for 10 minutes at 24°. Various amounts of pancreatic ribonuclease were added to the tubes and incubation was continued for an additional 5 minutes before the samples were assayed for cold trichloroacetic acid-precipitable radioactivity.

[14] S. Pestka, *J. Biol. Chem.* **242**, 4939 (1967).

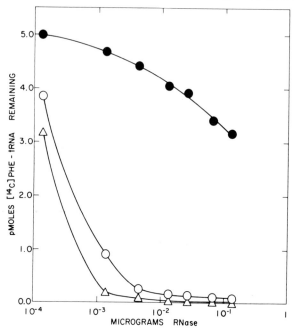

FIG. 1. The hydrolysis of [¹⁴C]Phe-tRNA by pancreatic RNase as a function of RNase concentration in the presence and in the absence of ribosomes and poly(U). Each 0.050-ml reaction mixture contained the components indicated in the text in addition to the following: 5.5 pmoles of [¹⁴C]Phe-tRNA (0.18 A_{260} unit); 2.3 A_{260} units of ribosomes where indicated; and 0.24 A_{260} unit of poly(U) where specified. The NH₄Cl-washed ribosomes were additionally washed and resuspended in 10 mM magnesium acetate and 10 mM TrisAc, pH 7.2, to remove NH₄Cl. The micrograms of RNase added (in a volume of 0.005 ml) in stage II of the incubation is given on the abscissa. ●, plus ribosomes and poly(U); ○, plus ribosomes, minus poly(U); △, minus ribosomes and poly(U).

As indicated in Fig. 1 in the presence of ribosomes and poly(U), [¹⁴C]Phe-tRNA is substantially protected from hydrolysis. At ribonuclease concentrations less than 0.004 μg per reaction mixture, the difference between the hydrolysis of free and bound [¹⁴C]Phe-tRNA diminished. Therefore, for most experiments, 0.013 μg of RNase was used to be sure that unbound [¹⁴C]Phe-tRNA was essentially completely digested. The addition of poly(U) in the absence of ribosomes produces no protection from digestion by RNase (data not shown). Under the specific conditions to be employed it is useful for each laboratory to determine a set of curves as illustrated in Fig. 1 in order to choose appropriate concentrations of RNase.

With the use of the RNase-TCA assay, the binding of [¹⁴C]Phe-tRNA

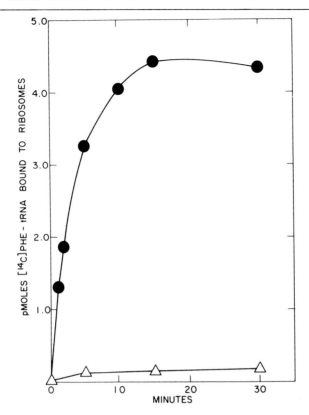

FIG. 2. Binding of [¹⁴C]Phe-tRNA to ribosomes as a function of time. Each 0.050-ml reaction mixture contained the components described in the text in addition to the following: 5.7 pmoles of [¹⁴C]Phe-tRNA (0.18 A_{260} unit); 2.3 A_{260} units of ribosomes; and 0.24 A_{260} unit of poly(U) where specified. [¹⁴C]Phe-tRNA bound to ribosomes was determined by resistance to hydrolysis by RNase as described in the text. The initial point (0 minute) was obtained by carrying out the incubation in the absence of ribosomes. For this experiment, the NH₄Cl-washed ribosomes were both washed and resuspended in 10 mM magnesium acetate and 10 mM TrisAc, pH 7.2, to remove NH₄Cl. ●, plus poly(U); △, minus poly(U).

to ribosomes as a function of time was examined (Fig. 2). The data indicate that the time course was similar to that studied with the nitrocellulose filter assay.[4,13] Ribonuclease T1 or micrococcal nuclease were used in some experiments, which suggested that they might be used instead of pancreatic RNase.

The binding of [¹⁴C]Phe-tRNA to ribosomes in response to poly(U) was evaluated by both the nitrocellulose filter and the RNase-TCA assays. As can be seen from the data of Table I, the binding of [¹⁴C]Phe-tRNA to ribosomes in response to poly(U) (△ pmoles) was identical for both

TABLE I
BINDING OF [¹⁴C]PHE-tRNA TO RIBOSOMES IN RESPONSE TO POLY(U):
COMPARISON OF ASSAYS FOR CODON RECOGNITION[a]

| Template | Ribosomes | [¹⁴C]Phe-tRNA | | | |
| | | Bound to filter | | Protected from RNase | |
		pMoles	ΔpMoles	pMoles	ΔpMoles
None	− Ribosomes	0.02	—	0.02	—
None	+ Ribosomes	0.73	—	0.18	—
Poly(U)	− Ribosomes	0.01	—	0.03	—
Poly(U)	+ Ribosomes	5.03	4.30	4.47	4.29

[a] Each 0.050-ml reaction mixture contained the components and was incubated as described in the text in addition to the following: 5.3 pmoles of [¹⁴C]Phe-tRNA (0.18 A_{260} unit); 0.24 A_{260} unit of poly(U) where designated. The Δpmoles represent the value obtained in the presence of ribosomes and template minus the value obtained with ribosomes alone.

assays. However, the binding of [¹⁴C]Phe-tRNA to ribosomes in the absence of template with the use of the RNase-TCA assay was one-fourth that of the nitrocellulose filter assay. A comparison of the assays as a function of ribosome concentration was made (Fig. 3, A and B). At most ribosome concentrations studied, the nitrocellulose filter assay appeared to yield slightly higher values (Fig. 3B). However, at the very highest ribosome concentrations studied, the difference between the two assays was small. The Phe-tRNA bound to ribosomes nonspecifically in the absence of poly(U) as a function of ribosome concentration increased more for the nitrocellulose filter assay than for the RNase-TCA assay.

The comparison of [¹⁴C]Phe-tRNA binding to ribosomes in response to UpUpU is given by the data of Table II. Since a higher Mg^{2+} concentration was used here (30 mM Mg^{2+}) than that used in the experiment summarized in Table I (20 mM Mg^{2+}) the nonspecific binding in the absence of template was higher than that in Table I. Although this nonspecific binding at 30 mM Mg^{2+} was four times that at 20 mM Mg^{2+} with the use of the RNase-TCA assay, it was still about half that of the nitrocellulose filter assay. The net stimulation of Phe-tRNA binding to ribosomes by UpUpU was possibly slightly greater with the RNase-TCA assay than with the other.

The binding of [¹⁴C]Lys-tRNA to ribosomes was studied in the presence and in the absence of template as indicated in Table III. The enhancement of Lys-tRNA binding in response to poly(A) or ApApA was greater

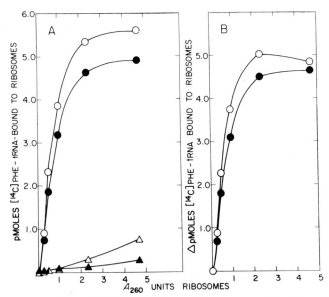

FIG. 3. (A) Binding of [¹⁴C]Phe-tRNA to ribosomes as a function of ribosome concentration. Reactions were performed as described in the text in addition to the following: 5.6 pmoles of [¹⁴C]Phe-tRNA (0.18 A_{260} unit); 0.24 A_{260} unit of poly(U) where noted; ribosome concentration as indicated. Ribosomes were prepared as described in the legend to Fig. 2. Incubations were performed for 15 minutes at 24°; at 15 minutes 1 × Standard Buffer or buffered RNase was added to appropriate tubes, and incubations were continued an additional 5 minutes at 24° prior to assay as described in the text. ○, △, Binding of [¹⁴C]Phe-tRNA to ribosomes in the presence (○) or absence (△) of poly(U) assayed by the nitrocellulose filter technique; ●, ▲, binding of [¹⁴C]Phe-tRNA to ribosomes in the presence (●) or absence (▲) of poly(U) determined by the RNase-TCA assay.

(B) The details of the experiment are given in the legend to Fig. 3A. The Δpmoles represent the stimulation of [¹⁴C]Phe-tRNA binding to ribosomes on addition of poly(U), namely, the difference between the appropriate curves of (A). ○, Δpmoles of [¹⁴C]Phe-tRNA bound to ribosomes assayed by the nitrocellulose filter technique; ●, Δpmoles [¹⁴C]Phe-tRNA bound to ribosomes determined by the RNase–TCA assay.

with the use of the RNase-TCA assay than with the nitrocellulose filter binding technique. Similar to the results with Phe-tRNA binding to ribosomes, the nonspecific binding of Lys-tRNA to ribosomes was less with the former than with the latter assay.

The specificity of codon recognition was evaluated with the use of the RNase-TCA assay. The binding of seven [¹⁴C]aminoacyl-tRNA preparations in response to six different polyribonucleotides was examined (Table

TABLE II

BINDING OF [^{14}C]PHE-tRNA TO RIBOSOMES IN RESPONSE TO UpUpU:
COMPARISON OF ASSAYS FOR CODON RECOGNITION[a]

Template	Ribosomes	[^{14}C]Phe-tRNA			
		Bound to filter		Protected from RNase	
		pMoles	ΔpMoles	pMoles	ΔpMoles
None	− Ribosomes	0.00	—	0.06	—
None	+ Ribosomes	1.63	—	0.70	—
UpUpU	− Ribosomes	0.01	—	0.07	—
UpUpU	+ Ribosomes	3.44	1.81	2.90	2.20

[a] Components and incubations of reaction mixtures are described in the legend to Table I except for the following: 30 mM magnesium acetate; 0.93 A_{260} unit of UpUpU where designated.

IV). As can be seen, [^{14}C]Phe-tRNA binds to ribosomes in response to all the uridylic acid containing polyribonucleotides; [^{14}C]Lys-tRNA in response poly(A) and poly(AC); [^{14}C]Val-tRNA, in response to poly(UG) only; [^{14}C]Ser-tRNA in response to poly(UC) and slightly to poly(U); [^{14}C]Thr-tRNA to poly(AC) chiefly and, perhaps, slightly to poly(UC); [^{14}C]Ile-tRNA, to poly(U), poly(UC), and poly(UA); and [^{14}C]Leu-tRNA, to poly(UC) slightly. These results are in agreement with the studies performed with the polyribonucleotides by the nitrocellulose filter

TABLE III

BINDING OF [^{14}C]LYS-tRNA TO RIBOSOMES: COMPARISON OF ASSAYS[a]

Template	Ribosomes	[^{14}C]Lys-tRNA			
		Bound to filter		Protected from RNase	
		pMoles	ΔpMoles	pMoles	ΔpMoles
None	− Ribosomes	0.03	—	0.09	—
None	+ Ribosomes	1.76	—	1.39	—
Poly(A)	− Ribosomes	0.07	—	0.11	—
Poly(A)	+ Ribosomes	4.06	2.30	3.85	2.46
ApApA	+ Ribosomes	3.32	1.56	3.40	2.01

[a] Each 0.050-ml reaction mixture contained the components and was incubated as described in the text in addition to the following: 0.22 and 0.13 A_{260} unit of poly(A) and ApApA, respectively, where designated; 5.1 pmoles of [^{14}C]Lys-tRNA (0.17 A_{260} unit). The Δpmoles represent the value obtained in the presence of ribosomes and template minus the value obtained with ribosomes alone.

TABLE IV

SPECIFICITY OF [¹⁴C]AMINOACYL-tRNA BINDING TO RIBOSOMES AS DETERMINED
BY RIBONUCLEASE RESISTANCE OF BOUND [¹⁴C]AMINOACYL-tRNA[a]

| | ΔpMoles [¹⁴C]Aminoacyl-tRNA Bound to Ribosomes Due to Addition of Polynucleotide | | | | | | |
[¹⁴C]Aminoacyl-tRNA	Poly (U)	Poly (A)	Poly (UA)	Poly (UG)	Poly (UC)	Poly (AC)	Minus polynu-cleotide (pMoles)
[¹⁴C]Phe-tRNA	**4.24**	0.02	**0.35**	**1.37**	**3.30**	−0.02	0.15
[¹⁴C]Lys-tRNA	−0.54	**2.56**	−0.06	−0.52	−0.58	**1.18**	1.58
[¹⁴C]Val-tRNA	0.00	0.00	0.02	**2.66**	−0.01	−0.04	0.09
[¹⁴C]Ser-tRNA	**0.36**	−0.10	−0.01	−0.18	**3.25**	0.12	0.69
[¹⁴C]Thr-tRNA	−0.04	−0.03	−0.03	0.01	**0.21**	**1.85**	0.19
[¹⁴C]Ile-tRNA	**1.21**	0.02	**0.22**	**0.19**	**0.67**	0.00	0.10
[¹⁴C]Leu-tRNA	−0.05	−0.12	−0.02	0.13	**0.52**	−0.27	1.27

[a] Each 0.050-ml reaction mixture contained the components and was incubated as described in the text in addition to the following: 5.3 pmoles of [¹⁴C]Phe-tRNA (0.18 A_{260} unit), 5.9 pmoles of [¹⁴C]Lys-tRNA (0.19 A_{260} unit), 8.0 pmoles of [¹⁴C]Val-tRNA (0.07 A_{260} unit), 14.6 pmoles of [¹⁴C]Ser-tRNA (0.36 A_{260} unit), 4.8 pmoles of [¹⁴C]Thr-tRNA (0.16 A_{260} unit), 8.6 pmoles of [¹⁴C]Ile-tRNA (0.14 A_{260} unit), and 15.9 pmoles of [¹⁴C]Leu-tRNA (0.23 A_{260} unit); 0.24 A_{260} unit of poly(U), 0.26 A_{260} unit of poly(A), 0.22 A_{260} unit of poly(UA) (1/1), 0.27 A_{260} unit of poly(UG) (1/1), 0.23 A_{260} unit of poly(UC) (1/1), and 0.25 A_{260} unit of poly(AC) (4/6). The background binding of [¹⁴C]aminoacyl-tRNA to ribosomes in the absence of polynucleotides is expressed in pmoles in the column designated "Minus polynucleotide." All other values (Δpmoles) were obtained by subtracting background binding of [¹⁴C]aminoacyl-tRNA from the binding obtained upon the addition of a polynucleotide preparation. Negative values indicate that the addition of polynucleotide decreased the binding of [¹⁴C]aminoacyl-tRNA to ribosomes from the background value. Reproducible stimulations of [¹⁴C]aminoacyl-tRNA due to the addition of polynucleotide are in boldface type.

assay.[15-17] The large extent of [¹⁴C]Ile-tRNA binding in response to poly(U) indicates substantial ambiguous codon recognition under these conditions.[13,16] The slight binding of [¹⁴C]Thr-tRNA to ribosomes in response to poly(UC) has been observed previously as well as a slight binding in response to polycytidylic acid and the copolymer of guanylic and cytidylic acids.[17]

[15] P. Leder and M. Nirenberg, *Proc. Nat. Acad. Sci. U.S.* **52**, 1521 (1964).
[16] S. Pestka, R. Marshall, and M. Nirenberg, *Proc. Nat. Acad. Sci. U.S.* **53**, 639 (1965).
[17] R. Marshall, S. Pestka, J. Trupin, C. O'Neal, and M. Nirenberg, unpublished studies.

TABLE V

REQUIREMENTS FOR PROTECTION OF [^{14}C]PHE-tRNA FROM HYDROLYSIS[a]

Deletions	pMoles [^{14}C]Phe-tRNA protected from hydrolysis by RNase
None	4.05
− Mg^{2+}	0.06
− K$^+$	3.40
− Poly(U)	0.13
− Ribosomes	0.05

[a] Reactions for following binding of [^{14}C]Phe-tRNA to ribosomes by the RNase-TCA assay were performed as described in the text in addition to the following: 4.9 pmoles of [^{14}C]Phe-tRNA (0.18 A_{260} unit); 2.3 A_{260} units of ribosomes where indicated; and 0.24 A_{260} unit of poly(U) where designated. The omission of individual components is indicated in the table. The NH$_4$Cl-washed ribosomes were washed and resuspended in 10 mM magnesium acetate and 10 mM TrisAc, pH 7.2, to remove monovalent cations.

Requirements for Protection of [^{14}C]Phe-tRNA from Hydrolysis by RNase

As can be seen from the data of Table V, maximum protection of [^{14}C]Phe-tRNA from hydrolysis required the presence of ribosomes, magnesium, and a template. Potassium was not required although it stimulated the binding. As indicated by the data of Fig. 4, the binding of [^{14}C]Phe-tRNA was a function of the magnesium concentration. Maximal binding of [^{14}C]Phe-tRNA to ribosomes occurred at slightly greater than 20 mM Mg^{2+}. All these results are very similar to those obtained with the nitrocellulose filter assay.[4,13]

Other enzymes can be used instead of ribonuclease in analogous assays. Reversal of the acylation reaction can be used. It was found that [^{14}C]Phe-tRNA is protected from enzymatic deacylation when bound to ribosomes.[11] In addition, protection of acetyl-[^{14}C]Phe-tRNA and [^{14}C]oligolysyl-tRNA from RNase digestion was observed with poly(U) and poly(A), respectively.[11] More recently, Vogel et al.[18] and Brun et al.[19] have shown that N-acylaminoacyl-tRNA bound to ribosomes is protected against cleavage by peptidyl-tRNA hydrolyase.

Protection of [^{14}C]Phe-tRNA from Hydrolysis with Ribosomal Subunits

With purified 30 and 50 S subunits, the protection of [^{14}C]Phe-tRNA from hydrolysis by RNase and binding of [^{14}C]Phe-tRNA to ribosomes

[18] Z. Vogel, T. Vogel, A. Zamir, and D. Elson, Eur. J. Biochem. 21, 582 (1971).
[19] G. Brun, D. Paulin, P. Yot, and F. Chapeville, Biochimie 53, 225 (1971).

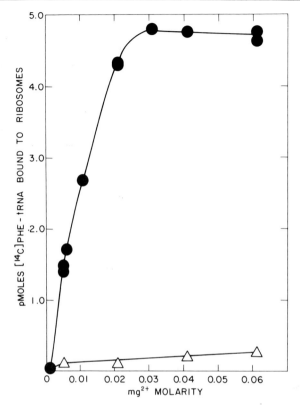

FIG. 4. Binding of [¹⁴C]Phe-tRNA to ribosomes as a function of magnesium concentration. Reactions were carried out as described in the legend to Fig. 2 with magnesium concentration as given. ●, plus ribosomes and poly(U); △, plus ribosomes, minus poly(U).

trapped by nitrocellulose filters were compared. As can be seen from the data of Table VI, although binding of [¹⁴C]Phe-tRNA to 30 S subunits in response to poly(U) was substantial, the [¹⁴C]Phe-tRNA was not protected from RNase digestion. Although little [¹⁴C]Phe-tRNA was bound to purified 50 S subunits, up to half of that bound may have been protected from digestion. The presence of both subunits was required for substantial protection of [¹⁴C]Phe-tRNA from RNase digestion.

To determine whether the [¹⁴C]Phe-tRNA bound to 30 S subunits could be converted to a state protected from RNase digestion, sensitivity of [¹⁴C]Phe-tRNA to digestion by RNase was followed as a function of time after the addition of 50 S subunits (Fig. 5). [¹⁴C]Phe-tRNA was incubated with 30 S subunits for 20 minutes. At 20 minutes an equimolar amount of 50 S subunits was added to some of the reactions, and incubation

TABLE VI

BINDING OF [¹⁴C]PHE-tRNA TO RIBOSOMAL SUBUNITS[a]

Subunits	pMoles [¹⁴C]Phe-tRNA bound to ribosomes	
	Nitrocellulose filter assay	RNase assay
30 S	2.52	0.12
50 S	0.22	0.13
30 S + 50 S	3.73	2.99

[a] Reactions were performed as described in the text except for the following: 72 pmoles of 30 S subunits (0.83 A_{260} unit) and 50 S subunits (1.96 A_{260} units) where indicated; 0.24 A_{260} unit of poly(U) in each reaction; and 4.9 pmoles of [¹⁴C]Phe-tRNA (0.18 A_{260} unit). The 30 S ribosome preparation contained 1.6% 50 S subunits; the 50 S subunit preparation was contaminated with less than 2% 30 S subunits.

was continued. Both the nitrocellulose filter assay (Fig. 5A) and the RNase–TCA assay (Fig. 5B) were used. As can be seen from Fig. 5A, in the presence of poly(U), [¹⁴C]Phe-tRNA was bound to 30 S particles in the presence or absence of 50 S subunits. However, only in the presence

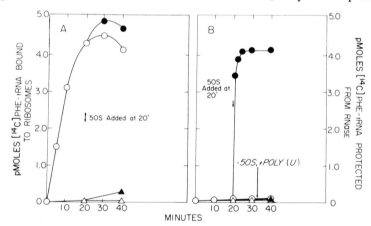

FIG. 5. Effect of 50 S subunits on protection and binding of [¹⁴C]Phe-tRNA. The procedures were performed as indicated in the legend to Fig. 3 with the following exceptions: 0.80 A_{260} unit of 30 S subunits (70 pmoles) were present in every reaction mixture; 1.9 A_{260} units of 50 S subunits (70 pmoles in a volume of 0.004 ml) were added at 20 minutes where indicated. The 30 S and 50 S ribosome preparations were contaminated with less than 1% 50 S and 5% 30 S particles, respectively. (A) Binding of [¹⁴C]Phe-tRNA to ribosomes was determined by the nitrocellulose filter assay. ●, plus 50 S subunits, plus poly(U); ○, no 50 S subunits, plus poly(U); ▲, plus 50 S subunits, no poly(U); △, no 50 S subunits nor poly(U). (B) [¹⁴C]Phe-tRNA protected from digestion by RNase was determined by the RNase-TCA assay. ●, plus 50 S subunits and poly(U); ○, no 50 S subunits, plus poly(U); ▲, plus 50 S subunits, no poly(U); △, no 50 S subunits nor poly(U).

of 50 S subunits was [¹⁴C]Phe-tRNA protected from RNase digestion (Fig. 5B). The development of the protected state was very rapid after the addition of 50 S particles. Greater than 80 and 90% of maximum protection was attained at 1 and 3 minutes after the addition of 50 S subunits, respectively. At 5 minutes, maximum protection was reached.

The conversion of [¹⁴C]Phe-tRNA from an RNase-sensitive to a RNase-resistant state on the addition of 50 S subunits to the [¹⁴C]Phe-tRNA · 30 S · poly(U) complex was used as an assay for the rate of attachment of 50 S subunits to the complex. By study of the rate of attachment at several temperatures an Arrhenius plot of the results yielded an activation energy of about 6 kcal/mole for the attachment of 50 S subunits to the complex. The rapid development of protection on addition of 50 S subunits to the [¹⁴C]Phe-tRNA · 30 S · poly(U) complexes indicates that the association of 50 S particles with the complex is rapid. The RNase-TCA assay can be used to study specifically the kinetics of association of 50 S particles with this complex. For more accurate kinetic studies, however, a higher concentration of RNase should be chosen with a shorter incubation time after the addition of RNase.

To examine the state of the [¹⁴C]Phe-tRNA bound to the ribosome · poly(U) complex after RNase exposure, the product was obtained and its size estimated by gel filtration through Sephadex G-75.[11] The coincidence of the tRNA absorbance and the [¹⁴C]Phe-tRNA product profile indicated that almost all the [¹⁴C]Phe-tRNA which was bound, and thus resistant to hydrolysis by RNase, remained substantially intact. Furthermore, for establishment of functional integrity of the Phe-tRNA product, it was shown that the [¹⁴C]phenylalanine could be incorporated into polyphenylalanine.[11] The results thus indicate that the [¹⁴C]Phe-tRNA which was bound to ribosomes and resistant to digestion by RNase remained intact. This phenomenon might be utilized as a relatively rapid method to obtain purified tRNA species in small amounts. Suggestions for obtaining purified tRNA species based on the specific binding of tRNA to ribosomes have been made by others.[3,4]

Concluding Remarks

The protection of aminoacyl-tRNA from ribonuclease digestion provides a useful assay for studying the binding of aminoacyl-tRNA to ribosomes. This assay is comparable in sensitivity to the nitrocellulose filter assay[4] for studying aminoacyl-tRNA binding to ribosomes stimulated by oligo- or polynucleotides. With the use of the RNase-TCA assay, the background binding of aminoacyl-tRNA to ribosomes in the absence of template is less than the background binding obtained with the nitrocellulose filter assay. This indicates that a substantial portion of the aminoacyl-tRNA contributing to the background binding is not protected from

RNase digestion and therefore is probably not in close association with the 50 S subunit. The specificity of the RNase–TCA assay was similar to that of the nitrocellulose filter assay. The high background values obtained with [14C]Lys- and [14C]Leu-tRNA were similar to those obtained with the nitrocellulose filter assay. Also, the negative values obtained for the stimulation of some aminoacyl-tRNA's on addition of polynucleotide which did not serve as templates for their binding to ribosomes were similar for both assays.

The assay described provides a method for study of the interaction of 50 S subunits with aminoacyl-tRNA and the association of 50 S subunits with aminoacyl-tRNA · 30 S · template complexes. Erdmann et al.[20] have employed the RNase–TCA assay to evaluate the functional binding of Phe-tRNA to ribosomes containing 50 S subunits lacking 5 S RNA. Also, the RNase–TCA assay provides an assay for those compounds which may prevent proper interaction of aminoacyl-tRNA with 50 S particles.

[20] V. A. Erdmann, S. Fahnestock, K. Higo, and M. Nomura, *Proc. Nat. Acad. Sci. U.S.* **68**, 2932 (1971).

[43] "Nonenzymatic" Translation

By L. P. GAVRILOVA and A. S. SPIRIN

The term "nonenzymatic" translation denotes that the reading out of the template and synthesis of peptide on the ribosome takes place without the participation of the transfer protein factors (T and G) and guanosine 5'-triphosphate (GTP). This designation is contrasted with the definition "enzymatic" translation signifying that the reading of the template and peptide synthesis on the ribosome occurs with the obligatory participation of the transfer protein factors (T and G) and is accompanied by the cleavage of GTP into GDP and inorganic phosphate.

The existence of "nonenzymatic" translation was first observed by Pestka[1,2] on purified ribosomes from *Escherichia coli*, which carried out polyuridylic acid [poly(U)]-directed synthesis of oligo- and polyphenylalanines in the absence of the transfer factors (T and G) and GTP. The required and sufficient components for the "nonenzymatic" translation are the ribosome, the template polynucleotide, and aminoacyl-tRNA, in the presence of Mg^{2+} and also K^+ or NH_4^+. The results obtained by Pestka

[1] S. Pestka, *J. Biol. Chem.* **243**, 2810 (1968).
[2] S. Pestka, *J. Biol. Chem.* **244**, 1533 (1969).

were reproduced and corroborated.[3] In contrast to "enzymatic" translation, the system of "nonenzymatic" translation is not inhibited by 5'-guanylyl-methylene diphosphonate, p-chloromercuribenzenesulfonate or p-chloro-mercuribenzoate, and fusidic acid.[1-3]

However, in studies of the system of "nonenzymatic" translation from *Escherichia coli* repeated experiments were, as a rule, badly reproducible. This occurred even when preparations of ribosomes, poly(U) and phenyl-alanyl-tRNA of the same lot were used. It was found that the factor determining the activity of ribosomes in the "nonenzymatic" translation may be the state of some SH-groups of ribosomes.[4-6] Thus, it was shown that under SH-group reducing conditions, i.e., in the presence of such SH-compounds as dithiothreitol (DTT) or β-mercaptoethanol (ME), the ability of ribosomes to carry out the "nonenzymatic" translation is suppressed. On the other hand, such an SH-reagent as p-chloromercuri-benzoate (PCMB) unexpectedly proved to be a strong stimulator of "nonenzymatic" translation.[4-6] The stimulation of "nonenzymatic" translation under the action of PCMB was found to be caused by a modification of SH-groups of the 30 S ribosomal subparticles.[6,7]

Similar to normal ("enzymatic") translation, the system of "nonenzymatic" translation is sensitive to specific inhibitors of protein synthesis such as tetracycline, chloramphenicol, and thiostrepton.[4,6,8,9]

The length of the peptide synthesized in the "nonenzymatic" system directly correlates with the length of the utilized template polynucleotide. Thus, in the presence of the triplet UpUpU or pUpUpU no synthesis of either di-, tri-, or oligophenylalanines takes place in the "nonenzymatic" system.[2,10] With $(Up)_6G$ or $(pU)_6$ the formation of diphenylalanine occurs, but there is no synthesis of tri- and oligophenylalanines. If $(Up)_9G$ or $(pU)_{9-11}$ is used as a template, then together with the formation of diphenylalanine the synthesis of triphenylalanine is observed.[2,10]

The presented chapter contains a description of a PCMB-activated

[3] L. P. Gavrilova and V. V. Smolyaninov, *Mol. Biol. (USSR)* **5**, 883 (1971).

[4] L. P. Gavrilova, *in* "Symposium on Molecular Mechanism of Antibiotic Action on Protein Biosynthesis and Membranes," June 1-4, Granada, Spain, 1971.

[5] L. P. Gavrilova and A. S. Spirin, *FEBS (Fed. Eur. Biochem. Soc.) Lett.* **17**, 324 (1971).

[6] L. P. Gavrilova and A. S. Spirin, *Mol. Biol. (USSR)* **6**, 311 (1972).

[7] L. P. Gavrilova and A. S. Spirin, *FEBS (Fed. Eur. Biochem. Soc.) Lett.* **22**, 91 (1972).

[8] S. Pestka, *Arch. Biochem. Biophys.* **136**, 89 (1970).

[9] S. Pestka and N. Brot, *J. Biol. Chem.* **246**, 7715 (1971).

[10] L. P. Gavrilova and V. E. Kotelyansky, unpublished.

[11] R. W. Erbe, M. M. Nau, and P. Leder, *J. Mol. Biol.* **39**, 441 (1969)

system of "nonenzymatic" synthesis of polyphenylalanine consisting of purified *Escherichia coli* ribosomes, polyuridylic acid [poly(U)] as a template and ^{14}C-phenylalanyl-tRNA (^{14}C-Phe-tRNA).

Isolation, Purification, and Storage of Ribosomes and Their Subparticles

Growth of Cells

Escherichia coli, strain MRE-600, were cultivated under vigorous aeration in an inorganic medium containing 0.5% glucose and 0.01% bactopeptone. At mid-log growth phase crushed frozen 0.14 M NaCl solution was added to the bacterial suspension and the cells were immediately collected in a continuous flow centrifuge at $+4°$. The bacterial paste was washed twice with a solution of 10 mM Tris·HCl, 10 mM MgCl$_2$, 22 mM NH$_4$Cl, and 1 mM mercaptoethanol or dithiothreitol, pH 7.8 ($0°$ to $+4°$). The washed bacteria were stored at $-20°$C.

Isolation, Purification, and Storage of 70 S Ribosomes

Solution I: 10 mM Tris·HCl, 10 mM MgCl$_2$, 50 mM NH$_4$Cl, 1 mM dithiothreitol, pH 7.1–7.2 at 25°

Solution II: 10 mM Tris·HCl, 10 mM MgCl$_2$, 1 M NH$_4$Cl, 1 mM dithiothreitol, pH 7.1–7.3 at 25°

Solution III: 10 mM Tris·HCl, 10 mM MgCl$_2$, pH 7.2 at 25°

The bacteria were broken in the cold by grinding in a mortar with aluminium oxide powder (2.5 parts of aluminium oxide to one part by weight of frozen bacterial paste); solution I was added to the ground bacteria and the suspension was freed from debris for 30 minutes at 16,000–20,000 rpm; DNase (0.5–1.0 μg per milliliter of the final mixture) was added to the supernatant. The ribosomes were pelleted for 90 minutes at 105,000 g.

A following purification of the ribosomes was done in 1 M NH$_4$Cl.[11] The ribosomal pellets were suspended in solution II. The suspension was left for 12 hours at $+4°$ under mild stirring. The aggregates were removed from the ribosome suspension by low-speed centrifugation for 30 minutes at 16,000 rpm. The ribosomes were pelleted for 3–4 hours in an ultracentrifuge at 105,000 g. The procedure of washing the ribosomes with solution II was repeated 4–6 times.

The pellets of the ribosomes washed 4–6 times with 1 M NH$_4$Cl were suspended in solution III; the suspension was diluted to 1 mg of ribosomes per milliliter, and the ribosomes were salted out at $0°$ to $+4°$ with ammonium sulfate by the addition of 42–49 g of dry salt per 100 ml of the initial suspension of ribosomes.[12] The ribosomal precipitate was stored under ammonium sulfate at $0°$ to $+4°$. A one-year storage of such salted out ribosomes did not display any noticeable decrease of their biological activity (specific binding of aminoacyl-tRNA in the presence of a template polynucleotide and incorporation of amino acids into peptide bonds in the "enzymatic" system) or change of their physical properties (sedimentation distribution and the dissociation-reassociation behavior).[3]

Separation of 30 S and 50 S Ribosomal Subparticles and
    Their Storage

Solution IV: 10 mM Tris·HCl, 20 mM MgCl$_2$, 100 mM NH$_4$Cl, pH 7.2 at $25°$

Solution V: 10 mM Tris·HCl, 2 mM MgCl$_2$, 500 mM NH$_4$Cl, pH 7.2 at $25°$. The sucrose gradient was 5–30% in solution V.

Ribosomal particles, 30 S and 50 S, were obtained using ribosomes twice pelleted from solution I by centrifugation for 90 minutes at 105,000 g (see "Isolation, Purification, and Storage of 70 S Ribosomes"). The pellets of such ribosomes were suspended in solution III, the suspension was diluted to 1 mg of ribosomes per milliliter, and the ribosomes were salted out at $0°$ to $+4°$ with ammonium sulfate adding 49 g of dry salt to 100 ml of the initial suspension of ribosomes. The ribosomes were stored under ammonium sulfate at $0°$ to $+4°$ (up to a year).

Before isolation of 30 S and 50 S subparticles, the precipitate of ribosomes in ammonium sulfate was pelleted for 15 minutes at 16,000 rpm; the pellet was suspended in solution IV and dialyzed against the same buffer for 1–2 hours at $+4°$ to remove the ammonium sulfate. Dialysis was then continued against buffer V for 2 hours with a change of solution after 1 hour. In these conditions a complete dissociation of ribosomes into ribosomal subparticles takes place. After dialysis, the ribosome suspension was cleared from aggregates by low-speed centrifugation. The separation of ribosomal subparticles was done in a 5–30% sucrose gradient prepared in buffer V using the zonal B15 rotor of the MSE Superspeed 65 preparative ultracentrifuge. The ribosomal subparticles were collected from fractions after the sucrose gradient centrifugation by precipitating with ammonium sulfate, adding 49 g of dry salt per 100 ml of the ribosomal 30 S or 50 S subparticle suspension; before the addition of salt the MgCl$_2$ concentration in the suspensions was adjusted to 10 mM. The preparations of precipitated 30 S and 50 S subparticles were stored under ammonium sulfate at $+4°$.[6]

---

[12] C. G. Kurland, *J. Mol. Biol.* **18**, 90 (1966).

The "Nonenzymatic" Ribosomal System of
Polyphenylalanine Synthesis

Solution VI: 10 mM Tris·HCl, 20 mM MgCl$_2$, 100 mM KCl, pH
7.1 at 25°
Solution VII: 10 mM Tris·HCl, 13 mM MgCl$_2$, 100 mM KCl, pH
7.1 at 25°

The "Nonenzymatic" System in the Presence of
p-Chloromercuribenzoate (PCMB)

Before the experiment, the suspension of 4–6 times washed with 1 M
NH$_4$Cl ribosomes or ribosomal subparticles stored under ammonium sulfate (see section "Isolation, Purification, and Storage of 70 S Ribosomes")
was centrifuged for 15 minutes at 16,000 rpm; the pellet was suspended
in solution VI and dialyzed against the same buffer for 2 hours at +4°
to remove the ammonium sulfate. Dialysis was then continued for 1–2
hours against the same buffer with the MgCl$_2$ concentration required for
the experiment.

The reaction mixture ("nonenzymatic" system) was prepared in buffer
VII. One 50-microliter sample of the mixture contained: 40–50 μg of ribosomes, 20 μg of poly(U) and 80–100 μg of ^{14}C-phenylalanine acylated
total tRNA from *E. coli* (130,000–180,000 cpm per milligram of tRNA).
Incubation was carried out at 25° for a few hours in the presence of

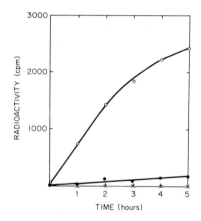

FIG. 1. Effect of PCMB and DTT on the poly(U)-directed synthesis of polyphenylalanine in the "nonenzymatic" system. The experiment was performed at 25°.
PCMB was added in a 10^{-4} M concentration, DTT in a 10^{-3} concentration. The
amount of hot TCA-precipitated ^{14}C-polyphenylalanine is plotted vs. incubation
time. ○, plus PCMB; ●, plus DTT; ×, plus PCMB without poly(U).

PCMB or DTT. PCMB was used in a concentration of $10^{-4} M$ and DTT in a concentration of $10^{-3} M$. The reaction was stopped by adding 3 ml of 5% trichloroacetic acid. About 50–100 μg of serum albumin as a carrier in a volume of 50–100 μl were added to each sample; the sample was hydrolyzed for 15 minutes at 90° and cooled; the precipitate was deposited on a nitrocellulose filter and washed with 5% trichloroacetic acid; the filters were dried at 90° and their radioactivity was assayed in a liquid scintillation counter in a standard toluene-PPO-POPOP mixture.

A typical experiment on "nonenzymatic" translation is represented in Fig. 1. It shows that the system consisting of ribosomes, poly(U) and ^{14}C-Phe-tRNA, without the T- and G-factors and GTP ("nonenzymatic" system), synthesizes TCA-insoluble ^{14}C-polyphenylalanine in the presence of PCMB. DTT inhibits such a system.

Polymerization of phenylalanine residues in a PCMB-activated "non-enzymatic" system is poly(U)-dependent; in the absence of poly(U) no incorporation of ^{14}C-phenylalanine into the TCA-insoluble product is observed (Fig. 1).

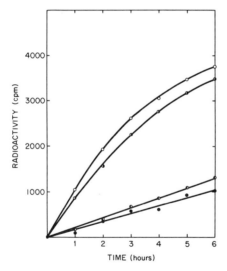

Fig. 2. Effect of preliminary PCMB-treatment of separate ribosomal subparticles (30 S and 50 S) on the poly(U)-directed synthesis of polyphenylalanine in the "nonenzymatic" system. The experiment was performed at 25° in the absence of PCMB. The amount of hot TCA-precipitated ^{14}C-phenylalanine is plotted vs. incubation time. ●, Both 30 S and 50 S subparticles pretreated with DTT; ○, both 30 S and 50 S subparticles pretreated with PCMB; ◑, the 30 S subparticle pretreated with PCMB and the 50 S subparticle pretreated with DTT; ◐, the 30 S subparticle pretreated with DTT and the 50 S subparticle pretreated with PCMB.

The "Nonenzymatic" System with PCMB-Pretreated Ribosomes
or Ribosomal 30 S Subparticles

Ribosomes and ribosomal subparticles were prepared for the experiment as described in the section "The 'Nonenzymatic' System in the Presence of PCMB."

Pretreatment of ribosomes or ribosomal subparticles with PCMB or DTT was done in solution VI or VII at 25° for 1–2 hours; the concentration of ribosomes was 4–6 mg/ml; the PCMB or DTT concentration was $10^{-4} M$ or $10^{-3} M$, correspondingly. After such treatment excess PCMB or DTT was removed by gel filtration through a G-50 Sephadex column (0.9×11 cm). 0.7 ml of the ribosome suspension (3–4 mg of ribosomes) was layered on the column. Usually elution from the column gave 1.5- to 2-fold dilution of the ribosome fraction. The reaction mixture was prepared in solution VII. One 50-μl sample of the mixture contained: 40 μg of PCMB-pretreated ribosomes, or 13 μg of PCMB-pretreated 30 S ribosomal subparticles and 26 μg of 50 S ribosomal subparticles. The other components were added in similar amounts as in the "nonenzymatic" system working in the presence of PCMB, namely: 20 μg of poly(U)

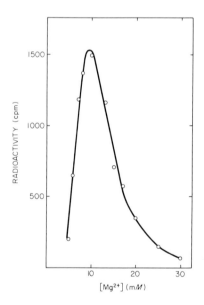

FIG. 3. Dependence of the poly(U)-directed polyphenylalanine synthesis rate in the "nonenzymatic" system on the Mg^{2+} concentration. The experiment was performed at 25° in the presence of PCMB. The amount of hot TCA-precipitated ^{14}C-polyphenylalanine synthesis during 3 hours of incubation is plotted vs. Mg^{2+} concentration.

and 80–100 μg of ^{14}C-phenylalanine acylated total tRNA from *E. coli* (130,000–180,000 cpm per milligram of tRNA).

A typical experiment on "nonenzymatic" translation with PCMB-pretreated 30 S ribosomal subparticles is represented in Fig. 2. It is seen that if only 30 S subparticles are treated with PCMB and the 50 S subparticles are treated with DTT, the system is practically just as active as in the case when both subparticles are treated with PCMB. On the contrary, treatment of 50 S subparticles with PCMB and the 30 S subparticles with DTT results in a low activity of the system. It follows from this that activation of the "nonenzymatic" translation system with PCMB depends on a modification (blocking of the SH-group or groups) only in the small, 30 S subparticle, not in the 50 S ribosomal subparticle.

Dependence of "Nonenzymatic" Translation on the
  Concentration of Cations and Temperature

Figure 3 shows the dependence of PCMB-activated "nonenzymatic" translation on the Mg^{2+} concentration. In this experiment the "nonenzymatic" system worked in the presence of PCMB at 100 mM K$^+$ and

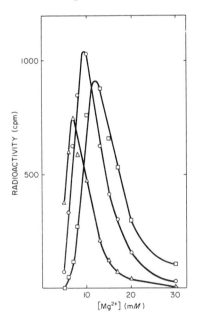

FIG. 4. Effect of K$^+$ concentration on the Mg^{2+}-dependence of poly(U)-directed polyphenylalanine synthesis in the "nonenzymatic" system. The experiment was performed at 25° in the presence of PCMB. The amount of hot TCA-precipitated ^{14}C-polyphenylalanine synthesized during 3 hours of incubation is plotted vs. Mg^{2+} concentration. △, 50 mM KCl; ○, 100 mM KCl; □, 200 mM KCl.

different concentrations of Mg^{2+} (see above "The 'Nonenzymatic' System in the Presence of PCMB"). Each experimental point means the radioactivity value of trichloroacetic acid-precipitated material synthesized in the sample during 3 hours incubation at 25°. It is seen that the dependence of the rate curve of "nonenzymatic" translation on the Mg^{2+} concentration has a distinct maximum.

Figure 4 demonstrates the effect of K^+ concentration on the dependence of PCMB-activated "nonenzymatic" translation on the Mg^{2+} concentration. Conditions of experiment are the same as for Fig. 3. It is seen in Fig. 4 that an increase in the K^+ concentration shifts the Mg^{2+} optimum of "nonenzymatic" translation to the region of higher Mg^{2+} concentrations, while a decrease in the concentration of K^+ leads to the shift of the Mg^{2+} optimum into the opposite direction, toward the region of lower Mg^{2+} concentrations.

K^+ ions in the PCMB-activated "nonenzymatic" translation system can be substituted by NH_4^+ ions. Such a substitution somewhat shifts the Mg^{2+} optimum of the system toward lower Mg^{2+} concentrations (Fig. 5).

"Nonenzymatic" translation is temperature-dependent: its activity increases with an increase of temperature (see Pestka[2] and Fig. 6). Tem-

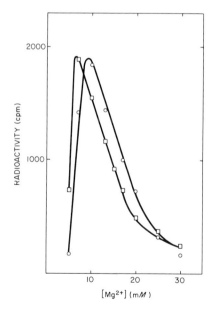

FIG. 5. Effect of substituting K^+ for NH_4^+ on the Mg^{2+}-dependence of poly(U)-directed polyphenylalanine synthesis in the "nonenzymatic" system. The experiment was performed at 25° in the presence of PCMB. The amount of hot TCA-precipitated ^{14}C-polyphenylalanine synthesized during 3 hours of incubation is plotted vs. Mg^{2+} concentration. $\bigcirc$, 100 mM KCl; $\square$, 100 mM NH$_4$Cl.

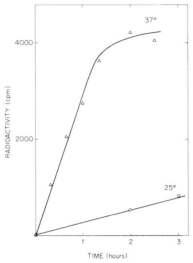

FIG. 6. Poly(U)-directed polyphenylalanine synthesis in the "nonenzymatic" system at two different temperatures. The experiment was performed at 25° and 37°, 15 mM Mg^{2+} in the presence of PCMB. The amount of hot TCA-precipitated ^{14}C-polyphenylalanine is plotted vs. incubation time. ○, 25°; △, 37°.

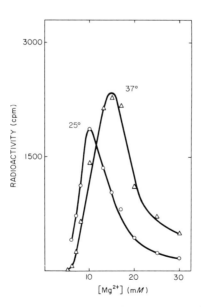

FIG. 7. Effect of temperature on the Mg^{2+}-dependence of poly(U)-directed polyphenylalanine synthesis in the "nonenzymatic" system. The experiment was performed at 25° or 37° in the presence of PCMB. The amount of hot TCA-precipitated ^{14}C-polyphenylalanine synthesis during incubation for 3 hours at 25° and for 50 minutes at 37° is plotted vs. Mg^{2+} concentration. ○, 25°; △, 37°.

perature affects the Mg^{2+} optimum of PCMB-activated "nonenzymatic" translation: an increase in temperature shifts the Mg^{2+} optimum toward the region of higher Mg^{2+} concentrations (Fig. 7).

The activating effect of PCMB on "nonenzymatic" translation can be due to either the stimulation of aminoacyl-tRNA binding or the stimulation of the peptidyltransferase center, or stimulation of translocation. It has been shown by some authors that in any case different SH-reagents do not stimulate binding of aminoacyl-tRNA to *Escherichia coli* ribosomes.[13-15] The peptidyltransferase center of the bacterial ribosome is indifferent to the effect of SH-reagents, including PCMB.[16,17] Hence, the stimulation of "nonenzymatic" translation in *Escherichia coli* ribosomes by PCMB can be apparently explained by the stimulation of translocation. This corroborates the idea that the capability for translocation in principle is inherent to the structural organization of the ribosome itself, not to the transfer protein factors.[4-7]

[13] R. R. Traut and A. L. Haenni, *Eur. J. Biochem.* **2**, 64 (1967).
[14] A. V. Furano, *Biochim. Biophys. Acta* **161**, 255 (1968).
[15] J. A. Retsema and T. W. Conway, *Biochim. Biophys. Acta* **179**, 369 (1969).
[16] R. R. Traut and R. E. Monro, *J. Mol. Biol.* **10**, 63 (1964).
[17] R. E. Monro, *J. Mol. Biol.* **26**, 147 (1967).

[44] Assay for Nonenzymatic and Enzymatic Translocation with *Escherichia coli* Ribosomes

By SIDNEY PESTKA

The movement of the ribosome with respect to the messenger RNA template has been termed translocation. Although the precise number and characteristics of all the events occurring during this process have not been delineated, the overall process can be measured. In the case of the enzymatic process, elongation factor G (EF G) has been implicated in the translocational events.[1-8] Presented below is a convenient assay

[1] Y. Nishizuka and F. Lipmann, *Arch. Biochem. Biophys.* **116**, 344 (1966).
[2] J. Lucas-Lenard and F. Lipmann, *Proc. Nat. Acad. Sci. U.S.* **55**, 1562 (1966).
[3] S. Pestka, *Proc. Nat. Acad. Sci. U.S.* **61**, 726 (1968).
[4] R. Ertel, N. Brot, B. Redfield, J. E. Allende, and H. Weissbach, *Proc. Nat. Acad. Sci. U.S.* **59**, 861 (1968).
[5] R. W. Erbe and P. Leder, *Biochem. Biophys. Res. Commun.* **31**, 798 (1968).
[6] A. Skoultchi, Y. Ono, J. Waterson, and P. Lengyel, *Cold Spring Harbor Symp. Quant. Biol.* **34**, 437 (1969).
[7] J. M. Ravel, *Proc. Nat. Acad. Sci. U.S.* **57**, 1811 (1967).
[8] J. W. Bodley, F. J. Zieve, L. Lin, and S. T. Zieve, *J. Biol. Chem.* **245**, 5656 (1970).

for elongation factor G.[3] In addition, during studies on translocation using poly(U) as a template, it was noted that a small quantity of oligophenylalanine was formed in the absence of elongation factor G and GTP.[9,10] This was investigated in order to determine if, indeed, the translocation which was observed could be accounted for by residual contaminating amounts of EF G remaining on the washed ribosomes.

For the following reasons it was felt that the nonenzymatic translocation could not be accounted for by residual amounts of EF G. Fusidic acid which inhibits elongation factor G did not inhibit nonenzymatic translocation.[9] The presence of the sulfhydryl inhibitor p-chloromercuribenzenesulfonate in the reaction inhibited oligophenylalanine synthesis dependent on EF G and GTP; however, the nonenzymatic reaction was not inhibited and in fact was stimulated approximately 20%.[9] Guanylyl 5'-methylene diphosphonate (GMPPCP), which inhibited translocation dependent on EF G and GTP, had no inhibitory effect on the nonenzymatic reaction in the absence of EF G and GTP. In the presence of EF G and GTP, the activation energy for translocation or oligophenylalanine synthesis is about 6.6 kcal/mole; in contrast, in the absence of EF G or GTP or both, the activation energy for the process is very significantly higher, being 23 kcal/mole.[9] In addition, in unpublished studies we have found that ribosomes which have been treated with 1 M NH$_4$Cl and ethanol to remove a ribosomal protein necessary for hydrolysis of GTP in the presence of EF G[11] can perform nonenzymatic translocation, but cannot function in the enzymatic process. Furthermore, with the use of a mutant of E. coli containing a temperature-sensitive EF G (E. coli G1B obtained from David Schlessinger), it was found that ribosomes from this strain can carry out nonenzymatic translocation at both permissive and nonpermissive temperatures. The observation of nonenzymatic translocation has been confirmed by Gavrilova and Spirin.[12]

The rationale for the assays described below is illustrated in Fig. 1. With the use of ribosomes washed in 1 M ammonium chloride, Phe-tRNA, magnesium, and potassium, poly(U) can be used to stimulate the incorporation of phenylalanine from Phe-tRNA into oligophenylalanine. For extensive incorporation, 100,000 g supernatant fraction and GTP are required.[13] The supernatant fraction contributes the transfer factors EF

[9] S. Pestka, J. Biol. Chem. 244, 1533 (1969).

[10] S. Pestka, Cold Spring Harbor Symp. Quant. Biol. 34, 395 (1969).

[11] E. Hamel and T. Nakamoto, Fed. Proc., Fed. Amer. Soc. Exp. Biol. 30, 1203 (1971).

[12] L. P. Gavrilova and A. S. Spirin, FEBS (Fed. Eur. Biochem. Soc.) Lett. 17, 324 (1971).

[13] M. W. Nirenberg and J. H. Matthaei, Proc. Nat. Acad. Sci. U.S. 47, 1588 (1961).

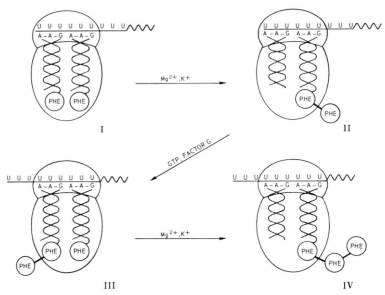

Fɪɢ. 1. Schematic illustration of sequence of events in oligophenylalanine synthesis.

Tu, EF Ts, and EF G.[2,14] With the use of a high magnesium concentration (20 mM), the requirement for T factors (EF Tu and EF Ts) for Phe-tRNA binding to ribosomes is circumvented. A high level of Phe-tRNA binding to ribosomes occurs at 20 mM Mg^{2+} in the absence of any additional transfer factors (I, Fig. 1).[15-17]

In the absence of K^+ or NH_4^+, no di- or oligopeptide bond formation is detectable.[18] Upon the addition of K^+ or NH_4^+, dipeptide bond formation occurs and, with the use of Phe-tRNA, diphenylalanine formation is substantial as shown in Panel II of Fig. 1.[18,19] In the absence of additional transfer factors, the reaction stops after the formation of Phe-Phe-tRNA.

Since the T factors are not required at high magnesium concentrations, it is relevant to inquire what factors are required to proceed beyond the accumulated Phe-Phe-tRNA. The translocational events are designated by the transition from II to III (Fig. 1). These events include the

[14] T. Nakamoto, T. W. Conway, J. E. Allende, G. J. Spyrides, and F. Lipmann, *Cold Spring Harbor Symp. Quant. Biol.* **28**, 227 (1963).

[15] M. Nirenberg and P. Leder, *Science* **145**, 1399 (1964).

[16] A. Kaji and H. Kaji, *Biochem. Biophys. Res. Commun.* **13**, 186 (1963).

[17] S. Pestka, *J. Biol. Chem.* **241**, 367 (1966).

[18] S. Pestka, *Arch. Biochem. Biophys.* **136**, 89 (1970).

[19] S. Pestka, *J. Biol. Chem.* **243**, 2810 (1968).

rejection of deacylated tRNA from the ribosome and translational movement of the polynucleotide template and ribosome relative to each other. Immediately after translocation and binding of a new Phe-tRNA, an additional peptide bond is synthesized to form triphenylalanine (IV, Fig. 1). By repetition of the translocation, binding, and peptide synthesis steps, chains of oligophenylalanine are constructed. In the presence of 20 mM Mg^{2+} and 50 mM K^+, binding of Phe-tRNA to ribosomes and peptide bond formation occur readily. Therefore, under these conditions, measurement of the requirements for oligophenylalanine formation reflect the requirements for the overall process of translocation, which must occur prior to elongation of the peptide. Thus, by measuring the quantity of phenylalanine incorporated into peptides of chain lengths 3 and greater, one can assay for translocation.

Materials

5 × Standard Buffer: 0.10 M $MgCl_2$, 0.25 M KCl, 0.25 M TrisAc, pH 7.2

1 M NH_4Cl washed ribosomes from *E. coli*[18]

[^{14}C]Phe-tRNA[19] (sp. act. [^{14}C]Phe $\geq$ 200 mCi/mmole; about 20 pmoles [^{14}C]Phe per A_{260} unit)

Poly(U) (52 A_{260} units/ml)

GTP, 0.5 mM

Elongation factor G[4] (1–5 mg/ml)

Reaction Mixtures

5 × Standard Buffer, 0.010 ml (final concentration of components is 20 mM Mg^{2+}, 50 mM K^+, and 50 mM TrisAc, pH 7.2)

Poly(U) solution, 0.005 ml (26 nmoles of base residues per reaction mixture)

GTP, 0.5 mM, 0.005 ml

EF G, 0.2–1.0 μg

Ribosomes, 0.3–0.5 A_{260} unit

[^{14}C]Phe-tRNA (0.25 A_{260} unit), 6 pmoles of [^{14}C]Phe

H_2O to 0.050 ml total volume

Each reaction mixture, therefore, contains in a final volume of 0.050 ml the following components: 20 mM Mg^{2+}, 50 mM K^+, and 50 mM Trisacetate, pH 7.2; 26 nmoles (0.26 A_{260} unit) of base residues of poly(U); 50 μM GTP; 0.3–0.5 A_{260} unit of ribosomes; 6 pmoles of [^{14}C]Phe-tRNA (0.25 A_{260} unit); and 0.2–1.0 μg of elongation factor G. Reactions are begun by addition of Phe-tRNA or ribosomes. Reaction mixtures are incubated at 24° for 20 minutes or 37° for 5 minutes for enzymatic translocation.[3,9] For nonenzymatic translocation elongation factor G and GTP are omitted and reactions are incubated at 37° for 30 minutes.[9]

A rapid and sensitive assay for separating oligopeptides of phenylalanine of chain length 3 and greater from mono- and diphenylalanine is by the use of small disposable columns of benzoylated diethylaminoethylcellulose. The procedure is described in Vol. 22.[20] If total recovery of all the oligophenylalanine peptides is not important, a rapid assay can be made by simply precipitating the larger peptides of phenylalanine with trichloroacetic acid.[9,21] In either case, the reactions are stopped by the addition of 0.1 ml of 0.5 M NaOH and incubating at 37° for 15 minutes to hydrolyze any aminoacyl- or peptidyl-tRNA present. The reaction mixtures are neutralized with 0.1 ml of 0.5 N HCl. The oligophenylalanine products are then precipitated with cold 10% trichloroacetic acid followed by filtration and washing with 5% trichloroacetic acid onto a nitrocellulose membrane filter[9]; or as indicated above, small disposable columns of benzolylated diethylaminoethylcellulose can be used to separate the oligophenylalanine from mono- and diphenylalanine.[20,21] Recovery of oligo-

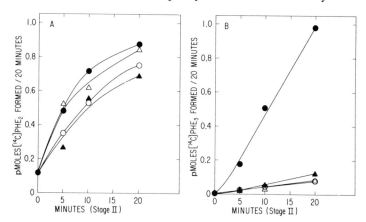

Fig. 2. Formation of di- and oligophenylalanine as a function of time. Each 0.050-ml stage I reaction mixture contained the components indicated under Assay Procedure in addition to: 0.38 A_{260} unit of ribosomes; 10 mM ammonium chloride; 6.8 pmoles of [^{14}C]Phe-tRNA (0.18 A_{260} unit); 46 nmoles of poly(U). Incubations were performed at 0° for 30 minutes (stage I). At the end of stage I, GTP (to a final concentration of 2 mM) and 1.2 μg of factor G were added to reaction mixtures where indicated. Reactions were then performed at 24° (stage II) and assayed at the times indicated on the abscissa. (A) Diphenylalanine formation; (B) oligophenylalanine formation. ●, Plus factor G, plus GTP; △, minus factor G, plus GTP; ○, minus factor G, minus GTP; ▲, plus factor G, minus GTP.

[20] S. Pestka, this series, Vol. 22, p. 508.
[21] S. Pestka, E. M. Scolnick, and B. H. Heck, *Anal. Biochem.* **28**, 376 (1969).

peptides is about one-third greater with the column technique than with the use of the trichloroacetic acid precipitation procedure where losses of the small oligopeptides occur.[9,21] The simultaneous measurement of mono-, di-, and oligophenylalanine requires the use of the column technique.

The time course of di- and oligophenylalanine synthesis as a function of time is shown in Fig. 2. As can be seen, maximal formation of oligo-phenylalanine (Phe $_{\geq 3}$) was dependent on the presence of both EF G and GTP. In the absence of EF G or GTP, much less Phe $_{\geq 3}$ was formed. In contrast, diphenylalanine formation is not dependent on the presence of GTP and EF G. Formation of Phe $_{\geq 3}$ is linear for at least 20 minutes at 24° under conditions of the assay. The data of Fig. 3 show the critical nature of the ribosome concentration, for a ribosome concentration greater than or less than optimal produces substantially less oligophenylalanine. This results simply from the ribosome:Phe-tRNA ratio. When ribosomes are in excess, the available Phe-tRNA is bound with little free Phe-tRNA remaining for di- and oligophenylalanine synthesis. It is therefore advis-able to perform a ribosome concentration curve for each ribosome prepara-tion under the specific conditions of the assay. Diphenylalanine formation as can be seen is not as critically dependent on ribosome concentration (Fig. 3). As shown in Fig. 4, these same characteristics are evident when nonenzymatic translocation is studied. In addition, a comparison of recov-ery of the oligophenylalanine products obtained by the column technique (Fig. 4A)[20] and by the trichloroacetic acid precipitation method (Fig. 4B) is also shown.

The table summarizes the methods of distinguishing enzymatic and

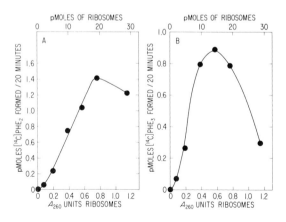

FIG. 3. Di- and oligophenylalanine formation as a function of ribosome concen-tration. Reaction conditions were similar to those described in the legend to Fig. 2 except that ribosome concentration was varied and stage II reactions were performed for 20 minutes. (A) Diphenylalanine formation; (B) oligophenylalanine formation.

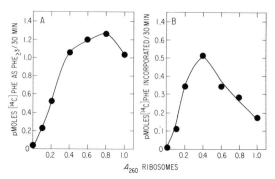

Fig. 4. The rate of oligophenylalanine formation in the absence of EF G and GTP as a function of ribosome concentration. Reactions were performed at 37° for 30 minutes under conditions given in the text. (A) Assays for oligophenylalanine performed by the column technique (S. Pestka, this series, Vol. 22, p. 508). (B) Oligopeptides of phenylalanine precipitated by trichloroacetic acid (TCA) as described in the text.

nonenzymatic translocation. The data in the table indicate that under these specific conditions for assay fusidic acid, p-chloromercuribenzenesulfonate, and GMPPCP inhibit enzymatic but not nonenzymatic translocation. The antibiotics thiostrepton, siomycin A, and micrococcin inhibit both enzymatic and nonenzymatic translocation (Figs. 5A and B; and table).[22,23]

INHIBITORS OF ENZYMATIC AND NONENZYMATIC TRANSLOCATION[a]

| | Percent of uninhibited control | |
Antibiotic	Enzymatic	Nonenzymatic
None	100	100
p-Chloromercuribenzenesulfonate ($3 \times 10^{-4}\,M$)	33	120
Guanylyl 5′-methylene diphosphonate ($1 \times 10^{-3}\,M$)	5	100
Fusidic acid ($6 \times 10^{-4}\,M$)	9	107
Thiostrepton ($1 \times 10^{-5}\,M$)	11	6
Siomycin A ($1 \times 10^{-5}\,M$)	6	10
Micrococcin ($1 \times 10^{-5}\,M$)	10	27

[a] The data are taken from S. Pestka, *Proc. Nat. Acad. Sci. U.S.* **61**, 726 (1968); S. Pestka, *J. Biol. Chem.* **244**, 1533 (1969); S. Pestka and N. Brot, *J. Biol. Chem.* **246**, 7715 (1971).

[22] S. Pestka and N. Brot, *J. Biol. Chem.* **246**, 7715 (1971).
[23] S. Pestka, *Biochem. Biophys. Res. Commun.* **40**, 667 (1970).

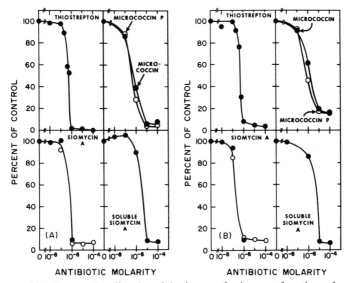

ANTIBIOTIC MOLARITY   ANTIBIOTIC MOLARITY

FIG. 5. (A) Enzymatic oligophenylalanine synthesis as a function of antibiotic concentration. (B) Nonenzymatic oligophenylalanine synthesis as a function of antibiotic concentration. Details of the reaction components are given by S. Pestka and N. Brot, *J. Biol. Chem.* **246**, 7715 (1971).

The steepness of the curves with respect to thiostrepton and siomycin suggest very tight binding of these antibiotics to ribosomes.

It is noteworthy that the localization of antibiotic action can often be readily made with the use of these assays for enzymatic and nonenzymatic translocation. For example, antibiotics which inhibit translocation through inhibition of EF G should inhibit enzymatic, but not nonenzymatic translocation. Fusidic acid and *p*-chloromercuribenzenesulfonate are agents whose actions were localized to inhibition of EF G activity in this manner (see table).[9] In contrast, antibiotics which inhibit translocation through interference with ribosome functions should inhibit both enzymatic and nonenzymatic translocation. The actions of thiostrepton, siomycin A, and micrococcin were thus localized as inhibitors of ribosome functions (see table and Figs. 5A and 5B).[22,23] A cautionary note is warranted here. Although thiostrepton and siomycin A inhibit translocation in cell-free assays, these antibiotics also inhibited Tu function.[24-26] Possibly in intact

[24] T. Kinoshita, Y.-F. Liou, and N. Tanaka, *Biochem. Biophys. Res. Commun.* **44**, 859 (1971).

[25] J. Modolell, B. Cabrer, A. Parmeggiani, and D. Vazquez, *Proc. Nat. Acad. Sci. U.S.* **68**, 1796 (1971).

[26] H. Weissbach, B. Redfield, E. Yamasaki, R. C. Davis, S. Pestka, and N. Brot, *Arch. Biochem. Biophys.* **149**, 110 (1972).

cells Tu function may be preferentially inhibited.[27,28] Of additional note is the effect of these inhibitors of translocation on EF G-dependent hydrolysis of GTP in the presence of ribosomes. Fusidic acid, thiostrepton, and siomycin A inhibit GTP hydrolysis[29]; micrococcin, however, does not.[22]

[27] E. Cundliffe, *Biochem. Biophys. Res. Commun.* **44**, 912 (1971).

[28] M. Cannon and K. Burns, *FEBS (Fed. Eur. Biochem. Soc.) Lett.* **18**, 1 (1971).

[29] S. Pestka, *Annu. Rev. Microbiol.* **25**, 487 (1971).

[45] Peptidyl-Puromycin Synthesis on Polyribosomes from *Escherichia coli*

By SIDNEY PESTKA

Several model systems have been used to study transpeptidation on ribosomes. The reaction of puromycin with numerous derivatives of aminoacyl-tRNA as synthetic donors has provided the basis for most model systems (Fig. 1). Common synthetic donors have included fMet-tRNA, acetyl-Phe-tRNA, polylysyl-tRNA, and fragments of tRNA such as C-A-C-C-A(Ac-Phe) and C-A-A-C-C-A(fMet).[1-10] The ribosomes used generally have been washed in 1 M NH$_4$Cl, although ribosomes washed in low salt have also been used. Because of apparent disparate results of inhibitors on peptide bond synthesis in model systems and intact cells,[11] it was useful to study transpeptidation in a more nearly physiological system, which behaved similarly to the process in intact cells. For this reason, native polyribosomes have been prepared and used as the enzyme–substrate complex containing peptidyl-tRNA. The transfer of the peptidyl group to [³H]puromycin as an acceptor was measured by determining the [³H]puromycin incorporated into trichloroacetic acid-precipitable material.

[1] B. E. H. Maden, R. R. Traut, and R. E. Monro, *J. Mol. Biol.* **35**, 333 (1968).

[2] M. E. Gottesman, *J. Biol. Chem.* **242**, 5564 (1967).

[3] I. Rychlik, *Biochim. Biophys. Acta* **114**, 425 (1966).

[4] M. S. Bretscher and K. A. Marcker, *Nature (London)* **211**, 380 (1966).

[5] A. Zamir, P. Leder, and D. Elson, *Proc. Nat. Acad. Sci. U.S.* **56**, 1794 (1966).

[6] J. Lucas-Lenard and F. Lipmann, *Proc. Nat. Acad. Sci. U.S.* **57**, 1050 (1967).

[7] H. Weissbach, B. Redfield, and N. Brot, *Arch. Biochem. Biophys.* **127**, 705 (1968).

[8] S. Pestka, *Arch. Biochem. Biophys.* **136**, 80 (1970).

[9] R. E. Monro and K. A. Marcker, *J. Mol. Biol.* **25**, 347 (1967).

[10] J. L. Lessard and S. Pestka, *J. Biol. Chem.* **247**, 690 (1972).

[11] S. Pestka, *Annu. Rev. Microbiol.* **25**, 487 (1971).

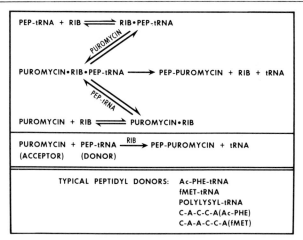

Fig. 1. Outline of the steps of the puromycin reaction. In the equations written, all starting components are considered to be free in solution. During protein synthesis, however, peptidyl-tRNA (PEP-tRNA) is a ribosomal bound intermediate and the usual acceptor is aminoacyl-tRNA. A list of several commonly used peptidyl donors is given in the figure.

Preparation of Polyribosome Extract

Reagents

Medium NK: 15 g nutrient broth in 1 liter of 0.1 M potassium phosphate, pH 7.7; 0.25% (w/v) glucose

Sucrose-Tris: 25% (w/v) sucrose in 40 mM Tris·HCl, pH 8.1

Lysozyme (muraminidase): 6.4 mg of egg white lysozyme (EC 3.2.1.17) in 1.0 ml of 0.25 M Tris·HCl, pH 8.1. This solution should be freshly made each time before use.

Na_2EDTA: 0.1 M

$MgSO_4$: 1.0 M

Lysing Medium: 0.5% (w/v) Brij 58; 50 mM NH_4Cl; 5 mM $MgSO_4$; 10 M Tris·HCl, pH 7.5

Deoxyribonuclease I (EC 3.1.4.5): 5 mg/ml (ribonuclease-free, Worthington)

Polysomes were prepared by modification of the method of Godson and Sinsheimer.[12] *E. coli* B was grown to about 300 Klett units (using the No. 42 blue filter), at which time cells were still in log phase in medium NK. Under these conditions 300 Klett units was approximately equivalent to 9×10^8 cells/ml. The cells were rapidly chilled to 0° to

---

[12] G. N. Godson and R. L. Sinsheimer, *Biochim. Biophys. Acta* **149**, 476, 489 (1967).

5° within 10 seconds by swirling in a stainless steel beaker surrounded by a dry ice–acetone bath. The suspension of cells was then centrifuged for 10 minutes at 10,000 rpm in a GSA rotor of a Sorvall RC2B centrifuge. The cells were resuspended in 9.0 ml sucrose-Tris solution; 1.0 ml of a freshly made solution of lysozyme was then added. To start the action of lysozyme, 0.5 ml of 0.1 M Na$_2$EDTA was added. The concentrated cell suspension was kept in an ice bath for 90 seconds with occasional shaking. To stop lysozyme action, 0.12 ml of 1 M MgSO$_4$ was added to the mixture. The cell protoplast suspension was sedimented at 10,000 rpm for 5 minutes in a SS-34 rotor of the Sorvall centrifuge. The supernatant portion was discarded and the inside of the tube carefully wiped to remove any remaining liquid. The pellet was resuspended in 9 ml of lysing medium by stirring with a glass rod: 0.1 mg of DNase in a volume of 0.02 ml was added to the suspension at this time. The suspension was kept in lysing medium 10 minutes at 0°–5°C. The viscous lysate was then centrifuged for 10 minutes in the SS-34 Sorvall rotor at 10,000 rpm. The supernatant was gently aspirated, and the pellet was discarded. The final solution contained 150–250 A_{260} units/ml. The absorbance at 260 nm of each milliliter as determined by sucrose gradient centrifugation was typically distributed as follows[13]: 45% in the fraction sedimenting at less than 30 S; 28%, 15%, 7%, and 5% of the absorbance was found in polyribosomes, 70 S, 50 S, and 30 S fractions, respectively. Washed polyribosomes free of supernatant fraction and free of ribosomal particles could be prepared by sedimenting the polysome extract through a sucrose gradient as described in the legend to Fig. 7. In order to obtain purified polysomes active in forming peptidylpuromycin, the sucrose gradients used must contain 0.01 to 0.1 M K$^+$. At lower or higher K$^+$ concentrations, the purified polysomes are inactivated. These purified polysomes behaved essentially identically to the polysome extracts with respect to peptidyl-puromycin synthesis.[14,15] Similar procedures were used to prepare polysomes from other *E. coli* strains such as MRE 600 and Q13.

Determination of Peptidyl-[³H]Puromycin Synthesis

Reagents

Polyribosome extracts (150–250 A_{260}/ml)
[³H]Puromycin: 40 μM; specific activity $\geq$ 700 mCi/mmole
5× Standard Buffer: 0.50 M KCl; 20 mM MgCl$_2$; 0.25 M Tris·acetate, pH 7.2

[13] S. Pestka and H. Hintikka, *J. Biol. Chem.* **246**, 7723 (1971).
[14] S. Pestka, *Proc. Nat. Acad. Sci. U.S.* **69**, 624 (1972).
[15] S. Pestka, *J. Biol. Chem.* **247**, 4669 (1972).

After addition of 0.010, 0.005, and 0.010 ml of the polyribosome extract, [^{3}H]puromycin solution, and 5× standard buffer, respectively, each 0.050-ml reaction mixture contained the following components: $0.10\ M$ KCl; 5 mM MgCl$_2$; 50 mM Tris·acetate, pH 7.2; 10 mM NH$_4$Cl and 0.1% (w/v) Brij 58 contributed by the polyribosome extract; about 2–4 A_{260} units of polyribosome extrac.; and 4 μM [^{3}H]puromycin. Unless otherwise noted, the polyribosome extract was added last to start reactions, and tubes were warmed at the temperature of incubation for 1 minute prior to addition of polysomes. Reactions were incubated for 1 minute at 24° or as otherwise indicated. [^{3}H]Puromycin incorporation into nascent polypeptides was measured as follows: reactions were stopped by addition of 2 ml of cold 10% (w/v) trichloroacetic acid and permitted to sit in an ice bath for at least 5 minutes; the contents of each tube were filtered through a polyvinylchloride Millipore filter (type BDWP, 0.6 μm pore size, 25 mm diameter); each tube and filter were washed three times with 3-ml portions of 5% (w/v) trichloroacetic acid; the filter was washed eight times with 3-ml portions of absolute ethanol at room temperature.[14] The filters were then dried under an infrared lamp and placed in a scintillation fluor; radioactivity was determined as previously described.[14] Extensive washing of these filters with absolute ethanol was necessary to remove completely unreacted [^{3}H]puromycin. In the absence of puromycin, under the conditions of the assay for peptidyl-puromycin synthesis, leucine incorporation into polypeptides was negligible.

Characteristics of Peptidyl-[^{3}H]Puromycin Formation

The kinetics of peptidyl-[^{3}H]puromycin formation at various temperatures is shown in Fig. 2. The initial rate of peptidyl-puromycin formation is nearly a linear function of temperature throughout the temperature range studied. The extent of peptidyl-puromycin synthesis at 15 minutes also increases with temperature. An Arrhenius plot provides an estimate of 5.6 kcal/mole for the activation energy of peptidyl-puromycin synthesis. The value for 0° (solid square) does not fall on the straight line as the other points representing temperatures higher than 0°.

Since the polysome extract contained 50 mM NH$_4$Cl, all reaction mixtures for evaluating the effect of monovalent cations contained 10 mM NH$_4$$^+$. Addition of K$^+$ beyond 10 mM NH$_4$$^+$ concentration increased the rate of peptidyl-puromycin synthesis slightly; nevertheless, peptidyl-puromycin synthesis occurred in the absence of additional K$^+$ or NH$_4$$^+$ ions (Fig. 3). Optimum NH$_4$$^+$ was found to be 50 mM to $0.1\ M$ NH$_4$$^+$. At higher concentrations of NH$_4$$^+$, the rate of peptidyl-puromycin synthesis was reduced. Addition of Na$^+$ or Li$^+$ was markedly inhibitory to peptidyl-puromycin synthesis.

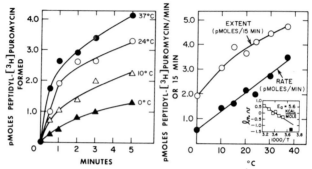

Fig. 2. Kinetics and extent of peptidyl-[³H]puromycin formation at various temperatures. Each 0.050-ml reaction mixture contained the components indicated under "Determination of Peptidyl-[³H]Puromycin Synthesis." Reaction mixtures were incubated at various temperatures and assayed at the times indicated on the abscissa of the left panel. For the determination of the rate of peptidyl-[³H]puromycin formation as a function of temperature (right panel, filled circles, ●), the picomoles of peptidyl-puromycin synthesized per minute was determined from the initial slope of time curves (not all of which are shown in the left panel); this rate is plotted as a function of time in the right panel as picomoles of peptidyl-[³H]puromycin formed per minute. In addition, the extent of peptidyl-puromycin synthesis at 15 minutes is also given at various temperatures (unfilled circles, ○). From the rate of peptidyl-puromycin synthesis at various temperatures (v) an Arrhenius plot is derived (inset of right panel); from the slope of the line, the activation energy (E_a) of peptidyl-[³H]puromycin synthesis is calculated to be 5.6 kcal/mole. The data for the Arrhenius plot are given as unfilled squares (□) for all temperatures except 0° which is plotted as a solid square (■).

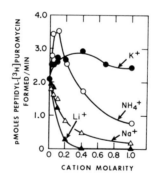

Fig. 3. Effect of monovalent cations on the rate of peptidyl-puromycin formation. Each 0.050-ml reaction mixture contained the following components: 50 mM Tris· acetate, pH 7.2; 5 mM MgCl₂; monovalent cation concentration as specified on the abscissa; 10 mM NH₄Cl was present in each reaction mixture as the contribution from the polysome suspension; 2.8 A_{260} units of the polysome preparation; and 4 μM [³H]puromycin (977 mCi/mmole). Reactions were incubated at 37° for 1 minute and the rate of peptidyl-[³H]puromycin formation was determined as described in the text.

Optimum magnesium concentration for peptidyl-puromycin synthesis was 4 mM (Fig. 4). At lower or higher concentrations of Mg^{2+}, the rate of peptidyl-puromycin synthesis was substantially reduced. In fact, at 20 mM Mg^{2+} the rate of synthesis was about half the rate at 4 mM Mg^{2+}. Maximal rate of peptidyl-puromycin formation occurred at the following cation concentrations: 4 mM Mg^{2+}, 1.3 mM Mn^{2+}, 1.2 mM Ca^{2+}, 10 mM putrescine, and 0.4 mM spermidine. The cations were effective in supporting peptidyl-puromycin synthesis in the following order: $Mg^{2+} > Mn^{2+} >$ putrescine $> Ca^{2+} >$ spermidine. All showed a stimulatory phase at low concentrations and an inhibitory phase at cation concentrations above the optimum. Although excess Mg^{2+} could only partially inhibit peptidyl-

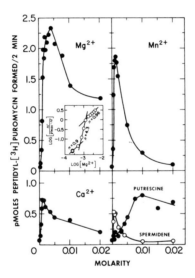

FIG. 4. Effect of divalent cations and oligoamines on rate of peptidyl-puromycin synthesis. Each reaction mixture contained the following in a total of 0.20 ml: divalent cation or oligoamine concentration as indicated on the abscissa; 0.10 M KCl; 50 mM Tris·acetate, pH 7.2; 2.5 A_{260} units of polyribosome preparation; and 1 μM [³H]puromycin. Polysomes were added last to start the incubations and reaction mixtures were incubated at 24° for 2 minutes. Formation of peptidyl-[³H]puromycin was determined as described in the text. Reaction volumes were 0.20 ml instead of the usual 0.05-ml so that the Mg^{2+} in the polyribosome preparation could be sufficiently diluted to determine Mg^{2+}, Mn^{2+}, Ca^{2+}, putrescine, and spermidine curves. The final concentration of Mg^{2+} due to Mg^{2+} in the polyribosome preparation was 0.25 mM. The cation concentrations on the abscissa refer to the concentrations of the cations in addition to this 0.25 mM Mg^{2+} present in each reaction mixture; however, in the case of Mg^{2+} the actual total Mg^{2+} concentration is plotted. A Hill plot of the data for the Mg^{2+} curve is also presented as an insert to the Mg^{2+} portion of the figure. In these experiments a polyribosome extract containing 250 A_{260} units/ml was used.

puromycin synthesis, excess Mn^{2+} or spermidine produced essentially total inhibition (Fig. 4).

The rate of peptidyl-puromycin formation as a function of pH is shown by the data of Fig. 5. At pH 5, the rate of the reaction is negligible. The optimum pH for the rate of peptidyl-puromycin formation was found to be 9.1. Above pH 9.1, the velocity of the reaction decreases.

Since it is known from independent experiments that one puromycin molecule reacts with each peptidyl-tRNA (see footnote 11 for a review), the value of the interaction coefficient as 1.0 is consistent with independent determination of this parameter (Fig. 6). The value for the K_m of 3 μM is about two orders of magnitude lower than the K_m determined for acetylphenylalanyl or formyl-methionyl-puromycin synthesis.[8,16]

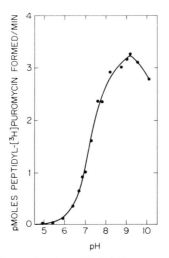

FIG. 5. Effect of pH on rate of peptidyl-[^{3}H]puromycin synthesis. Each 0.20-ml reaction mixture contained the components indicated in the text except that the buffer and volume were varied as follows: Tris concentration was 50 mM, and K$^+$ was present as potassium acetate (0.1 M); the appropriate pH was obtained by prior adjustment with acetic acid at the temperature of the reaction mixture. This Tris· acetate:potassium acetate combination was useful as a buffer throughout the pH range examined. To slow down the rate of peptidyl-puromycin synthesis, the volume of each reaction mixture was 0.20 ml with the following components present: 0.10 M potassium acetate; 2.5 mM NH$_4$Cl; 4.25 mM MgCl$_2$; 50 mM Tris·acetate, pH as given on the abscissa; 0.025% Brij 58 (w/v); 2.8 A_{260} units of polyribosome extract; and 1 μM [^{3}H]puromycin. Reaction mixtures were incubated at 15, 30, and 60 seconds at each pH. From this curve, the initial rate of peptidyl-puromycin synthesis was determined at each pH. The initial rate of peptidyl-puromycin synthesis as a function of pH is presented in the figure.

[16] S. Fahnestock, H. Neumann, V. Shashoua, and A. Rich, *Biochemistry* **9**, 2477 (1970).

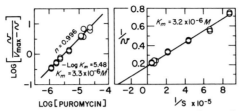

FIG. 6. Hill and double reciprocal plots for puromycin participation in peptidyl-puromycin synthesis. Each 0.050-ml reaction mixture contained the components described in the text except that the puromycin concentration was varied. The data were plotted in the left panel according to the Hill equation. In the right panel, a double reciprocal plot with respect to puromycin as a substrate is plotted. For substrates, $\log[v/(V_{max} - v)] = n \log [S] - \log K$; where v is reaction rate, V_{max} is maximal velocity, n is the interaction coefficient, [S] is substrate concentration and K is equal to K_m (when $n = 1$). For the Hill plot, K_m is 3.3 μM by extrapolation to the intercept of the ordinate; and 3.0 μM from the intercept of the abscissa (where $\log[v/(V_{max} - v)] = 0$). From the double reciprocal plot, $K_m = 3 \mu M$.

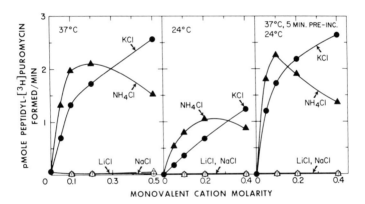

FIG. 7. Effect of monovalent cations on the rate of peptidyl-[³H]puromycin formation with washed polyribosomes. For these experiments, washed polyribosomes were used. These were prepared by layering 0.25 ml of the polyribosome preparation onto a 10% to 30% (w/v) sucrose gradient containing 5 mM MgCl₂ and 5 mM Tris·HCl, pH 7.2. The gradient was centrifuged in a Spinco SW 65 rotor at 60,000 rpm for 80 minutes at 0°. The brownish-yellow pellet of polyribosomes was resuspended in 5 mM MgCl₂ and 5 mM Tris·HCl, pH 7.2, at a concentration of 54 A_{260} units/ml. Each 0.050-ml reaction mixture contained the following components: 1.0 A_{260} unit of washed polyribosomes; 5 mM Tris·acetate, pH 7.2; 5 mM MgCl₂; no Brij 58 was present in reaction mixtures; 4 μM [³H]puromycin; and monovalent cation concentration as indicated on the abscissa. Reaction mixtures were incubated at 37° for 1 minute or 24° for 1 minute, the left and center panels, respectively. In the case of the right panel, reaction mixtures containing washed polyribosomes were incubated at 37° for 5 minutes prior to addition of [³H]puromycin; after addition of [³H]puromycin, incubations were performed at 24° for 1 minute.

When polyribosomes are centrifuged through a sucrose gradient containing no K^+ or NH_4^+, the resulting polyribosomes are inactive in synthesizing peptidyl-puromycin (Fig. 7). In contrast to the results with polyribosomes prepared and stored in the presence of 50 mM NH_4^+ (Fig. 3) where additional K^+ or NH_4^+ is only slightly stimulatory, these polyribosomes washed free of NH_4^+ are relatively inactive in forming peptidyl-puromycin unless they are activated by prior incubation with relatively high concentrations of K^+ or NH_4^+ (Fig. 7). That activation rather than a direct effect on the reaction rate is involved is suggested by comparison of the data of Figs. 3 and 7. If polyribosomes are incubated at 37° for 1 minute in the presence of K^+ or NH_4^+, substantial activity is seen (left panel, Fig. 7); in contrast, however, incubation at 24° for 1 minute produced a relatively small amount of peptidyl-puromycin synthesis (center panel, Fig. 7). If the polyribosomes are incubated at 37° for 5 minutes and then assayed for activity at 24° for 1 minute (right panel, Fig. 7), it can be seen that substantial activity was restored during the 5-minute 37° incubation. Under none of these conditions was Li^+ or Na^+ active.

ANTIBIOTIC	SYNTHETIC MODELS FOR PEPTIDE BOND SYNTHESIS	NATIVE PEPTIDE BOND SYNTHESIS
CHLORAMPHENICOL	●	●
SPARSOMYCIN	●	●
AMICETIN	●	●
GOUGEROTIN	●	●
BLASTICIDIN S	●	●
LINCOMYCIN	●	○
ERYTHROMYCIN	◐	○
NIDDAMYCIN	●	○
CARBOMYCIN	●	○
TYLOSIN	●	○
SPIRAMYCIN III	●	○
PA114 A	●	○
VERNAMYCIN A	●	○
ALTHIOMYCIN	◐	●

● , INHIBITION   ○ , NO EFFECT

FIG. 8. Summary of effects of antibiotics on assays. The column "model systems" refers to synthetic donors such as Ac-Phe-tRNA, fMet-tRNA, polylysyl-tRNA, C-A-C-C-A(Ac-Phe) or C-A-A-C-C-A(fMet). The column native peptide bond synthesis refers to peptidyl-[³H]puromycin synthesis on native polysomes. A filled circle, ●, designates inhibition by that antibiotic; an unfilled circle, ○, designates no effect. For a few of the antibiotics as erythromycin inhibition of peptide bond synthesis in model systems depends on the assay: polylysyl-puromycin synthesis is inhibited, but fMet-puromycin or Ac-Phe-puromycin synthesis is not. For details and a review, see S. Pestka, *Annu. Rev. Microbiol.* **25**, 487 (1971).

A comparison of the effects of antibiotics on model systems for peptide bond synthesis and on peptidyl-[³H]puromycin formation with native polysomes is shown in Fig. 8. Chloramphenicol, sparsomycin, amicetin, gougerotin, and blasticidin S inhibit peptide bond synthesis in both these assays. Lincomycin, the macrolides (erythromycin, niddamycin, carbomycin, tylosin, and spiramycin III), and the streptogramins A (PA114 A and vernamycin A) are potent inhibitors in appropriate model systems, but cannot inhibit peptidyl-puromycin synthesis on native polysomes.[11,15] One antibiotic, althiomycin, was found to inhibit transpeptidation on polyribosomes, but its effect on transpeptidation in model systems seemed to depend on the K^+ and Mg^{2+} concentrations. At high K^+ (0.4 M) or Mg^{2+} (0.04 M) concentrations the inhibitory effect of althiomycin on peptide bond formation appears to be reduced or abolished.[17] The correlation of the effects of antibiotics on native polyribosomes with their effects on intact cells or protoplasts is good.[11] Thus, studies with native polyribosomes can be used to predict the effects of antibiotics on intact cells; and these polyribosome extracts can be used to examine in detail the kinetics of transpeptidation.[14,15]

[17] S. Pestka, in "Antibiotics, Mode of Action" Vol. II (F. Hahn and J. Corcoran, eds.), in press.

[46] Peptidyl-Puromycin Synthesis on Polyribosomes from Rat Liver or Brain

By SIDNEY PESTKA

The reaction of puromycin with nascent peptides on polyribosomes was used to study transpeptidation on mammalian polysomes. This reaction on polyribosomes should be more nearly a model of transpeptidation in the intact cell than studies of peptide bond synthesis with synthetic donors and washed ribosomes. The incorporation of [³H]puromycin into nascent polypeptides provides a sensitive assay for transpeptidation. With the use of this assay, the characteristics and requirements of transpeptidation were evaluated as described below. The assay can be used as a sensitive method to evaluate the effect of antibiotics on peptide bond synthesis itself.

Preparation of Polyribosome Extract

Reagents

Solution A: 0.2 M sucrose, 0.1 M NH₄Cl, 5 mM MgCl₂, 1 mM dithiothreitol, and 20 mM Tris·HCl, pH 7.6

Solution B: 0.7 M sucrose with remaining components as solution A
Solution C: 2.0 M sucrose with remaining components as solution A
Solution D: 10 mM TrisAc, pH 7.2, 5 mM MgCl$_2$, and 50 mM KCl
Sodium deoxycholate: 10% (w/v) in 50 mM Tris·HCl, pH 8

Polyribosomes from rat liver and whole brain were prepared by minor modifications of methods previously reported.[1,2] Both rat liver and brain polyribosomes were prepared from Sprague-Dawley male rats 5–6 weeks old. All procedures were carried out at 0–5° unless otherwise noted. Rat livers were rinsed several times in 0.85% NaCl; 180 g of liver were minced, then homogenized in 3 volumes of solution A in a Teflon–glass tissue homogenizer with a clearance of 0.15–0.23 mm (TRI-R Instruments, No. S-37). The homogenate was centrifuged in a Sorvall SS-34 rotor at 12,000 rpm for 15 minutes. To this postmitochondrial supernatant fraction, 10% (w/v) sodium deoxycholate in 50 mM Tris·HCl, pH 8, was added to a final concentration of 1.3%; 25 ml of the suspension was then layered onto 25 ml of solution B which had been layered on 25 ml of solution C. Each 75 ml tube was then centrifuged in the Spinco Ty 42 rotor at 35,000 rpm for 8 hr. The supernatant fractions were discarded, and the pellets were rinsed and then resuspended in solution D. The polyribosome suspension was then centrifuged in the International PR-6 with the No. 253 rotor at 2500 rpm for 15 minutes to remove aggregated material. The final suspension of 6 ml contained 150–250 A_{260} units/ml and was stored in a liquid nitrogen refrigerator in small portions until used. Polyribosomes from rat brain were prepared similarly except that the final sodium deoxycholate concentration was 1.0% instead of 1.3%. It should be noted that rat liver and rat brain polyribosomes behaved almost identically with respect to peptidyl-puromycin synthesis, and so both could be used interchangeably. Approximately 1200–1500 A_{260} units of polyribosomes were obtained from 100 g of rat liver; and 1000 A_{260} units of polyribosomes from 100 g of whole rat brain.

Determination of Peptidyl-[³H]Puromycin Synthesis

Reagents

Polyribosomes: 150–250 A_{260} units/ml
[³H]Puromycin: 40 μM; specific activity $\geq$ 700 mCi/mmole

Each 0.050-ml reaction mixture contained the following components: 0.80 M KCl; 5 mM MgCl$_2$; 50 mM TrisAc, pH 7.2; about 1–2 A_{260} units

[1] A. K. Falvey and T. Staehelin, *J. Mol. Biol.* **53**, 1 (1970).
[2] C. E. Zomzely, S. Roberts, S. Peache, and D. M. Brown, *J. Biol. Chem.* **246**, 2097 (1971).

of polyribosomes; and 4 μM [³H]puromycin. Ordinarily, polyribosomes are added last to start reactions, which were performed at 24° for 1 minute. [³H]Puromycin incorporation into nascent polypeptides was determined as described in the previous section of this volume. The assay for [³H]puromycin incorporation into trichloroacetic acid-precipitable material is a modification of previously published procedures.[3,4] Reactions were stopped by addition of 2 ml of cold 10% (w/v) trichloroacetic acid to precipitate peptidyl-[³H]puromycin. The reaction mixtures were then filtered through a polyvinylchloride Millipore filter (type BDWP, 0.6 μm pore size, 25 mm diameter); each tube and filter were washed three times with 3-ml portions of 5% trichloroacetic acid; the filter was washed eight times with 3-ml portions of absolute ethanol at room temperature to remove unreacted puromycin.

Characteristics of Peptidyl-[³H]Puromycin Formation

The kinetics of peptidyl-[³H]puromycin formation at various temperatures is shown in Fig. 1.[5] The extent as well as the rate of peptidyl-puromycin formation increases with the temperature of the incubation. For

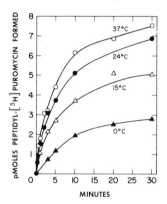

FIG. 1. Kinetics of peptidyl-puromycin synthesis. Each 0.050-ml reaction mixture contained the following components: 50 mM Tris·acetate, pH 7.2; 5 mM MgCl₂; 0.5 M KCl; 4 μM [³H]puromycin; and 1.4 A_{260} units of rat brain polyribosomes. Incubations were performed at the time and temperature indicated on the figure. Assays were performed as described in the text. Essentially identical results were obtained with the use of polyribosomes from rat liver.

[3] L. Skogerson and K. Moldave, *Arch. Biochem. Biophys.* **125**, 497 (1968).

[4] J. A. Schneider, S. Raeburn, and E. S. Maxwell, *Biochem. Biophys. Res. Commun.* **33**, 177 (1968).

[5] S. Pestka, R. Goorha, H. Rosenfeld, C. Neurath, and H. Hintikka, *J. Biol. Chem.* **247**, 4258 (1972).

measuring the rate of peptidyl-[³H]puromycin synthesis in most experiments reactions were incubated at 24° for 1 minute.

In contrast to polyribosomes from *E. coli*, mammalian polyribosomes appear to require high concentrations of K⁺ or Na⁺ in reaction mixtures for maximal transpeptidation (Fig. 2); NH₄⁺ was relatively ineffective and Li⁺ did not support peptidyl-puromycin synthesis. Although K⁺ was best, Na⁺ was also effective. Maximal transpeptidation occurred at 0.8 *M* K⁺ but was not reached even by 1.2 *M* Na⁺. This contrasts with the monovalent cation requirements for transpeptidation with bacterial polyribosomes; with bacterial polyribosomes, K⁺ and NH₄⁺ were effective, whereas both Na⁺ and Li⁺ were both ineffective and inhibitory.[6] Transpeptidation with both bacterial and mammalian polyribosomes was inhibited Li⁺. It should be noted that rat liver and rat brain polyribosomes gave essentially identical curves for monovalent cation effects (legend to Fig. 2). The requirement for high K⁺ has also been noted for acetyl-leucyl-puromycin synthesis.[7]

A double reciprocal plot with respect to K⁺ suggests that K⁺ may be participating in the reaction (Fig. 2). Furthermore, the fact that the Hill plot produces an interaction coefficient of 1 with respect to K⁺ suggests that a single critical K⁺ is required for transpeptidation similar to that found

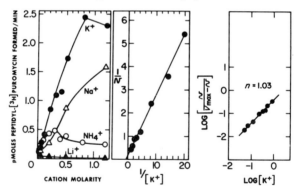

FIG. 2. Effect of monovalent cations on peptidyl-puromycin synthesis. Each 0.050-ml reaction mixture contained the following components: 50 m*M* Tris·acetate, pH 7.2; 5 m*M* MgCl₂; 4 μ*M* [³H]puromycin; 1.4 *A*₂₆₀ units of rat brain polyribosomes; and monovalent cation as indicated on the abscissa. Each reaction mixture contained 20 m*M* K⁺ which was the lowest concentration of K⁺ which could be attained due to the amount of K⁺ in the polyribosome preparation. Each reaction mixture was incubated at 24° for 1 minute and assayed as described in the text. Essentially identical curves were obtained with the use of polyribosomes from rat liver.

[6] S. Pestka, *Proc. Nat. Acad. Sci. U.S.* 69, 624 (1972).
[7] R. Neth, R. E. Monro, G. Heller, E. Battaner, and D. Vazquez, *FEBS (Fed. Eur. Biochem. Soc.) Lett.* 6, 198 (1970).

for polyribosomes from *E. coli.*[6] On the other hand, the high K[+] concentrations required may reflect that these polyribosome preparations were prepared in the presence of NH_4^+. It appears that with bacterial and mammalian polyribosomes, there is a single critical site which must be occupied by an appropriate monovalent cation for maximal transpeptidation. The sites on the mammalian ribosomes, however, must differ from those on bacterial polyribosomes since their ion specificities differ. These results suggest that, although the overall process of transpeptidation may be similar, the transpeptidase complex is not identical on bacterial and mammalian ribosomes. Similar conclusions are suggested from antibiotic effects on transpeptidation.[7]

Peptidyl-puromycin synthesis on native mammalian polyribosomes has additional characteristics and requirements different from peptidyl-puromycin formation with synthetic donors and washed ribosomes.[3,7,8] The optimum Mg^{2+} concentration for peptidyl-puromycin synthesis was some-

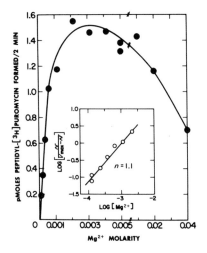

FIG. 3. Effect of Mg^{2+} concentration on peptidyl-puromycin synthesis. Each reaction was performed in a volume of 0.2 ml so that the Mg^{2+} present in the polyribosome suspension could be diluted out sufficiently to perform a Mg^{2+} curve. Each 0.20-ml reaction mixture contained the following components: 50 mM TrisAc, pH 7.8; 0.5 M KCl; Mg^{2+} concentration as indicated on the abscissa; 1 μM [³H]puromycin; 1.2 A_{260} units of rat liver polyribosomes. Reaction mixtures were incubated at 24° for 2 minutes and assayed as described in the text. The incubation time of 2 minutes was used because of the larger reaction volume. A double reciprocal plot of the data of this figure from 0.37 mM to 2.1 mM MgCl₂ produced a linear function which yielded a V_{max} of 2.27. From this value of the V_{max}, the Hill plot shown as an inset to the figure was calculated.

[8] D. Vazquez, E. Battaner, R. Neth, G. Heller, and R. E. Monro, *Cold Spring Harbor Symp. Quant. Biol.* 34, 369 (1969).

where between 2 mM and 5 mM (Fig. 3). This was lower than the 30 mM Mg^{2+} necessary when synthetic donors and washed ribosomes are used.[7] A double reciprocal plot of the reaction velocity with respect to Mg^{2+} also produced a straight line as in the case of K$^+$ from which the V_{max} could be obtained (data not shown). With this V_{max} value, a Hill plot (inset, Fig. 3) was obtained which suggested that one critical Mg^{2+} is necessary for transpeptidation on mammalian polyribosomes over the concentration range studied up to 5 mM. Concentrations of Mg^{2+} greater than 10 mM inhibited transpeptidation. This is analogous to the findings with *E. coli* ribosomes. It is likely, however, that additional Mg^{2+} is required for maintenance of ribosomal structure and perhaps other ribosomal functions as previously noted.[5,6] The K_m for Mg^{2+} is estimated to be about 1 mM. Mn^{2+} could substitute for Mg^{2+}; however, Ca^{2+}, spermidine, and putrescine were ineffective in supporting peptidyl-puromycin synthesis to any significant degree.[5]

The rate of peptidyl-puromycin formation as a function of pH is shown by the data of Fig. 4. At pH 6 and below, the rate of peptidyl-puromycin synthesis was negligible. The optimum pH for peptidyl-puromycin synthesis is in the vicinity of pH 8.3–8.8. Analysis of the effect of pH on peptidyl-puromycin synthesis was consistent with H$^+$ acting as both an inhibitor and substrate (Fig. 5). The data of Fig. 5 (left panels) indicate

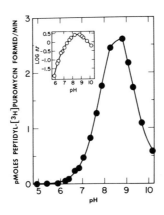

FIG. 4. Effect of pH on peptidyl-puromycin synthesis. Each 0.20-ml reaction mixture contained the following components: 50 mM TrisAc, pH as indicated on the abscissa; 5 mM MgCl$_2$; 0.5 M KCl; 1 μM [^{3}H]puromycin; 1.4 A_{260} units of rat brain polyribosomes. Reactions were incubated at 24° for various times and the initial rate of peptidyl-puromycin synthesis calculated from the initial slope of each curve. In order to obtain valid reaction velocities for pH values near the optimum, the reaction volume was increased to 0.2 ml rather than the usual 0.05 ml and [^{3}H]puromycin concentration reduced to 1 μM. Essentially an identical curve was obtained with rat liver polyribosomes.

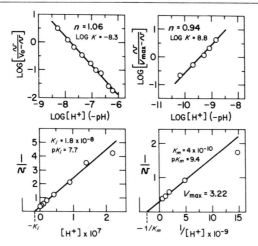

FIG. 5. Effect of H^+ on peptidyl-puromycin synthesis. The data of Fig. 4 were plotted in several ways. By plotting the data according to M. Dixon [*Biochem. J.* **55**, 170 (1963)] the data (lower left panel), indicate that the pK_i is equal to 7.7 for the ionizable group inhibiting peptidyl-puromycin synthesis in the protonated form [S. Pestka, *Proc. Nat. Acad. Sci. U.S.* **69**, 624 (1972)]. From the usual double reciprocal plot (lower right panel) for the pH range where peptidyl-puromycin synthesis was dependent on H^+ (pH 10 to 8.8), the pK_m was estimated to be 9.4. From the double reciprocal plot, the V_{max} was determined to be 3.22. This value for the V_{max} was used to calculate the Hill plots with respect to H^+ as inhibitor (left upper panel) and with respect to H^+ as a substrate (right upper panel). The intercepts are log $K' = -8.3$ and log $K = 8.8$ for the upper left and right panels, respectively; and the interaction coefficients were determined to be 1.06 and 0.94 for the upper left and right panels, respectively. It should be noted that the intercept of the Hill plots should be equivalent to the pK values obtained by the Dixon and double reciprocal plots. In the case of the inhibition by H^+, $-\log K'$ should be equal to the pK_a of the inhibitory ionizable group for noncompetitive inhibition by H^+ (Pestka, *loc. cit.*). The fact that $-\log K' = 8.3$ and $pK_i = 7.7$ may indicate some discrepancies in the extrapolation to obtain the intercept of the Hill plot on the ordinate. To obtain the intercept of the Hill plot, the extrapolation must be carried out far beyond the actual observable points, and thus small errors in the slope will produce a large error in the intercept of the ordinate. When v equals $V_0/2$, $\log[v/(V_0 - v)]$ equals 0; the $pI_{0.5}$ and pH equals 7.8 as determined from the intercept of the abscissa (upper left panel). This is close to the pK_i of 7.7 obtained from the Dixon plot. Similarly, in the case of H^+ participation in the reaction as a substrate (pH 10 to pH 8.8), the log K determined as 8.8 from the intercept of the ordinate is probably less accurate than the pK_m of 9.4 determined by the double reciprocal plot. However, when v equals $V_{max}/2$, $\log[v/(V_{max} - v)]$ equals 0; from the intercept of the abscissa the pH (and pK_m) is found to be 9.4 (upper right panel).

that a single H^+ is involved in inhibiting a rate-determining step in peptidyl-puromycin synthesis on mammalian polyribosomes. The pK_i of 7.7 (also pK_a) characterized protonation of the group involved in this inhibition. This dissociation constant is consistent with the possibility of a single imidazole residue (or conceivably an N-terminal amino group) being involved at the active center of the transpeptidase complex and in the rate-determining step for transpeptidation.[6] The protonated form of this group inactivated the complex.

By carrying out a kinetic analysis by the method of Dixon[9] throughout the appropriate pH range at several puromycin concentrations, it was also shown that inhibition by H^+ was noncompetitive (Fig. 6). The pK_i of 7.7 is consistent with the pK_a and pK_i as obtained above (Fig. 5). Since the pH curve is similar with hydroxypuromycin,[10] it is unlikely that protonation of the α-amino group of puromycin produces these pH effects. It should be noted that a plot of log v as a function of pH (Fig. 4, inset) according to Dixon and Webb[11] does not yield an indication of pK_a values for the protonated groups involved in the inhibitory or stimulatory phases. Thus, considering H^+ as an inhibitor or substrate may be useful as a simplified approach to determine the pK_a values of protonated groups involved in the reaction under appropriate circumstances.

For the pH range where peptidyl-puromycin synthesis was dependent

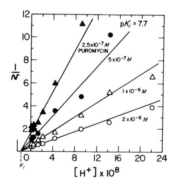

Fig. 6. Noncompetitive inhibition of peptidyl-puromycin synthesis by $[H^+]$. Each 0.20-ml reaction mixture contained the components and was assayed as indicated in the legend to Fig. 4 except that $[^3H]$puromycin concentration was varied: 0.25 μM (▲), 0.5 μM (●), 1 μM (△), 2 μM (○) and 1.2 A_{260} units of rat liver polyribosomes were present in each reaction mixture.

[9] M. Dixon, *Biochem. J.* **55**, 170 (1963).
[10] S. Fahnestock, H. Neumann, V. Shashoua, and A. Rich, *Biochemistry* **9**, 2477 (1970).
[11] M. Dixon and E. C. Webb, "Enzymes," 2nd ed. Academic Press, New York, 1964.

on pH (pH 10 to 8.8) apparently another single group with a pK_m of 9.4 is involved; protonation of this group activates the transpeptidase complex. The association constant of this group is consistent with it being either a phenolic hydroxyl (tyrosine) or an ϵ-ammonium group. An analysis of both inhibitory and stimulatory phases by Hill plots (Fig. 5, upper panels), suggested that a single H^+ is involved in each case ($n = 1.06$ and 0.94, respectively).

The pH optimum for transpeptidation on mammalian polyribosomes of 8.3 to 8.8 (Fig. 4) suggests that at physiological pH this reaction in the intact cell may be occurring at a substantially reduced rate compared to its potential capacity. However, the pH optimum measured for transfer of nascent peptides to puromycin may differ from the pH optimum for peptide bond synthesis with aminoacyl-tRNA as an acceptor. In addition, the analysis of hydrogen ion effects by considering H^+ as an inhibitor or substrate may under certain conditions be a simple and useful way of evaluating pH effects on reaction rates.[6]

Since it has been already reported that one puromycin molecule attaches to a single polypeptide chain of peptidyl-tRNA (see footnote 12 for a review), the determination of the interaction coefficient for puromycin was used as a control for these experiments (Fig. 7). The experimentally determined value of 1.03 for n is in good agreement with the expected value of 1. Thus, the methodology may be useful in understand-

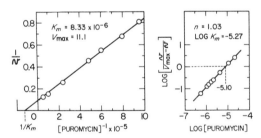

FIG. 7. Effect of puromycin on rate of peptidyl-puromycin synthesis. Each 0.050-ml reaction mixture contained the components indicated in the text except for the following: 0.05 TrisAc, pH 7.8; 1.2 A_{260} units of rat liver polyribosomes; and [³H]-puromycin concentration as indicated on the abscissa. The initial velocity of peptidyl-puromycin synthesis was determined from curves of peptidyl-puromycin synthesis as a function of time. In the left panel, the results are plotted in the form of a double reciprocal plot; and in the right panel, a Hill plot of the data is presented. From the Hill plot, log K_m was found to be equal to -5.27 as determined from the intercept of the ordinate; and -5.10 (8 μM) as determined from the intercept of the abscissa.

[12] S. Pestka, *Annu. Rev. Microbiol.* **25**, 487 (1971).

EFFECT OF ANTIBIOTICS ON PEPTIDYL-[³H]PUROMYCIN
SYNTHESIS WITH RAT POLYRIBOSOMES

Antibiotic	Molarity	Percent of control	Type of inhibition
None	—	100	—
Sparsomycin	$10^{-4} M$	1	Competitive
Anisomycin	$10^{-4} M$	16	Competitive
Amicetin	$10^{-3} M$	80	—
Gougerotin	$10^{-4} M$	26	Mixed
Blasticidin S	$10^{-4} M$	12	Mixed
Cycloheximide	$10^{-4} M$	54	Mixed
Acetoxycycloheximide	$10^{-4} M$	54	Mixed
Cryptopleurine	$10^{-3} M$	58	Mixed
Emetine	$10^{-3} M$	97	—
Aurintricarboxylic acid	$10^{-4} M$	138	—
Pactamycin	$10^{-4} M$	105	—
Fusidic acid	$10^{-3} M$	82	—
Edeine	$10^{-4} M$	97	—
Tetracycline	$10^{-4} M$	115	—
Chloramphenicol	$10^{-3} M$	102	—

[a] The data are taken from S. Pestka, H. Rosenfeld, R. Harris, and H. Hintikka, *J. Biol. Chem.* **247**, 6895 (1972).

ing the mechanisms involved in peptidyl-puromycin synthesis on native polyribosomes. The K_m for puromycin was determined as 8 μM.

Antibiotics that appear to inhibit peptidyl-puromycin synthesis on mammalian polyribosomes include sparsomycin, anisomycin, nucleoside antibiotics (gougerotin and blasticidin S), cycloheximide, and crypto-pleurine. The inhibitions by sparsomycin and anisomycin are competitive with respect to puromycin. The inhibitions by the other antibiotics appear to be of a mixed type by analysis with the double-reciprocal plots; however, analysis by the method of Dixon[9] suggests two phases of inhibition.[13] A summary of the effects of antibiotics on peptidyl-puromycin synthesis on mammalian ribosomes is presented in the table.

This assay for peptidyl-puromycin synthesis with the use of mammalian polyribosomes is a simple one and can be used readily to study the kinetics of transpeptidation as well as the effect of antibiotics on this reaction. As shown in studies with *E. coli* polyribosomes,[14] results obtained with this assay reflect results occurring in the intact cell.

[13] S. Pestka, H. Rosenfeld, R. Harris, and H. Hintikka, *J. Biol. Chem.* **247**, 6895 (1972).
[14] S. Pestka, *J. Biol. Chem.* **247**, 4669 (1972).

[47] Assay of Ester and Polyester Formation by the Ribosomal Peptidyltransferase

By STEPHEN FAHNESTOCK, HELMUT NEUMANN, and ALEXANDER RICH

The peptidyltransferase of the ribosome has several catalytic activities. One of these is the familiar function in normal peptide bond synthesis. In that process, a peptide group is transferred from one tRNA molecule to the α-amino group of an adjoining aminoacyl-tRNA, resulting in the formation of an amide or peptide linkage. Several assays have been developed for measuring the ability of the ribosomal peptidyltransferase to form peptide linkages. In this section, we describe an assay for the ability of the ribosomal peptidyltransferase to form ester linkages. In these experiments, the α-amino group of aminoacyl-tRNA is converted chemically to an α-hydroxy group. In this form, the deaminated aminoacyl-tRNA can be used to form ester linkages and polyester linkages, directed either by synthetic polynucleotides or by naturally occurring messenger RNA.[1,2] In a closely related reaction, an analog of puromycin is used in which the α-amino group has substituted for it an α-hydroxy group. In that form the α-hydroxy puromycin can be used in a modification of the fragment reaction. In the fragment reaction, formylmethionyl puromycin is formed.[3] When the analog is used, the formylmethionyl group is linked to the puromycin residue by an ester linkage.[4]

Both of these reactions can be used as an assay for the ability of the ribosomal peptidyltransferase to act catalytically in forming ester linkages as well as the normally occurring amide linkages. In the present section, we describe the synthesis of the α-hydroxy analogs both of puromycin and of aminoacyl-tRNA and describe the manner in which they can be used to assay ester formation.

Puromycin Analogs

Synthesis of Demethoxy-α-Hydroxypuromycin (Fig. 1)

6-Dimethylamino-9-(3′-(2-L-hydroxy-3-phenyl propionylamino)-3′-deoxy-β-D-ribofuranosyl)purine (III). A solution of 136 mg (0.66 mmole) of dicyclohexylcarbodiimide in 4.4 ml of acetonitrile was added to 97 mg (0.33 mmole) of puromycin aminonucleoside (I) and 110 mg (0.66 mmole) of L-(−)-3-phenyllactic acid (II) in a centrifuge tube, followed

[1] S. Fahnestock and A. Rich, *Science* **173**, 340 (1971).

[2] S. Fahnestock and A. Rich, *Nature (London) New Biol.* **229**, 8 (1971).

[3] R. E. Monro and K. E. Marcker, *J. Mol. Biol.* **25**, 347 (1967).

[4] S. Fahnestock, H. Neumann, V. Shashoua, and A. Rich, *Biochemistry* **9**, 2477 (1970).

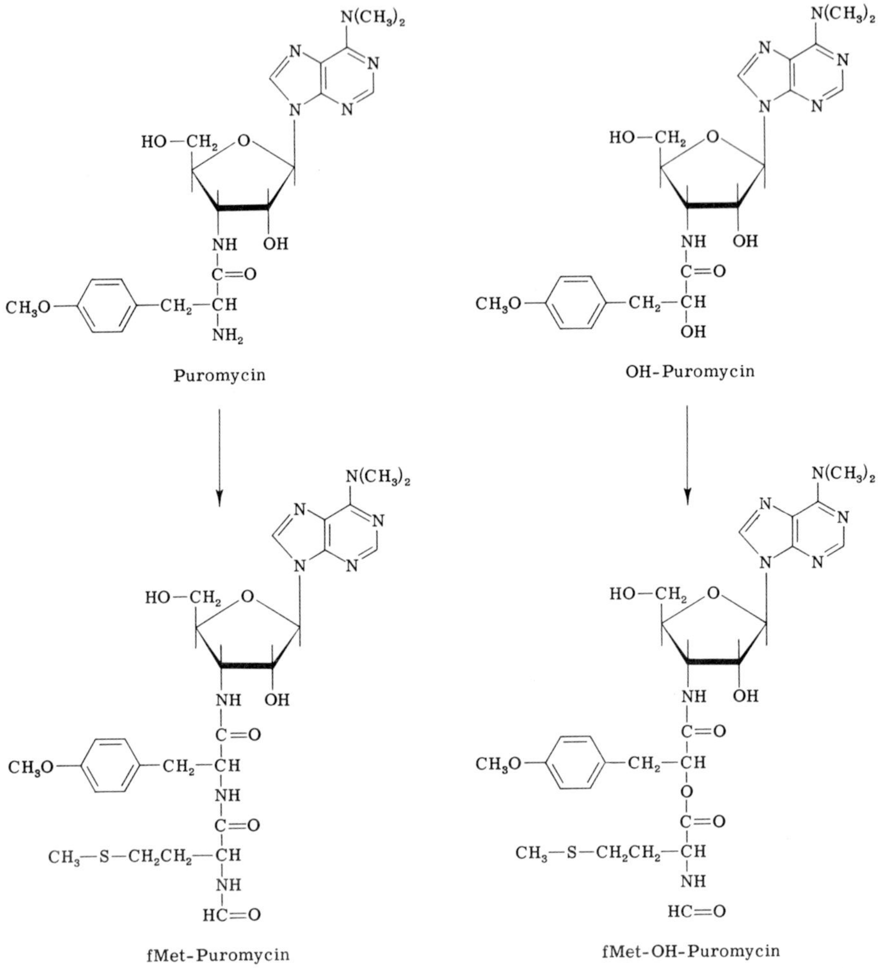

Fig. 1. Formulas of the products of the reaction of formylmethionyl-tRNA with either puromycin or demethoxy-α-hydroxypuromycin.

immediately by vigorous stirring for 15 minutes. The precipitated dicyclo-hexylurea was separated by centrifugation, and the supernatant was evaporated *in vacuo*. The colorless residue was dissolved in 2 ml of methanol and allowed to react with 0.5 ml of 2 N sodium hydroxide solution for 2 hours to hydrolyze any esters, neutralized to pH 5 with 1 N hydrochloric acid, and sodium bicarbonate was added until pH 8. The solvents were evaporated and the residue was extracted with 4 times 2 ml of acetonitrile. Evaporation of the acetonitrile solutions yielded 220 mg of white solid, which was purified by silica gel thin-layer chromatography (CCl₄:MeOH =

5:1) and recrystallized from methanol to give colorless needles in 80% yield; m.p. 209.5–210°, λ_{max} 272 nm ($\epsilon = 21{,}800$), $[\alpha]_D^{25} = -77°$ ($c = 0.1$ in ethanol).

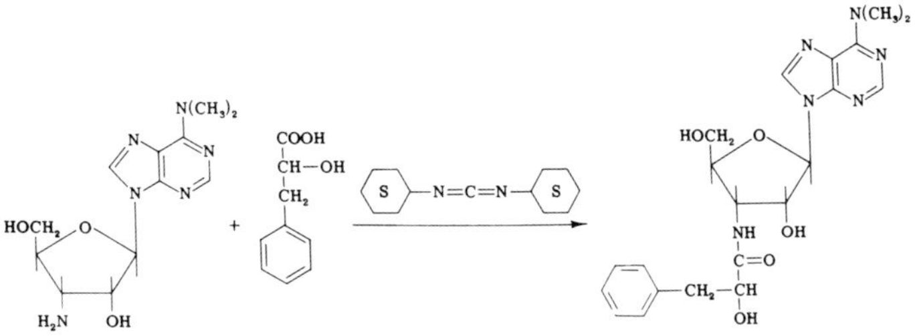

Synthesis of α-Hydroxypuromycin (Fig. 1)

α-Hydroxypuromycin was obtained by nitrous acid deamination of puromycin. One hundred milligrams of puromycin dihydrochloride was dissolved in 20 ml of cold (0°) 1 N HCl. To this was added, slowly with stirring, 20 ml of cold 1 N $NaNO_2$. The mixture was incubated overnight at 0°. Four milliliters of 5 N NaOH was then added, and the mixture was extracted with three 50-ml aliquots of ethyl acetate. Combined ethyl acetate layers were washed with 25 ml of H_2O and dried at room temperature on a Buchi flash evaporator. The material was dissolved in methanol and subjected to chromatography on a 2-mm thick layer of Silica Gel G with fluorescent indicator (Analtech) in carbon tetrachloride/methanol (10:1). Ultraviolet absorbing bands were located and eluted with acetone. Three bands were obtained, none of which corresponded to puromycin. The band pattern was not appreciably different if the deamination was carried out under conditions similar to those of Herve and Chapeville.[5] Only one of these bands proved to be active in the fragment system as described below. It was a minor component comprising about 10% of the total. The other components probably arise from rearrangements and racemization at the α-carbon.[6] They must be removed since they inhibit the activity of the active component.

Activity Assays

Aminoacyl-tRNA Synthesis

The standard system for aminoacyl-tRNA synthesis contained, per milliliter: 0.1 M HEPES pH 7.4, 20 mM magnesium acetate, 7.5 mM ATP, 0.2 mM CTP, 20 mM creatine phosphate, 70 μg of creatine phospho-

[5] G. Herve and F. Chapeville, *J. Mol. Biol.* 13, 757 (1965).
[6] J. D. Roberts and C. M. Regan, *J. Amer. Chem. Soc.* 75, 2069 (1953).

kinase, 10 mM glutathione, S100 extract[7] (2 mg of protein), 2–5 mg of stripped tRNA, and amino acids. In addition, mixtures for fMet charging contained 0.6 mM folinic acid. Incubation was for 15 minutes at 37°. The mixture was then chilled on ice and the pH was lowered to about 5 with acetic acid. Protein was extracted by shaking at 2° with an equal volume of vacuum-distilled, water-saturated phenol, followed by centrifugation for 15 minutes at 18,000 g in a Sorvall SS-1 rotor at 2°. RNA was precipitated from the aqueous phase by adding 0.1 volume 20% w/v potassium acetate, pH 5, and 2 volumes of cold ethanol, and after 30 minutes at −20° was collected by centrifugation.

[^{35}S]fMet-tRNA was prepared in the standard way using [^{35}S]methionine. It was then freed of aminoacyl-tRNA by treatment with 20 mM CuSO$_4$, 0.2 M sodium acetate, pH 5.5, for 60 minutes at 37°.[8] The fMet-containing hexanucleotide fragment was obtained from fMet-tRNA by digestion with ribonuclease T1, as described by Marcker.[9]

Incubation Systems

fMet Fragment Reaction.[3,10] Prior to methanol addition, the reaction mixture (0.1 ml) contained 60 mM Tris·HCl buffer (pH 8.1, measured at 0°), 0.4 M KCl, 20 mM magnesium acetate, 6 mM β-mercaptoethanol, 13.8 A_{260} units per milliliter of salt-washed ribosomes, and the formyl-methionyl T1 fragment or fMet-tRNA (50,000 dpm/ml).

The reaction was initiated with an equal volume of methanol (blank) or a methanolic solution of puromycin or puromycin analogs. Concentrations of puromycin and analogs were determined spectrophotometrically in 0.1 N HCl solution ($\lambda_{max} = 267.5$ nm; $\epsilon = 2.0 \times 10^4$ for all compounds). Incubation was at 0°. The reaction was terminated with 25 μl of 0.1 M BeCl$_2$, and 0.1 ml of 0.3 M sodium acetate (pH 5.5) saturated with MgSO$_4$, and then 1.5 ml of ethyl acetate were added. The mixture was shaken at room temperature for 15 seconds and centrifuged briefly. One milliliter of the ethyl acetate layer was counted in a liquid scintillation spectrometer.

An example of the reaction of [^{35}S]fMet hexanucleotide with the α-hydroxylpuromycin is shown in Fig. 2.[4] Puromycin is seen to react more rapidly than the analog.

mRNA Directed Reaction. The reaction can also be assayed with R17 viral RNA. The complete reaction mixture contained 0.1 M Tris·HCl (pH 7.4); 5 mM magnesium acetate; 50 mM KCl; 6 mM β-mercaptoethanol; 0.6 mM GTP; 0.93 A_{260} unit/ml [^{35}S]fMet-tRNA, containing 2.3 × 10^5

[7] M. Nirenberg, this series, Vol. 6, p. 17.
[8] P. Schofield and P. C. Zamecnik, *Biochim. Biophys. Acta* **155**, 410 (1968).
[9] K. A. Marcker, *J. Mol. Biol.* **14**, 63 (1965).
[10] B. E. H. Maden and R. E. Monro, *Eur. J. Biochem.* **6**, 309 (1968).

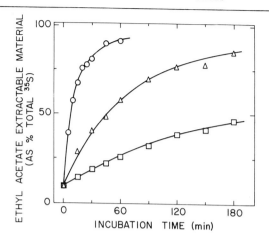

Fig. 2. Time course of the formylmethionyl T1 oligonucleotide fragment reaction with ψ-hydroxypuromycin and puromycin. The final concentration of puromycin was 0.25 mM (○) and the concentration of ψ-hydroxypuromycin was 1.9 mM (△). □, Control.

dpm/A_{260} unit and about 60% formylated; 41 A_{260} units per milliliter of ribosomes and 0.8 mg/ml of crude initiation factors[11]; 7.2 A_{260} units per milliliter of bacteriophage R17 RNA; and 1.0×10^{-3} M α-hydroxypuromycin. Aliquots (0.1 ml) were incubated for 15 minutes at 30°. The reaction was terminated with 25 μl of 0.1 M BeCl$_2$. The 0.1 ml of 0.3 M sodium acetate (pH 5.5) saturated with MgSO$_4$ and 1.5 ml of ethyl acetate were added and the mixture was shaken at room temperature for 15 seconds. After brief centrifugation, 1 ml of the ethyl acetate layer was counted in a liquid scintillation spectrometer.

α-Hydroxyacyl-tRNA's

In addition to the analogs of puromycin, the α-hydroxyacyl-tRNA's can also be used as peptidyltransferase substrates in polypeptide synthesizing systems directed by poly(U)[1] and by bacteriophage R17 RNA.[2] The α-hydroxyacyl-tRNA's are prepared by nitrous acid deamination of aminoacyl tRNA's.

[14C]Phelac-tRNA. The compound was prepared from [14C]Phe-tRNA^{Phe} by nitrous acid deamination under conditions which were shown by Carbon[12] to result in minimal alteration of the tRNA. Three milligrams of [14C]Phe-tRNA was dissolved in 1.5 ml of 0.25 M sodium acetate pH 4.3, 10 mM magnesium acetate, 1 M NaNO$_2$, and incubated at room tem-

[11] J. S. Anderson, M. S. Bretscher, B. F. C. Clark, and K. A. Marcker, *Nature* (*London*) 215, 490 (1967).

[12] J. A. Carbon, *Biochim. Biophys. Acta* 95, 550 (1965).

perature on a Radiometer pH-stat at pH 4.3 for 30 minutes. RNA was then precipitated with ethanol at $-20°$, the precipitate was redissolved in water and reprecipitated, and the pellet was washed with ethanol. Ethanol was removed under vacuum and the residue was dissolved in water and stored at $-20°$. An aliquot was incubated in 1 M triethylamine (pH 12.5) at 37° for 30 minutes, then subjected to paper chromatography [Whatman No. 1, solvent n-butanol/acetic acid/water (78:5:17)] and to silica gel thin-layer chromatography in the same solvent. The paper chromatogram was scanned on a Vanguard strip counter, and the silica gel plate was subjected to auto-radiography on Kodak Royal Blue X-ray film. Nonradioactive carrier Phe and Phelac were located by means of their fluorescence in liquid nitrogen under ultraviolet excitation. Paper chromatography demonstrates that less than 0.1% of the radioactivity is phenylalanine. About 20% remains at the origin and most of the radioactivity moves with carrier Phelac. Thin-layer chromatography reveals that most of the radioactivity cochromatographs with carrier Phelac, but there are two faster moving components. These may arise from rearrangements like those encountered in the nitrous acid deamination of puromycin. In that reaction the rearranged products comprised 90% of the yield. However, the presence of the p-methoxy group on the aminoacyl portion of puromycin would be expected to increase the amount of rearrangement observed,[6] so less rearrangement is expected in the deamination of Phe, and less is apparently observed. A significant portion of the radioactivity remains at the origin on silica gel also. This material may be unhydrolyzed Phelac-tRNA or some other rearrangement product. If it is unhydrolyzed tRNA, it must contain less than 0.1% Phe, since Phe-tRNA is more labile in alkali than is Phelac-tRNA.[5]

In the above charging of purified tRNA[Phe] a parallel incubation was carried out using tRNA which had previously been treated with nitrous acid in the standard way. Untreated tRNA accepted 643 pmoles Phe/A_{260} unit, while nitrous acid-treated tRNA accepted 454 pmoles/A_{260} unit. This 31% reduction in the amino acid acceptance of the tRNA after nitrous acid treatment reveals that there is some damage to the tRNA during such treatment. The treatment of Herve and Chapeville[5] was reported to decrease the Phe acceptance of tRNA by 70 to 80%.

Poly(U)-Directed System

The Incorporation System. In general the incorporation system contained 0.11 M Tris·HCl, pH 7.4; 0.096 M NH$_4$Cl; 16 mM magnesium acetate; 0.5 mM GTP; 4 mM PEP; 0.89 mg/ml S100 protein; 2.6 A_{260} units/ml ribosomes; 10 A_{260} units/ml [^{14}C]Phelac-tRNA (28 pmoles of Phelac/A_{260} unit, at 459 mCi/mmole); 24 μg/ml of poly(U). Components were generally mixed on ice in the order shown. The incorporation was

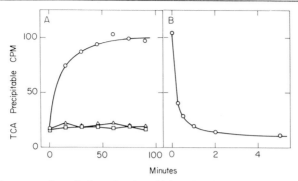

FIG. 3. Incorporation of phenyllactic acid into insoluble precipitate. (A) Kinetics of the incorporation. $\triangle$, Minus S100; $\square$, minus poly(U). (B) Alkaline digestion of acid-precipitable product in 0.67 M NaOH, 0°.

initiated by transfer to 30°, at which temperature they were incubated for various time periods.

The Assay. To each sample was added 0.5 volume of a solution of pancreatic ribonuclease (0.5 mg/ml) in 0.1 M EDTA, pH 7.4. They were then incubated for 15 minutes at 30°. Cold 10% TCA (40 times the original sample volume) containing 0.5% w/v casamino acids was added, and the samples were passed through Millipore type HAWP filters. The filters were washed three times with 3-ml aliquots of 5% TCA containing 0.25% w/v casamino acids, affixed to aluminum planchets by means of a drop of 1% ovalbumin, and dried under an infrared lamp. They were then counted at an efficiency of 23% in a Nuclear Chicago low background gas flow counter.

This assay is based on the ability of the EDTA-RNase treatment to degrade all tRNA both ribosome bound and unbound so that all unincorporated Phelac-tRNA is rendered TCA soluble. Control experiments show that the assay is sufficient to solubilize more than 99% of the Phelac-tRNA, whether it is bound to ribosomes or free.

An example of the kinetics of incorporation of phenyllactic acid into a polymer is shown in Fig. 3, together with its hydrolysis in alkali.[1]

R17 RNA-Directed System

An outline of the assay experiment and its results are shown in Fig. 4. The filled circle represents the ester linkage.[2]

The Incorporation System. The incubation mixture (1 ml) contained 50 mM Tris·HCl, pH 7.4, 7 mM magnesium acetate, 60 mM NH$_4$Cl, 0.2 mM GTP, 5.2 mM PEP, 6 mM β-mercaptoethanol, 8.0 A_{260} units/ml R17 amB$_2$ RNA, 50 A_{260} units/ml ribosomes (not salt washed, consisting of 60% native subunits)[13] 0.32 mg/ml RNA-free S100[13] 4.0 A_{260} units/ml

[13] E. Keuchler and A. Rich, *Nature (London)* **225**, 920 (1970).

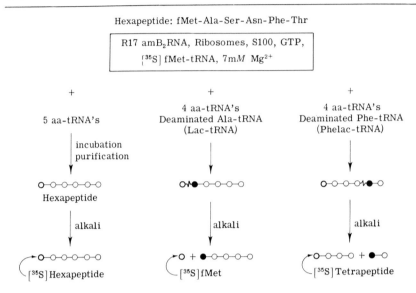

Fig. 4. Ester incorporation into R17 coat peptide. Outline of the experiment: open circles are used to indicate amino acids. The boldface open circles represent [35S]fMet, the only radioactive amino acid. The filled circles represent α-hydroxy acids.

tRNA charged with [35S]fMet (5.0×10^4 dpm/A_{260} unit) and 20 A_{260} units/ml tRNA charged with unlabeled amino acids and α-hydroxyacids as described below. The mixture was incubated at 34° for 30 minutes, then the TCA soluble, N-blocked peptides were isolated according to the procedure of Capecchi.[14] To 1 ml of incubation mixture (chilled) was added 4 ml of cold 5% w/w TCA, and the precipitate was removed by centrifugation. The supernatant was extracted by shaking with 5 ml of a mixture of m-cresol and ethyl acetate (0.7 ml:0.3 ml). The phases were separated by centrifugation at room temperature, and the upper (organic) phase was removed. To the organic phase was added 1 ml of H_2O and 10 ml of ether, and after shaking and centrifugation the ether phase was removed and the aqueous phase was washed with 10 ml of ether. The last traces of ether were removed in a stream of air, and the aqueous phase was passed through a column (0.5 cm × 2 cm) of Dowex 50-X2, equilibrated with H_2O. The material was eluted with H_2O and lyophilized.

The hexapeptide was further purified by electrophoresis on paper (Whatman 3 MM) in a buffer composed of 14 ml pyridine + 12.5 ml glacial acetic acid per liter (pH 4.8), at 28 V/cm for 5 hours. The hexapeptide region was eluted. An aliquot of this material was treated with 1 M triethylamine (pH about 12.5) at 35° for 20 minutes. Triethylamine

[14] M. R. Capecchi, *Biochem. Biophys. Res. Commun.* **28**, 773 (1967).

treated and untreated samples were then subjected to electrophoresis at 4°
on a cellulose thin layer (E. Merck) in the above pyridine-acetate buffer,
pH 4.8, for 105 minutes at 30 V/cm. Radioactivity was located by autoradiography on Kodak Royal Blue X-ray film.

   Preparation of α-Hydroxyacyl-tRNA's. tRNA was prepared from *E.
coli* B according to the procedure of Avital and Elson[15] described above.
Five mg of tRNA was charged with alanine or phenylalanine in the standard incubation mixture using S100 (2 mg/ml protein) chromatographed
on Sephadex G-25 immediately before use, and 10 mM phenylalanine or
alanine. After incubation at 30° for 30 minutes, tRNA was recovered by
phenol extraction and ethanol precipitation. The RNA was dissolved in
2 ml of 0.25 M sodium acetate, pH 4.3, 10 mM magnesium acetate, 1 M
NaNO$_2$, and incubated at room temperature (23°) on a pH-stat at pH 4.3
for 30 minutes. RNA was precipitated with ethanol and charged with a
mixture of amino acids. RNA-free S100 was used in this second charging
incubation, which was otherwise as described above.

   The products of the nitrous acid treatment of alanyl- and phenylalanyl-
tRNA are lactyl- and phenyllactyl-tRNA, respectively. This was verified
by chromatography of alkaline digests of similar preparations made with
the ^{14}C-labeled amino acids described above.

[15] S. Avital and D. Elson, *Biochim. Biophys. Acta* **179**, 297 (1969).

[48] Two-Dimensional Polyacrylamide Gel Electrophoresis
for Separation of Ribosomal Proteins

By H. G. WITTMANN

   Ribosomes from all organisms studied to date are very complex. They
consist of three RNA molecules and numerous proteins. A rapid, sensitive, and very reproducible method of determining the number of proteins in ribosomes is two-dimensional electrophoresis on slabs of polyacrylamide gel. This technique was first described for the separation of
the 55 ribosomal proteins of *Escherichia coli*[1,2] and was later somewhat
modified for other needs.[3,4] Because the original technique resulted in a
very good separation not only of ribosomal proteins from prokaryotes,[2]
but also of eukaryotes,[5-7] it will be described in this article in some detail.
Drawings with the technical details of the various parts of the apparatus

[1] E. Kaltschmidt and H. G. Wittmann, *Anal. Biochem.* **36**, 401 (1970).
[2] E. Kaltschmidt and H. G. Wittmann, *Proc. Nat. Acad. Sci. U.S.* **67**, 1276 (1970).
[3] H. Welfle, *Acta Biol. Med. Ger.* **27**, 547 (1971).
[4] O. H. W. Martini and H. J. Gould, *J. Mol. Biol.* **62**, 403 (1971).
[5] J. Delaunay and G. Schapira, *Biochim. Biophys. Acta* **259**, 243 (1972).
[6] C. C. Sherton and I. G. Wool, *J. Biol. Chem.* **247**, 4460 (1972).
[7] A. G. Lambertsson, *Mol. Gen. Genet.* **118**, 215 (1972).

were given elsewhere[1] in order to make possible the construction of the apparatus in an institute workshop. Where this is not feasible it might be of interest that the apparatus is commercially available from Firma Desaga, Heidelberg, Germany, or from C. A. Brinckmann, Westbury, New York.

Procedure

For separation in the first dimension, the protein sample is placed in the middle of a polyacrylamide rod. This is achieved in the following way: Close a glass tube (180 × 5 mm) at the lower end with a cap; fill cold acrylamide solution (buffer B of the table) up to the middle of the tube; overlayer with water; allow gel to polymerize; suck off water from the top of the polyacrylamide; add proteins dissolved in about 0.1–0.2 ml of buffer A (see table); overlayer with water; allow gel to polymerize with light; suck off water; fill the tube with cold acrylamide solution and allow gel to polymerize. After removal of the cap at the bottom, the filled tubes are placed into a rack which is constructed similarly to those used for one-dimensional disc electrophoresis but is high enough to take tubes of 18-cm length (Fig. 1).

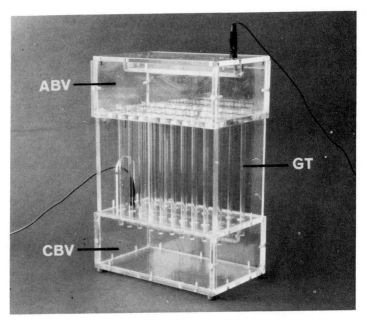

FIG. 1. Apparatus for protein separation in the first dimension. ABV: anode buffer vessel; CBV: cathode buffer vessel; GT: glass tube for polyacrylamide gel and protein sample. The apparatus can take 20 tubes.

Buffer A: for dissolving the protein sample

$$
\begin{array}{ll}
48.0 \text{ g} & \text{urea} \\
4.0 \text{ g} & \text{acrylamide} \\
0.2 \text{ g} & \text{bisacrylamide} \\
0.085 \text{ g} & \text{Na}_2\text{–EDTA} \\
0.32 \text{ g} & \text{boric acid} \\
0.06 \text{ ml} & \text{TEMED} \\
\end{array}
$$

H$_2$O to make 99 ml

Add 1 ml of solution in which 0.5 mg of riboflavin and 5 mg of ammonium peroxodisulfate are dissolved. Dissolve protein (10–50 μg per protein species) in about 0.1–0.2 ml of buffer A and apply onto the polyacrylamide rod for separation in the first dimension

Buffer B: for separation in the first dimension (pH 8.6)

$$
\begin{array}{ll}
36.0 \text{ g} & \text{urea} \\
8.0 \text{ g} & \text{acrylamide} \\
0.3 \text{ g} & \text{bisacrylamide} \\
0.8 \text{ g} & \text{Na}_2\text{–EDTA} \\
3.2 \text{ g} & \text{boric acid} \\
4.85 \text{ g} & \text{Tris} \\
0.3 \text{ ml} & \text{TEMED} \\
\end{array}
$$

H$_2$O to make 99.4 ml

Add 0.6 ml of a 7% ammonium peroxodisulfate solution for polymerization

Buffer C: electrode buffer (pH 8.6) for first dimension

$$
\begin{array}{ll}
360 \text{ g} & \text{urea} \\
2.4 \text{ g} & \text{Na}_2\text{–EDTA} \\
9.6 \text{ g} & \text{boric acid} \\
14.55 \text{ g} & \text{Tris} \\
\end{array}
$$

H$_2$O to make 1 liter.

The same buffer is used as anode or cathode buffer

Buffer D: for dialyzing between the first and second dimension

$$
\begin{array}{ll}
480 \text{ g} & \text{urea} \\
0.74 \text{ ml} & \text{glacial acetic acid} \\
2.4 \text{ ml} & 5\,N\text{ KOH} \\
\end{array}
$$

H$_2$O to make 1 liter

Buffer E: for separation in the second dimension (pH 4.6)

$$
\begin{array}{ll}
360 \text{ g} & \text{urea} \\
180 \text{ g} & \text{acrylamide} \\
5.0 \text{ g} & \text{bisacrylamide} \\
52.3 \text{ ml} & \text{glacial acetic acid} \\
9.6 \text{ ml} & 5\,N\text{ KOH} \\
5.8 \text{ ml} & \text{TEMED} \\
\end{array}
$$

H$_2$O to make 967 ml

Add 33 ml of a 5% ammonium peroxodisulfate solution for polymerization

Buffer F: electrode buffer for second dimension

$$
\begin{array}{ll}
140 \text{ g} & \text{glycine} \\
15 \text{ ml} & \text{glacial acetic acid} \\
\end{array}
$$

H$_2$O to make 10 liters

The same buffer is used as anode or cathode buffer

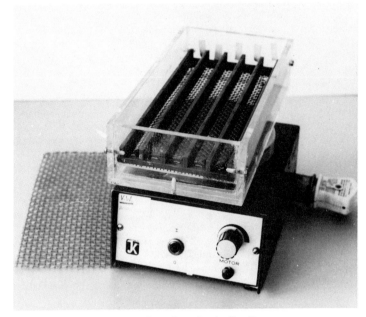

Fig. 2. Container for buffer D.

The vessels for the anode and cathode buffers are located at the top and the bottom of the apparatus, respectively. They are filled with buffer C (see table). The electrophoretic run is at 90 V for 36 hours at room temperature. The buffer in the two vessels is renewed after approximately 15 hours. At the end of the run the gels are removed from the glass rods by injecting glycerin with a syringe between the gel and the glass rod and then applying pressure with another syringe filled with water. The buffer in the gel rods is replaced by placing the gels in a container (Fig. 2) filled with buffer D (see table). After 45 and 90 minutes the buffer is replaced by a new one, and after additional 45 minutes the gel rods are placed on top of the chamber (Figs. 3 and 4) used for the electrophoresis in the second dimension. Before this is done, the chamber is closed at the bottom with polyacrylamide as follows: The same acrylamide solution that is used for the electrophoretic run in the second dimension (buffer E, see table) is filled into a flat container and overlayered with water. Before the acrylamide solution polymerizes the chamber shown in Fig. 3 is carefully placed into it (Fig. 5).

After polymerization of the acrylamide solution in the flat container the water is removed from the top of the gel and new cold acrylamide solution (buffer E of Table 1) is filled into the five vertical slots. Then

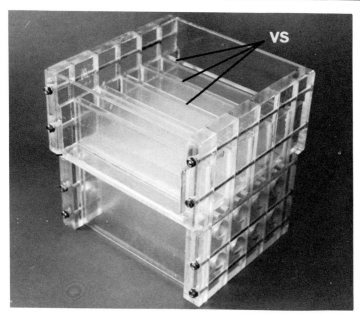

FIG. 3. Chamber for protein separation in the second dimension.

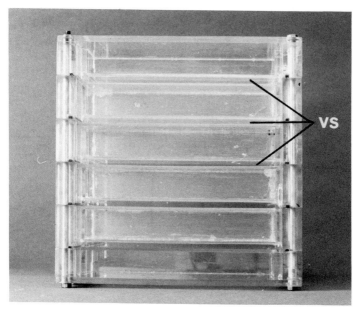

FIG. 4. Chamber (Fig. 3) seen from the top showing the vertical slots (VS) on which the polyacrylamide rods are placed.

FIG. 5. Chamber (Fig. 3) with flat container (FC) at the bottom.

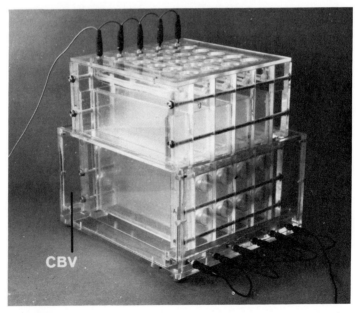

FIG. 6. Fully assembled apparatus during protein separation in the second dimension. CBV: cathode buffer vessel.

the gel rods which were dialyzed 3 × 45 minutes against buffer D of the table are placed on top of the vertical slots in such a way that they are almost embedded in the acrylamide solution. Care must be taken to remove air bubbles in the acrylamide solution. After polymerization the chamber is removed from the flat container, cleaned from residual polyacrylamide gel and placed into the vessel for the cathode buffer (Fig. 6). Then buffer F of the table is filled into the chamber, which serves as the anode buffer vessel and into the vessel for the cathode buffer. Finally a cover containing the platinum wires for the anode is put on top of the chamber. The electrophoretic run is done at room temperature for 24 hours at 100 V. Without voltage regulation the voltage increases during the run. Therefore the time for the run has to be shorter.

At the end of the run the chamber is taken out from the cathode container and dismantled. The gel slabs are removed, put into a rack (Fig. 7) and stained in solutions of Amido black or Coomassie blue. Destaining is done by removal of the stain with tap water for 1–2 hours and then with 1% acetic acid, which is either renewed every 12 hours or recycled through filters of charcoal. The destained and transparent polyacrylamide plates are photographed or stored wrapped in a very thin plastic sheet. The stained protein spots remain sharp for a period of several months if the plates are stored in the cold and dark.

FIG. 7. Rack for staining and destaining of five polyacrylamide slabs.

Applications

The two-dimensional polyacrylamide gel electrophoresis technique has been used for studying the following problems:

1. Number of ribosomal proteins in *E. coli* ribosomes (Fig. 8) and their subunits.[2]

2. Identification of isolated *E. coli* ribosomal proteins[8-10] and efficiency of various methods for extraction and fractionation of ribosomal proteins.[11,12]

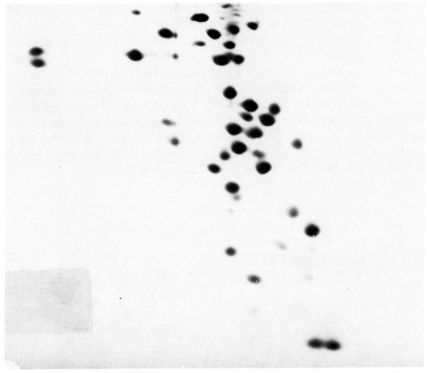

FIG. 8. Pattern of *Escherichia coli* ribosomal proteins (70 S) after two-dimensional polyacrylamide gel electrophoresis. For details see E. Kaltschmidt and H. G. Wittmann, *Proc. Nat. Acad. Sci. U.S.* **67**, 1276 (1970).

[8] I. Hindennach, G. Stöffler, and H. G. Wittmann, *Eur. J. Biochem.* **23**, 7 (1971).
[9] I. Hindennach, E. Kaltschmidt, and H. G. Wittmann, *Eur. J. Biochem.* **23**, 12 (1971).
[10] G. Funatsu, K. Nierhaus, and B. Wittmann-Liebold, *J. Mol. Biol.* **64**, 201 (1972).
[11] E. Kaltschmidt and H. G. Wittmann, *Biochemie* **54**, 167 (1972).
[12] E. Schwabe, *Hoppe-Seyler's Z. Physiol. Chem.* **353**, 1899 (1972).

3. Isoelectric points of *E. coli* ribosomal proteins.[13]

4. Identification of the proteins which form a complex with 5 S RNA.[14,15]

5. Determination of which proteins are present in the various precursors of *E. coli* ribosomes and their core particles obtained by salt treatment.[16,17]

6. Stoichiometry of proteins in *E. coli* ribosomes.[18]

7. Difference in ribosomal proteins from *E. coli* cells grown under different growth conditions.[19,20]

8. Topological studies on *E. coli* ribosomes.[21-24]

9. Determination which ribosomal proteins are different between *E. coli* wild type and mutants, e.g., revertants from streptomycin dependence to independence.[25-27]

10. Proteins different in ribosomes from various *E. coli* strains.[28]

11. Comparison of the ribosomal proteins isolated from bacteria belonging to the same or different families.[29]

12. Comparison of the proteins from cytoplasmic and chloroplast ribosomes of higher plants.[30]

13. Number of proteins from eukaryotic ribosomes and their subunits.[4-7,31]

[13] E. Kaltschmidt, *Anal. Biochem.* **43**, 25 (1971).

[14] P. N. Gray, R. A. Garrett, G. Stöffler, and R. Monier, *Eur. J. Biochem.* **28**, 412 (1972).

[15] J. R. Horne and V. A. Erdmann, *Mol. Gen. Genet.* **119**, 337 (1972).

[16] H. E. Homann and K. H. Nierhaus, *Eur. J. Biochem.* **20**, 249 (1971).

[17] K. H. Nierhaus, K. Bordasch, and H. E. Homann, *J. Mol. Biol.* **74**, 587 (1973).

[18] H. J. Weber, *Mol. Gen. Genet.* **119**, 233 (1972).

[19] E. Deusser and H. G. Wittmann, *Nature (London)* **238**, 269 (1972).

[20] E. Deusser, *Mol. Gen. Genet.* **119**, 249 (1972).

[21] R. R. Crichton and H. G. Wittmann, *Mol. Gen. Genet.* **114**, 95 (1971).

[22] P. Spitnik-Elson and A. Breimann, *Biochim. Biophys. Acta* **254**, 457 (1971).

[23] L. Kahan and E. Kaltschmidt, *Biochemistry* **11**, 2691 (1972).

[24] R. R. Crichton and H. G. Wittmann, *Proc. Nat. Acad. Sci. U.S.* **70**, 665 (1973).

[25] E. Deusser, G. Stöffler, H. G. Wittmann, and D. Apirion, *Mol. Gen. Genet.* **109**, 298 (1970).

[26] G. Stöffler, E. Deusser, H. G. Wittmann, and D. Apirion, *Mol. Gen. Genet.* **111**, 334 (1971).

[27] R. Hasenbank, C. Guthrie, G. Stöffler, H. G. Wittmann, L. Rosen, and D. Apirion, *Mol. Gen. Genet.* in press.

[28] E. Kaltschmidt, G. Stöffler, M. Dzionara, and H. G. Wittmann, *Mol. Gen. Genet.* **109**, 303 (1970).

[29] M. Geisser, G. Stöffler, and H. G. Wittmann, manuscript submitted.

[30] H. G. Janda, C. Gualerzi, H. Passow, G. Stöffler, and H. G. Wittmann, manuscript submitted.

[31] H. Welfle, J. Stahl, and H. Bielka, *Biochim. Biophys. Acta* **243**, 416 (1971).

[49] Two-Dimensional Polyacrylamide Gel Electrophoresis of Eukaryotic Ribosomal Proteins

By Corinne C. Sherton *and* Ira G. Wool

Most ribosomal proteins are basic and most lie in a narrow range of molecular sizes: they are, therefore, difficult to separate. Moreover, ribosomal proteins have no catalytic activity so their isolation, purification, and characterization cannot be followed by enzyme assay. The only feasible alternative is electrophoresis. A number of procedures have been used: two-dimensional polyacrylamide gel electrophoresis, developed by Kaltschmidt and Wittmann,[1] has expanded the resolving power of the technique and has proved valuable for analysis of prokaryotic[1-7] and eukaryotic[8-17] ribosomal proteins. For example, all 55 *Escherichia coli* ribosomal proteins can (with one exception) be displayed as individual spots on a single gel slab.[2]

We have adapted the Kaltschmidt and Wittmann procedure for our purposes and used it to analyze rat liver and skeletal muscle ribosomal proteins. We shall describe the technique in detail.

[1] E. Kaltschmidt and H. G. Wittmann, *Anal. Biochem.* **36**, 401 (1970).
[2] E. Kaltschmidt and H. G. Wittmann, *Proc. Nat. Acad. Sci. U.S.* **67**, 1276 (1970).
[3] M. Dzionara, E. Kaltschmidt, and H. G. Wittmann, *Proc. Nat. Acad. Sci. U.S.* **67**, 1909 (1970).
[4] E. Kaltschmidt, *Anal. Biochem.* **43**, 25 (1971).
[5] H. E. Homann and K. H. Nierhaus, *Eur. J. Biochem.* **20**, 249 (1971).
[6] E. Kaltschmidt and H. G. Wittmann, *Biochimie* **54**, 167 (1972).
[7] G. Stöffler and H. G. Wittmann, *J. Mol. Biol.* **62**, 407 (1971).
[8] O. H. W. Martini and H. J. Gould, *J. Mol. Biol.* **62**, 403 (1971).
[9] T. Hultin and A. Sjöqvist, *Anal. Biochem.* **46**, 342 (1972).
[10] B. L. Jones, N. Nagabhushan, A. Gulyas, and S. Zalik, *FEBS (Fed. Eur. Biochem. Soc.) Lett.* **23**, 167 (1972).
[11] H. Welfle, J. Stahl, and H. Bielka, *Biochim. Biophys. Acta* **243**, 416 (1971).
[12] H. Welfle, *Acta Biol. Med. Ger.* **27**, 547 (1971).
[13] H. Bielka, J. Stahl, and H. Welfle, *Arch. Geschwultsforsch.* **38**, 109 (1971).
[14] Huynh-Van-Tan, J. Delaunay, and G. Schapira, *FEBS (Fed. Eur. Biochem. Soc.) Lett.* **17**, 163 (1971).
[15] J. Delaunay and G. Schapira, *Biochim. Biophys. Acta* **259**, 243 (1972).
[16] C. C. Sherton and I. G. Wool, *J. Biol. Chem.* **247**, 4460 (1972).
[17] H. Welfle, J. Stahl, and H. Bielka, *FEBS (Fed. Eur. Biochem. Soc.) Lett.* **26**, 228 (1972).

Preparation of Ribosomes and Ribosomal Subunits

Media

Medium A: KCl, 100 mM; MgCl$_2$, 5 mM; EGTA, 5 mM; sodium pyrophosphate, 5 mM, pH 6.8. N.B.: The sodium pyrophosphate is added just before using.

Medium B: tris(hydroxymethyl)aminomethane (Tris)·HCl, 50 mM, pH 7.6; KCl, 250 mM; MgCl$_2$, 12.5 mM; EGTA, 5 mM; sucrose, 0.25 M

Medium C: Tris·HCl, 20 mM, pH 7.8; KCl, 80 mM; MgCl$_2$, 12.5 mM

Medium D: Tris·HCl, 20 mM, pH 7.8; KCl, 880 mM; MgCl$_2$, 12.5 mM; β-Mercaptoethanol (MSH), 20 mM, is included when the medium is used to dissociate ribosomes. The presence of a sulfhydryl-protecting agent is necessary to preserve the activity of subunits.

Medium E: Tris·HCl, 10 mM, pH 7.6; KCl, 80 mM; MgCl$_2$, 12 mM

Medium F: Tris·HCl, 10 mM, pH 7.6; KCl, 500 mM; MgCl$_2$, 5 mM

Medium G: Tris·HCl, 50 mM, pH 7.6

Medium H: Tris·HCl, 10 mM, pH 7.7; magnesium acetate, 100 mM

Reagents

Lubrol WX, 10% in 10 mM MgCl$_2$ (General Biochemicals)

Sodium deoxycholate (DOC), 10% in water (Nutritional Biochemicals Corp.)

Sodium dodecyl sulfate (SDS), 20% in water (Fisher)

Puromycin dihydrochloride (Nutritional Biochemicals Corp.)

Ethyleneglycol-bis(β-aminoethyl ether) N,N'-tetraacetic acid (EGTA), 50 mM (Sigma)

Sodium pyrophosphate, 50 mM (Fisher)

ATP, 0.1 M (Sigma)

GTP, 5 mM (Schwarz/Mann)

Solid phosphoenolpyruvate (Calbiochem Corporation)

Preparation of Ribosomes. We generally prepare muscle[18,19] and

[18] T. E. Martin, F. S. Rolleston, R. B. Low, and I. G. Wool, *J. Mol. Biol.* **43**, 135 (1969).

[19] W. S. Stirewalt, J. J. Castles, and I. G. Wool, *Biochemistry* **10**, 1594 (1971).

liver[18,20] ribosomes from male Sprague-Dawley rats of about 100–120 g; however, any of the standard procedures for the preparation of relatively uncontaminated particles (from any cell type) will do. One should choose conditions of centrifugation likely to yield the maximum number of pure particles.

A modification[21] that we have found valuable in the preparation of skeletal muscle ribosomes is the following: Finely minced skeletal muscle is stirred for 10 minutes in 2.5 volumes of medium A. The medium is drained through a stainless steel strainer and the muscle is stirred again for 10 minutes in 2.5 volumes of fresh medium A. The medium is removed, and the tissue is washed several times with medium B. EGTA chelates calcium ions and causes muscle to relax: the result is that the tissue is considerably softer (because of a lack of rigor) and hence easier to homogenize. The yield of ribosomes is almost doubled.

The muscle is now homogenized in 2 volumes of medium B in a Virtis 45 homogenizer—30 seconds, at a setting of 85. The homogenate is centrifuged (in a Sorvall model RC-2 centrifuge) in 250 ml polycarbonate bottles in a GSA rotor for 15 minutes at 13,000 g. The supernatant is filtered through glass wool and set aside; the pellet is resuspended in an amount of medium B equal to the original tissue volume and centrifuged at 13,000 g for 15 minutes. The supernatant is filtered through glass wool and retained. The pellet is discarded. The combined supernatant is centrifuged in a Spinco 30 rotor at 30,000 rpm for 2 hours in a Spinco Model L-2 centrifuge. The supernatant is decanted and discarded. The pellets are homogenized in medium C containing 0.25 M sucrose to which is added 6 ml of Lubrol WX (10% in 10 mM MgCl$_2$) and 12 ml of freshly prepared DOC (10% in water) for each 100 ml of medium C. Homogenization is with a motor-driven Teflon pestle and a size C glass tissue grinder. The homogenate is centrifuged in a Sorvall GSA rotor at 13,000 g for 15 minutes. The supernatant (7 ml) is layered on 5 ml of medium C in 0.5 M sucrose and centrifuged for 2.5 hours at 40,000 rpm in a Spinco 40 rotor in a Spinco L-2 centrifuge. The supernatant is carefully aspirated so that the DOC at the top does not contaminate the ribosomal pellet; the tubes are drained by inverting them, and the sides are dried. The ribosomal pellets may be stored at −70° for several months without loss of activity.

The particles used to prepare total ribosomal protein for electrophoresis should be as free from contamination as possible. It is helpful to incubate ribosome preparations to remove nascent peptide and endogenous mRNA.[20]

---

[20] T. E. Martin and I. G. Wool, *J. Mol. Biol.* **43**, 151 (1969).
[21] R. Zak, P. Etlinger, and D. A. Fischmann, *Res. Muscle Develop. and Muscle Spindle, Excerpta Med. Found. Int. Congr. Ser.* **240**, p. 163.

(Incubation is not necessary if the ribosomes are to be used to prepare subunits—see below.) Ribosomes are suspended by gentle homogenization in medium C, and incubated in medium C (7 mg of ribosomes per milliliter) containing: puromycin (0.1 mM); ATP (5 mM); GTP (0.5 mM); phosphoenolpyruvate (46.5 μg/ml); MSH (10 mM); and liver or muscle supernatant (G-25 fraction[22]—cf. this volume [18]). It is important that muscle supernatant be prepared in medium C (with 0.25 M sucrose) rather than medium B.[23] The higher concentration of KCl (250 mM) in medium B solubilizes myosin which interferes with the preparation of G-25 fraction. The ribosome suspension is incubated for 30 minutes at 37°. Ribosomal aggregates are removed by centrifugation at 12,000 g for 10 minutes. The ribosomes in the supernatant are collected by sedimentation (45,000 rpm for 5.5 hours at 4° in a Spinco Ti 50 rotor) through 4 ml of 0.5 M sucrose in medium D. The high concentration (880 mM) of KCl in medium D removes extraneous proteins as well as initiation and elongation factors bound to the ribosomes. The particles are referred to as stripped ribosomes.

Preparation of Ribosomal Subunits. Liver ribosomes (1–1.4 g is a convenient amount) that have not been stripped are dissociated by incubation for 15 minutes at 37° in medium D containing 0.1 mM puromycin and 20 mM MSH. The ribosomal subunits are separated by centrifugation at 13,500 rpm for 17 hours at 26° in a Spinco Ti 15 zonal rotor using a hyperbolic sucrose density gradient in medium D containing 20 mM MSH.[24] The gradient is formed over a cushion of 45% sucrose in the same medium; the concentration of sucrose in the gradient is 7.36% at 6.35 cm, and 37.2% at 8.66 cm from the rotor core. The sample, in a linear sucrose gradient of 0 to 7.4%, is layered onto the hyperbolic gradient. We have found that the yield of subunits from muscle ribosomes is improved if the concentration of potassium in the sucrose gradient is decreased from 0.88 to 0.5 M and that of magnesium from 12.5 to 3 mM.

After centrifugation the gradient is displaced with 60% sucrose and 10 ml fractions are collected. The fractions containing the subunits, either 40 S or 60 S (Fig. 1), are pooled and dialyzed against 8 liters of medium E for at least 36 hours with 3–4 changes of buffer. Sucrose and high concentrations of potassium interfere with precipitation of ribosomes by

[22] D. P. Leader, I. G. Wool, and J. J. Castles, *Proc. Nat. Acad. Sci. U.S.* **67**, 523 (1970).

[23] J. J. Castles and I. G. Wool, *in* "Protein Biosynthesis in Non-Bacterial Systems" (J. A. Last and A. I. Laskin, eds.), p. 1. Dekker, New York, 1972.

[24] E. F. Eikenberry, T. A. Bickle, R. R. Traut, and C. A. Price, *Eur. J. Biochem.* **12**, 113 (1970).

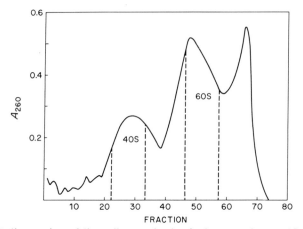

FIG. 1. Sedimentation of liver ribosomal subunits in a zonal rotor [C. C. Sherton and I. G. Wool, *J. Biol. Chem.* **247**, 4460 (1972)]. Liver ribosomal subunits (1.38 g) were separated by centrifugation for 17 hours at 13,500 rpm in a Spinco Ti 15 zonal rotor using a 7.4–38.0% hyperbolic sucrose density gradient [E. M. Eikenberry, T. A. Bickle, R. R. Traut, and C. A. Price, *Eur. J. Biochem.* **12**, 113 (1970)]. Fractions (10 ml) were collected, and those indicated by the interrupted lines were pooled; the other fractions were discarded; 117 mg of 40 S and 243 mg of 60 S subunits were recovered.

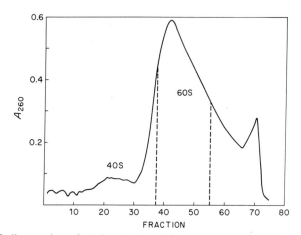

FIG. 2. Sedimentation of 60 S subunits of liver ribosomes in a zonal rotor [C. C. Sherton and I. G. Wool, *J. Biol. Chem.* **247**, 4460 (1972)]. Fractions containing 736 mg of 60 S subunits from three previous centrifugations in a zonal rotor were pooled and centrifuged again for 17 hours at 13,500 rpm in a Spinco Ti 15 zonal rotor using a 7.4–38.0% hyperbolic sucrose density gradient. [E. M. Eikenberry, T. A. Bickle, R. R. Traut, and C. A. Price, *Eur. J. Biochem.* **12**, 113 (1970)]. The fractions indicated by the interrupted lines were pooled; 324 mg of 60 S subunits were recovered.

ethanol.[25] Cold 95% ethanol (0.2 of a volume) is added. The suspension is kept at 0° for at least 1 hour (to ensure complete precipitation of the particles) and the subunits are collected by centrifugation at 10,000 g for 15 minutes. The ethanol-precipitated ribosomal pellets may be stored at −20°.

Purification of 60 S Subunits. The 60 S subunits isolated in this way are contaminated with variable amounts of 40 S subunits and must be resolved further to be suitable for a determination of the number of large subunit ribosomal proteins. The fractions containing 60 S subunits collected from centrifugation on three occasions in a zonal rotor (about 740 mg) are incubated again for 15 minutes at 37° in medium D containing 0.1 mM puromycin and 20 mM MSH. They are recentrifuged in a zonal rotor and fractions collected (Fig. 2), dialyzed, and precipitated with ethanol as described above.

Smaller amounts of 60 S subunit fractions can be purified by centrifugation in a Spinco SW 27 rotor. After suspension of the subunits in medium D, and centrifugation at 13,000 g for 10 minutes to remove

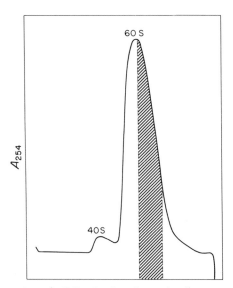

Fig. 3. Sedimentation of 60 S subunits of muscle ribosomes in a Spinco SW 27 rotor. Approximately 7.2 mg of 60 S subunits from a previous centrifugation on a zonal rotor were recentrifuged for 16 hours at 13,500 rpm using a 10–30% linear sucrose density gradient. The fractions indicated by the interrupted lines were pooled; 2.3 mg of 60 S subunits were recovered.

[25] M. S. Kaulenas, *Anal. Biochem.* **41,** 126 (1971).

aggregates, they are incubated for 5 minutes at 37° in 0.1 mM puromycin and 20 mM MSH. The suspension (0.8–1.0 ml containing 6.7–7.2 mg of ribosomes) is then layered onto a 37-ml 10 to 30% linear sucrose density gradient containing medium D and 20 mM MSH. Centrifugation is at 26° for 4 hours at 26,500 rpm, or 16 hours at 13,500 rpm. The distribution of the particles in the gradient is determined with an Instrument Specialties Co., Inc. (ISCO) Model D, density gradient fractionator, and the Model UA-2 UV analyzer,[18,23] and a portion of the 60 S peak is collected (cf. Fig. 3). The 60 S subunits are dialyzed and precipitated with ethanol as described above.

Analysis of the Purity of Ribosomal Subunits. The purity of the subunit fractions can be determined by zonal centrifugation (Fig. 4). The samples are dialyzed against medium E and layered on 5.2 ml of a 10–30% linear sucrose gradient in medium F. Centrifugation is at 60,000 rpm for 40 minutes at 26° in a Spinco SW 65 rotor, or at 48,000 rpm for 55 minutes in a Spinco SW 50.1 rotor. The distribution of particles

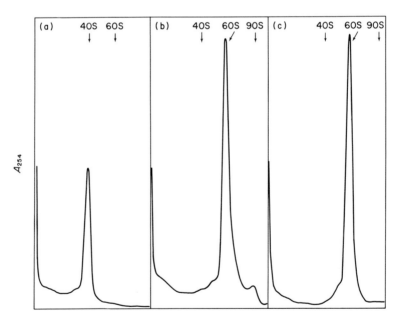

FIG. 4. Sedimentation of subunit fractions in sucrose gradients. [C. C. Sherton and I. G. Wool, *J. Biol. Chem.* **247**, 4460 (1972)]. The purity of subunit fractions was determined after dialyzing the samples against medium E. The samples were analyzed on 10–30% linear sucrose gradients in medium F; centrifugation was at 60,000 rpm for 40 minutes at 26° in a Spinco SW 65 rotor. (a) 40 S fraction (18 μg) from Fig. 1; (b) 60 S fraction (36 μg) from Fig. 1; (c) 60 S fraction (36 μg) from Fig. 2.

is determined with an ISCO density gradient fractionator and UV analyzer. The 40 S subunits from the first centrifugation in the zonal rotor are generally free of contamination with 60 S subunits (Fig. 4a); the 60 S subunits purified by either method described above are also generally free of contamination with 40 S subunits (Fig. 4c).

Another method of determining whether subunit preparations are contaminated is to analyze their constituent RNA.[18] Sufficient 20% SDS is added to a suspension of ribosomal particles in medium G so as to give a final concentration of 0.1%. The sample is incubated for 5 minutes at 37°, and then layered onto 5.2 ml of a 5 to 20% linear sucrose gradient in medium G. Centrifugation is at 4° for 2 hours at 60,000 rpm in a Spinco SW 65 rotor or for 2.75 hours at 48,000 rpm in a Spinco SW 50.1 rotor. The presence of 18 S (small subunit) and 28 S (large subunit) RNA is determined with an ISCO density gradient fractionator and UV analyzer (Fig. 5). Samples containing 60 S subunits that have not been purified contain 18 S RNA (Fig. 5b); whereas purified 60 S preparations do not (Fig. 5c).

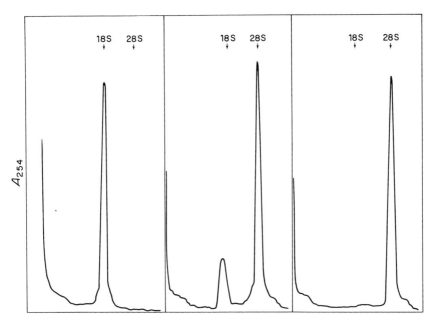

FIG. 5. Sedimentation in sucrose gradients of the RNA of subunit fractions. Subunit fractions were suspended in medium G with 0.1% SDS and incubated for 5 minutes at 37°. The samples were centrifuged in 5 to 20% linear sucrose density gradients in medium G at 60,000 rpm for 2 hours at 4° in a Spinco SW 65 rotor; (a) 40 S fraction (18 μg) from Fig. 1; (b) 60 S fraction (22 μg) from Fig. 1; (c) 60 S fraction (18 μg) from Fig. 3.

It is extremely important that ribosomal proteins to be used for two-dimensional electrophoresis be extracted from subunits that are intact and as free of contamination as possible. Subunits prepared by dissociation of ribosomes with high concentrations of KCl will recombine to form ribosomes active in protein synthesis,[18,26] whereas subparticles prepared by treatment with EDTA and other harsh methods are no longer active[18,24] and are not satisfactory for analysis of ribosomal proteins. The particles prepared by removal of magnesium from ribosomes unfold to varying degrees.[18,27–29] During zonal sedimentation, unfolded monomers may co-sediment with the large subunit, and unfolded large subunits contaminate the small subunit fraction. Contamination, caused in that way, has been responsible for erroneous reports that 60 S subunits (prepared with EDTA) contained all the proteins of the ribosome,[30] and that 40 S and 60 S subunits contained 15 proteins having identical mobilities.[11] In the latter case, cross contamination of less than 5% was reported. Obviously, even that is too much.

Extraction of Ribosomal Proteins

Reagents

Glacial acetic acid, and 1 N

Proteins are extracted from ribosomes and from ribosomal subunits with acetic acid by a modification of the procedure of Hardy et al.[31] Ribosomes are suspended (50–60 mg/ml) with gentle homogenization in medium H; 2 volumes of glacial acetic acid are added, and the mixture is stirred for 1 hour at 0°. Ribosomal RNA is removed by centrifugation at 15,000 g for 10 minutes; the pellet is washed with an equal volume of one-third medium H and two-thirds glacial acetic acid (v/v), and recentrifuged. The combined supernatant, containing the ribosomal proteins, is dialyzed against about 50 volumes of 1 N acetic acid for 48 hours with 4 changes of acid. The dialysis tubing is acetylated[32] to prevent the loss

[26] T. E. Martin and I. G. Wool, Proc. Nat. Acad. Sci. U.S. 60, 569 (1968).
[27] Y. Tashiro and T. Morimoto, Biochim. Biophys. Acta 123, 523 (1966).
[28] R. F. Gesteland, J. Mol. Biol. 18, 356 (1966).
[29] H. Lamfrom and E. R. Glowacki, J. Mol. Biol. 5, 97 (1962).
[30] M. DiGirolamo and P. Cammarano, Biochim. Biophys. Acta 168, 181 (1968).
[31] S. J. S. Hardy, C. G. Kurland, P. Voynow, and G. Mora, Biochemistry 8, 2897 (1969).
[32] L. C. Craig, this series, Vol. 11, p. 870.

of low molecular weight proteins. The dialyzed proteins are then lyophilized.

We have not extracted ribosomal proteins by other methods. Ford has reported[33] that acetic acid is not as efficient as lithium chloride-urea[34] or guanidine hydrochloride[35] in the extraction of proteins from *Xenopus* ribosomes. However, the method he used[36] did not include the modification (high concentration of magnesium) introduced by Hardy *et al.,*[31] which greatly increases the efficiency of the method. Kaltschmidt and Wittmann[6] extracted *E. coli* ribosomal proteins in various ways and evaluated the efficiency of the procedures by two-dimensional electrophoresis. The most complete extraction was with 2 parts glacial acetic acid and one part buffer containing 200 mM magnesium chloride. (The LiCl-urea and RNase methods were almost as effective.) Extraction was not complete if the concentration of magnesium in the buffer was 10 mM. Unfortunately, 100 mM magnesium (the concentration we have used in our buffer) was not tested.

If, for the moment, we take the number of ribosomal proteins as the criterion for the completeness of extraction (and there is a question whether that is an entirely valid standard), then it is obvious that the several procedures give remarkably similar results. Martini and Gould,[8] Hultin and Sjöqvist,[9] and Huynh-Van-Tan, *et al.*[14] all extracted mammalian ribosomes with LiCl-urea[34] and analyzed the proteins by two-dimensional electrophoresis: they found 62-63, 48, and 75 proteins, respectively. The number Hultin and Sjöqvist found is low, but they also found only 48 proteins when ribosomes were extracted with HCl, and from their electropherograms it is obvious that the proteins were incompletely resolved. Welfle *et al.* found 72 different proteins when they extracted ribosomes with HCl,[17] and 76 when they used 67% acetic acid and 50 mM magnesium chloride in the buffer.[11] With the exception of the findings of Hultin and Sjöqvist,[9] the several procedures give results quite comparable to the 68–72 different ribosomal proteins we found after extracting liver ribosomes with 67% acetic acid and 100 mM magnesium chloride in the buffer.[16] Although we cannot be certain that any procedure extracts all the eukaryotic ribosomal proteins, the acetic acid–magnesium chloride, urea–lithium chloride, and HCl methods seem equally efficient.

[33] P. J. Ford, *Biochem. J.* **125**, 1091 (1971).
[34] P. S. Leboy, E. C. Cox, and J. G. Flaks, *Proc. Nat. Acad. Sci. U.S.* **52**, 1367 (1964).
[35] R. A. Cox, this series, Vol. 12B, p. 120.
[36] P. B. Moore, R. R. Traut, H. Noller, P. Pearson, and H. Delius, *J. Mol. Biol.* **31**, 441 (1968).

At least that is a reasonable assumption until a systematic evaluation is carried out.

Two-Dimensional Polyacrylamide Gel Electrophoresis

Reagents

Acrylamide (Eastman Organic Chemicals)

N,N'-Methylene-bisacrylamide (Eastman Organic Chemicals)

N,N,N',N'-Tetramethylethylenediamine (TEMED) (Eastman Organic Chemicals)

Crystalline ammonium persulfate (APS), 0.6%; 1.5%; 6%; 8% (w/v) in water (Eastman Organic Chemicals)

Riboflavin (Eastman Organic Chemicals)

Naphthol Blue Black, 1% in 7.5% acetic acid (Eastman Organic Chemicals)

Solid boric acid (Sigma)

Solid disodium ethylenediamine tetraacetic acid (EDTA) (Sigma)

Solid Tris·HCl (Sigma)

Solid urea, and 8 M

NaOH, 2 N

KOH, 5 N

Acetic acid, 7.5%

Equipment

We have used the apparatus designed by Kaltschmidt and Wittmann and described by them in great detail in their first publication.[1] Comparable equipment is now available from several manufacturers.

General Procedures. We have adapted the technique of Kaltschmidt and Wittmann[1] for the electrophoresis of mammalian ribosomal proteins. The "standard" conditions for two-dimensional electrophoresis at room temperature are: 8% acrylamide gel, pH 8.6, for 40 hours at 90 V (initial current 2.7 mA/tube) in the first dimension; 18% acrylamide gel, pH 4.2, for 40 hours at 105 V (initial current 100 mA/gel) in the second dimension. If purified acrylamide solutions are used (see below) the voltage is reduced by 25%. The concentrations of the gels and the pH of the buffers are selected to maximize separation of the ribosomal proteins by charge in the first dimension and by size in the second, and to provide the best possible resolution of the largest number of proteins.

There is some leeway in the selection of the concentration of acrylamide, pH, voltage, etc., for the separation of mammalian ribosomal proteins.[8,9,11,14,17] Regardless of the conditions one finally selects as most suitable, it is important to perform electrophoresis with different per-

TABLE I

Buffers To Be Used for Electrophoresis of Ribosomal Proteins
in the First Dimension

Reagent[a]	pH			
	7.6	8.6	9.6	10.6
Boric acid (g)	37.1	28.8	6.3	1.86
Tris·HCl (g)	25.5	43.65	87.3	98.1
EDTA (disodium salt) (g)	7.05	7.2	11.7	7.05
2 N NaOH (ml)	—	55	52	32

[a] Sufficient water is added to give a final volume of 3 liters.

centages of acrylamide and at different pH's, in order to detect proteins
which may fortuitously coelectrophorese in the "standard" conditions.[2,16]
We generally vary the acrylamide percentage in the first dimension from
5% to 10% (while keeping the acrylamide:methylene–bisacrylamide ratio
constant), and vary the pH of the first dimension buffer from 7.6 to 10.6
(Table I).

First Dimension. Soft glass tubes (inner diameter 6 mm) are cut to
a length of 180 mm, the ends are fire polished, cleaned in chromic acid
solution, and dried. The acrylamide and methylene-bisacrylamide used in
the preparation of separation and sample gel solutions (Table II) should
first be recrystallized,[37] to reduce streaking and smudging in the electropherogram. The 8 M urea solution used in the preparation of the gels
is passed through a column of Bio-Rad AG 501-X8(D) mixed bed resin
to remove cyanate. Cyanate can lead to carbamylation of lysine residues,[38]

TABLE II

Preparation of Polyacrylamide Gel Solutions To Be Used for
Electrophoresis of Ribosomal Proteins in the First Dimension

Reagent[a]	Separation gel	Sample gel
Acrylamide (g)	40	10
Methylene-bisacrylamide (g)	1.5	0.5
Boric acid (g)	16	0.8
EDTA (disodium salt) (g)	4	0.21
Tris·HCl (g)	24.3	—
TEMED (ml)	1.5	0.15

[a] Sufficient water is added to give a final volume of 115 ml of separation gel or 60 ml
of sample gel.

[37] U. E. Loening, *Biochem. J.* **102**, 251 (1967).
[38] G. R. Stark, W. H. Stein, and S. Moore, *J. Biol. Chem.* **235**, 3177 (1960).

and thus increase the number of protein bands or spots obtained on electrophoresis.

For each 5 gel tubes to be filled, 18.75 ml of 8 M urea are combined with 5.75 ml of separation gel solution (Table II). The solution is deaerated (with vacuum suction) for 15 minutes to remove oxygen and reduce bubble formation, and APS (0.5 ml of 1.5%) is added. The final ratio of 8 M urea:separation gel solution:1.5% APS is 75:23:2. The solution is mixed, deaerated for 20 seconds, quickly pipetted into the gel tubes to a height of 100 mm, and overlayed with water (to ensure a level surface). At least 15 minutes is allowed for polymerization.

The sample gel is made by first combining 6.0 ml of 8 M urea with 1.92 ml of sample gel solution (Table II); 1.98 ml of the mixture are deaerated for 15 minutes. Next, 10 μl of 0.6% APS and 10 μl of ribo-flavin (1 mg/ml in 6 M urea) are added, mixed, and deaerated for 20 seconds. The final ratio is: 8 M urea (75):sample gel solution (24):0.6% APS (0.5):riboflavin (0.5). The water overlaying the polymerized gel

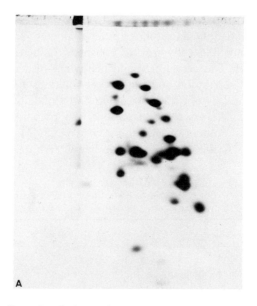

Fig. 6. Two-dimensional electropherograms of liver ribosomal proteins [C. C. Sherton and I. G. Wool, *J. Biol. Chem.* **247**, 4460 (1972)]. Standard conditions were used for the electrophoresis. The anode was on the left in the first dimension and at the top in the second. (a) 40 S ribosomal subunit protein (1 mg); (b) 60 S ribosomal subunit protein (1 mg); (c) 80 S ribosomal protein (1.6 mg). Some of the spots on the gels were clearly visible to the eye, but are barely discernible (S1) or cannot be seen (S12) in the photograph. Those proteins are included in the schematics (Figs. 7–9).

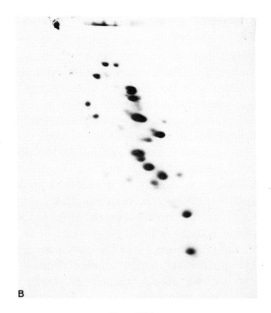

FIG. 6(b).

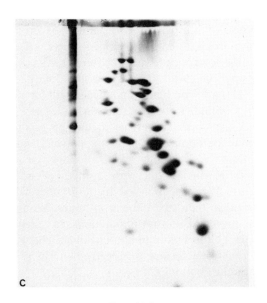

FIG. 6(c).

is removed and a 150-μl aliquot of the sample gel mixture is added to each tube (as spacer gel) and overlayed with water. Since the spacer gel contains only 4% acrylamide, it will, with an appropriate buffer, concentrate the proteins into thin layers before their entry into the 8% acrylamide separation gel. The spacer gel is polymerized in front of a fluorescent light for at least 20 minutes. A fluorescent light is necessary for polymerization when the gel contains riboflavin.

The ribosomal proteins are dissolved in sample gel: the concentration can be varied from 5 to 10 mg/ml and the volume from 50 to 400 μl. In general, the optimal amount for the analysis of 40 S and 60 S ribosomal proteins is 1.0 mg; and for 80 S, 1.6 mg (Fig. 6). To be able to record on film proteins which stain lightly, it is necessary to use at least twice the optimal amount of protein. On the other hand, proteins which stain intensely and which migrate closely can only be seen as separate when less than the optimal amount is used.

At the pH of the sample gel (8.2) the very basic ribosomal proteins are barely soluble. Incubating the gel solution containing the ribosomal protein sample at 37° for 10–15 minutes may help to solubilize the proteins, but with some preparations the solution remains translucent. To the sample gel solution containing ribosomal proteins are added 0.005 volume each of APS (0.6%) and riboflavin (1 mg/ml). The sample is deaerated for 20 seconds. An appropriate amount (see above) is layered onto the spacer gel and overlayed with water. Polymerization is carried out with a fluorescent light for at least 30 minutes.

Another portion of sample gel is deaerated, layered over the actual sample, and polymerized to form an upper spacer gel in the same way we described for the lower spacer gel. The upper separation gel is formed in the following way: 11.25 ml of 8 M urea and 3.45 ml of separation gel solution (Table II) are combined and deaerated for 15 minutes; APS (0.3 ml of 1.5%) is added and the solution is deaerated for 20 seconds and pipetted over the upper spacer gel to the top of the tubes. A small layer of water may be applied to facilitate polymerization, which is for at least 15 minutes.

Electrophoresis is carried out with 3 liters of whichever buffer is selected (Table I). We have omitted the 6 M urea which Kaltschmidt and Wittmann[1] used in their first dimension buffer. We find no discernible difference in the protein pattern without urea. In "standard" conditions (pH 8.6), the pH of the buffer in the chambers changes no more than 0.2 pH unit during electrophoresis. Electrophoresis is from the anode (top) to the cathode (bottom).

The first-dimension gels are removed from the tubes by first injecting glycerin to loosen the gels, which adhere to the tube wall, and then water

TABLE III
Preparation of Polyacrylamide Gel and Dialysis Buffer
To Be Used for Electrophoresis of Ribosomal Proteins
in the Second Dimension

Reagent[a]	Polyacrylamide gel	Dialysis buffer
Acrylamide (g)	360	—
Methylene-bisacrylamide (g)	10	—
Urea (g)	720	960
TEMED (ml)	11.6	—
Glacial acetic acid (ml)	104.6	1.48
5 N KOH (ml)	19.2	4.8

[a] Sufficient water is added to give a final volume of 1933 ml of polyacrylamide gel solution or 2 liters of dialysis buffer.

to force them out of the tube. (They break rather easily, but may be reassembled and placed atop the second-dimension gel slab.) On removal from the tubes the gels are dialyzed for a total of 45 minutes against 3 changes of 8 M urea, pH 7.2 buffer (Table III). Dialysis is to reduce the pH of the gels in order to adapt them to the second-dimension buffer.

Second Dimension. The 18% acrylamide gel solution (Table III) is prepared, deaerated for 30 minutes, and kept at 4° for at least 1 hour prior to use. Polymerization is performed at 4° to slow the process and to dissipate the large amount of heat generated. Gel solution, 483.3 ml, is combined with 16.7 ml of 6% APS (29:1) and poured into the base of the electrophoresis apparatus to seal off the bottoms of the slabs. At least 1 hour is allowed for polymerization. The first-dimension gels are positioned in the slots; 1160 ml of gel solution are combined with 40 ml of 8% APS and poured slowly to form slabs (20 cm × 20 cm × 0.5 cm). Bubbles under the first-dimension gels are removed. As polymerization proceeds, the gels shrink, and more gel solution is added to the top. The exact amount of APS that is required for polymerization to occur in a reasonable time varies with the age of the reagent.

Again, at least 1 hour must be allowed for complete polymerization. Electrophoresis is at room temperature. The 12 liters of pH 4.2 buffer to be used for electrophoresis in the second dimension contain 168 g of glycine and 18.0 ml of acetic acid (Table III). The pH will sometimes change as much as 2 units during the 40 hours required for electrophoresis. Electrophoresis is from the anode (top) to the cathode (bottom).

Staining and Destaining. The gels are stained for 20 minutes in 1% Naphthol Blue Black in 7.5% acetic acid. The dye is purified by electrophoresis through 10% acrylamide gel at 100 mA for 40 hours—the dye remaining on the cathode side of the gel is used. The gels are destained for

1 hour in running distilled water, and subsequently by continuous diffusion destaining[39] in 17 liters of 7.5% acetic acid. A simple, convenient method is to use an inexpensive model 425 "Dynaflo" aquarium pump to circulate the destaining solution through a filter of activated charcoal. The charcoal should be stirred for consecutive 24-hour periods with about 3 volumes each of 4 N NaOH, then 2 N HCl, and finally distilled water; the procedure removes colloidal carbon and soluble contaminants. We have found that Fisher Activated Cocoanut charcoal (6–14 mesh) works especially well.

Comments. When ribosomal proteins are analyzed by two-dimensional electrophoresis shadowed streaks occur in the upper portion of the gel slabs on the cationic side, a region where one would expect to find the proteins of greatest molecular size (Fig. 6). The streaks are probably caused by protein aggregates linked by disulfide bonds. Kaltschmidt and Wittmann[2] have shown that similar streaks, which occur when two-dimensional electrophoresis of *E. coli* ribosomal proteins is carried out, can be eliminated if the proteins are oxidized with performic acid or reduced and alkylated with iodoacetamide. We have not repeated that procedure. However, we have found that purification of acrylamide and methylenebisacrylamide by the procedure of Loening[37] diminishes the intensity of the streaks.

A much more serious difficulty is the variable amount of protein which remains at the origin during electrophoresis in the first dimension and causes the formation of bands, by migration directly down from the origin, during electrophoresis in the second dimension (Fig. 6). The formation of the bands is concentration dependent. We suspect the bands to be proteins that do not migrate in the first dimension because of the reduced solubility of ribosomal proteins at pH 8.6 (the pH in the first dimension). The proteins would be expected to be soluble at the pH (4.2) used in the second dimension, although a variable amount of material still remains at the origin. The bands appear to have mobilities in the second dimension equal to those of some of the ribosomal proteins. However, bands are not formed for all the proteins, and the bands present have different intensities. Therefore, the problem posed is whether variable amounts of each protein may be retarded at the origin, and whether, in fact, a particular protein might be lost in this way. As far as we know, we have never lost one of our proteins from the two-dimensional pattern,[16] although we have an instance in which a single protein (S9) was all but completely absent. Thus we cannot be certain that there do not exist

[39] L. J. Gathercole and L. Klein, *Anal. Biochem.* 44, 232 (1971).

ribosomal protein species which are insoluble at pH 7.6–10.6 and do not migrate from the origin.

Similar bands below the first dimension origin are apparent in all the other published electropherograms of mammalian ribosomal proteins

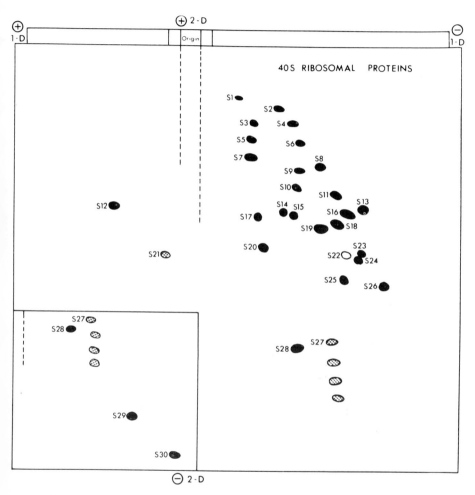

FIG. 7. Schematic of the two-dimensional electropherogram of the proteins of the 40 S subunit of liver ribosomes [C. C. Sherton and I. G. Wool, *J. Biol. Chem.* **247,** 4460 (1972)]. The solid spots were always seen; the crosshatched spots either varied in location (S21) or in intensity of staining (S27 and its three satellite spots); the open spot (S22) was seen only when the conditions of electrophoresis were changed. The diagram includes spots that are difficult to see in the photographs (Fig. 6). The intensity of the staining of the spots is not represented with fidelity.

analyzed by the Kaltschmidt and Wittmann procedure.[11-14,17] The two-dimensional electrophoresis method developed by Martini and Gould[8] (and also that used by Hultin and Sjöqvist[9]), has the great advantage of using a pH of 4.5 for electrophoresis in the first dimension. At this pH the ribosomal proteins are far more soluble, and there are no difficulties with proteins precipitating, or forming bands in the second dimension.

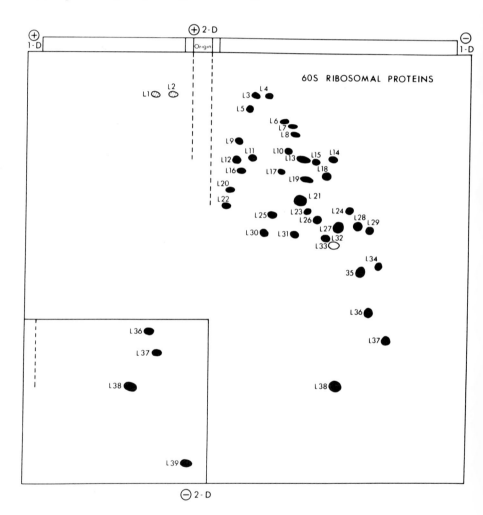

FIG. 8. Schematic of the two-dimensional electropherogram of the proteins of the 60 S subunit of liver ribosomes [C. C. Sherton and I. G. Wool, *J. Biol. Chem.* **247**, 4460 (1972)]. The solid spots were always seen; the crosshatched spots (L1 and L2) were seen only when an excess of protein was applied; the open spot (L33) was seen only when the conditions of electrophoresis were changed.

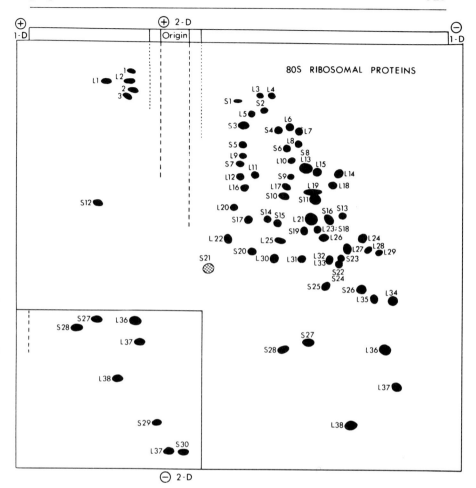

FIG. 9. Schematic of the two-dimensional electropherogram of the proteins of rat liver 80 S ribosome monomers [C. C. Sherton and I. G. Wool, *J. Biol. Chem.* **247,** 4460 (1972)]. The solid spots were always seen; the crosshatched spot (S21) varied in location; the open spots (1, 2, 3) were seen when ribosome monomer proteins were analyzed, but not when subunit proteins were electrophoresed.

Another advantage of that procedure is that electrophoresis in the second dimension is in 0.1% sodium dodecyl sulfate which enables one to determine the molecular weight of the ribosomal proteins.

Concluding Remarks

We have used two-dimensional electrophoresis to compile a catalogue of the proteins of rat liver ribosomes and ribosomal subparticles (Figs.

7–9). It is to be noted that the schematic drawings (which depict the proteins) do not represent with fidelity the intensity of staining of the individual spots; moreover, they are composites from a large number of electropherograms in which the amount of sample and the conditions of electrophoresis were varied. We have chosen to number the protein spots of each subunit separately along horizontal lines beginning at the upper left. The 60 S subunit proteins are designated by an L (large subunit), the 40 S subunit proteins by an S (small subunit).

There are 30 proteins in the 40 S subunit. Two are variable in position (S21, S27) or amount (S27), two can be seen only when the time of electrophoresis in the first dimension is decreased (S29, S30), and one is seen only when the conditions of electrophoresis are altered (S22). There are 39 proteins in the 60 S subunit. Two are present in very small amounts (L1, L2), one can be seen only at the shorter time (L39), and one is seen only in altered conditions (L33). The proteins of the two subunits, with one possible exception (S8, L13) are unique. Preparations of 80 S ribosomes contain 3 proteins not present in either subunit; thus, we estimate that mammalian ribosomes contain 68–72 different proteins.

Two-dimensional electrophoresis has proved a useful method for the resolution of mammalian ribosomal proteins. Electropherograms can be used to display and count the proteins of ribosomal subunits and monomers; and even to determine their molecular weights (using the SDS method[8]). In addition, two-dimensional electrophoresis can be employed to compare the protein composition of ribosomes from different species or tissues and to determine the isoelectric points of the proteins. The catalog of proteins should prove a valuable reference for the isolation, purification, and characterization of individual eukaryotic ribosomal proteins.

[50] A Modified Two-Dimensional Gel System for the Separation and Radioautography of Microgram Amounts of Ribosomal Proteins[1]

By G. A. Howard and R. R. Traut

Polyacrylamide gel electrophoresis has been a highly useful technique for separating complex mixtures of proteins. Recently a two-dimensional slab gel electrophoresis system was described by which all the ribosomal

[1] Supported by research grants from the U.S. Public Health Service (GM 17924) and the Damon Runyon Memorial Fund for Medical Research (DRG-1140). R.R.T. is an established Investigator of the American Heart Association.

proteins of *Escherichia coli* were separated into distinct spots.[2,3] This system and others have been used to separate eukaryotic as well as prokaryotic ribosomal proteins.[4-8] The system developed by Kaltschmidt and Wittmann[3] gives excellent resolution of both prokaryotic and eukaryotic ribosomal proteins moreover, standard and highly useful nomenclature of *E. coli* ribosomal proteins has been defined in terms of this system.[9]

We have adopted the basic system of Kaltschmidt and Wittmann to use thinner and smaller slabs which require smaller amounts of sample, and which can be dried for radioautography of radioactive proteins. The advantage of the smaller two-dimensional slab described here are summarized below:

	Miniature system	Kaltschmidt–Wittmann system[3]
Ribosomal protein required	100–200 μg	2–4 mg
Total electrophoresis time	12 hr	40–50 hr
Staining and destaining time	3–5 hr	30–40 hr
Radioautography	Very good—gel can be totally dried without distortion or shrinkage for accurate radioautograms	Must be done with wet slab
Cost of basic apparatus	Not yet commercially available (about $1150)	(Desaga-Brinkman) $1995.00

The resolution of *E. coli* ribosomal proteins with the miniature slab gel system described here is essentially identical to that reported by Kaltschmidt and Wittmann.[3] It has also been employed to separate eukaryotic ribosomal proteins and to identify by radioautography those phosphorylated by protein kinase. A two-dimensional gel system having many of the ad-

[2] H. G. Wittmann, this volume [48].

[3] E. Kaltschmidt and H. G. Wittmann, *Anal. Biochem.* **36**, 401 (1970).

[4] O. H. W. Martini and H. J. Gould, *J. Mol. Biol.* **62**, 403 (1971).

[5] T. Hultin and A. Sjöqvist, *Anal. Biochem.* **46**, 342 (1972).

[6] H. Van Tan, J. Delaunay, and G. Schapira, *FEBS* (*Fed. Eur. Biochem. Soc.*) *Lett.* **17**, 163 (1971).

[7] C. C. Sherton and I. G. Wool, *J. Biol. Chem.* **247**, 4460 (1972).

[8] H. Welfle, J. Stahl, and H. Bielka, *Biochim. Biophys. Acta* **243**, 416 (1971).

[9] H. G. Wittmann, G. Stöffler, I. Hindennach, C. G. Kurland, L. Randall-Hazelbauer, E. A. Birge, M. Nomura, E. Kaltschmidt, S. Mizushima, R. R. Traut, and T. A. Bickle, *Mol. Gen. Genet.* **111**, 327 (1971).

vantages of that described here has also been employed by Dr. David Elson (personal communication).

Methods

First Dimension by Disc Gel Electrophoresis

Solutions

Separating gel, pH 8.7 (modification of that reported by Kaltschmidt and Wittmann[3]):

> Urea, 6.0 M, 360.0 g/l
> Acrylamide, 4.0 wt. %, 40.0 g/l
> Bisacrylamide, 0.13 wt. %, 1.33 g/l
> EDTA-Na$_2$, 20 mM, 8.0 g/l
> Boric acid, 0.52 M, 32.0 g/l
> Tris, 0.4 M, 48.6 g/l
> TEMED, 0.45 ml

The above solution is filtered and may be stored at 4° for several weeks. It is degassed just before polymerization, catalyzed with ammonium persulfate: 5 μl of a freshly prepared 10% (w/v) solution per 1.0 ml of gel solution.

> Running buffer, pH 8.2
>> EDTA-Na$_2$, 10 mM, 2.4 g/l
>> Boric acid, 80 mM, 4.8 g/l
>> Tris, 60 mM, 7.25 g/l
> Tracking dyes
>> Pyronine G, 0.5% in H$_2$O (cationic)
>> Bromophenol blue, 0.1% in H$_2$O (anionic)

Acrylamide (technical grade), bisacrylamide, and TEMED were obtained from Eastman Chemicals; EDTA-Na$_2$ and Tris from Sigma; and the remaining reagents from Mallinckrodt. Recrystallization of the reagents was unnecessary and results were the same with crude or purified reagents.

Apparatus

Electrophoresis in the first dimension was performed in either of two ways:

(a) Standard disc gel electrophoresis apparatus, as described by Davis,[10] in which gels are run in glass tubes either 0.45 cm ×

---

[10] B. J. Davis, *Ann. N.Y. Acad. Sci.* **121**, 404 (1964).

9 cm, or 0.45 × 12.5 cm. Alternatively, tubes of smaller inner diameter (2–3 mm) can be employed in order to simplify insertion of the gel cylinder into the second dimension slab.

(b) Slab gel apparatus with preformed slots for individual samples as described by Reid and Bieleski,[11] and modified by Studier.[12,13] A slice of the slab containing the resolved sample in one of the slots is cut out and used for the second dimension. Further details of this method will not be presented here. The gel solution used is the same in (a) and (b).

Sample Application and Electrophoresis Conditions. Two methods of applying sample in the first dimension by method (a) above have been used in our laboratory. Both minimize the substantial loss of protein at the center-origin reported in other systems in which sample is polymerized in a sample gel containing acrylamide.[7,14]

METHOD 1. The sample (100–200 μg) is applied in agarose in the center of the first-dimensional gel; in this way proteins migrating both toward the anode and the cathode are resolved in the same run. For this method the longer disc gel tubes (0.45 cm × 12.5 cm) are used. The lower half of the separating gel is poured into the tube and overlayered with water according to standard methods. Agarose (1% in pH 8.2 running buffer) is liquefied in a boiling water bath, mixed in a 1:1 ratio with the protein sample and kept at 40° in a heating block until it is layered onto the flat surface of the polymerized lower gel. Water is again layered over the agarose-sample mixture until the agarose has hardened; it is then removed, and the tube is filled with more separating gel solution over the sample zone. Electrophoresis is carried out with the cathode above, with pyronine G as the tracking dye.

METHOD 2. Identical amounts of sample are applied at the top of each of two separate gels in the shorter tubes (0.45 cm × 9.0 cm). Electrophoresis of one of the gels is from the anode to the cathode with pyronine G as the tracking dye; electrophoresis of the other gel is from the cathode to the anode using bromphenol blue as the tracking dye.

By method 1 small losses of proteins occur as material immobilized in the agarose layer. These are much less, however, than those observed in other systems.[7,14] By method 2, essentially all the protein is recovered in the bands migrating from the origin. Thus, although ostensibly method 2 requires twice as much protein sample as method 1, in fact, because noth-

[11] M. S. Reid and R. L. Bieleski, *Anal. Biochem.* **22**, 374 (1968).
[12] F. W. Studier, *Science* **176**, 367 (1972).
[13] F. W. Studier, personal communication.
[14] E. Kaltschmidt and H. G. Wittmann, *Proc. Nat. Acad. Sci. U.S.* **67**, 1276 (1970).

ing is lost, the increased sample required is less than double that required when the sample is polymerized in the center of a single disc gel.

In both methods 1 and 2, electrophoresis is performed at 3 mA/gel tube for 30 minutes to facilitate the stacking of the protein bands, then the current is increased to 6 mA/gel and electrophoresis continued for 5–6 hours. The gels can be removed at once from the glass tubes in preparation for the second dimension, or stored in the tubes at 4° for 24–48 hours before use without significant diffusion of the protein bands or loss of resolution.

Preparations of First-Dimensional Gel for the Second-Dimension

Solutions

Dialysis buffer, pH 5.2 (Kaltschmidt and Wittmann,[3] "starting buffer")

'Urea, 8.0 M, 480.0 g/l
Acetic acid, glacial, 40 mM, 0.74 ml g/l
KOH, 10 mM, 0.67 g/l

Procedure. The first-dimensional disc gels of larger diameter are sliced in half longitudinally in order to fit between the glass plates in the second dimension. This can be done by placing the gel in a halved piece of tygon tubing matching the diameter of the gel, and then slicing the gel with uniform downward pressure of a thin knife blade. The small diameter gels can be used directly.

The first-dimensional gels are then dialyzed against the pH 5.2 buffer for a total of 60 minutes with at least two changes of buffer.

Second Dimension by Slab Gel Electrophoresis

Solutions

Separating gel solution, pH 4.5 (modification of that reported by Kaltschmidt and Wittmann[3]):

Urea, 6.0 M, 360.0 g/l
Acrylamide, 18.0 wt. %, 180.0 g/l
Bisacrylamide, 0.25 wt. %, 2.5 g/l
Acetic acid, glacial, 0.92 M, 53.0 ml
KOH, 0.048 N, 2.7 g/l
TEMED, 5.8 ml

The above solution is filtered and may be stored at 4° for several weeks. It is degassed just before polymerization is catalyzed with ammonium persulfate: 30 μl of a 10% (w/v) solution/1.0 ml of gel solution.

Running buffer, pH 4.0 (modification of that reported by Kaltschmidt and Wittmann[3]):

Glycine, 0.18 M, 14.0 g/l
Acetic acid, glacial, 6 mM, 1.5 ml

Tracking dye:
Pyronine G, 0.5% in H_2O containing 20% glycerol

Stains and destaining solution:
Coomassie brilliant blue, R-250, 0.1% in 7.5% glacial acetic acid:50% methanol:H_2O
Destaining solution, 50% methanol:7.5% glacial acetic acid: H_2O

Apparatus. The apparatus in the second dimension (Fig. 1) is an adaption of the thin-sheet gel apparatus referred to in first-dimensional method 1 above.[11-13] Two glass plates are sandwiched together with plexiglass spacers to allow the gel sheet of approximately 2 mm thickness to be formed. A thin coating of petroleum jelly on the spacers is used to prevent leakage of the gel solution.

Procedure. The first-dimensional slice is placed in position against the top spacer as shown (Fig. 1); any excess gel at the ends is trimmed off to allow the gel to fit in the defined width. When the first-dimension has been run as two separate short gels in opposite directions (Method 1), the two origins are placed adjacent to each other in the center as shown in Fig. 1. The two glass plates are clamped together at the sides and the top (where the first-dimensional slice is positioned) with foldback binder clips. The plates are inverted 180° and the cavity thus formed is filled with second-dimensional separating gel solution. After the gel has polymerized, the plates are again turned 180° so that the first-dimensional slice is on top. The spacer is removed, all excess petroleum jelly is wiped away, the binder clips are carefully removed, and the plates with the enclosed gel sheet are clamped to the apparatus previously described by Reid and Bieleski,[11] as modified by Studier.[12,13]

Electrophoresis Conditions. The tracking dye (0.1% pyronine G in 20% glycerol) is layered across the top of the first-dimensional slice, under the running buffer. Electrophoresis is carried out with the anode on top for 30–60 minutes at 40 V to allow stacking of the proteins in the first-dimensional gel slice, then the voltage is increased to 80–150 V and electrophoresis is continued for 6–12 hours. The exact voltage within the

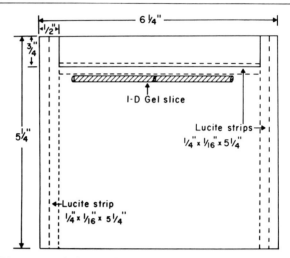

FIG. 1. Placement of the plexiglass spacers and the gel slice from the first dimension between the glass plates used for the second dimension. The two glass plates (one with a notch in the top as indicated) are made from ordinary double-strength window glass. The resulting "sandwich" is clamped in place in an apparatus like that previously described [M. S. Reid and R. L. Bieleski, *Anal. Biochem.* **22**, 374 (1968)], as modified by Studier [F. W. Studier, *Science* **176**, 367 (1972); F. W. Studier, personal communication].

limits tested makes no difference in the resulting separation of the proteins; thus convenience of running time is the major criterion for choosing a particular voltage. At 80 V the dye front moves 1 cm per hour.

Electrophoresis is stopped when the tracking dye is within 1 cm of the bottom of the gel. The top glass plate is loosened from the gel slab by carefully prying it upward with a wide spatula. The gel sheet is lifted off the lower plate and placed in a tray containing the stain solution, where it is left with occasional agitation for 1–4 hours. The gels are destained by slow shaking in a tray of the destaining solution on a mechanical shaking bath with several changes of the solution.

Treatment of Gel after Destaining. After destaining, the gel sheet can be photographed wet, or dried onto filter paper either for storage or for autoradiography. Drying is accomplished by using a vacuum and low heat as described by Maizel[15] (a modification of the procedure of Fairbanks *et al.*[16]). The thinness of the gels allows them to be totally dried in about 2 hours without distortion, shrinkage, or cracking.

[15] J. V. Maizel, Jr., *in* "Methods in Virology" (K. Maramorosch and H. Kaprowski, eds.), Vol. V, p. 179. Academic Press, New York, 1971.
[16] G. Fairbanks, C. Levinthal, and R. H. Reeder, *Biochem. Biophys. Res. Commun.* **20**, 393 (1965).

Determination of Molecular Weights of Proteins Separated by Two-Dimensional Electrophoresis

In addition to providing the molecular weight of the proteins contained in the individual spots resolved in the two-dimensional technique described, the analytical technique (SDS gel electrophoresis) which follows is also a criterion for the purity of the resolved spots.

Solutions

> Acrylamide disc gels, 10%, containing sodium dodecyl sulfate (SDS) as described by Bickle and Traut[17]
> Sample buffer, pH 7.2:
>
> $NaH_2PO_4 \cdot H_2O$, 2.9 mM, 0.4 g/l
> $Na_2HPO_4 \cdot 7H_2O$, 7.2 mM, 1.94 g/l
> Glycerol, 20 vol %, 200 ml
> β-Mercaptoethanol, 0.14 M, 10 ml
>
> Tracking dye: bromophenol blue, 0.1% in H_2O
> Stain and destaining solutions: same as for second dimension above

Apparatus

> Standard polyacrylamide disc gel electrophoresis apparatus, as described by Davis,[10] with gels formed in 0.45 cm × 9.0 cm glass tubes

Preparation of Sample. The center regions of stained spots on the two-dimensional slab are cut out, rinsed with water and macerated in small culture tubes with a glass rod. From 0.2 to 0.3 ml of a solution containing 1% SDS and 6 M urea is added to each tube, which are then left at room temperature for 24 hours. Then, 50 μl of sample buffer solution, 5 μl of 0.1% bromophenol blue (tracking dye), plus sufficient 10 N NaOH to adjust the pH to about 7.0 (about 10 μl) is added to each tube. The contents of each tube are mixed thoroughly, heated for 10 minutes at 65°, then applied to the top of the SDS gels as previously described.[17] To determine accurately the molecular weight of the proteins removed from the second-dimensional gel, protein standards of known molecular weight can be added directly to the macerated gel before the heating step, or run on separate gels.

[17] T. A. Bickle and R. R. Traut, *J. Biol. Chem.* **246**, 6828 (1971).

Electrophoresis Conditions. Electrophoresis is run from the cathode to the anode at 4–5 mA/gel until all the tracking dye has entered the SDS gel; the current is then increased to 10 mA/gel and electrophoresis is continued until the tracking dye is about 1 cm from the bottom of the gel. The gels are removed from the tubes, soaked about 20 minutes in 7.5% acetic acid to remove the SDS, then stained in 0.1% Coomassie brilliant blue overnight. The excess stain in the gels is then removed by transverse electrophoresis as previously described.[17]

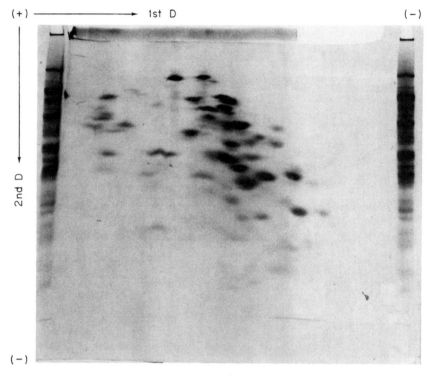

FIG. 2. Two-dimensional electrophoresis pattern of *Escherichia coli* MRE 600 50 S ribosomal subunit proteins extracted with 3 *M* LiCl–4 *M* urea [P. Spitnik-Elson, *Biochim. Biophys. Acta* **80**, 594 (1964); P. B. Moore, R. R. Traut, H. Noller, P. Pearson, and H. Delius, *J. Mol. Biol.* **31**, 441 (1968)]. The separated subunits were obtained by sucrose density gradient sedimentation of ribosomes washed once in 0.5 *M* NH$_4$Cl, 30 m*M* MgCl$_2$ [T. A. Bickle and R. R. Traut, *J. Biol. Chem.* **246**, 6828 (1971)]. Migration directions in each dimension are indicated by arrows. The O (top center) indicates the origin of the first-dimensional gel. In this case the sample was loaded in the center of the first-dimensional gel with agarose as described in the text. Gel and buffer conditions were as described in the text.

Specific Applications of the Method

Separation of Prokaryotic Ribosomal Proteins

As can be seen in Figs. 2 and 3, the methods described here give separation of the ribosomal proteins of *E. coli* 30 S and 50 S subunits comparable to that reported by Kaltschmidt and Wittmann.[2,3,14]

The two alternative methods described for applying the sample (methods 1 and 2) in the first dimension are illustrated with *E. coli* proteins. Both methods avoid the significant losses of protein at the center-origin as observed in the systems previously described.[7,14] In the first

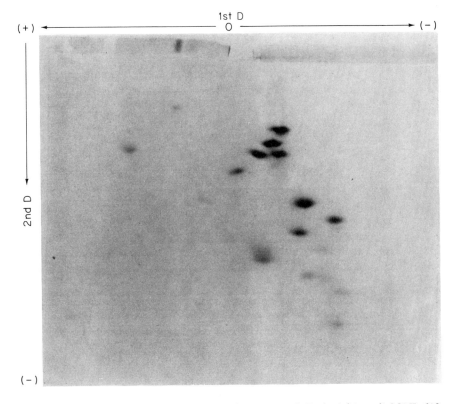

FIG. 3. Two-dimensional electrophoresis pattern of *Escherichia coli* MRE 600 30 S ribosomal subunit proteins extracted with 3 *M* LiCl–4 *M* urea. [P. Spitnik-Elson, *Biochim. Biophys. Acta* **80**, 594 (1964); P. B. Moore, R. R. Traut, H. Noller, P. Pearson, and H. Delius, *J. Mol. Biol.* **31**, 441 (1968)]. The separated subunits were obtained as in Fig. 2. In this case two short first-dimensional gels with identical top-loaded samples were run in opposite directions as described in the text. Gel and buffer conditions were as for Fig. 2.

method (Fig. 2) the sample was applied to the center of the first dimension in 0.5% agarose. Very little material remains stained at the origin. In the second method (Fig. 3) in which two separate, short first-dimensional gels were run in opposite directions no immobilized material appears at the origin. This method, although requiring twice as much material, is very useful in that it gives a pattern virtually identical to the single center-loaded gel and to that reported by Kaltschmidt and Wittmann,[3] while

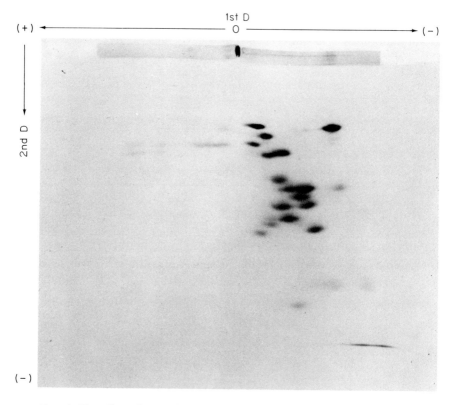

FIG. 4. Two-dimensional electrophoresis pattern of Novikoff hepatoma ascites cells 80 S ribosomal proteins extracted with 3 M LiCl–4 M urea [P. Spitnik-Elson, *Biochim. Biophys. Acta* **80**, 594 (1964); P. B. Moore, R. R. Traut, H. Noller, P. Pearson, and H. Delius, *J. Mol. Biol.* **31**, 411 (1968)]. The 80 S ribosomes were prepared as described by Busch [H. Busch, R. K. Busch, W. H. Spahn, J. Wikman, and Y. Daskal, *J. Exp. Biol. Med.* **137**, 1470 (1971)]. Migration direction in each dimension is indicated by an arrow. The first dimension was run in this case as a single top-loaded gel, thus the origin is at the upper left as indicated and the few acidic proteins in the mixture are not shown. The protein patterns at the sides of the second dimension were obtained by running samples of total protein loaded in wells at the top of the second dimension as described in the text. Gel and buffer conditions were as described in the text.

leaving essentially no protein remaining bound at the origin. Moreover, since our gel system requires only microgram amounts of ribosomal protein, the use of this "double" first dimension still uses only a fraction of the material required by the larger gel systems.

Separation of Eukaryotic Ribosomal Proteins

The method described here has been used in this laboratory to study various types of eukaryotic ribosomes. It has been especially useful in that it requires only small amounts of material. Figure 4 shows the separation of 69–70 ribosomal proteins from 80 S ribosomes of Novikoff hepatoma ascites cells. A further refinement to the gel system is also illustrated in Fig. 4. At the side of the gel slab samples of the total protein are run at the same time as the gel slice from the first dimension. This is done by using a plexiglass spacer at the top, when forming the second-dimensional gel, which has projections on it to make sample wells on each side of the first-dimensional slice. This has been very useful in doing comparative studies of ribosomal proteins from various sources, in that it makes it

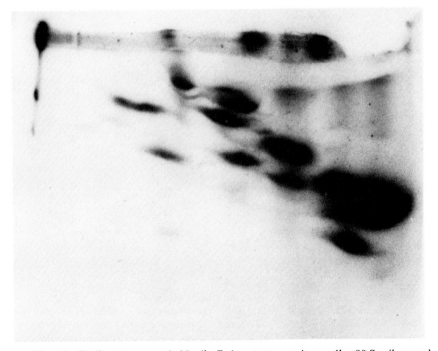

FIG. 5. Radioautogram of Novikoff hepatoma ascites cells 80 S ribosomal proteins labeled with [^{32}P]ATP and protein kinase. The two-dimensional gel slab was dried as described in the text and the autoradiogram was made by placing Kodak No-Screen Medical X-ray film directly against the dried gel.

easier to determine where to look for differences in the two-dimensional gel patterns. This is particularly important in the case of eukaryotic ribosomes where one may be looking for only 1 or 2 protein differences out of a total of 65–70 proteins.

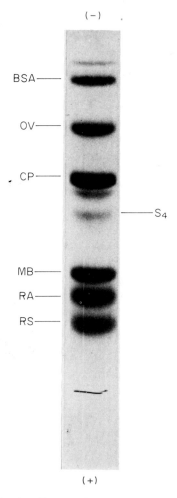

FIG. 6. Sodium dodecyl sulfate–10% acrylamide gel of *Escherichia coli* S4 protein. The stained protein spot taken from a two-dimensional gel slab like that in Fig. 3 was treated as described in the text before being applied onto the SDS gel. Standard proteins of known molecular weight were also applied onto the SDS gel for molecular weight calibration [K. Weber and M. Osborn, *J. Biol. Chem.* **244**, 404 (1964)]. BSA, bovine serum albumin; OV, ovalbumin; CP, carboxypeptidase A; MB, myoglobin; RA, ribonuclease A; RS, ribonuclease S.

Autoradiography

Figure 5 shows an autoradiogram of 80 S ribosomal proteins from Novikoff hepatoma ascites cells which were phosphorylated by protein kinase from rabbit skeletal muscle in a reaction in which the ^{32}P from [γ-^{32}P]ATP is transferred to the protein.[18] A comparison of Figs. 4 and 5 shows the specificity of the phosphorylation reaction. The totally dried gel was placed against the X-ray film. Although autoradiograms of ^{32}P can be made of wet gels,[19] those of ^{14}C or ^{35}S require drying of the gel.

Molecular Weight Determination of Separated Proteins

Figure 6 shows an example of the SDS gel analysis of one of the *E. coli* 30 S proteins separated on the two-dimensional gel. The S4 protein from *E. coli* 30 S subunits was cut from a sheet gel containing all 21 of the 30 S proteins, then treated as described above for the third dimension. Several proteins of known molecular weight were added as standards to give an accurate molecular weight value for the protein in question.[20] The molecular weight determined here of 27,000 for protein S4 compares quite favorably with that of 26,600 as reported by Traut *et al.*[21]

By means of this same method, molecular weights of other *E. coli* 30 S proteins have been determined which agree with previously reported values (Traut *et al.*[21]); see tabulation.

Protein	Determined molecular weight	Previously reported molecular weight
S1	68,000	68,000
S3	29,400	29,900
S5	21,000	20,200
S7	19,600	19,600
S9	16,200	17,200

[18] E. G. Krebs, *Curr. Top. Cell. Regul.* **5**, 99 (1972).

[19] J. Stahl, H. Welfle, and H. Bielka, *FEBS* (*Fed. Eur. Biochem. Soc.*) *Lett.* **26**, 233 (1972).

[20] K. Weber and M. Osborn, *J. Biol. Chem.* **244**, 4406 (1969).

[21] R. R. Traut, H. Delius, C. Ahmad-Zadeh, T. A. Bickle, P. Pearson, and A. Tissières, *Cold Spring Harbor Symp. Quant. Biol.* **34**, 25 (1969).

[51] Estimation of the Number and Size of Ribosomal Proteins by Two-Dimensional Gel Electrophoresis

By Mary G. Hamilton

Except for *Escherichia coli* ribosomes,[1] the stoichiometry of the proteins of ribosomes (r-proteins) is unknown. Moreover, because of the large quantities of material required, fractionation and characterization of the proteins may never be feasible for many kinds of eukaryotic ribosomes. Thus a need exists for simple analytical techniques that require no prior fractionation of the proteins.

Although Raymond[2] demonstrated the resolving potential of two-dimensional gel electrophoretic techniques for protein mixtures some years ago, our early efforts[3] with r-proteins did not yield clear maps. Ogata *et al.*[4] used a starch gel system with some success, but only recently have Kaltschmidt and Wittmann[5] perfected a method for polyacrylamide gels. In their method, resolution occurs primarily on the basis of charge. Although molecular sieving is utilized, the molecular weights of the proteins cannot be directly inferred from their positions. This article describes an electrophoretic mapping technique which combines charge separation in the first dimension with molecular sieving in the presence of the detergent sodium dodecyl sulfate (SDS)[6] in the second dimension. Molecular weights are obtained and the various species can be enumerated. Provided that the map can be quantitated and if the total protein content of the subunit is known, the number of copies of each species can also be calculated. At least two other groups[7,8] have concurrently developed methods almost identical to this one. Obviously, many variations of the conditions are possible, and new buffer systems[9] may lead to further improvements. Although the problem of quantitation has not been com-

[1] C. G. Kurland, *Annu. Rev. Biochem.* **41**, 377 (1972).

[2] S. Raymond, *Ann. N.Y. Acad. Sci.* **121**, 350 (1964).

[3] M. G. Hamilton and M. E. Ruth, *Abstracts, Tenth Annual Meeting of the Biophysical Society, Boston, Mass.* February, 1966.

[4] K. Ogata, K. Terao, T. Morita, and H. Sugano, *Biochim. Biophys. Acta* **129**, 217 (1966).

[5] E. Kaltschmidt and H. G. Wittmann, *Anal. Biochem.* **36**, 401 (1970).

[6] A. L. Shapiro, E. Viñuela, and J. V. Maizel, *Biochem. Biophys. Res. Commun.* **28**, 815 (1967).

[7] O. H. W. Martini and H. J. Gould, *J. Mol. Biol.* **62**, 403 (1971).

[8] T. Hultin and A. Sjöqvist, *Anal. Biochem.* **46**, 342 (1972).

[9] D. M. Neville, Jr., *J. Biol. Chem.* **246**, 6328 (1971).

pletely solved, the method should prove useful in accumulating data on r-proteins.

Procedures

The techniques and theory of gel electrophoresis are well described in numerous sources[10,11] and new developments are reported in almost every issue of the journal *Analytical Biochemistry*.

Sample Preparation. Rat liver r-proteins are extracted by the LiCl-urea method, i.e. an equal volume of 4 M LiCl, 8 M urea, 1 mM dithiothreitol, 30 mM sodium acetate, pH 5 is added to a solution of ribosomal subunits. After 1 or 2 days in the cold the precipitated RNA is removed by centrifugation for 30 minutes at 15,000 rpm in a Spinco No. 40 rotor. The supernatant solution is dialyzed exhaustively against 1% 2-mercaptoethanol and then lyophilized. The proteins are redissolved in 6 M urea, 1.2 mM methylamine, 0.5 mM NaH$_2$PO$_4$, pH 6.0, 1 mM dithiothreitol, and the concentration is estimated from the absorbance at 275 nm with an extinction coefficient, $E_{1\ cm}^{1\%} = 10$. On this basis, we use about 300 μg for the 2-D analysis. Other authors[7,8] have used less material, 100–200 μg. Since small RNA molecules like tRNA and 5 S RNA which are present on active subunits are soluble in 2 M LiCl and contribute disproportionately to the UV absorbancy, our estimates of the amount of protein are very rough. As with all analytical procedures, replicate analyses with various amounts of sample are essential.

Standard Proteins. The SDS gels are calibrated by proteins chosen for their ability to migrate to the cathode at pH 4.6 in the first dimension, e.g., ovalbumin (46,000), α-chymotrypsinogen A (25,700), lysozyme (14,100), and cytochrome c (12,700). About 20 μg of each is used. Alternatively, since prolonged exposure to urea may be deleterious, standards dissolved in 1% SDS, 1% 2-mercaptoethanol can be applied to cellulose acetate strips, which are then embedded in the second gel.

Apparatus. Resolution in the first dimension is carried out in 10 cm plastic tubes which have an inside diameter of 5 mm. Many commercial models are available for disc electrophoresis, and in fact, hand fabrication is simple. The second run is carried out in a gel slab apparatus designed by Raymond.[2] E-C Apparatus Corporation's survey gel model provides a gel slab, 23 cm wide × 17 cm long × 5 mm thick. It has cooling coils on both sides of the gel compartment and provides for recirculation of the buffer, which is essential to maintain constant pH. The bottom and sides must be sealed with gel plugs to permit layering of the resolving

[10] H. E. Whipple, ed., "Gel Electrophoresis." *Ann. N.Y. Acad. Sci.* **121**, Art. 2, 1964.
[11] A. Chrambach and D. Rodbard, *Science* **172**, 440 (1971).

gel in the vertical position. The apparatus is tilted while a small volume of gel solution is poured and allowed to polymerize first along the bottom, then each side in turn. The SDS gel is poured and overlayered with water.

Gel and Buffer Systems and Running Conditions. All gels contained 6 *M* urea. The acrylamide concentration is given as percent of A (w/v) and the *N,N'*-methylenebisacrylamide concentration as percent of B (w/v). A standard amount, 0.1% (v/v), of the accelerator, *N,N,N',N'*-tetramethylethylenediamine (TEMED) is used with ammonium persulfate as the catalyst.

For the first dimension, resolution at pH 4.6 where the r-proteins are positively charged, the standard system[12] for r-proteins is used. This is a two-gel, discontinuous-buffer system. The separation gel (7.5% A, 0.2% B in 60 mM KOH, 0.375 *M* acetic acid, pH 4.6) is about 10 cm long, and the upper or spacer gel (2.5% A, 0.6% B in 60 mM KOH, 20 mM acetic acid, pH 6.5) is about 1.5 cm long. The basic dye Pyronine Y is added with the sample to mark the front. The tank buffer is 70 mM β-alanine, 28 mM acetic acid, pH 4.6. The apparatus is placed in a cold room, and a current of 3 mA per tube is passed for about 5 hours, or until the marker dye has migrated close to the end of the gel. That position and the top of the resolving gel are marked by the injection of a 1% solution of bovine serum albumin dyed with Bromophenol blue.

The sample and the mixture of standards are run in duplicate. One gel of each is stained for 30 min with Amido black (0.1% in 7.5% acetic acid) which permits quick destaining by diffusion in several changes of 7.5% acetic acid. These gels serve as guides for trimming the duplicate gels for the second dimension and as a record of the first dimension analysis.

The duplicate gels are soaked in about 10 ml of 1% SDS, 1% 2-mercaptoethanol, and rocked gently for about 30 min at room temperature (see references cited in footnotes 7 and 8 for variations in this step). Then, the two gels, sample and standards, are positioned on the previously prepared gel slab and polymerized into place with a small volume of gel solution. The composition of the gel slab is 15% A, 0.5% B in 0.1% SDS, 0.1 *M* Na phosphate, pH 7.0. A 15% gel gives good resolution in the densest region of the map, but probably both 10% and 15% gels should be used. The tank buffer is 0.1% SDS, 10 mM sodium phosphate, pH 7.0. A current of 50 mA is passed for about 48 hr. Tap water is circulated through the cooling plates and the buffer is recirculated with an auxiliary pump. The Pyronine marker dye remains visible and conveniently migrates anodically in the second dimension.

Staining, Photography, and Evaluation of the Maps. The gels are

---

[12] P. S. Leboy, E. C. Cox, and J. G. Flaks, *Proc. Nat. Acad. Sci. U.S.* **52**, 1367 (1964).

soaked in 12.5% trichloroacetic acid for about an hour and then stained by soaking in 0.5% Coomassie blue in 10% acetic acid, 45% ethanol overnight or longer and destained by diffusion in several changes of 10% acetic acid, 25% ethanol. They are stored in the same solvent and improve with time. The gels are illuminated from below on a light box and photographed with a Polaroid Model 110B camera and a +4 closeup lens. An orange filter is used with projection film 46-L. The film is enlarged about 2.5× and traced on graph paper for evaluation. To normalize the maps so that separate analyses can be compared easily, each spot is identified by an R_f value and its molecular weight, M.

For R_f, the distance (x) of each spot from the origin $(x = 0)$, i.e., the top of the lower gel of the first dimension, is divided by the distance from the origin to the front. The origin and the front are identified in the SDS gel slab by the albumin injections described above (see spots indicated by arrows in Fig. 1).

For M, a separate plot is made on semilog graph paper of the distance

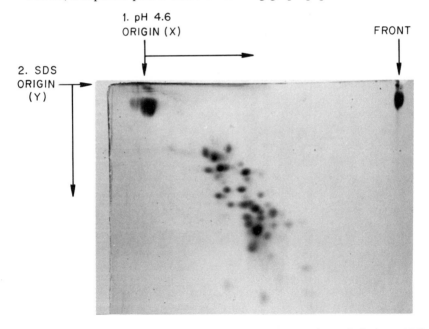

FIG. 1. Photograph of two-dimensional gel electrophoretic analysis in a 15% polyacrylamide gel slab of the proteins of the small subunit of rat liver ribosomes dissociated by treatment with EDTA. The subunits were separated by sucrose density gradient centrifugation in a B-15 batch-type zonal rotor, concentrated by ethanol precipitation and the proteins extracted as described in the text. The origin and front of the first analysis have been marked by bovine serum albumin (see arrows). The first dimension gel cylinder was embedded at the top of the second gel slab which contained sodium dodecyl sulfate.

(y) migrated by each standard protein from the edge ($y = 0$) of the SDS gel slab against its molecular weight. The resultant straight line is transposed to the first graph, and M for each r-protein spot is read off with the aid of a movable log scale fabricated from semilog paper. Alternatively, an equation, $\log M = my - b$, may be fitted to the line by first solving simultaneous equations containing the known M and y values of the standards for m and b, and then the unknown M values for each y.

Results

A map of the proteins of the small subunit derived from rat liver ribosomes by EDTA treatment is shown in Fig. 1 and the "normalized" map in Fig. 2. There obviously is some sieving in the first dimension; the large proteins generally have lower R_f values. If we assume that each spot represents a single protein, the sum of the molecular weights of the 25 prominent spots is 647,000. (In this analysis the 15 light spots will be ignored.) Although this number is quite close to the total protein content of the subunit, 600,000 daltons,[13] the variation in intensity of

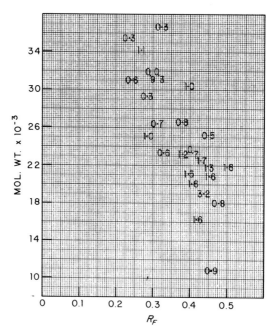

Fig. 2. Normalized "topographical" map of the r-proteins of the small subunit of rat liver ribosomes dissociated by EDTA treatment. The numerals locate the spots and give the number of copies of that component (see text).

[13] M. G. Hamilton and M. E. Ruth, *Biochemistry* **8**, 851 (1969).

staining suggests that the proteins are not present in equal amounts. We can estimate the amount of each species by densitometering a photograph of the map. For this purpose the gel was photographed at a higher magnification (about $2\times$) with orange light on GAF® Versapan Gafstar film type 2831. For this trial analysis, and since the spots are fairly symmetrical, we cut the film into strips and densitometered them in one dimension only. The "peak" areas were planimetered, the percentage of the total present in each calculated, and the number of copies computed. These numbers are plotted on Fig. 2 to present a 3-D analysis of the protein population.

It is interesting to examine these numbers. Of the proteins present in unitary (we included 0.7 and greater) and higher numbers, there are 18 species—a total of 25 molecules and 555,000 daltons. Five species occur in two copies, one in three copies, and these six account for 262,000 daltons of the 555,000. The "fractional" proteins comprise 7 species and total about 90,000 daltons. Presumably fractional species are not present on every subunit. Although it is possible that some subunits retain a protein that was extracted from others by the EDTA treatment, it is also possible that like the *E. coli* ribosomes the small subunit population of rat liver ribosomes is heterogeneous. Whether the detailed analysis we have attempted is correct, or even warranted, only time and more data will tell. It is clear that no one-dimensional analysis can adequately display the proteins of ribosomes. Our analysis assumes a constancy and reproducibility of R_f.

[52] Determination of Ribosomal Protein Molecular Weight Distributions from Stained Dodecyl Sulfate Gels[1]

By T. A. BICKLE and R. R. TRAUT

Many studies on the number and relative amounts of different proteins in complex, multiprotein structures, including ribosomes, have relied heavily on polyacrylamide gel electrophoresis as a means of resolving the proteins. Recently, gel electrophoresis in the presence of sodium dodecyl sulfate (SDS) has allowed the direct estimation of protein molecular weights; to a close approximation, the migration of polypeptide chains in SDS gels is a linear function of the chain molecular weight as illus-

[1] Supported by a research grant from the U.S. Public Health Service (GM 17924). R.R.T. is an Established Investigator of the American Heart Association.

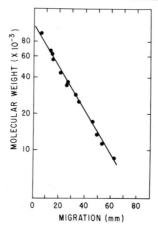

Fig. 1. Molecular weight scale determined with standard proteins. The proteins were obtained from commercial sources with the exception of proinsulin, which was a gift from Dr. D. Steiner. They were electrophoresed in 10% acrylamide 0.27% N,N-methylene bisacrylamide SDS gels as described in the text. The distances between the stained protein band and the origin of the gel was measured and plotted against the logarithm of the molecular weight. The proteins and their molecular weights are: phosphorylase a, 94,000; bovine serum albumin, 68,000; L-amino acid oxidase, 63,000; pyruvate kinase, 57,000; ovalbumin, 45,000; yeast alcohol dehydrogenase, 37,000; pepsin, 35,000; carbonic anhydrase, 29,000; chymotrypsinogen, 25,700; myoglobin, 17,200; RNase A, 13,600; RNase S, 11,400; bovine proinsulin, 8800. Taken from T. A. Bickle and R. R. Traut, J. Biol. Chem. 246, 6828 (1971) by permission of the copyright holders, the American Society of Biological Chemists.

trated in Fig. 1.[2-4] In principle, SDS gel electrophoresis can be used to determine the average molecular weights and molecular weight distributions of complex protein mixtures provided a method is available for quantitating the amount of protein at each position in the gel or of each component in the mixture.

Many previous studies have used radioactive proteins to determine the amounts of the different proteins in the mixtures analyzed. These have been separated electrophoretically and the amount of protein has been estimated by liquid scintillation counting or by densitometry of autoradiograms. There are two major disadvantages inherent in the use of radioactive protein. First, the protein is always expensive to prepare. Second, it is important for quantitative studies that the protein be uniformly labeled; that is, that the specific activity of all the protein species

[2] A. L. Shapiro, E. Viñuela, and J. V. Maizel, Biochem. Biophys. Res. Commun. 28, 815 (1966).
[3] A. K. Dunker and R. R. Rueckert, J. Biol. Chem. 244, 5074 (1969).
[4] K. Weber and M. Osborn, J. Biol. Chem. 244, 4406 (1969).

present in the mixture be the same. As it is generally necessary to chase the radioactivity out of precursors into mature structures by a chase with nonradioactive amino acids, either it must be demonstrated that the pool sizes and half-lives of each individual protein are the same, or these variables must be measured and allowed for in the calculations. Dice and Schimke[5] have shown that there may be significant variations in the specific activity of organelle proteins under different labeling conditions. They labeled rats with tritiated amino acids for 5 days and then with ^{14}C amino acids for the 12 hours preceding sacrifice of the animals and preparation of ribosomal proteins from the liver. The ribosomal proteins were then analyzed on SDS gels, and it was found that the pattern of labeling with tritium was quite different from that with ^{14}C; in this tissue at least, the ribosomal proteins turn over at varying rates. At neither time was the radioactivity a good guide to the actual amounts of the different proteins.

Here we describe a method for determining the quantitative distribution of protein mixtures in SDS gels as a function of molecular weight by the spectrophotometric scanning of gels stained with the protein dye Coomassie brilliant blue (R-250). The amount of this stain bound to protein has been found to be a linear function of the amount of protein up to at least 10 μg of protein per band.[6,7] The specific color yield varies over a range of $\pm 15\%$, depending on the protein, and the variations were found to be uncorrelated with molecular weight.[6] For a complex mixture of proteins separated according to molecular weight, the variations in color yield should be random and can be expected to cancel each other. We have used the method to determine the weight and number average molecular weights and the number of moles of protein in eukaryotic and prokaryotic ribosomes.[6,8]

SDS Gel Electrophoresis

Electrophoresis in 10% SDS gels is performed essentially according to Weber and Osborn.[4] For these gels the plot of log molecular weight against migration distance is linear over the range 10,000 to 100,000 (Fig. 1). Acrylamide and N,N'-methylene bisacrylamide (MBA) are from Eastman and are recrystallized before use. Acrylamide is dissolved in chloroform at 60° to saturation (about 60 g/liter), the solution is filtered hot and allowed to cool in a cold room at 4° overnight. The crystals are filtered out on a Buchner funnel, and washed once with cold chloroform;

[5] J. F. Dice and R. T. Schimke, *J. Biol. Chem.* **247**, 98 (1972).
[6] T. A. Bickle and R. R. Traut, *J. Biol. Chem.* **246**, 6828 (1971).
[7] J. Bennett and K. J. Scott, *Anal. Biochem.* **43**, 173 (1971).
[8] T. T. Sun, T. A. Bickle, and R. R. Traut, *J. Bacteriol.* **111**, 474 (1972).

air is sucked through them until they no longer smell of chloroform. A similar procedure is used to recrystallize MBA except that the solvent is acetone and the saturating concentration is about 10 g/liter.

Solutions

    I. Acrylamide, 22.2 g; MBA, 0.6 g; water to 100 ml
    II. $NaH_2PO_4 \cdot H_2O$, 7.9 g; $Na_2HPO_4 \cdot 7H_2O$, 38.9 g; SDS, 2 g; water to 1 liter
    III. Ammonium persulfate, 6 g in 2 ml of water; make fresh just before use
    IV. Bromophenol blue, 100 mg/100 ml

Buffers

    Sample buffer: 5 ml of II, 20 ml of glycerol, 1 g of SDS, 1 ml of 2-mercaptoethanol, water to 100 ml
    Reservoir buffer: 1 volume of II with 1 volume of water

Procedure. Mix 9 ml of solution I with 10 ml of solution II and add 15 μl of tetramethylethylenediamine (Eastman); degass and add 1 ml of solution III. Pipette the solution to a height of 7.5 cm in glass tubes 9 cm long with an internal diameter of 0.45 cm, overlay with degassed distilled water in order to obtain a flat meniscus, and allow to polymerize. Add samples containing either 25–75 μg of ribosomes or 2–5 μg of each of the standard proteins used for molecular weight calibration to 100 μl of the sample buffer, heat at 65° for 10 minutes to reduce disulfide bonds and allow the formation of the protein: SDS complex and cool in ice. (This procedure also dissociates the ribosomal protein from the RNA; the RNA does not interfere with the migration of the proteins and is not detected on the stained gels.) Add 1 μl of IV and apply the samples to the tops of the gels. Begin electrophoresis at 10 mA per tube and continue until the tracking dye is within 1 cm of the bottom of the tube (about 3.5 hours). Remove the gels from the tubes and insert a thin piece of copper wire through the center of the tracking dye band to mark its position. Soak the gels for 1 hour in 7.5% acetic acid to leach out the SDS; failure to remove SDS results in staining artifacts. Stain the gels for 8–16 hours in a solution of 0.1% Coomassie blue in 7.5% acetic acid, 50% methanol. Destaining is accomplished by soaking the gels for at least 2 hours in 3–5 changes of 7.5% acetic acid, 50% methanol, followed by transverse electrophoresis in 7.5% acetic acid, 5% methanol. These procedures give gels free from artifacts due to either loss of stain bound to protein or to high background coloration.

Scanning and Measurement of Gels

Gels are scanned at 600 nm; in our laboratory we use the scanning attachment for the Gilford 2400 spectrophotometer and scan the gels at 1 cm/minute with a chart speed of 0.5 minute per inch. This results in a scan sufficiently spread out for accurate integration. The distance of the peaks of standard proteins from the top of the scan is measured on the scan of the calibration gel. For gels containing ribosomal proteins, the absorbancy is measured every 0.1 inch on the tracing (i.e., corresponding to every 0.2 mm on the gel) between points corresponding to the top of the gel and the tracking dye.

Calculation of Weight and Number Average
  Molecular Weights

Programs are written for small desk-top computers to carry out the calculations. The program library supplied by the manufacturer of the computer contains programs for regression analysis and numerical integration, and the programs can be built around these.

Regression Analysis for the Determination of the Calibration Curve.
The measurements obtained from the scan of the calibration gel containing at least 5 standards are used to calculate the least mean squares line best fitting the plot of ln molecular weight vs. migration distance. The equation of the line has the form

$$\gamma = m \ln x + b$$

where γ is the migration distance, x is the protein molecular weight, and m and b are the constants determined by the analysis. These constants are stored and are used for calculating molecular weights later in the program. The program also calculates the regression coefficient r. This statistic has the value ± 1 when the calculated line is a perfect fit to the data points. Most SDS gels give values for r between -0.996 and -0.999; those falling outside these limits are not used.

Numerical Integration To Determine the Area of the Scan. A modification of Simpson's one-third rule for numerical integration is used to determine the area under the scans of ribosomal proteins using the trace heights read at 0.1 inch intervals. Equating stain intensity with protein concentration is equivalent to determining the total amount of protein on the gel. If c_i is the protein concentration at each point i in the gel, the total protein is Σc_i.

Calculation of the Weight and Number Average Molecular Weights. The molecular weight is calculated at each point along the gel with the constants determined from the calibration gel measurements. The molecular weights are multiplied by the corresponding trace heights to give $c_i m_i$

and are then divided into the trace heights to give c_i/m_i. The resulting figures are summed over the whole scan to give $\Sigma c_i m_i$ and $\Sigma(c_i/m_i)$. The weight average molecular weight is defined as[9]:

$$M_w = \Sigma c_i m_i/\Sigma c_i$$

and the number average molecular weight is:

$$M_n = \Sigma c_i/\Sigma(c_i/m_i)$$

Results

Analysis of Artificial Protein Mixtures

The entire procedure was tested with artificial mixtures of known amounts of pure proteins of known molecular weight. The average molecular weights of such mixtures can be calculated directly from the above formulas. The results, shown in Table I, demonstrate that (a) the calculated weight and number average molecular weights of single proteins are, as they should be, very nearly identical (to within about 0.5%) and (b) the averages calculated for the known protein mixtures are

TABLE I

WEIGHT AND NUMBER AVERAGE MOLECULAR WEIGHTS OF
PROTEIN MIXTURES OF KNOWN COMPOSITION[a]

Proteins	Calculated $\bar{M}_w$	From gel $\bar{M}_w$	Calculated $\bar{M}_n$	From gel $\bar{M}_n$
BSA	67,600	68,000	67,300	68,000
Ovalbumin	46,900	45,000	46,700	45,000
BSA + lactoglobin, equal weights	44,800	43,100	28,200	28,700
BSA + ovalbumin + equal weights	58,800	56,500	56,000	54,200
BSA + lactoglobin + ovalbumin, equal weights	45,300	43,700	32,300	32,700
BSA + ovalbumin + RNase A, equal weights	43,500	42,200	28,000	27,200
BSA + ovalbumin + RNase A, equal moles	53,800	53,400	39,800	42,200

[a] The pure, dry proteins listed in the table were accurately weighed and dissolved in SDS gel sample buffer. Equal weights or equal molar amounts of each were electrophoresed in 10% polyacrylamide-SDS gels, the gels were stained and scanned, and the weight and number average molecular weights were calculated as described. These experimentally determined values are compared to the theoretical, calculated from the known molecular weights and amounts of proteins.

[9] C. Tanford, "Physical Chemistry of Macromolecules," p. 147. Wiley, New York, 1951.

always within 5% and generally within 3% of the calculated theoretical values.

Analysis of Ribosomes

The method can be used to determine the weight and number average molecular weights and also the relative number of moles of material of each molecular weight for complex protein mixtures. For example, if the amount of protein contained in a structure is known (in daltons of protein per mole of particle), the number of moles of protein that the structure contains may be estimated by dividing this figure by the weight

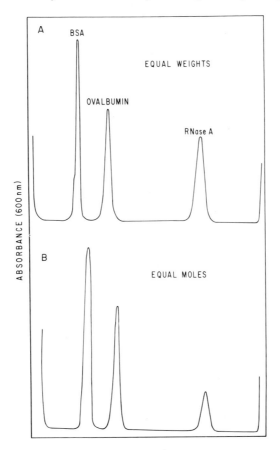

FIG. 2. Spectrophotometric scans of Coomassie blue-stained gels containing known amounts of standard proteins. (A) BSA, ovalbumin and RNase A, 5 µg each. (B) Equimolar amounts of BSA (10 µg), ovalbumin (6.6 µg) and RNase A (2 µg).

and number average molecular weights. The resulting figures represent extreme limits, and the true value falls somewhere between them.

In this way, the number of moles of protein contained by the ribosomes of a number of species of both prokaryotes and eukaryotes has been determined.[6,8] The number of moles of protein contained by the *Escherichia coli* ribosome is known by independent methods to be very close to 50.[10] The method described here gave a range of 45–56, average 50.5, a value in excellent agreement with that found independently by direct isolation of individual proteins. A similar method has been used by Lowey and Risby[11] and Osborn *et al.*[12] to determine the stoichiometry of, respectively, myosin subunits and R-17 bacteriophage proteins. These authors also found good agreement between results obtained by scanning stained gels and results obtained by chemical[11] or radiochemical means.[12]

Other Applications

The quantitative analysis of stained SDS gels can be useful in other related applications. For example, it was found that native 30 S ribosomal subunits isolated from lysates of *E. coli* contain a few nonribosomal proteins of high molecular weight that are not removed by washing the subunit with high salt.[13] In order to determine whether any of these proteins could be considered as being present in stoichiometric amounts on the subunit, the stained SDS gel shown on the left of Fig. 2 was scanned and the ribosomal protein and each of the nonribosomal protein bands were integrated separately. The area subtended by the ribosomal proteins was then set equal to 330,000, the number of daltons of protein on the

TABLE II

The Stoichiometry of the Extra Proteins on the Free 30 S Subunit[a]

Protein	MW	Daltons	Copies/subunit
1. (doublet)	88,000	5240	0.05
2.	78,000	32295	0.41
3.	63,000	41265	0.66
4.	52,000	60260	1.15

[a] The SDS gels of free 30 S ribosomal proteins shown in Fig. 3 was scanned, and the stoichiometry of the nonribosomal proteins (numbered in Fig. 3) was determined by means of the methods described in the text.

[10] R. R. Traut, H. Delius, C. Ahmad-Zadeh, T. A. Bickle, P. Pearson, and A. Tissières, *Cold Spring Harbor Symp. Quant. Biol.* 34, 25 (1969).
[11] S. Lowey and D. Risby, *Nature (London)* 234, 81 (1971).
[12] M. Osborn, A. M. Weiner, and K. Weber, *Eur. J. Biochem.* 17, 63 (1970).
[13] T. A. Bickle and R. R. Traut, unpublished observations.

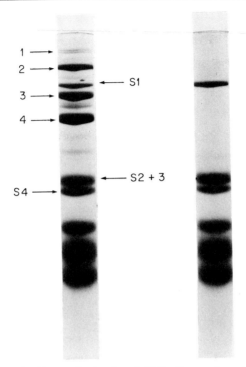

FIG. 3. Coomassie blue stained gels of 30 S ribosomal proteins derived from either the free 30 S subunit (left) or the 30 S subunit deprived from polysomes (right). The nonribosomal proteins associated with the free subunit are numbered from 1 to 4. The positions of the four largest ribosomal proteins, S1 to S4, are also indicated.

30 S particle,[14] and the number of daltons represented by the area of each of the nonribosomal components was then calculated. These values, together with the molecular weights of the proteins determined in the same analysis, allowed the number of copies of each of these proteins per 30 S subunit to be calculated. As is shown in Table II, only one of the proteins, that with a molecular weight of 52,000, could possibly be stoichiometric.

The methods described here should be useful for other studies on complex multiprotein structures. Among those to which they would seem suited are: virus, membranes, and other organized subcellular structures.

[14] A. Tissières, J. D. Watson, D. Schlessinger, and B. R. Hollingworth, *J. Mol. Biol.* **1**, 221 (1959).

[53] Reconstitution of 50 S Ribosomal Subunits from *Bacillus stearothermophilus*

By S. Fahnestock, V. Erdmann, and M. Nomura

Reconstitution of ribosomes from their dissociated molecular components is a valuable approach to the study of the roles of individual components in ribosome structure and function, as well as the process of assembly of the organelle. The complete reconstitution of the 30 S ribosomal subunit from a variety of bacterial sources including *Escherichia coli* has been demonstrated.[1] Using similar techniques, we have also been able to reconstitute the 50 S subunit from *Bacillus stearothermophilus*.[2] This chapter describes the techniques we have used for dissociation and reconstitution of 50 S subunits, and for functional characterization of the reconstituted particles.

Bacillus stearothermophilus, strain 799, is grown at 63° under forced aeration in a medium composed of 10 g of Bacto-tryptone (Difco), 5 g of yeast extract (Difco), 5 g of NaCl, 10 g of glucose, and 10^{-4} g of $MnCl_2$, per liter, pH 7.0. When the culture has reached early- to mid-log phase ($A_{550} = 0.500$), it is chilled below 10° by adding crushed ice, and the cells are harvested with a Sharples continuous-flow centrifuge. It is important not to allow the culture to become too dense, since 50 S subunits from older cultures cannot be reconstituted well. Cells are frozen in dry ice and can be stored at −80° for at least 1 year.

Preparation of Ribosomes

> Solution I: 10 mM Tris·HCl, pH 7.4 (23°), 30 mM NH$_4$Cl, 10 mM MgCl$_2$, 6 mM 2-mercaptoethanol (0.42 ml 2-mercaptoethanol per liter)
>
> Solution II: 1.1 M sucrose (RNase-free), 0.5 M NH$_4$Cl, 20 mM Tris·HCl, pH 7.4 (23°), 10 mM MgCl$_2$, 6 mM 2-mercaptoethanol
>
> Solution III: 10 mM Tris·HCl, pH 7.4 (23°), 30 mM NH$_4$Cl, 0.3 mM MgCl$_2$, 6 mM 2-mercaptoethanol

All steps are performed at 0–4°, unless otherwise stated. Freshly thawed cells are ground in an unglazed mortar with two times their weight of alumina (Norton Abrasives) until cell breakage is complete, i.e., when further grinding produces no further thinning of the paste. The paste is

[1] P. Traub and M. Nomura, *Proc. Nat. Acad. Sci. U.S.* **59**, 777 (1968).
[2] M. Nomura and V. A. Erdmann, *Nature* (*London*) **228**, 744 (1970).

then suspended in 3 volumes of solution I containing 2 μg of DNase (RNase-free) per milliliter. Alumina and debris are removed by centrifugation for 20 minutes at 8000 rpm in the Sorvall GSA rotor. The pellet is extracted with one-half the original volume of solution I. Remaining debris is removed from the combined supernatants by centrifugation for 30 minutes at 30,000 rpm in the Beckman Type 30 rotor.

The upper four-fifths of the supernatant (S30) is removed, and the pellet and lower supernatant are discarded.

Centrifuge tubes (Beckman Type 35 rotor) are two-thirds filled with S30 extract, then the tubes are filled by underlayering solution II. The ribosomes are pelleted through this sucrose–salt solution by centrifugation for 26 hours at 35,000 rpm, at 6°.

The resulting pellet (sucrose–salt washed 70 S) is resuspended (by stirring or with a loose-fitting Dounce homogenizer) in solution III. The pellet is generally nearly colorless at this stage, and the yield is about 300 A_{260} units per gram of cell paste, of which about 75% is 50 S subunits and 25% 30 S. If the ribosomes are to be frozen at this stage, the Mg^{2+} concentration is raised to 10 mM as soon as the pellet is completely resuspended.

Separation of Ribosomal Subunits

Sucrose–salt washed 70 S ribosomes, resuspended in solution III, can be subjected immediately to sucrose density gradient centrifugation for the purification of 50 S and 30 S subunits. Dialysis against buffer at low Mg^{2+} concentration is not necessary. We have adopted the procedure of Eikenberry et al.,[3] using the Beckman Ti 15 zonal rotor.

Solution IV: 38% (w/w) sucrose in solution III
Solution V: 7.4% (w/w) sucrose in solution III
Solution VI: 45% (w/w) sucrose in solution III
Solution VII: 60% (w/w) sucrose in solution III

All sucrose solutions are treated with diethyl pyrocarbonate (Calbiochem) to inactivate RNase. Sucrose is dissolved in H_2O and stirred for 24 hours at room temperature with 0.1% (v/v) diethyl pyrocarbonate. The solution is then thoroughly degassed under a water aspirator to remove CO_2 generated by decomposition of diethyl pyrocarbonate, and buffer solutions are added and final volumes adjusted.

The "6 cm" hyperbolic sucrose gradient described by Eikenberry et al.,[3] from 7.4% (w/w) to 38% (w/w) sucrose in solution III, is generated

[3] E. F. Eikenberry, T. A. Bickle, R. R. Traut, and C. A. Price, *Eur. J. Biochem.* **12**, 133 (1970).

by a Beckman Model 141 gradient pump. The two reservoirs contain solutions III and IV. The gradient is pumped into the rotor edge, followed by solution IV until solution is displaced from the rotor center. The sucrose concentration of the displaced material is monitored by refractometer, and the pump is stopped when the effluent contains 7.4% w/w sucrose. The sample is applied at the rotor center as an inverse linear gradient (vs. solution V) in a total of 100 ml, followed by a 750-ml overlay of solution III. After sample and overlay are in place, a 50-ml cushion of solution VI is inserted at the rotor edge.

Centrifugation is at 35,000 rpm for 7 hours at 4°. The gradient is displaced with solution VII, and, after the first 1000 ml is discarded, 15-ml fractions are collected. The A_{260} profile is determined, appropriate fractions are pooled, and the Mg^{2+} concentration is adjusted to 10 mM. Subunits are recovered by centrifugation. (Dilute with an equal volume of solution I and spin 24 hours at 35,000 rpm in a Beckman type 35 rotor at 6°.)

Ribosomes can be frozen in a dry ice–ethanol bath in small aliquots and stored at $-80°$. Best reconstitution results are obtained with freshly prepared ribosomes which have never been frozen, but freezing once is permissible. Once thawed, however, they should not be refrozen. Furthermore, 50 S reconstitution activity seems to deteriorate during storage at $-80°$, so ribosomes should not be stored longer than 2 months.

Urea–LiCl Dissociation

Both sucrose–salt-washed 70 S ribosomes and purified 50 S subunits have been used interchangeably in the following procedure. In general, 50 S activity after reconstitution is higher if components are derived from 70 S ribosomes, so 70 S ribosomes are preferable for many experiments. The difference in activities is not due to the presence of 30 S subunits or 16 S RNA and 30 S proteins, since addition of these to the 50 S-derived system does not affect the reconstitution of 50 S activity.[4] The difference may be due to the prolonged exposure of the ribosomes to low concentrations of Mg^{2+} during purification of subunits.

Solution VIII: 8 M urea, 4 M LiCl

Ribosomes are adjusted to $A_{260} = 300–500$ in solution I. An equal volume of solution VIII is added, and the mixture is placed on ice for 36–48 hours. The urea solution should be prepared from freshly dissolved urea to minimize isocyanate which is formed in urea solutions on storage.

---

[4] S. Fahnestock, W. Held, and M. Nomura, in "Generation of Subcellular Structures," Proceedings of the First John Innes Symposium. Norwich, England, July 1972.

This treatment dissociates most of the ribosomal proteins and causes the 23 S ($+16$ S) RNA to precipitate. The RNA precipitate is recovered by low speed centrifugation. In some cases it is advantageous to wash the RNA pellet with a 1:1 mixture of solution I and solution VIII. At this point, the pellet contains the 23 S ($+16$ S) RNA, about 30% of the 5 S RNA, and some 50 S proteins, still bound to the 23 S RNA (see below). The supernatant contains all the ribosomal proteins (except one 50 S protein) and about 70% of the 5 S RNA.

5 S RNA-Dependent System

To obtain a reconstitution system which is dependent on added 5 S RNA, the urea–LiCl fractions are freed of 5 S RNA—the supernatant protein fraction by chromatography on DEAE-cellulose, and the RNA pellet by repeated precipitation with 5 M LiCl.[5]

Removal of 5 S RNA from Urea–LiCl Supernatant

Solution IX: 6 M urea, 10 mM Tris·HCl, pH 7.4 (23°), 1 mM MgCl$_2$, 6 mM 2-mercaptoethanol, 0.25 M KCl
Solution X: Same as solution IX, but 0.30 M KCl
Solution XI: Same as solution IX, but 1.00 M KCl

All solutions are prepared from freshly dissolved urea. Urea–LiCl supernatant is dialyzed against 4×100 volumes of solution IX for 4 hours at 6°, then applied to a small DEAE-cellulose column (5 cm $\times$ 1.2 cm is sufficient for protein derived from 1000 A_{260} units of 50 S subunits) equilibrated with solution IX. The protein is eluted with solution X, monitoring A_{280}. The first peak, including the pass-through, contains all of the ribosomal proteins and should be free of 5 S RNA. 5 S RNA can be eluted from the column with solution XI. We have had best results using a preparation of DEAE-cellulose obtained from Sigma Chemical Co.; the separation has been unsuccessful with some other preparations from other sources.

Removal of 5 S RNA from Urea–LiCl Pellet

5 S RNA is soluble in 5 M LiCl, while 23 S (and 16 S) RNA is not. Thus, 5 S RNA can be removed from the urea–LiCl pellet by repeated LiCl precipitation of 23 S RNA. The urea–LiCl RNA pellet is redissolved in H$_2$O (to $A_{260} = 200$ to 300) by stirring briefly at room temperature. An equal volume of 10 M LiCl is added, and the precipitate is collected

---

[5] V. A. Erdmann, S. Fahnestock, K. Higo, and M. Nomura, *Proc. Nat. Acad. Sci. U.S.* **68**, 2932 (1971).

immediately by low speed centrifugation. This procedure is repeated 3 times to obtain 5 S RNA-free 23 S (and 16 S) RNA.

Removal of Residual Protein from Urea–LiCl Pellet

Solution XII: 2 M Mg(Ac)$_2$, adjusted with HCl to pH 2.0

The urea–LiCl RNA pellet contains 2–4 50 S ribosomal proteins which are bound to 23 S RNA.[6,7] One protein, which we have designated L3, is mostly (90%) found in the pellet, being present only in small amounts (10%) in the supernatant. The other proteins are present in the pellet in much smaller amounts and are mostly in the supernatant. These proteins can be removed from 23 S RNA by precipitation at low pH. RNA is redissolved in 6 M urea to A_{260} = 200 to 400. To this solution is added one-third volume of solution XII. After 1 hour on ice, the precipitated RNA is collected by low speed centrifugation, suspended in solution III and dialyzed until dissolved (4 hours) against solution III. Any 5 S RNA which is present is entirely precipitated along with 23 S and 16 S RNA by this treatment. The resulting pellet is essentially protein-free; protein L3 and traces of other proteins are found in the supernatant.

Reconstitution of 50 S Subunits

Solution XIII: 30 mM Tris·HCl, pH 7.4 (23°), 20 mM MgCl$_2$, 1 M KCl, 6 mM 2-mercaptoethanol

Solution XIV: Same as solution XIII, but no KCl

Protein fractions are dialyzed for 6 hours against solution XIII. RNA fractions are dissolved in H$_2$O and can be dialyzed against Solution XIV, but if they are sufficiently concentrated (A_{260} > 300) dialysis is not necessary. Protein concentrations are expressed as A_{260} equivalents per milliliter, one A_{260} equivalent being the amount of material obtained from one A_{260} unit of ribosomes assuming no losses (and considering changes in volume during dialysis). RNA concentrations can be determined spectrophotometrically and, to a good approximation, 1 A_{260} unit of 23 S RNA (or 23 S + 16 S) is one 50 S (or 70 S) A_{260} equivalent.

The basic reconstitution mixture contains, per milliliter, 10 A_{260} units of urea–LiCl 23 S RNA and 12.5 50 S A_{260} equivalents of urea–LiCl supernatant fraction. If protein-free 23 S RNA is used, the RNA and urea–LiCl protein concentrations are the same, and 20 50 S equivalent units per milliliter of protein L3 (or pH 2 Mg^{2+} supernatant fraction) are added. If 5 S RNA-free components are used, the RNA concentra-

[6] V. A. Erdmann, S. Fahnestock, and M. Nomura, *Fed. Proc., Fed. Amer. Soc. Exp. Biol.* 30, 1203 (Abstract), (1971).
[7] S. Fahnestock, V. Erdmann, and M. Nomura, *Biochemistry* 12, 220 (1973).

tion is the same, the protein concentration is fifteen 50 S A_{260} equivalents per milliliter and the 5 S RNA concentration 0.35 A_{260} unit/ml or ten 50 S A_{260} equivalents per milliliter. Solutions XIII and XIV are mixed in such proportions that the final ratio is 2 parts solution XIV to 1 part solution XIII, assuming that all protein fractions are in solution XIII and all RNA fractions are in solution XIV. The final concentration of KCl is 0.33 M. Components are mixed in the following order: (1) buffer components, (2) RNA fractions, (3) protein fractions.

The mixture is incubated at 60°. Activity sometimes continues to increase for up to 5 hours, but at a continually decreasing rate. We have generally limited our incubations to 2 hours.

The activity of the reconstituted particles can be assayed directly (see below) or the particles can be concentrated by centrifugation (2.5 hours at 60,000 rpm or equivalent at 4°) and, if necessary, purified by sucrose density gradient centrifugation.

Functional Characterization of Reconstituted Particles

Poly(U)-Dependent Polyphenylalanine Synthesis

The activity of reconstituted ribosomes has generally been assayed by polypeptide synthesis directed by poly(U). Activity of 50 S is determined in the presence of an excess of added 30 S subunits. We use 30 S subunits prepared as described by Traub et al.[8] from E. coli Q13, since they are more active than those from B. stearothermophilus, even with 50 S subunits from B. stearothermophilus. The enzyme fraction required for the assay is also prepared from E. coli Q13, as described by Traub et al.[8]

Each incubation mixture contains (mixed at 0° in the order shown) 45 μl of Mix-1, 4–12 μl of enzyme (the optimum must be determined for each preparation of enzyme), 3 μl of 1 mg/ml pyruvate kinase, 10 μl of 4 mg/ml poly(U), 5 μl of [^{14}C]Phe (15 μCi/ml at a specific activity of 10 μCi/μmole), 1 A_{260} unit of 30 S subunits, 1 A_{260} unit of 50 S subunits or reconstituted particles, and enough solution I to make the final volume 150 μl. Mix-1 contains 21.5 mM Tris·HCl, pH 7.4 (23°), 64.5 mM NH$_4$Cl, 21.5 mM Mg(Ac)$_2$, 4.3 mM ATP·Na$_2$, 0.13 mM GTP·Na$_2$, 13.5 mM 2-mercaptoethanol, 4.3 mM dithiothreitol, 21.5 mM phosphoenol pyruvate·Na$_3$, and 4.3 mg/ml tRNA from E. coli K12 (General Biochemicals Co.).

For direct assay of reconstitution mixtures, the total volume can be increased to 0.2 ml to accommodate a 0.1-ml aliquot of the sample to

---

[8] P. Traub, S. Mizushima, C. V. Lowry, and M. Nomura, this series, Vol. 20, p. 391.

be assayed, in reconstitution buffer. In this case, the amounts of components listed above need not be increased, and the dilution and change in ionic conditions due to the addition of 0.1 ml of reconstitution buffer do not reduce the activity of the system.

When particles are assayed directly, in most cases no reconstitution of active 50 S subunits takes place during the assay because the temperature (37°) is below the minimum required. However, in some cases inactive particles can be activated during assay if 50 S ribosomal proteins are present. For example, particles reconstituted without 5 S RNA show significant activity if assayed directly, probably because in the presence of the unbound 50 S proteins,[5] 5 S RNA which is present in the tRNA preparation can be incorporated and the particles activated to some extent. In cases like this, free ribosomal proteins must be removed by pelleting the reconstituted particles. (Activation by 5 S RNA takes place only in the presence of 50 S proteins.)

Samples are incubated at 37° for 30 minutes. The reaction is terminated by adding 2 ml of 5% (w/v) trichloroacetic acid (TCA), and the precipitates are heated in a boiling water bath for 15 minutes, chilled on ice and filtered onto Reeve Angel glass fiber filters, type 934AH. The filters are washed with 5% TCA and dried, and the radioactivity is measured in a gas-flow or liquid-scintillation counter.

Peptidyltransferase Assay

The peptide bond-forming enzyme, peptidyltransferase, which is an integral part of the 50 S ribosomal subunit,[9] catalyzes a reaction between N-formylmethionyl-tRNA and puromycin.[10,11] This simple reaction, which requires only 50 S subunits, Mg^{2+} and K^+ ions, and some alcohol, provides a convenient and specific assay for peptidyltransferase activity.

Solution XV: 110 mM Tris·HCl, pH 7.4 (23°), 30 mM $MgCl_2$, 800 mM KCl

Each incubation mixture contains 50 μl of solution XV, [^{35}S]fMet-tRNA (30,000 cpm in 1 μl, see below), the sample to be assayed (1 A_{260} unit 50 S) in solution I, and solution I to make the total volume 100 μl. The reaction is initiated by adding 50 μl of 1 mM puromycin in methanol, and allowed to proceed for 10 minutes at 0°. It is terminated by adding 10 μl of 10 N KOH and transferring the samples to a 37° bath, where

[9] R. E. Monro, T. Staehelin, M. L. Celma, and D. Vazquez, Cold Spring Harbor Symp. Quant. Biol. 34, 357 (1969).
[10] R. E. Monro, J. Cerna, and K. Marcker, Proc. Nat. Acad. Sci. U.S. 61, 1042 (1968).
[11] R. Miskin, A. Zamir, and D. Elson, J. Mol. Biol. 54, 355 (1970).

they are incubated for 10 minutes. This alkaline treatment hydrolyzes unreacted fMet-tRNA and fMet-methylester, which is formed by nonenzymatic methanolysis.[11,12] One ml of 1 M sodium phosphate buffer, pH 7.0, is added, with 1.5 ml ethyl acetate, the samples are mixed vigorously with a vortex mixer and centrifuged briefly at 1000 g. An aliquot (1 ml) of the upper (ethyl acetate) phase is removed for radioactivity determination in a liquid scintillation counter.

Synthesis of fMet-tRNA. Either unfractionated tRNA or purified tRNA$_F^{Met}$ can be used. The incubation mixture contains, per milliliter, 100 mM sodium cacodylate, pH 7.0, 10 mM MgCl$_2$, 10 mM KCl, 6 mM 2-mercaptoethanol, 7.5 mM ATP, 0.2 mM CTP, 100 mM phosphoenol pyruvate, 0.02 mg of pyruvate kinase, 0.6 mM folinic acid, 1 mg "enzyme" [the same preparation used for the poly(U) assay, described above], tRNA (5 mg unfractionated tRNA or 1 mg tRNA$_F^{Met}$), and 1 mCi [^{35}S]Met at about 20 Ci/mmole. The mixture is incubated for 30 minutes at 37°, then the pH is lowered to about 5 with acetic acid and the mixture is extracted at 4° with an equal volume of phenol saturated with H$_2$O. The organic phase is reextracted with 0.5 volume of H$_2$O and the combined aqueous phases are brought to 2% KAc, pH 5, and the RNA is precipitated with 2 volumes of ethanol at −20°.

The pellet is washed with ethanol and redissolved at 3 mg/ml in 20 mM CuSO$_4$, 200 mM NaAc, pH 5.5. This solution is incubated at 37° for 1 hour to hydrolyze unformylated met-tRNA and other aminoacyl-tRNA's,[13] then the RNA is again precipitated with 2 volumes of ethanol at −20° and redissolved in 1 ml of 0.2 M EDTA, pH 5. The tRNA is then purified by filtration through a 1.5 × 20 cm column of Sephadex G-25 in H$_2$O to remove nucleoside triphosphates and concentrated by ethanol precipitation if necessary. It can be stored frozen in 6 mM 2-mercaptoethanol solution.

Aminoacyl-tRNA Binding Assay

The 50 S subunit participates in the ribosome binding of aminoacyl tRNA by associating with the 30 S–aminoacyl-tRNA complex to form a 70 S couple. This association results in the protection of bound tRNA from degradation by low concentrations of ribonuclease.[14] This protection from RNase is the basis of a convenient assay for the ability of the 50 S subunit to associate with the 30 S subunit.[14]

[12] S. Fahnestock, H. Neumann, V. Shashoua, and A. Rich, *Biochemistry* **9**, 2477 (1970).
[13] P. Schofield and P. C. Zamecnik, *Biochim. Biophys. Acta* **155**, 410 (1968).
[14] S. Pestka, *J. Biol. Chem.* **243**, 4038 (1968).

Solution XVI: 500 mM Tris·HCl, pH 7.4 (23°), 100 mM MgCl$_2$, 250 mM KCl

Each incubation mixture contains 10 μl of solution XVI, 10 μl of 4 mg/ml poly(U), 0.5 A_{260} unit of 30 S subunits from *E. coli* Q13 in 5 μl of solution I, 0.005 μCi [^{14}C]Phe-tRNA at 385 Ci/mole of Phe (New England Nuclear) in 10 μl of H$_2$O, and the sample to be assayed in 15 μl of solution I, containing 1 A_{260} unit of 50 S subunits. Incubation is at 30° for 20 minutes. Pancreatic RNase (0.014 μg in 5 μl) is then added, and the samples are incubated again at 30° for 7 minutes. They are then chilled, and 2 ml of cold (0°) TCA is added. Precipitates are collected on Reeve Angel glass fiber filters (type 934AH), washed with cold 5% TCA, and dried; radioactivity is determined in a gas-flow counter or liquid-scintillation counter.

EF G-GTP Binding Assay

The 50 S subunit has a site to which EF G and GTP bind as part of the translocation function.[15] The antibiotic fusidic acid stabilizes the 50 S subunit-EF-G-GTP complex so that it can be collected on a nitrocellulose filter.[16] This is the basis of a convenient assay, developed by Bodley, for the EF-G-GTP binding function of the 50 S subunit.

Each incubation mixture contains 2 μl of 75 mM fusidic acid, 50 pmoles of [^{3}H]GTP in 10 μl of solution I and the sample to be assayed in 33 μl of solution I. Purified EF G (5 μl, 1 mg/ml in solution I) is added last to initiate the reaction, which is allowed to proceed for 5 minutes at 0°. It is terminated by diluting each sample with 3 ml of solution I containing 0.1 μM fusidic acid, and the sample is immediately passed through a Millipore filter (type HA, 0.45 μm), which is washed with solution I plus 0.1 μM fusidic acid, dried, and counted in a liquid scintillation counter.

[15] J. W. Bodley and L. Lin, *Nature (London)* **227**, 60 (1970).
[16] J. W. Bodley, F. J. Zieve, L. Lin, and S. T. Zieve, *Biochem. Biophys. Res. Commun.* **37**, 437 (1969).

[54] Isotopic Labeling and Analysis of Phosphoproteins from Mammalian Ribosomes

By LAWRENCE BITTE and DAVID KABAT

I. Introduction and Scope

It has been recently shown that at least five polypeptide chains from ribosomes of rabbit reticulocytes,[1,2] rat liver cells,[3] and mouse sarcoma 180 tumor cells[4] are phosphoproteins rather than simple proteins. The phosphoryl groups in these proteins are present in o-phosphoseryl and in o-phosphothreonyl residues. Since these phosphoryl groups turn over intracellularly, the proteins become radioactive when the cells are incubated with [^{32}P]orthophosphate. Furthermore, these phosphoproteins from rabbit reticulocyte and from mouse sarcoma 180 ribosomes have the same molecular weights and may be homologous proteins related by evolution.[4] Intracellular phosphorylation of these polypeptides is markedly elevated by hormonal stimuli or by cyclic AMP.[5-7] In vitro phosphorylation of ribosomal proteins by cyclic AMP-dependent protein kinases (ATP:protein phosphotransferase EC 2.7.1.37) has also been reported by several groups.[3,8-10] In addition, a protein kinase which may be the physiologically active enzyme has been isolated in association with reticulocyte[11] and with fibroblast[10] ribosomes.

This is the only enzymatic modification of ribosomal proteins known to occur in mammalian cells. Although a similar metabolism is apparently absent from Escherichia coli,[12] cyclic AMP-dependent protein phosphorylation reactions have been increasingly implicated in the control of metabolism and growth of mammalian cells and in the regulation of physiological responsiveness. Frequently, the phosphoproteins occur in complex assem-

[1] D. Kabat, Biochemistry 9, 4160 (1970).
[2] D. Kabat, J. Biol. Chem. 247, 5338 (1972).
[3] J. E. Loeb and C. Blat, FEBS (Fed. Eur. Biochem. Soc.) Lett. 10, 105 (1970).
[4] L. Bitte and D. Kabat, J. Biol. Chem. 247, 5345 (1972).
[5] L. Bitte and D. Kabat, manuscript in preparation, 1972.
[6] M. L. Cawthon, L. Bitte, A. Krystosek, and D. Kabat, J. Biol. Chem. (in press) 1973.
[7] C. Blat and J. E. Loeb, FEBS (Fed. Eur. Biochem. Soc.) Lett. 18, 124 (1971).
[8] C. Eil and I. G. Wool, Biochem. Biophys. Res. Commun. 43, 1001 (1971).
[9] G. M. Walton, G. N. Gill, I. B. Abrass, and L. D. Garren, Proc. Nat. Acad. Sci. U.S. 68, 880 (1971).
[10] C. Li and H. Amos, Biochem. Biophys. Res. Commun. 45, 1398 (1971).
[11] D. Kabat, Biochemistry 10, 197 (1971).
[12] J. Gordon, Biochem. Biophys. Res. Commun. 44, 579 (1971).

blages of macromolecules, such as in chromosomes,[13] in viruses,[14] in secretory granules,[15] in microtubules,[16] and in rhodopsin on photoreceptor membranes.[17]

The purpose of this chapter is to describe the methods which we have found most useful for the isotopic labeling and characterization of phosphoproteins from mammalian ribosomes. We describe methods for removing contaminating phosphoproteins and for establishing the reality of this ribosomal modification. These same methods should be applicable to analysis of other phosphoproteins, especially those which occur in complex subcellular organelles.

II. Labeling of Ribosomal Phosphoproteins with [^{32}P]-Orthophosphate (Methods for Reticulocytes and for Sarcoma 180 Cells)

A. Principles

Nearly all phosphorus in mammalian ribosomes is present in the constituent RNA molecules (approximately 6.5×10^3 P-atoms), with only a small number occurring in the proteins (approximately 7–11 P-atoms).[2] Consequently, it is desirable to ^{32}P-label the phosphoryl groups on the proteins in conditions in which ribosomal RNA synthesis is absent. This has been accomplished by utilizing reticulocytes (which lack a nucleus and are accordingly inactive in RNA synthesis),[1,2] by utilizing tumor cells cultured in the presence of actinomycin D (an inhibitor of RNA synthesis),[4] or by labeling purified ribosomes in vitro with [γ-^{32}P]ATP in the presence of protein kinase.[3,8–11]

B. Solutions Used

 a. Physiological salt solution (0.13 M NaCl, 5 mM KCl, 1.5 mM MgCl$_2$), used for washing mammalian cells
 b. Buffer A (10 mM KCl, 1.5 mM MgCl$_2$, 10 mM Tris·HCl, pH 7.4), a low ionic strength buffer in which ribosomes are stable
 c. Buffer B (0.25 M KCl, 10 mM MgCl$_2$, 10 mM Tris·HCl, pH 7.4), a high ionic strength buffer in which single ribosomes may partially dissociate into subunits

[13] T. A. Langan, "Regulatory Mechanisms for Protein Synthesis in Mammalian Cells," p. 101. Academic Press, New York, 1968.
[14] M. Strand and J. T. August, Nature (London) New Biol. 233, 137 (1971).
[15] F. Labrie, S. LeMaire, G. Poirer, G. Pelletier, and R. Boucher, J. Biol. Chem. 246, 7311 (1971).
[16] D. B. P. Goodman, H. Rasmussen, and F. DiBella, Proc. Nat. Acad. Sci. U.S. 67, 652 (1970).
[17] D. Bownds, J. Dawes, J. Miller, and M. Stahlman, Nature (London) New Biol. 237, 639 (1972).

d. Buffer C (0.10 M KCl, 40 mM NaCl, 5 mM Mg acetate, 20 mM Tris·HCl, pH 7.6), an intermediate ionic strength buffer in which ribosomes are stable

e. Sodium deoxycholate. A 10% solution is stored at room temperature

f. New methylene blue stain: 5.0 g Methylene blue NN (Allied Chemical, Morristown, New Jersey), 8.5 g of NaCl and 4.0 g of sodium citrate·2H$_2$O, mixed with 1 liter of H$_2$O followed by filtration

g. Toluidine blue stain: 1% solution in physiological salt solution

h. Nutritional medium for reticulocytes (modified from Hori and Rabinovitz[18]). The modified medium contains physiological salt solution supplemented with 1 mg/ml glucose, 0.4 mg/ml NaHCO$_3$, 0.2 mM ferrous ammonium sulfate, and 5% fetal calf serum. Amino acids are present in the following concentrations, expressed as millimoles per liter: L-glutamine 0.096, L-histidine 0.116, L-leucine 0.20, L-lysine 0.090, L-phenylalanine 0.080, L-serine 0.086, L-tryptophan 0.015, L-tyrosine 0.042, L-valine 0.154, L-methionine 0.040, L-arginine 0.040 and L-isoleucine 0.031. L-Cysteine, L-alanine, L-asparagine, glycine, L-proline, and L-threonine are present at 10 μM

i. Nutritional medium for sarcoma 180 cells. This medium contains Krebs bicarbonate buffer lacking inorganic phosphate (0.12 M NaCl, 5 mM KCl, 2.5 mM CaCl$_2$, 1 mM MgSO$_4$, 0.2% NaHCO$_3$, pH 7.8–8.0) supplemented with 1 mg/ml glucose, with 10% fetal calf serum, and with L-amino acids in the final concentrations recommended by Lee et al.[19]

C. Procedures for Reticulocytes

Rabbits are made anemic by subcutaneous injection for at least 7 days with 10 mg per kilogram of body weight of phenylhydrazine hydrochloride. A neutralized solution containing 25 mg/ml is prepared daily just before use. We generally make a new rabbit anemic each week and use it throughout the following week. The blood is collected into heparinized beakers which are kept chilled in ice. The cells are sedimented by centrifugation at 800 g for 6 minutes and are washed three times with physiological salt solution. Blood cells are routinely stained with new methylene blue stain in order to determine the proportion of reticulocytes. The washed cells are resuspended in one volume of physiological salt solution. Four

[18] M. Hori and M. Rabinovitz, *Proc. Nat. Acad. Sci. U.S.* **59**, 1349 (1968).
[19] S. Y. Lee, V. Krsmanovic, and G. Brawerman, *Biochemistry* **10**, 895 (1971).

drops of cell suspension and three drops of stain are mixed in a tube
and are incubated at 37° for 1 hour. A smear of cells is then made on
a slide, which is examined in the microscope. We routinely obtain by
these methods at least 95% reticulocytes.

The washed cells are resuspended at 7×10^8 cells/ml in the nutritional medium for reticulocytes. Generally, we add 5 μg of actinomycin
D per milliliter to suppress any RNA synthesis in contaminating leukocytes. The cell suspension is swirled in a water bath at 37° for 15 minutes
before addition of 50 μCi/ml [^{32}P]orthophosphoric acid (New England
Nuclear Corp., Boston, Massachusetts). After incorporation for various
time periods, the cell suspension is diluted 5-fold with ice-cooled physiological salt solution and the cells are sedimented at 800 g for 6 minutes.
All subsequent procedures are at 2°. The packed cells are lysed by the
addition of 4 volumes of buffer A. After centrifugation at 10,000 g for
10 minutes, the supernatant is collected into an ice-cooled beaker. A

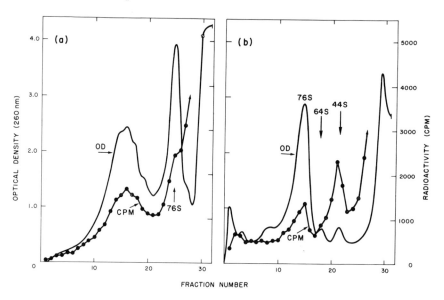

FRACTION NUMBER

FIG. 1. Sucrose gradient sedimentation of ^{32}P-labeled reticulocyte ribosomes.
Reticulocytes were incubated for 60 minutes with [^{32}P]orthophosphate. The extract
containing ribosomes was layered onto 29-ml linear gradients of 15–30% sucrose
dissolved in buffer A in tubes for the SW 25.1 Spinco rotor. In (a), centrifugation
was at 25,000 rpm for 2.5 hours. In (b) centrifugation was at 17,500 rpm for 20
hours; however, 18,000 rpm for 19 hours gives better results for preparative purposes. The gradient fractions were supplemented with 0.050 mg bovine serum
albumin and were then precipitated with 10% trichloroacetic acid at 0° for 30
minutes. The precipitates were filtered onto 0.45 μm Millipore filters for radioactivity measurement from D. Kabat, *Biochemistry* 9, 4160 (1970).

solution of 0.1 M acetic acid is added dropwise with swirling until the pH is reduced to pH 5.1; this causes the ribosomes and many cellular enzymes to coprecipitate, leaving the hemoglobin in solution. The precipitate is collected by centrifugation at 10,000 g for 10 minutes. The precipitate obtained from 1 ml of packed cells is redissolved in 1 ml of buffer A, B, or C (the choice depends on the experiment, as is described below); a few drops of 1 M Tris·HCl, pH 7.5, are added to the samples in order to facilitate their resolution; and the solutions are then clarified by centrifugation at 10,000 g for 10 minutes.

The resulting solutions contain ^{32}P-labeled ribosomes and are used for further ribosome purification (see Section III). When such solutions are sedimented into sucrose gradients, results such as those in Fig. 1 are reproducibly obtained. It can be seen that radioactivity is associated with the rapidly sedimenting polyribosomes, with the single ribosomes (76 S), and with the large (66 S) and small (44 S) subribosomal particles. This radioactivity is precipitable with 10% trichloroacetic acid, suggesting that it is associated with macromolecules.

D. Procedures for Sarcoma-180 Cells

The sarcoma 180 cell culture is maintained by transferring 0.2 ml of ascites fluid (1 × 10⁷ cells) from the peritoneum of infected mice into the peritoneum of new 28–35 g female Swiss Webster mice. Cells are harvested for experimentation between 5 and 7 days postinoculation and are then washed three times by centrifugation with cold physiological salt solution. Generally, we obtain between 2 and 6 ml of ascites fluid (with 0.1–0.4 ml of packed cells per milliliter of ascites fluid) from each mouse. Cultures of ascites cells occasionally change their growth characteristics or karyotype. Consequently, in beginning work with such cultures it is desirable to store aliquots of the cells in a liquid N_2 freezer. We routinely use the procedure of Hauschka et al.[20] for sarcoma 180 cells except that dimethyl sulfoxide (10% of the final cell suspension volume) is used instead of glycerol.

For isotopic incorporation, the cells are suspended in the nutritional medium for sarcoma 180 cells in the ratio of 1 ml of packed cells to 29 ml of medium; the flasks are incubated at 37° in a rotary shaking water bath. Actinomycin D (10 μg/ml) is added to the cell suspension to inhibit RNA synthesis 10 minutes prior to addition of [32P]orthophosphoric acid (50 μCi/ml of cell suspension). After incorporation for various time periods, the cells are chilled by dilution with four volumes of cold physiological salt solution and are collected by centrifugation at 600 g for 5

---

[20] T. S. Hauschka, J. T. Mitchell, and D. J. Niederpruem, *Cancer Res.* 19, 643 (1959).

minutes at 2–4°. The cells are then washed twice by centrifugation in cold physiological salt solution.

Cell lysis and ribosome preparation are performed at 2°. Washed cells are suspended in three volumes of buffer C. Triton X-100 (Sigma Chemical Co., St. Louis, Missouri) is added at a concentration of 0.25% and lysis is accomplished by drawing the cell suspension into a Pasteur pipette six times. A small sample of the cell lysate is stained with toluidine blue and is observed with a phase contrast microscope in order to ensure that cell lysis is complete with no disruption of nuclei. The cell lysate is then centrifuged at 12,000 g for 10 minutes to produce a supernatant fraction which we shall refer to as the cell extract. Ribosomes are precipitated from this cell extract by titration to pH 5.1 and are further treated as is described above for reticulocytes. Sedimentation of the resulting solutions containing [32P]-labeled sarcoma 180 ribosomes in sucrose gradients is

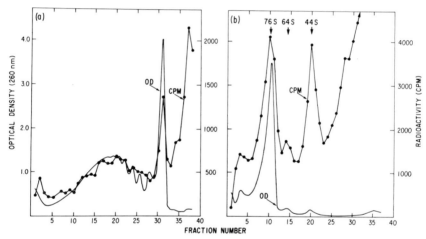

FIG. 2. Sucrose gradient sedimentation of [32P]-labeled sarcoma 180 ribosomes in a Spinco SW 27 rotor. Sarcoma 180 cells were incubated for 60 minutes with [32P]-orthophosphate in a nutrient medium containing 10 μg/ml actinomycin D. In (a), the ribosomes were concentrated by pH 5.1 precipitation and were then further purified by reprecipitation with 50 mM MgCl₂ [M. Takanami, *Biochim. Biophys. Acta* **39**, 318 (1960)]. Forty OD₂₆₀ units of ribosomes were then sedimented at 25,000 rpm for 3 hours in a linear 38 ml 15–40% sucrose gradient in buffer C. In (b), the preparative procedures were the same except that centrifugation was in buffer A at 25,000 rpm for 17 hours. In addition, the MgCl₂ precipitation step was eliminated because it causes a loss of subribosomal particles. Analytical procedures were as in Fig. 1, except that precipitates in 10% trichloroacetic acid were heated to 90° for 20 minutes before filtration onto membranes in order to remove nucleic acids. The experiments in frames (a) and (b) were done independently with different mice. Adapted from L. Bitte and D. Kabat, *J. Biol. Chem.* **247**, 5345 (1972).

shown in Fig. 2. The labeling results appear to be very similar to those obtained with reticulocytes (Fig. 1).

This choice of conditions for lysis of sarcoma 180 cells with Triton X-100 deserves some comment. Lysis of rat liver or rabbit muscle cells in the presence of 1% Triton X-100 is known to solubilize certain membrane constituents[21-23] and to cause leakage of ribonucleoproteins from nuclei[24]; these can contaminate the resulting ribosome preparations. Furthermore, when this lysis is carried out in low ionic strength buffers, there occurs an artifactural adsorption of polyribosomes onto the modified membranes and released membranous proteins, resulting in a severe lowering of polysome yields. Such adsorption does not occur if the monovalent cation concentration is elevated to 150 mM.[21] For these reasons, we have used as low a Triton X-100 concentration as possible for complete cell lysis (0.25%) and a lysis buffer 140 mM in monovalent cations. Higher salt concentrations were avoided because S-180 single ribosomes are partially dissociated into subunits in 150 mM KCl.[19] When isolated in our conditions, polysomes are routinely obtained in high yields as compared with previous reports[19,23,25-27] and are undegraded as judged by the absence of nascent polypeptide chains from single ribosomes. We obtain between 4–7 mg of ribosomes from each milliliter of sedimented sarcoma 180 cells. Furthermore, the polysome size distribution is unaffected by dissolution of membranes in the cell extract with 0.5% sodium deoxycholate. This implies that the polysomal clusters in the extracts are not significantly aggregated onto membranes.[21]

III. Purification of ^{32}P-Labeled Mammalian Ribosomes

A. Principles

The above methods yield solutions containing partially purified ^{32}P-labeled reticulocyte and sarcoma 180 ribosomes. Further purification is needed in order to obtain preparations suitable for analysis of phosphorylated ribosomal components. However, there are no rigorous criteria for defining the purity of mammalian ribosomal preparations. Some

[21] S. Olsnes, Biochim. Biophys. Acta 232, 705 (1971).
[22] B. H. McFarland and G. Inesi, Arch. Biochem. Biophys. 145, 456 (1971).
[23] A. L. J. Gielkens, T. J. M. Berns, and H. Bloemendal, Eur. J. Biochem. 22, 478 (1971).
[24] S. Olsnes, Biochim. Biophys. Acta 213, 149 (1970).
[25] B. L. M. Hogan and A. Korner, Biochim. Biophys. Acta 169, 139 (1968).
[26] R. K. Morse, H. Hermann, and S. M. Heywood, Biochim. Biophys. Acta 232, 403 (1971).
[27] I. Faiferman, L. Cornudella, and A. O. Pogo, Nature (London) 233, 234 (1971).

true intracellular constituents may be only loosely bound and may be lost readily; conversely, many contaminants adsorb strongly to ribosomes. It has been our experience that multiple cycles of ribosome sedimentation and resolution is an ineffective means of removing such adsorbed contaminants. Furthermore, it is generally recognized by workers in this field that highly purified mammalian ribosomes are very unstable, especially when they have been repeatedly sedimented from solution and dissolved in fresh buffers.

Accordingly, we have used preparative procedures that do not require multiple cycles of ribosome sedimentation and resolution. The ribosomes are sedimented through sucrose solutions in buffers of differing ionic strengths. Although the phosphoprotein content of the preparations is highly dependent on the ionic strength used during the centrifugation, certain of the phosphoproteins cannot be extracted from the ribosomes with high ionic strength buffers, even in the presence of 0.5% sodium deoxycholate. Since these firmly bound components in reticulocyte and in sarcoma 180 ribosomes are very similar (see below), and since they occur in ribosomes in a reasonable stoichiometry,[2] we have concluded[4] that they are true ribosome constituents rather than contaminants.

B. Sucrose Gradient Purification

Solutions containing ribosomes dissolved in buffers A, B, or C (see Section II, B) are fractionated by centrifugation in sucrose gradients made in the same buffers. Frequently, ribosomes in buffer B are adjusted to 0.5% sodium deoxycholate 5 minutes before layering onto the sucrose gradients. The sedimentation conditions for the Spinco SW 25.1 and SW 27 rotors are described in the legends to Figs. 1 and 2, respectively. Polyribosomes are obtained from gradients like those in Figs. 1a and 2a, whereas single ribosomes and subribosomal particles are prepared from gradients like those in Figs. 1b and 2b. After centrifugation, the gradients are pumped directly through a flow-cell in a Gilford spectrophotometer and the absorbance at 260 nm is plotted on a recorder. The appropriate regions of the sucrose gradient are collected directly into ice-cooled flasks and the ribosome fractions are then pelleted by centrifugation at 65,000 rpm for 2 hours in the Spinco 65 rotor or at 40,000 rpm for 4 hours in the Spinco 40 rotor. The ^{32}P-labeled ribosome pellets are stored in a freezer.

C. Direct Sedimentation of Ribosomes

Partially purified ribosomes dissolved in buffers A, B, or C are diluted to 7 ml and are layered carefully over 2 ml of 15% sucrose dissolved in the same buffers. Ribosomes in buffer B are often adjusted to 0.5% sodium deoxycholate before they are layered over the sucrose solutions. The ribosomes are then pelleted by centrifugation at 65,000 rpm for 45 minutes in the Spinco 65 rotor or at 40,000 rpm for 90 minutes in the Spinco 40 rotor. The supernatants are carefully decanted, and the ribosome pellets are stored in a freezer.

D. Uses of Ionic Strength in Preparation of ^{32}P-Labeled Ribosomes

Ribosomes prepared by the above two sedimentation methods (Sections B and C) are equally pure as judged by their phosphoprotein compositions. In other words, sedimentation of ribosomes through a small 2-ml cushion of sucrose solution removes the same ^{32}P-labeled contaminants which are removed by sedimentation of the ribosomes through sucrose gradients. The more important variable which influences ^{32}P-phosphoprotein content of isolated ribosomes is the ionic strength used

TABLE I

INFLUENCE OF IONIC STRENGTH OF BUFFERS ON ^{32}P-CONTENT
OF RIBOSOMAL PREPARATIONS

Experiment	Source of ribosomes[a]	Buffer used[b]	Specific activity of ribosome preparations[c] (cpm/OD$_{260}$)	Percentage of radioactivity remaining
1.	Reticulocytes	A	2,750	100
		C	1,760	64
		B	1,180	43
		B+	450	16
2.	Reticulocytes	A	3,470	100
		C	1,820	54
		B+	880	26
3.	Sarcoma 180	A	25,400	100
		C	22,800	90
		B+	12,000	47

[a] Ribosomes were all prepared by the direct sedimentation method from cells labeled for 60 minutes with [^{32}P]orthophosphate. The cells were labeled as described in Section II, except that the sarcoma 180 cell suspension contained 0.25 mCi/ml of [^{32}P]orthophosphate rather than 0.050 mCi/ml.

[b] Formulas of these buffers are given in Section II, B. Buffer "B+" is buffer B supplemented with 0.5% sodium deoxycholate.

[c] Radioactivity in ribosome preparations was measured after precipitation with 10% trichloroacetic acid. The precipitated samples in 10% trichloroacetic acid were heated at 90° for 20 minutes to hydrolyze nucleic acids and were then filtered onto 0.45 μm Millipore membranes. The membranes were rinsed thoroughly with 5% trichloroacetic acid.

for their preparation. The ionic strengths of buffers A, C, and B are
0.02, 0.17, and 0.29, respectively. Table I shows that a substantial portion of the nonnucleic acid ^{32}P-labeled material is extracted from ribosomes by the higher ionic strength buffers.

As we will document more fully below, certain ^{32}P-phosphoproteins
remain associated with ribosomes regardless of the buffers employed for
their preparation, whereas other phosphoproteins are more readily extracted. The extracted phosphoproteins may be contaminants or weakly
bound ribosome constituents. Figure 3 shows an electrophoretic comparison on polyacrylamide gels of the ^{32}P-labeled materials from reticulocyte
ribosomes prepared in either buffers A or B; the radioactive components
are visualized by autoradiography. Clearly, the effect of raising the ionic
strength is to cause a highly selective extraction of phosphorylated
components.

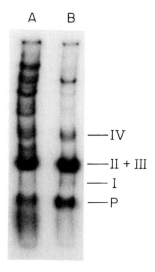

Fig. 3. Role of ionic strength in preparation of ^{32}P-labeled ribosomes. The figure
shows an electrophoretic fractionation of ^{32}P-labeled components from reticulocyte
ribosomes which had been purified either in low ionic strength buffer A or in high
ionic strength buffer B. The techniques of electrophoresis in sodium dodecyl sulfate–
polyacrylamide gels and of autoradiographic visualization of radioactive components
are described in Section IV. Migration is toward the bottom. All resolved components
have been identified as phosphoproteins except for the band labeled "P" which migrates faster than any ribosomal proteins and is extracted from ribosomes with 0.5%
sodium deoxycholate. The nomenclature used to identify the phosphoproteins which
remain bound to ribosomes washed with buffer B is described below (Section IV).
Gel concentration is 4%. Components Ia and Ib are not separated in these 4%
gels, and component III is not resolved from the more heavily labeled component
II. Adapted from D. Kabat, Biochemistry 9, 4160 (1970).

IV. Electrophoresis of ^{32}P-Labeled Ribosomal Components
in Sodium Dodecyl Sulfate-Polyacrylamide Gels

A. Principles

The ^{32}P-labeled constituents of mammalian ribosomes can be conveniently analyzed by electrophoresis in polyacrylamide gels in the presence of sodium dodecyl sulfate (SDS). Many of the basic techniques have been recently described.[28] The advantages of this method are the following: (a) Ribosomes and most other subcellular assemblages of macromolecules are fully dissolved and dissociated into their components in the presence of the SDS buffer. The sample can be layered directly onto the gel without prior separation of the RNA and protein constituents. The method is therefore very convenient and simple. (b) The fractionation of proteins in SDS-gels is on the basis of their molecular weights; a plot of the mobility of proteins versus the logarithm of their molecular weights falls approximately on a straight line.[29,30] (c) The resolution of constituents is excellent; it is at least as good as other electrophoretic methods of fractionating mammalian ribosomal proteins.[31]

B. Reagents and Solutions

a. Acrylamide-bisacrylamide stock solutions. These are stable at 4° for several months. The acrylamide and *N,N'*-methylenebisacrylamide are obtained from Eastman Organic Chemicals (Rochester, New York) and are recrystallized before use.[32] Acrylamide is dissolved in $CHCl_3$ at 50°, and the solution is filtered. Crystallization occurs during storage overnight at $-20°$. The crystals are washed with heptane. Methylenebisacrylamide is recrystallized from water and then from acetone. After dissolving in H_2O at 90° and filtering, the solution is stored overnight at 4° and the crystals are collected by filtration. 10 g of crystals are then dissolved in acetone at 50° and are recrystallized at $-20°$.

Stock 1 (for gels containing less than 5% acrylamide). This contains 15 g acrylamide and 0.75 g of *N,N'*-methylenebisacrylamide in 100 g of solution.

[28] J. V. Maizel, Jr., *in* "Methods in Virology" (K. Maramorosch and H. Koprowski, eds.), Vol. V. Academic Press, New York, 1971.
[29] A. L. Shapiro, E. Viñuela, and J. V. Maizel, Jr., *Biochem. Biophys. Res. Commun.* **28**, 815 (1967).
[30] A. K. Dunker and R. R. Reuckert, *J. Biol. Chem.* **244**, 5047 (1969).
[31] H. W. S. King, H. J. Gould, and J. J. Shearman, *J. Mol. Biol.* **61**, 143 (1971).
[32] U. E. Loening, *Biochem. J.* **102**, 251 (1967).

Stock 2. This contains 20 g of acrylamide and 0.5 g of bisacrylamide in 100 g of solution.

b. Ammonium persulfate. A 1.6% solution is kept in the refrigerator and is prepared freshly each week.

c. Stock electrophoresis buffer (10×). It is 0.36 M Tris, 0.30 M NaH_2PO_4, and 10 mM EDTA. The actual electrophoresis buffer contains 100 ml of stock buffer, 900 ml of H_2O, and 6.0 g of sodium dodecyl sulfate (Matheson, Coleman and Bell).

d. TEMED: N,N,N',N'-Tetramethylethylenediamine (Eastman Organic Chemicals, Rochester, New York).

e. Coomassie brilliant blue stain (Schwarz/Mann, Orangeburg, New York). This contains 0.2% in destaining solution.

f. Destaining solution. This contains 100 ml of acetic acid, 200 ml of methanol, and 700 ml of H_2O.

g. Gel dissolving solution.[33] It must be freshly made. It contains 1% NH_4OH dissolved in 30% H_2O_2.

h. Scintillation fluid. This is made by mixing 3 g of 2,5-diphenyloxazole (PPO), 0.3 g of 1,4-bis[2(4-methyl-5-phenyloxazolyl)]-benzene(dimethyl POPOP), 400 ml of toluene, and 800 ml of Triton X-100.

C. Gel Preparation

Eight-centimeter gels are made in 6 mm (i.d.) plastic tubes. For preparation of 4% gels, 6.67 ml of stock 1 acrylamide–bisacrylamide and 1.25 ml of 1.6% ammonium persulfate are diluted to 25 ml with electrophoresis buffer lacking SDS; 20 μl of TEMED is added. The solution is then quickly mixed and placed into the plastic tubes, and the menisci are overlain with 100 μl of H_2O. Polymerization is complete in 30 minutes at room temperature. 8% Gels are made identically, except that we use 10 ml of the stock 2 acrylamide–bisacrylamide solution. Before use, the gels in the plastic tubes are soaked overnight at room temperature in electrophoresis buffer containing 0.6% SDS.

D. Sample Preparation

The pellets containing purified ribosomes are dissolved at room temperature in a few drops of 0.6% SDS–electrophoresis buffer adjusted to 5% sucrose. Of the resulting solution, 10 μl, is added to 1 ml of H_2O and the absorbance at 260 nm is measured. The absorbance should be between 0.5 and 1.0 units/ml. The solution is then adjusted to 1%

---

[33] D. Goodman and H. Matzura, *Anal. Biochem.* **42**, 481 (1971).

2-mercaptoethanol and is heated at 60° for 30 minutes to reduce disulfide bonds.

E. Electrophoresis

The gels are prerun at 5 mA/gel for 1 hour before 1 OD_{260} unit of the ribosome sample in a volume of 10–30 μl is layered onto the upper surface of the gel. Electrophoresis is at room temperature at 5 mA/gel for approximately 2 hours.

F. Staining of Proteins in the Gels

Proteins in the gel can be stained with Coomassie brilliant blue. The gels are removed from the plastic tubes and are washed overnight at room temperature with 12.5% trichloroacetic acid. This is required for removing SDS from the gel, since the detergent coprecipitates with the stain. The gels are then placed into the stain for 8–16 hours and are then washed overnight with several changes of destaining solution. Destaining is continued until the gel background is clear.

G. Autoradiographic Detection of ^{32}P-Labeled Constituents

Radioactive components in the gels can be visualized by autoradiography. The cylindrical gels are cut into 4 longitudinal sections with the apparatus described by Fairbanks et al.[34] They are then dried onto high-wet strength paper (Schleicher and Schuell, No. 497) or onto Whatman 3 MM paper. Gels with 4% polyacrylamide can be dried without special apparatus on the high-wet strength paper; the gel slices dry smoothly without shrinking or cracking in an oven at 75°. However, more concentrated gels require a special drying technique.[28,34] The dried gels on the paper backing are then pressed together with X-ray film (Kodak, single coated type SB-54) for varying time periods before the film is developed.

H. Transverse Sectioning of Gels for Direct Measurement of Radioactivity

We frequently transection the gels with a commercially available sectioning apparatus (Brinkman Instruments, Westbury, New York). Gels for this purpose are made as described above except that they also contain 10% glycerol; this facilitates the accurate sectioning of frozen gels.[35] The 1 mm gel sections are dissolved by incubation at room temperature in 0.8 ml of gel dissolving solution. The incubation is for 24–48 hours in

[34] G. Fairbanks, Jr., C. Levinthal, and R. H. Reeder, *Biochem. Biophys. Res. Commun.* **20**, 343 (1965).
[35] R. A. Weinberg, U. E. Loening, M. Willems, and S. Penman, *Proc. Nat. Acad. Sci. U.S.* **58**, 1088 (1967).

tightly stoppered scintillation vials. After addition of 15 ml of scintillation fluid, radioactivity is measured in a liquid scintillation spectrometer.

I. Uses of Electrophoresis for Analyzing ^{32}P-Labeled Ribosomes

We present here some representative electrophoretic analyses of the radioactive components from ^{32}P-labeled mammalian ribosomes. On the one hand, these data illustrate the reproducibility and some uses of this method. Second, the ^{32}P-labeled phosphoproteins from rabbit reticulocyte and from mouse sarcoma 180 ribosomes coelectrophorese in the SDS-polyacrylamide gels, suggesting that they have the same molecular weights.[4] Based on this apparent similarity between different species and tissues, we anticipate that these electrophoretic patterns may be generally obtained for mammalian cells. Accordingly, this data should serve as a useful prototype for comparison with future analyses made in other laboratories.

Figure 4 shows a typical electrophoretic fractionation in an 8% poly-

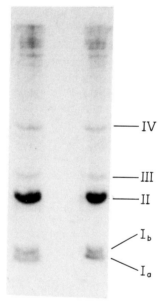

FIG. 4. Electrophoresis of ^{32}P-labeled components from reticulocyte ribosomes in a sodium dodecyl sulfate–polyacrylamide gel. The gel concentration is 8%; visualization of radioactive components is by autoradiography; and migration is toward the bottom. The reticulocytes were labeled with [^{32}P]orthophosphate for 60 minutes (Section II), and the ribosomes were then purified by the direct sedimentation method in high ionic strength buffer B supplemented with 0.5% sodium deoxycholate (Section III). The two internal sections of the gel were dried next to each other on the paper backing and used for autoradiography. Both of the sections are presented to illustrate the reproducibility obtained.

acrylamide gel of the ^{32}P-labeled components from reticulocyte ribosomes, the radioactive components being visualized by autoradiography. All the resolved components are phosphoproteins, as determined by the criteria described below, and they are numbered in the order of their mobilities in the gel. Components Ia and Ib migrate closely together on 4% gels and are resolved slightly on 8% gels. In addition to the five numbered components (Ia, Ib, II, III, and IV) there are several more slowly migrating phosphoproteins which occur in ribosome preparations in variable amounts. The five components assigned numerals in Fig. 4 are reproducibly obtained in ribosome preparations. We have estimated their molecular weights[28–30] as 18,200, 19,500, 27,500, 33,000, and 53,000 for components Ia, Ib, II, III, and IV, respectively.

However, polyribosomes and single ribosomes contain different proportions of these radioactive phosphoproteins. Figure 5 shows a time course for the ^{32}P-labeling of these constituents in reticulocyte poly- and single

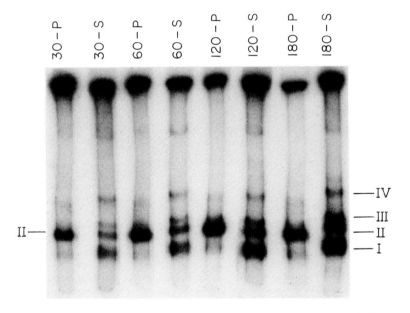

FIG. 5. Comparison of ^{32}P-labeled components from reticulocyte polyribosomes and single ribosomes by electrophoresis in sodium dodecyl sulfate–polyacrylamide gels. Gel concentration is 4%, and the visualization of radioactive components is by autoradiography. Ribosomes were purified in buffer B containing 0.5% sodium deoxycholate (see Section III) from cells labeled for 30, 60, 120, or 180 minutes with [^{32}P]orthophosphate. S is single ribosomes; P is polyribosomes. These types of ribosomes have distinctive and reproducibly obtained patterns of phosphorylation. Adapted from D. Kabat, *J. Biol. Chem.* **247**, 5338 (1972).

ribosomes. Single ribosomes contain relatively more of components I and III, whereas polysomes contain relatively more radioactivity in component II. Analysis of this difference has indicated[2] that mammalian ribosomes are heterogeneous in their molecular constitution.

A comparison of the ^{32}P-labeled components from rabbit reticulocyte and mouse sarcoma 180 polyribosomes and single ribosomes is shown in

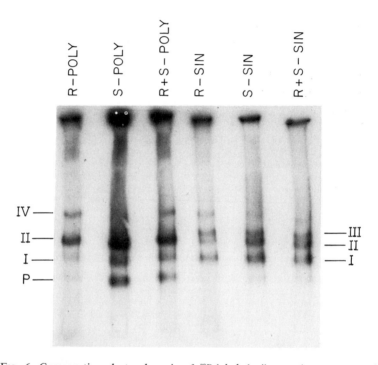

FIG. 6. Comparative electrophoresis of ^{32}P-labeled ribosomal components from rabbit reticulocytes and from mouse sarcoma 180 cells. The cells were labeled for 60 minutes and their ribosome fractions were then prepared from sucrose gradients containing buffer C. Gel concentrations are 4%; migration is downward; and the radioactive components are visualized by autoradiography. The nomenclature of the phosphoproteins I–IV is described in the text. The "P" component occurs in polysomes from reticulocytes and from sarcoma 180 cells in variable amounts but may not be a protein (see Fig. 3). S-SIN and S-POLY are sarcoma 180 single ribosomes and polyribosomes, respectively. R-SIN and R-POLY are reticulocyte single ribosomes and polyribosomes, respectively. R + S-SIN is mixture of reticulocyte and of sarcoma 180 single ribosomes. R + S-POLY is a mixture of reticulocyte and of sarcoma 180 polyribosomes. The sarcoma 180 preparations are usually more contaminated than those from reticulocytes, as evidenced by the higher background of radioactivity on the sarcoma 180 gels. From L. Bitte and D. Kabat, *J. Biol. Chem.* **247**, 5345 (1972).

Fig. 6. All the labeled components except P have been identified as phosphoproteins. The reticulocyte phosphoproteins are identified by Roman numerals as was described above. Clearly, the phosphoprotein patterns from reticulocyte and from sarcoma 180 ribosomes are very similar. No electrophoretic components appear only in sarcoma 180 or only in reticulocyte ribosomes since mixtures of ribosomes from these two sources exhibit the same number of bands as do the unmixed samples. Furthermore, the distribution of these components between polysomes and single ribosomes is similar in both cells; however, significant differences between these distributions have been noted.[4]

V. Membrane Filter Assay of Radioactivity in the Phosphoprotein Components of ^{32}P-Labeled Ribosomes

A. Principles

The quantity of radioactivity incorporated into protein components of ^{32}P-labeled ribosomes can be analyzed following protein precipitation with 10% trichloroacetic acid (TCA). The acidified precipitates are heated to 90° for 20 minutes and are then collected onto membrane filters. Methods are described for removing three types of radioactive contaminants from the precipitated proteins.

B. Removal of RNA

Two procedures are employed to eliminate contamination with radioactive RNA. First, the incorporation of [^{32}P]orthophosphate into ribosomes occurs in conditions in which RNA synthesis is greatly inhibited (Section II). Second, nucleic acids are removed by heating in 10% TCA at 90° for 20 minutes.[36,37] However, it should be recognized that this treatment also results in a 10–20% loss of protein from the precipitates.[37]

The amount of radioactivity lost from TCA-precipitated ^{32}P-labeled ribosomes by heating at 90° for 20 minutes is presented in Table II. Reproducibly, this treatment removes only a small fraction of radioactivity from the precipitates.

That this treatment with hot TCA causes quantitative extraction of ^{32}P-labeled RNA was verified using purified radioactive rRNA. In addition, we have analyzed reticulocyte ribosomes which were labeled *in vivo*, in conditions favoring the labeling of RNA. An anemic rabbit was bled 38 hours after it had been injected with 1 mCi of [^{32}P]orthophosphate. The

---

[36] W. C. Schneider, *J. Biol. Chem.* **161**, 293 (1945).
[37] D. Kennell, this series, Vol. 12, p. 686.

TABLE II

Hot Trichloroacetic Acid Labile Radioactivity in Precipitates
OF ^{32}P-Labeled Ribosomes

Experiment	Source of ribosomes[a]	Filtering method[b]	Percentage of radioactivity lost to hot TCA treatment
1.	Reticulocytes	A	14
2.	Sarcoma 180	A	25
		B	27
		C	23

[a] Ribosomes were ^{32}P-labeled and isolated from buffer B as described in the text (Sections II, B and C, and III, C). The resulting ribosomal pellets were dissolved in cold water and were used for 10% TCA precipitation at 0°.

[b] Filtration Method A. All ribosome samples were precipitated in cold 10% TCA in test tubes. Control samples were chilled for 60 minutes, were collected by filtration onto Millipore filters (0.45 μm), and were then rapidly rinsed 3 times with 5% TCA at room temperature. The remaining samples were allowed to stand in ice for only 10 minutes, were heated at 90° for 20 minutes and then chilled to 0° for 30 minutes, and were then collected onto Millipore filters as described above.

　　Filtration Method B. This method is identical to method A except that fiberglass filters (Reeve Angel, 934A-A) were substituted for Millipore filters.

　　Filtration Method C. The ribosomal preparations were spotted onto the center of fiberglass filters (50 μl samples) and were dried at room temperature. The filters were then placed into 50 ml of cold 10% TCA for 30 minutes. Control filters remained in the cold acid for an additional 50 minutes before they were placed in a Millipore filtration apparatus for rinsing with 5% TCA as in Methods A and B. The remaining filters were heated to 90° for 20 minutes, were chilled on ice for 30 minutes, and were then rinsed on the filtration apparatus as described above.

polysomes had a specific activity of 31,000 cpm/mg protein, as measured after precipitation with 10% TCA at 0°. However, the specific activity was reduced to 430 cpm/mg protein after heating in 10% TCA to 90° for 20 minutes. This residual radioactivity was not further reduced by a second treatment with hot acid, suggesting that it was not in unhydrolyzed nucleic acids. Thus, the hot TCA method is very efficient for removing RNA from ^{32}P-labeled ribosomal proteins.

C. Extraction of Phospholipids

　　The quantity of radioactivity in contaminating phospholipids is estimated by measuring the loss of radioactivity when the TCA-precipitate is extracted with a series of lipid solvents. The solvents and sequence of extractions used were suggested by Davidson et al.[38] The precipitate is ex-

[38] J. N. Davidson, S. C. Fraser, and W. C. Hutchison, Biochem. J. 49, 311 (1951).

tracted sequentially once with acetone, ethanol, and chloroform; twice with ethanol–ethyl ether (3:1); and then once with ethyl ether.

One method we have used[1,4] is to extract the TCA-precipitated ribosomal proteins in test tubes following the hot acid treatment (see Section B). The precipitate is sedimented by centrifugation, and the solvents are removed by aspiration. However, this procedure produces variable losses of protein from the precipitate and is especially difficult for the chloroform extraction in which the precipitate floats on the solvent surface.

Consequently, we have more recently used a method in which the TCA precipitation is done in a glass fiber filter (Table II, Method C) and the filter is then extracted with the series of nonaqueous solvents. We observe no loss of radioactivity when this procedure is used for samples of ^{32}P-labeled ribosomes from reticulocytes and from sarcoma 180 cells. This is true regardless of the buffers used for ribosome preparation (see Table I). Hence, we conclude that phospholipid contamination does not contribute significantly to the radioactivity which remains in the ribosomal precipitates after the hot acid treatment.

D. Other Contaminants

Greenaway[39] has shown that Mg–ATP complexes can contaminate proteins precipitated with cold 10% TCA. Dissolving the precipitates in cold 0.1 M NaOH and then reprecipitating with TCA was suggested as a method for removing this source of contamination. Other workers[40] have also used a 0.1 M NaOH resolution step as a means to remove contaminants from phosphoproteins precipitated with cold 10% TCA. However, such contamination is apparently not significant when the precipitates of ^{32}P-labeled ribosomal phosphoproteins are heated in 10% TCA at 90° for 15–20 minutes. Resolution of such treated precipitates in cold 0.1 M NaOH does not cause any further reduction in measured radioactivity.

VI. Proteolytic Digestion of Ribosomal Phosphoproteins

A. Principles

Pronase (*Streptomyces griseus* protease) is highly active in solutions containing 0.6% sodium dodecyl sulfate, whereas ribonuclease is markedly inhibited in these conditions.[41] This forms the basis of a simple procedure for proteolysis of phosphoproteins in the absence of any degradation of

[39] P. J. Greenaway, *Biochem. Biophys. Res. Commun.* **47**, 639 (1972).
[40] R. J. DeLange, R. Kemp, W. D. Riley, R. A. Cooper, and E. G. Krebs, *J. Biol. Chem.* **243**, 2200 (1968).
[41] D. Kabat, *Anal. Biochem.* **39**, 228 (1971).

nucleic acids. The electrophoretic mobilities of the ^{32}P-labeled ribosomal constituents are compared before and after the proteolysis.

B. Proteolysis

Sedimented pellets of ^{32}P-labeled ribosomes are dissolved in electrophoresis buffer containing 0.6% sodium dodecyl sulfate as described above (Section IV, D). After the absorbance of the solution at 260 nm is adjusted to 50 units/ml, an aliquot containing 20 μl is placed into a second tube. Two microliters of a freshly prepared Pronase solution [5 mg/ml Pronase (Calbiochem, Los Angeles, California) dissolved in the 0.6% SDS-electrophoresis buffer] is added to this tube. After 30 minutes at room temperature, the Pronase-treated sample is layered onto the surface of a polyacrylamide gel for electrophoresis. Twenty microliters of the untreated control sample is layered onto another gel. Electrophoresis is then performed as described above (Section IV).

Typical electrophoretic results with Pronase digested ^{32}P-labeled reticulocyte ribosomes have been described.[1] All the radioactive components are converted by pronase into more rapidly migrating materials which have a mobility approximately the same as the fast moving P component (see Fig. 3). Because polypeptide mobilities in SDS-gels are increased when their molecular weights are reduced,[28-30] such data suggest that the labeled components are proteins. Furthermore, it has been shown[41] that RNA is not degraded by this proteolysis procedure.

VII. Methods for Selective Cleavage of Phosphate–Protein Bonds

A. Principles

Information concerning the mode of binding of phosphate with protein can be obtained by studying the conditions which result in bond cleavage.

B. Alkaline Phosphatase

Escherichia coli and intestinal alkaline phosphatases (orthophosphoric monoester phosphohydrolase EC 3.1.3.1) are relatively nonspecific phosphomonoesterases which also catalyze hydrolysis of acyl phosphates.[42] The phosphate which is bonded to ribosomal proteins is efficiently cleaved by the *E. coli* enzyme,[1] in conditions in which ribonucleic acid remains stable.

^{32}P-labeled ribosomes are dissolved in buffer B (see Section II, B) at a concentration of 8 mg/ml and the solution is separated into two portions. After warming to 37°, a small volume of *E. coli* alkaline phosphatase

[42] J. W. Sperow and L. G. Butler, *Biochim. Biophys. Acta* 146, 175 (1971).

(ribonuclease-free type BAPF, Worthington Biochemical Corp., Freehold, New Jersey) is added to one portion of the ribosome solution to give a final enzyme concentration of 0.4 mg/ml (13 U/ml); 0.1-ml aliquots of the solutions are periodically removed into tubes containing 1.9 ml of 5% trichloroacetic acid. These tubes are warmed to 90° for 15 minutes and are then chilled to 0° for at least 30 minutes before the precipitates are filtered onto Millipore membranes for radioactivity measurement. Other samples of the ribosome solutions are cooled in ice at various times after enzyme addition and the ribosomes are subsequently pelleted by centrifugation (Section III, C). These sedimented ribosomes are then analyzed by electrophoresis in polyacrylamide gels (Section IV).

This enzymatic treatment quantitatively removes the ^{32}P-labeled phosphate from the mammalian ribosomal proteins I–IV (Fig. 4). The half-time for removal of the ribosome-associated radioactivity is 5 minutes in these conditions. Components Ia and Ib are dephosphorylated more rapidly than is component II. Furthermore, the integrity of polyribosomal structure is not destroyed by dephosphorylation of the ribosomal proteins.

C. Hydroxylamine

One common mode of bonding of phosphate with proteins is by acyl linkage to carboxyl groups. Such acyl phosphates are susceptible to rapid aminolysis at pH 5.5 with hydroxylamine. On the contrary, phosphate esters are stable in these conditions.

^{32}P-labeled ribosomes are dissolved in water at a concentration of 10 mg/ml. One aliquot is diluted into 9 volumes of 1 M succinic acid–1 M hydroxylamine, pH 5.5; whereas a second aliquot is diluted into 9 volumes of 1 M succinic acid, pH 5.5 lacking any hydroxylamine. The resulting ribosome solutions are incubated at 37°, and 0.1-ml aliquots are periodically taken into tubes containing 1.9 ml of 5% trichloroacetic acid. These tubes are heated to 90° for 15 minutes and are then chilled for at least 30 minutes before the precipitates are filtered onto Millipore membranes for radioactivity measurement. Acyl phosphate bonds in proteins are very rapidly cleaved in these conditions.[43,44] However, the bonds of phosphate with ribosomal proteins are resistant to this treatment.

D. The pH Stability of Phosphate–Protein Bonds

The stability of phosphate–protein bonding is highly dependent upon the pH. Furthermore, the shape of the curve relating bond stability with pH is characteristic of the type of bonding.[44] For example, acyl phosphates

[43] A. Martonosi, J. Biol. Chem. 244, 613 (1969).
[44] R. S. Anthony and L. B. Spector, J. Biol. Chem. 247, 2120 (1972).

are rapidly hydrolyzed at both extremes of pH, a fact that imposes a U shape on the stability curves. On the other hand, phosphate monoesterified to hydroxyl groups of seryl or threonyl residues are very stable at low pH and unstable at high pH. If the phosphoryl group is linked to a nitrogen of a histidyl or lysyl residue, the hydrolysis curve is high at low pH and low at high pH. Phosphoryl groups attached to the oxygen of a tyrosyl residue or to a sulfur atom of a cysteinyl residue also exhibit distinctive hydrolytic behaviors.[44]

[32]P-labeled ribosomes are dissolved in water at a concentration of 5 mg/ml. 0.1-ml aliquots are then diluted into tubes containing 1.9 ml of various buffers. Two tubes are prepared for each buffer. The buffers (0.10 M unless otherwise specified) are pH 1.2 and 2.4 (HCl and KCl), pH 3.1 (KH_2PO_4), pH 5.9 (K_2HPO_4), pH 7.2 and 8.4 (Tris·HCl), pH 9.8 ($KHCO_3$–K_2CO_3), pH 13.1 (0.20 N KOH) and pH 13.5 (1.0 N KOH). After incubation at 55° for 120 minutes, the samples are chilled to 0° and adjusted to 20% trichloroacetic acid. The resulting precipitates are collected by centrifugation and resuspended in cold 5% trichloroacetic acid. Of the two samples incubated in each of the buffers, one is filtered directly onto a Millipore membrane for radioactivity measurement. The other sample is heated to 90° for 20 minutes and is then chilled to 0° for 30 minutes before filtration. The latter procedure removes ^{32}P-labeled nucleic acids which may be present in the ribosome preparations. Radioac-

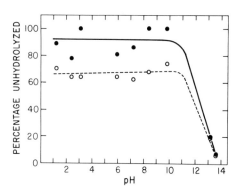

FIG. 7. Effect of pH on the stability of ribosomal protein-phosphate bonds. The ^{32}P-labeled sarcoma 180 ribosomes were isolated by the direct sedimentation method. After incubation at 55° for 2 hours at each pH, samples were precipitated with cold 5% trichloroacetic acid and were filtered onto membranes. Other samples were heated in 5% trichloroacetic acid at 90° for 20 minutes before filtration; this removes nucleic acids and 15–20% of proteins from the precipitates. This latter portion of the data shows that the radioactivity analyzed here is not in nucleic acids. The percentage of radioactivity is measured with respect to a control sample at pH 7.2 which was kept chilled at 0°. ●—●, Cold trichloroacetic acid; ○--○, treated with hot trichloroacetic acid.

tivity on the membranes is then measured in a low background gas flow counter. The pH stability profile of sarcoma 180 ribosomal phosphoproteins is shown in Fig. 7. The shape of this profile is that expected of o-phosphoseryl and o-phosphothreonyl residues.

VIII. Detection of o-Phosphoserine and o-Phosphothreonine in Ribosomal Hydrolyzates

A. Principles

Radioactive o-phosphoserine and o-phosphothreonine can be detected in acid hydrolyzates of ^{32}P-labeled ribosomal proteins. This is possible because the hydrolysis of the phosphomonoester bonds is considerably slower in acid than the hydrolysis of peptide bonds. After hydrolysis of the protein, the amino acids are fractionated by high voltage paper electrophoresis at pH 1.85; only the phosphorylated amino acids migrate toward the cathode in these conditions.[13]

B. Hydrolysis

Solutions containing ^{32}P-labeled ribosomes are precipitated with 5% trichloroacetic acid at 0°. After heating at 90° for 15 minutes to hydrolyze nucleic acids and subsequent chilling at 0° for 30 minutes, the radioactive precipitate is collected by centrifugation. It is washed with ethanol–ethyl ether (3:1) and then with acetone. The precipitate from 1 mg of ribosomes is suspended in 2 ml of 6 M HCl in a sealed hydrolysis tube, which is then heated to 105° for 7 hours. The hydrolyzate is evaporated to dryness in a vacuum desiccator and the residue is dissolved in 100 µl of paper

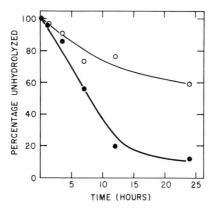

FIG. 8. Kinetics of hydrolysis of o-phosphoserine (●) and of o-phosphothreonine (○) in 6 M HCl at 105°. These amino acids were purified from hydrolyzates by paper electrophoresis, were staining with Cd-ninhydrin, and were then quantitatively assayed (see text).

electrophoresis buffer (see below). The quantity of o-phosphoserine and of o-phosphothreonine is analyzed following electrophoretic separation of the amino acids.

The rates of hydrolysis of o-phosphoserine and of o-phosphothreonine in these hydrolysis conditions are shown in Fig. 8. Phosphothreonine is relatively resistant to hydrolysis. We selected 7 hours as the optimal time for phosphoprotein hydrolysis in order to allow complete polypeptide hydrolysis with only minimal loss of these phosphorylated amino acids.

C. Paper Electrophoresis

Electrophoresis is in a water-cooled flat plate apparatus on Whatman 3 MM paper.[13] The paper strip (10 inches wide by 20 inches long) is wet with paper electrophoresis buffer (2.5% formic acid, 7.8% acetic acid) and is blotted so that it is only slightly damp. Radioactive sample (10 μl) is applied at the origin; 10-μl control samples containing 2 μg of o-phosphoserine and of o-phosphothreonine are applied to adjacent positions on the origin. A control sample containing [^{32}P]orthophosphate is also analyzed. A voltage of 3000 V is then applied across the electrodes for 120 minutes. The paper is dried in an oven at 75°, then sprayed thoroughly with a Cd-ninhydrin stain[45] (made by mixing 200 ml of acetone, 20 ml of H_2O, 4.0 ml of acetic acid, 200 mg of Cd acetate and 2.0 g of ninhydrin; this stain is stable in dark bottles at 4°); it is then heated at 75° for 3 hours. Positions of the marker phosphoserine and phosphothreonine are circled with a pencil and the paper is placed together with X-ray film for autoradiographic localization of radioactive compounds. Such regions of the paper can be cut out and analyzed quantitatively for radioactivity in a liquid scintillation spectrometer.

The color which develops on the paper after Cd-ninhydrin staining can be quantitatively eluted with methanol. Its absorbance at 500 nm is proportional to the quantity of the amino acid.[45]

Figure 9 shows a typical electrophoretic analysis of an acid hydrolyzate of ^{32}P-labeled reticulocyte ribosomes. The radioactive spots are visualized by autoradiography; and the positions of electrophoresis of o-phosphoserine, o-phosphothreonine, and orthophosphate are also indicated. Of the total radioactivity in the hydrolyzates, we routinely observe that approximately 13% coelectrophoreses with o-phosphoserine, 2% with o-phosphothreonine, and the remainder with orthophosphate. These data support other evidence that these amino acids are present in mammalian ribosomes; however, the data do not exclude the possibility that other phosphorylated amino acids might occur in ribosomes. Furthermore, the actual percentages

---

[45] W. J. Dreyer and E. Bynum, this series, Vol. 11, p. 32.

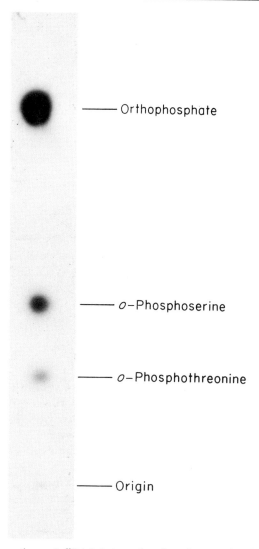

FIG. 9. Separation of ^{32}P-labeled *o*-phosphoserine, *o*-phosphothreonine, and orthophosphate by paper electrophoresis. The radioactive hydrolyzate (12,000 dpm) is from ^{32}P-labeled reticulocyte polyribosomes, and the compounds are visualized by autoradiography. The positions of electrophoresis of standard marker compounds are indicated.

of phosphoserine and phosphothreonine cannot be determined by this method because the rates of phosphomonoester hydrolyses are likely to be different (from Fig. 8) when the amino acids are present in peptide chains.

IX. Quantitative Analysis of Phosphate in Ribosomal Proteins

A. *Principles*

Nucleic acids are removed from ribosomes by heating in 5% trichloroacetic acid (TCA) at 90° for 30 minutes and from phospholipids by extraction. The residual protein fraction contains phosphoseryl and phosphothreonyl residues. Such phosphate is released by incubation with 1 M NaOH at 37° for 16 hours, and it is analyzed by a chemical method. Such analysis indicates that the quantity of phosphate bonded to ribosomal proteins is appreciable (approximately 14 phosphates per ribosome).

B. *Extraction of Nonprotein Phosphate*

Solutions containing 10 mg of ribosomes are adjusted to 10% TCA, and the precipitates are collected by centrifugation. They are then washed twice by centrifugation with 5 ml of cold 5% TCA. The precipitates are resuspended in 5 ml of 5% TCA, then the suspensions are heated to 90° for 15 minutes; after chilling to 2° for 30 minutes, the precipitates are resedimented. The latter step is repeated, and the resulting precipitates are washed once again with 5 ml of cold 5% TCA. The precipitates are then extracted with ethanol–ethyl ether (3:1) at 60° for 15 minutes and finally with cold acetone.

C. *Alkaline Hydrolysis of Protein–Phosphate Bonds*

The resulting protein precipitates are transferred into plastic tubes and are dissolved in 0.50 ml of 1 M NaOH (reagent grade, low carbonate type from Merck and Company, Inc., Rahway, New Jersey). Use of plastic tubes is necessary because NaOH causes a leaching of phosphate from glass. A control tube is incubated without any protein. The solutions are incubated at 37° for 16 hours, and the protein is reprecipitated by addition of 0.33 ml of 50% trichloroacetic acid. After chilling to 0° for 1 hour, the precipitates are removed by sedimentation and are saved for protein determination. The supernatants are transferred into separate tubes and are analyzed for orthophosphate concentration.

D. *Phosphate Determination*

Phosphate is analyzed by a modification of Sumner's method.[46] To the phosphate-containing solutions resulting from step C (see above) are added 0.10 ml of 7.5 N H_2SO_4, 0.1 ml of 6.6% ammonium molybdate [$(NH_4)_6Mo_7O_{24} \cdot 4H_2O$], and 0.08 ml of freshly prepared ferrous sulfate solution (made by mixing 5 g of $FeSO_4 \cdot 7H_2O$, 50 ml H_2O, and 1 ml of 7.5 N H_2SO_4). The quantity of phosphate is determined by comparing the

[46] J. B. Sumner, *Science* 100, 413 (1944).

TABLE III

QUANTITATIVE ANALYSIS OF PHOSPHATE IN RIBOSOMAL PROTEINS

Cells used	Proteins analyzed[a]	Nanomoles P_i per milligram protein
Reticulocytes	A-ribosomes	9.6
	C-ribosomes	9.1
	B+-ribosomes	7.1
Sarcoma 180 Cells	C-ribosomes	17
	B-ribosomes	12.5
	Cell-extract proteins[b]	14
Sea urchin eggs[c]	Cell-extract proteins	2.8[d]
Sea urchin Blastulae[c]	Cell-extract proteins	1.8[d]

[a] Ribosomes were prepared by the direct sedimentation method, A-ribosomes from buffer A, B-ribosomes from buffer B, C-ribosomes from buffer C, B+-ribosomes from buffer B supplemented with 0.5% sodium deoxycholate. Formulas of these buffers are given in Section II, B.

[b] This fraction is the supernatant remaining after the cell lysate is centrifuged at 12,000 g for 10 minutes (see Section II, D).

[c] Generously provided by Lawrence Kedes, Stanford University Medical School, Palo Alto, California.

[d] These relatively low values for sea urchin proteins are maximal estimates of their P_i content and may reflect the presence of some contaminating material which contributes to the P_i assay.

absorbance at 700 nm of these latter solutions with similar solutions prepared with known amounts of orthophosphate. If the ribosomal proteins were ^{32}P-labeled, the specific activity of orthophosphate can be determined by measuring the radioactivity in an aliquot of the solutions after their absorbance has been recorded.

E. Protein Determination

The protein precipitates from step C are dissolved in 20 ml of 1 M NaOH, and the quantity of protein is determined by the procedure of Lowry et al.,[47] using bovine serum albumin as the standard.

F. Results with Mammalian Ribosomes

Some representative data for mammalian ribosomes are presented in Table III. The quantities of P_i in proteins from ribosomes prepared with high ionic strength buffer B are slightly lower than the value of 16.1 nmoles of P_i per milligram of protein measured[3] for ribosomes from rat liver cells; however, the purity of these latter ribosomes was not demonstrated. Our

[47] O. H. Lowry, N. J. Rosebrough, A. L. Farr, and R. J. Randall, *J. Biol. Chem.* **193**, 265 (1951).

data suggest that there are an average of approximately 14 phosphoryl groups in the proteins of reticulocyte ribosomes prepared from buffer B containing 0.5% sodium deoxycholate. [This calculation assumes that these ribosomes are 50% protein and have a molecular weight of 4.0 × 10^6. Thus, each nanomole of ribosomes contains 2.0 mg of protein and, accordingly, 14.2 nmoles of P_i.] Furthermore, the sarcoma 180 ribosome preparations contain even more P_i; however, they are also more contaminated with nonribosomal proteins. These analyses of phosphate content are consistent with the previous conclusion, based on another technique, that there are 7–11 protein-associated phosphoryl groups which are turning over in each reticulocyte ribosome.[2]

Acknowledgments

Report of work supported by U.S. Public Health Service Grants CA-11347 and HL-CA-14960-04. We thank M. Laurence Cawthon and Janet Ploss for their assistance and advice.

[55] Phosphorylation of Ribosomal Proteins: Preparation of Rat Liver Ribosomal Protein Kinases. Assay of the Phosphorylation of Ribosomal Subunit Proteins. Assay of the Function of Phosphorylated Ribosomes

By CHARLES EIL and IRA G. WOOL

The proteins of eukaryotic ribosomes are phosphorylated *in vivo* and *in vitro*.[1-8] In the latter case the reaction is catalyzed by protein kinases present in the cytosol[1,4,5,8] or associated with ribosomes.[3,5,7] Two such enzymes, both stimulated by cyclic adenosine 3′,5′-monophosphate (cyclic AMP), can be isolated from rat liver cytosol; the protein kinases transfer

[1] J. E. Loeb and C. Blat, *FEBS (Fed. Eur. Biochem. Soc.) Lett.* **10**, 105 (1970).
[2] D. Kabat, *Biochemistry* **9**, 4160 (1970).
[3] D. Kabat, *Biochemistry* **10**, 197 (1971).
[4] G. M. Walton, G. N. Gill, I. B. Abrass, and L. D. Garren, *Proc. Nat. Acad. Sci. U.S.* **68**, 880 (1971).
[5] C. Eil and I. G. Wool, *Biochem. Biophys. Res. Commun.* **43**, 1001 (1971).
[6] C. Blat and J. E. Loeb, *FEBS (Fed. Eur. Biochem. Soc.) Lett.* **18**, 124 (1971).
[7] C. C. Li and H. Amos, *Biochem. Biophys. Res. Commun.* **45**, 1398 (1971).
[8] H. Yamamura, Y. Inoue, R. Shimomura, and Y. Nishizuka, *Biochem. Biophys. Res. Commun.* **46**, 589 (1972).

the terminal phosphate from ATP[5] to serine and threonine residues of specific proteins of rat liver ribosomal subunits.[9]

We shall describe techniques we have devised (or adapted) for the preparation of rat liver ribosomal protein kinases, for the assay of the phosphorylation of the proteins of ribosomal subunits, for analysis of the product of the reaction, for assay of ribosomal phosphoprotein phosphatase, and for the comparison of the function of phosphorylated and nonphosphorylated ribosomes.

Preparation of Rat Liver Cytosol Ribosomal Protein Kinases I and II

Reagents

β-Mercaptoethanol (MSH), 1 M
Glacial acetic acid
Glycerol
KOH (0.5 M)
KH_2PO_4, 1 M
K_2HPO_4, 1 M
Solid $(NH_4)_2SO_4$
Solid NaCl
Aged calcium phosphate gel (10.5% solids) (Sigma)
DEAE-cellulose (DE-52 resin) (Whatman)
Hydroxylapatite (Bio-Rad)

Buffers

TM: tris(hydroxymethyl)aminomethane (Tris)·HCl, 10 mM, pH 7.7; MSH, 10 mM
TMG: Tris·HCl, 10 mM, pH 7.7; MSH, 10 mM; glycerol, 10%
PMG: potassium phosphate buffer, pH 7.5, containing: MSH, 10 mM; glycerol, 10%; and potassium phosphate at concentrations of 0.05 and 0.25 M
Potassium phosphate, 250 mM, pH 8.1; MSH, 10 mM

Separation of Protein Kinases I and II. Rat liver ribosomal protein kinases are prepared by a modification of the method Gill and Garren[10] used to purify a histone kinase from bovine adrenal supernatant. All procedures are carried out at 4°. The 100,000 g post-microsomal supernatant from rat liver (prepared as described by Martin and Wool[11]) is made

[9] E. Eil and I. G. Wool, *J. Biol. Chem.* **248**, 5123 (1973).
[10] G. N. Gill and L. D. Garren, *Biochem. Biophys. Res. Commun.* **39**, 335 (1970).
[11] T. E. Martin and I. G. Wool, *J. Mol. Biol.* **43**, 151 (1969).

10 mM in MSH with the 1 M stock solution. The pH is adjusted to 5.0 with glacial acetic acid and the precipitate is removed by centrifugation at 10,000 g for 20 minutes. The pH of the supernatant is made 7.0 with 0.5 M KOH and brought to 45% saturation by the slow addition of solid ammonium sulfate. The precipitate is collected by centrifugation at 27,000 g for 20 minutes, dissolved in TM, and dialyzed against the same buffer for 16 hours.

Aged calcium phosphate gel, adjusted to 3% in TM (w/v), is added slowly to the protein solution (about 12 mg/ml) and stirred for 35 minutes. The ratio of gel to protein should be 1:1 (w/w). The gel is collected by centrifugation at 3000 g for 5 minutes and washed twice with 75 ml of TM. The protein kinase activity is eluted three times with 50 ml each of 0.25 M potassium phosphate buffer (pH 8.1). The combined eluates are dialyzed against TMG for 16 hours.

Generally about 550 mg of dialyzed protein is applied to a DEAE-cellulose column (22 × 3 cm) that has been equilibrated with TMG. The contents of the column are washed with two column volumes of TMG at a flow rate of 30 ml/hour and the protein kinase eluted (at the same rate) with 1 liter of a linear gradient of 0.04 to 0.40 M NaCl in TMG. Fractions (10 ml) are collected and the absorbance at 280 nm is determined. The protein kinase activity of the fractions are assayed (see below). Generally,

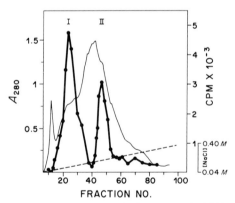

FIG. 1. The separation of rat liver protein kinases by DEAE-cellulose chromatography [C. Eil and I. G. Wool, *Biochem. Biophys. Res. Commun.* **43**, 1001 (1971)]. Selected fractions were assayed for ribosomal protein kinase activity in the presence of cyclic AMP using 60 S ribosomal subunits as substrate. The radioactivity incorporated into 60 S ribosomes in the absence of added protein fractions was 375 cpm. Each 1000 cpm incorporated into ribosomal protein is equivalent to 9.1 pmoles of phosphate. The continuous thin line is the A_{280} of the eluate; the thicker line is the protein kinase activity of selected fractions (cpm of ^{32}P incorporated into protein of 60 S ribosomal subunits); the interrupted line is the concentration of NaCl.

protein kinase I eluates at 0.1 M and protein kinase II at 0.2 M NaCl (Fig. 1).[5]

Purification of Protein Kinase I. Protein kinase I can be further purified by a modification of the procedure of Yamamura *et al.*[12] The peak fractions of protein kinase I from the DEAE-cellulose eluate are pooled and brought to 65% saturation by slow addition of solid ammonium sulfate. The precipitate is collected by centrifugation, dissolved in TMG, and dialyzed overnight against 50 mM PMG.

Approximately 25 mg of dialyzed protein is applied to a hydroxyapatite column (2.3 × 2.5 cm) that has been equilibrated with 50 mM PMG. The column is washed with 40 ml of 50 mM PMG at a flow rate of 12 ml/hour and the protein kinase activity eluted (at the same rate) with 200 ml of a linear gradient of 50 mM to 0.25 M potassium phosphate buffer (pH 7.5). Fractions (4 ml) are collected and the absorbance at 280 nm is determined. The protein kinase activity of the fractions are assayed (see below). The fractions from the peak containing the cyclic AMP-dependent enzyme activity (generally eluted at 0.1 M potassium phosphate) are pooled and the protein precipitated by the slow addition of solid ammonium sulfate to 65% saturation. After centrifugation, the precipitate is dissolved in TMG and dialyzed overnight against TMG to remove ammonium sulfate and potassium phosphate. We designate the sample purified protein kinase I.

Preparation of Ribosomes and Ribosomal Subunits

Reagents

Puromycin dihydrochloride, 1 mM (Nutritional Biochemicals Corporation)
Ethanol

Media

Medium A: Tris·HCl, 50 mM, pH 7.6; KCl, 80 mM; MgCl$_2$, 12.5 mM; MSH, 20 mM
Medium B: Tris·HCl, 50 mM, pH 7.6; KCl, 880 mM; MgCl$_2$, 12.5 mM; MSH, 10 mM

We generally prepare liver ribosomes from male Sprague-Dawley rats of about 100–120 g,[11] however, any of the standard procedures for preparation of relatively uncontaminated particles (from any eukaryotic cell type) will do. The ribosomes are suspended in medium B and incubated for 15 minutes at 37° with 0.1 mM puromycin—to remove nascent peptide

[12] H. Yamamura, M. Takeda, A. Kumon, and Y. Nishizuka, *Biochem. Biophys. Res. Commun.* **40**, 675 (1970).

and dissociate the ribosomes into subparticles.[13,14] After incubation the ribosomal suspension is layered on a 10–30% linear sucrose gradient and centrifuged for 4 hours at 25° in a Spinco SW 27 rotor at 27,000 rpm. The gradients are displaced with 50% sucrose using an Instrument Specialities Co., Inc. (ISCO) Model D density gradient fractionator, and the effluent is analyzed at 254 nm with an ISCO model UA-2 UV analyzer. Subunit fractions are collected[13] and dialyzed overnight against medium A. (Sucrose and high concentrations of potassium interfere with precipitation of ribosomes by ethanol.[15]) Cold 95% ethanol (0.2 of a volume) is added to the subunit fractions and the suspension is kept at 0° for at least 1 hour (to ensure complete precipitation of the particles). The subunits are collected by centrifugation at 10,000 g for 10 minutes, carefully drained free of residual ethanol, and suspended in medium suitable for assay of ribosomal protein kinase activity (see below). The concentration of ribosomes and ribosomal subunits is calculated from their absorption at 260 nm[16]; 1 A_{260} unit is taken to be the equivalent of 45 μg of ribosomal RNA (rRNA).

Assay of the Phosphorylation of the Proteins of 40 S and 60 S Ribosomal Subunits by Protein Kinase

Reagents

Cyclic AMP, 10^{-5} M (Sigma)
Caffeine, 10^{-3} M (Nutritional Biochemicals Corporation)
[γ-^{32}P]ATP, 0.1 mM, 2–10 $\times$ 10^{-4} cpm/nmole; prepared by the method of Glynn and Chappell[17]
Trichloroacetic acid (TCA), 5 and 10%
Formic acid, 88%
BBOT scintillation fluid: 5 g of 2,5-bis[2-(5-*tert*-butylbenzoxazolyl)]thiophene (Packard) in 500 ml of toluene and 500 ml of Methyl Cellosolve[16]

Media

Medium C: Tris·HCl, 40 mM, pH 7.7; MgCl$_2$, 5 mM; MSH, 10 mM; glycerol, 5%

Assay. The reaction requires only a substrate protein that will accept phosphate, ATP, magnesium, and protein kinase. Variations in the condi-

[13] T. E. Martin, I. G. Wool, and J. J. Castles, this series, Vol. 20, p. 417 (1971).
[14] W. S. Stirewalt, J. J. Castles, and I. G. Wool, *Biochemistry* 10, 1594 (1971).
[15] M. S. Kaulenas, *Anal. Biochem.* 41, 126 (1971).
[16] I. G. Wool and P. Cavicchi, *Proc. Nat. Acad. Sci. U.S.* 56, 991 (1966).
[17] I. M. Glynn and J. B. Chappell, *Biochem. J.* 90, 147 (1964).

tions are frequently necessary, but we describe a typical assay. Ribosomal subunits—generally 28 μg (rRNA) of 40 S or 70 μg of 60 S subunits— are incubated in 0.1 ml of medium C containing a preparation of protein kinase (the amount varies with the purity and activity of the enzyme: 10 μg of purified protein kinase I would be adequate) and [γ-^{32}P]ATP (0.1 mM; 2 to 10 × 10^4 cpm/nmole). When cyclic AMP is added, the concentration is 10^{-5} M. If the enzyme preparation has not been purified by chromatography on DEAE-cellulose, and if cyclic AMP is included, it is best to add caffeine (1 mM) to inhibit contaminating phosphodiesterase.[18] The reaction is started by adding [γ-^{32}P]ATP and the tubes are stirred (on a Vortex mixer) and incubated for 15 minutes at 37°. The reaction is stopped by adding 1 ml of 10% TCA and mixing.

Determination of the Incorporation of Radioactivity from [γ-^{32}P]ATP into Ribosomal Proteins. A convenient procedure is to collect the radioactive phosphorylated ribosomal proteins on glass fiber filters (grade 934 AH, 2.4 cm, Reeve Angel). The samples (in TCA) are heated in a boiling-water bath for 15 minutes, to hydrolyze nucleic acids, and then cooled at 0°. The protein is collected on glass fiber disks held in a Millipore multiple sample manifold. The filters are washed with 30 ml of 5% TCA and transferred to scintillation vials containing 0.5 ml of 88% formic acid to dissolve the protein. BBOT scintillation fluid (10 ml) is added, the samples are stirred on a Vortex mixer and the radioactivity is determined (with an efficiency of 95%) in Packard Tri-Carb liquid scintillation spectrometer.

Characterization of Ribosomal Phosphoproteins by
    Electrophoresis in Polyacrylamide Gels and
    by Radioautography

Reagents

MgCl$_2$, 0.5 M
KCl, 2.5 M
Magnesium acetate, 1 M
Acetic acid, glacial and 7%
Formic acid, 88%
Urea, 7 M containing 0.1 M MSH
Solid Tris

Media

Medium D: Tris·HCl, 40 mM, pH 7.7; MSH, 10 mM; glycerol, 5%
Medium E: Tris·HCl, 10 mM, pH 7.7; magnesium acetate, 100 mM

---

[18] T. W. Rall and E. W. Sutherland, *J. Biol. Chem.* **232**, 1077 (1958).

Autoradiography of Phosphorylated Ribosomal Proteins. The identity of the individual ribosomal proteins phosphorylated by protein kinase can be determined if the proteins are extracted from the ribosomal subunits,[19] separated by electrophoresis in polyacrylamide gels,[20,21] and autoradiographs of the gels made.[5]

Ribosomal subunits, approximately 400 μg (rRNA) of 40 S or 1.0 mg of 60 S, are phosphorylated in 12-ml conical Corex tubes in 1.6 ml of medium D containing 0.1 mM [γ-^{32}P]ATP and protein kinase, with or without cyclic AMP. Neither magnesium nor potassium are essential for phosphorylation of ribosomal proteins,[9] but it is at times important to add the cations (see below) and that can be done with appropriate amounts of concentrated solutions. Incubation is for 30 minutes at 37°. The reaction is stopped by the addition of 0.18 ml of ice-cold 1 M magnesium acetate (to precipitate subunits) and the samples kept at 0° for at least 15 minutes. The ribosomes are collected by centrifugation at 10,000 g for 10 minutes, resuspended in approximately 1 ml of medium A, and reprecipitated by the addition of 1/9 volume of 1 M magnesium acetate. After centrifugation the ribosomes are suspended with a glass rod in 0.1–0.2 ml of medium E; 2 volumes of glacial acetic acid are added, and the mixture is stirred with a glass rod for 1 hour at 0° to extract the ribosomal proteins.[19] The ribosomal RNA is removed by centrifugation at 15,000 g for 10 minutes; the pellet is washed with approximately the same volume of 1/3 medium C and 2/3 glacial acetic acid and recentrifuged. The combined supernatant, containing the ribosomal proteins, is dialyzed against 1000 volumes of 7% acetic acid for 3–4 hours. Dialysis is in acetylated tubing[22] to prevent loss of low molecular weight proteins. The concentration[23] and the radioactivity (see above) of the protein in the extract is determined. The dialyzate is lyophilized (to reduce the volume and to remove acetic acid) and the protein dissolved in 0.1–0.2 ml of 7 M urea containing 0.1 M MSH. The pH of the samples is adjusted to 8 with a small amount of solid Tris and incubated for 3 hours at 37° to assure that the proteins are reduced.[21] The ribosomal proteins (80 μg of 40 S or 120 μg of 60 S) are separated by electrophoresis in discontinuous polyacrylamide gels at pH 4.5 in 6 M urea.[20,21] The gels are stained, destained, and sliced into several 1 mm

[19] S. J. S. Hardy, C. G. Kurland, P. Voynow, and G. Mora, *Biochemistry* **8**, 2897 (1969).

[20] P. S. Leboy, E. C. Cox, and J. G. Flaks, *Proc. Nat. Acad. Sci. U.S.* **52**, 1367 (1964).

[21] R. B. Low and I. G. Wool, *Science* **155**, 330 (1967).

[22] L. C. Craig, this series, Vol. XI, p. 870 (1967).

[23] O. H. Lowry, J. T. Rosebrough, A. L. Farr, and R. J. Randall, *J. Biol. Chem.* **193**, 265 (1951).

longitudinal sections. The slices are dried under strong vacuum on filter paper on a Buchner funnel; drying is speeded by covering the slices on the funnel with Saran Wrap and heating with an infrared lamp. X-ray film (Kodak, NS-54T) is placed over the dried gels on filter paper and exposed 1–8 days.

Determination of Phosphoserine and Phosphothreonine in
   Hydrolyzed Phosphorylated Ribosomal Proteins

   Reagents

   HCl, 6 N
   Perchloric acid, 70%
   Solid ammonium molybdate
   Phosphoserine, 2 mg/ml H_2O
   Phosphothreonine, 2 mg/ml H_2O
   KH_2PO_4, 2 mg/ml H_2O

It is important to establish that the phosphate incorporated into ribosomal protein is covalently bound to amino acid residues, for ^{32}P-labeled RNA can contaminate ribosome proteins. Since the RNA might be far more radioactive than the protein a small amount of contamination could be very misleading.

Phosphorylation of ribosomal proteins by protein kinase leads to the formation of phosphoserine and phosphothreonine.[1-4,9] To detect those amino acids ribosomal subunits are phosphorylated with [γ-^{32}P]ATP and the ribosomal proteins are extracted. The proteins are then hydrolyzed in 6 N HCl and the hydrolyzate is subjected to paper electrophoresis at pH 1.9. At that pH, phosphoserine and phosphothreonine migrate to the anode whereas the other amino acids migrate to the cathode. The radioactivity is located by radioautography of the electrophoretogram.

Ribosomal subunits, 40 S and 60 S, are phosphorylated, and the proteins are extracted as described above for the preparation of radioactive ribosomal proteins for gel electrophoresis. Lyophilized 40 S proteins (90 μg, about 25,000 cpm) and 170 μg of lyophilized 60 S proteins (about 75,000 cpm) are dissolved separately in 0.5–1.0 ml of 6 N HCl. The tubes are evacuated, sealed, and heated at 110° for approximately 7 hours. Hydrolysis of the proteins for longer periods leads to extensive breakdown of the ester bonds in phosphoserine and phosphothreonine.[3] The samples are dried in a rotary evaporator. The hydrolyzates are dissolved in 100 μl of 2.5% formic acid–7.8% acetic acid, pH 1.9. The samples to be analyzed, authentic phosphoserine (50 μl), phosphothreonine (50 μl), and phosphate (50 μl), are spotted onto Whatman 3 MM paper (46 × 57 cm)

10 cm from the end to be attached to the cathode. The sample is applied in several small portions, with drying after each application, to prevent spreading. Drying is speeded with a hair dryer. Both ends of the paper are moistened with the formic acid–acetic acid mixture (pH 1.9) and then applied to the electrophoresis apparatus (Savant; Flat Plate). Electrophoresis is for 2 hours at 2750 V. The paper is dried in an oven at 60° and cut to separate the radioactive samples from the authentic markers. The radioactive samples on the paper are covered with X-ray film (Kodak NS-2T) and exposed for 2 weeks. The portion of the paper with the authentic samples of phosphoserine, phosphothreonine, and inorganic phosphate is sprayed with a mixture of 3% perchloric acid, 0.1 N HCl, and 1% w/v ammonium molybdate to locate inorganic and organic phosphate.[24] The paper is dried in an oven and developed with strong ultraviolet light.[24]

Comment

Effect of Cations. Magnesium is generally required for kinase reactions having ATP as a substrate; nonetheless, maximum phosphorylation of ribosomal proteins occurs when subunits are incubated without addition of magnesium. Magnesium is bound to the ribosome and no doubt is the source of sufficient cation to support catalytic activity of the enzyme. If ribosomes are treated with EDTA to remove the bound magnesium and then assayed without added magnesium, no phosphorylation occurs. With ordinary preparations of ribosomal subunits increasing concentrations of magnesium (up to 20 mM) inhibit protein phosphorylation.

Potassium also inhibits ribosomal protein phosphorylation, though to a lesser extent than magnesium. Increments in the concentration of potassium, to about 50 mM, decrease phosphorylation of ribosomal subunits. The effects may be a consequence of the influence of the cation on the structure of the ribosomal subparticles. However, potassium also inhibits the phosphorylation of histone by protein kinase.[9] The structure of histones is not known to be affected by potassium; hence, the effect of the cation may be in part on the enzyme.

Assay of Ribosomal Phosphoprotein Phosphatase

A prerequisite for the study of the function of phosphorylated ribosomes, is assurance that the preparation of factors used to measure ribosome activity does not contain ribosomal phosphoprotein phosphatases. If phosphorylated ribosomes are incubated with unresolved cytosol (from ascites cells or from rat liver) in circumstances used to assay protein syn-

---

[24] R. M. C. Dawson, *in* "Lipid Chromatography Techniques" (G. V. Marinetti, ed.), Vol. I, p. 179. Dekker, New York, 1967.

thesis, as much as 70% of the phosphate is released from the particles.[25] It is obvious that an inhibitor of ribosomal phosphoprotein phosphatase would be a great convenience (provided, of course, that it did not interfere with ribosome function). Unfortunately, we have not discovered such a reagent, although we have tried a number of inhibitors of other phosphatases. What is required then is that the factor preparations be sufficiently resolved so as to be relatively free of ribosomal phosphoprotein phosphatase. We describe an assay that allows one to assess the extent of contamination of factor preparations with ribosomal phosphoprotein phosphatase.

Reagents

ATP, 0.1 M (Pabst)
GTP, 5 mM (Pabst)
Creatine phosphate (Boehringer Mannheim Corporation)
Creatine phosphokinase (Boehringer Mannheim Corporation)
Aminoacyl-tRNA–$E. coli$ B tRNA aminoacylated[26,27] with 20 different amino acids
Polyuridylic acid (Miles Laboratories Incorporated)

Medium

Medium F: Tris·HCl, 20 mM, pH 7.5; KCl, 125 mM, magnesium acetate, 5 mM; MSH, 10 mM

Preparation of Radioactive Phosphorylated Ribosomes. The substrate for assay of ribosomal phosphoprotein phosphatase is ^{32}P-labeled ribosomes. They are prepared by incubating ten times the usual amount of ribosomal subunits—280 μg (rRNA) of 40 S or 720 μg (rRNA) of 60 S subparticles—in 1 ml of medium C in Corex centrifuge tubes with saturating amounts of protein kinase, 10^{-5} M cyclic AMP, and 0.1 mM [γ-^{32}P] ATP. Incubation is for 25 minutes at 37°. The concentration of MgCl$_2$ is raised to 12.5 mM (by addition of sufficient 0.5 MgCl$_2$) and potassium to 80 mM (with sufficient 2.5 M KCl). The ribosomes are precipitated by the addition of 0.2 volume of cold 95% ethanol (see above) and collected by centrifugation at 10,000 g for 10 minutes. The phosphorylated ribosomes are suspended in 0.5 ml of medium C and insoluble material is removed by centrifugation at 3000 g for 5 minutes.

Ribosomal Phosphoprotein Phosphatase Assay. The ribosomal phosphoprotein phosphatase activity of various factor preparations can be determined by incubating them with ^{32}P-labeled subparticles—2.5 μg (rRNA)

[25] C. Eil and I. G. Wool, *J. Biol. Chem.* **248**, 5130 (1973).
[26] G. von Ehrenstein and F. Lipmann, *Proc. Nat. Acad. Sci. U.S.* **47**, 941 (1961).
[27] I. G. Wool and P. Cavicchi, *Biochemistry* **6**, 1231 (1966).

of 40 S and 6.25 μg (rRNA) of 60 S subunits—in 0.1 ml of medium C. However, our purpose has generally been to test the function of phosphorylated ribosomes, so we have assayed phosphatase in the same circumstances in which we measure protein synthesis. The ^{32}P-labeled subunits (in the amount specified above) are incubated in 0.1 ml of medium F containing: ATP, 1 mM; GTP, 0.1 mM; polyuridylic acid, 10 μg; creatine phosphate, 5 mM; creatine phosphokinase, 20 μg; and aminoacyl-tRNA, 50 μg. The reaction is started by adding the radioactive ribosomes. Incubation is at 37° for 10–45 minutes depending on the amount of ribosome phosphoprotein phosphatase. The reaction is stopped by adding 1 ml of 10% TCA, and the mixture is heated at 90–95° for 15 minutes. The precipitate is collected on glass fiber disks, and the radioactivity is measured as described for the protein kinase assay (see above). Ribosome phosphoprotein phosphatase activity of protein fractions is calculated from the loss of radioactivity from ^{32}P-labeled ribosomes after allowance is made for loss of radioactivity from ribosomes incubated without the protein fractions. The loss of radioactivity in the latter case does not exceed 15% of the amount originally present.

Assay of the Function of Phosphorylated Ribosomes

Medium

Medium G: Tris·HCl, 40 mM, pH 7.7; KCl, 10 mM; MgCl$_2$, 5 mM; MSH, 10 mM; glycerol, 5%

There are two requisites for the study of the function of phosphorylated ribosomes: the particles should be fully phosphorylated at the start of the experiment and the loss of phosphate during the assay should be kept to a minimum. The first requirement is met by choosing optimal conditions (concentrations of ions, of enzymes, etc.) for the phosphorylation of ribosomal proteins by protein kinase. For ribosomes to retain activity, phosphorylation of the subunits must be carried out in medium containing magnesium and potassium. Since these ions also inhibit protein kinase, they should be kept to the minimal concentration consonant with maintenance of ribosome function. If subunits are phosphorylated in medium containing 5 mM MgCl$_2$ and 10 mM KCl, protein kinase activity is not appreciably depressed, and the capacity of the subunits to synthesize protein (during which assay the potassium is usually 75–125 mM) is hardly decreased. No synthesis of protein occurs if potassium is omitted from the preincubation medium.[9,28] Although protein kinase I does not require a reducing agent for its activity, MSH is included in

[28] P. H. Näslund and T. Hultin, *Biochim. Biophys. Acta* 204, 237 (1970).

the protein kinase reaction to preserve the function of the ribosomes.[29,30]

Ribosomal subunits, 40 S (20–30 μg of rRNA) and 60 S (50–75 μg of rRNA), are preincubated separately in 0.1 ml of medium G containing: ATP, 1 mM—the ATP ordinarily need not be radioactive; cyclic AMP, 10 μM; and protein kinase. Incubation is for 25 minutes at 37°. The reaction is stopped by cooling the samples on ice, and aliquots of the subunits are used to assay their function.

We have used this procedure to compare the ability of phosphorylated and nonphosphorylated ribosomal subunits to synthesize polyphenylalanine, to translate encephalomyocarditis virus RNA, to bind aminoacyl-tRNA, and to reassociate to form 80 S monomers.[25] While we have not examined all the possibilities, we have not yet discovered an appreciable difference in the activity of ribosomal subunits as a result of phosphorylation.[25]

Remarks

A reliable assay of the activity of ribosomal protein kinase requires that the enzyme be resolved from inhibitors present in crude rat liver cytosol. The eluate from calcium phosphate gel contains detectable ribosomal protein kinase activity, but catalyzes considerable phosphorylation in the absence of ribosomes. The protein in the enzyme preparation that is phosphorylated is removed by DEAE-cellulose column chromatography. That procedure also separates two protein kinases: neither shows a substrate preference; both phosphorylate 40 S subunits, 60 S subunits, and calf histone.[5] Moreover, protein kinase I also phosphorylates casein and protamine sulfate as well as 80 S ribosomes.[9] However, the 80 S ribosomes accept less phosphate than subunits and are contaminated with protein kinase since they can be phosphorylated in the absence of added enzyme. Apparently, the enzyme is removed when subunits are prepared by treating ribosomes with 0.88 M KCl and centrifuging them through a sucrose gradient.

At least three proteins of the 40 S subunit of liver ribosomes and nine of the 60 S subunit are phosphorylated by ribosomal protein kinases.[5] The extent of phosphorylation varies with the kinase preparation used, the presence of cyclic AMP, and the ionic conditions, although none of these variations leads to the phosphorylation of different ribosomal proteins.[9] The reaction involves the transfer of the terminal phosphate of ATP and not the exchange of phosphates already present in ribosomal proteins.[9] GTP will not serve as a phosphate donor.[9]

[29] T. E. Martin, F. S. Rolleston, R. B. Low, and I. G. Wool, *J. Mol. Biol.* 43, 135 (1969).
[30] H. C. McCallister and R. S. Schweet, *J. Mol. Biol.* 34, 519 (1968).

Section V

Messenger RNA and Protein Synthesizing Systems

[56] The Isolation of Messenger RNA from Mammalian Cells[1]

By George Brawerman

Criteria for the Identification of mRNA

Recent advances in our knowledge of mammalian messenger RNA have permitted a rational approach to its identification and isolation. Until recently, the only criteria generally available for the detection of mRNA in animal cells were: (a) presence in polysomes; (b) rapid labeling; (c) heterogeneity in size distribution. DNA-like base composition is used sometimes as an additional criterion. In order to distinguish between polysomal RNA and other RNA species that might cosediment with the polysomes, some investigators have dissociated the polysomes with EDTA and utilized only those RNA components released as more slowly sedimenting material by this treatment.[2] The preferential inhibition of ribosomal RNA synthesis by actinomycin D has provided a convenient means for avoiding the presence of ribosomal label in the polysomes. In many mammalian cell cultures, actinomycin levels of 0.04 μg/ml are sufficient to block rRNA synthesis, but have little effect on the labeling of the presumed mRNA.[2,3]

The above criteria provide a clear distinction between ribosomal and nonribosomal RNA, but fail to discriminate effectively between mRNA and other RNA components that might contaminate the polysome preparations. Such contaminants could cause difficulties in cases where highly labeled nuclear RNA has leaked from the nuclei during cell disruption.[2,4]

The development of mammalian cell-free systems that will synthesize defined proteins in response to exogenous RNA has provided a precise means for identifying unique mRNA species. The messenger for hemoglobin, for instance, has been well characterized by its sedimentation coefficient of 9–10 S and its capacity to induce globin synthesis in heterologous cell-free systems.[5,6] Other mRNA species have been assayed in a similar fashion.

[1] Supported by a research grant from the U.S. Public Health Service (GM 17973).
[2] S. Penman, C. Vesco, and M. Penman, J. Mol. Biol. 34, 49 (1968).
[3] R. Perry, Exp. Cell Res. 29, 400 (1963).
[4] R. P. Perry and D. E. Kelley, J. Mol. Biol. 35, 37 (1968).
[5] R. E. Lockard and J. B. Lingrel, Biochem. Biophys. Res. Commun. 37, 204 (1969).
[6] D. Housman, R. Pemberton, and R. Taber, Proc. Nat. Acad. Sci. U.S. 68, 2716 (1972).

Presence of Poly(A) in mRNA

It has been found recently that most of the heterodisperse nonribosomal polysomal RNA of mammalian cells contains a polyadenylate sequence at the 3' end of the molecules.[7–10] The occurrence of poly(A) in mRNA may be a characteristic of all eukaryotic cells. The poly(A) segment has been detected in most of the mRNA species active as templates for synthesis of specific proteins. The mRNA for histones, however, has been shown to lack the poly(A) sequence.[11] The presence of poly(A) in most mRNA, and its absence from ribosomal and transfer RNA, provides a precise criterion for the identification of mRNA in polysomes. Poly(A) possesses unique properties, such as capacity to form complementary base-paired structures with poly(U) and poly(dT) and to bind to cellulose nitrate membrane filters (Millipore filters) at high ionic strength. These properties provide simple means for the separation of mRNA from other RNA species. Radioactive mRNA can be assayed most conveniently by its capacity to bind to Millipore filters,[7] and the labeling need not be done in the presence of actinomycin, since labeled RNA components other than mRNA are not retained on the filters.

Poly(A) also exhibits a unique behavior during deproteinization with phenol. It appears to bind to denatured proteins in the presence of monovalent cations, and, as a result, is carried to the nonaqueous phase.[12] Because of this property of the poly(A) segment, mRNA may not be readily extracted from biological preparations, even under conditions that favor the recovery of the bulk of the cellular RNA. It has been possible, however, to define some of the parameters that control the apparent protein–poly(A) interaction in the phenol mixture, and to establish optimal conditions for the extraction of mRNA.[12]

Extraction of mRNA

Resistance of mRNA to Phenol Extraction

It had been known for some time that mammalian cells contain an RNA fraction refractory to phenol extraction. Aqueous phenol treatment

[7] S. Y. Lee, J. Mendecki, and G. Brawerman, *Proc. Nat. Acad. Sci. U.S.* **68**, 1331 (1971).
[8] M. Edmonds, M. H. Vaughan, and H. Nakazato, *Proc. Nat. Acad. Sci. U.S.* **68**, 1336 (1971).
[9] J. E. Darnell, R. Wall, and R. J. Tushinski, *Proc. Nat. Acad. Sci. U.S.* **68**, 1321 (1971).
[10] J. Mendecki, S. Y. Lee, and G. Brawerman, *Biochemistry* **11**, 792 (1972).
[11] M. Adesnik and J. E. Darnell, *J. Mol. Biol.* **67**, 397 (1972).
[12] G. Brawerman, J. Mendecki, and S. Y. Lee, *Biochemistry* **11**, 637 (1972).

TABLE I
PHENOL FRACTIONATION OF SARCOMA 180 POLYSOMES[a]

RNA fraction	Total RNA in fraction (μg)	Radioactivity (cpm)	Millipore-bound radioactivity (% of total in fraction)
pH 7.6	1152	54,500	2
pH 9	38	79,250	100

[a] Polysomes from cells labeled with uridine in presence of 0.04 μg/ml actinomycin D were subjected to phenol extraction in presence of 0.1 M Tris·HCl (pH 7.6). Non-aqueous residue reextracted first with 0.1 M Tris (pH 7.6) then twice with 0.1 M Tris (pH 9.0). RNA purified from pooled neutral and alkaline extracts. Data derived from G. Brawerman, J. Mendecki, and S. Y. Lee, *Biochemistry* **11**, 637 (1972).

at neutral pH in the cold of homogenized liver and of disrupted ascites cells was shown to result in the release of the ribosomal RNA into the aqueous phase, but to leave a minor RNA fraction in the nonaqueous phase.[13] The latter could be recovered by reextraction at high temperature. It was shown to be DNA-like in base composition. A similar RNA fraction could be recovered from the nonaqueous phase by reextraction with slightly alkaline Tris buffer.[14] The latter fraction shall be referred to as pH 9 RNA (RNA not extracted into the aqueous phase in the presence of Tris·HCl pH 7.6, but recoverable with Tris pH 9). This RNA fraction was found to be present in purified polysomes from mouse sarcoma 180 ascites cells.[7] It was further observed that the poly(A) sequence was present in the polysomal pH 9 RNA, and absent from the RNA extractable at neutral pH. Table I shows the distribution in the two phenol fractions of the bulk polysomal RNA and of the material labeled in the presence of a low level of actinomycin. Only the labeled RNA extracted at pH 9 contains poly(A), as indicated by its capacity to bind to Millipore filters.

Conditions for the Phenol Extraction of mRNA

The behavior of the pH 9 RNA during phenol extraction can be accounted for by the properties of the poly(A) segment. It was observed that synthetic poly(A), when mixed with polysomes and subjected to the phenol treatment under the conditions used during RNA extraction

[13] G. P. Georgiev and V. L. Mantieva, *Biochim. Biophys. Acta* **61**, 153 (1962).
[14] G. Brawerman, L. Gold, and J. Eisenstadt, *Proc. Nat. Acad. Sci. U.S.* **50**, 630 (1963).

TABLE II

EFFECT OF POLYSOMES AND OF METHYLATED ALBUMIN ON THE
BEHAVIOR OF POLY(A) DURING PHENOL EXTRACTION[a]

	Labeled RNA remaining in aqueous phase (% of total)		
Labeled RNA preparation	No addition	Polysomes added	Methylated albumin added
Poly(A)	96	3	6
Polysomal RNA, pH 9	98	19	35
Polysomal RNA, pH 7.6	—	68	80

[a] Labeled RNA preparations mixed with unlabeled polysomes or methylated albumin and subjected to phenol treatment in 0.1 M Tris·HCl (pH 7.6). Data derived from G. Brawerman, J. Mendecki, and S. Y. Lee, *Biochemistry* **11**, 637 (1972).

at pH 7.6, is removed from the aqueous phase (Table II). This appears to be caused by interaction between the poly(A) and denatured ribosomal proteins. The same effect is observed when methylated albumin is substituted for the ribosomal proteins. Deproteinized pH 9 polysomal RNA shows a similar behavior, in contrast to the pH 7.6 RNA fraction.[12]

The above interaction between poly(A) and protein is promoted by monovalent cations. Since Tris is nearly completely protonated at neutral pH, a 0.1 M solution of this buffer contributes a concentration of monovalent cations close to 0.1 M. K^+ and Mg^{2+}, included in the medium used to suspend polysomes, are also present. Reextraction of the pH 9 RNA from the nonaqueous residue after pH 7.6 extraction as in Table I, takes place in a considerably reduced ionic environment, since Tris at pH 9.0 occurs primarily in the nonionized form. Inclusion of Na^+ or K^+ at this stage interferes with the pH 9 extraction. It appears that the pH 9 RNA can be extracted at neutral pH in the absence of cations, but Tris buffer of pH 9.0 seems more effective and has been used consistently in this laboratory.

Procedure for RNA Extraction[12]

Reagents

Tris buffer, pH 9.0: 1 M solution of tris(hydroxymethyl)aminomethane (Tris) neutralized with HCl to pH 9.0 (measured at room temperature and diluted as required)

Aqueous phenol: redistilled phenol diluted with H_2O to a concentration of 80% (v/v); stored in the cold in a brown bottle

SDS: 5% solution of sodium dodecyl sulfate; kept at room temperature

Procedure. All operations are carried out at 0–5°. The polysome suspension is diluted with water to a concentration no greater than 100 A_{260} units per milliliter. One-tenth volumes of 5% SDS and of 1 M Tris, pH 9.0, are added, followed by 1 volume of aqueous phenol. The mixture is stirred vigorously for about 5 minutes, then centrifuged at 12,000 g for 10 minutes to separate the phases, and the aqueous phase is removed. The nonaqueous residue (phenol phase plus gel interphase) is reextracted with an equal volume of 0.1 M Tris, pH 9.0, and 0.5% SDS by vigorous stirring as before, and the aqueous phase is removed after centrifugation. The pooled aqueous phases contain most of the polysomal rRNA and mRNA. They are extracted three times with an equal volume of fresh aqueous phenol by brief vigorous stirring followed by centrifugation and removal from the phenol phase. These last extractions with phenol serve to remove residual protein material, as well as most of the SDS. The RNA solution can be used without precipitation, but must be then extracted at least four times with ether to remove dissolved phenol. The ether is subsequently removed by blowing air over the solution. The RNA can also be precipitated from the aqueous phase after the last phenol extraction by addition of 0.1 volume of 1 M NaCl and 2.5 volumes of ethanol. After storage overnight at 4°, the precipitate is collected by centrifugation and washed twice with cold 66% ethanol in 0.1 M NaCl, to remove residual phenol. The precipitate can be dissolved in water and stored at $-20°$.

Extraction of Nuclear RNA

Poly(A) segments are also present in the nucleoplasmic RNA. These poly(A)-containing nuclear RNA molecules are presumed to represent the precursors to cytoplasmic mRNA.[10,15] The pH 9 phenol procedure is effective for their extraction from the nuclei, but DNA is also extracted. The high viscosity of the extracts makes their handling very difficult. Moreover, the RNA cannot be separated easily from the contaminating DNA. Incubation with DNase is necessary to hydrolyze the DNA, and contamination of the enzyme preparation by traces of RNase could easily lead to fragmentation of the RNA molecules during this treatment.

The hot SDS-phenol procedure,[16] which does not lead to extraction of DNA, appears to be as effective as the pH 9 phenol procedure for the recovery of the poly(A)-containing RNA molecules from nuclei.

[15] J. E. Darnell, L. Philipson, R. Wall, and M. Adesnik, *Science* **174**, 507 (1971).
[16] M. Girard, this series, Vol. 12A, p. 581.

Other Extraction Procedures for Polysomal mRNA

It has been reported that addition of $CHCl_3$ to the phenol leads to the release of poly(A)-containing RNA into the aqueous phase, even when conditions tend to favor the protein–poly(A) interaction.[17] The procedure was found to be effective in this laboratory, with yields of mRNA nearly as high as those produced by the pH 9 phenol treatment.

The hot SDS-phenol procedure is also used for the extraction of polysomal mRNA.[8,9] The relative efficiencies of the hot phenol and the cold pH 9 phenol treatments have not been compared on polysomal mRNA.

Purification of mRNA

Binding Properties of Poly(A)

A variety of techniques are available for the selective absorption of poly(A)-containing RNA molecules and their effective separation from other types of RNA. In addition to its capacity to form complementary base-paired structures with poly(U) and poly(dT), the poly(A) segment is capable of binding to cellulose nitrate membrane filters (Millipore filters) at high ionic strength.[7] In this respect the behavior of poly(A) is similar to that of single-stranded DNA. Thus most of the labeled polysomal RNA from cells incubated in the presence of a low level of actinomycin will bind to Millipore in the presence of 0.5 M KCl. Ribosomal RNA is not retained under these conditions. Quantities of RNA up to 60 μg have been adsorbed on a single filter.[12] The maximum capacity of the filters has not been determined. The adsorbed RNA can be eluted with SDS in Tris buffer pH 9.0. The RNA cannot be adsorbed on the filters in the presence of the detergent.

The unique behavior of poly(A) is not limited to binding to Millipore. As mentioned above, it will also bind to denatured proteins when the concentration of monovalent cation is sufficiently high. The precise conditions for selective binding to proteins, however, are not well understood, and this property does not appear to be useful for the purification of mRNA. With diluted suspensions of polysomes from mouse sarcoma 180 cells, the fractionation based on sequential phenol extraction with Tris buffers of pH 7.6 and 9.0 is quite effective (see Table I). With more concentrated polysome suspensions, the pH 9 fraction remains heavily contaminated with ribosomal RNA, and the mRNA is obtained in lower yields.[12]

The size of the poly(A) segment may play an important role in its

[17] R. P. Perry, J. LaTorre, D. E. Kelley, and J. R. Greenberg, *Biochim. Biophys. Acta* **262**, 220 (1972).

binding characteristics to Millipore. The newly formed mRNA of mammalian cells contains a poly(A) segment 100–200 nucleotides long.[10] Poly(A) of this size will bind effectively under the conditions described below. A substantially lower poly(A) size has been reported in the case of rabbit reticulocyte hemoglobin mRNA labeled for a long period of time.[18] It is possible that poly(A) segments with less than 50 nucleotides are not effectively retained on Millipore.[19] A substantial portion of the rabbit hemoglobin mRNA, however, is capable of binding to Millipore.[12] The procedures using complementary base pairing of the poly(A) appear to be effective with shorter polynucleotide chains.

The Millipore-binding techniques provides a convenient analytical procedure for the routine assay of newly formed mRNA, since the adsorbed material can be counted directly on the filter. It permits the study of mRNA metabolism even in the absence of inhibitors of rRNA synthesis, since the labeled rRNA is not retained on the filters. This technique, however, does not appear to be effective for the selective adsorption of the nuclear poly(A)-containing RNA molecules.

Procedure for the Adsorption of mRNA on Millipore Filters[12]

Reagents

KCl buffer: 0.5 M KCl, 1 mM MgCl$_2$, and 10 mM Tris·HCl, pH 7.6; SDS must be absent from solution
Eluting buffer: 0.5% SDS in 0.1 M Tris·HCl, pH 9.0

Procedure. The adsorption can be carried out either in the cold or at room temperature. The latter condition appears to be more effective in reducing contamination by residual ribosomal RNA. The RNA solution is diluted 10- to 20-fold with KCl buffer. The final RNA concentration should be no greater than 0.3 mg/ml. The solution is passed through a Millipore filter that had been presoaked for 30 minutes in KCl buffer. A rate of filtration of approximately 0.5 ml per minute is adequate. The filter is next washed several times with KCl buffer.

The filter with the adsorbed RNA is either dried for scintillation counting or used for elution of the adsorbed material as follows. The filter is placed in 0.5–1 ml of eluting buffer and kept in ice for about 30 minutes with occasional shaking. Because of the residual KCl on the filter, a heavy precipitate of potassium dodecyl sulfate appears rapidly. This does not interfere with the elution of the RNA. The precipitate can be removed by centrifugation at 12,000 rpm for 10 minutes, and does not carry along any adsorbed RNA.

[18] H. Burr and J. B. Lingrel, *Nature (London) New Biol.* 233, 41 (1971).
[19] D. Sheiness and J. E. Darnell, personal communication.

The material recovered from the Millipore filters can be still contaminated with rRNA to an extent of approximately 50%. Zone sedimentation can be used to remove some of the contaminating RNA.

Other Separation Procedures for mRNA

Several procedures have been developed for the adsorption of mRNA using complementary base pairing. They involve coupling either poly(U) or poly(dT) to an insoluble matrix and adsorbing the RNA preparation under conditions that favor base pairing. Poly(U) has been coupled to glass fiber filters by subjecting the filters coated with poly(U) to UV irradiation.[20] The RNA solution is adsorbed to the treated filters in the presence of 0.12 M NaCl in 10 mM Tris·HCl (pH 7.5). Poly(U)-glass fiber columns have been used in a similar fashion.

Poly(dT) coupled to cellulose is an effective adsorbent for poly(A)-containing molecules. It had been used successfully for the isolation of hemoglobin mRNA on a relatively large scale.[21] The cyclohexylamine salt of thymidylic acid in pyridine is first polymerized in the presence of N,N'-dicyclohexylcarbodiimide (DCC) under strictly anhydrous conditions; the polymer is next coupled to thoroughly dehydrated cellulose, also in the presence of DCC.[22] The poly(dT)-cellulose is used as a small column, and the RNA solution is applied in the presence of 0.5 M KCl in 0.01 M Tris·HCl pH 7.5. After thorough washing, the mRNA is eluted with 10 mM Tris, pH 7.5.[21]

It has been reported that untreated cellulose can adsorb poly(A)-containing RNA under the conditions used for the poly(dT)-cellulose procedure.[23] These conditions are also similar to those for adsorption on Millipore filters. It is possible, however, that not all cellulose preparations exhibit this binding property.

Poly(U) coupled to agarose can be used as a selective adsorbent for poly(A)-containing RNA molecules.[24] Also, messenger RNA hybridized with poly(U) can be isolated by chromatography on hydroxyapatite.[25] In this latter procedure, the complementary poly(A)–poly(U) segment on the RNA causes it to bind more tightly to hydroxyapatite. The mRNA, however, is recovered with contaminating poly(U).

---

[20] R. Sheldon, C. Jurale, and J. Kates, *Proc. Nat. Acad. Sci. U.S.* **69**, 417 (1972).

[21] H. Aviv and P. Leder, *Proc. Nat. Acad. Sci. U.S.* **69**, 1408 (1972).

[22] P. Gilham, *J. Amer. Chem. Soc.* **86**, 4982 (1964).

[23] P. A. Kitos, G. Saxon, and H. Amos, *Biochem. Biophys. Res. Commun.* **47**, 1426 (1972).

[24] M. Adesnik, M. Salditt, W. Thomas, and J. E. Darnell, *J. Mol. Biol.* **71**, 21 (1972).

[25] J. R. Greenberg and R. P. Perry, *J. Mol. Biol.* **72**, 3 (1972).

[57] Isolation of Rabbit Reticulocyte 9 S mRNA

By BERNARD LEBLEU

A fraction of rabbit reticulocyte polyribosomal RNA with some of the expected properties of globin mRNA (messenger RNA) was isolated several years ago by Marbaix and Burny.[1] Indirect evidence suggesting that this particular RNA fraction (which sediments at 9 S when centrifuged through a sucrose gradient) essentially consists of globin mRNA has accumulated over the next few years in several laboratories.[2] More recently a mouse reticulocyte 9 S RNA species isolated in a similar way has been shown unequivocally to direct the synthesis of mouse β-globin chains in a rabbit reticulocyte cell-free system.[3] Equivalent fractions of RNA isolated from reticulocyte polyribosomes have been prepared from different species and shown to code for both α and β globin chains in various in vitro[4,5] as well as in vivo[6] amino acid incorporation systems.

In the method described here, rabbit reticulocyte polyribosomes are dissociated by EDTA and the globin mRNA, which is liberated as a 15 S messenger ribonucleoprotein complex (mRNP),[7] is separated from the ribosomal subunits and tRNA by sucrose gradient centrifugation. This 15 S mRNP has been well characterized[8] and directs the synthesis of globin chains in an in vitro amino acid incorporation system to the same extent as the free mRNA.[9] Globin 9 S mRNA is then separated from the associated proteins and further purified by a sucrose gradient centrifugation. This method, which has been extensively used for the isolation of different mRNA species, conveniently allows the preparation of up to 0.5 mg of 9 S mRNA with a high degree of purity as judged by sucrose

[1] G. Marbaix and A. Burny, Biochem. Biophys. Res. Commun. 16, 522 (1964).
[2] H. Chantrenne, A. Burny, and G. Marbaix, Progr. Nucl. Acid Res. Mol. Biol. 7, 173 (1967).
[3] R. E. Lockard and J. B. Lingrel, Biochem. Biophys. Res. Commun. 37, 203 (1969).
[4] M. B. Mathews, M. Osborn, and J. B. Lingrel, Nature (London) New Biol. 233, 206 (1971).
[5] P. M. Prichard, D. J. Picciano, D. G. Laycock, and W. F. Anderson, Proc. Nat. Acad. Sci. U.S. 68, 2752 (1971).
[6] C. D. Lane, G. Marbaix, and J. B. Gurdon, J. Mol. Biol. 61, 73 (1971).
[7] G. Huez, A. Burny, G. Marbaix, and B. Lebleu, Biochim. Biophys. Acta 145, 629 (1967).
[8] B. Lebleu, G. Marbaix, G. Huez, A. Burny, J. Temmerman, and H. Chantrenne, Eur. J. Biochem. 19, 264 (1971).
[9] U. Nudel, B. Lebleu, T. Zehavi-Willner, and M. Revel, Eur. J. Biochem. 33, 314 (1973).

gradient or polyacrylamide gel electrophoresis analysis. Scaling up the preparation can be achieved by using zonal rotor centrifugation.[10] Globin mRNA prepared in such a way is active in directing the synthesis of globin chains in different cell-free systems including a Krebs ascites cell-free system which has been modified for increasing its efficiency of translation,[11] as will be described briefly.

Alternative procedures for the preparation of globin mRNA have been proposed on the basis of the poly(A) content of mRNA's.[12]

Preliminary Remarks

All reagents are made from sterilized stock solutions kept at 2°. Sucrose containing solutions are made from a sterile 50% (w/v) aqueous solution and kept at $-20°$ in order to prevent any bacterial growth.

All glassware and related equipment is either heated at 200° for 2 hr, autoclaved, or treated with sulfochromic acid.

All manipulations are carried out at low temperature (0–4°) unless otherwise specified.

Reagents

Phenylhydrazine, 2.5% neutralized aqueous solution (phenylhydrazine hydrochloride, Fluka AG) stored in sealed glass ampules in the cold

Cresyl brilliant blue (Serva, Germany), 1% (w/v) in isotonic solution

Anesthesic solution made up from 5 parts Nembutal (Nembutal veterinary anesthetic, Abbot Laboratories Ltd., Queenborough, England; 60 mg of Pentobarbitone sodium/ml), one part heparin (5000 units/ml, Boots Pure Drug Co., Nottingham, England), and four parts of isotonic solution

Isotonic solution: 140 mM NaCl, 5 mM MgCl$_2$, 5 mM KCl

Hypotonic solution: 10 mM Tris·HCl, pH 7.4; 2 mM MgCl$_2$; 10 mM KCl

Tris-EDTA buffer: 20 mM Tris·HCl, pH 7.4; 33 mM EDTA, pH 7.0

NETS buffer: 10 mM Tris·HCl, pH 7.4; 10 mM EDTA; 10 mM NaCl; 0.2% SDS

SDS buffer: 300 mM NaCl; 10 mM Tris·HCl, pH 7.4; 0.5% SDS

[10] R. Williamson, M. Morrison, G. Lanyon, R. Eason, and J. Paul, *Biochemistry* **10**, 3014 (1971).

[11] B. Lebleu, U. Nudel, E. Falcoff, C. Prives, and M. Revel, *FEBS* (*Fed. Eur. Biochem. Soc.*) *Lett.* **25**, 97 (1972).

[12] H. Aviv and P. Leder, *Proc. Nat. Acad. Sci. U.S.* **69**, 1408 (1972).

SDS (sodium lauryl sulfate, BDH Chemical Ltd., Pools, England), eventually recrystallized from ethanol.[13]

Procedures

Reticulocytosis

New Zealand white rabbits (weighing 3–4 kg) are made anemic by 4 daily subcutaneous injections of phenylhydrazine (0.3 ml/kg). The blood is collected 2 days after the last injection providing the reticulocyte level is high enough. This can be checked easily as follows: Some drops of blood are removed from the ear and mixed with twice their volume of an isotonic solution of cresyl brilliant blue. This mixture is shaken intermittently over a period of 15 minutes. One drop is layered onto a microscope slide, and the preparation is checked using an immersion objective. Erythrocytes are stained yellow while reticulocytes become violet. The rabbits are considered anemic when reticulocytes comprise more than 50% of the total cell population.

Preparation of Reticulocytes

The rabbits are killed by injection of a heparin-containing anesthetic solution (2 ml/kg) in the marginal vein of the ear. An incision is made at the level of the xyphoid appendix and the thoracic cavity is opened along the sternum. The aorta is severed, and blood is collected in an ice-cold beaker. Additional blood is then obtained by making a large incision in the heart. The yield of blood is around 80–100 ml per rabbit. The blood is filtered through cheese cloth and centrifuged at 2° at 1000 g for 20 minutes in an International swinging-bucket rotor. The plasma is carefully eliminated by aspiration. The cells are gently resuspended in isotonic solution and centrifuged as described above. This washing procedure is repeated twice.

Preparation of Polyribosomes

The packed cells are lysed with twice their volume of hypotonic solution. They are shaken vigorously for 2 minutes in the cold, and the isotonicity is reestablished by the addition of NaCl to a final concentration of 0.15 M. This minimizes contamination of the cell lysate by leukocytes, which are less sensitive to hypotonic shock than red cells. The crude lysate is centrifuged for 30 minutes at 2° at 30,000 g in a Sorvall centrifuge to pellet the stroma. The supernatant is carefully removed and layered on a 5-ml cushion of 25% (w/v) sucrose in hypotonic solution. The lysate is

[13] A. M. Crestfield, K. Smith, and F. Allen, J. Biol. Chem. 216, 185 (1955).

then centrifuged for 210 minutes at 76,000 g at 2° in the No. 30 rotor of a Spinco ultracentrifuge. The pellet consists essentially of polyribosomes and monomers with relatively few ribosomal subunits. The presence of the sucrose layer minimizes the contamination of ribosomal material by hemoglobin. The ribosomal pellets are rinsed 2 to 3 times with hypotonic solution before further use or storage.

Isolation of mRNP

mRNP is generally prepared from freshly isolated polyribosomes although storage in liquid air does not appear to promote any degradation. The ribosomal pellets are gently detached from the centrifugation tubes

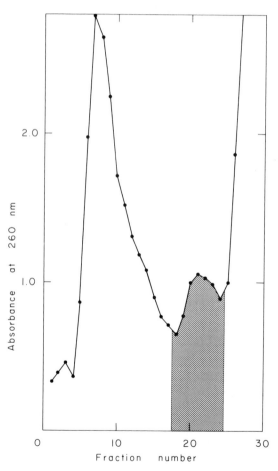

FIG. 1. Sucrose gradient centrifugation profile of EDTA dissociated polyribosomes. Centrifugation is from left to right.

with a glass rod and resuspended at an approximate concentration of 20 mg/ml in Tris·EDTA buffer, using a Potter homogenizer with a Teflon pestle. Up to 20 mg of material is layered carefully onto a 15–30% (w/v) linear sucrose gradient in 20 mM Tris·HCl (pH 7.4). Centrifugation is carried out for 42 hours at 25,000 rpm at 6° in the SW 27 rotor of a Spinco ultracentrifuge. For collection, the gradients are displaced upward with a 50% (w/v) sucrose aqueous solution through a Teflon conical cork. Around 30 fractions are collected with an Isco fraction collector. Their optical density at 260 nm is determined manually with a Zeiss spectrophotometer. This way of collecting minimizes contamination of the 14 S mRNP by ribosomal material.

A typical fractionation is illustrated in Fig. 1. Messenger RNA sediments at 14 S as a ribonucleoprotein[7,8] and is well resolved from the ribosomal subunits which are pelleted to the bottom of the centrifugation tubes. The material which sediments at the top of the gradient essentially consists of proteins (mainly hemoglobin), tRNA, and 5 S RNA. At that stage, mRNP is contaminated with 5 S rRNA which is released in the form of an 8 S nucleoprotein complex from the polyribosomes when they dissociate. The fractions corresponding to the mRNP region of the gradient (hatched zone) are pooled from several gradients. If to be used as such, the top fractions of the mRNP only are pooled in order to reduce contamination by 5 S rRNA, which essentially takes place on the light side of the peak. These fractions are dialyzed several hours against 20 mM Tris·HCl (pH 7.4) and stored in small aliquots in liquid air. This mRNA material directs the synthesis of globin chains in a cell-free system derived from Krebs ascites cells.[9]

Preparation of mRNA

Sodium dodecyl sulfate (SDS) is added to a final concentration of 1% to the pooled mRNP material. NaCl is then added to a final concentration of 0.3 M and the mRNP is precipitated with 2 volumes of ethanol overnight at −20°. The precipitate is collected by centrifugation at 1500 g at 0° in conical tubes for 30 minutes. The pellet is resuspended in 1% SDS, layered onto 5–20% (w/v) linear sucrose gradients in NETS buffer and centrifuged for 20 hours at 12° at 26,000 rpm in the SW 27 rotor of a Spinco ultracentrifuge. (Occasionally there is an SDS precipitate at the bottom of the gradient which does not affect the separation.) The gradients are collected as described above.

Two main components are separated which correspond to 9 S mRNA and 5 S rRNA[7] (Fig. 2). The material which appears as a shoulder on the heavy side of the 9 S peak mainly consists of 12 S RNA, function and

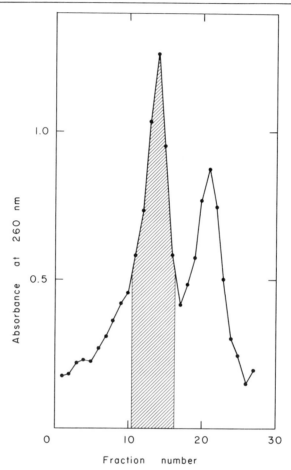

FIG. 2. Sucrose gradient centrifugation profile of RNA extracted from the 14 S mRNP (corresponding to the hatched zone of Fig. 1). Centrifugation is from right to left.

origin of which is still in debate.[10,14] The hatched zone is collected and precipitated with 2 volumes of alcohol in the presence of 0.3 M NaCl. The precipitate is collected by centrifugation at 1500 g at 0° in conical tubes for 30 minutes.

Deproteinization of mRNA

This step is performed to eliminate any traces of residual proteins that may still contaminate the 9 S region of the gradient or stick to the RNA.

[14] G. Spohr and K. Scherrer, *Cell Differentiation* 1, 53 (1972).

The alcohol pellet is resuspended in SDS buffer at a concentration of 0.1–0.3 mg/ml. An equal volume of isoamyl alcohol–chloroform (1:24, v/v) is added and the mixture is shaken for 20 minutes at room temperature. This is conveniently achieved using a rotatory shaker at 250 rpm. The phases are separated by centrifugation for 10 minutes at 600 g in an International clinical centrifuge. The aqueous phase is carefully removed and stored. The organic phase is reextracted with half its volume of SDS buffer for 10 minutes as above. The aqueous phases are combined and precipitated with 2 volumes of ethanol overnight at $-20°$. The alcohol precipitate is collected by centrifugation as above and redissolved in water at a concentration of 0.3–1 mg/ml. The ethanol precipitation is repeated one more time. The alcohol precipitate is washed twice with an alcohol–water (65:35, v/v) mixture containing 0.3 M NaCl in the aqueous phase. The precipitate is dried under vacuum, resuspended in sterile water at a concentration of 0.5–1.0 mg/ml and stored in small aliquots in liquid air. It is stable for several months in those conditions.

Test of Purity

Sucrose Gradient Analysis. Globin mRNA, 1–2 μg, in 0.1 ml of NETS buffer is carefully layered on 5 to 20% (w/v) sucrose gradients in NETS buffer and centrifuged for 5 hours at 45,000 rpm at 10° in the SW 50.1 rotor of a Spinco ultracentrifuge. The gradients are displaced upward with a 50% (w/v) aqueous solution of sucrose and recorded

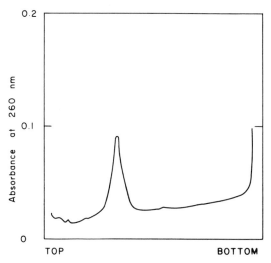

Fig. 3. Sucrose gradient centrifugation analysis of purified 9 S mRNA. Centrifugation is from left to right.

automatically at 260 nm on the expended scale of a Gilford spectro-photometer through a wide path flow cell. This material behaves as a homogeneous symmetrical peak (Fig. 3).

Polyacrylamide Gel Electrophoresis. Polyacrylamide gel electrophoresis provides a more sensitive test of purity. Purified 9 S mRNA migrates as a single homogeneous species when applied to 3.5% polyacrylamide gels prepared according to Loening.[15] It splits into two main components, however, and several contaminants of lower molecular weight on composite agarose–acrylamide gels[16] as well as on 3.5% polyacrylamide gels containing 7 M urea.[17] Mouse 9 S RNA is also resolved in several distinct bands on 6% polyacrylamide gels.[10,18] Whether or not the two main components correspond to α- and β-globin chains is still under investigation. The nature of the low molecular weight contaminants is unknown. They do not represent more than 5–10% of the total material.

Test of Activity

Various cell-free systems have been recently described in which globin mRNA can be efficiently translated such as reticulocyte lysate,[3] L cells[19] and Krebs ascites cell-free extracts.[4,11] We have routinely used a Krebs ascites cell-free system for characterizing the products made *in vitro* under the direction of globin 9 S RNA. It is derived from ascites tumor cells propagated intraperitoneally in unbred mice. The cell S30 fraction is isolated, preincubated, and passed on a Sephadex G-25 column as described by Mathews and Korner.[20]

For the amino acid incorporation assay, 0.05 ml incubation mixtures are generally used containing 0.012 ml of preincubated S30 (50 A_{260}/ml), 30 mM Tris·HCl (pH 7.6), 3.5 mM $MgCl_2$, 70 mM KCl, 7 mM β-mercaptoethanol, 1 mM ATP, 0.5 mM GTP, 5 mM creatine phosphate, 7.5 μg creatine kinase, 0.1 mM of each amino acid except the radioactive one and 0.15 μCi of ^{14}C-leucine (351 μCi/μmole).

After 30 minutes of incubation at 37°, 40-μl aliquots are spotted on Whatman No. 3 filter paper disks and processed for measurement of amino acid incorporation into protein.[21]

The ionic conditions (both KCl and $MgCl_2$ concentrations) are rather critical for optimal efficiency of translation. The conditions given here

[15] U. E. Loening, *Biochem. J.* **102**, 251 (1967).
[16] A. C. Peacock and C. W. Dingman, *Biochemistry* **7**, 668 (1968).
[17] J. Beckmann and B. Lebleu, unpublished observations.
[18] R. E. Lockard and J. B. Lingrel, *J. Biol. Chem.* **247**, 4174 (1972).
[19] E. Falcoff, B. Lebleu, R. Falcoff, and M. Revel, *J. Virol.* **12**, 421 (1973).
[20] M. B. Mathews and A. Korner, *Eur. J. Biochem.* **17**, 328 (1970).
[21] M. Revel, H. Greenshpan, and M. Herzberg, this series, Vol. 20, p. 261.

TRANSLATION OF GLOBIN mRNA IN KREBS ASCITES CELL-FREE SYSTEM[a]

mRNA	No addition	Globin mRNA, 1.5 pmoles[b]
Additions		
None	0.93[c]	3.2
Reticulocyte RWF	1.6	40.0

[a] Reaction mixtures, 0.05 ml, were incubated as described in the text; 0.025 ml of reticulocyte ribosomal wash fluid (RWF) (75 μg protein) was added when indicated.

[b] Endogenous activity has been subtracted.

[c] Results are expressed as picomoles of [14C]leucine incorporated per 0.012 ml of Krebs S30.

are only indicative as some variations have been observed from one preparation to another as well as for the translation of different mRNA species.

The sensitivity of the assay is greatly improved by the addition of a 0.5 M KCl ribosomal wash fluid[11] prepared from reticulocyte polyribosomes as described by Shafritz and Anderson[22] (see table) (3–4 globin chains are made per each globin mRNA chain provided that each of them is active). Metafora et al.[23] have recently reported similar findings.

The cell-free product is in fact globin; this has been demonstrated by several methods including chromatography on carboxymethyl cellulose columns,[24] SDS polyacrylamide gel electrophoresis,[25] and high voltage ionophoresis at pH 3.5 of the [35S]methionine-labeled tryptic peptides.[26]

[22] D. A. Shafritz and W. F. Anderson, J. Biol. Chem. 2245, 5533 (1970).

[23] S. Metafora, L. Terada, L. W. Dow, P. A. Marks, and A. Bank, Proc. Nat. Acad. Sci. U.S. 65, 1299 (1972).

[24] H. M. Dintzis, Proc. Nat. Acad. Sci. U.S. 47, 247 (1961).

[25] U. K. Laemmli and J. V. Maizel, Nature (London) New Biol. 227, 690 (1970).

[26] D. Housman, R. Pemberton, and R. Taber, Proc. Nat. Acad. Sci. U.S. 68, 2716 (1971).

[58] Preparation of Globin Messenger RNA

By ARTHUR W. NIENHUIS, AMY K. FALVEY, and W. FRENCH ANDERSON

Procedures for isolating globin messenger RNA (mRNA) from reticulocytes have been utilized in this laboratory in the course of our investigations into the mechanism of initiation of protein synthesis in mammalian cells and, specifically, in characterizing the molecular defect in globin synthesis in cells of patients with homozygous beta thalassemia and other human blood diseases.[1-3] In order to examine the role of globin mRNA during erythropoiesis and also to investigate the genetic mechanism of the anemia-induced switch in hemoglobin type in sheep and goats,[4] a procedure has been developed for preparing active globin mRNA from nucleated bone marrow cells. These methods will be described in detail.

Procedure 1: Isolation of Globin mRNA from Reticulocyte Polysomes

This procedure is most effective when the reticulocyte count is markedly elevated (greater than 30%). When using blood with a lower reticulocyte count, Procedure 2 is recommended. A convenient method for obtaining a high reticulocyte count in rabbits, sheep, or goats is by inducing anemia by means of daily injections with phenylhydrazine. Polysomes can then be obtained in high yield from the lysate of this blood; the polysomes serve as starting material for the preparation of globin mRNA. The method described is an adaption of the technique described by Lingrel and his co-workers.[5,6]

Solutions and Reagents

Rich's saline: 0.14 M NaCl, 50 mM KCl, and 5 mM MgCl$_2$

Lysing Solution A: 5 mM MgCl$_2$, 1 mM dithiothreitol, 0.1 mM EDTA (titrated to pH 7 with NaOH)

Solutions for discontinuous sucrose gradients: 24% and 60% sucrose (Mann ultrapure, ribonuclease-free) containing 20 mM Tris·HCl, pH 7.5, 50 mM KCl, and 5 mM MgCl$_2$

Polysome buffer: 20 mM Tris·HCl, pH 7.5, 10 mM KCl, 5 mM MgCl$_2$, 0.5 mM EDTA (titrated to pH 7 with NaOH)

Sodium dodecyl sulfate (SDS), 5% (w/v), (Pfaltz-Bauer)

Solutions for zonal sucrose gradients: 15% sucrose and 30% sucrose (Mann ultrapure, ribonuclease-free) containing 5 mM Tris·HCl pH 7.5

Tris·HCl, 5 mM, pH 7.5

Tris·HCl, 10 mM, pH 7.5

Sucrose, 35%

Sucrose, 75%

[1] A. W. Nienhuis, D. G. Laycock, and W. F. Anderson, *Nature (London) New Biol.* **231**, 205 (1971).

[2] A. W. Nienhuis and W. F. Anderson, *J. Clin. Invest.* **50**, 2458 (1971).

[3] A. W. Nienhuis, P. H. Canfield, and W. F. Anderson, *J. Clin. Invest.* **52**, 1735 (1973).

[4] A. W. Nienhuis and W. F. Anderson, *Proc. Nat. Acad. Sci. U.S.* **69**, 2184 (1972).

[5] M. J. Evans and J. B. Lingrel, *Biochemistry* **8**, 3000 (1969).

[6] R. E. Lockard and J. B. Lingrel, *Nature (London) New Biol.* **233**, 204 (1971).

NaCl, 4 M
Ethanol, 95%, USP grade

Preparation of Lysate

The detailed procedure for producing a phenylhydrazine-induced anemia in rabbits and for preparing a lysate of the reticulocyte-rich blood is described in the chapter by Crystal *et al.*[7] The cells from approximately 400 ml of heparinized blood (obtained from eight rabbits) are pelleted by centrifugation at 5000 g for 10 minutes at 4°. The cells are washed two times with three volumes of Rich's saline and are lysed by the addition of 2 volumes of Lysing Solution A. The lysate (approximately 240 ml) is clarified by centrifugation at 15,000 g for 20 minutes at 4°.

Isolation of Polysomes

Starting with the lysate obtained from eight rabbits, 40-ml aliquots of lysate are layered onto each of six discontinuous sucrose gradients [15 ml of 60% sucrose (lower layer) and 15 ml of 24% sucrose (upper layer), each containing 20 mM Tris·HCl, pH 7.5, 50 mM KCl, and 5 mM MgCl$_2$] in polycarbonate centrifuge tubes for the Beckman type 35 rotor. The gradients are centrifuged at 35,000 rpm for 14–16 hours at 2°. After centrifugation, the supernatant fractions are discarded; the sides of the tubes and the pellets are rinsed 3 times with 5 ml each of polysome buffer. Each pellet is then suspended in 3–5 ml of polysome buffer. This can be conveniently accomplished over approximately a 30 minute period by breaking the pellet up in small fragments with a glass stirring rod after the buffer is added and redispersing them in the solution every 4–5 minutes; or the pellets can be dissolved by homogenizing (2–3 strokes) with a very loose Teflon pestle. It is important that the pestle be loose-fitting (thus reducing the shearing stress on the polysomes). The resulting polysome suspension is clarified by centrifugation at 5000 g for 5 minutes at 2°. The pellets are discarded, and the supernatant fraction (containing the polysomes) is used to prepare globin mRNA immediately, or it can be stored in liquid nitrogen prior to subsequent use. The polysome suspension is opalescent and has a faint yellow color; the concentration in the final solution is 100–200 A_{260} units per milliliter. Total recovery from 240 ml of lysate (8 rabbits) is approximately 6000 A_{260} units.

Isolation and Purification of Globin mRNA

A 1665-ml exponential sucrose gradient (15% to 30% sucrose containing 5 mM Tris·HCl, pH 7.5) is generated at 15° in a Beckman Ti 15

[7] R. G. Crystal, N. A. Elson, and W. F. Anderson, this volume [12].

zonal rotor. Twenty milliliters of polysome solution (2500–3000 A_{260} units) is made up to final concentrations of 50–60 A_{260} units of RNA per milliliter, 5 mM Tris·HCl, pH 7.5, 5% sucrose (by addition of 75% sucrose), and 0.5% SDS (by addition of 5% SDS). After incubation at 37° for 5 minutes, the SDS-disrupted polysomes are loaded onto the gradient at room temperature, followed by 60 ml of 5 mM Tris·HCl, pH 7.5 (the latter layers easily over the sample because the polysome-SDS solution contains 5% sucrose). The gradient is centrifuged in the Ti 15 zonal rotor at 30,000 rpm at 15° for 1.5–2 hours. The temperature is then lowered to 8°, and centrifugation is continued at 30,000 rpm for 60 hours. These temperatures were chosen because they provide the best resolution of the RNA peaks and the highest yield of active mRNA. If the entire centrifugation is carried out at 2° as suggested by Lingrel,[5,6] we have found less RNA is recovered in the 9 S region of the gradient and the RNA in this region is contaminated with protein. At 15° the SDS does not precipitate, and an initial centrifugation at this temperature separates the RNA from the protein. Any SDS precipitating later during centrifugation at 8° does not appear to disturb the gradient. A 15% to 30% sucrose gradient is used because it is stable at 15° and also because it is more stable to the sedimenting SDS precipitate.

At the end of the centrifugation, the rotor is slowed to 2500 rpm without the use of the brake. The gradient is then displaced with 35% sucrose at approximately 25 ml per minute while the absorbance is continuously monitored at 260 nm. Absorbance peaks corresponding to tRNA (4 S), mRNA (9 S), and ribosomal RNA (18 S and 28 S) are readily discernible. Fractions corresponding to sedimentation values of 9–10 S are pooled, and the RNA is then concentrated by centrifugation at 35,000 rpm for 50 hours at 2° in a Beckman type rotor. After this length of centrifugation, the bottom 10 ml of the tube contains all the RNA. This fraction is made 0.25 M in NaCl by addition of 4 M NaCl; the RNA is precipitated by addition of 2.5 volumes of cold 95% ethanol. After standing for 12–18 hours at −20°, the RNA precipitate is collected by centrifugation at 15,000 g for 45 minutes at −20°. The RNA is then dissolved in approximately 1 ml of 10 mM Tris·HCl, pH 7.5, dialyzed against this buffer for 6–8 hours at 4°, and is stored in liquid nitrogen. The RNA does not lose activity after repeated freezing and thawing. On 3.5% or 5% polyacrylamide gel electrophoresis 90–95% of the RNA in this preparation runs as a single band.

The mRNA activity of these preparations is assayed in an mRNA-dependent cell-free system derived from rabbit reticulocytes and is described in detail by Crystal et al.[7] Even though RNA with activity is found not only in the 9–10 S region but also in the RNA which sediments as a light shoulder of the 18 S RNA, a second sucrose gradient fractionation indicates that it is the 9–10 S RNA present in both fractions that has the mRNA activity for globin synthesis. Ribosomal RNA freed of this 9–10 S RNA does not stimulate the mRNA-dependent cell-free system.

Procedure 2: Isolation of Globin mRNA from Reticulocyte Lysates

We have obtained highly active mRNA from the polysomes of animals with phenylhydrazine-induced reticulocytosis. However, we have had only modest success when using Procedure 1 in isolating mRNA from human or animal blood having low reticulocyte counts. Poor yield of polysomes, contamination of these polysomes with denatured hemoglobin, and loss of mRNA activity presumably due to RNase, all led us to adopt a phenol extraction procedure starting with red cell lysate.[8] Active mRNA preparations are obtained routinely from 100–150 ml of blood with a reticulocyte count of 5% to 30%.

Solutions and Reagents

Lysing Solution B: 2 mM $MgCl_2$, 1 mM dithiothreitol, 0.1 mM EDTA (titrated to pH 7.0 with NaOH)
SDS, 1%, containing 0.4 M NaCl
Tris·HCl, 0.1 M, pH 9.0
Phenol solution: 500 ml of phenol (Fisher, reagent grade), 70 ml of metacresol (reagent grade), and 0.5 g of 8-hydroxyquinolone; 80 ml of water are added to saturate the solution
Solutions for sucrose gradients: 5% and 35% sucrose (Mann ultrapure, ribonuclease-free) each containing 50 mM Tris·HCl, pH 7.5

Isolation of Total RNA from Red Blood Cells

Lysate is prepared from whole blood as described under Procedure 1 except that the red blood cells are lysed with 3–4 volumes of Lysing Solution B. The lysate is diluted 1:1 with 1% SDS containing 0.4 M NaCl. One volume of the phenol–metacresol–hydroxyquinolone solution is added, and the mixture is shaken in a mechanical shaker in 250-ml centrifuge bottles for 10 minutes at room temperature. The phenol and aqueous phases are separated by centrifugation at 6000 g for 5 minutes. The aqueous layer is then removed and the phenol and interphase layers are reextracted with an equal volume of 0.1 M Tris·HCl, pH 9.0. Again the mixture is shaken for 10 minutes at room temperature, the aqueous and phenol layers are separated by centrifugation and the upper aqueous phase is removed and combined with the original aqueous phase. The pooled aqueous phases are reextracted with an equal volume of phenol–metacresol–hydroxyquinolone

[8] J. H. Parish and K. S. Kirby, *Biochim. Biophys. Acta* **129**, 554 (1966).

as described above. The aqueous phase at this stage should be rather clear, but occasionally it contains some SDS and denatured protein, which make it somewhat cloudy. These contaminants are removed during subsequent sucrose gradient fractionation.

The RNA in the final aqueous phase is precipitated by the addition of two volumes of cold 95% ethanol, and the suspension is allowed to stand at $-20°$ for a minimum of 6 hours. The RNA precipitate is then recovered by centrifugation at 15,000 g for 45 minutes at $-20°$. The RNA precipitate generally will be found as a thin layer of very fine, flaky material on the upper side of the centrifuge bottle; occasionally the precipitate will form a tight pellet at the corner of the bottle. If the phenol extractions have not completely removed the protein, the precipitate will be larger and more compact. After the centrifugation, the ethanol is decanted, the sides of the bottle are allowed to drain, and the residual ethanol is carefully removed with a Pasteur pipette. The RNA precipitate is dissolved in a small volume (3–6 ml) of 5 mM Tris·HCl, pH 7.5. Pure RNA dissolves easily, but denatured protein is relatively insoluble. The sides of the bottle are washed carefully to obtain any RNA that may be present but not readily visible. The RNA may be frozen and stored in liquid nitrogen at this stage, or sucrose gradient fractionation of the RNA may be performed immediately.

Isolation of Globin mRNA from the Total RNA of Red Blood Cells

Sucrose gradients are prepared for the Beckman SW 41 rotor using the Technicon autoanalyzer.[9] Six identical gradients are prepared as follows: The mixing flask contains 78 ml of 5% sucrose containing 50 mM Tris·HCl, pH 7.5, and the reservoir contains 100 ml of 35% sucrose containing 50 mM Tris·HCl, pH 7.5. Six 12-ml exponential sucrose gradients are simultaneously generated. The resulting gradients are 5% to about 22% sucrose; these conditions are nearly isokinetic for the sedimentation of RNA.[9] A sample volume of 0.2–1.0 ml containing 15–20 A_{260} units of RNA can be fractionated on these gradients. (Note: the RNA at this stage of purification is contaminated with phenol which has a strong absorbance maximum at 270 nm. The contribution of phenol to the absorbance at 260 nm may be estimated by comparing the absorbance at 270 nm with that at 260 nm, or the sample can be dialyzed or passed down a small Sephadex G-25 column to remove the phenol.) The RNA sample is made 0.5% in SDS and incubated for 5 minutes at 37°. The sample is then carefully layered onto the gradients and centrifugation is performed in the Beckman SW 41 rotor at 41,000 rpm for 15 hours at 2°. After this length

---

[9] H. Noll, in "Techniques in Protein Biosynthesis" (P. N. Campbell and J. R. Sargent, eds.), Vol. 2, p. 101. Academic Press, New York, 1969.

of centrifugation, the 28 S RNA has pelleted and the 18 S RNA and 4 S RNA peaks are well separated. A distinct 9–10 S RNA peak is not always seen, therefore a fraction ranging from the 7 S region to the 14 S region is taken. The absorption profile of the gradients is continuously monitored at 260 nm, and the desired fractions are pooled. These fractions are made 0.2 M in KCl by addition of 4 M KCl. The RNA is then precipitated by addition of two volumes of cold 95% ethanol. The suspension is allowed to stand at −20° for 12–18 hours, and the RNA is recovered by centrifugation at 15,000 g for 45 minutes at −20°. At the end of the centrifugation, the ethanol is decanted, and the tubes are drained. The RNA precipitate may be nearly invisible at this stage, so care should be taken not to disturb the thin layer of precipitate on the upper side of the tubes. The RNA is dissolved in 2 mM KCl and the residual ethanol is removed by dialysis against 2 mM KCl for 6–8 hours at 4°. Alternatively, simultaneous removal of ethanol and concentration of the RNA can be accomplished by lyophilization. The RNA is stored in liquid nitrogen and is stable for at least two years.

Procedure 3: Isolation of Globin mRNA from Bone Marrow Cells

Active globin mRNA from rabbit, sheep, goat, or human bone marrow cells can be isolated by the following method. This method includes a hot phenol extraction[10,11] under conditions which are thought to aid in the recovery of poly A-containing RNA.[12] Intact nucleated cells are directly extracted with phenol. Attempts to isolate active mRNA from bone marrow cytoplasmic lysates or polysomes have been unsuccessful.

Solutions and Reagents

> Acetate Buffer 1: 50 mM sodium acetate, pH 5.0, 0.1 M NaCl, 10 mM EDTA (titrated to pH 5.0 with NaOH), 0.5% SDS
> Acetate Buffer 2: 50 mM sodium acetate, pH 5.0, 0.1 M NaCl, 10 mM EDTA (titrated to pH 5.0 with NaOH)
> Phenol (Fisher, reagent grade) saturated with Acetate Buffer 1
> Phenol (Fisher, reagent grade) saturated with Acetate Buffer 2
> Minimal Essential Eagle's Medium (MEM)[13]

Collection of Tissue

Animal bone marrow cells are recovered from the proximal long bones (femurs and humeri). The bones from 2–3 immature New Zealand white rabbits (or a single long bone from an immature sheep or goat) provide

[10] K. Scherrer and J. E. Darnell, *Biochem. Biophys. Res. Commun.* 7, 486 (1962).
[11] M. Girard, this series, Vol. 12A, p. 581.
[12] G. Brawerman, J. Mendecki, and S. Y. Lee, *Biochemistry* 11, 637 (1972).
[13] H. Eagle, *Science* 130, 432 (1959).

sufficient marrow for the isolation of mRNA. The rabbits are anesthesized by intraperitoneal injection of sodium barbital (30 mg per kilogram of body weight), and the large animals are anesthesized by intravenous injection of sodium barbital. The animals are exsanguinated by cardiac puncture for the rabbits or by means of a femoral artery catheter for sheep and goats. The proximal long bones are removed, and the marrow cavities are flushed with MEM. Under sterile precautions the marrow is dispersed by serial aspiration through Nos. 16-, 19-, and 21-gauge needles. The cells are then collected by centrifugation at 3000 g for 5 minutes at 4°, washed once with MEM, and then suspended in sufficient MEM to give a final cell concentration of 5×10^7 cells/ml.

Human bone marrow is collected from the posterior iliac spine of the patient by standard clinical techniques. Three or four aspirates are obtained, and the marrow from each aspirate is collected in a syringe containing 10 ml of MEM (with 400 units of heparin per milliliter). The aspirates are pooled and the nucleated cell count is adjusted to 5×10^7 per milliliter.

Isolation of Total RNA from Bone Marrow Cells

This description assumes an initial volume of the cell suspension of 40 ml. Any multiple of that volume can be extracted by simply increasing the phenol and buffer proportionately. The cell concentration in the original suspension should not exceed 5×10^7 cells/ml, since otherwise the concentration of protein and DNA in the phenol extract will be so great that a gel may form thus making separation of the phases very difficult.

Eighty milliliters of phenol saturated with buffer are mixed with 40 ml of Acetate Buffer 1 at room temperature. Forty milliliters of cell suspension are slowly dripped into the phenol–buffer–SDS solution with constant stirring. After 10 minutes, the mixture is transferred to a 250-ml polyethylene centrifuge tube; cooled in ice for 10 minutes, and then centrifuged at 6000 g for 5 minutes at 4° in order to separate the aqueous phase from the phenol phase. After centrifugation, the aqueous phase is removed and kept on ice; 80 ml of Acetate Buffer 2 are added to the phenol phase and interphase layer with continuous stirring. The suspension is heated to 55° in a water bath with continuous stirring and is maintained at that temperature for an additional 10 minutes. It is then cooled in ice until the temperature is less than 10°, and the aqueous phase is recovered by centrifugation as described above. This second aqueous phase is combined with the first aqueous phase.

The phenol phase and interphase layer remaining after the second extraction are again extracted with 80 ml of Acetate Buffer 2 as described above. The mixture, while being stirred, is heated to 55° and maintained for 10 minutes at this temperature; it is then transferred to ice, cooled, and centrifuged to separate the aqueous and phenol phases. The third aqueous phase is combined with the first two aqueous phases; the phenol phase is discarded.

An equal volume of phenol saturated with Acetate Buffer 2 is added to the three pooled aqueous phases and the suspension is stirred for 10 minutes at room temperature. The layers are separated by centrifugation at 8000 g for 10 minutes at 4°, and the final aqueous phase is recovered.

The RNA in the final aqueous phase is precipitated by the addition of two volumes of cold 95% ethanol. The suspension is allowed to stand at $-20°$ for a minimum of 6 hours. The RNA precipitate is recovered by centrifugation at 15,000 g for 45 minutes at $-20°$. The ethanol is decanted, the sides of the bottles are allowed to drain, and the remaining ethanol is removed with a Pasteur pipette. The precipitated RNA may be barely visible at this stage. The RNA is washed from the sides of the bottles with a small volume of 5 mM Tris·HCl, pH 7.5. The volume in which the RNA is dissolved should not exceed 6 ml since the RNA in the sample must now be fractionated on sucrose gradients and it is advisable to layer as small a sample volume as possible onto each gradient.

Isolation of Globin mRNA from the Total RNA of Bone Marrow Cells

Sucrose gradient fractionation of the total RNA is performed as described under Procedure 2. Generally, the RNA from 1×10^9 cells is fractionated on 3 to 5 12-ml gradients. We have found the highest concentration of mRNA activity in the 7 S to 14 S region of the gradients. However, there is detectable globin mRNA activity in the 28 S region as well. The significance of this observation is not clear.

Discussion

Messenger RNA prepared by each of the three procedures described above is highly active in the mRNA-dependent rabbit reticulocyte cell-free system. The protein synthetic activity of the system is linearly related to the amount of mRNA added to the system until saturation is achieved. Generally, the preparations of mRNA from reticulocytes have had somewhat greater activity than those from bone marrow cells. This difference is thought to relate to the relative amount of globin mRNA in these partially purified RNA preparations. We have found that the ratio of α- to β-globin synthesized in the cell-free system is a function of the amount of mRNA added to the system. The ratio of α-globin to β-globin synthesized in the system decreases as the amount of mRNA added is increased (Fig. 1). While Fig. 1 provides data for human reticulocyte mRNA prepared by

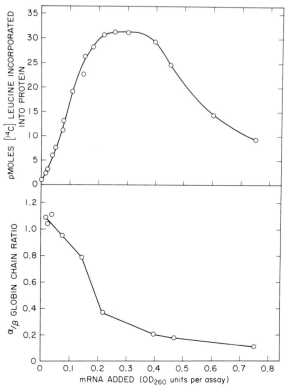

FIG. 1. Translation of normal human globin mRNA in the mRNA-dependent cell-free protein-synthesizing system derived from rabbit reticulocytes. For details, see A. W. Nienhuis, P. H. Canfield, and W. F. Anderson, *J. Clin. Invest.* **52**, 1735 (1973).

Upper panel: [^{14}C]leucine incorporation as a function of mRNA added to the assay system (see this volume [12]). Lower panel: ratio of α globin chains to β globin chains synthesized in the cell-free system as a function of mRNA added to the assay. Globin chains are fractionated by carboxymethyl-cellulose chromatography in 8 *M* urea as originally described by J. B. Clegg, M. A. Naughton, and D. J. Weatherall, *J. Mol. Biol.* **19**, 91 (1966).

Procedure 2, very similar curves have been obtained for rabbit reticulocyte mRNA prepared by Procedure 1 and human bone marrow mRNA prepared by Procedure 3. Only when the mRNA is rate-limiting in the system is there balanced synthesis of globin chains.

[59] Immunoadsorption of Ovalbumin Synthesizing Polysomes and Partial Purification of Ovalbumin Messenger RNA

By ROBERT T. SCHIMKE, RAFAEL PALACIOS, DREW SULLIVAN, MARY L. KIELY, CARMEN GONZALES, and JOHN M. TAYLOR

An understanding of the mechanisms underlying the regulation of specific protein synthesis requires the ability to isolate and quantitate the various elements that may be rate limiting in specific protein synthesis, including specific mRNA. In special instances a specific mRNA may be isolated because it constitutes such a large portion of the total mRNA of a cell type, and is of a unique size such that it can be separated from other RNA species by size; e.g., globin mRNA in red blood cells,[1-3] or by virtue of unique physical properties resulting from an unusual base composition, e.g., silk fibroin mRNA.[4] However the majority of mRNA's will require other methods for isolation. Therefore, we have developed techniques for the isolation of specific polysomes and mRNA based on the specificity of antibody reaction with nascent polypeptide chains. The methods described below were developed for the isolation of ovalbumin mRNA. We believe from our experience with other mRNA's, that these methods are generally applicable to any protein.

The methodology involves (a) the binding of anti-ovalbumin antibody to polysomes,[5] (b) binding of the polysome–antibody complex to an ovalbumin matrix, (c) washing the adsorbed polysomes to remove nonspecifically trapped polysomes, (d) release of RNA from the matrix, and (e) selective enrichment of mRNA by Millipore filtration.[6] Each step is described separately, together with the appropriate controls required to indicate the specificity of the techniques employed.

Chemicals

Ovalbumin, five times recrystallized, is obtained from Nutritional Biochemical Corporation, bovine serum albumin (BSA)[7] from Miles Labora-

[1] H. Chantrenne, A. Burny, and G. Marbaix, *Progr. Nucl. Acid Res. Mol. Biol.* **7**, 173 (1967).

[2] F. Labrie, *Nature (London)* **221**, 1217 (1969).

[3] R. E. Lockard and J. B. Lingrel, *Biochem. Biophys. Res. Commun.* **37**, 204 (1969).

[4] Y. Suzuki and D. D. Brown, *J. Mol. Biol.* **63**, 409 (1972).

[5] R. Palacios, R. D. Palmiter, and R. T. Schimke, *J. Biol. Chem.* **247**, 3216 (1972).

[6] R. Palacios, D. Sullivan, N. M. Summers, M. L. Kiely, and R. T. Schimke, *J. Biol. Chem.* **248**, 540 (1973).

tory, Inc., ribonuclease and deoxyribonuclease from Worthington, and Pronase and lactoperoxidase from Calbiochem, glutaraldehyde (Grade V) from Sigma Chemical Corporation, heparin from Fisher Scientific, Triton X-100 from J. T. Baker, and SDS from BDH Chemicals Ltd. Sucrose (ribonuclease free), CsCl, and DOC are from Schwarz/Mann, and sarkosyl from CIBA-Geigy.

Animals

White Leghorn laying hens are obtained from Kimber Farms, Fremont, California. Oviduct preparations are made within the first 3 hours after obtaining the hens. To ensure that all the hens are in an actively laying state, only those that have laid an egg or that have an egg in the oviduct the day the tissue is prepared are used. Medium-sized, male, neutered goats (*Capra hircus*) are used for anti-OV preparation.

Radioactivity Measurements

Samples are precipitated in 5% trichloroacetic acid with 400 μg of carrier BSA, collected on Whatman GF/C filters, dissolved in 0.5 ml of NCS (Amersham-Searle), and counted in 10 ml of a toluene-based scintillator fluid {4 g of 2,5-diphenyloxazole and 50 mg of 1,4-bis[2-(4-methyl-5-phenyloxazole)] in 1000 ml}. For measuring radioactivity in the pellets of sucrose gradients, the bottom of the tube is cut and treated the same as the filters.

Sterile Technique

In order to minimize the possibility of RNase contamination, all glassware is sterilized by dry heat or autoclaving, and all buffers and sucrose solutions with which polysomes are in contact are sterilized by autoclaving. All reactions with polysomes are carried out in sterile plastic tubes, and all solutions are handled with sterile plastic pipettes. Suspensions of DEAE- and CM-cellulose and Sephadex G-100 in equilibrium buffers are sterilized by autoclaving. Dialysis tubing is boiled in 5% NaHCO$_3$ containing 0.1 *M* EDTA, washed in distilled water, and autoclaved.

Oviduct Preparation

Hens are killed by decapitation, and the magnum portion of the oviduct is removed and placed in a Petri dish on ice. The oviduct is opened longi-

---

[7] Abbreviations used include: BSA, bovine serum albumin; SDS, sodium dodecyl sulfate; DOC, deoxycholate; EDTA, ethylenediaminetetraacetic acid; anti-OV, antibody against ovalbumin; anti-BSA, antibody against bovine serum albumin; anti-CON, antibody against conalbumin; OV, ovalbumin.

tudinally with scissors and scraped with a scalpel, leaving behind the serous membrane. About 15 g of tissue is obtained from each oviduct. The oviduct preparation is either used immediately or divided into portions which are stored in liquid nitrogen.

Homogenization

Homogenate is prepared with either fresh tissue or with tissue that has been frozen for up to 4 weeks. All procedures are performed at 0–4°. When frozen tissue is used, homogenization begins while the tissue is thawing. It is homogenized with 7 volumes (w/v) of polysome buffer (25 mM Tris·HCl, pH 7.6, at 4°, 25 mM NaCl, 5 mM MgCl$_2$ 0.14 M sucrose and 500 μg of sodium heparin per milliliter, using a Dounce homogenizer (Kontes Glass). Six strokes are made with a loose pestle, then 2 volumes of a freshly prepared detergent solution containing 5% sodium deoxycholate and 5% Triton X-100 is added; three more strokes with a loose pestle is followed by two with the tight pestle. The homogenate is centrifuged for 5 minutes at 27,000 g in a Sorvall refrigerated centrifuge, and the supernatant is used for polysome isolation.

Isolation of Polysomes

The polysomes used for immunoadsorption must not be degraded, aggregated or contaminated with supernatant proteins. We have therefore used a method that involves an initial centrifugation of polysomes through 1.0 M sucrose to a cushion of 2.5 M sucrose. The polysomes can be obtained free of monosomes and small polysomes in large amounts. Polysomes isolated by the "cushion" method are undegraded, as judged both on the similarity of their sedimentation profile to polysomes obtained directly from oviduct homogenates, and on their content of ovalbumin mRNA.

Aliquots of the supernatant (6 ml) are layered over a discontinuous sucrose gradient containing 2 ml of 2.5 M sucrose and 4 ml of 1.0 M sucrose (both sucrose solutions are prepared in polysome buffer containing 100 μg of sodium heparin per milliliter). The tubes are centrifuged at 4° for 90 minutes at 41,000 rpm in the Spinco SW 41 rotor. After centrifugation, an opalescent polysome band at the boundary of the two sucrose layers is extracted in about 0.7 ml of sucrose solution into a sterile syringe by puncturing the side of the tube approximately 5 mm below the band. The polysomes are dialyzed at least 6 hours against polysome buffer at 0°. After dialysis the preparation is centrifuged for 5 minutes at 27,000 g and stored at 0°. In this condition polysomes show no signs of degradation for at least 24 hours as determined by sucrose gradient sedimentation.

Polysome profiles are determined by placing 1 ml of homogenate or polysomes (approximately 10 A_{260} units) on the top of a 11.5 ml 0.5–

1.5 M sucrose gradient in polysome buffer containing 100 μg of sodium heparin per milliliter. Centrifugation is at 40,000 rpm in a Spinco SW 41 rotor for 100 minutes at 2°. Polysome profiles are analyzed with a Gilford recording spectrophotometer equipped with a continuous-flow cell.

Purification of γ-Globulin

The commercially obtained ovalbumin is purified further on a column (2.5 × 40 cm) of DEAE-cellulose using the gradient system of Mandeles.[8] The ovalbumin fractions of the eluted protein are checked by acrylamide gel electrophoresis using two different systems: Tris-glycine buffer, pH 8.9, and phosphate buffer pH 7.1 containing SDS.

Goats are immunized with the electrophoretically homogeneous ovalbumin. Five milligrams of ovalbumin, mixed with complete Freund's adjuvant, is injected intramuscularly. After 3 weeks a second injection of the same mixture is made, and 2 weeks later goats are bled by cannulation of the carotid artery. Serum is separated and precipitated twice by addition of solid ammonium sulfate to 40% saturation at 4°. The crude γ-globulin fraction obtained is dialyzed against 10 mM sodium phosphate buffer, pH 7.2 containing 15 mM NaCl, and passed through a 5-cm diameter sterile column containing 10 cm of CM-cellulose over 10 cm of DEAE-cellulose, equilibrated with the same buffer. This column removes RNase activity.

Goats are used routinely so that a large amount of the antibody can be obtained at one time. In order to make a single characterization of the antibody preparation, we collect serum from two goats (about 1.7–2.0 liters), characterize a sample of the serum, and store the remainder at −20° for subsequent workup. The antibody preparations are titered and checked for specificity by their capacity to precipitate ovalbumin from a homogenate of oviduct fragments whose proteins have been labeled by incorporation of an amino acid mixture for 30 minutes (see section on isolation of [¹⁴C]ovalbumin below).

The absence of ribonuclease activity is determined by incubating the polysomes with antibody 45–60 minutes at 4° (the temperature and time used for specific polysome isolation), and subsequently isolating the RNA and determining its capacity for directing ovalbumin synthesis.

Preparation of Ovalbumin, BSA, and Anti-OV Matrices

The proteins are covalently cross-linked with glutaraldehyde according to the method of Avrameas and Ternynck.[9] Commercial ovalbumin or BSA is dissolved in 0.1 M sodium phosphate buffer, pH 7.0, to a final concentration of 50 mg/ml. For each milliliter of protein solution, 0.1 ml of

[8] S. J. Mandeles, *J. Chromotogr.* 3, 256 (1960).
[9] S. Avrameas and T. Ternynck, *Immunochemistry* 6, 53 (1969).

12.5% glutaraldehyde is added dropwise with stirring. The mixture is allowed to gel for 4 hours at room temperature and homogenized three times in a Teflon pestle glass homogenizer in 0.2 M sodium phosphate buffer, pH 7.3, containing 0.15 M sodium chloride. When the gels are used for isolating polysomes, they are washed three times with polysome buffer (see above).

The crude γ-globulin fraction (40% ammonium sulfate fraction) is used to prepare the anti-OV matrix. To each milliliter of the γ-globulin (45 mg/ml) 0.1 ml of 1.0 M sodium phosphate buffer, pH 7.0, and 0.05 ml of 12.5% glutaraldehyde is added. The mixture is allowed to gel for 30 minutes at room temperature and is homogenized and washed as described above. All gels are stored at 4° for up to 1 week before use.

Purification of [¹⁴C]Ovalbumin with Anti-OV Matrix

A ¹⁴C-labeled oviduct supernatant is prepared according to Palmiter et al.[10] from chicks stimulated with estrogen for 5 days. Essentially this procedure involves cutting oviduct tissue into small fragments (1–2 mm in diameter) and incubating them in small plastic petri dishes in Hank's basal salt medium with ¹⁴C-labeled amino acid(s) for 5–8 hours in a 95% O_2–5% CO_2 atmosphere at 37°. The tissue is then homogenized in 10 mM sodium phosphate, pH 7.5, containing 15 mM NaCl, further disrupted by sonication, and then centrifuged at 100,000 g. Five milliliters of the labeled supernatant are incubated with 1 g (wet weight) of anti-ovalbumin matrix for 45 minutes at room temperature. The mixture is centrifuged for 10 minutes at 6000 rpm and the supernatant is discarded. The matrix is washed with 10 ml of 10 mM sodium phosphate buffer, pH 7.3, 0.15 M NaCl, centrifuged, and the supernatant discarded. The washing is repeated three more times. The absorbed protein is eluted with 0.1 M glycine·HCl buffer, pH 2.8, as described by Avrameas and Ternynck.[9] The matrix is continually mixed for 10 minutes on a magnetic stirrer at room temperature with 2 ml of the glycine buffer. The mixture is then centrifuged 10 minutes at 6000 rpm and the supernatant saved. The elution is repeated two more times, and the supernatants are pooled and dialyzed overnight against 0.15 M NaCl, 10 mM sodium phosphate buffer, pH 7.3. The material obtained migrates as a single peak of radioactivity in an SDS acrylamide gel.

Purification of Anti-OV with Ovalbumin Matrix

The anti-ovalbumin γ-globulin fraction isolated from goat serum is incubated with an ovalbumin matrix (20 ml of γ-globulin to 1 g wet weight of matrix) for 45 minutes at room temperature with continuous stirring.

---

[10] R. D. Palmiter, T. Oka, and R. T. Schimke, *J. Biol. Chem.* **246**, 724 (1971).

The mixture is centrifuged, washed four times with 0.15 M NaCl, 10 mM sodium phosphate buffer, pH 7.3, and eluted with 0.1 M glycine·HCl buffer, pH 2.8, as described above. The sample is dialyzed overnight against 0.15 M NaCl, 10 mM sodium phosphate buffer, pH 7.3, and frozen in small aliquots. The titer of the γ-globulin preparation is increased approximately 7-fold after purification. One milligram of the purified preparation typically precipitates 100 μg of ovalbumin at the equivalence point. The antibody is more than 90% pure as indicated by the amount of γ-globulin precipitated at the equivalence point.

Iodination of γ-Globulin

For every 10 ml of purified γ-globulin (8–12 mg/ml in the phosphate-NaCl buffer described above but with pH adjusted to 7.8 with NaOH), 0.5 ml of lactoperoxidase (0.8 mg/ml in the same pH 7.8 buffer) and 0.1 ml of K^{125}I (New England Nuclear, 10 mCi/ml, 2.4 μmoles/mCi) are added. The reaction is started by addition of 0.5 ml H$_2$O$_2$ (90 μM, freshly prepared) and continued for 15 minutes at room temperature with occasional shaking. The reaction mixture is then passed through a sterile 0.8-cm diameter disposable column containing 1 cm of DEAE-cellulose overlaid with 1 cm of CM-cellulose prepared as described under "Purification of γ-Globulin." This step is essential to remove the large amount of ribonuclease present in the lactoperoxidase preparation. The γ-globulin is then concentrated by addition of ammonium sulfate to 40% saturation. The resultant precipitate is collected by centrifugation and dissolved in 10 mM sodium phosphate–0.1 mM EDTA, pH 7.8. The protein (approximately 30 mg in 2 ml) is applied to a Sephadex G-100 column (2.5 × 90 cm) equilibrated at 4° with the same buffer. Radioactive fractions which constitute a single peak are pooled and frozen at −20°.

Reticulocyte Lysate Assay of Ovalbumin mRNA

Ovalbumin synthesis is measured by immunological precipitation of ovalbumin from a rabbit reticulocyte lysate protein-synthesizing system (see this volume [64]).

Isolation of Ovalbumin-Synthesizing Polysomes

Polysome preparations at a concentration of 10 A_{260} units/ml in polysome buffer (see "Homogenization") containing 100 μg of heparin per milliliter are incubated with pure anti-OV (1 mg of antibody per 20 A_{260} units of polysomes) at 4° for 45 minutes. The reaction mixture is transferred to a Corex centrifuge tube containing ovalbumin matrix (400 mg per 20 A_{260} units of polysomes) and incubated with constant stirring at 4° for 45 minutes. The preparation is centrifuged in a swinging-bucket rotor for 10 minutes at 30,000 g_{max} and the supernatant saved. The matrix is washed with 0.5 M sucrose, 0.15 M NaCl, 1% Triton X-100, 1% DOC in polysome buffer with 100 μg/ml of heparin (4 ml of buffer per 400 mg wet weight of matrix), centrifuged as before and the supernatant saved in a separate tube. This washing procedure is repeated twice more. Detergents and sucrose are removed in a final washing with polysome buffer alone. To elute the adsorbed polysomes, a buffer containing 10 mM Tris·HCl, 50 mM EDTA, pH 7.5, and 100 μg/ml of heparin is added (2 ml/400 mg of wet weight matrix) and the suspension is stirred constantly at 4° for 15 minutes. The preparation is centrifuged as before and washed one additional time. The supernatants are saved in separate tubes and precipitated with ethanol as described below.

Notes on the Immunoadsorption Techniques

Binding of Anti-OV. We have found that this reaction occurs as well at 4° as at 37°. The binding is saturable. Approximately 50% of the binding occurs at relatively low anti-OV input, but to approach saturation requires approximately 10 times as much antibody. For bulk polysome immunoadsorption necessary for mRNA isolation, we have found it impractical to obtain quantitative recovery of specific polysomes. In part this is due to the quantity of antibody required.

Washing Procedures. It is imperative to employ detergents (1% DOC and 1% Triton X-100) as well as using 0.5 M sucrose in the washes. Without their use, extensive nonspecific trapping of polysomes occurs to a degree which reduces the specificity markedly. The amount of trapped polysomes will also vary with the protein matrix used. For instance, when using BSA-matrices in control experiments, considerably more nonspecific binding of polysomes occurs than with the OV-matrix.

Elution. To obtain mRNA and ribosomal subunits, but not the nascent chains, elution only with EDTA is necessary. To obtain the nascent peptide chains, as well as the RNA, elution with 1% SDS is necessary. Treatment of the OV-matrix with SDS partially destroys its immunologic reactivity.

Extension of This Method to Other Systems

We have employed this general method successfully with hen oviduct conalbumin and rat serum albumin mRNA's. There is no reason to believe that the methodology cannot be applied to any protein or protein fragment (peptide) that is antigenic, and for which an antibody will react with nascent chains. Since our experience has been with proteins that constitute relatively large proportions of the protein synthesized by a tissue (ovalbumin, 60% conalbumin, 10%, serum albumin, 3%), the extension to proteins synthesized in a relatively small amount remains to be determined.

However, we have been successful in immunoprecipitation of ovalbumin synthesizing polysomes from a rabbit reticulocyte lysate programmed with ovalbumin mRNA, when ovalbumin constitutes only 3% of the protein synthesized. The RNA obtained from such reticulocyte polysomes, contains active ovalbumin mRNA as determined by a "second" translation in the rabbit reticulocyte lysate system.[11]

A matrix of antigen, as in the above case of ovalbumin, cannot be used with proteins difficult to isolate in large amounts. We have also employed a matrix made of rabbit antigoat γ-globulin. This latter technique should find more universal use with proteins not readily available in large amounts for use as a matrix.

Controls for the Specificity of Immunoadsorption

(a) Binding of anti-OV to specific polysomes. The specificity of binding can be demonstrated by use of anti-OV labeled with [125]I. Figure 1 shows a representative experiment in which polysomes from hen oviduct are incubated with anti-OV (panel A), showing binding in the polysome region approximating 12 ribosomes per mRNA. That this process is saturable is shown in panel B, where a large excess of unlabeled anti-OV is first added,

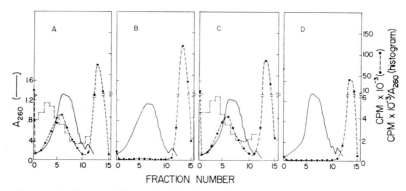

FIG. 1. Binding of [125I]anti-ovalbumin and [125I]anti-BSA to hen oviduct polysomes. Polysomes (10 A_{260} units in 1.0 ml) were incubated at 4° with (A) 30 µg of [125I]anti-ovalbumin for 30 minutes; (B) 500 µg of unlabeled anti-ovalbumin for 30 minutes followed by 30 µg of [125I]anti-ovalbumin for 30 minutes more; (C) 500 µg of unlabeled anti-BSA for 30 minutes followed by 30 pg of [125I]anti-ovalbumin for 30 minutes; and (D) 30 µg of ^{125}I labeled anti-BSA for 30 minutes. After the incubation the polysomes were layered over a continuous sucrose gradient and centrifuged. Fractions (1.0 ml) were collected to measure specific activity and radioactivity. From R. Palacios, R. D. Palmiter, and R. T. Schimke, *J. Biol. Chem.* **247**, 3216 (1972).

[11] R. Palacios and R. T. Schimke, *J. Biol. Chem.* **248**, 1424 (1973).

followed 30 minutes later by the same amount of [^{125}I]anti-OV as in panel A. Panel C shows that the addition of an amount of anti-BSA antibody comparable to that added in panel B does not block the binding of anti [^{125}I]anti-OV. Panel D shows that [^{125}I]anti-BSA does not bind to polysomes.

Figure 2 shows further specificity of the binding reactions by comparing the binding of [^{131}I]anti-OV and [^{125}I]anti-CON in the same polysome preparation. The anti-CON binds to polysomes larger than anti-OV, a finding that is consistent with molecular weights of the two proteins (conalbumin, 72,000; ovalbumin, 43,000).

Other types of controls not shown include the demonstration that anti-CON does not displace binding of [^{125}I]anti-OV (and vice versa), and that anti-OV does not bind to polysomes from tissues that are not synthesizing ovalbumin.

Figure 3 shows another test of specificity, and indicates that the

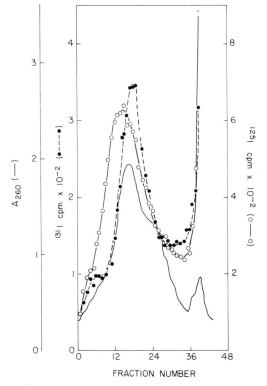

FRACTION NUMBER

FIG. 2. Binding of [^{135}I]anti-OV and [^{121}I]anti-CON to hen oviduct polysomes. The techniques are similar to those described in Fig. 1; 6.0 μg of each of the two antibodies were used in the reactions.

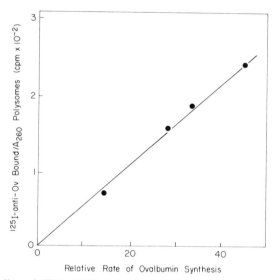

FIG. 3. Binding of [^{125}I]anti-OV to oviduct ribosomes from chicks given 1–4 days of secondary stimulation with estrogens. Chicks were treated with estrogen (2 mg per day, secondary stimulation). The relative rate of ovalbumin synthesis was measured in explants of tissue in culture and is expressed as percent of the total protein synthesis. Polysomes were prepared from each tissue, and an aliquot used to measure the binding of [^{125}I]anti-OV. From R. D. Palmiter, R. Palacios, and R. T. Schimke, *J. Biol. Chem.* **247**, 3296 (1972).

amount of binding (at saturating amounts) of anti-OV is proportional to the amount of ovalbumin synthesized by the tissue from which the polysomes are isolated. Thus the binding reaction can be used to quantitate the number of polysomes synthesizing specific protein.

Another necessary control is the demonstration that the antibody does not react with cytoplasmic protein bound nonspecifically to the polysomes during initial homogenization. To test for this possibility, hen liver homogenate is mixed with an equal volume of hen oviduct 100,000 *g* supernatant, and the polysomes are subsequently isolated and tested for binding of [^{125}I]anti-OV. No binding of anti-OV to polysomes is detected in this control experiment.

(b) *Degree of contamination with nonspecifically trapped polysomes.* The degree of contamination can be evaluated by adding to the oviduct polysomes, polysomes from a tissue that does not synthesize ovalbumin and which have been labeled *in vivo* by an appropriate precursor, e.g., uridine. This is shown in Fig. 4, where oviduct polysomes plus labeled polysomes from chick liver were combined, and the immunoadsorption was performed. This figure shows that most of the radioactivity is removed by

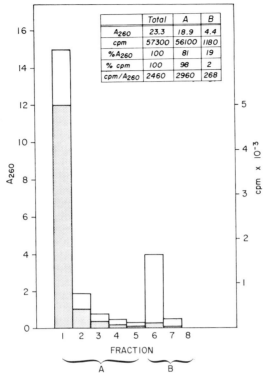

	Total	A	B
A_{260}	23.3	18.9	4.4
cpm	57300	56100	1180
%A_{260}	100	81	19
% cpm	100	98	2
cpm/A_{260}	2460	2960	268

FIG. 4. Immunoadsorption of ovalbumin synthesizing polysomes from a mixture of labeled liver and nonlabeled oviduct polysomes. A 2-week-old chick received an intraperitoneal injection of [³H]uridine (2 mCi). After 20 hours it was killed, the liver was removed, and a polysome preparation was made. The final preparation had a specific activity of 15,000 cpm/A_{260} of polysomes. Hen oviduct polysomes (20 A_{260} units), and labeled liver polysomes (4 A_{260} units, 6×10^4 cpm) were mixed and treated for isolation of ovalbumin polysomes. Fractions collected from the different steps were precipitated with ethanol, and A_{260} was measured. To measure radioactivity, an aliquot of the sample was precipitated with 5% TCA and collected on a glass fiber filter. Open bars, A_{260}; lined bars, ³H cpm. The A_{260} and radioactivity recovered from the fractions in A (washings) B (elution) or total (washing and elution) are tabulated below. From R. Palacios, D. Sullivan, N. M. Summers, M. L. Kiely, and R. T. Schimke, *J. Biol. Chem.* **248**, 540 (1973).

	Total	A	B
A_{260}	23.3	18.9	4.4
Cpm	57,300	56,100	1180
% A_{260}	100	81	19
% Cpm	100	98	2
Cpm: A_{260}	2460	2960	268

the washing procedures. Upon elution of the immunoadsorbed polysomes with EDTA, approximately 20% of the A_{260} (oviduct polysomes) is recovered with only 2% of the radioactivity (liver polysomes).

(c) Size distribution of nascent peptide chains obtained after immunoprecipitation. Figure 5 shows the size distribution of nascent chains ob-

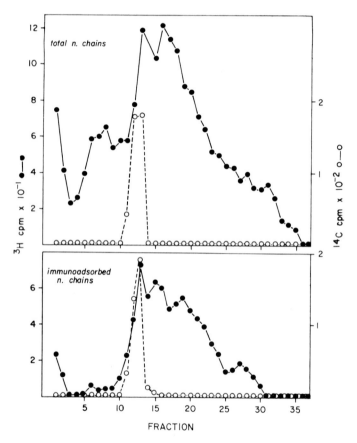

FIG. 5. SDS-acrylamide gel electrophoresis of nascent polypeptide chains from total (A) and immunoadsorbed polysomes (B). Explants of hen oviduct were incubated in Hanks medium with ^{3}H-labeled amino acids for 5 minutes. Polysomes were prepared and treated for immunoadsorption except the elution was made with 1% SDS at 20°C. Aliquots from total and eluted polysomes were mixed with [^{14}C]ovalbumin and precipitated with 5% TCA. The precipitates were dissolved in boiling SDS buffer and applied to 6 × 90 mm 10% acrylamide gels. Electrophoresis was as described by R. D. Palmiter, T. Oka, R. T. Schimke [*J. Biol. Chem.* **246**, 724 (1971)]. Gel slices were shaken overnight at 38° in 0.7 ml of NCS and counted. From R. Palacios, D. Sullivan, N. M. Summers, M. L. Kiely, and R. T. Schimke, *J. Biol. Chem.* **248**, 540 (1973).

tained from total polysomes and from immunoadsorbed polysomes. Whereas nascent chains from total polysomes are widely different in size, those obtained from the immunoadsorbed polysomes are of the size of ovalbumin and smaller.

(d) Purification of ovalbumin mRNA activity. The selective purification of ovalbumin mRNA activity is shown in Fig. 6. In this experiment polysomes from estrogen-stimulated chicks, where 17% of total protein synthesis was ovalbumin, were immunoadsorbed, and the original immunoadsorbed, and nonadsorbed polysomal RNA was assayed for ovalbumin mRNA activity. This figure shows an approximately 7-fold purification of

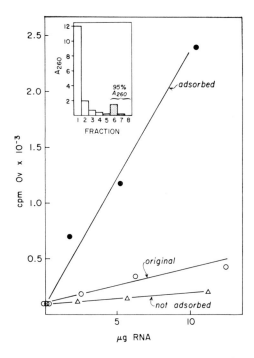

FIG. 6. *In vitro* synthesis of ovalbumin by RNA extracted from immunoadsorbed chick oviduct polysomes. Chicks received a secondary stimulation with estrogen for 18 hours [R. D. Palmiter, T. Oka, and R. T. Schimke, *J. Biol. Chem.* **246**, 724 (1971)]. Oviduct magnum was isolated and the relative rate of ovalbumin synthesis was measured by incubating explants of the oviduct in Hanks medium with ^{3}H-labeled amino acids and then determining the percentage of the total protein synthesized that was precipitated by anti-OV (Palmiter *et al. loc. cit.*). Ovalbumin was 17% of total protein synthesized. Polysomes prepared from the same oviduct (20 A_{260} units) were treated for immunoadsorption. RNA was extracted from total (○—○), nonadsorbed (△—△), and adsorbed (●—●) polysomes and assayed at different concentrations in the reticulocyte lysate system. From R. Palacios, D. Sullivan, N. M. Summers, M. L. Kiely, and R. T. Schimke, *J. Biol. Chem.* **248**, 540 (1973).

mRNA activity in the immunoadsorbed polysomes with little activity remaining in the nonadsorbed polysomes. This is the predicted degree of purification. Recovery of total mRNA activity was essentially 100%.

Ethanol Precipitation of RNA

RNA samples are made 0.1 M in NaCl and adjusted, if necessary, to neutral pH by addition of 0.2 volume of 1 M Tris·HCl, pH 7.0. Two volumes of ethanol are added and the RNA precipitated a minimum of 6 hours at $-20°$. The precipitated RNA is pelleted out of the ethanol at 14,000 g for 20 minutes at $0°$.

Deproteinization of RNA

Three methods of deproteinizing RNA from polysomes and removing SDS and heparin are used alternately or in combination. These steps are necessary to obtain RNA preparations that can be assayed for mRNA content. In particular heparin at a concentration as low as 1 μg/ml is a potent inhibitor of protein synthesis, presumably initiation. In one method, polysomes are made 50 mM in EDTA, 1% in SDS, and 0.1 M in NaCl and precipitated with 2 volumes of ethanol as described above for RNA. The precipitate is dissolved in 1 ml of 0.5% SDS–acetate-EDTA buffer (20 mM sodium acetate, 5 mM EDTA, pH 5.0). The sample is then layered on a 11.5 ml 5–20% sucrose gradient in 0.5% SDS acetate EDTA buffer and centrifuged at 40,000 rpm for 6 hours at 20° in a Spinco SW 41 rotor. The gradient is pumped through a flow cell in a Gilford recording spectrophotometer and the material in the lower part of the gradient (including all the 18 S RNA peak) is collected and precipitated with ethanol as described above. The material in the upper part of the gradient, consisting of the tRNA, SDS-treated proteins, and heparin (added as a nuclease inhibitor during polysome isolation) is discarded.

Alternatively polysomes are made 1% with Sarkosyl, NaCl, and ethanol added as indicated above, precipitated at least 6 hours at $-20°$C, and the precipitate is dissolved in 1% Sarkosyl, 10 mM EDTA, 25 mM Tris·acetate, pH 6.5. Dry heat-sterilized CsCl is then added to this solution to a density of 1.723–1.739 g/cc (about 105 g CsCl/80 ml of solution). Aliquots of the sample (8.8 ml) are dispensed into Beckman polyallomer tubes, overlayered with mineral oil, and centrifuged 60 hours at 21°, 33,000 rpm, in a Beckman type 40 rotor. Under these conditions the larger RNA species pellet while the detergent-treated proteins, any DNA present, and tRNA, which does not come to equilibrium because of its small size, remained in the CsCl solution. At the end of the centrifugation oil and water are drawn off and the RNA pellet is dissolved in 10 mM Tris·HCl, pH 7.5, and ethanol precipitated several times. Up to 10 mg of RNA can be loaded on one such gradient.

Both of these methods require considerable time with ultracentrifuges and limited capacity. An alternative method which can be used for large preparations involves the use of lithium chloride, which precipitates the RNA, whereas the protein and heparin remain in solution.

Total polysome or immunoadsorbed and eluted polysomal RNA is adjusted to 15 mM EDTA, pH 7.5, and 1% SDS (stock solution 20% w/v) and mixed until the suspension is no longer turbid. Sodium chloride is added to a final concentration of 0.1 M. The RNA is precipitated by ethanol as described above. The RNA pellet is resuspended in 10 mM EDTA, pH 7.3, an equal volume of 4.0 M LiCl added, and the solution placed at 2° for at least 12 hours. After centrifugation at 10,000 g for 15 minutes at 2°, the supernatant is again discarded. The pellet is resuspended in ice-cold 2.0 M LiCl containing 10 mM EDTA, pH 7.3. After thorough mixing, the solution is again centrifuged as above. This washing step is repeated two additional times to ensure complete removal of heparin and protein. The resulting pellet is now suspended in distilled water, and adjusted to 0.1 M NaCl with 4.0 ml stock solution, and precipitated with two volumes of ethanol. The ethanol precipitation is repeated two additional times to ensure removal of residual LiCl and SDS.

The lithium chloride technique is the method of choice for bulk isolation of polysomal RNA and gives mRNA activity as good, or better, than the two techniques outlined above. Although relatively time-consuming, the actual amount of time required for manipulations is minimal.

A final $A_{260}:A_{280}$ ratio is greater than 2.0 indicating essentially pure RNA.

Millipore Adsorption Procedure for Purification of Ovalbumin mRNA

This procedure was essentially that of Brawerman et al.[12] and is based on the existence of poly(A)-rich regions in the mRNA. A 47-mm Millipore HAWP filter is soaked overnight in KTM buffer (0.5 M KCl, 10 mM Tris·HCl, 1 mM MgCl$_2$, pH 7.6, at the temperature of RNA passage, either 0° or 20° as indicated). The filter is placed in a Millipore fritted glass collection apparatus and washed 10 times under vacuum with 10 ml of KTM buffer. The RNA sample, in KTM buffer, is allowed to flow through the filter by gravity. Up to 10 mg of RNA in 100 ml of buffer have been passed through one filter. After the sample passes through, the filter is washed 10 times with 10-ml aliquots of KTM buffer under vacuum. The procedure is done at 0° and 20° with no apparent differences in the RNA

---

[12] G. Brawerman, J. Mendecki, and S. Y. Lee, *Biochemistry* 11, 637 (1972).

fractions obtained. The filter is then removed and placed in a sterile, covered petri dish. Three milliliters of KTM buffer are added, the dish is agitated briefly, and the buffer is removed and discarded. This step is designed to remove any RNA trapped on the filter under the edges of the glass chimney. Three milliliters of low salt buffer (10 mM EDTA, 0.5% SDS, adjusted to pH 5.0 with NaOH) are added, the filter agitated for 15 minutes at 20°, and the solution is removed and saved. The low salt wash is repeated 4 times, and the combined fractions are ethanol pre-cipitated 3 times.

As shown in Fig. 7, the Millipore purification procedure results in an

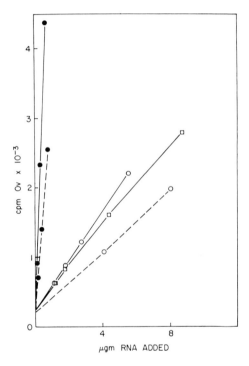

FIG. 7. Ovalbumin synthesizing activity after Millipore treatment of RNA extracted from immunoadsorbed ovalbumin polysomes. Hen oviduct polysomes (600 A_{260} units) were treated with anti-OV and OV-matrix, washed, and eluted with EDTA. Aliquots from the adsorbed and nonadsorbed polysomes were deproteinized and the RNA was fractionated by Millipore filters. Aliquots of the RNA were assayed at different concentrations in the reticulocyte lysate system. Original polysomes (□—□); immunoadsorbed polysomes before Millipore (○—○); nonimmunoad-sorbed polysomes before Millipore (○- - - -○); immunoadsorbed polysomes after Millipore (●—●); nonimmunoadsorbed polysomes after Millipore (●- - - -●). From R. Palacios, D. Sullivan, N. M. Summers, M. L. Kiely, and R. T. Schimke, *J. Biol. Chem.* **248**, 540 (1973).

approximately 25-fold purification of ovalbumin mRNA relative to starting total polysome RNA. In this experiment only 50% of the specific oval-bumin synthesizing polysomes were immunoadsorbed (see "Notes on Immunoadsorption Techniques"). Hence the difference in specific activity of the Millipore-bound RNA between immunoadsorbed polysomes and those remaining after immunoadsorption represents the difference in oval-bumin mRNA content of the two polysome fractions. Approximately 30–40% of ovalbumin mRNA activity is retained by Millipore filtration. The mRNA obtained from Millipore adsorption is not pure. On the basis of competition hybridization experiments with labeled 28 S and 18 S RNA and ribosomal DNA using the methods of Brown and Weber,[13] we estimate that approximately 25% of the Millipore-bound RNA is ovalbumin mRNA, whereas the remainder of the RNA is predominantly 28 S RNA. Aviv and Leder[14] have been successful in the use of poly(dT) cellulose columns to partially purify hemoglobin mRNA. This method is under investigation on our laboratory and may prove to be an additional step for obtaining oval-bumin mRNA, or other mRNA's containing poly(A) sequences, free of other contaminating RNA species.

Criteria for Specificity of Ovalbumin mRNA Isolation

The criteria for purity of the ovalbumin mRNA theoretically should consist of the following:

1. The presence of a single band of RNA on acrylamide gel electro-phoresis which when extracted codes for ovalbumin. We have not yet accomplished this.

2. The directing of synthesis of only ovalbumin in a protein-synthesiz-ing system that is totally dependent of exogenous mRNA. The currently available system amenable to this analysis is the Krebs ascites tumor sys-tem of Mathews and Korner.[15]

3. Hybridization kinetics. Another and equally stringent criterion of purity relates to hybridization properties of the complimentary DNA syn-thesized by RNA viral RNA-dependent DNA polymerase.[16] The Millipore-bound immunoadsorbed RNA (see Fig. 6) is capable of catalyzing the incorporation of dGTP into DNA in the presence of the other deoxy-nucleoside triphosphates, in a system that is totally dependent on addition of primer, poly(dT), RNA, and Rous sarcoma virus enzyme. Purified ribo-somal RNA does not serve as a template for this enzyme, and hence with

[13] D. D. Brown and C. S. Weber, *J. Mol. Biol.* **34**, 661 (1968).
[14] H. Aviv and P. Leder, *Proc. Nat. Acad. Sci. U.S.* **69**, 1408 (1972).
[15] M. B. Mathews and A. J. Korner, *Biochemistry* **17**, 328 (1970).
[16] A. J. Faras, J. M. Taylor, J. P. McDonnell, W. E. Levinson, and J. M. Bishop, *Biochemistry* **11**, 2334 (1972).

mRNA fractions containing ribosomal RNA contamination, an assessment of the homogeneity of the mRNA species is possible assuming that all mRNA's are transcribed equally well. Theoretically, if only a single mRNA species is present, and used as template for the enzyme, then the two complementary strands of DNA synthesized should hybridize 100% with kinetics along a CoT curve consistent with a single reacting species of complementary DNA.[17]

The DNA synthesized with the ovalbumin mRNA fraction hybridizes to an extent of approximately 85–90%, and with kinetics indicating a single reacting species of DNA.[18]

[17] R. J. Britten and D. E. Kohne, *Science* 161, 529 (1968).
[18] D. Sullivan, R. Palacios, J. Stavneger, J. M. Taylor, A. J. Faras, M. L. Kiely, M. N. Summers, J. M. Bishop, and R. T. Schimke, *J. Biol. Chem.* 248, Nov. 10 (1973).

[60] The Purification of the Messenger RNA for Silk Fibroin

By DONALD D. BROWN and YOSHIAKI SUZUKI

The messenger RNA (mRNA) which codes for the protein silk fibroin can be purified from the posterior gland of the silkworm *Bombyx mori.*[1] The purification involves three sucrose gradient sedimentation steps, which make use of the very large size of the mRNA. Identification of the mRNA is established by virtue of its unusual base composition and simple repetitive nucleotide sequence.[1] The mRNA is about 60% G + C with 40% G residues. The protein consists mainly of alternating glycine and alanine residues[2] so that the mRNA has a simple repetitious nucleotide sequence. The purification of mRNA involves two SDS-sucrose gradients in which the mRNA is the principal component sedimenting between 40 S and 60 S. Two such gradients are enough to give 40 μg of a 90% pure mRNA from a pair of posterior glands. This apparent large size is due in part to aggregation of the mRNA, probably as a result of its high G content. Traces of ribosomal RNA are present in the mRNA aggregates and are removed by a third sucrose centrifugation in formamide. Secondary interactions are eliminated in the mRNA under the denaturing conditions, and the rRNA contaminant is released. Under these conditions the mRNA sediments at about 32 S while both 28 S and 18 S rRNA sediment together at 18 S. The 28 S RNA of *B. mori* like that of other insects[3] appears to have a specific

[1] Y. Suzuki and D. D. Brown, *J. Mol. Biol.* 63, 409 (1972).
[2] F. Lucas and K. M. Rudall, *Comp. Biochem.* 26, 475 (1968).
[3] S. W. Applebaum, R. P. Epstein, and G. R. Wyatt, *J. Mol. Biol.* 21, 28 (1966).

knick so that the molecule falls into two 18 S pieces when it is denatured. The resultant mRNA preparation is pure enough to carry out specific hybridization with *B. mori* DNA under conditions of RNA excess.[4]

Raising of *Bombyx mori*

Eggs of the silkworm are commercially available throughout the year from Dahl Company (Berkeley, California) or Turtox General Biological Supply House (Chicago, Illinois). The eggs have been treated so that they will not undergo their usual diapause of 6 months. They can be stored in the refrigerator at 5°C for a few months, after which their viability begins to decrease. When placed at room temperature, they hatch in about 10 days. The worms undergo 5 larval instars in 4 weeks before spinning a cocoon and changing into a pupa. The fifth-instar lasts about 8 days. During the first 4 days DNA synthesis occurs and very little fibroin is synthesized.[5] The exact period that the mRNA is synthesized has not been determined, but it can be labeled *in vivo* with $^{32}PO_4$ or nucleoside precursors from day 2 through day 8. By day 6 to day 7, the mRNA comprises about 1% of the total RNA in the posterior gland,[1] and almost all the protein synthesized by this gland after day 5 is fibroin.

Eggs are sterilized by soaking for 10 minutes in 70% alcohol followed by 10 minutes in 3% formaldehyde. Finally the eggs are washed with 100% alcohol for 10 seconds, quickly blotted and air dried. They are incubated at room temperature in 80–90% humidity until hatching. During summer months the larvae are fed mulberry leaves. These leaves are washed after collection in a penicillin–streptomycin solution containing 100 μg/ml of each antibiotic for 20–30 minutes, and can be stored in air-tight bags at 4° for up to 3 weeks. The animals are fed daily or twice daily. A number of artificial diets have been tried for *B. mori*. We have benefited from the generosity of Dr. K. Hayashiya of the Kyoto University of Industrial Arts and Textile Fibres whose complex, artificial diet can be found described in detail.[1,6] An essential ingredient is dried mulberry leaves. About 25–30 animals at different stages of development can easily be maintained throughout the year with several kilograms of this diet. The powder is mixed with water (11 g/25 ml) and autoclaved for 5 minutes at 120°. The final paste is kept at 5° and has been used for more than 2 weeks. Animals become progressively less interested in diet which has been autoclaved for increasingly longer periods. The animals are fed fresh diet daily. Drying of the food is retarded by raising the animals in dishes within a closed plastic

[4] Y. Suzuki, L. P. Gage, and D. D. Brown, *J. Mol. Biol.* **70**, 637 (1972).
[5] Y. Tashiro, T. Morimoto, S. Matsurvia, and S. Nagata, *J. Cell Biol.* **38**, 574 (1968).
[6] F. Matsubara, Y. Ohtsuki, and K. Hayashiya, *Annu. Rep. Kyoto Univ. Arts Textile Fibres* **5**, 146 (1968). (In Japanese.)

crisper. An open beaker of water is sufficient to maintain the right humidity.

A similar diet is now commercially available from Nippon Formula Feed Manufacturing Co., Ltd., Chiyoda-ku, Tokyo, Japan.

Larvae shed their skins at each molt and careful timing is begun as they enter the fifth instar. Animals are sacrificed for mRNA isolation on day 5 or day 6 of the fifth instar. If radioactive mRNA is desired, the animal is injected with isotope 24 hours before sacrifice. As much as 2 mCi of $^{32}PO_4$ or 10 mCi of [^3H]uridine have been injected in about 0.1 ml into the coelom of a single animal. The $^{32}PO_4$ solutions are neutralized before use while tritiated nucleosides need to have the ethanol removed (*in vacuo*) before injection. A No. 26 disposable needle is introduced just under the skin about in the middle of the ventral side and slipped about 0.5 inch under the skin. The intracoelomic pressure is reduced by the loss of hemolymph through the injection hole which takes place almost immediately. Then the solution is injected. Care should be taken to avoid injecting the isotope into the large filled intestine.

The animal is killed after being immobilized in ice water. It is convenient to remove the gut by slitting the animal along its dorsal length, cutting the intestine at its posterior end and gently lifting it and teasing it free in an anterior direction. The gut can be removed usually without spilling any of its contents, which are full of nucleases, and the pair of glands are now easily visible. There are three obvious parts of the gland: the thin anterior part is cut and the large colon-shaped middle gland is lying free and can be easily identified and discarded. The middle gland narrows abruptly and the posterior gland begins. Each of the two long tortuous intestine-like posterior glands is attached to the surface laterally by filamentous trachea. The gland is dissected free along its length and can be frozen at $-70°$ for later processing or homogenized and extracted immediately.

The Extraction of RNA

Reagents

Homogenization buffer: 30 mM Tris (pH 7.4), 0.1 M NaCl, 1 mM EDTA, 10 μg/ml polyvinylsulfate (Eastman Organic Chemicals No. 8587) and 1% sodium dodecyl sulfate (SDS)

Pronase (CalBiochem): freshly dissolved stock solution of 3 mg/ml in H_2O, preincubated 1 hour at 35° just before use

Phenol, water saturated, redistilled, stored at 4°

DNase (RNase-free, Worthington): freshly made up stock solution of 200 μg/ml in 0.01 M sodium acetate pH 5.0. DNase digestion medium is 10 mM sodium acetate, 5 mM $MgCl_2$, and 10 μg/ml polyvinylsulfate

Procedure. The posterior glands are homogenized with 5–10 strokes of a loose-fitting Dounce homogenizer in about 5 ml of buffer for each pair of posterior glands. The homogenate is incubated with 300 μg/ml of preincubated Pronase at 35° for 30 minutes, then cooled to 0° and shaken vigorously with an equal volume of water-saturated phenol. The aqueous layer is saved, and the first phenol layer is reextracted with 0.5 volume of buffer. The aqueous layers are combined and reextracted with phenol, and nucleic acids are precipitated twice with 0.3 M NaCl and 2 volumes of absolute ethanol. The precipitate is dissolved in the sodium acetate buffer and incubated with 10 μg of DNase per milliliter at 0° for 30 minutes. The nucleic acids are precipitated with 0.3 M NaCl and 2 volumes of ethanol.

Purification of Fibroin mRNA from Total RNA of *B. mori*[1,4]

Reagents

Buffer for sucrose gradients: 20 mM Tris (pH 7.9), 0.1 M NaCl, 1 mM EDTA, and 0.5% SDS; 15% w/w sucrose dissolved in buffer; 30% w/w sucrose dissolved in buffer; 50% w/w sucrose dissolved in buffer for the gradient cushion.

Buffer for formamide sucrose gradients: 3 mM Tris, 3 mM EDTA, 70% formamide, 4.6% and 22% w/w sucrose dissolved in formamide buffer.

Procedure. The purified RNA precipitate is dissolved in the buffer for the first sucrose gradient. About 2–4 mg/ml RNA concentration is satisfactory. A pair of posterior glands contain about 4 mg of total RNA; about 1% of this RNA is fibroin mRNA. The RNA purified from the posterior glands of one animal can be centrifuged in two sucrose gradients of 38.5 ml in the large tubes of the Spinco SW 27 rotor. The gradients are 15 to 30% sucrose in the same SDS buffer in which the RNA is dissolved. The gradient is prepared over a 1.5 ml 50% sucrose cushion. The first sucrose gradient (Fig. 1) is centrifuged at 25° for 13 to 16 hours at 20,000 rpm. The gradients are fractionated conveniently by pumping out the contents through a Gilford recording spectrophotometer. The absorbance at 260 nm is monitored and the fractions are collected automatically. Sometimes a broad peak of radioactivity and/or absorbance can be seen between 40 S to 60 S. This region of the gradient is pooled and the RNA is precipitated with 0.3 M NaCl and 2 volumes of ethanol. The pooled RNA from six of these gradients can be rerun in a single sucrose gradient.

The second gradient is a 17-ml linear gradient of 15 to 30% sucrose in SDS-buffer prepared over a 0.5 ml cushion of 50% sucrose. The RNA

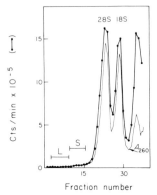

FIG. 1. First-round SDS-sucrose gradient of [³²P]RNA from the posterior silk gland.

is centrifuged for the same time as the first run and fractionated in the same manner. Once again the RNA is collected by ethanol precipitation. At this point the mRNA is 90% pure (Fig. 2).

The RNA is dissolved in the formamide buffer and centrifuged at

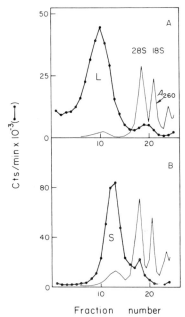

FIG. 2. Second-round SDS-sucrose gradient of [³²P]RNA fractions "L" and "S" (see Fig. 1). Nonradioactive 18 S and 28 S RNA were added as optical density markers. Absorbance in the regions of "L" and "S" is due to mRNA.

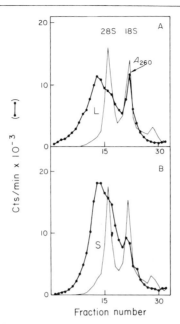

Fig. 3. Third-round formamide-sucrose gradients of [^{32}P]RNA fractions "L" and "S" (see Figs. 1 and 2). Nonradioactive 18 S and 28 S RNA purified from the frog *Xenopus laevis* were added as optical density markers. *Bombyx mori* 28 S RNA sediments at 18 S under these conditions.

25° in a 4.6 to 22% w/w sucrose gradient for 20 hours at 39,000 rpm in an SW 41 Spinco rotor. At room temperature the mRNA is denatured by the high concentration of formamide and bands at 32 S with suggestion of degradation (Fig. 3). The mRNA is precipitated with 2 volumes of ethanol and 0.2 *M* NaCl and can be redissolved in the desired buffer.

Characteristics of the mRNA

The size of fibroin mRNA has not been determined accurately. The mRNA bands broadly in the formamide gradients (Fig. 3) with a peak at 32 S. This sedimentation value is estimated for an RNA of about 2×10^6 daltons. The molecular weight of fibroin is a subject of some controversy. The latest value in the literature estimates that the principal polypeptide is 3×10^5 daltons,[7] which should be encoded by a mRNA of about 4×10^6 daltons. The broad sedimentation of the mRNA in SDS-sucrose gradients does not accurately reflect its heterogeneity because of the aggregation of the mRNA. The RNA sedimenting more rapidly in SDS-sucrose gradients (labeled "L" in Figs. 1–3) sediments at the same

[7] Y. Tashiro, E. Otsuki, and T. Shimadzu, *Biochim. Biophys. Acta* **257**, 198 (1972).

rate as the slower banding fractions (labeled "S") when the two are compared in formamide gradients (Fig. 3). The two fractions have the same base composition and oligonucleotide profiles after RNase digestion.[1]

The base composition of fibroin mRNA is about 60% G + C containing 40% G, 22% U, 19% A, and 19% C. The predominant nucleotide sequence is UCA-GG$_A$^{U}-GCU-GG$_A$^{U}-GCU-GG$_A$^{U}- which codes for the polypeptide sequence -Ser-Gly-Ala-Gly-Ala-Gly-.[1] Purity of the mRNA is estimated by its base composition,[1] its oligonucleotide profile,[1] and the specificity with which highly radioactive mRNA hybridizes with the fibroin genes.[4]

[61] DNA- and RNA-Directed Synthesis *in Vitro* of Phage Enzymes

By PETER HERRLICH and MANFRED SCHWEIGER

Due to the recent rapid development of techniques of *in vitro* protein synthesis and their extensive application to the study of regulation of protein synthesis after phage infection, it appears necessary to supplement the description of the techniques given in Volume 20 of "Methods in Enzymology."[1]

In Vitro Synthesis of Proteins

The standard technique of the "DEAE-system" was modified and adapted to suit the various purposes. To indicate the basic variations, the preparation of the standard system is described briefly, points which have been found to be of special importance are indicated, and the steps are discussed which proved to be less critical. Incubation mixtures, mixes, and buffers are described in the Appendix.

The following abbreviations are used: S100, 100,000 g supernatant (ribosome-free); S30, 30,000 g supernatant (free of debris); PEP, phosphoenolpyruvate; TCA, trichloroacetic acid; SDS, sodium dodecyl sulfate; EDTA, ethylenediaminotetraacetic acid; SAMase, S-adenosylmethionine hydrolase; OD$_x$, optical density at x nm (1 cm light pathway).

The Standard DEAE-System

All strains of *Escherichia coli* K12, that have been tested, gave active systems. The growth media are not critical. Cells from early logarithmic

---

[1] L. M. Gold and M. Schweiger, this series, Vol. 20, 537.

and late logarithmic growth phase might be used. However, for routine systems, we grow cells in a supplemented rich medium[1] and harvest at log phase by rapid cooling. The cells are washed twice. They can be disrupted by stirring with glass beads in a Sorvall Omni-Mixer for 3 minutes (6 g of cells, 6 ml of TMA buffer, 24 g of beads) or by shaking with glass beads in a Bühler homogenizer[2,3] for 2 minutes; in any case, the container must be cooled with ice water and shaking is interrupted several times to keep the batch cool. Disruption by a French press or by the alumina technique also yields extracts suitable for protein synthesis. When glass beads have been used, they are removed by filtering through a sintered glass filter or by low speed centrifugation. Cell debris is removed by spinning at 35,000 rpm (80,000 g) for 20 minutes, and ribosomes are then pelleted at 50,000 rpm (200,000 g) for 80 minutes.

Protein Fraction

The supernatant including the flocculent layer of DNA and polymerase on top of the ribosomes (critical) is adsorbed to 2 g DEAE cellulose (Serva, Heidelberg. 45051, mval/g, dry weight, which has previously been suspended in TMA buffer several times to remove superfine particles that slow down chromatography). The DEAE-cellulose slurry is washed with TMA buffer batchwise on a sintered glass filter (6–10 cm diameter) 3–4 times, then filled into a glass column (2.5 cm diameter) and proteins are eluted with TMA buffer, containing additional 0.25 M ammonium chloride.[2,3] Fractions, 1 ml, are collected, and the peak fractions are pooled (fractions which contain more than 1 OD_{280} units/ml). This protein fraction is stored, in aliquots, in liquid nitrogen.

Ribosomes

The pelleted ribosomes (from 6 g cells) are suspended in 3 ml of TMA buffer, 0.3 ml of a solution of ATP (10 mM), and 3-phosphoglycerate (100 mM) in TMA are added and incubated at 37° for 90 minutes to remove endogenous messenger RNA. The suspension is clarified by low speed centrifugation and the ribosomes pelleted at 50,000 rpm (200,000 g) for 90 minutes. The pellet is dissolved in TMA buffer at 100 mg/ml (1500 OD_{260}/ml) and stored in liquid nitrogen (in aliquots).

Conditions for Protein Synthesis

Protein synthesis is performed by mixing at 0° the following components: 5 μl of phage DNA ($>$500 μg/ml in Tris·HCl 10 mM pH 8.0),

[2] P. Traub and W. Zillig, *Hoppe-Seyler's Z. Physiol. Chem.* **343**, 246 (1966).

[3] P. Traub, W. Zillig, R. L. Millette, and M. Schweiger, *Hoppe-Seyler's Z. Physiol. Chem.* **343**, 261 (1966).

5 μl of ribosomes (100 mg/ml in TMA), 20 μl of protein fraction (7–8 mg/ml), 10 μl of protein synthesis mix, 2.5 μl of tRNA from *E. coli* K12 or *E. coli* B (10 mg/ml in H_2O, stripped, General Biochemicals, Chagrin Falls, Ohio), 2.5 μl of $MgCl_2$ solution (20 × the final concentration desired), 5 μl of H_2O. Normally we prepare a mixture of several components which is then equally distributed between several tubes. DNA, however, is always pipetted separately. The incubation temperature is 37°, and enzyme synthesis proceeds linearly until about 40 minutes. The magnesium concentration is critical, and optimal magnesium has to be determined individually for each enzyme to be synthesized. Spermidine and putrescine can, to some extent, substitute for magnesium (e.g., 8 m*M* Mg^{2+}, 2 m*M* spermidine).[4]

Variations

In the Preparation of the Ribosomes

Crude Ribosomes. Crude ribosomes are obtained by dissolving the first ribosomal pellet in TMA buffer at 100 mg/ml (1500 OD_{260}/ml). They are stored in liquid nitrogen. Crude ribosomes mediate the synthesis of about three times as much enzymes as do preincubated ribosomes under otherwise comparable conditions. Ribosomes are preincubated only when amino acid incorporation and enzyme synthesis are to be observed in parallel.

Ammonium Chloride-Washed Ribosomes and Preparation of Initiation Factors.[5] Six grams of *E. coli* K12 are disrupted as above. After removal of the debris, the ribosomes are extensively pelleted (50,000 rpm, 165,000 *g* for 5.5 hours). The upper four-fifths of the supernatant are chromatographed on DEAE-cellulose as described above. The remainder is thoroughly suspended and recentrifuged under the same conditions. The resulting supernatant is discarded, the ribosomal pellet is dissolved in 3.5 ml TMA buffer. The total is mixed with 35 ml "ribosome washing buffer" and stirred at 0° for 3 hours. The "washed" ribosomes are pelleted at 50,000 rpm (165,000 *g*) for 6–8 hours. The ribosomes are dissolved in TMA buffer at 100 mg/ml. The supernatant is brought to an ionic strength of 0.3 (dilution 1 : 3.9) and allowed to flow through 1 g of DEAE-cellulose (dry weight, presoaked in 1 : 3.9 diluted "ribosome washing buffer"). RNA and ribosomes adsorb to DEAE cellulose, initiation factors pass through. $(NH_4)_2SO_4$ is added to the flow-through volume (16.7 g/100 ml). The precipitate formed at 0° after 30 minutes is dis-

---

[4] P. Herrlich, E. Scherzinger, and M. Schweiger, *Mol. Gen. Genet.* **114**, 31 (1971).
[5] We thank C. Gualerci for advice in preparing initiation factors.

carded. Again $(NH_4)_2SO_4$ is added (26 g/100 ml). The precipitate, this time, is dissolved in 1 ml TMA buffer, dialyzed against TMA buffer and used as source of initiation factors.

For protein synthesis, the initiation factors (5 μl) are added to the other components described above (instead of the 5 μl of H_2O), "washed ribosomes" are substituted for crude ribosomes and the amount of magnesium present in the initiation factor fraction is taken into account (increases the total by 1 mM). These initiation factors stimulate "washed ribosomes" 4-fold in enzyme synthesis while crude ribosomes are not stimulated at all.

Disintegration and in Vitro Reconstitution of 30 S Ribosomes. The 30 S ribosomes can be isolated, disintegrated, and reconstituted from 16 S RNA and ribosomal proteins. These reconstituted 30 S ribosomes are active together with 50 S subunits in enzyme synthesis *in vitro.*[6]

In the Preparation of the Protein Fraction

DEAE-cellulose chromatography of the S100 protein removes most of the nucleic acids and nucleases. It may, in some cases, suffice to work with cruder systems. The content in template nucleic acids is not disturbing when phage genes are added that are not present in the system and when specific phage enzyme synthesis is observed. Nucleases reduce the yield of enzymes synthesized, but not dramatically, probably because transcription and translation proceed synchronously the RNA being protected.

S100. The S100 is used instead of DEAE-cellulose protein by simply replacing—in the protein synthesis incubation mixture—the 20-μl DEAE fraction by 20 μl of S100. Because the S100 contains mainly salt equivalent to TMA buffer, the ionic strength is increased by additional KCl or NH_4Cl.

S30. An S30 fraction is also active in enzyme synthesis without further treatment when the appropriate salt conditions are established. Since the actually needed Mg^{2+} or salt concentrations depend on the constituents (for instance PEP, total amount of RNA, ribosomes, etc.) of the protein synthesis reaction, it is advisable to optimize salt and Mg^{2+} concentration for any newly prepared system. A dialyzed S30, a method developed for synthesis *in vitro* of β-galactosidase,[7] is active in phage enzyme synthesis as well.

---

[6] E. Egberts, P. Traub, P. Herrlich, and M. Schweiger, *Biochim. Biophys. Acta* **277**, 681 (1972).

[7] C. Chambers, L. Cheong, and G. Zubay, *in* "The Lac Operon" (D. Zipser and J. Beckwith, eds.), p. 375. Cold Spring Harbor Laboratory, Cold Spring Harbor, New York, 1970.

"Brij Extracts." Even very crude extracts obtained by the Brij-lysozyme method, can be used for *in vitro* protein synthesis.[8]

Addition of Purified RNA Polymerase to the in Vitro System

E. coli RNA Polymerase. RNA polymerase may be a limiting factor in the *E. coli* system. *E. coli* RNA polymerase is purified[9] and added to the system without changing the standard conditions.

T3 or T7 Phage RNA Polymerase. T3 or T7 polymerase serve as an example where conditions have to be changed in order to meet transcription and translation requirements. The principle is as follows: *E. coli* RNA polymerase, constituent of the system, is inhibited by rifampicin while phage RNA polymerase is not. In the presence of rifampicin, only the action of phage RNA polymerase will be observed. T3 and T7 phage RNA polymerase, however, need salt and magnesium concentrations different from those optimal for *E. coli* RNA polymerase and different from those actually used in the *E. coli* protein synthesizing system.[10,11]

The components of such an *in vitro* system vary from those described as standard system: 5 μl DNA, 5 μl T3 or T7 RNA polymerase (1–2 units[12]), 10 μl protein fraction, 5 μl ribosomes, 10 μl modified protein synthesis mix (omitting K acetate), 2.5 μl tRNA, 2.5 μl MgCl$_2$ solution, 5 μl rifampicin (500 μg/ml in H$_2$O), 5 μl H$_2$O. The addition of glycerol with the polymerase (used for polymerase storage) increases the magnesium requirement. It is advisable to test the optimal magnesium concentration each time.

It should be noted that most of the *E. coli* RNA polymerase can be removed by centrifugation (S100 after prolonged centrifugation). This may be used when other polymerizing enzymes are added that cannot be distinguished by their rifampicin resistance.

Uncoupled Transcription and Translation[13]

Transcription is allowed to proceed in the absence of translation. Transcription is then interrupted by the addition of actinomycin and translation is started by the addition of some missing component or of the whole translatory machinery.

[8] M. Schweiger, P. Herrlich, E. Scherzinger, and H. J. Rahmsdorf, *Proc. Nat. Acad. Sci. U.S.* 69, 2203 (1972).

[9] R. R. Burgess, *J. Biol. Chem.* 244, 6160 (1969).

[10] M. Chamberlin, J. McGrath, and L. Waskell, *Nature (London)* 228, 227 (1970).

[11] E. Scherzinger, P. Herrlich, and M. Schweiger, *Mol. Gen. Genet.* 118, 67 (1972).

[12] W. Zillig, E. Fuchs, and R. L. Millette, *in* "Procedures in Nucleic Acid Research" (G. L. Cantoni and D. R. Davies, eds.), p. 323. Harper, New York, 1966.

[13] M. Schweiger, P. Herrlich, and R. L. Millette, *J. Biol. Chem.* 246, 6707 (1971).

We routinely presynthesize RNA in volumes of 0.025 ml containing 5 μl DNA, 5 μl *E. coli* RNA polymerase (2 units[12]), 10 μl of protein synthesis mix, 5 μl MgCl₂-solution (e.g., 5 μl 80 mM MgCl₂ = 16 mM final concentration). After 15 minutes at 37°, RNA synthesis is stopped by 20 μg/ml actinomycin D or 50 μg/ml DNase I (electrophoretically purified, Worthington). The volume is increased to 0.05 ml by addition of 25 μl of a prepared mixture of the remaining components: 18 μl of protein fraction, 5 μl of ribosomes, 1 μl of tRNA, 1 μl of magnesium chloride (no additional magnesium = 12.6 mM, 1 μl of 0.1 M MgCl₂ leads to 14.6 mM MgCl₂ and so forth). Incubation is continued for another 15 minutes. The method allows the variation of the salt and magnesium concentrations separately, in the transcriptional and the translational processes. It is also possible to study transcription with one or another component of the translational apparatus present, but without concomitant translation. Still the specific examination of the RNA products by translation into enzymes is possible.

Phage RNA-Dependent Enzyme Synthesis in Vitro

Isolation of RNA from Phage-Infected Cells.[14] Aliquots of a culture with phage-infected cells (about 5 × 10⁹ cells per aliquot) are harvested on frozen 10 mM Tris·HCl pH 7.6, 10 mM KCl. The cells are pelleted and resuspended in 0.8 ml of the same buffer. Egg-white lysozyme and EDTA are added to give 200 μg/ml and 1 mM, respectively. After 10 minutes at 4°, the mixture is treated twice by freeze-thawing. Pancreatic DNase I (electrophoretically purified, Worthington) and MgCl₂ are added at 20 μg/ml and 20 mM. A clear nonviscous lysate develops when, after 5 minutes at 20°, the mixture is treated with 1% sodium dodecyl sulfate. The lysates are brought to pH 5.2 and 0.2 M sodium acetate (by 1/10 volume of 2 M sodium acetate) and extracted twice with redistilled water-saturated phenol. To the aqueous phase (approximately 1 ml) another 0.1 ml of 2 M sodium acetate is added plus 2 volumes of ethanol. The precipitate is dried and dissolved in 0.2 ml H₂O, yielding an RNA solution of 80 OD₂₆₀ units/ml.

In Vitro Synthesized Phage Messenger RNA. A transcription mixture for *in vitro* synthesis of RNA by *E. coli* RNA polymerase consists of 50–100 μg of DNA and 20 units of *E. coli* RNA polymerase (holoenzyme) in 1 ml with 0.5 mM each of ATP, CTP, GTP, and UTP, 1 mg/ml bovine serum albumin, 1 mM dithiothreitol, 130 mM NH₄Cl, 70 mM MgCl₂, 30 mM Tris·HCl, pH 7.9, 0.1 mM EDTA. The ionic strength is

[14] E. T. Young and G. van Houwe, *J. Mol. Biol.* **51,** 605 (1970).

approximately 0.37.[11,15] The mixture for RNA synthesis by T3 or T7 RNA polymerase contains: DNA, 100 μg/ml; polymerase, 20 units/ml; each of the nucleotide triphosphates, 0.5 mM; bovine serum albumin, 1 mg/ml; dithiothreitol, 1 mM; NH$_4$Cl, 40 mM; MgCl$_2$, 15 mM; Tris·HCl, 40 mM, pH 7.9; EDTA, 0.1 mM, and rifampicin, 20 μg/ml.[11] After incubation at 37° for 30 minutes, the mixtures are made to 0.2% in sodium dodecyl sulfate, the RNA is isolated by centrifugation through CsCl,[16] and two subsequent ethanol precipitations. To ascertain complete release from the DNA·enzyme complex, an incubation with 20 μg/ml pancreatic DNase I (electrophoretically purified, Worthington) can be included before the treatment with sodium dodecyl sulfate. In this case, RNA can also be extracted by phenolization at pH 5.2 and 0.2 M sodium acetate, followed by ethanol precipitation. The precipitate is either dissolved in 50–100 μl H$_2$O and used as template RNA, or dissolved in SDS-containing buffer for further purification by sucrose gradient centrifugation.[13]

RNA-Dependent Enzyme Synthesis. The synthesis is performed in any of the described systems without changing the conditions, except that DNA is replaced by RNA. It is advisable to test the optimal magnesium concentration again. In the standard system the optimal concentration for RNA-dependent synthesis is 11–13 mM.

An Enzyme Synthesizing System from Yeast Mitochondria[17]

The isolation of a mitochondrial enzyme fraction and of mitochondrial ribosomes, and the problem of bacterial contamination have been discussed elsewhere.[17] In principle; the yeast cells (*Saccharomyces cerevisiae*) are transferred into spheroplasts by gluculase; treatment with Triton X-100 and differential centrifugation leads to a mitochondrial fraction which is passed through a French press. The mitochondrial extract is treated similarly to the *E. coli* extract. The S100 protein is adsorbed to DEAE-cellulose, eluted in one step by 0.35 M KCl, and dialyzed. Crude ribosomes are obtained by high speed centrifugation. The method yields a solution of mitochondrial protein of 20 mg/ml and mitochondrial ribosomes of 65 mg/ml. The mitochondrial fractions substitute for the corresponding *E. coli* fractions. However, only 15 μl of protein fraction is used, and, because the mitochondrial protein does not contain NH$_4$Cl or KCl, a corresponding amount of NH$_4$Cl is added.

[15] R. L. Millette and C. D. Trotter, *Proc. Nat. Acad. Sci. U.S.* **66**, 701 (1970).
[16] W. Sauerbier and A. R. Bräutegam, *Biochim. Biophys. Acta* **199**, 36 (1970).
[17] D. Richter, P. Herrlich, and M. Schweiger, *Nature (London) New Biol.* **238**, 74 (1972).

Detection of Synthesized Proteins

Enzyme Assays

The cell-free incubations for protein synthesis under the direction of T3 DNA and T7 DNA can be assayed for the following enzymes:

S-Adenosylmethionine Hydrolase.[18] The assay measures the conversion of [methyl-14C]S-adenosylmethionine to [14C]thiomethyladenosine and homoserine (unlabeled). Aliquots of the protein synthesis incubation mixture (usually less than 1 μl) are diluted with SAMase dilution buffer to give a volume of 10 μl. These are mixed with 10 μl of SAMase mix and incubated at 30° for 30 minutes. Then 5-μl aliquots are spotted on cellulose thin-layer sheets (Macherey-Nagel, Düren, Polygram Cel 300 UV$_{254}$). The spots are allowed to air-dry, and the chromatograms are developed with 64% ethanol in water. The solvent is allowed to move to the 4.0 cm mark, the sheets are dried, and cut into three parts: The front 1.5 cm contain thiomethyladenosine, the strip of 2 cm with the start (1.5 cm above, 0.5 cm below the start) contains S-adenosylmethionine. The radioactivity is determined in a scintillation spectrometer. SAMase activity is expressed in percent conversion of S-adenosylmethionine to thiomethyladenosine. Protein synthesis in the presence of T3 DNA leads to sufficient SAMase synthesis that 0.2 μl converts 60% of the original amount of S-adenosylmethionine to thiomethyladenosine. The assay responds linearly to enzyme concentrations between 0 and 90% conversion. The precision of the assay allows detection of about 0.5% conversion. For convenience we define 1 unit as the amount of enzyme converting 130 pmoles of S-adenosylmethionine to thiomethyladenosine under the assay conditions; 130 pmoles correspond to 1% conversion.

T3 and T7 RNA Polymerase.[19] The principle of the assay is the incorporation of [3H]UTP (as [3H]UMP) into TCA-precipitable material in the presence of rifampicin. Usually the extracts to be tested contain complexes of RNA polymerase and DNA in the process of elongation of RNA synthesis. Therefore, a preincubation period is needed in the presence of rifampicin. In addition, in DNA-dependent protein synthesis the salt conditions are different from those optimal for phage polymerase.[11] The ionic strength has to be lowered.

We routinely take 30-μl aliquots of the protein synthesis mixture, add 10 μl of 500 μg/ml rifampicin (in water) and preincubate at 30° for 5 minutes. Then, 10 μl DNA is added, 20 μl of polymerase mix plus the necessary amount of MgCl$_2$. The incubation volume is 0.1 ml. When a

[18] P. Herrlich and M. Schweiger, *J. Virol.* 6, 750 (1970).
[19] P. Herrlich and M. Schweiger, *Mol. Gen. Genet.* 110, 31 (1971).

DNA-dependent protein synthesis mixture is used as source of enzyme, the final concentrations of components are as follows: DNA, 50 μg/ml (T3 or, T7); rifampicin, 50 μg/ml; $MgCl_2$, 15 mM; NH_4Cl, 33 mM; K-acetate, 15 mM; Tris·HCl, 16.5 mM, pH 8.0; dithiothreitol, 1 mM; bovine serum albumin, 400 μg/ml; and, in addition to the triphosphates remaining in the aliquot of the protein synthesis mixture, 0.5 mM each of ATP, UTP, GTP, and CTP, and 80 mM phosphoenolpyruvate; ribosomes and *E. coli* proteins are present. After 30 minutes at 30°, the incubation is terminated by 1 ml of ice water; 1 ml of cold 10% TCA is added under vigorous stirring (by whirl-mixing for a few instants). The precipitate is collected and washed with cold 5% TCA by filtering through a glass filter (Whatman GF/C). The glass filters are dried, and the radioactivity is determined in a scintillation spectrometer. One unit of RNA polymerase activity is defined as the amount of enzyme incorporating 1 nmole of UMP into TCA-precipitable material within 10 minutes at 37°[12] or 20 minutes at 30°. T7 DNA-directed protein synthesis in the *E. coli* system yields usually 20 units of phage polymerase per milliliter. Assay linearity has been observed up to incorporation of 50 nmoles of UMP per milliliter, and there is probably still linearity of the assay beyond 50 nmoles.

DNA Ligase. T3 or T7 ligase catalyzes, in the presence of ATP, the formation of covalently closed circular molecules from [d(A-T)]$_n$ oligomers. The circles cannot be digested by exonuclease III while the linear molecules are digested. The circles are precipitated and represent a measure of ligase activity.

The method, as originally published,[20,21] is as follows: Protein synthesis incubation mixtures are diluted in ligase dilution buffer. Aliquots, 10 μl, are mixed with 100 μl of ligase mix and incubated at 30° for 30 minutes. The reaction is terminated by treatment at 100° for 2 minutes. The reaction mixture is made 10 mM in 2-mercaptoethanol, and exonuclease III[22] is added in an amount that would digest 99.5% of an equivalent amount of linear [d(A-T)]$_n$ oligomers under identical conditions. Incubation is continued at 37° for 45 minutes. The samples are collected on ice, the volume is increased to 0.4 ml by a solution of 10 mM Tris·HCl, pH 8.0, 10 mM EDTA, 0.1 mg/ml of calf thymus DNA; 0.4 ml of 10% TCA is added. The precipitate is collected on glass filters (Whatman GF/C), and the radioactivity is determined. The assay is linear up to 4500 cpm (about 7000 dmp). In T7 DNA-dependent

[20] P. Modrich and I. R. Lehman, *J. Biol. Chem.* **245**, 3626 (1970).
[21] E. Scherzinger, P. Herrlich, M. Schweiger, and H. Schuster, *Eur. J. Biochem.* **25**, 341 (1972).
[22] T. M. Jovin, P. T. Englund, and L. L. Bertsch, *J. Biol. Chem.* **244**, 2996 (1969).

protein synthesis, 0.01 μl usually contains sufficient ligase to produce 3000–4000 cpm circles.

SYNTHESIS OF SUBSTRATE. In principle ^{3}H-[d(A-T)]$_n$ is synthesized with *E. coli* DNA polymerase[22] and primer d(A-T). The radioactive substrate is then cut into fragments optimal for ligase dependent circle formation, by limited endonucleolytic cleavage.

Preparation of primer d(A-T): 2.5 OD$_{260}$ units d(A-T) (Miles, Kankakee, Illinois) in 0.5 ml of 10 mM MgCl$_2$, 0.1 M Tris·HCl pH 8.0, 1 mM EDTA, are incubated with 2.5 ng of DNAase I (electrophoretically purified, Worthington, DNase to be diluted in 1 μg/ml albumin) at 37° for 35 minutes. The reaction is stopped by heating to 75° for 30 minutes (I).

Synthesis of ^{3}H-[d(A-T)]$_n$: In a total volume of 40 ml, the following components are incubated at 37°.

Potassium phosphate, 60 mM, pH 7.5
MgCl$_2$, 6 mM
2-Mercaptoethanol, 2 mM
dATP, 0.5 mM
^{3}H-dTTP, 0.5 mM (100 Ci/mole)
Primer d(A-T), 0.4 ml (mixture I)
E. coli polymerase I, 1.5 ml (350 units/ml)[22]

The reaction is followed by measuring the OD at 260 nm in a 0.1 cm cuvette against a thymine solution of OD$_{260}$ = 0.8. Under these conditions, a minimum of OD$_{260}$ is reached after 4 hours. The reaction is stopped by addition of 2.34 g of NaCl and heating to 70° for 30 minutes. After extensive dialysis against 1 M NaCl, 1 mM EDTA, the ^{3}H-[d(A-T)]$_n$ solution is concentrated by a factor of 3 (e.g., rotatory evaporation) and dialyzed against 0.1 M Tris·HCl, pH 8.0, 1 mM EDTA. The yield is 16–18 ml of OD$_{260}$ 3.5–4.0 (II).

Aliquots of II are mixed with nonradioactive [d(A-T)]$_n$ and subjected to the incubation conditions I. However, 1.5 ng/ml proved to yield optimal length [d(A-T)]$_n$ substrate for T7 and T3 DNA ligase under these conditions. After heating to 75° for 30 minutes, the solution is used directly as ligase substrate.

Lysozyme.[23] The assay is based on measuring the release of radioactivity from filter paper disks that contain adsorbed *E. coli* C grown in medium with ^{3}H-diaminopimelic acid. To an exponentially growing culture (at 0.05 OD$_{600}$) of *E. coli* C in tryptone medium (1% tryptone Difco,

---

[23] M. Schweiger and L. M. Gold, *Proc. Nat. Acad. Sci. U.S.* **63**, 1351 (1969). We thank W. Leutgeb and U. Schwarz for communicating to us their original method.

0.5% NaCl), is added 10 μCi/ml ^{3}H-diaminopimelic acid (Amersham), and the culture is allowed to grow to stationary phase under vigorous agitation. To Whatman 3 MM filter disks (2.5 cm) are added 0.1-ml aliquots which are dropped into cold 10% TCA. After 4 hours, the disks are washed twice with 5% TCA and once with ethanol-ether (v/v = 1:1). The filters are then boiled in 0.1 M ammonium acetate for 10 minutes, rinsed with cold 0.1 M ammonium acetate and stored in the same buffer. Each filter contains approximately 100,000 cpm (determined in Bray's solution[24] by liquid scintillation spectrometry), of which about 50,000 cpm can be released by lysozyme. The wet filters, containing an estimated constant volume of 0.1 ml buffer, are placed in scintillation vials, 0.5 ml of 0.1 M ammonium acetate plus the enzyme sample are added, and the vials are placed in a 37° water bath. Aliquots (0.05 ml) of the supernatant fluid are removed and the radioactivity is determined with 5 ml of Bray's counting solution. The assay follows linear kinetics until a total excorporation of almost 50,000 cpm. One unit of lysozyme activity is defined as the amount of enzyme excorporating 1 cpm per minute at 37° under these conditions.[11] After T7 DNA-directed *in vitro* enzyme synthesis, the incubation mixture contains normally about 50–80 units/μl. The test has been used for detection of *in vitro* synthesized enzymes under the direction of the following DNA's: T4, T3, T7, ϕ29, SPP-1, SP82, T2, SP 5C, λ, and ϕ80.

Assays for dCMP Deaminase and α- and β-Glucosyltransferase. These have been described in a previous volume.[1]

Polyacrylamide Gel Electrophoresis

For the synthesis of radioactive protein that can be detected by polyacrylamide gel electrophoresis, the ribosomes must be preincubated. The standard system with preincubated ribosomes still incorporates amino acids without added template. But the material made is of small size. Upon addition of template, the incorporation of amino acids is stimulated by a factor of 5–10, whereas a system with ribosomes that had not been preincubated will be stimulated by a factor of only two. (These factors, however, are no measure for the specificity of synthesis. In the presence of intact template, a disproportionately high amount of "senseful" protein is synthesized.) Protein synthesis *in vitro* is allowed to proceed under the standard conditions, but with a mixture of ^{14}C-amino acids added (New England Nuclear NEC 445; 5 μCi per sample). It is desirable to apply a quantity equivalent to 300,000 to 500,000 dpm incorporated amino acids, per gel or well of the slab gel. After synthesis, the mixture

---

[24] G. Bray, *Anal. Biochem.* 1, 249 (1960).

is precipitated by 10% TCA; the precipitate is washed by acetone and dissolved in solubilizing buffer (usually 100 μl). Additional 0.1 M NaOH may be necessary to dissolve the proteins. The samples are heated to 100° for 2 minutes and divided for analysis on two concentrations of polyacrylamide (7.5 and 15% or 10 and 20%).

Electrophoresis. Gel electrophoresis is done in glass tubes of 10 cm length and 0.5–0.6 cm internal diameter.[25,26] Gels are formed from solutions containing 0.1 M sodium phosphate, pH 6.8, 0.1% SDS, 0.05% (v/v) tetramethylethylenediamine, 10% ammonium persulfate and the desired concentration of acrylamide and N,N'-bismethylene acrylamide (weight ratio 30:0.8). The ammonium persulfate solution should be freshly prepared and added as the last component. All other ingredients can be stored in solution, the combined acrylamide-N,N'-bismethylene acrylamide stock solution usually as 30%, or 0.8% respectively, in H_2O. Nine centimeters of each tube is filled with the separating gel. After polymerization, about 0.5 cm of 5% spacer gel are placed on top. The tubes are then brought into the correct position in the electrophoresis apparatus and the *in vitro* samples are layered on top of the spacer gel. Electrophoresis buffer (0.1 M sodium phosphate, pH 6.8, 0.1% SDS) is filled into the chambers and a defined voltage is maintained over several hours (for example, 10% gel, 35 V, 16 hours; 20% gel, 80 V, 16 hours; the time can, however, be decreased by electrophoresis at higher voltage). The gels are removed from the tubes by carefully introducing glycerol with a syringe and needle alongside the tube walls. The gels are sliced by some cutting device into slices of equal size. Each slice is placed in a scintillation vial containing 0.5 ml of 0.1% SDS. The vials are usually heated to 60° overnight and the solubilized radioactivity determined by the addition of 5 ml of Aquasol (New England Nuclear)[27] and counted in a scintillation spectrometer. An alternative method is drying the gel slices at 60° in a vacuum oven, adding 0.1 ml of H_2O, shaking for several hours and counting after addition of 1 ml of NCS (Amersham) and 10 ml of Toluene-scintillator fluid by scintillation spectrometry.

Slab Gel Electrophoresis. This powerful technique will be described.[28] Details can be obtained from the authors. We have adapted their technique according to our needs. In principle, gels are formed between glass plates, kept apart by plexiglass strips of 1.6 mm thickness. The upper-

[25] F. W. Studier and J. V. Maizel, Jr., *Virology* 39, 575 (1969).
[26] J. V. Maizel, Jr., *in* "Fundamental Techniques in Virology" (K. Habel and N. P. Salzman, eds.), p. 334. Academic Press, New York, 1969.
[27] T. Platt, J. H. Miller, and K. Weber, *Nature (London)* 228, 1154 (1970).
[28] J. V. Maizel, Jr. and F. W. Studier, *in* "Methods in Virology" (K. Maramarosch and H. Koprowski, eds.), Vol. 6, 1973.

most part is kept free of gel by a plexiglass blank. Spacer gel is filled into this remaining part, and a comb forces the gel to form wells into which the samples can be placed. After electrophoresis, the gels are placed on wet Whatman 3 MM filter paper on one side and a household plastic sheet on the other side. The sandwich is placed between two sheets of silicone rubber. The center of one sheet (at the filter paper side of the sandwich) carries a connection to a vacuum pump. By inserting at the same filter paper side a sheet of V2A steel wire mesh, a vacuum chamber is formed. After vacuum has been established, the whole device is heated over a steam bath. It takes 10–15 minutes to dry the gels completely. They shrink only to the extent the filter paper does. There are no distortions. The dried gels are pressed to X-ray film (Kodak RP, R54). Bands are visible after exposure for several hours (overnight). The bands are quantitatively measured by microdensitometry.

Immunotechniques

If newly synthesized proteins cannot be detected by enzyme assays, immunotechniques are useful. The first DNA-dependent *in vitro* synthesis of a specific protein was demonstrated by immunoprecipitation.[3]

Using immunotechniques in protein synthesis *in vitro,* one should be aware of the fact that the methods detect intact antigenic determinants and are therefore limited in their information concerning fidelity of *in vitro* synthesis.

Immunoprecipitation.[3] Reaction mixtures, including radioactive amino acids (as described in the paragraph on gel methods), are incubated under standard conditions for protein synthesis and then treated with pancreatic ribonuclease (100 μg), DNase I (10 μg) and carrier antigen in 2 ml of precipitation buffer at 37° for 10 minutes. The sediment is pelleted at 45,000 rpm (170,000 g) for 90 minutes. Antiserum is added to the supernatant and the mixture is kept at 0°C for 12 hours. The optimal carrier to antibody ratio has to be determined in a pilot experiment. The immunoprecipitate is collected by centrifugation at 38,000 g for 10 minutes. The pellet is washed three times with TMA buffer and, for counting of radioactivity, dissolved in 0.5 ml of soluene-100 (Packard) and toluene scintillator.

"Sandwich" Technique.[11] If carrier antigen is not available, the "sandwich"-technique should be applied. Fifteen microliters of labeled *in vitro* synthesized proteins, treated with RNase and DNase, are added to 1 ml of precipitation buffer with 2 μl of specific antiserum from rabbits or control serum from rabbits. After incubation at 30°C for 30 minutes, antirabbit serum (from sheep) is added. In a pilot experiment the amount of antirabbit serum for optimal precipitation of the rabbit serum has to

be determined. For the precipitation of the rabbit serum proteins, incubation at 30° is continued for another 30 minutes. The precipitate is kept at 4° for 2 days and then collected at 30,000 g for 15 minutes, washed three times vigorously with 0.14 M NaCl, and finally dissolved in 0.5 ml of soluene-100 and toluene scintillator for counting radioactivity. The use of polyallomer tubes and the presence of albumin is believed to reduce the unspecific precipitation.

Immunoelectrophoresis.[29,30] *In vitro* synthesized proteins are incubated with RNase and DNase (as described under immunoprecipitation). Fifteen microliters of labeled proteins with more than 200,000 dpm of ^{14}C are used for immunoelectrophoresis: 2 ml of a 1% agarose gel in barbital buffer (pH 8.2, ionic strength below 0.05) on glass slides (26 × 76 mm), 90 minutes at a voltage gradient of 6 V/cm. After electrophoresis, 50 μl of an appropriate antiserum are used to develop the immunodiffusion pattern in a moisture chamber (24 hours, room temperature). The slides are extensively washed with buffer to remove radioactive amino acids and dried for autoradiography or stained for proteins, respectively. By tracing the bands obtained by autoradiography with a microdensitometer, the method can be used as semiquantitative assay of individual proteins. It is beyond the scope of this paper to go further into the description of immunoelectrophoresis. For details it is referred to the literature.[30]

Immunodiffusion Test (Ouchterlony). More sample can be analyzed for individual antigens with less expense by the immunodiffusion test. DNase- and RNase-treated protein synthesis reaction mixtures, 2–50-μl aliquots, are developed using 50 μl of antiserum. The agarose gels are washed extensively as above, dried, and used for autoradiography or stained for proteins.

Appendix

Buffers and Mixes

TMA-buffer: Tris·HCl, 10 mM, pH 7.5
MgCl$_2$, 10 mM
NH$_4$Cl, 22 mM
Glycerol, 5% (vol)
Dithiothreitol, 1 mM
Protein synthesis mix: to be diluted into protein synthesis incubation mixtures 1:5
Tris·HCl, 0.25 M, pH 8.0
Potassium acetate, 0.25 M
20 Amino acids, 1 mM

[29] J. J. Scheidegger, *Int. Arch. Allergy Appl. Immunol.* **7**, 103 (1955).
[30] C. A. Williams, *in* "Methods in Immunology and Immunochemistry" (C. A. Williams and M. W. Chase, eds.), p. 234. Academic Press, New York, 1971.

Incubation Mixtures for Protein Synthesis (final concentrations of components)

Component[a]	Standard system	Protein synthesis in the presence of T7 RNA polymerase	Uncoupled system, 1st part	Mito-chondrial system
Tris·HCl (mM)	56	55	104	56
K-acetate (mM)	50	—	100	50
Dithiothreitol (mM)	2.5	2.5	4	2.5
20 Amino acids (mM)	0.2	0.2	0.4	0.2
CTP, UTP, GTP (mM)	0.5	0.5	1.0	0.5
ATP (mM)	2.0	2.0	4.0	2.0
Phosphoenolpyruvate (mM)	20	20	40	20
NH_4Cl (mM)	111	56.6	—	80
tRNA (μg/ml)	500	500	—	500
$MgCl_2$ (mM)	9–16	9–16	10	12
S100 protein (mg/ml)	about 4	about 2	—	3–6
Ribosomes (mg/ml)	10	10	—	6–10
Template nucleic acid (μg/ml)	50	50	100	50
Approximate ionic strength (μ)	0.270	0.165	0.274	0.242

[a] Other additions include T7 polymerase and *E. coli* polymerase.

    Dithiothreitol, 10 mM
    ATP, 10 mM
    CTP, UTP, GTP, 25 mM each
    Phosphoenolpyruvate, 0.1 M
Ribosome washing buffer: Tris·HCl, 100 mM, pH 7.8
    KCl, 50 mM
    Magnesium acetate, 10 mM
    Glycerol, 5% (vol)
    NH_4Cl, 1 M
    Dithiothreitol, 1 mM
SAMase mix: *S*-Adenosylmethionine, 1.3 mM (1.4 μCi/μmole)
    Tris·HCl, 10 mM, pH 6.3
    Dithiothreitol, 1 mM
    Bovine serum albumin, 10 μg/ml
SAMase dilution buffer: Tris·HCl, 10 mM, pH 6.3
    Dithiothreitol, 10 mM
Polymerase mix: 1:1 mixture of the following two solutions:
    I: ATP, CTP, UTP, GTP (sodium salts) 5 mM each ^{3}H-UTP, 50 μCi/ml
    Phosphoenolpyruvate (potassium salt) 0.4 M
    Tris·HCl, 20 mM, pH 8.0
    Dithiothreitol, 10 mM
    II: Bovine serum albumin, 4 mg/ml
    The polymerase mix (I + II complete) is diluted 1:5 into the final assay volume.
DNA ligase mix: Tris·HCl, 50 mM, pH 8.0

MgCl$_2$, 4 mM
EDTA, 1 mM
ATP, 0.2 mM
Dithiothreitol, 5 mM
Bovine serum albumin, 50 μg/ml
^{3}H-Labeled [d(A − T)]$_n$, 0.15 mM (nucleotide equivalents) (1850 cpm/nmole = about 2700 dpm/nmole)
DNA ligase dilution buffer: Tris·HCl, 50 mM, pH 8.0
MgCl$_2$, 4 mM
EDTA, 1 mM
Dithiothreitol, 1 mM
Bovine serum albumin, 500 μg/ml
Immunoprecipitation buffer: Tris·HCl, 50 mM, pH 8.2
Mg acetate 14 mM
KCl, 60 mM
EDTA, 0.25 M
Bovine serum albumin, 1 mg/ml
Solubilizing buffer:
SDS, 1%
2-Mercaptoethanol, 1%
Glycerol or sucrose, 10%
Tris·HCl, 0.5 M, pH 6.8

[62] Cell-Free Synthesis of Myosin

By Stuart M. Heywood and Arthur W. Rourke

Muscle myosin is composed of two identical large subunits (200,000 daltons) and three smaller subunits (ranging in size from 16,000 to 27,000 daltons).[1,2] The large molecular weight subunits, or myosin heavy chains (MHC), have been shown to be synthesized on large polysomes consisting of 55–60 ribosomes.[3] These polysomes can be isolated by sucrose density gradient centrifugation, and a preparation of messenger RNA coding for MHC can subsequently be obtained. The procedures for isolating myosin mRNA, synthesizing MHC on heterologous ribosomes, and identifying the final product are detailed below:

Solutions

Buffer A: 0.25 M KCl (NH$_4$Cl may be substituted), 10 mM MgCl$_2$, 20 mM Tris·HCl (pH 7.4)

[1] A. G. Weeds and S. Lowey, *J. Mol. Biol.* **61**, 701 (1971).
[2] J. Dow and A. Stracher, *Proc. Nat. Acad. Sci. U.S.* **68**, 1107 (1971).
[3] S. M. Heywood, R. Dowben, and A. Rich, *Proc. Nat. Acad. Sci. U.S.* **57**, 1002 (1967).

Buffer B: 5 mM EDTA (acid form), 0.5% SDS, 50 mM Tris·
acetate, pH 7.4

Buffer C: 0.15 M KCl, 5 mM MgCl$_2$, 6 mM β-mercaptoethanol,
10% glycerol, 20 mM Tris·HCl, pH 7.6

Buffer D: 5 mM MgCl$_2$, 6 mM β-mercaptoethanol, 10% glycerol,
20 mM Tris·HCl, pH 7.4

Buffer E: 1.0 M KCl, 2.5 mM MgCl$_2$, 6 mM β-mercaptoethanol,
20 mM Tris·HCl, pH 7.4

Cell-Free Amino Acid Incorporating System

In vitro amino acid incorporation is carried out in Buffer C. The
following are added to each 0.5 ml of final volume (a smaller incubation
volume may be used with corresponding adjustment in the concentration
of the constituents): (a) 1 mg of KCl-washed ribosomes (11 OD units/
mg per milliliter at 260 nm); (b) 0.2 mg of unfractionated muscle initiation factor preparation or 0.04 mg each of DEAE-cellulose fractionated
muscle initiation factors; (c) 2.5 mg of an 240,000 g muscle supernatant
protein fraction; (d) 2 μmoles of ATP, 0.5 μmoles of GTP, 7.5 μmoles
of creatine phosphate, 100 μg of creatine kinase; (e) 50 g of tRNA[4];
(f) 0.25 nmole each of 20 amino acids containing 2–3 μCi of radioactively
labeled amino acid(s); (g) myosin mRNA obtained from polysomes using
12–18 sucrose gradients (approximately 18 g of 14-day embryonic chick
muscle).

All the constituents of the reaction mixture, with the exception of
initiation factors and mRNA, are assembled at 2°. The proper volume of
this mixture is added directly to the ethanol-precipitated mRNA preparation in the cold. As soon as the RNA is in solution, the mixture is transferred to plastic incubation tubes and the initiation factors added. Care
is taken to exclude RNase activity in all procedures. The reaction mixtures
are incubated at 35° for 30–45 minutes. Short incubation times yield low
amounts of radioactivity (8 minutes is required for completion of the
HMC chain[5]), and during longer incubation times results have been
obtained that suggest that proteolysis of synthesized myosin may occur.

Special Procedures

Myosin Synthesizing Polysome Preparation. Polysomes are prepared
from 14 to 15-day embryonic chick leg muscle. For each sucrose gradient,
1 g of muscle (dissected free of bone) is homogenized at 0° in 0.5 ml
Buffer A with 3 strokes of a loose-fitting Dounce type homogenizer (clear-

---

[4] G. Von Ehrenstein, this series, Vol. 12A, 588.

[5] G. E. Morris, E. A. Buzash, A. W. Rourke, K. Tepperman, W. C. Thompson, and
S. M. Heywood, *Cold Spring Harbor Symp. Quant. Biol.* 37, 535 (1972).

ance of 0.005 inch). This allows a 70–80% recovery of ribosomes and leaves the large polysomes intact. The homogenate is then centrifuged at 10,000 g for 10 minutes to remove mitochondria, nuclei, and cell debris. The mitochondrial free supernatant (equivalent to 1 g of tissue or approximately 1 ml) is layered on a 27 ml, 15–40% (w/w) linear sucrose gradient made in buffer A. After centrifuging for 1.5 hours at 25,000 rpm (I.E.C. Rotor Type SB-110) the large myosin synthesizing polysomes may be collected directly from the bottom 4.5–5 ml of the sucrose gradients (see Heywood et al.[3] for typical 14-day muscle polysome profile).

Myosin mRNA Preparation. The large myosin synthesizing polysomes have been concentrated successfully by either of two procedures: (1) The fractions from the sucrose density gradients containing the myosin-synthesizing polysomes are pooled and the polysomes are pelleted by centrifugation at 60,000 rpm for 2 hours (I.E.C. Rotor Type A-321); (2) the myosin synthesizing polysome fractions from the sucrose gradients are filtered at 0° through SDS-washed, nitrocellulose membrane filters (0.45 μ pore) (Millipore Corp.). One filter is used for each sucrose gradient and care is taken to assure that the surface of the filter is maintained with a thin layer of buffer A. This latter procedure is rapid and yields the same amount of RNA as the centrifugation procedure.

The polysome pellets (from 1 above) are gently resuspended in 1 ml of buffer B. Alternatively, the nitrocellulose filters containing the bound polysomes are immediately immersed in 1 ml of buffer B at 4°. The filters are washed for 5 minutes and then pulled up on the side of the beaker and allowed to drain completely. Either the resuspended polysome pellets or the nitrocellulose filter wash is then layered on a 27 ml, 10–30% sucrose density gradient made in buffer B and centrifuged for 18–20 hours at 5° at 25,000 rpm (I.E.C. Rotor Type SB-110). The gradient is monitored on a continuous-flow spectrophotometer at 260 nm and the RNA sedimenting from 25 S to 27 S is collected. 20 μg of carrier tRNA (prepared from embryonic chicks) is added and the RNA is precipitated with 2 volumes of ethanol at $-20°$. The mRNA preparation may be stored as an ethanol precipitate for at least 2 weeks without loss of activity. Prior to use, the mRNA is centrifuged at 10,000 g for 10 minutes; the pellets are rinsed in cold ethanol and recentrifuged, and the tubes are allowed to drain at 0° for 30 minutes. Myosin mRNA prepared in this manner contains a considerable amount of 28 S rRNA. The activity of the mRNA does not seem to be affected by the rRNA contamination.[5] An mRNA preparation from 12 polysome gradients will normally direct the incorporation of 500–2000 dpm when a uniformly labeled amino acid mixture (New England Nuclear NEC-445) is used in the cell-free system.

S240 and Ribosome Preparation. The enzyme fraction (S240) is prepared fresh for each experiment. Approximately 8 g of 14-day embryonic chick muscle is dissected and allowed to equilibrate for 30 minutes at 2° in buffer D. The tissue is then homogenized with 15 strokes of a loose-fitting Dounce type homogenizer without additional buffer. Care is taken to avoid bubble formation. The homogenate is centrifuged at 15,000 g for 20 minutes and the subsequent supernatant is filtered through nylon cloth and centrifuged for 2 hr at 60,000 rpm (I.E.C. Rotor Type A-310). The middle one-third of the supernatant is then dialyzed against buffer C at 0°. This enzyme fraction normally contains 15–20 mg/ml of protein and contains relatively little myosin.

Salt-washed ribosomes are prepared from either embryonic muscle or adult chicken erythroblasts. Muscle ribosomes are prepared by dissecting 40 g of tissue in cold buffer C, except that the K^+ concentration is increased to 0.25 M. The tissue is homogenized with 15 strokes of a loose-fitting Dounce type homogenizer and centrifuged at 10,000 g for 20 minutes; the mitochondrial free supernatant is then centrifuged at 60,000 rpm for 2 hours. A thin buffy white layer is removed from the surface of the pellet, and the remaining ribosome pellet is used for preparing both muscle initiation factors and salt washed ribosomes.

Chicken erythroblasts are prepared from phenylhydrazine and acetyl-phenylhydrazine-treated adult chickens by the method of Laver et al.[6] This procedure produces a cell preparation consisting of 80–85% erythroblasts. After bleeding, the cells are precipitated and washed four times with 0.15 M NaCl, 10 mM KCl, 5 mM MgCl$_2$. A bluish layer of cells is removed from the surface of the cell pellet after each centrifugation. Finally the cells are lyzed in 4 volumes of cold 12 mM β-mercaptoethanol. After 1–2 minutes, salts are rapidly added to the lysate to a final concentration of 0.15 M KCl, 5 mM MgCl$_2$, and 20 mM Tris·HCl (pH 7.4). This rapid addition of salts prevents DNA leakage from the nuclei. The lysate is then centrifuged at 30,000 g for 30 minutes, and the supernatant is carefully decanted to avoid nuclear contamination. The mitochondrial supernatant is then centrifuged at 60,000 rpm for 2 hours in order to pellet the ribosomes. These ribosomes are subsequently used for preparation of salt-washed erythroblast ribosomes and erythroblast initiation factors.

Either the muscle or erythroblast ribosomes are resuspended in buffer E to give a final concentration of 15–20 mg per milliliter of ribosomes. The ribosomes are washed in the high-salt buffer at 2° for 5 minutes and subsequently centrifuged at 60,000 rpm (maximum g force of 320,000 g) for 1 hour. The total volume of the ribosome suspension is normally be-

[6] W. Laver, A. Neuberger, and S. Udenfriend, *Biochem. J.* **70**, 4 (1958).

tween 0.5 ml and 1.0 ml. The supernatant is carefully removed and used for the preparation of initiation factors (IF). The ribosomes are resuspended in 1 ml of buffer C and dialyzed against buffer C. They may then be used directly in the cell-free system or stored up to 2 weeks at $-70°$.

Initiation Factor Preparation. The 1 M KCl wash from muscle ribosomes (see above) is precipitated at 80% saturation of $(NH_4)_2SO_4$ prior to dialysis against buffer C. The precipitate which occasionally forms during dialysis against low salt buffer may be ignored. Once the factor preparation is in a low salt buffer, care is taken to avoid the use of glass, for IF2 and IF3 have a tendency to bind to glass under these conditions. Therefore, all manipulation of initiation factors under low-salt conditions is performed using plasticware.

If the KCl ribosome wash is to be fractionated into the individual factors on DEAE-cellulose, the crude factors are dialyzed against 1 mM EDTA, 50 mM KCl, 20 mM Tris·HCl (pH 7.4), 12 mM β-mercaptoethanol, and 10% glycerol. A sample, containing 8–12 mg protein is applied to a 1 × 12 cm DEAE-23 (Whatman) cellulose column equilibrated with the dialysis buffer. Initiation factor 1 (IF1) is eluted with 50 mM KCl buffer, IF3 is eluted with 27 mM KCl buffer, and IF2 is eluted with 0.35 M KCl buffer. These fractions are eluted in a stepwise manner. The DEAE-fractionated factors are precipitated at 80% saturation of $(NH_4)_2SO_4$ and dialyzed against buffer C.

Identification of Synthesized Myosin Heavy Chain

After incubation, the reaction mixture is rapidly chilled and made 0.5 M KCl and 20 mM $K_4P_2O_7$ (pH 7.5). Finally 0.5 mg of carrier myosin, prepared by the method of Patterson and Strohman (1970),[7] is added to the reaction mixture and centrifuged at 320,000 g for 15 minutes. Myosin is then precipitated by dialysis against 20 mM Tris·HCl (pH 7.4), 6 mM β-mercaptoethanol. The precipitate is solubilized in 20 mM $K_4P_2O_7$ (pH 7.5) and applied to a 0.5 × 8.0 cm DEAE-cellulose column equilibrated with the same buffer. After washing the column with 20 ml buffer, myosin is eluted with 0.35 M KCl, 20 mM $K_2P_2O_7$ (pH 7.5). The myosin-containing fraction is made 5 M guanidine hydrochloride, 4 mM N-ethylmaleimide and finally concentrated to 0.5 ml against Sephadex G-200. All of the above operations are performed at 2°.

Rapid analysis of the amount of radioactivity incorporated into myosin heavy chain can be performed by dialyzing the concentrated, denatured

[7] B. Patterson and R. C. Strohman, *Biochemistry* **9**, 4094 (1970).

myosin against 4 M urea, 6 mM β-mercaptoethanol, 0.1 M Tris, 50 mM acetic acid, 0.5% SDS, and 25% sucrose. The sample is then applied to a 5% acrylamide gel containing Tris·acetate and SDS, and electrophoresis is performed for 1.5–2 hours. The radioactivity migrating with myosin heavy chain can then be assayed by standard procedures.

To assure the fidelity of translation of myosin heavy chain in the cell-free system, peptide analysis of synthesized product is performed. The concentrated, denatured myosin (see above) is added to 3–5 mg of DEAE-Sephadex A-50 purified myosin which has been denatured in 5 M guanidine hydrochloride and 4 mM N-ethylmaleimide.[8] The sample is then applied to a 0.9 × 60.0 cm column of Sepharose 4B equilibrated with 5 M guanidine hydrochloride. The column is poured and run with a hydrostatic head pressure of 12–15 cm and washed with guanidine hydrochloride until the optical density at 280 nm is less than 0.03. Each column is used only once, and the separation of myosin heavy chain from the low molecular weight components of myosin is monitored at 280 nm (see Rourke and Heywood[8] for column elution profile and subsequent peptide fingerprint analysis). Fingerprint analysis of Sepharose 4B purified myosin heavy chains is performed by digesting the protein [equilibrated against 1% NH$_4$HCO$_3$ (pH 8.4)] at 35° with trypsin at an enzyme:substrate ratio of 1:100. Toluene is added to the reaction mixture to a final concentration of 0.1% after 3.5 hours of incubation trypsin is added again at the same ratio. Following 6 additional hours of digestion, x-chymotrypsin is added at an enzyme:substrate ratio of 1.5:100. After 8 more hours the peptide mixture is diluted 3-fold with 1% NH$_4$HCO$_3$ (pH 8.4), and insoluble material is removed by centrifugation. The supernatant is lyophilized and the peptides dissolved in H$_2$O at 60°. The peptides are subsequently analyzed either by two-dimensional electrophoresis on a 30.5 × 30.5 cm Whatman No. 3 MM paper or by elution through an ion exchange resin (Aminex A-5, Bio-Rad Laboratories) of 0.9 × 9 cm column dimensions. The electrophoretic analysis is performed for 50 minutes at 2000 V at pH 2.0 (7.5% acetic acid–2.5% formic acid) and the second dimension for 60 minutes at 2000 V at pH 6.4 (pyridine–acetic acid–water, 10:0.4:90, v/v). The peptide map is developed using ninhydrin-collidine stain. The ion exchange chromatography is performed by elution of the peptides using a 160 ml linear gradient (0.2 M pyridine acetate, pH 2.5, to 1.6 pyridine acetate, pH 5.6, at 50° under 25 psi). Either the peptide map or the fractions eluted from the ion exchange column are assayed for radioactivity using standard procedures.

[8] A. W. Rourke and S. M. Heywood, *Biochemistry* **11**, 2061 (1972).

[63] Translation of mRNA from Vertebrate Eye Lenses

By ANTON J. M. BERNS and HANS BLOEMENDAL[1]

The vertebrate eye lens belongs to the few tissues that manufacture a limited number of proteins in rather large quantities. These proteins are called crystallins. A rough classification of crystallins can be given on the basis of their molecular weights: α-crystallin, 400,000–1,000,000 daltons; β-crystallin, 50,000–200,000 daltons; and γ-crystallin, 25,000–30,000 daltons.

The wide range of high molecular weights of α-crystallin has recently been related to the existence of size classes with identical polypeptide composition.[2] For biosynthetic studies α-crystallin is the most attractive protein as its composition is quite simple and it is synthesized in considerably higher amounts than the other lens proteins.[3]

By means of 7 M urea or other detergents, α-crystallin is dissociated into the component polypeptide chains, which can be separated either by polyacrylamide gel electrophoresis at alkaline pH,[4] by isoelectric focusing,[5] or by column chromatography on DEAE-cellulose.[6]

The electrophoretic behavior, nomenclature, and a number of relevant properties of the α-crystallin chains, the knowledge of which is essential not only for the characterization of the native protein but also for the analysis of newly synthesized chains *in vitro,* are summarized in Fig. 1 and in the table.

Reagents

Medium A: Medium B + 0.35 M sucrose
Medium B: 5 mM MgCl$_2$, 25 mM KCl, 50 mM Tris·HCl, pH 7.6
Medium C: 50 mM Tris·HCl, pH 7.4, 6% sucrose

---

[1] The present investigations have been carried out in part under the auspices of the Netherlands Foundation for Chemical Research (SON) and with financial aid from the Netherlands Organization for the advancement of Pure Research (ZWO).

[2] H. Bloemendal, A. J. M. Berns, A. Zweers, H. Hoenders, and E. L. Benedetti, *Eur. J. Biochem.* **24**, 401 (1972).

[3] J. Delcour and J. Papaconstantinou, *J. Biol. Chem.* **247**, 3289 (1972).

[4] J. G. G. Schoenmakers, R. Matze, M. van Poppel, and H. Bloemendal, *Int. J. Protein Res.* **1**, 19 (1969).

[5] J. G. G. Schoenmakers and H. Bloemendal, *Nature* (*London*) **220**, 790 (1968).

[6] G. J. van Kamp, H. J. Hoenders, and H. Bloemendal, *Biochim. Biophys. Acta* **243**, 149 (1971).

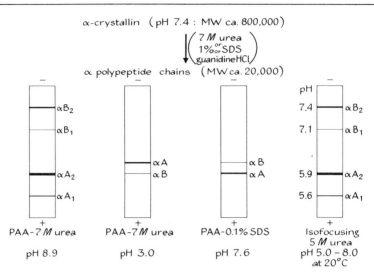

FIG. 1. Schematic representation of the electrophoretic behavior of the poly-peptide chains of the lens protein α-crystallin under different conditions (PAA = polyacrylamide).

Methods

A. Preparation of Lens Polysomes (Procedure of Bloemendal et al.[7])

This procedure is described as used routinely in our laboratory. Poly-ribosomes are prepared from lenses of 1-day-old calves. The isolated eyes may be stored on ice up to 10 hours. All further handling was performed at 0°–4°. The lenses are taken out. The lens capsules containing the epithelial cells are removed with the aid of forceps, and the outer cortex is obtained by punching out a layer 1–2 mm thick from the equator with

NH$_2$ TERMINUS OF α-CRYSTALLIN[a]

Enzyme	Amino acid sequence
Pronase	Acetyl-Met-Asp[b]
Subtilisin	Acetyl-Met-Asp-Ile-Ala[b]
Trypsin	Acetyl-Met-Asp-Ile-Ala-Ile-Gln-His-Pro-Trp-Phe-Lys[c]

[a] First column: The enzymes which release the peptides enlisted in the second column.
[b] From αA and αB chains.
[c] From αA chains.

[7] H. Bloemendal, J. Schoenmakers, A. Zweers, R. Matze, and E. L. Benedetti, Biochim. Biophys. Acta 123, 217 (1966).

the aid of a glass trephine. The isolated material is homogenized in 1 volume (v/w) of ice-cold medium A. A Teflon–glass homogenizer with a clearance of 0.15–0.2 mm is used; five strokes at 2000 rpm are applied. Nuclei, mitochondria, and cell debris are removed by centrifugation at 12,000 g for 15 minutes in a rotor GSA of the Sorvall RC2-B preparative centrifuge. To the supernatant one-tenth volume of 10% (w/v) sodium deoxycholate in 50 mM Tris·HCl buffer, pH 7.6, is added, and the solution is layered on top of a discontinuous density gradient consisting of 10 ml of 1.5 M sucrose in medium B and 10 ml of 2 M sucrose in medium B. Centrifugation is carried out in a rotor 30 of the Spinco L-50 ultracentrifuge at 78,000 g for 17 hours at 2°. For 15 ml of postmitochondrial supernatant 100–120 calf lenses are required.

B. Preparation of Lens mRNA (Procedure of Berns et al.[8])

For the isolation of RNA components the polyribosomes are pelleted at 150,000 g (average) for 2 hours at 2° in a Spinco Ti 50 rotor, suspended in medium C containing 1% sodium dodecyl sulfate (SDS) to a final concentration of about 6 mg/ml, incubated at 37° for 5 minutes and diluted twice with the same medium lacking SDS.

In order to separate the polysomal RNA components zonal centrifugation with exponential gradients is applied. These gradients are constructed by pumping in a 37% sucrose solution (w/w), boiled with 0.02% diethylpyrocarbonate for 30 minutes and containing 50 mM Tris·HCl, pH 7.4, in a mixing chamber with 300 ml of the same Tris buffer only; 50 ml of the gradient together with 100 ml of buffer is used as overlayer. The density of the sample (15–20 ml) is adjusted by addition of either sucrose or buffer in order to stabilize the sample zone between the highest density of the overlayer and the lowest density of the gradient, from which the first 10 ml is discarded. The sample and the overlayer are forced into the rotor with air pressure up to 0.2 atm. After centrifugation at 4° in a zonal rotor type B XXX of an IEC ultracentrifuge for 15 hours at 50,000 rpm the gradient is displaced from the edge of the rotor at about 10–15 ml/min by pumping in a 50% (w/w) sucrose solution. The optical density is monitored at 260 nm using a Gilford spectrophotometer equipped with a 2-mm flow cell of IEC. The eluate is fractionated by hand. The separation shown is obtained with approximately 25 mg of polysomal RNA (Fig. 2). Larger quantities can be handled in rotor type BXXIX of the IEC ultracentrifuge. The gradient for this centrifugation is as follows: 40% sucrose solution containing 50 mM Tris·HCl, pH 7.4, which has

[8] A. J. M. Berns, R. A. de Abreu, M. van Kraaikamp, E. L. Benedetti, and H. Bloemendal, *FEBS* (*Fed. Eur. Biochem. Soc.*) *Lett.* **18**, 159 (1971).

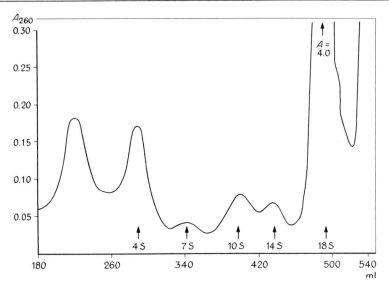

FIG. 2. Sedimentation profile of calf lens polysomal RNA. Pattern obtained after 15 hours of zonal centrifugation at 50,000 rpm in rotor BXXX of the IEC. The sample contained 25 mg of RNA. (Reproduced from *FEBS Lett.*, with permission.)

been boiled with 0.02% diethylpyrocarbonate for 30 minutes is pumped in the mixing chamber, filled with 800 ml of the same Tris buffer; 180 ml of the gradient and 300 ml of the Tris buffer is used as overlayer. The sample volume is about 40 ml. Adjustment of the density of the sample is as described above. The first 20 ml of the gradient is discarded. The air pressure is 0.4 atm. Centrifugation time is 24–30 hours at 35,000 rpm. The gradient is displaced with the aid of 50% (w/w) sucrose solution at a speed of about 20–30 per minute. A typical separation is shown in Fig. 3. Fractions of each "messenger" peak are assayed separately. The different RNA fractions are precipitated by adding 1/10 volume of 2 *M* potassium acetate at pH 5.0 and 2.5 volumes of cold ethanol. The solution is allowed to stand for 16 hours at −25°, then the precipitates are collected by centrifugation and stored under ethanol at −25°. By this procedure, 2 µg/ml RNA can be precipitated quantitatively. The approximate yield from 1000 calf lenses is 5 mg of polysomal RNA. From this, 50 µg of messenger RNA can be obtained after zonal centrifugation. The messenger preparations can further be purified on oligo-thymidylic acid-cellulose columns as described by Aviv and Leder.[9] However, this is no prerequisite for the assays described in Section E.

[9] H. Aviv and P. Leder, *Proc. Nat. Acad. Sci. U.S.* **69**, 1408 (1972).

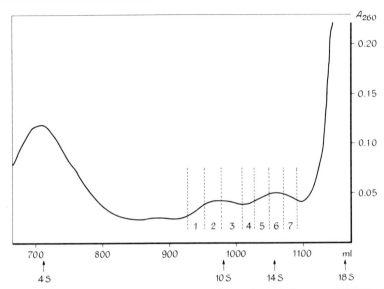

FIG. 3. Zonal centrifugation of calf lens polysomal RNA in rotor BXXIX of the IEC. Only part of the pattern obtained after 25 hours of centrifugation at 35,000 rpm is depicted. The numbers 1–7 indicate that the two messenger regions (8–15 S) have been divided into seven fractions to be tested separately in the cell-free system. The sample contained 40 mg of RNA. (Reproduced from *Eur. J. Biochem.*, with permission.)

C. Preparation of Lens RNP Particles (Procedure of Berns[10,11])

For the isolation of RNP particles the polyribosomes are pelleted as described above and suspended in medium C containing 10 mM EDTA to a final concentration of about 3 mg/ml. To separate the RNP particles obtained in this way, zonal centrifugation as described above is applied with the following modifications. The sucrose concentration is 28% (w/w). Thirty-five milliliters of the gradient and 160 ml of the buffer is used as overlayer. Centrifugation at 50,000 rpm is carried out for 2–4 hours at 4° in the BXXX rotor of the IEC centrifuge. The separation profile is depicted in Fig. 4. After the zonal run the mRNP region (compare Fig. 4a–c) is divided into three fractions. These fractions are run in rotor SB 283 of the IEC ultracentrifuge for 15 hours at 40,000 rpm in an exponential sucrose gradient (10–26%). The profile of the individual RNP fractions is shown in Fig. 5a,b. RNA can be extracted from the two RNP fractions a and b with SDS as described for the extraction of the polyribo-

[10] A. J. M. Berns, Thesis, Nijmegen, 1972.

[11] H. Bloemendal, A. Berns, G. Strous, M. Mathews, and C. D. Lane, *Proc. 8th FEBS Meeting Amsterdam*. North-Holland Publ., Amsterdam, **27**, 237, 1972.

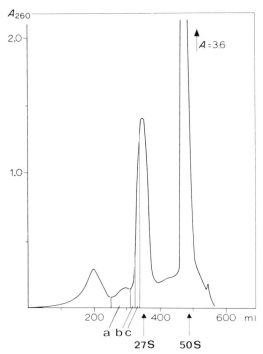

Fig. 4. Zonal centrifugation of calf lens RNP particles. The pattern is obtained after 135 minutes of spinning at 50,000 rpm in rotor BXXX of the IEC. The mRNP region is divided into three fractions as indicated.

somes. The sedimentation profile of the RNA fractions is shown in Fig. 5a′,b′. For comparison a similar experiment has been carried out with globin mRNP (g and g′).

D. Assay Systems

1. Reticulocyte Lysate (Procedure of Lockard and Lingrel[12]). Rabbits weighing 2–3 kg are made anemic by daily injection with 1 ml of 2.5% neutralized phenylhydrazine solution for 6 days. Routinely the phenyl-hydrazine solution is stored at −25° in the presence of 1 mM glutathione. On day 8, 1 ml of heparin (5000 IU) and 0.8 ml of 10% Nembutal are injected. The animals are bled by heart puncture. The blood, about 70 ml per rabbit, is centrifuged at 2500 g for 10 minutes. The cells are washed three times with a medium containing 0.13 M NaCl, 5 mM KCl, and 7.4 mM magnesium acetate. To the washed cells one volume of cold distilled water is added and the lysed cells are centrifuged at 30,000 g

[12] R. E. Lockard, and J. B. Lingrel, *Biochim. Biophys. Res. Commun.* **37**, 204 (1969).

A_{260}

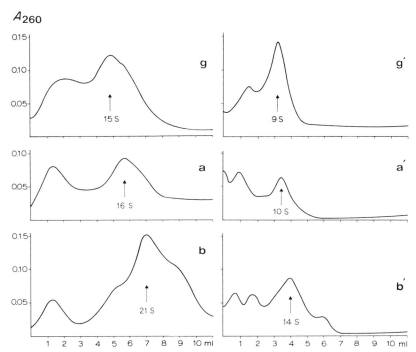

FIG. 5. Ultracentrifugation pattern of reticulocyte 15 S RNP, lens 16 S and 21 S mRNP particles and corresponding messenger fractions. (a and b) Fractions from the gradient shown in Fig. 4 after spinning in rotor SB 283 of the IEC for 15 hours at 40,000 rpm. (g) For comparison a control experiment with reticulocyte RNP particles is shown. In the right part of this figure the sedimentation profiles of RNA fractions extracted from the different RNP particles are shown.

for 10 minutes at 2°. The supernatant fraction is recentrifuged. The lysate is divided into 0.3–1.0 ml fractions, frozen in liquid nitrogen, and stored at −70° until use. The lysate containing about 150 mg protein/ml keeps its activity during several months.

2. *Krebs Ascites S30 Lysate (Procedure of Mathews and Korner[13])*. Ascitic fluid is obtained from mice bearing Krebs II ascites tumors. The material is suspended in a buffer containing 35 mM Tris·HCl, pH 7.6, and 146 mM NaCl. The cells are centrifuged for 6 minutes at low speed (60 g), and the pellet is repeatedly washed in the same buffer to remove contaminating red cells. Finally the cells are centrifuged at 250 g for 5 minutes. The washed cells are resuspended in 2 volumes of a buffer containing 10 mM KCl, 1.5 mM Mg acetate, and 10 mM Tris·HCl, pH 7.5. The mixture is transferred to an all glass homogenizer with tight-

[13] M. B. Mathews and A. Korner, *Eur. J. Biochem.* **17**, 328 (1970).

fitting pestle. Two strokes are applied and the suspension is left in ice for 5 minutes. Homogenization is performed by 20–30 strokes. Immediately thereafter one-third volume of a medium containing 0.3 M Tris·HCl pH 7.5, 1.25 M KCl, 50 mM magnesium acetate, and 60 mM 2-mercaptoethanol is added to restore the tonicity. The homogenate is centrifuged at 30,000 g for 10 minutes. Preincubation is performed for 30 to 40 minutes at 37° as described for incubation (see Section E, 2). The preparation is ready for use after passage through a column of Sephadex G-25. The soluble enzymes are obtained from untreated S30 after centrifugation for 2.5 hours at 50,000 g in a Spinco L50 preparative ultracentrifuge.

E. Incubation Conditions

1. *For the Reticulocyte System.* Assays for messenger translation are performed at 26° for 120 minutes in a mixture containing: 200 μl of reticulocyte lysate, 100 mM ammonium acetate, 10 mM Tris·HCl, pH 7.4, 2 mM magnesium acetate (in addition to the concentrations of salts present in the lysate), 0.25 mM heme, 1 mM ATP, 0.2 mM GTP, 15 mM creatine phosphate, 30 μg of creatine phosphokinase, 100 μM amino acids, all except methionine, 5 μCi [^{35}S]Met (sp. act., 10–25 Ci/mmole) or 0.5 μCi formyl-[^{35}S]Met-tRNA$_f^{Met}$ and 20 μg of lens mRNA. Final volume: 0.5 ml. After incubation, 25 μl of 0.25 M EDTA, 10 μg of pancreatic RNase, and 150 μg of carrier α-crystallin is added and the mixture is kept for 30 minutes at 37°. An aliquot of the mixture is precipitated by adding 1 volume of 10% TCA for measurement of the radioactivity incorporated. The major part of the incubation mixture is precipitated by adding dropwise 15 volumes of 2.5% oxalic acid in acetone at 0° under vigorous shaking. The precipitate is centrifuged at 3000 g for 5 minutes, washed 3 times with cold acetone and washed once with ether. The material obtained is used for gel electrophoresis.

2. *For the Ascites System.* The incubation mixture contains, per 50 μl: 20 μl of preincubated ascites S30, 2.5 mM magnesium acetate, 100 mM KCl, 25 mM Tris·HCl, pH 7.5, 6 mM 2-mercaptoethanol, 1 mM ATP, 0.1 mM GTP, 5 mM creatine phosphate, 0.2 mg/ml creatine phosphokinase, ^{14}C-labeled amino acid mixture: 5 μCi/ml (54 mCi/matom, Radioachemical Centre, Amersham, Great Britain), supplemented with the six missing amino acids (asparagine, cysteine, glutamine, histidine, methionine, and tryptophan) at 20 μM each or [^{35}S]methionine: 10 μM, 10 Ci/mmole, plus the remaining nineteen unlabeled amino acids, 20 μM each; lens messenger is added at 60 μg/ml. After 60 minutes at 37°, reactions are terminated by incubation for 15 minutes at 37° with 20 μl of pancreatic RNase and 10 mM EDTA. This mixture is either used directly or precipitated with 5% TCA and washed twice in 5% TCA,

thereafter in ethanol, ethanol:ether (1:1 v/v), and ether. The product is analyzed on polyacrylamide gels.

F. Polyacrylamide Gel Electrophoresis

1. SDS-Gel Analysis. SDS-gel electrophoresis according to Weber and Osborn[14] is performed in 8.0 × 0.5 cm cylindrical glass tubes at 5 mA/gel for 12 hours. The gels contain: 12.5% acrylamide, 0.333% N,N'-methylenebisacrylamide, 0.05% TEMED, 0.4 mg/ml ammonium persulfate, 0.1% SDS, 0.1 M sodium phosphate, pH 7.0. The electrode buffer consists of 0.1 M sodium phosphate, pH 7.0, 0.1% SDS, and 1% 2-mercaptoethanol. The gels are preelectrophoresed for 1 hour at 5 mA/gel. The sample is solubilized in 1% SDS, 1% 2-mercaptoethanol and 10 mM sodium phosphate, pH 7.0, and heated at 100° for 3 minutes. After electrophoresis the gels are stained with Coomassie blue for 2 hours. The Coomassie blue solution is prepared by adding 1.25 g of Coomassie blue to a mixture of 454 ml of 50% methanol and 46 ml of acetic acid. The gels are destained at 37° for several hours in a mixture containing 250 ml of methanol and 75 ml of acetic acid per liter. If staining is too faint, the staining procedure can be repeated.

An alternative electrophoretic procedure is performed as described by Laemmli.[15] These gels which are 9 cm long contain 15% acrylamide, 0.4% methylenebisacrylamide, and 0.1% SDS. In this method a stacking gel is applied. Staining and destaining is as described by Weber and Osborn. The separation pattern on these gels is shown in Fig. 6. The result indicates that 14 S lens messenger directs the synthesis of αA chains whereas 10 S gives rise to αB chains, and some additional chains coinciding with β-crystallin polypeptides.

2. Electrophoresis on Urea Gels at Acid pH. Electrophoresis on acidic urea gels is performed according to Schoenmakers et al.[4] in 8.0 × 0.5 cm cylindrical glass tubes at 4 mA/gel for 3.5 hours. The gels contain: 10% acrylamide, 0.2% N,N'-methylenebisacrylamide, 0.006% potassium metabisulfite, 0.015% ammonium persulfate, 6 M urea, and 0.25 M formic acid. Polymerization is performed at 40° for 2 hours. The electrode buffer contains: 0.25 M formic acid, 6 M urea, and 1% 2-mercaptoethanol. Preelectrophoresis is carried out at 5 mA/gel for 1 hour. Samples are solubilized in electrode buffer, containing 10% sucrose and applied to the gel in amounts up to 400 μg. After electrophoresis the gels are stained with Amido black for 30 minutes. The Amido black solution is prepared by adding 2.5 g of Amido black to 250 ml of methanol, 35 ml of acetic acid,

---

[14] K. Weber and M. Osborn, *J. Biol. Chem.* **244**, 4406 (1969).
[15] U. K. Laemmli, *Nature (London)* **227**, 680 (1970).

STAINED GELS   AUTORADIOGRAPHS

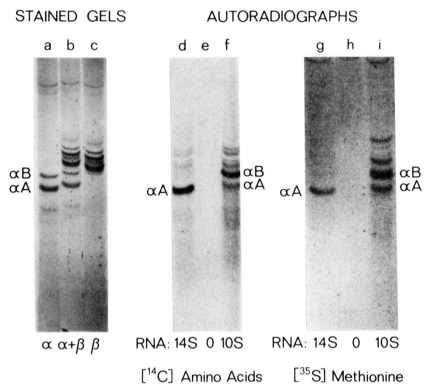

FIG. 6. SDS-gel electrophoretic analysis of products synthesized in the Krebs ascites cell-free system under direction of lens messengers. (a, b, c) Protein staining. (d, e, f) A mixture of ^{14}C amino acids was added during incubation; (d) plus 14 S lens mRNA; (e) minus lens mRNA; (f) plus 10 S lens mRNA. (g, h, i) as (d, e, f) but instead of the ^{14}C amino acid mixture [^{35}S]methionine was used as radioactive precursor in the assay. From this type of experiment it is seen that lens 14 S mRNA directs the synthesis of αA chains almost exclusively. It cannot be concluded from this type of experiment whether the αA$_2$ or the αA$_1$ chain is synthesized. (Reproduced from *Nature,* with permission.)

and 215 ml of water. The gels are destained electrophoretically in 2% acetic acid. A typical pattern is shown in Fig. 7. It can be seen that the addition of 14 S lens messenger to the reticulocyte lysate gives rise predominantly to radioactive material in the αA region.

3. Electrophoresis on Urea Gels at Alkaline pH. Electrophoresis on basic urea gels is performed according to Schoenmakers *et al.*[4] in 8.0 × 0.5 cm cylindrical glass tubes at 4 mA/gel for 3.5 hours. The gels contain: 10% acrylamide, 0.2% *N,N′*-methylenebisacrylamide, 0.006% TEMED, 0.004% potassium ferricyanide, 0.07% ammonium persulfate, 75 m*M* Tris–1.8 m*M* EDTA–8.4 m*M* boric acid, pH 8.9, and 6 *M* urea. The

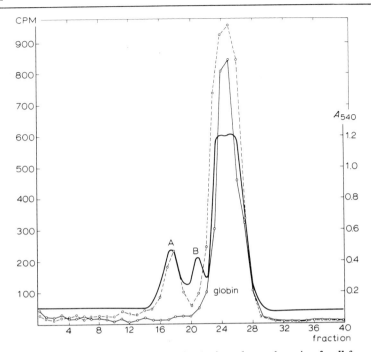

FIG. 7. Tracing of a gel pattern obtained after electrophoresis of cell-free products in the reticulocyte system after addition of lens 14 S messenger. —————, Absorbance at 540 nm; ○- -○, radioactivity in the presence of 14 S; ○—○, radioactivity in the absence of the messenger. Gel type: 6 M urea-polyacrylamide, pH 3.0; sample: 400 μg of protein from the incubation mixture and 100 μg of carrier α-crystallin. From this type of experiment it is seen that lens 14 S mRNA directs the synthesis of αA chains. Whether αA$_2$ or αA$_1$ chains are made can be concluded neither from the experiment shown in Fig. 6 nor from this figure. (Reproduced from *Nature,* with permission.)

electrode buffer contains: 75 mM Tris, 1.8 mM EDTA, 8.4 mM boric acid, pH 8.9, 6 M urea, and 1% 2-mercaptoethanol. The electrophoresis buffer may be replaced by 25 mM Tris·glycine, pH 8.5. Gels and buffers are prepared shortly before use. Preelectrophoresis is performed at 5 mA/gel for 1 hour. Samples are applied in electrode buffer, containing 10% sucrose in amounts up to 400 μg. Staining and destaining is carried out as described for the acidic urea gels. The pattern obtained by this procedure allows the conclusion that under direction of the 14 S lens messenger exclusively αA$_2$ chains are synthesized (Fig. 8).

4. *Preparative Electrophoresis on Polyacrylamide Gels.* Preparative gel electrophoresis is performed in 9.0 × 1.3 cm cylindrical glass tubes at 10 V/cm for 14 hours. The gels are prepared as described above. Samples

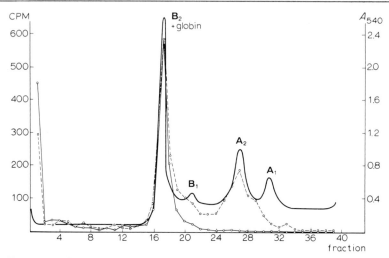

FIG. 8. Tracing of a gel pattern obtained after electrophoresis of cell-free products in the reticulocyte system after addition of 14 S lens messenger. —————,
Absorbance at 540 nm; O--O, radioactivity in the presence of 14 S mRNA; O—O,
radioactivity in the absence of lens messenger. Gel type: 6 M urea-polyacrylamide,
pH 8.9. Sample: 400 μg of protein from the incubation mixture plus 100 μg of carrier
α-crystallin. This experiment reveals that αA_2 chains have been synthesized almost
exclusively. (Reproduced from *Nature,* with permission.)

containing 5 mg of protein can be applied to these gels. After electrophoresis, a small slice of the gel is stained. The area containing the
desired polypeptide is cut out, minced, stirred for 16 hours with 10
volumes of a 6 M urea solution, the gel particles are removed by filtration,
the filtrate is diluted three times, dialyzed against distilled water containing
5 mM 2-mercaptoethanol, and lyophilized.

G. Determination of the Radioactivity on the Gels

The radioactivity on polyacrylamide gels can be determined in two
ways: (1) After monitoring at 540 nm in a Gilford spectrophotometer,
adapted with a gel scanner, the gels are cut into 1.5 mm slices, which are
solubilized with 200 μl of NCS (Nuclear Chicago solubilizer) at 60° for
16 hours, and, after addition of a toluene-based scintillator, counted in a
Packard liquid scintillation counter. (2) The gels are sliced longitudinally,
dried down under vacuum on filter paper and autoradiographed with a
Kodak X-ray film, type RHP-Royal X-omat.

The processing of gels for autoradiography is most conveniently performed with the aid of the apparatus shown in Fig. 9. First the gel rod
is fixed between the two Perspex plates in the gel holder (Fig. 9A) and

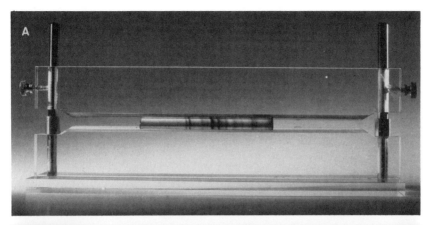

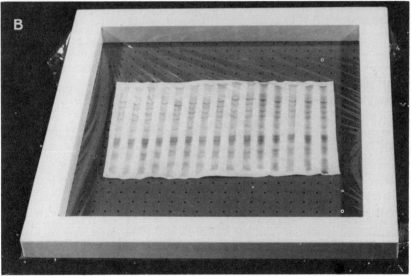

FIG. 9. Device for drying of gel slices to be autoradiographed. (A) Gel holder. (B) Hollow base plate which can be connected to a vacuum pump. The film cover is held in place by a metal frame.

sliced longitudinally with a very thin thread (e.g., thread as used for fishing). A piece of filter paper cut to size is then put on a plate in which a great number of holes have been drilled. This plate is hollow and can be evacuated (Fig. 9B). The sliced gels are carefully laid down on the wetted filter paper and covered with a thin sheet of plastic film (e.g., Saran Wrap, Dow Chemical Co.). The white metal frame holds the cover in place. The outlet of the base plate is connected to a vacuum pump. A hair-dryer placed at a distance of about 50 cm from the surface of the plate is used to provide a stream of warm air. In this way the gel slices are dried in 4–5 hours.

H. Analysis of Newly Synthesized Peptides after Various Enzymatic Digestions

1. Pronase Digestion. Pronase digestion on the newly synthesized crystallin chains is performed in a solution of 0.1 M ammonium bicarbonate and 1 mM CaCl$_2$ for 6–24 hours at 37°[16] (the enzyme:substrate ratio is 1:50 w/w). If no complete solubilization is achieved, a second treatment is performed. After digestion the peptides are either lyophilized or treated with Dowex 50 in order to remove the unblocked peptides. For this treatment, 3/4 volume of a Dowex 50 suspension is added, mixed thoroughly, and centrifuged. The supernatant is removed and the resin is washed with 1 volume of distilled water. After centrifugation the two supernatants are combined and lyophilized.

The residual peptides are dissolved in distilled water and subjected to paper electrophoresis on Whatman 3 MM paper in acetic acid-pyridine-water (6:200:794 v/v), pH 6.5. Radioactivity is determined after drying of the electropherogram which is cut into 1 cm strips. The paper pieces are immersed in toluene-based scintillator and counted in a liquid scintillation counter. Reference peptides are stained for methionine using platinic iodide. A typical electropherogram is shown in Fig. 10. In our original experiments fmet tRNA$_f^{Met}$ labeled in methionine was used as radioactive precursor. This enables further characterization of the N-terminal dipeptide. Treatment of the eluted blocked dipeptide with 0.5 N HCl at 90° for 30 minutes removes the formyl group. The radioactivity shifts to the position of Met-Asp (Fig. 11). If radioactive methionine is used as label, the acetylated N terminus is formed as in the native α-crystallin chains. This means that in the reticulocyte system the necessary machinery is present for N-terminal acetylation of, at least, lens crystallins.[17]

2. Subtilisin Digestion. Subtilisin digestion and subsequent paper electrophoretic analysis are performed as described for Pronase digestion except that calcium chloride is omitted from the solution.[18]

3. Tryptic Digestion. Tryptic digestion is carried out on the polypep-

[16] H. J. Hoenders, J. G. G. Schoenmakers, J. J. T. Gerding, G. I. Tesser, and H. Bloemendal, *Exp. Eye Res.* **7**, 291 (1968).

[17] G. J. A. M. Strous, T. J. M. Berns, H. van Westreenen, and H. Bloemendal, *Eur. J. Biochem.* **30**, 48 (1972).

[18] H. J. Hoenders, J. van Tol, and H. Bloemendal, *Biochim. Biophys. Acta* **160**, 283 (1968).

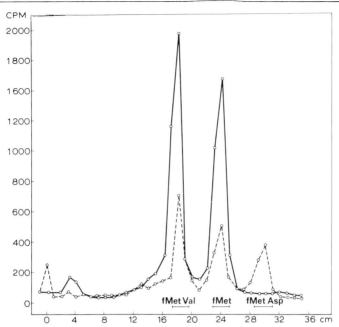

FIG. 10. Paper electrophoretic analysis of the N-terminal dipeptide(s) derived from the cell-free products synthesized under direction of 14 S lens mRNA in the reticulocyte system. — — —, 14 S RNA added; ——, without lens messenger. Electrophoresis at 45 V/cm for 2.5 hours. The reference peptides were detected by staining with platinic iodide. In this experiment formyl[^{35}S]met-tRNA$_f^{Met}$ was used as radioactive precursor. The experiment reveals that the formyl group is not split from the N terminus during incubation. [^{35}S]Methionine may also be used as radioactive precursor. In that case acetyl-Met-Asp will be found in this type of experiment since the reticulocyte lysate contains the machinery to acetylate the N terminus [G. J. A. M. Strous, T. J. M. Berns, H. van Westreenen, and H. Bloemendal, *Eur. J. Biochem.* **30**, 48 (1972)].

tide chains separated by polyacrylamide gel electrophoresis on basic urea gels. The analysis of the αA_2 chain is here described. The gel segment containing the αA_2 chains is cut out and handled as described in Section F, 4. The polypeptides are aminoethylated by the method of Raftery and Cole[19]: The polypeptides are solubilized in a solution containing 8 M deionized urea, 0.5 M Tris·HCl, pH 8.6, 2 mg/ml EDTA, and 1.3% 2-mercaptoethanol. Before the protein is added, this solution is fluxed with argon during 5 minutes. The dissolved protein (about 15 mg/ml) is left for 2 hours at room temperature, and 25 μl of ethylenimine per milliliter is added. After aminoethylation for 2 hours at room temperature,

[19] M. A. Raftery and R. D. Cole, *J. Biol. Chem.* **241**, 3557 (1966).

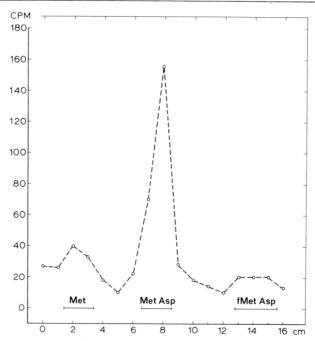

FIG. 11. Paper electrophoretic analysis of the products obtained after deformyla-
tion of the material eluted from the fMet-Asp region detected on the preceding
electropherogram (Fig. 10). The paper segments on which formyl-Met-Asp was
found were taken from the counting vials and washed with toluene. Radioactivity
was recovered by elution with 2 ml of bidistilled water. The material was treated
with 0.5 M HCl at 90°C for 30 minutes in order to remove the formyl group. After
lyophilization the product was again subjected to high-voltage paper electrophoresis
at 45 V/cm for 1 hour.

the polypeptides are precipitated with 15% TCA. The TCA is removed
by washing with acetone and the material is dissolved in 0.1 M ammonium
bicarbonate, adjusted to pH 8.9 with ammonia, and digested with trypsin
(1:50 w/w) for 5 hours at 37°. The resulting peptides are lyophilized,
dissolved in water, and subjected to descending paper chromatography on
Whatman 3 MM paper eluted with n-butanol–acetic-acid–pyridine–water
mixture (60:12:48:40, v/v). Tryptic peptides from [35S]Met-labeled αA_2
crystallin, prepared by $in\ vitro$ incubation of a lysate from calf lens are
used as markers. After chromatography for 18 hours, the chromatogram
is dried and cut into 1-cm strips which are counted in a liquid scintillation counter a toluene-based scintillator, being used. The native methionine
containing peptides of αA_2 have the same chromatographic behavior as
the markers (compare Fig. 12).

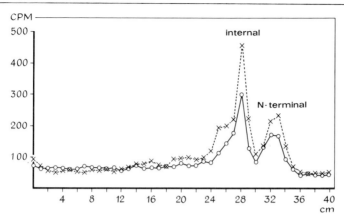

Fig. 12. Paper chromatography of the methionine-containing tryptic peptides from αA_2 chains synthesized under direction of 14 S lens mRNA. ×--×, radioactivity derived from the newly synthesized products; ○—○, radioactivity derived from the tryptic peptides of native α-crystallin polypeptide chains. The shoulder of the peak corresponding to the internal tryptic methionine peptide presumably represents the oxidized form of this peptide. Identical results are obtained with newly synthesized αA_2 chains in the oocyte system [A. J. Berns, M. van Kraaikamp, H. Bloemendal, and C. D. Lane, *Proc. Nat. Acad. Sci. U.S.* **69**, 1606 (1972)].

Analysis of all tryptic peptides can be performed on Bio-Rad Aminex A-5 columns (0.4 × 17 cm). After lyophilization the tryptic digest is redissolved in 2 ml of a 0.2 M pyridine–acetate buffer, pH 3. Elution is started with 15 ml of the same buffer at 50° under a pressure of 12 atm. Thereafter a 2 × 125 ml linear gradient is applied ranging from pH 3.1 to pH 5.0 (the concentration of the pH 5.0 is 2.0 M pyridine acetate). Fractions of 1.5 ml are collected. An aliquot of 50 µl is mixed with 5 ml of Instagel (Packard) and counted in a liquid scintillation counter. Another aliquot of 50 µl is hydrolyzed in 3 M KOH. After neutralization the fractions are stained with ninhydrin, and the absorbance is measured at 570 nm (Fig. 13). The individual fractions from the Aminex A-5 column can further be analyzed by descending paper chromatography.[20] After evaporation under reduced pressure at room temperature the fractions are redissolved in 50 µl of distilled water and applied at distances of 0.6 cm on a sheet of Whatman 3 MM filter paper. Elution is performed with a mixture of *n*-butanol–glacial acetic acid–pyridine–water (60:12:48:40 v/v) for 12 hours. The chromatogram is stained with a solution of 0.5% ninhydrin and, after drying, autoradiographed for 2 weeks.

[20] A. J. M. Berns, V. Schreurs, M. van Kraaikamp, and H. Bloemendal, *Eur. J. Biochem.* **33**, 551 (1973).

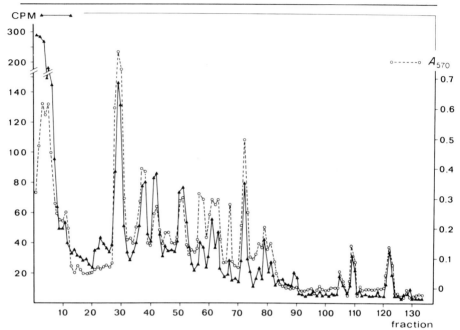

FIG. 13. Column chromatography of tryptic peptides from newly synthesized αA_2 chains in the reticulocyte cell-free system under direction of 14 S lens messenger. ○----○, absorbance at 570 nm. ▲—▲, radioactivity. In this experiment the incubation medium contained a ^{14}C-labeled amino acid mixture (20 μM each) missing Asp, Glu, Met, Trp, Cys, and His. These amino acids were added unlabeled in a concentration of 50 μM each. The other incubation conditions were as described in the text. Before tryptic digestion, 30 mg of carrier α-crystallin was added per 0.5 μCi of newly synthesized αA_2 chain. Trypsin (1:50 w/w) was added at zero time and after 2 hours. Thereafter digestion was continued for 3 hours. (Reproduced from *Eur. J. Biochem.*, with permission.)

4. Chymotryptic Digestion. Frequently incubation of a protein with trypsin does not result in complete digestion. The residual insoluble core can be digested with chymotrypsin. In order to verify the fidelity of translation of messenger RNA in a cross system, one should not neglect this additional analytical step. Also, in case of lens messenger translation, chymotryptic digestion is included. The enzyme substrate ratio is 1:50 w/w. Digestion at 37° is carried out for 16 hours in 0.1 M ammonium bicarbonate adjusted to pH 8.9. Thereafter the material is lyophilized and redissolved in 20–50 μl of electrophoresis buffer. The sample is applied on a 30 × 100 cm sheet of Whatman 3 MM paper which has been dipped in pyridine–glacial acetic acid–water (25:1:225 v/v), pH 6.5. Electrophoresis in a Savant High Voltage apparatus with Varsol-cooling is carried out at 40 V/cm for 90 minutes. Then, after drying for 1 hour at room temperature and at 80° for 2 hours, the sheet is cut longitudinally in a number of strips depending on the amount of sample applied (50 nmoles of digested protein is convenient). Using a sewing machine, each strip is stitched on a 46 × 58 cm Whatman 3 MM paper at a distance of approximately 8 cm from the short side. Descending chromatography in *n*-butanol–glacial acetic acid–pyridine–water is performed as described in Section H, 3 for tryptic peptides.

Comment

Analysis of the products synthesized under direction of lens messenger added to different cell-free cross systems reveals that the translation occurs faithfully. The same holds for an *in vivo* system derived from living oocytes of *Xenopus laevis,* which has been described in detail elsewhere.[21] The accuracy of translation is even extended to the posttranslational process of *N*-acetylation. Both the reticulocyte lysate[17] and the oocytes[21] have the machinery to acetylate the newly synthesized α-crystallin chains in N-terminal position, yielding biosynthetic products indistinguishable from the native α-chains. As no addition of messenger specific protein factors is required in all systems assayed, it is strongly suggested that such recognition factors are no prerequisite for reliable translation. Recent results show that the addition of a 0.5 *M* KCl wash from reticulocyte ribosomes to a Krebs ascites lysate[22] programmed with lens messenger enhances the formation of crystallin chains. Also these observations render the necessity of tissue- or species-specific factors at least questionable, since the lysate may be derived from mice, the factors from rabbits and the messenger from calves. The argument that the messenger may contain a small amount of attached protein is not tenable, as after thorough protein extraction from the mRNA preparations the messenger activity is maintained.

An intriguing problem is the length of the 14 S messenger. This messenger, with a molecular weight of about 360,000, directs the synthesis of a polypeptide with a molecular weight of only 20,000, whereas the lens 10 S messenger population with an average molecular weight of 260,000 directs the synthesis of polypeptide chains with significantly higher molecular weights. Although poly(A) tracks are present in lens 14 S messenger,[23] they cannot account for the calculated excess of nucleotides. This problem awaits elucidation.

[21] A. J. M. Berns, M. van Kraaikamp, H. Bloemendal, and C. D. Lane, *Proc. Nat. Acad. Sci. U.S.* 69, 1606 (1972).

[22] M. B. Mathews, I. B. Pragnell, M. Osborn, and H. R. V. Arnstein, *Biochim. Biophys. Acta* 287, 113 (1972).

[23] U. Bertazzoni, G. Piperno, A. J. M. Berns, and H. Bloemendal, in preparation.

Lens mRNA, like other poly A containing messengers can be transcribed into complementary DNA using purified DNA polymerase from AMV. The product though smaller in size than the template is a faithful transcript of this template.[24]

[24] A. J. M. Berns, H. Bloemendal, S. J. Kaufman, and I. M. Verma, *Biochem. Biophys. Res. Commun.* **52**, 1013 (1973).

[64] Assay of Ovalbumin mRNA in Reticulocyte Lysate

By ROBERT T. SCHIMKE, ROBERT E. RHOADS, and G. STANLEY MCKNIGHT

For studies of the regulation of protein synthesis in higher organisms and for the isolation of specific mRNA's, it is necessary to have sensitive and quantitative assays of mRNA content of tissues. We have found the crude rabbit reticulocyte lysate system to be an adequate system for such assays. The techniques are simple, and specific protein synthesis directed by exogenous RNA can be monitored readily by use of highly specific immunologic techniques. Although the procedures described below are for the synthesis of ovalbumin, these techniques work equally well for conalbumin and rat serum albumin mRNA's. Thus the reticulocyte lysate system may provide a general method for the quantitative measure of mRNA's from animal tissue.

Rabbit Reticulocyte Lysate

New Zealand white rabbits weighing 2–3 kg are made anemic by 5 daily subcutaneous injections of 0.3 ml/kg of neutralized 2.5% phenylhydrazine·HCl. On day 7 the rabbits are bled from the marginal ear vein into heparinized tubes (approximately 60 ml from each animal). The blood is centrifuged at 600 g for 10 minutes at 2°, and the plasma is discarded. The cells are suspended gently in 3 volumes of a solution containing 0.13 M NaCl, 5 mM KCl, 7.5 mM MgCl$_2$ and centrifuged at 600 g for 10 minutes. This washing procedure is repeated once more. The buffy coat is partially removed by aspiration during these washings. Finally the packed cells are lysed by resuspension in an equal volume of water and allowed to stand at 0° with occasional stirring for 5 minutes, after which cell debris and unlysed cells are removed by centrifugation at 15,000 g for 10 minutes at 2°. The lysate is stored in aliquots of 2–3 ml in liquid nitrogen.

Components of the Protein Synthesizing System

Chemicals. Creatine phosphate and creatine kinase are from Sigma. [4,5-³H]leucine (40 Ci/mmole) is from Schwarz/Mann. Other components are from various suppliers.

Assay. Incubations are performed at 25° for 1 hour in disposable, sterile 5-ml tubes (Falcon). The final volume is 0.250 ml and contains 100 μl of lysate, 50 μl of the mix (see the table), and various amounts of exogenous nucleic acid preparation. The volume is made to 250 μl with water. The mixture of supplemental components is constituted just before use, since it loses activity at a rate of approximately 20% per hour. After incubation, 30 μl of a solution containing 0.5 mg/ml of ovalbumin and 0.1 M L-leucine are added, and the reaction mixtures are transferred to Beckman Model 152 microfuge tubes and centrifuged at 15,000 g in a Sorvall HB rotor with specially made adaptors.

Incubation Mixture

ATP, GTP, creatine phosphate, and amino acids are neutralized to pH 7.2 with 1.0 M KOH. The [³H]leucine is neutralized by adding 15 μl of 1 N KOH to 1 ml of the acid solution as obtained from the supplier. The amino acid mixture consists of all 19 amino acids without leucine. The stock solution of amino acids is sterilized by passage through a HA Millipore filter after preparation. All stock solutions are stored at −20°.

The concentration of nucleic acid assayed varies with the purity of the mRNA preparation. Nucleic acid preparations generally inhibit total incorporation by the reticulocyte lysate system at high concentrations. Consequently it is best to assay mRNA preparations at as low a concentration as possible. Generally three concentrations of nucleic acid less than

COMPONENTS FOR OVALBUMIN mRNA ASSAY

Component	Concentration of stock solution	Stock solutions/20 assays (ml)	Final concentration in incubation
KCl	3.76 M	0.1	75 mM
MG acetate	0.1 M	0.1	2 mM
ATP	50 mM	0.1	1 mM
GTP	10 mM	0.1	0.2 mM
L-Amino acids, 19	2 mM	0.1	40 μM
L-Leucine	250 μM	0.1	5 μM
[4,5-³H]L-Leucine	0.5 mCi/ml	0.25	25 μCi/ml
Creatine phosphate	0.75 M	0.1	15 mM
Creatine kinase	4000 U/ml	00.05	40 U/ml

25 μg/250 ml are assayed. This must to a certain extent be determined empirically for each preparation to be assayed. The nonlinearity of the assay can be markedly improved by expressing ovalbumin radioactivity as a percent of total incorporation.

Immunoprecipitation of Labeled Ovalbumin

The ovalbumin and anti-ovalbumin antibody preparations have been described (see this volume [59]). The anti-ovalbumin used is the γ-globulin fraction obtained by ammonium sulfate precipitation of goat serum. The immunoprecipitation is performed in duplicate. To 100 μl of the centrifuged lysate reaction mixture is added 100 μl of a solution containing (a) sufficient antibody to quantitatively precipitate 5 μg of ovalbumin, (b) 30 μl of a solution containing 10% Triton-X-100, and 10% DOC, and (c) 10 mM sodium phosphate, pH 7.5, and 0.15 M NaCl. After mixing, the antibody precipitate is allowed to form for at least 30 minutes at room temperature. Each reaction mixture is then layered on top of 200 μl of a solution previously centrifuged at 15,000 g for 5 minutes containing 1.0 M sucrose, 10 mM L-leucine, 1% (w/v) each of Triton X-100 and DOC in a Beckman microfuge tube. The prior centrifugation of the sucrose detergent solution removes sediment that otherwise traps radioactivity nonspecifically, resulting in high blank values. The precipitate is separated from nonspecific radioactivity by centrifuging through the sucrose-detergent solution for 5 minutes at 15,000 g_{max}, in the Sorvall HB rotor. The contents of each microfuge tube are then frozen by brief immersion in dry ice-acetone, and the tip containing the antibody precipitate is cut off and placed in a scintillation vial. The precipitate is dissolved by shaking with 0.5 ml NCS (Amersham/Searle) at 37° for several hours. Ten milliliters of toulune scintillation fluid containing 5 g of 2,5-diphenyloxazole and 0.3 g of 1,4-bis(4 methyl-5-phenyloxazole-2-yl)benzene per liter are added, and the radioactivity is determined by liquid scintillation spectrometry.

Total reticulocyte protein synthesis is determined by spotting duplicate 10-μl samples of the centrifuged lysate reaction mixture on Whatman No. 540 filter disks and placing the disks in a beaker containing approximately 200 ml of 10% trichloroacetic acid.[1] The disks are washed with four 200-ml portions of 10% trichloroacetic acid and then one of 5% trichloroacetic acid. More 5% trichloroacetic acid is added, and the beaker is placed in a 90° water bath for 15 minutes. After three more washes with 5% trichloroacetic acid, the filters are dried for 30 minutes under a heat lamp. They are then placed in scintillation vials, 0.7 ml of

[1] F. J. Bollum, this series, Vol. 12B, p. 169.

NCS is added, and protein is dissolved by shaking at 37° for several hours. Radioactivity is determined as with the immunoprecipitates.

The use of detergents and sucrose in the washing procedure is imperative in ensuring low backgrounds (50–120 cpm). Routinely, immunoprecipitates from reactions with added mRNA contain 1000–4000 cpm. Control immunoprecipitates (bovine serum albumin antibody plus serum albumin) equivalent in size to ovalbumin immunoprecipitates similarly have low radioactivity.

Optimal conditions for the translation of ovalbumin mRNA appear to be the same as for endogenous mRNA.[2] The pH optimum is that of the unadjusted lysate, pH 7.2, the Mg^{2+} optimum is 2.0–2.4 mM, and the K^+ optimum is 75–100 mM. Incorporating activity per milliliter of reaction mixture can be increased by using more lysate per assay.[3] A greater specific activity of leucine could be achieved if endogenous amino acids were removed from the lysate, e.g., by Sephadex G-25 gel filtration, since the lysate contributes the major proportion of nonradioactive leucine, (14 nmoles per milliliter of reaction mixture).

We do not routinely add hemin to the reaction mixture, as employed by several groups.[4] It stimulates ovalbumin synthesis as well as hemoglobin synthesis at 35°, but the degree of stimulation is sharply dependent on hemin concentration, and the optimal concentration varies from one lysate preparation to another. Based on the results of Maxwell et al.,[5] that protein synthesis is inhibited less at 25° in the absence of hemin, this temperature is used for assay, and hemin is omitted.

Preparation of RNA for Assay

For quantitative measure of the ovalbumin mRNA content of a tissue, it is necessary to ensure optimal recovery of mRNA activity. The yield of ovalbumin mRNA activity was used as a criterion in choosing optimal conditions for the isolation of undegraded nucleic acid from oviduct. Nearly all the mRNA activity can be recovered after fractionation of oviduct into subcellular fractions (crude nuclear pellet, 15,000 g supernatant, and polysomes).

Oviduct, either fresh or previously frozen in liquid nitrogen, is homogenized at 2° in eight volumes (v/w) of polysome buffer (25 mm Tris· HCl, 25 mM NaCl, 5 mM $MgCl_2$, pH 7.5) containing 5% sucrose and

[2] S. D. Adamson, E. Herbert, and W. Godchaux, *Arch. Biochem. Biophys.* **125**, 671 (1968).
[3] S. D. Adamson, G. A. Howard, and E. Herbert, *Cold Spring Harbor Symp. Quant. Biol.* **34**, 547 (1969).
[4] T. Hunt, G. Vanderhoff, and I. M. London, *J. Mol. Biol.* **66**, 471 (1972).
[5] C. R. Maxwell, C. S. Kamper, and M. Rabinovitz, *J. Mol. Biol.* **58**, 317 (1971).

500 μg/ml sodium heparin in a Dounce homogenizer with five to eight strokes with a loose pestle. One volume of a solution containing 10% each of Triton X-100 and DOC is added, and the tissue is further homogenized with three strokes with a tight pestle. For the preparation of total nucleic acid, an aliquot of the homogenate is added to an equal volume of buffer at room temperature containing 2% (w/v) SDS, 10 mM EDTA, 40 mM sodium acetate 80 mM Tris·HCl, pH 7.0 (two times concentrated SDS buffer). The mixture is then homogenized at room temperature in a Sorvall Omnimixer for 30 seconds at top speed and centrifuged at 27,000 g_{max} for 5 minutes. One milliliter of the supernatant is layered over 10.4 ml of a 5 to 20% linear sucrose gradient in 1 × concentrated SDS buffer with a 1.0 ml cushion of 40% sucrose in SDS buffer. Nucleic acid is separated from protein, heparin, and other low molecular weight substances by centrifugation at 40,000 rpm in the Spinco SW 41 rotor at 20° for 6 hours. Gradients are collected from the bottom of the tube, and OD_{260} is monitored continuously with a Gilford spectrophotometer and flow cell. The tubing and flow cell are previously exposed briefly to 0.1% diethylpyrocarbonate to destroy ribonuclease activity. All nucleic acid sedimenting faster than the midpoint of the trough between 18 S rRNA and the supernatant proteins is collected, made 0.2 M in NaCl, and precipitated by addition of 2 volumes of cold (−20°) ethanol. After standing overnight at −20°, the nucleic acid precipitate is collected by centrifugation for 20 minutes at 3000 g, redissolved in 5 mM Tris·HCl, pH 7.5, made 0.2 M in NaCl and reprecipitated with ethanol. The precipitate is collected by centrifugation and

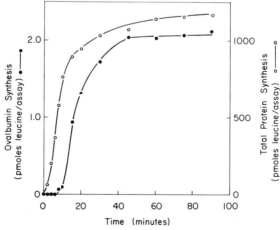

FIG. 1. Time course of ovalbumin and endogenous protein synthesis. Reactions contained 11 μg of hen oviduct polysomal RNA per milliliter.

all traces of ethanol are removed by lyophilization. Immediately before assay, nucleic acid fractions are dissolved in water.

Alternative methods for isolation of RNA, including various purified fractions, are described in this volume [59]. In purified nucleic acid fractions not containing Mg^{2+}, one OD_{260} unit corresponds to 50 μg of nucleic acid per milliliter. The use of the SDS method for mRNA isolation is superior to the use of phenol extraction in the presence of 0.5% SDS, 25 mM EDTA, and 75mM NaCl at pH 8.0. In our hands phenol extraction leads to a 60–70% decrease in total tissue ovalbumin mRNA activity. This may result from the trapping of mRNA in the large protein containing interface. The use of 500 μg/ml of heparin is necessary for

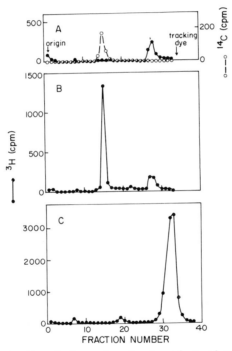

FIG. 2. SDS-acrylamide gel electrophoresis of products of protein synthesis in rabbit reticulocyte lysates. (A) Reaction mixture contained no oviduct RNA. Radioactivity is that precipitated with ovalbumin antibody and electrophoresed after dissociation in SDS and dithiothreitol (●—●). [^{14}C]Ovalbumin prepared in oviduct explants and added after incubation as a marker (○- - -○). (B) Reaction mixture contained 40 μg/ml of oviduct polysomal RNA. Radioactivity is that present in the ovalbumin immunoprecipitate. (C) Radioactivity incorporated into total protein in the absence of added oviduct RNA and precipitated with TCA before solution in SDS. Details are given by R. E. Rhoads, G. S. McKnight, and R. T. Schimke, *J. Biol. Chem.* **246**, 7407 (1971).

inhibition of ribonuclease activity. At lower concentrations, i.e., 40–100 μg/ml, recovery of mRNA activity is only 40–50% of that at 500 μg/ml. Higher heparin concentrations are without effect on mRNA recovery. It is essential to remove the heparin from RNA solutions since it is a potent inhibitor of protein synthesis initiation. Heparin removal is accomplished either by the sucrose centrifugation as described above or by the use of lithium chloride precipitation of RNA as described in this volume [59]. Figure 1 shows the time course of incorporation of isotope into total protein and into immunoprecipitated ovalbumin. Incorporation into total protein is rapid for 10 minutes, slows to a new rate, and essentially ceases after 60 minutes. There is a lag of 8–10 minutes in the appearance of ovalbumin, and then incorporation continues until 40 minutes.

That the assay is quantitative in indicated by the linear relationship between counts incorporated into ovalbumin and the amount of added RNA (see Figs. 4 and 6 from this volume [59]).

Proof That Immunoprecipitated Product Is Ovalbumin

1. RNA extracted from hen liver, which does not synthesize ovalbumin does not stimulate amino acid incorporation into a form that is immunoprecipitable.

2. The immunoprecipitated material is specific for anti-ovalbumin since an immunologic reaction of added bovine serum albumin and anti-bovine serum albumin does not precipitate any radioactivity.

3. The immunoprecipitated material is the same size as ovalbumin.

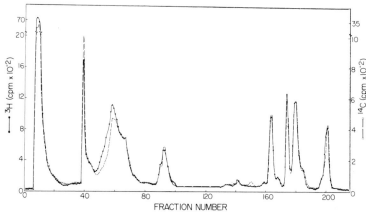

FIG. 3. Ion exchange chromatography of peptides produced by tryptic digestion of anti-ovalbumin precipitated [³H]leucine-labeled reaction product (●—●) and [¹⁴C]leucine-labeled ovalbumin (○----○). Details are given by R. E. Rhoads, G. S. McKnight, and R. T. Schimke, *J. Biol. Chem.* **246**, 7407 (1971).

Figure 2 shows that only on addition of oviduct RNA is there the appearance of a band of radioactivity with migratory properties of ovalbumin, when the immunoprecipitate is subjected to SDS acrylamide gel electrophoresis.

4. The tryptic peptides of the immunopreciptated ovalbumin are the same as those of authentic ovalbumin. This is shown in Fig. 3.

[65] Isolation on Cellulose of Ovalbumin and Globin mRNA and Their Translation in an Ascites Cell-Free System

By GUNTHER SCHUTZ, MIGUEL BEATO, and PHILIP FEIGELSON

The procedures described below have been developed for studying changes in the messenger RNA activities during cell differentiation and in the course of steroid hormone action. Exploiting the poly(A) content of cellular and viral mRNA's,[1] eukaryotic messenger RNA's can be isolated by chromatography on nucleotide-free cellulose. A modified cell-free protein synthesis system, derived from Krebs II–ascites cells allows the efficient and faithful translation of these mRNA's.

Isolation of Messenger RNA from Rabbit Reticulocytes and Chicken Oviduct

Extraction of RNA. (a) The polysomal-ribosomal fraction of reticulocytes was prepared from blood of 10 phenylhydrazine-treated rabbits as described by Anderson.[2] The polysomes were suspended in 40 ml of cold 50 mM Tris·HCl (pH 8.3)–5 mM EDTA–75 mM NaCl–0.5% sodium dodecyl sulfate (SDS) and an equal volume of cold buffer saturated phenol, adjusted to pH 8.3, was added. The mixture was shaken for 5 minutes at 4° and the phases were separated by centrifugation at 10,000 g for 10 minutes. The aqueous phase was twice reextracted with 0.75 and 0.5 volume, respectively, of buffer saturated phenol, made 0.3 M with respect to LiCl, and the RNA was precipitated with 2 volumes of ethanol at −20°. The ethanol precipitate was dissolved in H₂O.

(b) Chicken oviduct RNA was obtained by homogenizing 20 g of the magnum portion of the oviduct from laying hens in 130 ml of cold 50 mM Tris·HCl (pH 8.3)–5 mM EDTA–75 mM NaCl–0.5% SDS plus 150 ml of cold buffer saturated phenol in a Waring Blendor for 1 minute. The

[1] J. E. Darnell, L. Philipson, R. Wall, and L. M. Adesnik, *Science* **174**, 507 (1971).
[2] J. M. Gilbert and W. F. Anderson, *J. Biol. Chem.* **245**, 2342 (1970).

RNA was obtained as above except that the phenol extraction was repeated five times.

Cellulose Chromatographic Isolation of Poly(A)Containing Messenger RNA

The ability of poly(A) and poly(A)-containing RNA to bind to nitrocellulose filters,[3-5] to oligo-(dT)-cellulose[6,7] and to poly(U)-cellulose[8] has been established. More recently, poly(A) has been reported to bind to certain preparations of nucleotide-free cellulose.[9] We have found that microgranular cellulose (SIGMACELL type 38 from Sigma Co.) was

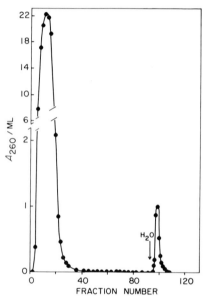

FIG. 1. Chromatography on cellulose of rabbit reticulocyte polysomal RNA. 260 A_{260} units were applied in 20 ml of 10 mM Tris·HCl (pH 7.6), 500 mM KCl, and 0.2 mM MgCl, and extensively washed with the same buffer. Elution with water is indicated by arrow.

[3] S. Y. Lee, J. Mendecki, and G. Brawerman, *Proc. Nat. Acad. Sci. U.S.* **68**, 1331 (1971).

[4] G. Brawerman, J. Mendecki, and S. Y. Lee, *Biochemistry* **11**, 637 (1972).

[5] G. C. Rosenfeld, J. P. Comstock, A. R. Means, and B. W. O'Malley, *Biochem. Biophys. Res. Commun.* **47**, 387 (1972).

[6] M. Edmonds and M. G. Caramela, *J. Biol. Chem.* **244**, 1314 (1969).

[7] H. Aviv and P. Leder, *Proc. Nat. Acad. Sci. U.S.* **69**, 1408 (1972).

[8] J. Kates, *Cold Spring Harbor Symp. Quant. Biol.* **35**, 743, (1970).

[9] P. A. Kitos, G. Saxon, and H. Amos, *Biochem. Biophys. Res. Commun.* **47**, 1426 (1972).

over twenty times more effective than fibrillar cellulose (Cellex-N-1, BioRad Laboratories) in selectively binding poly(A). Furthermore, at high ionic strength, poly(A)-rich RNA was bound and ribosomal and transfer RNA did not bind to the Sigma microgranular cellulose. The bound poly(A)-rich RNA species were elutable by low ionic strength media.

Thus, 100–400 A_{260} units of total oviduct RNA or reticulocyte polysomal RNA, dissolved in 10 mM Tris·HCl (pH 7.6)–500 mM KCl–0.2 mM MgCl$_2$, were applied at room temperature to a 2–4 ml cellulose column (dry weight 0.5–1 g) which had been equilibrated with the same buffer. The column was washed with this buffer eluting the unbound 28 S and 18 S ribosomal RNA species. The washing was continued until the eluant A_{260} level had dropped below 0.01 (Fig. 1). The poly(A) containing adsorbed RNA (messenger RNA) was eluted with neutralized water then adjusted to be 0.3 M in LiCl and precipitated with two volumes of ethanol at $-20°$. The precipitated RNA was washed two times with ethanol and dissolved in H$_2$O to a concentration of 1 mg/ml (assuming 25 A_{260}/ml equals 1 mg/ml). Two to three percent of the applied total oviductal RNA and 1.2% of reticulocyte polysomal RNA was retained by the column at high ionic strength, eluted by water, and hence presumably represents poly(A)-rich messenger RNA.

Cell-Free System

Preparation of Cell-Free System. Except for some minor modifications, the procedure of Mathews and Korner was followed.[10]

Krebs II–ascites tumor cells were transplanted at weekly intervals by intraperitoneal injection of 0.1 ml of the 1:1 diluted ascitic fluid into 6-week-old female CD-1 mice, obtained from Charles River Breeding Laboratories, Boston, Massachusetts. Only nonbloody tumors were used for propagation of the cell line and the preparation of the cell-free system.

The ascitic fluid was removed by aspiration, and the cells were obtained by a 5-minute centrifugation at 70 g. The cells were washed three times in 35 mM Tris·HCl, pH 7.5–0.14 M NaCl. After packing the cells by centrifugation at 200 g for 5 minutes, 2 volumes of 10 mM Tris·HCl, pH 7.5–1.5 mM magnesium acetate–10 mM KCl–1 mM dithiothreitol were added, and the cells allowed to swell for 5 minutes. They were then homogenized with 15 strokes of a tight-fitting glass homogenizer. One-tenth volume of 300 mM Tris·HCl, pH 7.5–1.2 M KCl–50 mM magnesium acetate–10 mM dithiothreitol was added and the homogenate was centrifuged for 10 minutes at 30,000 g. The supernatant was brought to 1 mM

[10] M. B. Mathews and A. Korner, *Eur. J. Biochem.* **17**, 328 (1970).

ATP, 0.1 mM GTP, 0.6 mM CTP, 5 mM creatine phosphate, 0.16 mg/ml creatine kinase, and 20 amino acids, each 40 μM; the extract was then incubated for 45 minutes at 37° and again centrifuged for 10 minutes at 30,000 g. The supernatant was dialyzed against 3 changes of 30 mM Tris·HCl, pH 7.5–120 mM KCl–5 mM magnesium acetate–1 mM dithiothreitol and stored in small aliquots after quick freezing at −80°.

The "salt wash fraction" from rabbit reticulocyte ribosomes was found to markedly enhance the rate, and extend the period of rapid amino acid incorporation. This fraction was prepared from reticulocyte ribosomes obtained from phenylhydrazine-treated rabbits, according to Gilbert and Anderson.[2] To a suspension of ribosomes ($\simeq 100$ A_{260}/ml) in 0.25 M sucrose–0.5 mM EDTA–1 mM DTT adjusted to pH 7.0, KCl was slowly added from a 3.0 M solution to yield a final concentration of 0.5 M. The solution was stirred for 1 hour at 4°, followed by a centrifugation for 90 minutes at 200,000 g. The supernatant was diluted with 0.25 M sucrose–0.5 mM EDTA–1 mM DTT to a KCl concentration of 0.3 M and to remove any polynucleotide contaminant was then passed through a DEAE-cellulose column which had been equilibrated in the same buffer. The protein containing eluate, as indicated by the presence of hemoglobin, was pooled and diluted to 0.1 M final KCl concentration and centrifuged for 13 hours at 58,000 rpm in a 60 Ti rotor (240,000 g average). The bottom milliliter was aspirated by a Pasteur pipette and after quick freezing, stored at −80° in small aliquots.

The tRNA was prepared from livers of laying hens following the procedure of Beck et al.[11]

Incubation Conditions for Protein Synthesis

Routine assays contained in 50 μl: 30 mM Tris·HCl (pH 7.5), 95 mM KCl, 3 mM magnesium acetate, 1 mM dithiothreitol, 1 mM ATP, 0.1 mM GTP, 0.6 mM CTP, 5 mM creatine phosphate, 0.16 mg/ml creatine kinase, 0.04 mM of 19 unlabeled amino acids, 5 μCi of [³H]leucine (31.6 Ci/mmole), 0.15 A_{260} of laying hen hepatic tRNA, 0.3 A_{280} of a 0.5 M KCl wash fraction from rabbit reticulocyte ribosomes, 0.3–0.4 A_{260} unit of ascites lysate, and 0.1–2.5 μg of the various mRNA fractions.

Incubation was performed at 37° and the incorporated radioactivity was determined on filter paper disks as described by Bollum.[12] Time-course studies indicated a rapid extensive [³H]leucine incorporation, at a slightly declining rate of up to 150 minutes. The optimal KCl concentration in the absence of the reticulocyte ribosomal salt wash fraction was found to be

[11] G. Beck, D. Hentus, and J. P. Egel, *Biochim. Biophys. Acta* **213**, 55 (1970).
[12] F. J. Bollum, this series, Vol. 12, p. 169.

Cellulose-RNA added (μg/50 μl)	RSW	KCl (mM)	[³H]Leucine incorporated (cpm/50 μl)
None	−	65	760
Oviduct (2.5)	−	65	1,770
Reticulocyte (0.3)	−	65	3,055
None	+	95	10,740
Oviduct (2.5)	+	95	258,100
Reticulocyte (0.3)	+	95	234,200

65 mM whereas in its presence the optimal concentration shifted to 95 mM; 3 mM Mg acetate was optimal in both instances. The system shows a requirement for tRNA, the optimal concentration of 0.15 A_{260} unit permitting 2.5-fold higher incorporation rates. As seen in the table, the efficiency of translation of globin and oviduct mRNA is dramatically increased by supplementing the system with initiation factors from reticulocyte ribosomes, which make the system's response sensitive enough to easily detect fractions of micrograms mRNA.[13] The incorporation of [³H]leucine into protein is proportional to the amount of added mRNA up to about 1 μg per 50 μl of assay volume.[14]

Cell-Free Synthesis of Globin and Ovalbumin

To establish that faithful translation of the rabbit globin and the hen oviductal ovalbumin messenger RNA's was occurring, the proteins synthesized *in vitro* in response to these respective mRNA were characterized.

Globin Synthesis. Following the incubation period, 50 μl of water were added and the ribosomes were removed by centrifugation of the incubation mixture at 150,000 g for 60 minutes; 0.2 ml of 0.1 M KOH was added to the supernatant containing the "released chains" and incubated for 20 minutes at 37°C. One milliliter of 10% Cl₃CCOOH was added, and the precipitate was washed two times with cold 5% Cl₃CCOOH and twice with acetone. The precipitate was dissolved in 10 mM sodium phosphate (pH 7.0)–1% SDS–1% β–mercaptoethanol and electrophoresed according to Weber and Osborn.[15] The gels were cut into 1.6-mm slices and dissolved in 1 ml of H₂O₂ by incubation overnight at 60°. The samples were

[13] S. Metafora, M. Terada, L. W. Dow, P. A. Marks, and A. Bank, *Proc. Nat. Acad. Sci. U.S.* **69**, 1299 (1972).
[14] G. Schutz, M. Beato, and P. Feigelson, *Biochem. Biophys. Res. Commun.* **49**, 680 (1972).
[15] K. Weber and M. Osborn, *J. Biol. Chem.* **244**, 4406 (1969).

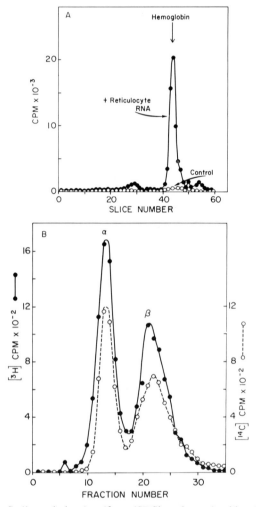

FIG. 2. (A) Sodium dodecyl sulfate (SDS)–polyacrylamide electrophoresis of proteins synthesized and released *in vitro* in response to reticulocyte mRNA. The chains released after 100 minutes incubation from a 30-μl reaction mixture containing 0.72 μg of cellulose-purified reticulocyte RNA were analyzed on a SDS-polyacrylamide gel. The arrow indicates the position of hemoglobin used as internal marker. The direction of anodal migration is from left to right. (B) Separation of *in vitro* synthesized α- and β-globin chains on CM-cellulose. The globin chains released after 60 min incubation from a 100-μl reaction mixture containing 2.5 μg of cellulose-purified reticulocyte RNA were prepared after addition of 2 mg of ^{14}C-labeled globin as described.[14] The α- and β-chains were separated on CM-cellulose by the method described.[16]

counted in Aquasol. As seen in Fig. 2A, where reticulocyte mRNA was utilized, the majority of the released polypeptide chains comigrated with authentic rabbit globin. To determine the relative rates of synthesis of the α- and β-globin chains, an aliquot of the [³H]leucine labeled newly synthesized released chains was mixed with [¹⁴C]labeled globin (specific activity 20,000 cpm/mg). Heme was removed by acetone extraction and the α- and β-globin chains separated on CM-cellulose according to Rabinowitz and Fischer.[16] Approximately symmetrical synthesis of α- and β-globin chains was demonstrated by CM-cellulose chromatography of the proteins newly synthesized in response to reticulocyte mRNA (Fig. 2B).

Ovalbumin Synthesis. Ovalbumin synthesis directed by the fraction of chicken oviduct RNA retained by cellulose was established by specific

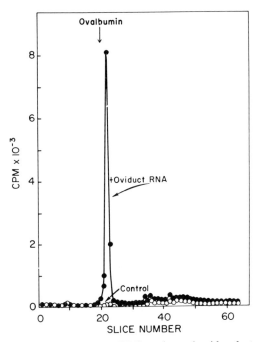

FIG. 3. Sodium dodecyl sulfate, (SDS)–polyacrylamide electrophoresis of the proteins synthesized *in vitro* in the presence of oviduct mRNA and immunologically isolated. The released polypeptide chains after 60 minutes incubation period of a 90-μl reaction system containing 4.5 μg oviductal mRNA were isolated immunochemically with rabbit antiovalbumin γ-globulin. The immunoprecipitate was subjected to SDS-polyacrylamide electrophoresis. The arrow indicates the position of authentic ovalbumin used as an internal marker. The direction of migration is from left to right.

[16] M. Rabinowitz and J. M. Fischer, *Biochim. Biophys. Acta* 91, 313 (1964).

immunoprecipitation with anti-ovalbumin γ-globulin and subsequent SDS-polyacrylamide electrophoresis of the immunoprecipitate. Aliquots (100 μl) of the released chains were incubated in a total volume of 0.35 ml containing 10 mM sodium phosphate, pH 7.0, 0.14 M NaCl, 2% Triton X-100 with 10 μg of authentic ovalbumin as carrier and 50 μl of anti-ovalbumin γ-globulin,[17] for 30 minutes at room temperature and then overnight at 4°. The immunoprecipitate was collected by centrifugation and then washed three times with 0.35 ml of 10 mM sodium phosphate (pH 7.0)–0.14 M NaCl–2% Triton X-100. The immunoprecipitate was dissolved in 10 mM sodium phosphate (pH 7.0)–1% SDS–1% β-mercapto-ethanol and electrophoresed on SDS gels as described for the globin chains. An incubation mixture without added mRNA was similarly processed and served as control. The translation of the ovalbumin message yields a polypeptide product that reacts with anti-ovalbumin γ-globulin and electrophoreses upon SDS gel slightly faster than authentic ovalbumin, as shown in Fig. 3. We consider it likely that the newly synthesized "ovalbumin" lacks the 1500 molecular weight carbohydrate moiety, which is a component of the authentic ovalbumin, and hence migrates slightly faster on the SDS-acrylamide gel.[18]

Addendum Added in Proof

Recently, these procedures have enabled (a) the partial purification of the rat hepatic mRNA coding for tryptophan oxygenase, (b) its faithful translation *in vitro* involving initiation, elongation, and release of the completed enzymic protomeric units, and (c) the assay of the liver level of the mRNA for tryptophan oxygenase during hormonal induction of this enzyme.[19]

[17] D. M. Weir, "Handbook of Experimental Immunology," pp. 3–10. Davis, Philadelphia, 1967.

[18] R. E. Rhoads, G. S. McKnight, and R. T. Schimke, *J. Biol. Chem.* **246**, 7407 (1971).

[19] G. Schutz, M. Beato, and P. Feigelson, *Proc. Nat. Acad. Sci.* **70**, 1218 (1973).

[66] Translation of Reovirus mRNA, Poliovirus RNA, and Bacteriophage Qβ RNA in Cell-Free Extracts of Mammalian Cells

By LYDIA VILLA-KOMAROFF, MAXSON MCDOWELL, DAVID BALTIMORE, and HARVEY F. LODISH

The translation of viral messenger RNA in cell-free extracts from mammalian cells is one of the most powerful tools available for elucidation both of regulation of virus gene function and of the more general problem of the mechanism and regulation of polypeptide chain initiation and termination. In this chapter we discuss the techniques currently in use in our laboratory for the study of translation of three viral mRNA's: poliovirus RNA, reovirus mRNA, and bacteriophage Qβ RNA. We will discuss the techniques for growing the cells, and the viruses and for purifying the mRNA's. We have utilized two types of cell-free systems. First, preincubated extracts from several types of cultured mammalian cells.[1-3] In this case incorporation of amino acids into protein is completely dependent on the addition of exogenous mRNA. Second, a non-preincubated lysate prepared from rabbit reticulocytes.[3-5] In this case the addition of viral mRNA does not stimulate—and often inhibits—overall protein synthesis. However, since the predominant protein made by these lysates is globin, it is often easy to identify virus-specific proteins. The advantages of the lysate system, compared to the preincubated extracts are: (1) the amount of virus-specific protein is considerably greater, (2) the system is free of nucleases, and (3) longer completed virus polypeptides are made.[3]

Cell Culture

HeLa, L, and CHO cell lines are maintained in suspension cultures at 37° in Eagle's Minimal Essential Medium (Joklik's modification, Grand Island Biological Co.) supplemented with 7% serum. Horse serum is used for the HeLa cells; fetal calf serum for the CHO and L cells. Medium for the CHO cells is further supplemented with nonessential amino acids (1:100 dilution of 100 × NEAA from Grand Island Biological Co.).

HeLa cells should be maintained between 2 and 4×10^5 cells/ml;

[1] I. M. Kerr and E. M. Martin, *J. Virol.* **7**, 438 (1971).
[2] M. B. Mathews and A. Korner, *Eur. J. Biochem.* **17**, 328, 339 (1970).
[3] M. J. McDowell, W. K. Joklik, L. Villa-Komaroff, and H. F. Lodish, *Proc. Nat. Acad. Sci. U.S.* **69**, 2649 (1972).
[4] R. E. Rhoads, G. S. McKnight, and R. T. Schimke, *J. Biol. Chem.* **246**, 7407 (1971).
[5] S. Stavnezer and R. C. Huang, *Nature (London)* **230**, 172 (1971).

L cells between 5 and 10×10^5 cells/ml; and CHO cells between 2 and 6×10^5 cells/ml.

Krebs II ascites cells (courtesy of A. Burness, Sloan-Kettering Institute) are maintained by passaging in the peritoneal cavity of genetically heterogeneous female white mice (strain CD-1, Charles River Breeding Laboratory, Wilmington, Massachusetts). These cells should be harvested and passaged every 7 days, using the following procedure.[3,6]

Using a blunt tool break the neck of a mouse that had been injected 7 days previously. Open the peritoneal cavity with a pair of clean scissors and drain the cell suspension into a small, clean, sterile beaker. One mouse will provide 2–10 ml of fluid with about 10^8 cells/ml. Bright yellow fluid, which is probably contaminated with bacteria, or very bloody fluid should not be used for passaging.

Take up the fluid in a sterile syringe. Inject 0.1 to 0.2 ml of fluid into a mouse—be sure the needle has penetrated into the peritoneal cavity. Cells injected under the skin usually do not form tumors.

Preparation of Cell-Free Extracts from Cultured Cells[3,6]

Reagents

> Isotonic Buffer: 35 mM Tris·HCl, pH 7.5; 146 mM NaCl; 11 mM glucose
> Hypotonic Buffer: 15 mM KCl; 10 mM Tris·HCl, pH 7.5; 1.5 mM MgAc$_2$; 6 mM 2-mercaptoethanol
> 10X Incubation Buffer: 200 mM HEPES, pH 7.5; 1200 mM KCl; 50 mM MgAc$_2$; 60 mM 2-mercaptoethanol
> Column Buffer: 20 mM HEPES, pH 7.5; 120 mM KCl; 1.5 mM MgAc$_2$; 6 mM 2-mercaptoethanol

All procedures should be carried out at 0–4° unless otherwise specified.

Harvest the liquid tumors of ascites cells from five mice. Filter the fluid through one or two layers of cheesecloth into about 400 ml of isotonic buffer. All fluids, except bright yellow ones, can be used for the extracts. Wash the cells five times by differential centrifugation (80 g for 5 minutes) in isotonic buffer to remove reticulocytes. Collect by centrifugation L cells, HeLa cells, and CHO cells from 2 or more liters of suspension culture. Wash three times in isotonic buffer. After this point the procedure is identical for all four types of cells.

After the last wash, resuspend the cells in a small volume of isotonic

---

[6] D. Housman, R. Pemberton, and R. Taber, *Proc. Nat. Acad. Sci. U.S.* **68**, 2716 (1971).

buffer and transfer them to a graduated conical centrifuge tube. Pack the cells by centrifugation at 2000 rpm for 10 minutes in the International PR-6 centrifuge. Note the cell volume. Drain off as much buffer as possible. Resuspend one volume of cells in 3 volumes of hypotonic buffer. After 10 minutes at 0°, break the cells with 30–50 strokes of a tight-fitting dounce homogenizer. Add 0.1 volume of 10X incubation buffer and centrifuge the homogenate at 30,000 g for 20 minutes. Discard the pellet. Add ATP to a final concentration of 1 mM, GTP to 0.2 mM, creatine phosphate to 8 mM, and creatine phosphokinase to 0.2 mg/ml. Incubate the extract for 30–60 minutes at 37°. If the extract is turbid, centrifuge at 5000 rpm in the Sorvall No. SS-34 rotor for 10 minutes and remove any fluocculent material with a sterile glass tube or by passing the extract through several layers of cheesecloth. Pass the extract through a G-25 column (3 × 30 cm) that has been equilibrated with column buffer and elute with the same buffer. Pool the opalescent fractions and freeze in small aliquots at −80° or in liquid nitrogen.

Preparation of Reticulocyte Lysates[7,8]

Reagents

Saline: 0.140 M NaCl; 1.5 mM MgAc$_2$; 5 mM KCl
Acetylphenylhydrazine (Sigma Chemical Co.) 1.2% solution
Neutralized to pH 7.5 with 1 M HEPES solution (pH 7.0)

Six New Zealand White rabbits weighing 4–6 pounds are made anemic by subcutaneous injection of 1.2% acetylphenylhydrazine according to the following schedule: 2 ml on day 1, 1.6 ml on day 2, 1.2 ml on day 3, 1.6 ml on day 4, and 2 ml on day 5. On days 7 and 8 the rabbits are bled: One ear is swabbed with cotton saturated with xylene, and a single incision, using new razor blades, is made in the posterior ear vein about midway along the length of the ear. Each rabbit should yield 50–60 ml of blood, which is collected into 50 ml of chilled saline containing 0.001% heparin. The blood is filtered through cheesecloth, then centrifuged at 3500 rpm for 5 minutes. The cells are washed by centrifugation three times, with the last centrifugation at 7000 rpm. Packed cells are lysed at 0° with an equal volume of cold water. After 1 minute, the lysate is centrifuged at 15,000 rpm for 20 minutes. Aliquots (usually 0.5 ml) of the supernatant are frozen at −80°, at which temperature activity is stable for several months.

[7] D. Housman, M. Jacobs-Lorena, U. L. RajBhandary, and H. F. Lodish, *Nature* (*London*) **227**, 913 (1970).
[8] S. D. Adamson, W. Godchaux, and E. Herbert, *Arch. Biochem. Biophys.* **125**, 671 (1968).

Preparation of Poliovirus[9,10]

Poliovirus, like all picornaviruses, contains a single-stranded RNA which is the template for the synthesis of viral proteins in virus-infected cells.[9] The RNA contains about 7500 nucleotides, and it sediments at about 35 S in SDS-gradients.[9] *In vivo,* protein synthesis is initiated only at the 5′ end and the viral proteins are generated by cleavage from one long precursor protein.[11-13] The order of the synthesis of the viral polypeptides has been determined.[14,15] Poliovirus Moloney type I is used, and the virus is grown in HeLa cells.

The preparation of EMC virus and EMC viral RNA, another picornavirus used widely in the ascites system, has been described elsewhere.[1,2,16,17,18]

Reagents

Earle's Saline: 0.47 M NaCl; 20 mM KCl; 0.87 mM MgCl$_2$; 20 mM glucose; 0.1 M NaHCO$_3$; 10 mM NaH$_2$PO$_4$; pH 7.4
RSB: 10 mM Tris, pH 7.35; 10 mM NaCl; 15 mM MgCl$_2$
SDS Buffer: 0.1 M NaCl; 10 mM Tris, pH 7.5; 1 mM EDTA
Sucrose Solutions: 15% and 30% sucrose in SDS-buffer with 0.5% SDS

The HeLa cells from 6 liters of culture (4×10^5 cells/ml) are harvested by centrifugation in 250-ml glass centrifuge bottles at 1000 rpm for 5 minutes in the International PR-6 centrifuge. The cells are washed once with Earle's saline and resuspended at 4×10^8 cells/ml in a solution of virus containing about 4×10^9 PFU. The virus is allowed to adsorb at room temperature for 30 minutes and the infection is then initiated by diluting the cells to 4×10^6 cells/ml with warm (37°) medium containing 5% horse serum. Six and one half to 7 hours after the beginning of the infection, harvest the cells by centrifugation at 1000 rpm for 5 minutes. Re-

[9] D. Baltimore, *in* "The Biochemistry of Viruses" (H. B. Levy, ed.), p. 101. Dekker, New York, 1969.
[10] D. Baltimore, M. Girard, and J. E. Darnell, *Virology* 29, 179 (1966).
[11] D. Baltimore, *in* "From Molecules to Man" (M. Pollard, ed.), p. 1. Academic Press, New York, 1971.
[12] M. F. Jacobson and D. Baltimore, *Proc. Nat. Acad. Sci. U.S.* 61, 77 (1968).
[13] M. F. Jacobson, J. Asso, and D. Baltimore, *J. Mol. Biol.* 49, 657 (1970).
[14] R. Taber, D. Rekosh, and D. Baltimore, *J. Virol.* 8, 395 (1971).
[15] D. Rekosh, *J. Virol.* 9, 479 (1972).
[16] H. Aviv, I. Boime, and P. Leder, *Proc. Nat. Acad. Sci. U.S.* 68, 2303 (1971).
[17] A. G. Smith, K. A. Marcker, and M. B. Mathews, *Nature (London)* 225, 184 (1970).
[18] A. T. H. Burness, *J. Gen. Virol.* 5, 291 (1969).

suspend the cells to a concentration of 4×10^7 cells/ml in RSB. Freeze and thaw the preparation three times, then remove the cellular debris by centrifugation at 1500 rpm for 10 minutes. Recentrifuge the supernatant at 1500 rpm for 3 minutes and discard any pellet. At this point production of virus can be quantitated by the plaque assay.[19]

Preparation of Polio Messenger RNA[20]

Add sodium dodecyl sulfate to 1%, then recover the virus by centrifuging the solution at 28,000 rpm in the 30 rotor of a Spinco ultracentrifuge for 3 hours at 22°. Resuspend the virus pellet in SDS buffer containing 0.5% SDS and layer on a linear 15–30% sucrose gradient in SDS buffer also containing 0.5% SDS. Centrifuge the gradient in the Spinco SW 27 rotor at 27,000 rpm for 2.5 hours. Collect the gradient through a recording spectrophotometer, and pool the virus fractions. Centrifuge the pooled fractions for four hours in the Spinco 30 rotor at 30,000 rpm. Resuspend the virus pellet in a small volume of RSB, add SDS to 1% and disrupt the virions by adding 0.1 M sodium acetate pH 3.5 to the virus until the solution is pH 3.5. Layer the solution onto a linear 15–30% sucrose gradient in SDS buffer containing 0.5% SDS. Centrifuge the RNA in the SW 27 Spinco rotor for 15 hours at 22,000 rpm. Pool the RNA fractions, and add NaCl to a concentration of 0.4 M. Add 2 volumes of ethanol and store the RNA overnight at $-20°$. Pellet the RNA by centrifugation at 10,000 rpm for 15 minutes in the Sorvall centrifuge. Wash the RNA by centrifugation twice with ethanol and once with ether. Resuspend the RNA in H_2O. The yield of poliovirus RNA from 6 liters of cells in about 11 A_{260} units.

Preparation of Reovirus and Reovirus mRNA

Reovirus particles contain a double-stranded-RNA dependent RNA polymerase which transcribes messenger RNA from the viral genome. This enzyme becomes capable of synthesizing messenger RNA *in vitro* when the virus is treated with chymotrypsin.[21–23] Ten species of monocistronic messenger RNA are synthesized *in vitro* in nearly equal amounts by weight.[23] They can be divided into three groups on the basis of their size; their molecular weights are in the order of 1.3×10^6 daltons for the three largest species, 0.8×10^6 for the three medium-sized species, and $0.4 \times$

[19] P. D. Cooper, *Virology* 13, 153 (1961).
[20] N. Granboulan and M. Girard, *J. Virology* 4, 475 (1969).
[21] J. Borsa and A. F. Graham, *Biochem. Biophys. Res. Commun.* 33, 895 (1968).
[22] A. J. Shatkin and J. D. Sipe, *Proc. Nat. Acad. Sci. U.S.* 61, 1462 (1968).
[23] J. J. Skehel and W. K. Joklik, *Virology* 39, 822 (1969).

10^6 for the four smallest species.[24-26] That these are functional messenger RNA molecules is shown by the fact that they are able to code for the *in vitro* synthesis of complete viral polypeptides.[3,27]

Reagents

Earle's Saline: NaCl, 0.47 M; KCl, 20 mM; MgCl$_2$, 0.87 mM; glucose, 20 mM; NaHCO$_3$, 0.1 M; NaH$_2$PO$_4$, 10 mM; pH 7.2

Homogenization Buffer: NaCl, 0.25 M; 2-mercaptoethanol, 10 mM; Tris, 10 mM; pH 8

SSC: NaCl, 0.15 M; sodium citrate, 15 mM; pH 7.5

Sucrose Gradient Solutions: sucrose, 20% and 40%; NaH$_2$PO$_4$, 1 mM; pH 7.2

Loening's Buffer: NaH$_2$PO$_4$, 30 mM; EDTA, 1 mM; Tris, 36 mM; pH 7.8

Methylene Blue Solution: methylene blue, 0.2%; sodium acetate, 0.4 M; pH 5

Preparation of Virus[28]

Six liters of an actively growing suspension culture of L cells containing 12×10^5 cells/ml are infected by adding purified virus at a multiplicity of 500 virus particles per cell. The Dearing strain of Reovirus type 3 is used. One-third volume of fresh medium is added and the cells are incubated at 34° for 20 hours.

All subsequent steps are carried out at 4°. The cell culture is allowed to settle, excess medium is siphoned off and the cells, which contain the progeny virus, are collected by centrifugation at 500 g for 5 minutes. The cells are gently resuspended in Earle's saline, centrifuged again, and then resuspended in homogenization buffer at a concentration of 0.5×10^8 cells/ml and homogenized in a Virtis 23 homogenizer for 2 minutes at full speed. One-third volume of Freon 113 (trichlorotrifluoroethane) is added, the homogenization is repeated, and the phases are then separated by centrifuging at 2000 g for 15 minutes. The organic phase is reextracted with an equal volume of homogenization buffer, after which the aqueous phases are combined and reextracted with one-third volume of Freon 113. Virus is collected from the final aqueous phase by centrifuging at 25,000 rpm

[24] Y. Watanabe, L. Prevec, and A. F. Graham, *Proc. Nat. Acad. Sci. U.S.* **58**, 1040 (1967).

[25] H. J. Zweerink, M. J. McDowell, and W. K. Joklik, *Virology* **45**, 716 (1971).

[26] A. R. Bellamy and W. K. Joklik, *J. Mol. Biol.* **29**, 19 (1967).

[27] W. D. Graziadei and P. Lengyel, *Biochem. Biophys. Res. Commun.* **46**, 1816 (1972).

[28] R. E. Smith, H. J. Zweerink, and W. K. Joklik, *Virology* **39**, 791 (1969).

for 60 minutes in the SW 27 rotor of a Spinco ultracentrifuge. The virus is resuspended from the pellet by stirring overnight in SSC, then sonicated briefly and layered onto six 20–40% linear sucrose density gradients which are centrifuged in the SW 27 rotor at 25,000 rpm for 60 minutes. The visible bands of virus are collected, and the pellets from these gradients are resuspended and recentrifuged through sucrose gradients as before, yielding further visible bands of virus, which are also collected. [The band of top component,[28] which is less dense than virus and sediments more slowly, can be separated from the virus band during collection from the sucrose gradients or more conveniently during collection from the subsequent CsCl gradient.] Virus from the combined bands is diluted with three volumes of SSC and centrifuged at 25,000 rpm for 60 minutes. The pellet is resuspended into SSC and the virus suspension is layered onto six preformed CsCl gradients of density 1.2–1.4 g/ml in the SW 41 rotor of a Spinco ultracentrifuge. These gradients are centrifuged at 25,000 for 120 minutes, and the resultant bands of virus (density 1.36 g/ml) are collected and dialyzed overnight against several changes of SSC. The yield from 6 liters of cells is normally greater than 100 mg of virus. Virus concentration is estimated from the relationship: 5.4 $OD_{260} \simeq 1$ mg of virus $\simeq 1.1 \times 10^{13}$ virus particles.

Preparation of Subviral Particles[28]

Freshly purified virus (50 mg) in 10 ml of SSC is incubated with stirring at 37° for 90 minutes in the presence of 0.1 mg of chymotrypsin per milliliter. The subviral particles thus formed are layered onto four preformed CsCl gradients of density 1.3–1.5 g/ml in the SW 41 rotor and centrifuged at 36,000 rpm for 100 minutes. The visible bands of subviral particles (density 1.44 g/ml) are collected and dialyzed against SSC overnight. With this procedure there is a nearly quantitative yield of purified subviral particles from virus.

Preparation of Messenger RNA[23]

Stringent precautions should be taken to keep all solutions and glassware free of RNase for this preparation. Messenger RNA is synthesized in 15 ml of reaction mixture which contains freshly purified subviral particles derived from 50 mg of virus and the following components in the indicated concentrations: ATP, GTP, UTP, CTP, 1.25 mg/ml each; phosphoenolpyruvate, 5 mg/ml; pyruvate kinase, 0.05 mg/ml; Macaloid (Baroid Division of National Lead Co.), 0.5 mg/ml; $MgCl_2$, 10 mM; Tris, pH 8, 100 mM. The reaction is allowed to proceed with stirring for 6 hours at 37°; the reaction mixture is then centrifuged at 35,000 rpm for 60 minutes in the SW 50 rotor of a Spinco ultracentrifuge to remove subviral particles.

The supernatant, which contains completed molecules of messenger RNA, is extracted with one-third volume of water-saturated phenol to remove residual proteins, mixed with two volumes of ethanol and kept at $-20°$ for several hours. The dense precipitate which is formed contains RNA and an excess of nucleoside phosphates. This material is twice redissolved in distilled water and reprecipitated in the presence of 2 M lithium chloride at 4°, then twice redissolved in distilled water and reprecipitated in the presence of 0.2 M potassium acetate and two volumes of ethanol at $-20°$. The final precipitate which contains purified messenger RNA is rinsed with ethanol, dissolved in distilled water, and stored at $-20°$. The yield is approximately 40 OD_{260} units from 15 ml of reaction mixture.

Assay for the Integrity of the Messenger RNA

It is necessary to check that the RNA which has been synthesized is not significantly degraded, and for this purpose a sample of the RNA is electrophoresed on polyacrylamide gels[29] after being denatured with DMSO (dimethyl sulfoxide) and then allowed to renature. Although the messenger RNA cannot be separated into ten species by electrophoresis, the resolution which is obtained is quite sufficient for this assay.

Acrylamide gels (3%) with a diameter of 0.6 cm and a length of 12 cm are prepared from a solution with the following composition: 30% acrylamide, 0.8% N,N'-methylenebisacrylamide, 3 ml; 10% sodium dodecyl sulfate, 0.3 ml; N,N,N',N'-tetramethylethylenediamine, 0.03 ml; fresh ammonium persulfate (10 mg/ml), 0.3 ml; Loening's buffer, 26.4 ml. The electrophoresis buffer is Loening's buffer supplemented with 0.1% SDS.

Messenger RNA (30 μg) is dissolved in 0.1 ml of water, 0.9 ml of DMSO is added, and the mixture is incubated at 45° for 5 minutes. Transfer RNA (50 μg) is added as carrier, and the RNA is precipitated in the presence of 0.2 M potassium acetate and two volumes of ethanol at $-20°$. The RNA precipitate is then collected by centrifugation, washed in ethanol and dissolved in distilled water. This RNA sample is electrophoresed for 5 hours at a constant current of 5 mA per gel in parallel with a control sample of 15 μg of messenger RNA which has not been treated with DMSO. The gels are washed overnight in water at 37° to remove all SDS, stained for 60 minutes with methylene blue at room temperature and destained in water for 2 days at 37°. Under these conditions the large class of messenger RNA forms a single band, the medium-sized class is usually separated into two bands and the small class is separated into three bands. Each band is rather diffuse. If these bands are observed in both the control

---

[29] U. E. Loening, *Biochem. J.* 113, 131 (1969).

gel and in the gel containing DMSO-treated RNA, then this indicates that the messenger RNA which has been prepared is largely intact.

Preparation of Qβ Phage and Qβ RNA

Qβ RNA, like f2 RNA, has been widely utilized in studies of cell-free protein synthesis in bacterial extracts. Recently, we have shown that Qβ RNA, in extracts from any of the four cultured cells described above, will direct synthesis of coat protein.[30] Since Qβ RNA can be prepared in large amounts, and since amber mutants in the coat gene are available,[31] this system should be extremely useful for study of the mechanism and regulation of initiation and termination of protein synthesis in mammalian cells, and of possible amber suppressor strains.

Reagents

Medium: Per liter: Tryptone (Difco), 32 g; yeast extract (Difco, 20 g; NaCl, 5 g; sodium hydroxide, 5 ml of 1 M solution
Tris Buffer: 0.1 M Tris·HCl, pH 8
Bacteria: Any male (Hfr) strain of E. coli will do

We have used with success derivatives of both Hfr C (Caralli) and HfrH (Hayes), for instance, strain K38 or strain CA264.

Preparation of Phage[32]

An inoculum of cells is grown overnight in the above medium, and then diluted 1:100 into fresh medium. For small-scale preparations (as of mutants) 500 ml medium is used in a 1-liter bottle; alternatively, one can use 8 liters in a 20-liter carboy. Aerate cells at 37° as vigorously as possible, using one sintered-glass filter per one-half liter of medium. Cells will grow logarithmically until an A_{550} of 2 is reached; saturation is reached at about $A_{550} = 8$. At $A_{550} = 2$ add $CaCl_2$ to 2 mM and 5–10 PFU per cell of Qβ. Continue aeration at 37° for at least 2–3 hours; add antifoam (Fisher Scientific Co.-Antifoan type A) as needed. Phage titer should be about 5×10^{12} PFU/ml, measured after lysis of an aliquot with $CHCl_3$ and lysozyme (50 μg/ml).

Add, per liter of lysate: 115 g of polyethylene glycol (MW, 6000) and 35 g NaCl. Mix well to dissolve, and keep overnight at 4°. Centrifuge in the GSA Sorvall rotor (250-ml bottles) for 30 minutes at 8000 rpm. Resuspend the pellet (containing cells and phage) in about 1/30 the initial volume of 0.1 M Tris pH 8. It is best to use a motor-driven stirrer to dis-

[30] T. Morrison and H. F. Lodish, Proc. Nat. Acad. Sci. U.S. 70, 315 (1973).
[31] K. Horiuchi and S. Matsuhashi, Virology 42, 60 (1970).
[32] E. Goldman and H. F. Lodish, J. Virol. 8, 417 (1971).

perse the cells. Keep at 4° for 1–2 hours. Add $CHCl_3$ (1:1000); EDTA to 5 mM; and lysozyme (50 mg/liter). Stir gently until very viscous—at least 1–2 hours. Then add Mg^{2+} to 20 mM and DNase (10 mg/liter). Stir until viscosity drops. Then centrifuge in the Sorvall GSA rotor 1 hour at 8000 rpm. Discard the pellet, which contains bacterial debris. To the supernatant, add polyethylene glycol and NaCl in the above proportions. Again keep overnight at 4°, then centrifuge 1 hour at 8000 rpm. Resuspend the pellet in 10–50 ml of Tris buffer (0.1 M, pH 8), and centrifuge 10 minutes at 8000 rpm to remove debris. The phage is recovered by centrifugation either 5 hours at 28,000 rpm in the Beckman No. 30 rotor, or 2.5 hours at 40,000 rpm in the type 40 rotor. The pellet is dissolved in Tris buffer and diluted to about 10 mg of phage/ml (1 mg/ml = 7 A_{260}). Add, per milliliter of solution, 0.625 g of CsCl, centrifuge for 24 hours at 4° either at 28,000 rpm in the No. 30 rotor or at 35,000 rpm in the No. 40 rotor. Phage should form an opalescent band about the middle of the tube; it can be recovered easiest with a long needle attached to a syringe. The phage is diluted at least 3-fold with water, then recovered by centrifugation in the ultracentrifuge as detailed above. Finally it is dissolved in Tris buffer. Yield is between 20 and 100 mg of phage per liter of culture.

Preparation of Messenger RNA[33]

To extract RNA, phage is diluted with water to about 3 mg/ml, then shaken for 3 minutes at room temperature with an equal volume of water-saturated phenol. After centrifugation at 4° for 5 minutes at 4000 rpm, the aqueous layer is set aside and the phenol layer is extracted with one-fifth the volume of water. After centrifugation the two aqueous layers are pooled, made 0.2 M in sodium acetate (pH 4.7) and reextracted with phenol. To the aqueous phase is added 2 volumes of ethanol. After at least 20 minutes at −20°, the RNA is recovered by centrifugation (5000 rpm for 10 minutes) and dissolved in 0.2 M sodium acetate (pH 4.7). Again the RNA is precipitated with 2 volumes of ethanol. The precipitate is washed by centrifugation once in ethanol, then once with ether. Finally, the RNA is dissolved in water; a gentle stream of nitrogen gas is used to remove traces of ether. The RNA is stored frozen at −20°.

Cell-Free Protein Synthesis

General. Extracts are generally frozen in 0.5 ml aliquots, an amount generally sufficient for one experiment. They are never refrozen. All solutions, except salts and buffers, are stored frozen at −20°. The optimum concentrations of K^+ and Mg^{2+} ions (and, for reticulocyte extracts, hemin)

[33] H. F. Lodish, *J. Mol. Biol.* **50**, 689 (1970).

are different for each extract and each mRNA and it is essential, for each extract and each mRNA used, to vary systematically the concentrations of each. For instance, in reactions using an ascites extract, with a final K⁺ concentration of 83 mM, the optimal final concentration of magnesium ion was different for different mRNA's: Reo mRNA, 1.5 mM; globin mRNA, 2 mM; Qβ RNA, 1.0 mM.[34]

Reagents

ATP: 50 mM; neutralized to pH 7.0 with NH₄OH
GTP: 10 mM; neutralized to pH 7.0 with NH₄OH
Creatine phosphate: 0.3 M; neutralized to pH 7.0 with NH₄OH
Creatine phosphokinase
Dithiothreitol: 1 M
HEPES (Calbiochem): pH 7.5, 1 M
KCl: 3 M

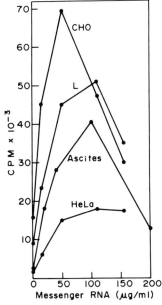

FIG. 1. Stimulation in the preincubated systems of *in vitro* protein synthesis by varied concentrations of added Reovirus messenger RNA. Fifty-microliter aliquots of reaction mixture were incubated for 100 minutes. The label was [³H]leucine. Cell-free systems were prepared from the indicated cell types [M. J. McDowell, W. K. Joklik, L. Villa-Komaroff, and H. F. Lodish, *Proc. Nat. Acad. Sci. U.S.* **69**, 2649 (1972)].

---

[34] L. Villa-Komaroff and H. F. Lodish, unpublished observations (1972).

Magnesium acetate: 0.1 M

Nineteen amino acids, all except leucine: 5 mM

Radioactive amino acid—either [³H]leucine, 40,000 mCi/mmole, 1 mC/ml (for ascites reactions) or [¹⁴C]leucine, 400 mCi/mmole, 100 μCi/ml (for reticulocyte reactions) both from New England Nuclear Corp.

Hemin: 4 mg/ml (for reticulocyte reactions)

NaOH: 0.1 M

Trichloroacetic acid: 10% and 5% (TCA)

Difco vitamin-free casamino acids: 3%

To prepare hemin, dissolve 20 mg in 0.5 ml 0.1 M NaOH. Add 0.1 ml Tris pH 7.8, 1 M; then neutralize to pH 7.8 to 8.0 with HCl. Dilute to 1.0 ml with water. Add 4 ml of ethylene glycol. Centrifuge at 10,000 rpm for 5 minutes, and store supernatant at $-20°$.

Reactions with Ascites, HeLa, CHO, or L-Cell Extracts[3]

A mix is constructed of all ingredients common to all tubes in the reaction. It usually contains:

Reagent	Volume (μl)	Final concentration in reaction (includes components in the extract)
Cell extract	500	
ATP	15	0.9 mM
GTP	15	0.18 mM
Creatine phosphate	30	10.8 mM
Creatine phosphokinase	(1 mg)	1.2 mg/ml
Dithiothreitol	8	1.0 mM
HEPES	15	18 mM
KCl	5	90 mM
Magnesium acetate	5	1.5 mM
19 Amino acids	20	0.12 mM, each
[³H]leucine	40	5 μCi/ml
Water	20	
	668	

Final reactions generally contain 40 μl of mix, other additions especially RNA's, and sufficient water to make the final volume 50 μl. Reactions are incubated at 30° for variable periods of time, generally 40 minutes. One-half milliliter of NaOH (0.1 M) is added, and the reactions

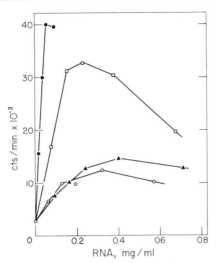

FIG. 2. Translation of bacteriophage Qβ RNA in extracts from mouse ascites cells. Reactions (50 μl) contained [³H]leucine and were incubated at 30° for 40 minutes. Globin mRNA (●—●) was prepared as previously described [D. Housman, R. Pemberton, and R. Taber, *Proc. Nat. Acad. Sci. U.S.* **68**, 2716 (1971)], and f2 RNA (○—○) and the amber mutant Sus-3 (▲—▲) [D. L. Engelhardt, R. E. Webster, and N. D. Zinder, *J. Mol. Biol.* **29**, 45 (1967)] were prepared as described in the text for Qβ RNA (□—□) [taken from T. G. Morrison and H. F. Lodish, *Proc. Nat. Acad. Sci. U.S.* **70**, 315 (1973)].

are incubated an additional 15 minutes in order to hydrolyze peptidyl-tRNA. They are chilled in a bath of ice-water, and 0.5 ml of casamino acids are added. After mixing, 1.0 ml of 10% TCA is added. After mixing, the reactions are kept on ice for at least 10 minutes, then filtered through a fiberglass filter (Whatman GF/A) and washed well with 5% TCA. Counting of the radioactive precipitates can be by any established procedure.

Results of typical reactions using Reo mRNA or Qβ mRNA are shown in Figs. 1 and 2. Further work showed that Reo mRNA directed synthesis of between 6 and 8 of the 8 known Reo proteins.[3] Qβ RNA directed synthesis mainly of Qβ coat protein.[30]

Reactions with Reticulocyte Extracts[3,35]

A mix is constructed of all ingredients common to all samples. Generally it contains:

[35] H. F. Lodish, D. Housman, and M. Jacobsen, *Biochemistry* **10**, 2348 (1971).

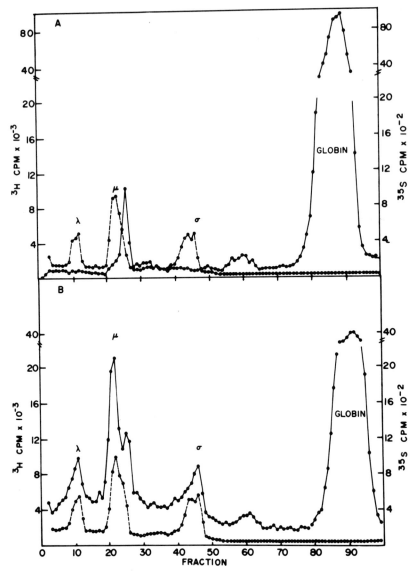

FIG. 3. Sodium dodecyl sulfate gel electrophoretic analysis of total polypeptides synthesized in the reticulocyte cell-free system. Incubation was at 25° for 100 minutes. Polypeptides were labeled *in vitro* with [^{35}S]methionine (150 Ci/mmole, 250 μCi/ml) (●——●) and coelectrophoresed with viral polypeptides labeled *in vivo* with ^{3}H labeled protein hydrolyzate (○- - -○). Conditions for gel electrophoresis and counting gel slices were detailed previously [M. J. McDowell, W. K. Joklik, L. Villa-Komaroff, and H. F. Lodish, *Proc. Nat. Acad. Sci. U.S.* 69, 2649 (1972); M. J. McDowell and W. K. Joklik, *Virology* 45, 724 (1971)]. Migration was from left to right. (A) Polypeptides synthesized due to endogenous message. (B) Polypeptides synthesized in the presence of 200 μg/l of Reovirus message. The volume of reaction mixture analyzed on gel B was 3.5 times that in gel A.

Reagent	Volume (μl)	Final concentration in reaction
Reticulocyte extract	500	
ATP	15	0.86 mM
GTP	15	0.17 mM
Creatine phosphate	30	11.0 mM
Creatine phosphokinase	(0.1 mg)	0.12 mg/ml
Dithiothreitol	4	5 mM
HEPES	10	12 mM
KCl	15	56 mM
Magnesium acetate	15	1.8 mM
19 Amino acids	15	92 μM, each
[^{14}C]leucine	50	6 μCi/ml
Hemin	4	20 μg/ml
	673	

Final reaction generally contains 50 μl of mix, and other components to a final volume of 60 μl. Incubation is at 30°. Aliquots of 10 μl are taken at intervals into 0.5 M NaOH solution. Further work-up is as in the preceding section.

Figure 3 demonstrates synthesis of Reovirus proteins by the reticulocyte lysate.[36,37] Proteins labeled *in vitro* and labeled with [^{35}S]methionine (20 Ci/mmole; 50 μCi/ml) are mixed with viral polypeptides labeled *in vivo* with [^{3}H]protein hydrolyzate; the mixture is analyzed by gel electrophoresis in a buffer containing urea and sodium dodecyl sulfate. The top panel shows that the reticulocyte lysate, without added mRNA, synthesizes mainly globin and one other protein. Upon addition of Reovirus mRNA, there is inhibition of synthesis of globin, but also synthesis of all three classes of Reovirus proteins: γ (average MW, 30,000); μ (MW, 80,000), and λ (MW, 130,000). Higher resolution gels resolved all eight Reovirus proteins.[3]

Acknowledgments

This work was supported by grants Nos. AI-8909, AI-08814, AI-08388, and AM-15929 from the U.S. Public Health Service and Contract No. AT-(40-1)-3857 from the Atomic Energy Commission. L.V.-K. is supported by a Cell Biology training grant from the National Institutes of Health, M.J.McD. is a James B. Duke Fellow. D.B. is a recipient of a Faculty Research Associate Award from the American Cancer Society No. PRA-50. H.F.L. is a recipient of a Research Career Development Award from the National Institutes of Health No. GM-50, 175.

[36] D. L. Engelhardt, R. E. Webster, and N. D. Zinder, *J. Mol. Biol.* **29**, 45 (1967).
[37] M. J. McDowell and W. K. Joklik, *Virology* **45**, 724 (1971).

[67a] Protein Synthesis with Rabbit Reticulocyte Preparations[1]

By WILLIAM R. WOODWARD, JOEL L. IVEY, and EDWARD HERBERT

Two reticulocyte cell-free systems are described for the study of hemoglobin synthesis. The first is a crude lysate preparation (CFS I)[2] obtained by lysing the cells and removing the membrane fraction. The other preparation (CFS II) is a partially fractionated lysate system consisting of a postribosomal supernatant, salt-treated ribosomes, and a KCl extract of ribosomes. One advantage of these systems over more fractionated systems[3,4] is that they synthesize *de novo* much more hemoglobin per ribosome. For example, CFS I is capable of synthesizing almost as much hemoglobin as the intact cell for approximately 20 minutes, initiating 10 to 15 chains per ribosome. This makes it possible to demonstrate phenomena that require multiple chain turnover, such as hemin control of globin synthesis. In the absence of added hemin, globin synthesis ceases within a few minutes at 37°; whereas, in the presence of added hemin, globin synthesis is maintained for much longer periods. The more highly fractionated systems do not respond to hemin addition.

The preparation, storage, and use of these systems in the study of a variety of phenomena is described. Data are also presented to illustrate the degree of variability that can be expected in the synthetic activity and in the response to added hemin in lysates prepared from different rabbits.

Reagents

Solution A (in grams per liter): NaCl, 70; KCl, 4.5; $CaCl_2 \cdot 2H_2O$, 1.5; $MgCl_2 \cdot 6H_2O$, 3.0; glucose, 20. This solution can be stored frozen in 100-ml aliquots.

Solution B: Solution A diluted 1:10 (v/v) with water

Amino Acid Mixture I (in mg per liter): leucine, 1063; valine, 925; alanine, 450; and phenylalanine, 650

[1] This work was supported by a United States Public Health Service research grant CA-07373 from the National Cancer Institute and a National Science Foundation grant GB-4063. One of us (E.H.) is the recipient of a Career Development Award from the United States Public Health Service (1-K6-CA-2101).

[2] Abbreviations used: CFS I and CFS II, cell-free systems I and II; PRS, postribosomal supernatant; TCA, trichloroacetic acid; RSU, ribosomal subunit; A_{260} units, the quantity that has an A_{260} when dissolved in 1 ml of solvent.

[3] R. L. Miller and R. Schweet, *Arch. Biochem. Biophys.* **125**, 632 (1968).

[4] H. R. V. Arnstein, R. A. Cox, and J. A. Hunt, *Biochem. J.* **92**, 648 (1964).

Amino Acid Mixture II (in mg per liter): arginine hydrochloride, 250; aspartate, 950; glycine, 1000; histidine, 1250; isoleucine, 100; lysine hydrochloride, 825; methionine, 125; proline, 400; serine, 440; threonine, 500; tryptophan, 150; tyrosine, 375; cysteine hydrochloride, 125

Glutamine: 4 mg/ml

Master Mix:

ATP, 50 mM, pH 7.2, 20 μl

GTP, 50 mM, pH 7.2, 20 μl

Glutamine, 20 μl

Amino Acid Mixture I, 50 μl

Amino Acid Mixture II, 50 μl

KCl, 4 M, 30 μl

L-[^{14}C]Leucine, 25 μl (0.1 μCi/μl; the specific activity of the final incubation will be about 5 μCi/μmole)

Creatine phosphate, 1 M, 65 μl

Creatine phosphokinase, 20 μl: either 15 mg/ml Boehringer Mannheim or 30 mg/ml Schwarz/Mann

Water, 50 μl

Hemin: 6.0 mg/ml made up as follows: 60 mg of hemin is dissolved in 1.0 ml 1 N NaOH, 1.0 ml of 0.5 M Tris base, and 7.2 ml of water. The solution is stirred while 0.8 ml of 1 N HCl is slowly added. The solution is centrifuged at 20,000 g for 10 minutes. The clarified supernatant is stored frozen in 50-μl aliquots and is diluted with an equal volume of water for use.

Bleaching Solution: 2 parts 10% TCA; 1 part 88% formic acid, 1 part 30% hydrogen peroxide

Lysis and Storage of Rabbit Reticulocytes

Reticulocytes are obtained from New Zealand white rabbits (female) which have been made anemic by daily subcutaneous injections of a neutralized 2.5% (w/v) solution of phenylhydrazine hydrochloride (see regimen in Godchaux et al.[5]). The rabbits are bled by cardiac puncture on day 10. The cells are sedimented, washed in solution B and then lysed in an equal volume of 4 mM MgCl$_2$. The cell stroma are removed by centrifugation at 20,000 g, and 4.0 ml aliquots of the supernatant fraction (lysate) are stored in polyethylene vials in liquid nitrogen. This preparation is the lysate and is either used as CFS I or further fractionated to give CFS II. No loss in capacity to respond to hemin addition is detected after storage in liquid nitrogen for more than a year.

[5] W. Godchaux, S. Adamson, and E. Herbert, *J. Mol. Biol.* **27**, 57 (1967).

Incubation Conditions for CFS I

CFS I Activity Assays

The purpose of these assays is to determine the synthetic activity of CFS I in the presence and absence of added hemin. Lysates are incubated at hemin concentrations of 0, 50 μM and 100 μM, respectively. These hemin concentrations are in the range that gives maximal hemin response.[6] Master mix and lysate are mixed in proportions 25:200 (master mix: lysate, v/v). Hemin, the last component to be added, is diluted with an equal volume of water, and 2-μl and 4-μl aliquots are added to the appropriate incubation mixtures. The reaction mixtures and all components are kept on ice until the start of the incubation.

The reaction mixtures are incubated, with shaking, at 37° for 60 minutes. Synthesis of hemoglobin generally is linear up to 20 minutes at 37° and almost ceases by 60 minutes. Less than a 10% increase in incorporation is observed between 60 and 120 minutes.

Aliquots (20 μl) are spotted on filter paper disks (Whatman No. 1, 24 mm). After the sample has been absorbed by the paper, the disk is immersed in cold 10% TCA for 10 minutes. The disks are treated twice with hot 10% TCA for 10 minutes and then bleached with 200 ml of bleaching solution for at least 2 hours. The bleached disks are washed twice with absolute ethanol, once with ethanol:ethyl ether (1:1, v/v), once with ethyl ether and dried. The filter papers are counted in a liquid scintillation counter using 5 g of Omniflor (New England Nuclear) per liter toluene.

Substitution of creatine phosphate–creatine phosphokinase energy-generating system (CP system) for the phosphoenol pyruvate–pyruvate kinase system (PEP system) results in a 2.5-fold increase in the rate of protein synthesis at 37° in CFS I. The enhancement is even more pronounced at 26° (approximately 3.5-fold).[6] Also, the number of chains completed per ribosome increases from less than 2 with the PEP system to greater than 6 with the CP system. The explanation for this difference is not known, but our studies indicate that the lower incorporation with the PEP system is not due to contaminants in the phosphoenolpyruvate or pyruvate kinase.

Variability of Hemin Effects

The level of synthesis in CFS I in the presence and absence of added hemin (37° for 60 minutes) varies from rabbit to rabbit. This variability

[6] S. Adamson, E. Herbert, and W. Godchaux, *Arch. Biochem. Biophys.* **125**, 671 (1968).

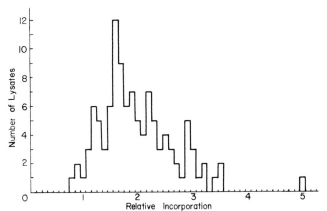

FIG. 1. The distribution of 50 μM hemin stimulation in Class I lysates (see Table I) relative to no added hemin. The ordinate values are calculated as follows:

Relative incorporation

$$= \frac{\text{cpm incorporated in 50 } \mu M \text{ hemin}}{\text{cpm incorporated with no added hemin}} \text{ (standard assay conditions)}$$

The incorporation levels in the absence of added hemin ranged from 1100 to 17,900 cpm per 20 μl (data from 110 lysates).

is seen in data collected from assays on 157 different lysate preparations performed in this laboratory over a period of about one year. Figure 1 shows the magnitude of the stimulation by added hemin. Only lysates with maximum incorporation of greater than 5000 cpm/20 μl (2.2 nmoles of globin per milliliter of lysate) were retained for use in cell-free system experiments. Based on the assumption that 1 ml of lysate contains approximately 1.5 mg of ribosomes (MW $\simeq 4.5 \times 10^6$),[7] incorporation above this cutoff point represents synthesis of more than 6.5 globin chains per ribosome. (The maximum synthesis observed was slightly more than 27 globin chains per ribosome.) Less active lysates are used in the preparation of inhibitor fraction or tRNA.[8,9]

The lysate response to added hemin falls into three categories: inhibited (those lysates in which added hemin depresses synthesis); moderately stimulated (those lysates in which the added hemin stimulates synthesis, but less than 2-fold); and highly stimulated (those lysates in which the added hemin stimulates greater than 2-fold). Only highly stimulated lysates were used for studies of phenomena that require multiple turnover of nascent chains.

[7] W. R. Woodward and E. Herbert, *Science* **177**, 1197 (1972).
[8] W. R. Woodward and E. Herbert, this volume [67b].
[9] W. R. Woodward, P. Wilairat, and E. Herbert, this volume [67c].

TABLE I
VARIATIONS IN HEMIN RESPONSES[a]

Classification of lysate[b]	Magnitude of hemin effect[c]	Number of lysates	Percent of total Class I lysates	Percent of total lysates
Class I	>2-fold	71	63.5	45.2
	<2-fold	38	33.9	24.2
	Inhibited	3	2.6	1.9
Total Class I		112		71.3
Class II	>2-fold	8		5.1
	<2-fold	30		19.1
	Inhibited	7		4.5
Total Class II		45		28.7

[a] These data are based on 157 lysates prepared over a one-year period (1971–1972). The lysates were incubated at 37° for 1 hour as described (CFS I Activity Assay).

[b] The lysates were arbitrarily divided into two categories: Class I, those lysates that incorporated greater than 2.5×10^5 cpm per milliliter of lysate (L-[14C]leucine, 5 μCi/μmole, counting efficiency ~60%) or greater than 25% of the amount synthesized by intact cells (37° for 30 minutes, see Table II); and Class II, those lysates that incorporated less than 2.5×10^5 cpm per milliliter of lysate. The maximal incorporation levels whether in the presence (those lysates stimulated by hemin) or absence (those lysates inhibited by hemin) of added hemin were used to determine classification of lysates.

[c] The optimal stimulation observed by hemin addition (either 50 or 100 μM) compared to no hemin addition.

The data in Table I illustrate the degree of variability encountered in working with lysates from different animals. Pooling lysates from different animals gives more uniform responses. Whereas this approach has some advantages, it masks differences among animals that might provide a clue to understanding the physiology of the phenomena being studied.

The Effect of Temperature on Hemin Response

The incubation temperature has an effect on the magnitude of the hemin response. When CFS I is incubated at 26° the initial rate of amino acid incorporation is slightly less than half that at 37°. However, synthesis continues for longer periods of time at 26°, and the total incorporation is approximately the same at both temperatures. At temperatures below 26° the rate of synthesis declines more rapidly with decreasing temperature, and total synthesis is significantly reduced.[6] Concomitant with the temperature-dependent decrease in the rate of synthesis is a loss of the requirement for added hemin. Lysates which show greater than 2-fold hemin stimulation at 37° show either no hemin stimulation or a slight hemin inhibition (up to 10%) at 26°. The hemin requirement at 37° is apparently unrelated to the assembly of globin chains into hemoglobin, since approximately the same amount of tetrameric hemoglobin is synthesized at 37° (in the presence of optimum hemin) and at 26° (no added hemin). The lack of hemin response at lower temperatures in CFS I is utilized in assaying for an inhibitor fraction which can be inactivated by hemin under certain conditions.[8]

Preparation and Incubation Conditions for CFS II

Preparation of CFS II

CFS II is a more fractionated system than CFS I and differs from the latter in that the lysate is separated into a supernatant fraction (PRS) and a ribosome fraction. The ribosome fraction is treated with 0.5 M KCl which removes from the ribosomes a protein fraction, designated I fraction or initiation fraction,[3] and nascent chains in the form of peptidyl-tRNA. Addition of I fraction to the salt-treated ribosomes, PRS, and master mix stimulates *de novo* synthesis of globin. This system is only slightly less active than CFS I, and normally synthesizes greater than 2 globin chains per ribosome (1 hour at 37°) when prepared from a Class I lysate. The CFS II system shows the same hemin stimulation observed in CFS I.

A lysate is centrifuged for 3 hours at 105,000 g (4°), and the PRS fraction is carefully removed without disturbing the pelleted ribosomes. All subsequent operations are carried out at 4°. The PRS may be frozen in liquid nitrogen for later use. The ribosomes are gently rinsed (2 times) with 0.25 M sucrose, then homogenized with 0.25 M sucrose (100:6.4, v/v; lysate: 0.25 M sucrose). The homogenate is then made 0.5 M in KCl by addition of 4 M KCl and kept on ice for 60 minutes. The ribosomes (in the form of ribosomal subunits), RSU, are sedimented by centrifugation for 3 hours at 105,000 g (4°). The supernatant (I fraction) is carefully removed and dialyzed against 0.375 M KCl for 12 hours.[10] The RSU are gently rinsed with 0.25 M sucrose then homogenized with 0.25 M sucrose such that the final concentration of RSU is about 100 A_{260} units per ml.[3,11] The RSU and I fraction may be frozen in liquid nitrogen for later use.

---

[10] A high KCl concentration is used in the dialysis to prevent inactivation that occurs at low salt concentrations.

[11] S. D. Adamson, P. M. P. Yau, E. Herbert, and W. V. Zucker, *J. Mol. Biol.* **63**, 247 (1972).

Incubation Conditions for CFS II

The incubation (250 μl) consists of the following components: 100 μl of PRS, 100 μg of salt-treated ribosomes,[12] up to 50 μl of I fraction,[13] sufficient 0.375 M KCl to make final incubation 70 mM KCl, and 25 μl of a modified master mix. The master mix is modified by omitting KCl, reducing the amino acid mixture I by one half (effectively doubles the specific activity of the leucine), and adding 25 μl of 0.2 M MgCl$_2$. The volume of the master mix is maintained by adjusting the amount of water added.

TABLE II

COMPARISON OF THE SYNTHETIC ACTIVITY OF RETICULOCYTE
INTACT CELL AND CELL-FREE PREPARATIONS[a]

System	Nmoles leucine per milligram ribosomes	Chains of globin per ribosome
Intact cell[b]	100–120 (30 min)	26–31
CFS I	25–107	6.5–27.6
CFS II	10–20	2.6–5.2
Ammonium sulfate[c]	3.3	0.86
pH 5 enzyme[d]	0.85	0.22

[a] Incorporation was measured after incubation for 1 hour at 37° except for incorporation by the intact cell which was measured after 30 minutes at 37°. The intact cell has been observed to sustain a linear rate of protein synthesis for periods greater than 1 hour. The CFS I sustains a linear rate of protein synthesis for only about 20–30 minutes (the other cell-free systems are linear for shorter periods). Therefore, incorporation by the intact cell at 30 minutes serves as an upper limit for comparing the activities of various cell-free systems. It should be noted that incorporation by a cell-free system may be less than that in the intact cell because of early cessation of synthesis or a lower initial rate of synthesis, or both.

[b] S. D. Adamson, G. A. Howard, and E. Herbert, *Cold Spring Harbor Symp. Quant. Biol.* **34**, 547 (1969).

[c] R. L. Miller and R. Schweet, *Arch. Biochem. Biophys.* **125**, 632 (1968). Their values are reported as nanomoles of valine. This was converted into nanomoles of leucine by assuming 17.5 residues of leucine and 14.5 residues of valine per chain ["Atlas of Protein Sequence and Structure" (M. O. Dayhoff and R. V. Eck, eds.), pp. 128 and 142. National Biomedical Research Foundation, Silver Spring, Maryland, 1967–1968].

[d] H. R. V. Arnstein, R. A. Cox, and J. A. Hunt, *Biochem. J.* **92**, 648 (1964).

[12] It is assumed that 1.0 mg of ribosomes in 1.0 ml of 0.25 M sucrose gives an A_{260} of 11.2 [P.O.P. Ts'o and J. Vinograd, *Biochim. Biophys. Acta* 49, 113 (1961)].

[13] The levels of I fraction for optimum incorporation and the levels at which hemin effects are observable may be determined by prior assay. The levels of I fraction which show optimum incorporation may be higher than those at which hemin effects are observable, and optimum I fraction levels may not show any hemin stimulation.

The reaction mixture is incubated, with shaking, at 37° for 1 hour. Aliquots are spotted on filter disks and processed for counting. All synthesis occurs *de novo* in this system, making it extremely useful in the study of initiation effects.

Comparison of Synthetic Activity of Reticulocyte Intact Cell and Cell-Free Preparations

Table II presents a comparison of the synthetic activity of CFS I and CFS II with two other commonly used reticulocyte cell-free systems and with the intact cell. CFS I synthesizes hemoglobin at the same rate as the intact cell[14] for 15–30 minutes. The more highly fractionated systems (ammonium sulfate[3] and pH 5 enzyme[4]) are much less active than either CFS I or II, completing only a fraction of a globin chain per ribosome.

[14] S. D. Adamson, G. A. Howard, and E. Herbert, *Cold Spring Harbor Symp. Quant. Biol.* **34**, 547 (1969).

[67b] Preparation and Characterization of an Inhibitor of Globin Synthesis from the Postribosomal Supernatant of Reticulocyte Lysates[1]

By WILLIAM R. WOODWARD and EDWARD HERBERT

When the postribosomal supernatant, PRS,[2] from reticulocytes is incubated at temperatures above 30° in the absence of added hemin, an inhibitor of globin synthesis forms which we call Q fraction.[3] Hemin retards the formation of Q fraction and under some circumstances inactivates it.[4] An inhibitor with similar properties can also be isolated from

[1] This work was supported by a United States Public Health Service research grant CA-07373 from the National Cancer Institute and a National Science Foundation grant GB-4063. One of us (E.H.) is the recipient of a Career Development Award from the United States Public Health Service (1-K6-CA-2101).

[2] Abbreviations used: DEAE-cellulose, diethylaminoethyl cellulose; EDTA, ethylenediaminetetraacetic acid. The following abbreviations are defined more completely elsewhere (W. R. Woodward, J. L. Ivey, and E. Herbert, this volume [67a]); CFS I and CFS II, cell-free systems I and II; PRS, postribosomal supernatant.

[3] S. D. Adamson, P. M. P. Yau, E. Herbert, and W. V. Zucker, *J. Mol. Biol.* **63**, 247 (1972).

[4] The inhibition shown by this fraction at short incubation periods (less than 30 minutes at 37°) can be overcome by the addition of hemin, whereas later addition of hemin has no effect on the inhibitory activity. The minimum concentrations of hemin that overcome the early inhibition are in the same range as those

mature erythrocytes, and in this case no prior incubation is required for full activity.

The Q fraction acts at the site of chain initiation. Therefore, in order to detect the inhibitor and to study its mode of action, it is necessary to use cell-free systems, such as CFS I and CFS II, that initiate several rounds of synthesis during the course of an incubation. We present several assays for Q fraction activity, each designed to measure a different property of the inhibitor.

CFS I is more sensitive to the Q fraction than CFS II but exhibits a delay in responding to the inhibitor. Since chain completion is insensitive to Q fraction,[3] it is reasonable to postulate that the delayed response of CFS I to Q fraction is the time it takes to complete nascent chains already present on the ribosomes at the beginning of the incubation. This idea is supported by the fact that no delay in the expression of inhibitory activity is seen in CFS II, in which the ribosomes contain no nascent chains and all hemoglobin synthesis occurs *de novo*. Therefore the CFS I assay is used to measure the effect of Q fraction on the total amount of hemoglobin synthesized, and CFS II is used to measure the effect of Q fraction on initial rates of synthesis of hemoglobin.

The CFS I assay is incubated at 26° to avoid the formation of endogenous Q fraction in the assay system.[5] This obviates the need for adding hemin (necessary at 37°),[6] thus avoiding the complication of hemin inactivation of the inhibitor. On the other hand it is necessary to incubate the CFS II assay at 37°, because synthesis is greatly depressed at the lower temperatures. However, activation of Q fraction in the PRS of CFS II is a slow process and does not interfere with initial rate measurements which take only a few minutes.

Assays for Q Fraction Activity

Assay A

The CFS I and master mix are prepared as described previously[7] except that the amount of Amino Acid Mixture I added to the master mix is reduced by half, thereby doubling the final specific activity of the

---

that maximally stimulate CFS I incubations at 37° (W. R. Woodward, J. L. Ivey, and E. Herbert, this volume [67a]). Except where otherwise noted, Q fraction will refer to the inhibitory activity which is unaffected by addition of hemin ("irreversible inhibitor" in the terminology of Rabinovitz [M. Gross and M. Rabinovitz, *Proc. Nat. Acad. Sci. U.S.* **69**, 1565 (1972)]).

[5] C. R. Maxwell, C. S. Kemper, and M. Rabinovitz, *J. Mol. Biol.* **58**, 317 (1971).
[6] S. D. Adamson, E. Herbert, and W. Godchaux, III, *Arch. Biochem. Biophys.* **125**, 671 (1968).
[7] W. R. Woodward, J. L. Ivey, and E. Herbert, this volume [67a].

leucine (10 μCi/μmole). A 200-μl incubation mixture [CFS I:master mix; 200:25 (v/v)] is combined with up to 50 μl of Q fraction and the final volume is adjusted to 250 μl. The reaction mixture is incubated with shaking for 2 hours at 26°. At the end of the incubation, aliquots are spotted on filter paper disks and processed for counting.

Due to the variability of the cell-free system response to Q fraction inhibition and the variability of the levels of Q fraction present in different PRS preparations, it is necessary to define inhibitory units for each assay. A unit of Q fraction is that level necessary for half-maximal inhibition of globin synthesis in the standard assay. The amount of inhibition is determined by comparing the incorporation of radioactivity into protein in the presence of Q fraction to that in the presence of the appropriate water or buffer blank. The percentage inhibition relative to a mixture containing water can be calculated by the following relationship[8]:

$$\% \text{ Incorporation} = \frac{\text{cpm in inhibitor assay}}{\text{cpm in water blank}} \times 100 \tag{1}$$

$$\% \text{ Inhibition} = 1 - \% \text{ incorporation} \tag{2}$$

When the logarithm of percentage incorporation is plotted against increasing concentration of Q fraction, a straight line is obtained, and the level of Q fraction yielding 50% inhibition may easily be determined (Fig. 1).[5]

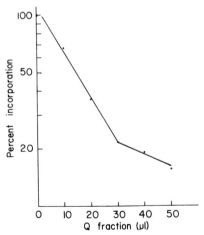

FIG. 1. Effect of Q fraction addition on incorporation in CFS I (Assay A, 26° for 2 hours, final volume 250 μl). The percent incorporation is calculated from Eq. (1). A unit of Q fraction is defined as that volume of Q fraction which reduces incorporation by 50% in the standard assay.

[8] G. A. Howard, S. D. Adamson, and E. Herbert, *Biochim. Biophys. Acta* **213**, 237 (1970).

Assay B

The CFS II components and master mix are prepared as previously described[7] except that the final specific activity of the leucine is doubled as above. Up to 100 μl of Q fraction or appropriate control mixture are combined with 500 μl of CFS II reaction mixture and incubated at 37° with shaking. Aliquots (20 μl) are removed at 1-minute intervals for 10–15 minutes and spotted on filter paper disks for counting. Incorporation will continue at a linear rate for at least 10 minutes in this system. The period of linear incorporation can be extended for as long as 30 minutes by addition of hemin.[3]

I fraction may be purified on DEAE-cellulose[9] before use in this system, but the resulting I fraction is less stable and, hence, more difficult to work with than crude I fraction. Moreover, the results obtained are the same with either fraction.

Properties of Q Fraction

Two basic properties of Q fraction help to distinguish it from other kinds of inhibitors of globin synthesis: (1) Q fraction requires activation, and the rate of activation is temperature dependent and is affected by addition of hemin. The activation does not appear to result from the loss of some smaller fragment (see section on purification of Q fraction), but may involve a conformational change in the material. The requirement for activation distinguishes Q fraction from many other types of inhibitors which require no prior activation. (2) In CFS II the initial rate of synthesis increases with increasing levels of I fraction. Q fraction inhibits initial rates in this system in a manner that suggests competition between a component of Q fraction and I fraction for some essential site in the system.[3] Other evidence indicates that I fraction actually titrates Q fraction.

These properties distinguish Q fraction from other inhibitors of globin synthesis, such as apomyoglobin, which requires no prior activation, and which affect only the duration of synthesis in CFS II.

Each of these properties has been observed in experiments using both a crude and a partially purified Q fraction preparation. Although these properties may be modified by subsequent purification procedures, they serve as a basis for defining the inhibitory fraction arising in the cytoplasm of reticulocytes.

[9] R. L. Miller and R. Schweet, *Arch. Biochem. Biophys.* **125**, 632 (1968). The I fraction activity elutes from a DEAE-cellulose column developed with a linear concentration gradient of KCl in 0.1 mM EDTA and 20 mM Tris, pH 7.5, at about 0.2 M KCl. The purified I fraction prepared by this method is free of hemoglobin which is not retained by the column and RNA (including peptidyl-tRNA) which elutes at higher KCl concentrations.

Activation of Q Fraction

Formation of maximum Q fraction activity generally takes place in 4–5 hours at 37° (Fig. 2).[10] Formation of Q fraction is accelerated as temperature is raised from 30° to 45° and is slowed as temperature is reduced below 30°. At 26° Q formation is barely detectable after 2 hours (the incubation time for Assay A). The presence of other components of the cell-free system, including ribosomes or master mix components, does not affect the formation of Q fraction activity.

Because of its greater sensitivity, Assay A (CFS I) is used routinely rather than Assay B (CFS II) to detect Q fraction and to define units of Q fraction activity. This assay system is prepared as described above and incubated for 2 hours at 26°. The amount of inhibition is determined

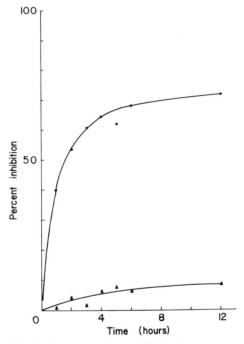

FIG. 2. Rate of Q fraction activation at 37° in the presence and in the absence of added hemin (50 μM hemin). Aliquots of PRS (50 μl) are incubated at 37° in the presence (▲—▲) and absence (●—●) of added hemin, then assayed for inhibitory activity by Assay A (26° for 2 hours). The percentage inhibition is calculated (relative to unincubated PRS addition) from Eq. (2).

[10] Although Maxwell *et al.*[5] report formation of maximum inhibitory activity only after 10–12 hours at 37°, results in this laboratory indicate maximum Q fraction formation after 4–5 hours at 37°.

by comparing incorporation in the presence of Q fraction with that in the presence of unincubated PRS or water. Note that as Q fraction is purified, the incubation time required for activation at 37° increases; therefore, it is important to determine the kinetics of the Q fraction activation at each stage of purification.

Assay A may also be used to study the effect of hemin on Q fraction activation. It has been observed that when hemin is added at early times to the Q fraction activation system (before 30 minutes at 37°) the Q fraction is inactivated (designated "reversible inhibitor" by Rabinovitz[4]). Later addition of hemin has no effect on Q fraction activation. The effect of hemin on activation in partially purified Q fraction has not been thoroughly investigated.

Effect of Q Fraction on Initial Rates of Synthesis

Several inhibitors of globin synthesis have been found which appear to function by removing hemin from the reaction mixture. Of these, apomyoglobin is the most potent (Fig. 3A). However, at low levels of apomyoglobin the inhibition can be prevented or reversed by addition of hemin to the assay system. Human globin chains stripped of heme also have an inhibitory effect that can be overcome by hemin addition. Human globin chains prepared without stripping off heme are not inhibitory in this system.

It was mentioned earlier that the initial rate of synthesis in Assay B increases as the I fraction level is increased. Addition of Q fraction to an assay system containing a low level of I fraction causes a depression of the initial rate, whereas the addition of apomyoglobin, a heme scavenger, depresses only the duration of hemoglobin synthesis (Fig. 3B). This result clearly establishes that the inhibitory action of Q fraction differs from that of apomyoglobin. Related experiments show that rabbit globin behaves similarly to apomyoglobin.

Titration of Q Fraction Inhibition by I Fraction

Addition of excess I fraction can completely titrate the effect of added Q fraction. This may be examined either by investigating the initial rates of globin synthesis in Assay B at varying levels of I fraction in the presence or absence of Q fraction[11] or by measuring the total incorporation at increasing levels of Q fraction in the presence of various levels of

---

[11] At high levels of I fraction the initial rate of synthesis is the same in the presence or in the absence of added Q fraction. This effect should be noted when investigating the initial rates of globin synthesis in the presence of Q fraction (see "Effect of Q Fraction on Initial Rates of Synthesis").

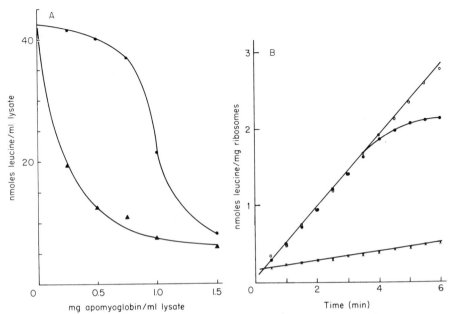

FIG. 3. A comparison of the inhibition by apomyoglobin and Q fraction. (A) Effect of hemin on inhibition by apomyoglobin. Varying concentrations of apomyoglobin are assayed in the presence (40 μM hemin, ●—●) and absence (▲—▲) of added hemin by Assay A (26° for 2 hours). (B) Effect of Q fraction and apomyoglobin on the initial rate of globin synthesis. Three Assay B incubations (1 ml) were prepared and 200 μl of either water (○—○), 5 mg/ml apomyoglobin (●—●) or Q fraction (×—×) were added. The reaction mixtures were incubated at 37°.

I fraction in either Assay A (26° for 2 hours) or Assay B (37° for 2 hours). As shown in Fig. 4, Q fraction inhibits protein synthesis at low levels of I fraction; whereas at high I fraction levels added Q fraction shows no effect. I fraction partially purified by DEAE-cellulose chromatography shows the same protective action against Q fraction.

Purification of Q Fraction and Isolation of a Pre-Q Fraction

Two general procedures have been developed for the purification of Q fraction; gel filtration, and weak ion exchange. Fractionation of crude Q fraction on Sepharose 4B (Pharmacia, Piscataway, New Jersey) shows inhibitory activity eluting ahead of hemoglobin. The estimated molecular weight of Q fraction is 5×10^5. When an unincubated PRS (containing no inhibitory activity) is fractionated on Sepharose 4B column under the same conditions, no inhibitory activity is detected in any of the column fractions until they are incubated at 37° (10 hours in the absence of

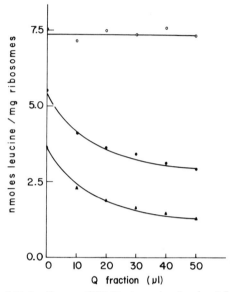

FIG. 4. Effect of Q fraction on CFS II at varying levels of I fraction. Incorporation at increasing levels of Q fraction is measured in Assay B (37° for 2 hours) with either no added I fraction (▲—▲), 50 μl of I fraction per milliliter (●—●), or 200 μl of I fraction per milliliter (○—○). Q fraction reduces incorporation to the level of chain completion when no I fraction is added.

added hemin). Then, inhibitory activity appears in the same region of the column as Q fraction. Therefore, a pre-Q fraction exists which has essentially the same molecular weight as Q fraction.

Gel Filtration Chromatography of Either Pre-Q or Q Fraction

A Sepharose 4B column (0.9 × 110 cm) is equilibrated with 0.1 M ammonium acetate, pH 7.0, and 1 mM glutathione (reduced). A 1–3 ml sample of either PRS or Q fraction is placed on the column and eluted with the above buffer; 1 ml fractions are collected. When an unincubated PRS is fractionated on the column, 0.5-ml aliquots from each fraction are incubated for 10 hours at 37° (conditions required for optimal activation of Q fraction). Assay A is used to detect inhibitor activity. Aliquots of 50 μl from each fraction from the column are added to 200 μl of Assay A and incubated at 26° for 2 hours. In the case of gel filtration of Q fraction, the column buffer serves as a control, and for pre-Q fractionation, the unincubated portion of each fraction serves as a control. As can be seen in Fig. 5, pre-Q and Q fraction elute at the same position on the column.

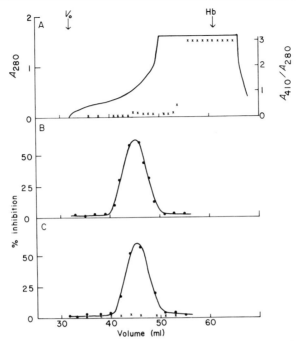

Fig. 5. Chromatography of Q fraction and unincubated PRS on Sepharose 4B. (A) A_{280} (——) and $A_{410} : A_{280}$ (×—×, hemoglobin has a ratio of approximately 3) for a 3-ml sample of either Q fraction or unincubated PRS. (B) Q fraction elution pattern; 50-μl aliquots are assayed for inhibitory activity in Assay A (26° for 2 hours). (C) Unincubated PRS elution pattern; 50-μl aliquots are assayed for inhibitory activity in Assay A (26° for 2 hours) before (×—×) and after (●—●) incubation at 37° for 10 hours.

DEAE Chromatography of Pre-Q or Q Fraction

DEAE-cellulose resin is washed and converted to the hydroxide form. A column is prepared (0.9 × 15 cm) and washed with 10 mM ammonium acetate, pH 7.0. The sample, up to 20 ml of unincubated PRS or Q fraction, is loaded, and the column is washed with 10 mM ammonium acetate, pH 7.0. When the A_{280} falls to less than 0.1, the column is eluted with 0.3 M ammonium acetate, pH 7.0, 3-ml fractions being collected. In the case of unincubated PRS fractionation, small aliquots from each fraction are incubated at 37° for 10 hours, and these are assayed for inhibitory activity in Assay A. The unincubated portion of each fraction serves as a control. The Q fraction activity and the inhibitory activity produced by incubating the PRS fractions at 37° for 10 hours are found in the 0.3 M ammonium acetate fraction from the DEAE-cellulose column.

Other Properties of Q Fraction

Q fraction activity is destroyed by trypsin, but not affected by treatment with pancreatic ribonuclease or phospholipase C. It contains no detectable hemin as determined by spectral measurements at 410 nm (see Fig. 5).

[67c] Preparation of Reticulocyte Aminoacyl-tRNA and the Assay of Codon Recognition Properties of Isoacceptor tRNA's in a Reticulocyte Cell-Free System[1]

By WILLIAM R. WOODWARD, PRAPON WILAIRAT, and EDWARD HERBERT

Investigation of the codon recognition properties of isoacceptor tRNA's under conditions of protein synthesis provides a more reliable index of codon–anticodon interactions than trinucleotide induced binding of tRNA to ribosomes. In a cell-free system that is synthesizing several rounds of a well defined protein per ribosome, it is possible to measure the efficiency of recognition at particular amino acid sites in the protein. The reticulocyte cell-free system (CFS I)[2] provides such a system.

Preparation of Reticulocyte Aminoacyl-tRNA

Principle. This section presents methods for rapidly separating crude reticulocyte tRNA and for acylating the tRNA for subsequent fractionation or addition to the cell-free system. These methods take advantage of the high binding capacity of B-D cellulose[3,4] for tRNA. B-D cellulose is used (1) to purify the crude RNA fraction (ethanol precipitable material from the aqueous phase of a phenol extract of a reticulocyte postribosomal supernatant[5]), and (2) to separate the aminoacyl-tRNA from other com-

[1] This work was supported by a United States Public Health Service research grant CA-07373 from the National Cancer Institute and a National Science Foundation grant GB-4063. One of us (E.H.) is the recipient of a Career Development Award from the United States Public Health Service (1-K6-CA-2101). One of us (P.W.) is the recipient of a Rockefeller Foundation Fellowship.

[2] Abbreviations used; A_{260} units, the quantity that has an A_{260} when dissolved in 1 ml of solvent; B-D cellulose, benzoylated DEAE-cellulose. CFS I, cell-free system I; and PRS, postribosomal supernatant are defined elsewhere (W. R. Woodward, J. L. Ivey, and E. Herbert, this volume [67a]).

[3] I. Gillam, S. Millward, D. Blew, M. von Tigerstrom, E. Wimmer, and G. Tener, *Biochemistry* 6, 3043 (1967).

[4] I. C. Gillam and G. M. Tener, this series, Vol. 20 [6].

[5] J. M. Gilbert and W. F. Anderson, this series, Vol. 20 [59].

ponents of an acylation reaction. In the latter case, the reaction mixture is placed directly onto a B-D cellulose column, thereby eliminating the necessity for phenol extraction of the aminoacyl-tRNA. The principle advantages of this method are its speed, mild conditions for handling the tRNA, and high yield.

Reagents

B-D cellulose: either from commercial sources or prepared according to Tener[4]

Buffer A: 0.3 M NaCl; 10 mM sodium acetate, pH 4.5; and 10 mM MgCl$_2$

Buffer B: 1.0 M NaCl; 10 mM sodium acetate, pH 4.5: 10 mM MgCl$_2$ and 10% ethanol

Buffer C: 2.0 M NaCl and 40% ethanol

Acylation reaction containing the following components (final concentrations):

Glycyl-glycine, 0.1 M, pH 8.0

MgCl$_2$, 15 mM

ATP, 1–5 mM (neutralized)

Radioactive amino acid, 10 μM, at convenient specific activity

A_{280} units, 0.25–1.0 per milliliter of crude synthetase

A_{260} units, 1.0–20 per milliliter of tRNA

The optimum levels of synthetase and tRNA should be determined for each amino acid. It is important to select a tRNA level that is in the linear portion of the incorporation versus tRNA concentration curve.

Aminoethylation Conditions. The sample to be aminoethylated is dissolved in 7.5 ml of the following: 8 M urea; 50 mM Tris·HCl, pH 8.6; and 0.025% EDTA. β-Mercaptoethanol is added to 0.2 M (0.1 ml), and the mixture is incubated at room temperature for 30 minutes. The reduced product is aminoethylated by incubation with 0.5 M ethylenimine (0.2 ml) at room temperature for 30 minutes.[6] The aminoethylated product is desalted on Sephadex G-25 equilibrated with 0.2 M acetic acid.

Preparation of Reticulocyte tRNA

Procedure. All procedures are carried out at 0–4° unless otherwise noted. The postribosomal supernatant is extracted with phenol and precipitated with ethanol as previously described.[5] The ethanol precipitate is dried *in vacuo*, then dissolved in buffer A to 20 A_{260} units per milliliter. The solution is centrifuged to remove any insoluble material and placed

---

[6] R. T. Jones, *Cold Spring Harbor Symp. Quant. Biol.* **29**, 297 (1964).

on a B-D cellulose column equilibrated with buffer A.[7] The column is washed with buffer A until the A_{260} falls below 1% of the peak value. This treatment removes mainly small molecular weight components. The amino acid acceptor tRNA is eluted with buffer B. The column is washed with buffer B until the A_{260} falls below 1% of the peak value. This fraction is precipitated with 2.5 volumes of absolute ethanol at $-20°$ for at least 4 hours. The column is regenerated with buffer C and reequilibrated with buffer A for reuse.

The yield of tRNA by this procedure is approximately 0.25 mg tRNA per milliliter of packed reticulocytes (1 mg of tRNA dissolved in 1 ml of buffer gives an A_{260} of 24). The tRNA in this fraction contained acceptor activity for valine, lysine, phenylalanine, leucine, serine, histidine, cysteine, and tyrosine (the only amino acids assayed). Approximately 10–15% of the A_{260} units placed on the column remained tightly bound even after washing with buffer C.[8] Although this material did not interfere with subsequent use of the column, the resins were discarded after 5 usages.

Preparation of Aminoacyl-tRNA-Synthetases and Acylation of tRNA

Procedure. Crude reticulocyte synthetases were prepared by a modification of the method of Yang and Novelli.[9] A fresh or frozen PRS ($\sim$40 ml, the amount obtained from a single rabbit) is fractionated by gel-filtration (Sephadex G-100) and step elution chromatography on DEAE-cellulose. The crude synthetases are eluted from the DEAE-cellulose column with a buffer containing the following: 10 mM KPO$_4$, pH 6.5; 20 mM β-mercaptoethanol; 0.25 M KCl; and 10% glycerol. The peak tubes of synthetase activity are pooled and concentrated in dialysis tubing to 10% of the original volume by treatment with Ficoll (Pharmacia). The concentrated fraction is then dialyzed for at least 6 hours against 500 volumes of 50% glycerol. The final protein concentration is approximately 25 A_{280} units/ml. Preparations of the crude synthetase lost no more than 20% of their activity in 4 months of storage at $-20°$.

Before the tRNA (either crude or fractionated) can be acylated, it is necessary to determine the optimal levels of tRNA and enzyme for the acylation reaction. This may be accomplished by assaying varying

---

[7] Best results are obtained if the sample does not exceed 200 A_{260} units per milliliter of resin bed.

[8] The fraction retained on the B-D cellulose resin is not thought to contain any tRNA activity. For those aminoacyl-tRNA's investigated (valine, phenylalanine, and lysine), essentially all of the tRNA (as measured by recovery of radioactivity of [^{14}C] aminoacyl-tRNA) placed on the resin was eluted with buffer B.

[9] W.-K. Yang and G. D. Novelli, this series, Vol. 20 [5].

levels of synthetase and tRNA in 30-μl reaction mixtures containing the components described above. The reaction mixtures are incubated at 37° for 15 minutes, and 25-μl aliquots are spotted on filter paper disks (Whatman No. 1, 24 mm). The aminoacyl-tRNA is precipitated on the filter disks as described by Bollum,[10] dried, and counted.

When tRNA is acylated on a large scale the reaction mixture is incubated at 37° for 15 minutes, and the reaction is halted by adding 0.1 volume of 1.0 M sodium acetate, pH 4.5, and chilling the mixture on ice.[11] The reaction mixture is placed on a small B-D cellulose column (3 × 30 mm, equilibrated with buffer A) and washed with about 5 ml of buffer A. The washing is complete when the radioactivity nears background levels. The tRNA is then eluted with buffer B in 6-drop fractions. The first three fractions usually contain 85% of the radioactivity eluted with this buffer, and the recovery of radioactivity in ethanol-precipitable material is between 80 and 90% of the radioactivity in TCA precipitable material placed on the column. When phenylalanyl-tRNA is prepared, buffer B is made 20% in ethanol to ensure complete recovery of the acylated tRNA. Alternatively, the reaction mixture may be placed directly on a B-D cellulose column and the aminoacyl-tRNA fractionated using a linear gradient of increasing salt concentration. The aminoacyl-tRNA fractions to be used in the cell-free system are precipitated with ethanol and subsequently dissolved in an appropriate volume of water.

Assay of the Codon Recognition Properties of Isoacceptor tRNA's in a Reticulocyte Cell-Free System

The Effect of Added tRNA on Protein Synthesis

Principle. The efficiency of codon recognition on hemoglobin mRNA is measured by quantitating the amount of amino acid transferred from a given species of aminoacyl-tRNA into specific sites in hemoglobin. There are two requirements that should be met before one can effectively achieve this goal.[12] (1) The system should be carrying out chain initiation as well as chain elongation and termination. If the system is initiation limited, there will be an increasing gradient of radioactivity from the NH_2 terminal to the COOH terminal of the soluble protein produced. This gradient will make comparison of levels of radioactivity in different amino acid sites difficult. (2) The added tRNA should not significantly perturb protein syn-

---

[10] F. J. Bollum, this series, Vol. 12B, p. 171.

[11] If greater than 0.75 A_{280} units of enzyme per milliliter incubation are used, then 1.0 M sodium acetate, pH 5.4, should be used to prevent undue precipitation of the protein.

[12] W. R. Woodward and E. Herbert, *Science* **177**, 1197 (1972).

thesis. If the added tRNA inhibits protein synthesis or alters the product synthesized, one can no longer be certain that the codon recognition observed bears any relationship to *in vivo* codon recognition.

In order to satisfy the above criteria, it is necessary to examine protein synthesis as a function of added tRNA. A compromise must be reached between maximal sensitivity (achieved by increasing the levels of acylated tRNA) and minimal perturbation of protein synthesis (obtained by decreasing the levels of added tRNA). In this laboratory we try to use levels of tRNA that inhibit incorporation no more than 15%. For reticulocyte tRNA this level is about 10 A_{260} units per milliliter of cell-free system; whereas for nonmammalian tRNA's such as yeast, this level is about 5 A_{260} units/ml. The use of tRNA enriched for a particular species, reduces the level of tRNA that satisfies the above requirement.

Transfer of Amino Acid from Aminoacyl-tRNA into Hemoglobin

Several procedures for the fractionation of isoacceptor tRNA's have been presented elsewhere[4,13,14] and, therefore, will not be discussed here. The aminoacylated isoacceptor tRNA is dissolved in distilled water. The lysate and master mix (CFS I) are prepared and mixed as previously described,[15] except that water replaces added radioactive amino acid in the master mix. A small aliquot of the tRNA is added to a proportionate amount of the cell-free reaction mix to serve as a control. Two controls are run as follows: Two parallel samples are incubated along with the main transfer reaction, one with and one without added tRNA (an equal volume of water replaces added tRNA). The rate and extent of synthesis are monitored by measuring incorporation of free amino acid (0.5 μCi L-[^{14}C]-leucine added per 200 μl of incubation). The amount of radioactivity incorporated from the added tRNA will be insignificant compared to the amount incorporated from the free amino acid. The samples are incubated for 1 hour at 37° and then aliquots are spotted on filter paper disks for counting.[15]

In order to obtain sufficient sensitivity to perform the analysis of amino acid transfer into individual sites described below, we have found that 0.5 to 1.0 × 10^4 dpm per residue of amino acid in the protein is desirable. Since greater than 85% of the radioactivity initially present in the form of acylated tRNA appears in hemoglobin when CFS I is used, the amount of radioactive aminoacyl-tRNA needed for the experiment can be readily

[13] A. D. Kelmers, H. O. Weeren, J. F. Weiss, R. L. Pearson, M. P. Stulberg, and G. D. Novelli, this series, Vol. 20 [2].
[14] R. Stern and U. Z. Littauer, this series, Vol. 20 [8].
[15] W. R. Woodward, J. L. Ivey, and E. Herbert, this volume [67a].

determined. A 2 ml CFS I incubation is generally sufficient to incorporate enough radioactive amino acid into hemoglobin for analysis.

Incorporation of amino acid from the aminoacyl-tRNA proceeds rapidly even at 4°; therefore it is important that the tRNA be the last component added. Although no difference in the product synthesized has been detected, we prefer to start the incubation before adding the tRNA to ensure that all synthetic reactions have reached a steady state rate. Deacylation of aminoacyl-tRNA's (valyl-tRNA, lysyl-tRNA) before incorporation has not been a problem.[12] There are at least three reasons for this: (1) The amino acid is rapidly incorporated into protein (less than 5 minutes for complete incorporation), (2) concentration of substrates for the aminoacyl-tRNA-synthetase is high at all times, favoring maintenance of the acylated state of tRNA, and (3) any amino acid released from the tRNA would be so diluted by the large pool of free amino acid in the incubation that its subsequent incorporation would be below the limits of detection in the system.

Analysis of the Product Synthesized in the Cell-Free System

Following the incubation the ribosomes are sedimented by centrifugation at 105,000 g for 2 hours, and the resulting postribosomal supernatant is diluted with 2 parts water and slowly added to 25 volumes of cold 0.1% acid-acetone (1 ml HCl per liter acetone) to remove the heme and precipitate the globin. The precipitated globin is combined with an equal number of disintegrations per minute of uniformly labeled globin prepared by incubation of intact cells with a radioactive amino acid[16] ([³H]amino acid if [¹⁴C] is used for the acylation of the tRNA and vice versa). The uniformly labeled globin is prepared by passing the postribosomal supernatant fraction from the intact cell incubation through a DEAE-cellulose column, after equilibration of the column with 10 mM Tris·HCl, pH 7.2, and 25 mM KCl. Under these conditions hemoglobin is not bound to the resin. Globin is prepared from the partially purified hemoglobin by precipitation in cold acid-acetone as described above. The α and β chains are then separated on CM cellulose,[17] and the purified subunits are aminoethylated and digested with trypsin (trypsin:protein; 1:100 (w/w) at 37° pH 8.0).

The tryptic peptides are chromatographed by a modification of the procedure of Jones[6] on a Dowex-50 column (Aminex A-5 resin, 0.9 × 16 cm), jacketed at 50°. The peptides are eluted with a 500-ml linear gradient of pyridine acetate. Fractions of approximately 3 ml are collected,

[16] W. Godchaux, S. D. Adamson, and E. Herbert, *J. Mol. Biol.* **27**, 57 (1967).
[17] H. M. Dintzis, *Proc. Nat. Acad. Sci. U.S.* **47**, 247 (1961).

and the radioactivity is monitored. The uniformly labeled globin serves to mark the peptides of interest and to measure recovery of peptides. The peptides are rechromatographed on Dowex 50-X2 (0.9 × 50 cm), jacketed at 50° using the same pyridine acetate gradient. The purity of the peptides is then checked by cellulose thin-layer chromatography (n-butanol, pyridine, acetic acid, and water, 15:10:3:12); cellulose thin-layer electrophoresis at pH 1.9 (88% formic acid, acetic acid, and water, 50:150:800), and pH 5.6 (pyridine, acetic acid, and water, 4:1:995). If a peptide shows more than a single ninhydrin spot in any of the above systems, it is purified by that system on a batch scale, and its purity is rechecked. The peptides are then hydrolyzed with 6 N HCl for 24 hours at 110° and subjected to cellulose thin-layer chromatography (n-butanol, acetic acid, and water, 143.5:13:43.5, upper phase) and the identity of each peptide verified by its composition. The ratio of isotope transferred by the aminoacyl-tRNA to isotope from the uniformly labeled material for the purified peptides can then be used as a measure of the relative codon recognition for an isoacceptor tRNA at a particular site in a peptide.[12] Peptides containing more than one residue of amino acid may require further digestion and purification to determine the specific site recognized by an isoacceptor tRNA, although this might not be necessary in cases where no transfer occurs into any site or where all sites are recognized.

Results obtained for lysine isoacceptor tRNA have been published and may be consulted for further details.[12]

[67d] Preparation and Analysis of Nascent Chains on Reticulocyte Membrane-Bound Ribosomes[1]

By WILLIAM R. WOODWARD and EDWARD HERBERT

Membrane-bound ribosomes from reticulocytes have been shown to synthesize globin.[2] Methods are described for preparing nascent peptides labeled with radioactive amino acid and for analyzing the tryptic digestion products from the nascent peptides.

[1] This work was supported by a United States Public Health Service research grant CA-07373 from the National Cancer Institute and a National Science Foundation grant GB-4063. One of us (E.H.) is the recipient of a Career Development Award from the United States Public Health Service (1-K6-CA-2101).
[2] W. R. Woodward, S. D. Adamson, H. M. McQueen, J. W. Larson, S. M. Estvanik, P. Wilairat, and E. Herbert, J. Biol. Chem. 248, 1556 (1973).

Reagents

Master Mix (intact cell incubation):
   Amino Acid Mixture I,[3] 1 ml
   Amino Acid Mixture II, 1 ml
   Glutamine, 4 mg/ml, 0.25 ml
   Solution A,[3] 0.5 ml
   Sodium bicarbonate 2.2 g/100 ml, 0.5 ml
   [3H]amino acid[4] dried under an air stream in the incubation flask
   Water, 1.75 ml
Ferrous ammonium sulfate, 70 mg/ml
Buffer A: 1 mM Tris·HCl, pH 7.4; 10 mM KCl; 1.5 mM MgCl$_2$
Buffer B: 10 mM Tris·HCl, pH 7.4; 10 mM ammonium acetate; and
   1.5 mM MgCl$_2$
Sucrose pad: 55 g of sucrose in 100 ml of buffer B

Preparation of 3H Amino Acid-Labeled Nascent Chains on Reticulocyte Membrane-Bound Ribosomes

The nascent chains on the bound ribosomes are labeled with [3H]amino acid in an intact cell incubation. Phenol red (1 μl) is added to the master mix, and carbon dioxide is bubbled through the solution until the indicator turns colorless.[5] Master mix solution is combined with an equal volume of reticulocytes,[6] and 20 μl of the ferrous ammonium sulfate solution are added. The reaction mixture is incubated for 10 minutes at 35°, chilled, and centrifuged to sediment the cells. The cells are lysed by adding 2.4 ml of buffer A to 1.0 ml of cells. The membrane fraction is sedimented by centrifugation. The supernatant may be used for the preparation of nascent chains on free ribosomes. The membrane pellet is washed twice with buffer A, suspended in 20 ml of buffer A which has been made 0.2% in DOC,[7] placed on ice for 30 minutes, and centrifuged to sediment membranes. The supernatant fraction, containing the released membrane bound ribosomes

---

[3] Amino Acid Mixtures I and II and Solution A are prepared as previously described (W. R. Woodward, J. L. Ivey, and E. Herbert, this volume [67a]) except that the radioactive amino acid used in the incubation is omitted from the appropriate Amino Acid Mixture.

[4] The [3H]amino acid (highest specific activity available) is dried in the incubation flask to remove any ethanol present.

[5] Carbon dioxide can be conveniently generated in a sidearm vacuum flask by adding 1 N HCl to sodium bicarbonate. Tubing connected to the sidearm can be used to direct the carbon dioxide generated into the reaction vessel. After the master mix has been flushed, it should be tightly stoppered.

[6] W. R. Woodward, J. L. Ivey, and E. Herbert, this volume [67a].

[7] Abbreviations used; DOC, deoxycholate; A_{260} units, the quantity that has an A_{260} when dissolved in 1 ml of solvent.

is centrifuged at 105,000 g for 150 minutes over a 1.0 ml sucrose pad. The sedimented ribosomes are dissolved in 5 ml of buffer B and centrifuged again over a 1.0-ml sucrose pad. The second centrifugation is necessary to completely remove hemoglobin adsorbed to the ribosomes. The bound ribosomes are dissolved in 1.0 ml of buffer B, and the A_{260} units and radioactivity are determined.

Analysis of Nascent Chains on Membrane-Bound Ribosomes by Tryptic Peptide Mapping

One important criterion that can be used to establish that the radioactivity associated with the membrane bound ribosomes is in the form of nascent chains is to show that the specific activity of the amino acid decreases from the NH_2-terminal to COOH-terminal end of the protein. Therefore, in selecting the amino acid to be used to label the nascent chains, it is important to select one that is distributed throughout the globin chain. Moreover, since the nascent peptides cannot be separated into α and β chains it is necessary to select an amino acid such as tyrosine whose α and β tryptic peptides (3 tyrosine residues in each chain) separate well on Dowex-50 chromatography.[8]

An equal number of dpm of ^{14}C-labeled uniformly labeled globin[9] is combined with the ^{3}H-labeled nascent chains. The mixture is aminoethylated as previously described[9] except that the desalting step is carried out on a Sephadex G-15 column in 0.2 M acetic acid. The aminoethylated product is digested with trypsin and chromatographed on Dowex 50 (Aminex A-5) and Dowex 50-X2. The amount of radioactivity in the peptides is carefully determined, and the ratio of nascent peptides to uniformly labeled peptides so obtained are plotted against the position of the amino acid in the globin chain.

Uniformly labeled globin is used to locate the globin peptides of interest and to quantify recovery of the peptides. This is especially important for the analysis of nascent chains because peptides near the COOH-terminal of the protein contain considerably less radioactivity than peptides near the NH_2-terminal and may be difficult to locate. A plot of the ratio of radioactivity in the nascent chain peptides to that in the uniformly labeled peptides against the amino acid position should yield a linear plot with a negative slope that intercepts the abscissa at the COOH-terminal amino acid. If significant amounts of completed chains contaminate the ribosomal preparation, the intercept will be considerably greater than the expected chain length. This can be used as an indication of the levels of contamination.

[8] R. T. Jones, *Cold Spring Harbor Symp. Quant. Biol.* **29**, 297 (1964).
[9] W. R. Woodward, P. Wilairat, and E. Herbert, this volume [67c].

[68] The Wheat Embryo Cell-Free System[1]

By Abraham Marcus, Dov Efron, and Donald P. Weeks

Cell-free extracts of wheat embryos catalyze the incorporation of radioactive amino acids into protein in response to several mRNA's.[2-4] The translational fidelity of the system has been established by the analysis of the products of reactions catalyzed by STNV-RNA,[4] by BMV-RNA[5] and by rabbit globin mRNA.[6] The system is most active with eucaryotic mRNAs but Qβ-RNA can also be translated in the wheat system yielding authentic phage coat protein (J. W. Davies and P. Kaesberg, unpublished observations). The following sections describe the methods used for isolating viable wheat embryos and for carrying out a typical TMV-RNA-catalyzed amino acid incorporation reaction.

Isolation of Wheat Embryos

The method described here for the isolation of wheat embryos is a modification of the procedure of Johnston and Stern.[7] The major features of the method are: (1) removal of the loosely attached embryo from the seed by mechanical agitation in a Waring Blendor; and (2) separation of embryos from endosperm fragments by a series of sieving, blowing, and a selective organic solvent flotation procedure based on differences in bouyant density between embryos and endosperm.

Procedure. Wheat seeds, 250 g, are blended for 5 seconds at full speed in a Waring Blendor. Freezing of seeds in dry ice prior to blending, as suggested by Johnston and Stern,[7] does not appear to be necessary. The contents of the quart-sized glass blending container are emptied onto a set of three sieves (numbers 10, 14, and 28 of the Tyler scale). The sieves are vibrated for approximately 30 seconds at a setting of 40 on a "Syntron" automatic vibrator.[8] Material in the top sieve (No. 10) is pooled to be

[1] Supported by USPHS grant GM-15122 and NSF grant GB-23041 as well as by NIH grants CA-06927 and RR-05539 awarded to this institute and by an appropriation from the Commonwealth of Pennsylvania.
[2] A. Marcus, B. Luginbill, and J. Feeley, *Proc. Nat. Acad. Sci. U.S.* 59, 1243 (1968).
[3] A. Marcus, *J. Biol. Chem.* 245, 955 (1970).
[4] W. H. Klein, C. Nolan, J. M. Lazar, and J. M. Clark, Jr., *Biochemistry* 11, 2009 (1972).
[5] D. S. Shih and P. Kaesberg, *Proc. Nat. Acad. Sci. U.S.* 70, 1799 (1973).
[6] D. Efron and A. Marcus, *FEBS Lett.* 33, 23 (1973).
[7] F. B. Johnston and H. Stern, *Nature* (*London*) 179, 160 (1957).
[8] Model J-2C, Syntron Company, Homer City, Pennsylvania.

blended and sieved once more. The embryo fraction retained by the number 28 sieve is freed of bran by processing[9] in a seed blower.[10]

A major portion of contaminating endosperm fragments is then removed by adding approximately 50 g of the bran-free embryo fraction to a 350-ml mixture of cyclohexane and carbon tetrachloride (both reagent grade). The optimal ratio of cyclohexane to carbon tetrachloride varies from 1:1.75 to 1:2.5 for different lots of seed. The mixture is stirred with a glass rod, and the endosperm fragments are allowed to settle away from the floating embryos (exposure time to organic solvent should be kept to a minimum). Embryos are then carefully poured into a large sintered glass funnel (coarse) mounted on a 1000-ml Buchner funnel and dried under vacuum. The organic solvent mixture may be reused for 4–5 flotations. Embryos should be spread on absorbent paper and allowed to air dry for 15–30 minutes.

Following flotation, embryos are resieved using sieve numbers 10, 14, 16, and 28. The embryo preparation is first vibrated gently at a setting of 5, until all particles pass through the top sieve. The top two sieves are then removed and the bottom sieves are again vibrated until all the particles retained by the number 16 sieve are freely moving. The number 28 sieve is then covered and vibrated at setting 40 for 10–12 seconds to remove fragments of embryo and endosperm. Embryos are further fractionated by differential blowing, with fragments first being removed by gentle blowing,[11] and intact embryos then separated from larger endosperm fragments in a more forceful air stream.[12]

Both the flotation and the blowing steps require individual judgments as to the optimal settings and procedures for a given lot of seed. A second passage of embryos through the flotation procedure has been useful with certain seed batches.

Embryos may be stored in sealed glass jars at 4° for several years with only a minor loss of the activity of an *in vitro* amino acid incorporating system derived from these embryos. Different varieties of seed have shown some variation in the activity obtained from the embryos. Preparations with high activities have been consistently obtained from the variety Fortuna.[13] Seed lots contaminated by small weed seeds should be avoided.

[9] Setting of 25 with Port No. 2 open; 30 g of material per 30 second blowing.

[10] Model L, New Brunswick General Sheet Metal Works, New Brunswick, New Jersey.

[11] Embryo, 15 g per blowing, Port No. 1 open; initial setting of 15 is raised to 21 over a 20-second period, and the blowing is continued for 1 minute.

[12] Embryo, 7 g per blowing; all ports closed; setting of 25; retain embryos that are blown from the sample chamber into the collecting tray.

[13] Obtained from Mr. Glenn Hartman, Montana State University, Agriculture Experimental Station, P.O. Box 393, Sidney, Montana 59270.

Preparation of TMV-RNA

Isolation of the Virus. TMV infected tobacco leaves (*Tabacum* var. *Samsun*)[14] (200 g) are blended at high speed (6 blendings of 30 seconds each at 1-minute intervals) with 300 ml of the following solution: 105 ml of 1 M K_2HPO_4, 45 ml of 1 M KH_2PO_4, 3.0 ml of 0.5 M EDTA (disodium brought to pH 7.4 with KOH), 0.3 ml of 2-mercaptoethanol, and 150 ml of H_2O. The slurry is then squeezed through a double layer of cheesecloth and cleared by centrifugation for 10 minutes at 23,500 g. To the supernatant 8/100 volume of *n*-butanol is added dropwise at room temperature with stirring, with the stirring continued for an additional 15 minutes after addition of the butanol. The suspension is centrifuged for 10 minutes at 12,000 g, and small fragments are removed from the supernatant by filtration through Whatman No. 1 filter paper. The virus isolation is then carried out by high speed centrifugation; 40 minutes at 28,000 rpm (Spinco rotor No. 30). The viral pellet is gently suspended in 5 mM potassium phosphate pH 7.4 (5 ml for each No. 30 rotor tube),[15] and each tube is rinsed twice with 2.5 ml of the phosphate solution. The combined suspension is clarified by low speed centrifugation (10 minutes at 23,000 g), and the virus is repelleted by 40 minutes centrifugation at 28,000 rpm in the Spinco No. 30 rotor. Several such cycles of high and low speed centrifugation are repeated until the viral suspension achieves a ratio of optical densities at 260/280 nm of 1.2. This preparation is then suspended in 5 mM phosphate buffer containing 20 mM EDTA pH 7.4 at a viral concentration of 15–20 mg/ml (1 A_{260} unit of virus = 0.3 mg/ml) and incubated overnight at 30°. The virus is repelleted by a 1 hour centrifugation in the Spinco No. 30 rotor, and resuspended in 30 ml H_2O giving a concentration of 17–20 mg/ml. This suspension is stored at 2–4°.

Extraction of TMV-RNA. TMV suspension, 8–10 ml containing 180–200 mg of virus, is brought to 5.0 mM EDTA and 0.5% sodium dodecyl sulfate (SDS) by adding 0.6 ml of 0.1 M EDTA, 0.6 ml of 10% SDS and H_2O to a final volume of 12 ml; 14 ml of EDTA-washed phenol[16] are

---

[14] Infected leaves are obtained by painting a virus solution (150 μg/ml in water) on leaves of plants, 4–6 inches in height, the leaves having just previously been rubbed with Carborundum. About 1 minute after the virus inoculation, the leaves are carefully rinsed so that all the Carborundum is removed. The plants are maintained for 18–23 days and the leaves are then harvested and stored at −20°.

[15] A convenient procedure is to allow the buffer to sit over the virus pellet for several hours (or overnight) with periodic swirling.

[16] To a 1 pound bottle of reagent-grade phenol (Matheson, Coleman and Bell), 260 ml of H_2O are added and the mixture is melted at 30°. Twelve milliliters of 0.5 M EDTA pH 7.4, and 8 ml of 1 N KOH are added; the suspension is stirred for 10 minutes at room temperature and transferred to two 600-ml beakers. As soon as

added, and the suspension is stirred for 5–10 minutes in a 125-ml Erlenmeyer flask at room temperature. The emulsion is centrifuged for 10 minutes at 23,500 g at 2–4°, and the upper layer is reextracted twice with 14 ml of EDTA-washed phenol.[16] Then, 0.05 volume of 2 M KAc, pH 5.5, and 2.5 volumes of cold ethanol are added. The suspension is mixed and kept overnight at −15°. The RNA is then collected by centrifugation, dissolved in 6.0 ml of 0.1 M KAc, pH 5.5, and reprecipitated with 2.5 volumes of cold ethanol. After at least 2 hours at −15°, the pellet is collected by centrifugation, dissolved in 2.5 ml H₂O and dialyzed[17] for 2 hours against distilled H₂O. The RNA yield from 200 mg of virus is generally 2.5–3.0 ml containing 3.3–3.8 mg/ml.

Amino Acid Incorporation

Reagents[18]

GTP, 0.01 M
Dithiothreitol, 0.3 M
Wheat germ tRNA,[19] 0.6 mg/ml
TMV-RNA, 0.33 mg/ml
Mix 1: mix 4.0 ml of 0.32 M disodium creatine phosphate, 4.0 ml of 1.0 M Tris·acetate, pH 8.0, 2.0 ml of 0.08 M disodium ATP adjusted to pH 6.7 with KOH, 6.4 mg creatine kinase and 2.0 ml of H₂O. This solution is stable indefinitely when stored at −20°.

---

the two layers have separated, the top layer is removed by aspiration and discarded. The bottom layer is transferred to a 1-liter beaker and 400 ml of H₂O are added. The suspension is stirred for 2 minutes at room temperature, and the top layer is again removed, this time after standing in the dark for 30 minutes. The water wash is repeated. Subsequently, 4 ml of 0.5 M EDTA pH 7.4 and 34 ml of H₂O are added; the suspension is stirred for 10 minutes at room temperature and then kept in the dark for 30 minutes. The top layer is removed by aspiration, and the bottom layer is stored in a brown bottle at 2–4° as EDTA-washed phenol. Occasionally, the layers do not separate well during the second water wash. This situation may be remedied by adding 1 ml of 0.5 M EDTA pH 7.4 to the suspension.

[17] The dialysis tubing used in 0.25-inch diameter (A. H. Thomas) and is prewashed with EDTA in the following manner. A 100-foot roll cut into lengths of 8–10 inches is kept for 1 hour in 1 liter of boiling 10 mM EDTA, pH 7.4. The procedure is repeated for 0.5 hour with a fresh EDTA solution. The tubing is then rinsed 10× with double distilled water and stored at 2–4° (under double-distilled water).

[18] The trilithium salt of GTP and creatine kinase may be obtained from Boehringer Mannheim Corp., creatine phosphate from Calbiochem, and ATP from Sigma Chemical Co.

[19] A. Marcus, S. N. Seal, and D. P. Weeks, this volume [11].

Preparation of S23

Dry wheat embryos, 300 mg, are ground thoroughly with a small amount of sand in a precooled mortar with pestle in a total volume of 3.3 ml of 90 mM KCl, 2 mM CaCl$_2$, 1 mM Mg(Ac)$_2$, 6 mM KHCO$_3$.[20] The embryos are initially ground in 1.0 ml with 0.5- and 1.8-ml increments added subsequently. The slurry is then centrifuged for 10 minutes at 23,500 g and the supernatant is removed with a Pasteur pipette, taking care to leave behind as much as possible of the upper lipid layer. Just prior to use, 0.5–3.0 ml are dialyzed[17] against 500 ml of 1 mM Tris·acetate, pH 7.6, 50 mM KCl, 2 mM Mg(Ac)$_2$, 4 mM 2-mercaptoethanol for 1.75 hours. This preparation is termed S23. It may be kept at $-20°$ (undialyzed) with only moderate loss in activity. It is important, however, that dialysis be carried out just prior to use. A typical preparation tested in this manner after 8 days of storage had 75–80% of the original activity.

Assay

To tubes in an ice bath are added 0.28 ml of an incubation mixture containing 30 μl Mix 1, 1 μl of GTP, 3 μl of dithiothreitol, 20 μl tRNA, 12 μl of 0.1 M Mg(Ac)$_2$, 27 μl of 0.5 M KCl, 10 μl of 1.25 mM of each of 19 unlabeled amino acids,[21] 30 μl of TMV-RNA, 0.125 μCi of [14C]leucine (0.4 nmole), and 0.12 ml of freshly dialyzed S23. Reaction mixtures are then incubated for 30 minutes at 30°, and the reaction is stopped by adding 0.2 ml of 16% TCA containing 13 μmoles of [12C]leucine followed by 4 ml of 5% TCA. The suspension is cooled on ice for 10 minutes and the insoluble material is collected by centrifugation. The pellet is then resuspended in 4.0 ml of 5% TCA containing 6.5 μmoles of [12C]leucine, and heated for 15 minutes at 90°. After 8 minutes on ice, the insoluble material is collected by filtration on glassfiber (GF/C) disks. The disks are washed twice with 5% TCA, dried under an infrared lamp, and counted for radioactivity in 10 ml of a toluene-based scintillator (5 g of PPO and 300 mg of dimethyl POPOP per liter).

The incorporation obtained for such an incubation is approximately 50,000 cpm. Increasing the TMV-RNA component to 20 μg (the level of saturation) increases incorporation by 75%, while extending the incubation

---

[20] The pH of the homogenate should be 6.4–6.8 (as checked on the supernatant obtained after the 23,500 g centrifugation). If the pH is lower than 6.3, preparations of reduced activity are obtained. A higher pH results in release of the endogenous messenger [D. P. Weeks and A. Marcus, *Biochim. Biophys. Acta* **232**, 671 (1971)], giving correspondingly less dependence on the exogenous mRNA. Adjustment of the pH, if necessary, is done by increasing or decreasing the KHCO$_3$ molarity in the grinding medium.

[21] A. Marcus and J. Feeley, *Proc. Nat. Acad. Sci. U.S.* **51**, 1075 (1964).

time to 1 hour increases incorporation by an additional 50%. In a typical 60 minutes incubation with 20 μg TMV-RNA, an incorporation of 125,000 cpm is obtained. If TMV-RNA is omitted, the incorporation is 4200. Presence of chloramphenicol (10 μg/ml) is without effect on the reaction and its addition is recommended for reaction times longer than 30 minutes.

[69] *In Vitro* Translation of STNV-RNA

By JOHN M. CLARK, JR. and WILLIAM H. KLEIN

Satellite tobacco necrosis virus (STNV) is a degenerate virus requiring *in vivo* coinfection with tobacco necrosis virus (TNV) for growth.[1] This unique dependence is assumed to arise from the fact that STNV-RNA contains information for only STNV coat protein.[2] STNV replication, *in vivo,* is then assumed to require TNV dependent enzymes (e.g., TNV dependent RNA replicase).

The full details of this *in vivo* dependence are yet to be worked out. Yet various *in vitro* studies with STNV and STNV-RNA emphasize the uniqueness of the STNV system. STNV is a small virus (approximately 2×10^6 daltons) containing one piece of 400,000 molecular weight single-stranded RNA.[2] This STNV-RNA serves as an *in vitro* messenger-RNA in both prokaryotic and eukaryotic systems.[3] Further, in both systems, such translation results in *in vitro* synthesis of STNV coat protein.[3,4] Thus STNV-RNA is a readily available, monocistronic, messenger-RNA capable of directing the synthesis of a single defined protein. Such an RNA has many uses in *in vitro* protein biosynthesis studies.

Virus Growth and Isolation

Principles. There are several strains of STNV and TNV.[1] Further, these various virus strains can be grown on a variety of plants (e.g., tobacco, beans, tulips). To date, the STNV-RNA used in *in vitro* translation studies has been derived from the Rothamsted strain of STNV. This procedure features growth of Rothamsted STNV on mung beans in combination with the Rothamsted *b* strain of TNV (TNV-*b*). These choices pro-

[1] B. Kassanis and H. L. Nixon, *J. Gen. Microbiol.* **25**, 459 (1961).
[2] M. E. Reichman, *Proc. Nat. Acad. Sci. U.S.* **52**, 1009 (1964).
[3] W. H. Klein, C. Nolan, J. M. Lazar, and J. M.Clark, Jr., *Biochemistry* **11**, 2009 (1972).
[4] R. E. Lundquist, J. M. Lazar, W. H. Klein, and J. M. Clark, Jr., *Biochemistry* **11**, 2014 (1972).

vide two distinct advantages. First, mung beans can be grown in trays, so cumbersome greenhouse operations can be avoided. Second, STNV can be isolated relatively free of TNV due to the cold-sensitive character of TNV-*b*; i.e., −15° storage of beans infected with TNV-*b* results in the disruption of TNV-*b*.

The detailed instructions and scheduling of STNV growth and isolation are flexible. For example, slight variations in bean and virus growth occur and result in varied times of harvest for infected beans. Further, the virus isolation may be scheduled over a 1-, 2-, or 3-day period as convenient. Yet, particular care must be taken to maintain the specific temperatures called for in the bean growth, inoculation, and washings specified. Failure to maintain these temperatures results in lowered virus growth and may also result in bean damage and rot.

Growth of Virus on Mung Beans.[5] As indicated, the Rothamsted *b* strain of TNV is cold sensitive. Thus one inoculates germinating mung beans with an inoculum prepared from STNV-TNV-*b* infected, unfrozen, beans (i.e., previously stored at 0–2°). Infected beans decay upon prolonged storage at 0–2°. One must therefore repeat the 8-day virus growth at least once every 3–4 weeks in order to maintain a virus inoculum. The best virus yields require use of freshly harvested infected beans as inoculum.

Late in the afternoon of the first day, submerge 350 g of dry mung beans (variety Jumbo, Palecek Mills, Enid, Oklahoma) in a bath of 0.5% sodium hypochlorite (Clorox: H_2O, 1:10) and stir for 5 minutes. Then pour off the fluid and wash the beans 10 times with tap water (i.e., submersion followed by decantation). Finally, place the washed beans in a 2-liter beaker; add 1 liter of water; cover the beaker with plastic (e.g., parafilm or polyethylene) and a dark cloth; and incubate overnight (approximately 12 hours) at 30°.

On the morning of the second day, wash (submersion and decantation) the swelled beans a few times to remove germination waste materials. Then divide the lot of beans into two enamel 16-inch × 10-inch trays lined with double layers of blotter paper. Spread the beans as a monolayer on the blotter paper and then add tap water so as to cover one-third the height of the beans (*most of the beans' surfaces should be open to the air*). Cover the tray with plastic and a dark cloth and incubate 24 hours at 30°, adding more water after approximately 8 hours so as to retain the original water level in the trays.

At the start of the third day, transfer the beans to a large beaker and repeatedly rinse and decant the beans in a stream of 25–30° tap water. This will allow removal of many of the undesirable green seed coats that

[5] H. Liu, F. O. Holmes, and M. E. Reichmann, *Phytopathology* **59**, 883 (1969).

will float above the white seeds. Then return the beans to two 16-inch × 10-inch enamel trays lined with a double layer of fresh blotter paper, add 25–30° tap water and cover as before. Incubate these trays of beans at 37° for 5–6 hours.

Toward the end of the 37° incubation period, briefly homogenize (Waring Blendor) 100 g of infected beans with 100 ml of tap water so as to obtain an even slurry. Squeeze this preparation through 4 layers of cheesecloth; centrifuge (0–2°) the resultant fluid at 8500 g for 20 minutes; after decantation, divide this clarified solution into 2 equal portions (approximately 75 ml each) and store at 0–2° until used as inoculum for the 37° preconditional beans.

Immediately as the inoculum solution is ready, decant the water from one tray of 37° beans and place the beans in a large glass container (e.g., a 3-liter, partially baffled, wide-mouthed, round-bottom flask). Quickly raise the temperature of one of the two batches of inoculum solution to 30–37°, add 0.5 g of Celite (Hyflo-Supercel) to the warm inoculum solution, and pour the entire slurry onto the beans. Then haltingly rotate the glass container for 15 minutes so as to tumble the beans and inoculum.

The exact extent of this inoculation process will depend upon the nature of the container, the harshness of the tumbling, the grade of Celite, etc., and must be worked out by each investigator. Most people are not harsh enough when they first attempt this process, yet one can be too harsh. If the infected beans average fewer than 10 virus lesions per bean, the inoculation period should be harsher or extended. If one detects limited bean growth and extensive bean rot, the process is probably too harsh.

At the end of the timed inoculation process, wash the residual fluid, Celite, and most of the remaining seed coats from the beans by repetitive washes and decantations with 30–37° tap water. Then decant off the water and distribute the beans evenly as a monolayer on a double layer of moist blotter paper lined in a 20-inch × 17-inch tray. Cover the tray with a layer of plastic punctured with small holes spaced approximately 3 inches apart, cover the plastic with a dark cloth, and incubate the tray at 19–20°. Finally, repeat the same inoculation process with the second batch of inoculum and the second tray of 37° preconditioned beans.

During the next 5 days of 19–20° incubation, add water to the trays one or two times so as to maintain moisture in the blotters (no standing water should be evident). Virus lesions will develop 2–3 days after inoculation. Harvest the infected beans once (most of the virus lesions are 1 mm in diameter, usually 5 days after inoculation), and store at 0–2° for inoculum or −15° for virus isolation. Best virus yields are obtained from beans that have contiguous virus lesions (i.e., >20 lesions per seed).

Isolation of STNV from Mung Beans. As indicated earlier, the scheduling of the following STNV isolation is flexible. Slightly higher yields are obtained by shortening the overall isolation process. Most of the operations are performed at room temperature with appropriate use of ice buckets, refrigerated centrifuges, and refrigerated dialysis facilities as indicated. General details of this isolation have been published elsewhere.[5]

Working in small lots, blend (Waring Blendor) 400–600 g of infected, −15° prestored, mung beans with an equal weight of room temperature 25 mM Na_2EDTA, 1% K_2HPO_4, pH 7.0 (9.25 g $Na_2EDTA \cdot 2H_2O$ + 10 g K_2HPO_4 + water up to 1 liter) until an even slurry is formed. Transfer the slurry to ice-chilled centrifuge bottles and centrifuge (e.g., Sorvall, GSA head) for 30 minutes, at 0–2° and 8000 g. Decant the supernatant through 4 layers of cheesecloth and, using gloves, finally gently squeeze the residual fluid from the cheesecloth. Over a 1-minute period, gently stir in 20 ml of room-temperature absolute ethanol per each 100 ml of 8000 g supernatant and then immediately transfer this cool alcoholic fluid to ice-chilled centrifuge bottles and centrifuge (0–2°) at 5000 g for 30 minutes. Immediately after this centrifugation, decant and measure this 5000 g supernatant. Then over a 5-minute period, gently stir in an equal volume of room-temperature saturated $(NH_4)_2SO_4$ solution. Allow this mixture to stand at room temperature. If flocculation does not occur after 30 minutes, stir in a 2% volume (of total volume) of saturated $(NH_4)_2SO_4$ solution, and wait 15 minutes for flocculation. Repeat additional 2% additions of the $(NH_4)_2SO_4$ until flocculation occurs. (This period of waiting is a convenient time to make the sucrose gradients required later in this procedure.)

One hour after flocculation is detected in the $(NH_4)_2SO_4$ containing solution, centrifuge (0–2°) the flocculent preparation for 30 minutes at 5000 g. Immediately decant the supernatant and aspirate off any residual fluid above the *loose* pellets (prolonged pellet contact with $(NH_4)_2SO_4$ aggregates STNV). Then using a sterile glass rod, rapidly resuspend the pellets in a total volume of 200–300 ml of cold (0–2°) distilled water. Usually this solution is then dialyzed (0–2°) overnight, with one change, against 3–4 liters of cold (0–2°) distilled water. Alternately, if a continuous preparation is envisioned, 2–3 hours of dialysis, as above, will suffice.

Clarify the dialyzed material by centrifuging (0–2°) the solution for 20 minutes at 30,000 g. Gently decant the supernatant from the *loose* pellet. Then further centrifuge (0–2°) this 30,000 g supernatant for 4 hours at 80,000 g (e.g., Spinco, No. 30 head). Immediately after centrifugation, decant the supernatant fluid from the resultant yellow-brown pellets, and finally aspirate off any residual fluid over the pellets. Using an ice bucket to chill the tubes, resuspend the pellets (prechilled sterile glass rod) in sufficient cold distilled water so as to obtain a final volume of 15 ml. Allow this preparation to sit (0–2°) for 1 hour to assure solubilization of the virus before applying it as three 5-ml aliquots to the tops of three 48-ml, cold (0–2°), linear, 4–20% sucrose gradients in 0.1 M KCl (e.g., in tubes used in a Spinco SW 25.2 rotor). Centrifuge (0–2°) in a Spinco SW 25.2 rotor at 22,500 rpm (62,000 g) for 4 hours. After centrifugation, remove the gradient tubes and using a cold room, mount one tube in a ring stand arrangement featuring a flashlight aimed directly down the length of the gradient tube. Then in the dark with the flashlight on, locate any band(s) of light-scattering material. STNV usually appears as a milky band approximately one-third of the way down the tube. Occasionally, additional bands of milky appearing residual TNV and/or STNV aggregates (dimer, trimer, etc.) appear beneath the STNV bond. Further, one sees a layer of brownish soluble protein near the top of the gradient. Using a fine-tipped pipette (e.g., Pasteur pipette) and a propipette, aspirate off just the STNV containing band (the top milky band) from the tube. Repeat this process with all the gradient tubes, combining the STNV solutions in a single cooled (0–2°) vessel.

Two alternatives exist at this stage in the procedure. One may proceed with the isolation as below or one may dialyze (0–2°) the STNV solution overnight, with one change, against 10 volumes of cold distilled water. In either case, next pellet the collected STNV by centrifuging (0–2°) for 4–6 hours at 85,000 g. Decant off the supernatant solution and then while chilling (0–2°) the centrifuge tubes, use a fine-tipped pipette (e.g., Pasteur pipette) to aspirate off the last of the fluid above the gelatinous pellets. Resuspend (0–2°) (sterile glass rod) the pellets in a total of 0.5 ml of distilled water and assay spectrophotometrically for STNV concentration (A_{260} of 6.8 = 1 mg STNV/ml when in 1 cm deep cuvette). Make up to a final concentration of 10–20 mg/ml with cold distilled water and then freeze away the virus in a dry-ice bath before storing at −80°. STNV yield will be 15–60 mg STNV per 600 g of well infected beans.

One may assay the homogeneity of STNV preparations (as follows) during and/or after the final virus isolation. Add 0.75 ml of the original extract (i.e., 8000 g supernatant) or 1 mg of STNV from the final STNV solution to a Spinco Model E analytical ultracentrifuge rotor. Then after appropriate evacuation of the rotor chamber, observe (schlieren optics) the rate of migration of the component(s) detected at 35,600 rpm, ambient temperature, 60° bar angle, in pictures taken at 4-minute intervals. STNV will migrate as a single peak of 48 S material that will appear at the midpoint of the sixth picture frame. The original extract will often also reveal the presence of residual TNV (122 S) and aggregated STNV (222 S).

Isolation of STNV-RNA

The single-stranded RNA of STNV can be isolated by rather conventional phenol extraction procedures. Yet various *in vitro* assays of STNV-

RNA structure and function indicate that STNV-RNA is particularly susceptible to nuclease mediated disruption. Thus, in the following procedure, one should utilize reagents and containers known to be free of nuclease contamination (e.g., glassware pretreated with hot 0.1 M KOH and then exhaustively washed with distilled water).

Mix a 1–2 ml sample (called 1 volume) of STNV solution (10–20 mg STNV/ml), a 0.1 volume of 2 M $(NH_4)_2CO_3$, and a 1.2 volume of predistilled, room temperature water-saturated, phenol. Shake (Vortex mixer) this mixture to obtain an even suspension and then incubate (room temperature) for 15 minutes, reshaking the sample every 5 minutes. Centrifuge (0–2°) this phenol suspension for 15 minutes at 5000 g. Carefully aspirate off the aqueous (top) layer being careful to avoid the phenol and the denatured protein at the interface. Add 5 ml of 0–2° ether to the aqueous phase and then cool it (0–2°). Mix (Vortex mixer) 1 volume of distilled water into the phenol phase. Recentrifuge (0–2°, 15 minutes, 5000 g) the phenol–water mixture; remove the aqueous (top) layer as before; and combine this aqueous phase with the other cooled, ether containing, aqueous phase. Using a Vortex mixer, aspiration with a Pasteur pipette, and where possible using an ice bucket to cool the sample, remove the residual phenol from the combined aqueous phases by extracting 8 times with 5-ml batches of cold (0–2°) ether. After aspirating off the last ether phase, evaporate the residual ether from the aqueous phase by blowing over the fluid surface for 5 minutes with a gentle stream of N_2 gas.

Start the precipitation of the RNA from the ether-free aqueous phase by mixing in 6 drops of 3 M sodium acetate followed by a volume of −15° absolute ethanol twice that of the aqueous sample, and then storing the solution at −15°. If no turbidity appears after 5 minutes at −15°, mix in a minimum (up to 10 drops) of 3 M sodium acetate and continue −15° storage until turbidity appears. Twenty minutes after turbidity is detected in the −15° sample, centrifuge (10 minutes, 5000 g, −15 to 0°) down the precipitate, decant off most of the supernatant, and then using an ice bucket to cool the precipitate in its centrifuge tube, evaporate off the last of the supernatant with a gentle stream of N_2 gas. Continue the N_2 drying until the RNA pellet begins to crack and the pellet edges lighten in color. (This step is critical. The ethanol must be removed to facilitate future use of the RNA, yet too much drying destroys the RNA.)

Using a sterile glass rod, redissolve the moist RNA pellet in a minimum of cold distilled water. (A small amount of residual turbid protein will probably be present. Do not overdilute in attempting to dissolve this insoluble material.) Remove a 2-μl sample and assay spectrophotometrically for STNV-RNA concentration (A_{260} of 25 = 1 mg STNV-RNA/ml when in a 1 cm deep cuvette). Make up to a final concentration of 4 mg/ml with cold distilled water and then freeze away desired aliquots of the RNA in a dry-ice bath before storing at $-80°$. STNV-RNA yield will be 4–6 mg STNV-RNA per 40 mg of original virus.

If STNV coat protein is desired, mix 10 ml of acetone into the residual phenol phase to precipitate the viral coat protein. Then wash the precipitated protein 5 times with 10 ml washes of acetone by use of repeated centrifugations (5 minutes, 5000 g) and decantations. Finally dry the white coat protein under vacuum and store it in a dry place. This rather insoluble protein can be solubilized, if desired, by 24-hour dialysis $(0–2°)$ into 0.5 M NaOH followed by dialysis into 1% $(NH_4)_2CO_3$ or $5 × 10^{-4} M$ KOH.

In Vitro Message Potential of STNV-RNA

Translation by a Prokaryotic (E. coli) System. Our initial attempts at *in vitro* translation of STNV-RNA using the S30 system of Nirenberg and Matthaei[6] yielded low levels of protein synthesis. We assumed this low level of translation resulted from nucleases released from the cells by the alumina grinding step of the Nirenberg and Matthaei procedure. This theory has never really been tested. It is highly probably that a variety of cell-free systems from various bacteria will carry out the *in vitro* translation of STNV-RNA. It is known that STNV-RNA can be readily translated in a well documented[7] *in vitro* system from *E. coli* that is uniquely low in endogenous nucleases and proteases. Such translation, as described below, leads to the predominant synthesis of STNV coat protein.[3,4,7]

STNV-RNA translation is conveniently measured by assay of the incorporation of one or more ³H- or ¹⁴C-labeled amino acids into protein. In an optimal experiment, a 50-μl system contains 40 mM Tris·HCl, pH 7.8, 10 mM Mg(OAc)$_2$, 30 mM KCl, 8 mM β-mercaptoethanol, 6 mM PEP as K⁺ salt, 1.5 mM ATP as K⁺ salt, 300 μM GTP as K⁺ salt, 40 μM concentrations of each of the 20 common amino acids (i.e., 40 μM for each labeled and unlabeled amino acid), 40 μg of STNV-RNA, 2.5 μg of pyruvic kinase, and 10 μl of preincubated, dialyzed *E. coli* S30 extract prepared as previously reported.[7] Start the reaction by adding the STNV-RNA and then the preincubated S30. Incubate at 37° for 20 minutes to obtain essentially complete synthesis of 2–4 pmoles of STNV coat protein.

Subsequent usages of the labeled protein product dictate the method of stopping the reaction(s). If a simple assay of protein synthesis is desired, stop the reaction with an excess of 5% trichloroacetic acid and then measure incorporation of radioactivity into hot trichloroacetic acid-

[6] M. Nirenberg and J. H. Matthaei, *Proc. Nat. Acad. Sci. U.S.* **47**, 1588 (1961).
[7] J. M. Clark, Jr., A. Y. Chang, S. Spiegelman, and M. E. Reichmann, *Proc. Nat. Acad. Sci. U.S.* **54**, 1193 (1965).

insoluble material by conventional procedures.[8] If some sort of protein analysis or fingerprinting is desired, stop the reaction by adding 20–100 μg of pancreatic RNase, and proceed to remove unincorporated amino acids, etc., as desired.

Translation by Eukaryotic (Wheat Embryo) System. Optimum *in vitro* translation of STNV-RNA by the S23 preparation from wheat embryos of Marcus[9] requires assay conditions different from those most optimal for this wheat embryo system during the *in vitro* translation of other messenger-RNA's. The following optimal conditions for STNV-RNA translation result in the predominant synthesis of STNV coat protein.[3,4]

Prepare a 0.36 ml reaction containing 30 mM Tris·HCl, pH 7.6, 3 mM Mg(OAc)$_2$, 45 mM KCl, 4.6 mM β-mercaptoethanol, 4.2 mM PEP as K$^+$ salt, 1 mM ATP as K$^+$ salt, 200 μM GTP as K$^+$ salt, 1–2 μM concentrations of ^{3}H- or ^{14}C-labeled amino acids, and 30 μM concentrations of the nonradioactive amino acids needed to provide the other amino acids common to proteins, 40 μg of STNV-RNA, 35 μg of pyruvic kinase, and 0.15 ml of S23 extract from wheat embryos.[9] Start the reaction by adding the STNV-RNA, and then the S23 extract. Incubate at 30° for 30 minutes to obtain essentially complete synthesis of 0.3–0.6 pmoles of STNV coat protein.

As with the prokaryotic translation of STNV-RNA, the purpose of the experiment dictates the method for stopping the reaction(s). Consult the section on the *in vitro* prokaryotic translation of STNV-RNA for details.

[8] J. E. Allende, R. Monroe, and F. Lipmann, *Proc. Nat. Acad. Sci. U.S.* **51**, 1211 (1964).

[9] A. Marcus, D. Efron, and D. P. Weeks, this volume [68].

Author Index

Numbers in parentheses are reference numbers and indicate that an author's work is referred to although his name is not cited in the text.

Subject Index